T0358909

The Optical Communications Reference

Note from the Publisher

This book has been compiled using extracts from the following books within the range of Optical Communications books in the Elsevier collection:
Okamoto, (2006) Fundamentals of Optical Waveguides, 9780125250962
Menn, (2004) Practical Optics, 9780124909519
DeCusatis, (2006) Fiber Optic Essentials, 9780122084317
Pal, (2006) Guided Wave Optical Components and Devices, 9780120884810
Lam, (2007) Passive Optical Networks, 9780123738530
Kaminow, Li and Willner, (2008) Optical Fiber Telecommunications V B, 9780123741721
DeCusatis, (2008) Handbook of Fiber Optic Data Communication, 9780123742162
DeCusatis, (2002) Fiber Optic Data Communication, 9780122078927

The extracts have been taken directly from the above source books, with some small editorial changes. These changes have entailed the re-numbering of Sections and Figures. In view of the breadth of content and style of the source books, there is some overlap and repetition of material between chapters and significant differences in style, but these features have been left in order to retain the flavour and readability of the individual chapters.

End of chapter questions

Within the book, several chapters end with a set of questions; please note that these questions are for reference only. Solutions are not always provided for these questions.

Units of measure

Units are provided in either SI or IP units. A conversion table for these units is provided at the end of the book.

The Optical Communications Reference

Edited by Casimer DeCusatis and Ivan Kaminow

Amsterdam · Boston · Heidelberg · London · New York · Oxford
Paris · San Diego · San Francisco · Sydney · Tokyo
Academic Press is an imprint of Elsevier

Academic Press is an imprint of Elsevier
The Boulevard, Langford Lane, Kidlington, Oxford OX5 1GB, UK
30 Corporate Drive, Suite 400, Burlington, MA 01803, USA

First edition 2010

British Library Cataloguing in Publication Data
A catalogue record for this book is available from the British Library

Library of Congress Cataloging-in-Publication Data
A catalog record for this book is availabe from the Library of Congress

ISBN–13: 978-0-12-375163-8

For information on all Academic Press publications
visit our web site at books.elsevier.com

Printed and bound in the United Kingdom

Transferred to digital print 2012

Contents

Section One

Optical theory

Chapter 1.1

1.1

Geometrical optics

Menn

1.1.1. Ray optics conventions and practical rules. Real and virtual objects and images

Electro-optical systems are intended for the transfer and transformation of radiant energy. They consist of active and passive elements and sub-systems. In active elements, like radiation sources and radiation sensors, conversion of energy takes place (radiant energy is converted into electrical energy and vice versa, chemical energy is converted in radiation and vice versa, etc.). Passive elements (like mirrors, lenses, prisms, etc.) do not convert energy, but affect the spatial distribution of radiation. Passive elements of electro-optical systems are frequently termed optical systems.

Following this terminology, an optical system itself does not perform any transformation of radiation into other kinds of energy, but is aimed primarily at changing the spatial distribution of radiant energy propagated in space. Sometimes only concentration of radiation somewhere in space is required (like in the systems for medical treatment of tissues or systems for material processing of fabricated parts). In other cases the ability of optics to create light distribution similar in some way to the light intensity profile of an "object" is exploited. Such a procedure is called imaging and the corresponding optical system is addressed as an imaging optical system.

Of all the passive optical elements (prisms, mirrors, filters, lenses, etc.) lenses are usually our main concern. It is lenses that allow one to concentrate optical energy or to get a specific distribution of light energy at different points in space (in other words, to create an "image"). In most cases experienced in practice, imaging systems are based on lenses (exceptions are the imaging systems with curved mirrors).

The functioning of any optical element, as well as the whole system, can be described either in terms of ray optics or in terms of wave optics. The first case is usually called the geometrical optics approach while the second is called physical optics. In reality there are many situations when we need both (for example, in image quality evaluation, see Chapter 2). But, since each approach has advantages and disadvantages in practical use, it is important to know where and how to exploit each one in order to minimize the complexity of consideration and to avoid wasting time and effort.

This chapter is related to geometrical optics, or, more specifically, to ray optics. Actually an optical ray is a mathematical simplification: it is a line with no thickness. In reality optical beams which consist of an endless quantity of optical rays are created and transferred by electro-optical systems. Naturally, there exist three kinds of optical beams: parallel, divergent, and convergent (see Fig. 1.1.1). If a beam, either divergent or convergent, has a single point of intersection of all optical rays it is called a homocentric beam (Fig. 1.1.1b,c). An example of a non-homocentric beam is shown in Fig. 1.1.1 d. Such

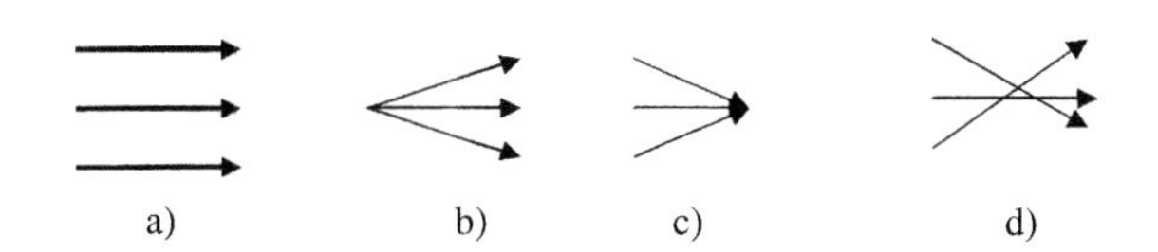

Fig. 1.1.1 Optical beams: (a) parallel, (b,c) homocentric and (d) non-homocentric.

Practical Optics; ISBN: 9780124909519

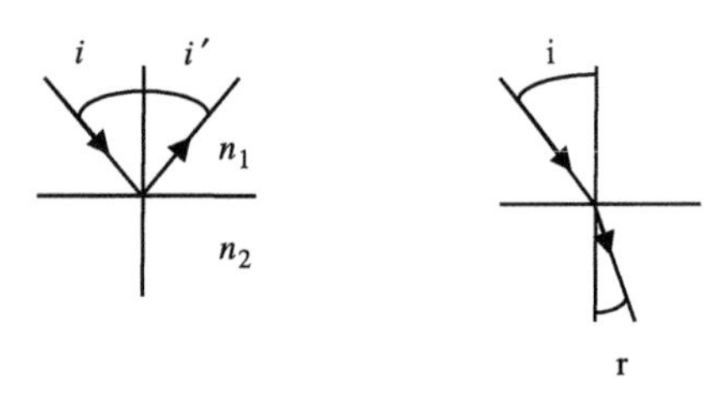

Fig. 1.1.2 Reflection and refraction of radiation.

a convergent beam could be the result of different phenomena occurring in optical systems (see Chapter 2 for more details).

Ray optics is primarily based on two simple physical laws: the law of reflection and the law of refraction. Both are applicable when a light beam is incident on a surface separating two optical media, with two different indexes of refraction, n_1 and n_2 (see Fig. 1.1.2). The first law is just a statement that the incident angle, i, is equal to the reflection angle, i'. The second law defines the relation between the incident angle and the angle of refraction, r:

$$\sin(i)/\sin(r) = n_2/n_1. \tag{1.1.1}$$

It is important to mention that all angles are measured from the vertical line perpendicular to the surface at the point of incidence (so that the normal incidence of light means that $i = i' = r = 0$).

In the geometrical optics approach the following assumptions are conventionally accepted:

(a) radiation is propagated along a straight line trajectory (this means that diffraction effects are not taken into account);

(b) if two beams intersect each other in space there is no interaction between them and each one is propagated as if the second one does not appear (this means that interference effects are not taken into account);

(c) ray tracing is invertable; in other words, if the ray trajectory is found while the ray is propagated through the system from input to output (say, from the left to the right) and then a new ray comes to the same system along the outgoing line of the first ray, but propagates in the reverse direction (from the right to the left), the trajectory of the second ray inside and outside of the system is identical to that of the first ray and it goes out of the system along the incident line of the first ray.

Normally an optical system is assumed to be axisymmetrical, with the optical axis going along OX in the horizontal direction. Objects and images are usually located in the planes perpendicular to the optical axes, meaning that they are along the OY (vertical) axis. Ray tracing is a procedure of calculating the trajectory of optical rays propagating through the system. Radiation propagates from the left to the right and, consequently, the object space (part of space where the light sources or the objects are located) is to the left of the system. The image space (part of space where the light detectors or images are located) is to the right of the system.

All relevant values describing optical systems can be positive or negative and obey the following sign conventions and rules:

ray angles are calculated relative to the optical axis; the angle of a ray is positive if the ray should be rotated counterclockwise in order to coincide with OX, otherwise the angle is negative;

vertical segments are positive above OX and negative below OX;

horizontal segments should start from the optical system and end at the relevant point according to the segment definition. If going from the starting point to the end we move left (against propagated radiation), the segment is negative; if we should move right (in the direction of propagated radiation), the corresponding segment is positive.

Examples are demonstrated in Fig. 1.1.3. The angle u is negative (clockwise rotation of the ray to OX) whereas u' is positive. The object Y is positive and its image Y′ is negative. The segment S defines the object distance. It starts from the point O (from the system) and ends at the object (at Y). Since we move from O to Y against the light, this segment is negative ($S < 0$). Accordingly, the segment S' (distance to the image) starts from the system (point O′) and ends at the image Y′. Since in this case we move in the direction of propagated light (from left to right) this segment is positive ($S' > 0$).

The procedure of imaging is based on the basic assumption that any object is considered as a collection of separate points, each one being the center of a homocentric divergent beam coming to the optical system. The optical system transfers all these beams, converting each one to a convergent beam concentrated in a small spot (ideally a point) which is considered as an image of the corresponding point of the object. The collection of such "point images" creates an image of the whole object (see Fig. 1.1.4).

An ideal imaging is a procedure when all homocentric optical beams remain homocentric after traveling

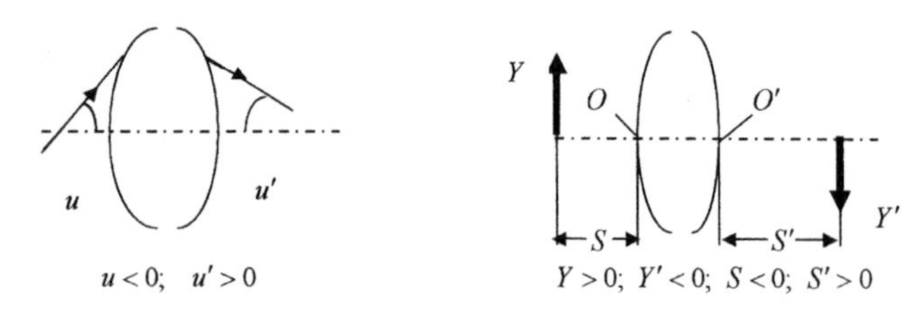

Fig. 1.1.3 Sign conventions.

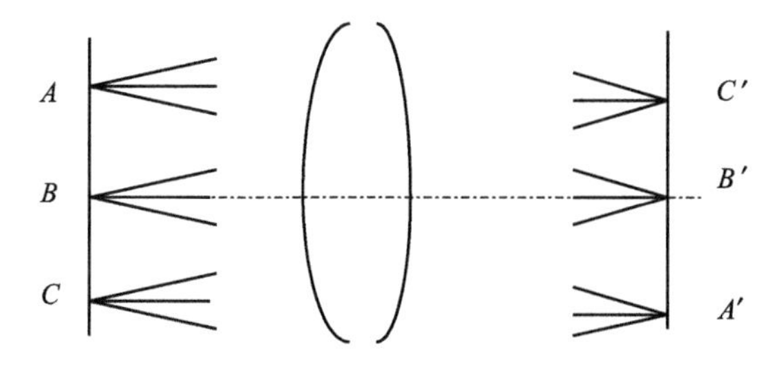

Fig. 1.1.4 Concept of image formation.

through the optical system, up to the image plane (this case is demonstrated in Fig. 1.1.4). Unfortunately, in real imaging the outgoing beams become non-homocentric which, of course, "spoils" the images and makes it impossible to reproduce the finest details of the object (this is like a situation when we try to draw a picture using a pencil which is not sharp enough and makes only thick lines – obviously we fail to draw the small and fine details on the picture). The reasons for such degradation in image quality lie partially in geometrical optics (then they are termed optical aberrations) and partially are due to the principal limitations of wave optics (diffraction limit). We consider this situation in detail in Chapter 1.2. Here we restrict ourselves to the simple case of ideal imaging.

In performing ray tracing one should be aware that doing it rigorously means going step by step from one optical surface to another and calculating at each step the incident and refraction angles using Eq. (1.1.1). Since many rays should be calculated, it is a time-consuming procedure which today is obviously done with the aid of computers and special programs for optical design. However, analytical consideration remains very difficult (if possible at all). The complexity of the procedure is caused mainly by the nonlinearity of the trigonometrical functions included in Eq. (1.1.1). The situation can be simplified drastically if we restrict ourselves to considering small angles of incidence and refraction. Then $\sin(i) \approx i$; $\sin(r) \approx r$; $r = i/n$ and all relations become linear. Geometrically this approximation is valid only if the rays are propagated close to the optical axis of the system, and this is the reason why such an approximation is called paraxial. A paraxial consideration enables one to treat optical systems analytically. Because of this, it is very fruitful and usually is exploited as the first approximation at the early stage of design of an optical system.

Even in the paraxial approach we can further simplify the problem by neglecting the thickness of optical lenses. Each lens consisting of two refractive surfaces (spherical in most cases, but sometimes they could be aspherical) separated by glass (or other material) of thickness t is considered as a single "plane element" having no thickness, but still characterized by its ability to concentrate an incident parallel beam in a single point (called the focal point or just focus). In such a case the only parameter of the lens is its focal length, f', measured as the distance between the lens plane and the focus, F′. Each lens has two focuses: the back (F′) and the front (F), the first being the point where the rays belonging to a parallel beam incident on the lens from the left are concentrated and the second being the center of the concentrated rays when a parallel beam comes to the lens from the right. Obviously, if the mediums at both sides of the lens are identical (for example, air on both sides or the lens being in water) then $f' = -f$. In the case when the mediums are different (having refractive index n and n' correspondingly) the relation should be

$$nf' = -n'f. \tag{1.1.2}$$

The optical power of a lens, defined as

$$\phi = 1/f', \tag{1.1.3}$$

is used sometimes in system analysis, as we shall see later.

Imaging with a simple thin lens obeys the two following equations:

$$\frac{1}{S'} - \frac{1}{S} = \frac{1}{f'}, \tag{1.1.4}$$

$$V = S'/S = y'/y, \tag{1.1.5}$$

where V is defined as the optical magnification. These two formulas enable one to calculate the positions and sizes of images created by any thin lens, either positive or negative, if all values are defined according to the sign conventions and rules described earlier in this section. A number of thin lenses which form a single system can also be treated using expressions (1.1.4) and (1.1.5) step-by-step for each component separately, the image of element i being considered as a virtual object for element $(i + 1)$. An example of such a consideration with details for a two-lens system is presented in Problem P.1.1.7.

The next step in approaching the real configuration of an optical system is to take into account the thickness of its optical elements. Still remaining in the paraxial range one can describe the behavior of a single spherical surface (see Fig. 1.1.5) by the Abbe invariant (r is the radius of the surface):

$$n\left(\frac{1}{r} - \frac{1}{S}\right) = n'\left(\frac{1}{r} - \frac{1}{S'}\right). \tag{1.1.6}$$

Then, the ray tracing for an arbitrary number of surfaces can be performed with the aid of the following two simple relations (see also Fig. 1.1.6):

$$u_{k+1} = \frac{n_k}{n_{k+1}} u_k + \frac{h_k}{r_k n_{k+1}} (n_{k+1} - n_k). \tag{1.1.7}$$

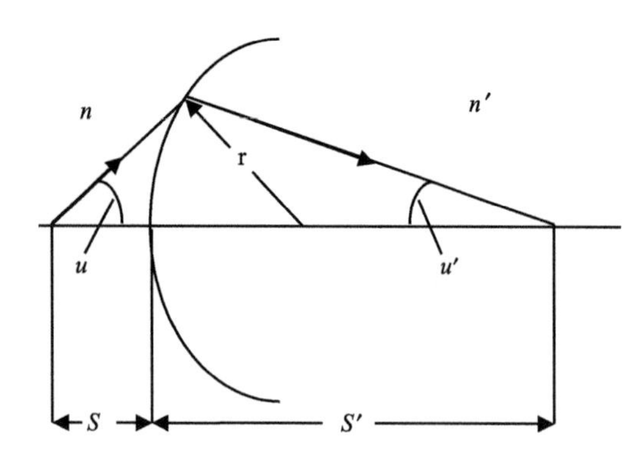

Fig. 1.1.5 Refraction of rays at a single spherical surface.

$$h_{k+1} = h_k - u_{k+1}d_k \qquad (k = 1,2...,N). \quad (1.1.8)$$

Given the radii of the spherical surfaces, the refraction indexes on both sides, and the distances between them, all angles, u_k, and heights, h_k, can be easily found, starting from initial values u_1, h_1.

To apply Eqs. (1.1.7) and (1.1.8) to a single lens defined by two spherical surfaces of radii r_1 and r_2 separated by the segment d, we first have to remind ourselves of the definition of the principal planes, H, H′ and the cardinal points. As is seen from Fig. 1.1.7, the real ray trajectory ABCD can be replaced by ABMM'CD in such a way that they are identical outside the lens, but inside the lens the rays intersect two virtual planes H and H′ at the same height (OM = O′M′). Actually these principal planes, H, H′, can represent the lens as far as ray tracing is considered. Furthermore, the focal distances, f, f', are measured from the cardinal points O, O′ to the front and back focuses, F and F′. The terms "back focal length" (BFL) and "front focal length" (FFL) are related to the segments S_F, S_F from the back and front real surfaces to F' and F, respectively (see Fig. 1.1.7). Calculation of BFL and FFL enables one to determine the location of both principal planes with regard to the lens surfaces. Leaving the details of calculation to Problem P.1.1.5 we just indicate here the final results:

$$S_{F'} = f'\left[1 - \frac{d}{r_1 n}(n-1)\right], \qquad (1.1.9)$$

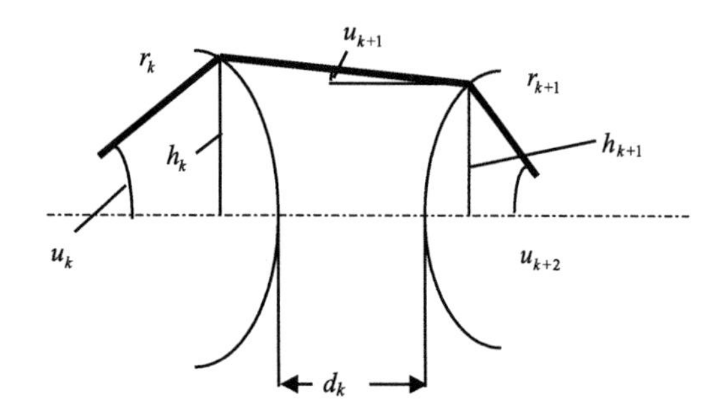

Fig. 1.1.6 Ray tracing between two spherical surfaces.

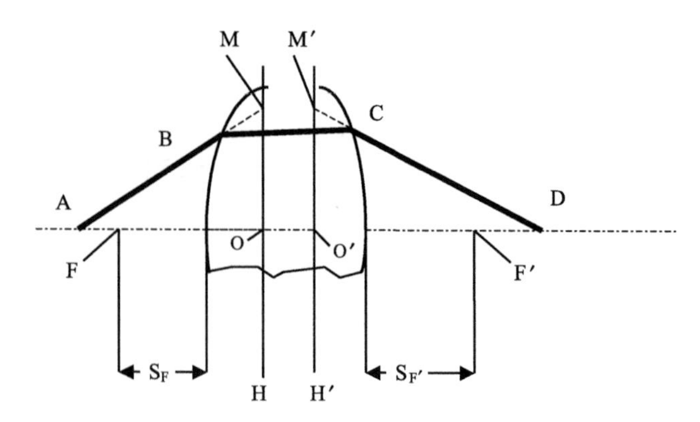

Fig. 1.1.7 Principal planes of a thick lens.

$$S_F = -f'\left[1 + \frac{d}{r_2 n}(n-1)\right], \qquad (1.1.10)$$

and for the focal distance

$$\frac{1}{f'} = (n-1)\left(\frac{1}{r_1} - \frac{1}{r_2}\right) + \frac{d(n-1)^2}{r_1 r_2 n}. \qquad (1.1.11)$$

In many cases the second term of the last formula can be neglected since it is much smaller than the first one.

Problems

P.1.1.1. Find the image of the object OA in Fig. 1.1.8 using the graphical method.

P.1.1.2. Find the image of the point source A and direction of the ray AB after the positive lens L_1 (Fig. 1.1.9a) and the negative lens L_2 (Fig. 1.1.9b).

P.1.1.3. *Ray tracing in a system of thin lenses.* Find the final image of a point source A after an optical system consisting of thin lenses L_1, L_2, and L_3 ($f_1' = f_2' = 15$ mm; $f_3' = 20$ mm) if A is located on the optical axis 30 mm left of the lens L_1 and the distances between the lenses are $d_{12} = 40$ mm, $d_{23} = 60$ mm. [Note: Do this by ray tracing based on Eq. (1.1.4).]

P.1.1.4. *Method of measurement of focal length of a positive lens.* An image of an object AB created by a lens is displayed on a screen P distant from AB at $L = 135$ mm (Fig. 1.1.10). Then the lens is moved from the initial position, 1, where the sharp image is observed at

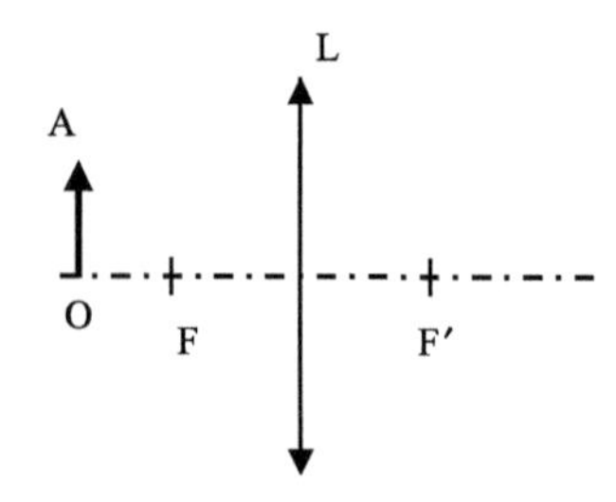

Fig. 1.1.8 Problem P.1.1.1 – Imaging by the graphical method.

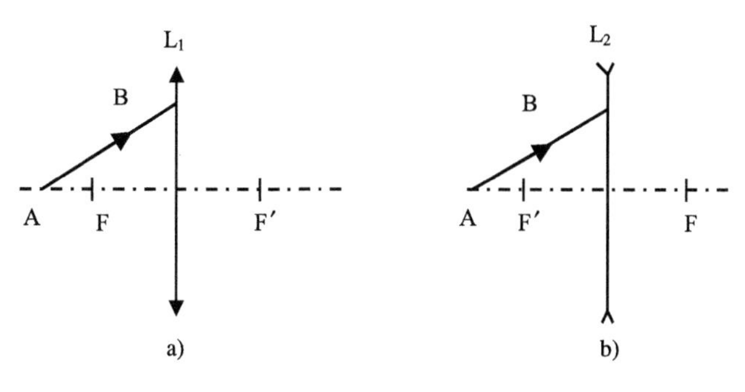

Fig. 1.1.9 Problem P.1.1.2 – Imaging by the graphical method: (a) with a positive lens; (b) with a negative lens.

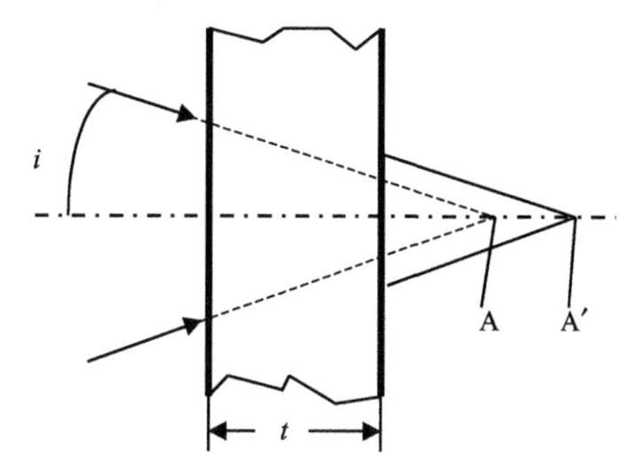

Fig. 1.1.11 Problem P.1.1.6 – Consideration of a parallel plate.

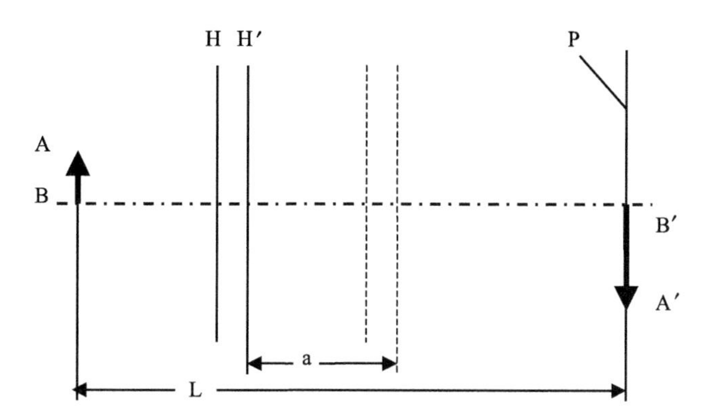

Fig. 1.1.10 Problem P.1.1.4 – Method of focal length measurement.

magnification V_1, to the position 2 where again the sharp image is observed on the same screen, but at magnification $V_2 = 1/V_1$. The distance between positions 1 and 2 is $a = 45$ mm. Find the focal length of the lens and estimate the uncertainty of the measured value if the lens thickness, t, is about 5–6 mm.

P.1.1.5. *Location of the principal planes of a thick lens.* Find the positions of two principal planes H and H′, BFL, and FFL of a lens made of glass BK-7($n = 1.5163$) having two spherical surfaces of radii $R_1 = 50$ mm and $R_2 = -75$ mm and thickness $t = 6$ mm.

P.1.1.6. *Violation of homocentricity of a beam passed through a flat slab.* A flat slab of glass is illuminated by a homocentric beam which fills the solid angle $\omega = 1.5$ sr with the center at point A, 30 mm behind the slab (Fig. 1.1.11). The thickness of the slab $t = 5$ mm and refractive index $n = 1.5$. Find the location of the point A′ after the slab as a function of incident angle, i, and estimate the deviation of the outgoing beam from homocentricity.

P.1.1.7. *A two-lens system in the paraxial range.* A lens L_1 of 100 mm focal length is followed by a lens L_2 of 75 mm focal length located 30 mm behind it. Considering both lenses as a unified system find the equivalent optical power and position of the focal plane.

P.1.1.8. *A ball lens.* Find the location of the principal planes of a ball lens (a full sphere) of radius $r = 3$ mm and its BFL.

1.1.2. Thin lenses layout. Microscope and telescope optical configurations

We will consider here the following basic configurations: (i) magnifier; (ii) microscope; and (iii) telescope. All three can be ended either by a human eye or by an electro-optical sensor (like a CCD or other area sensor).

1.1.2.1. The human eye

Although the details of physiological optics are beyond the scope of this book, we have to consider some important features of the human eye (for further details, see Hopkins, 1962) as well as eye-related characteristics of optical devices. Usually the "standard eye" (normal eye of an adult person) is described in terms of a simplified model (so-called "reduced eye") as a single lens surrounded by the air from the outside, and by the optically transparent medium (vitreous humor) of refractive index 1.336 from the inside. As a result, the front focal length of the eye, f, differs from the back focal length, f' (see Eq. (1.1.2)). The front focal length is usually estimated as 17.1 mm where as f' is equal to 22.9 mm. The pupil of the eye varies from 2 mm (minimum size) to 8 mm (maximum size) according to the scene illumination level (adaptation). The lens creates images on the retina which consists of huge numbers of photosensitive cells. The average size of the retina cells dictates the angular resolution of the eye (ability of seeing two small details of the object separately). The limiting situation is that the images of two points are created at two adjacent cells of the retina. This renders the angular resolution of a normal eye to be 1 arcminute (3×10^{-4} rad). The lens curvature is controlled by the eye muscles in such a way that the best (sharp) image is always created on the retina, whether the object is far or close to the eye (accommodation process). The distance of best vision is estimated as 250 mm, which means that the eye focused on objects at distances of 250 mm is not fatigued during long visual operation and can still differentiate small details.

Three kinds of abnormality of eye optics are usually considered: myopia, hyperopia, and astigmatism. The first one (also called near-sightedness) occurs when a distant object image is not created on the retina but in front of it. A corrective negative lens is required in such a situation. In the second case (called far-sightedness) the opposite situation takes place: the images are formed behind the retina and, obviously, the corrective lens should be positive. Astigmatism means that the lens curvature is not the same in different directions which results in differences in focal lengths, say in the horizontal and vertical planes. Correction is done by spectacles with appropriately oriented cylindrical lenses.

The other properties of the eye related to visual perception are considered in Chapter 10.

1.1.2.2. Magnifications in optical systems

Generally, four adjacent magnifications can be defined for any optical system: (i) linear magnification, V, for objects and images perpendicular to the optical axis; (ii) angular magnification, W; (iii) longitudinal magnification, Q (magnification in the direction of the optical axis); and (iv) visible magnification, Γ (used only for systems working with the human eye).

Linear magnification, defined earlier for a single lens by Eq. (1.1.5), is still applicable for any complete optical system. Angular magnification can be defined for any separate ray or for a whole beam incident on a system. For example, for the tilted ray shown in Fig. 1.1.12 W is calculated as follows:

$$W = \tan(u')/\tan(u). \tag{1.1.12}$$

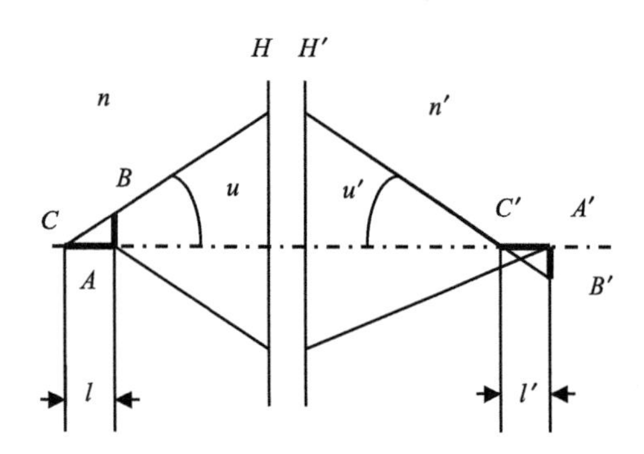

Fig. 1.1.12 Imaging of vertical and horizontal segments.

As can be shown, the product VW is a system invariant: it does not depend on the position of the object and image, but is determined by the refractive indexes on both sides of the optical system (n and n'). If $n = n'$ then $VW = 1$.

Considering the segment l along the optical axis and two pairs of conjugate points, A and A′, C and C′ (Fig. 1.1.12), we can find the longitudinal magnification, Q:

$$Q = l'/l. \tag{1.1.13}$$

It can be shown that for small segments l, i' one can use the formula $Q = V^2$.

Finally, visible magnification is related to the size of images on the retina of an eye. It is defined as the ratio of the image created by the optical system to the image of the same object observed by the naked eye directly. Since the image size is proportional to the observation angle (see Fig. 1.1.13), Γ is determined as follows:

$$\Gamma = \tan(\gamma')/\tan(\gamma). \tag{1.1.14}$$

1.1.2.3. A simple magnifier

This is usually operated with the eye. While observing through a magnifying glass an object is positioned between the front focus of the lens and the lens itself (Fig. 1.1.14). The image is virtual and its position corresponds to the distance of best vision of the eye (250 mm). The closer the object to F, the higher the magnification. Therefore, approximately, we can define that

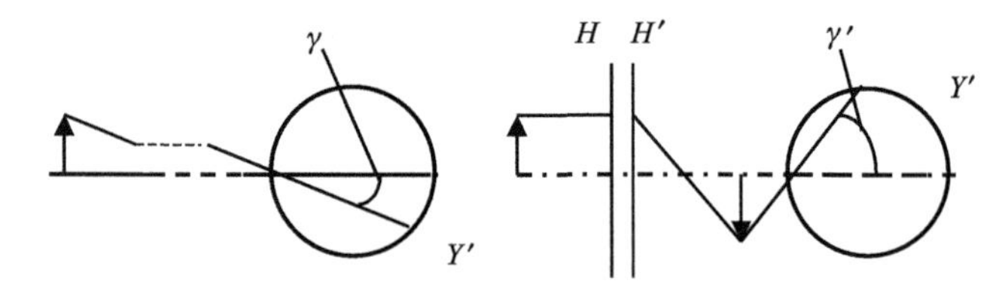

Fig. 1.1.13 Explanation of visible magnification.

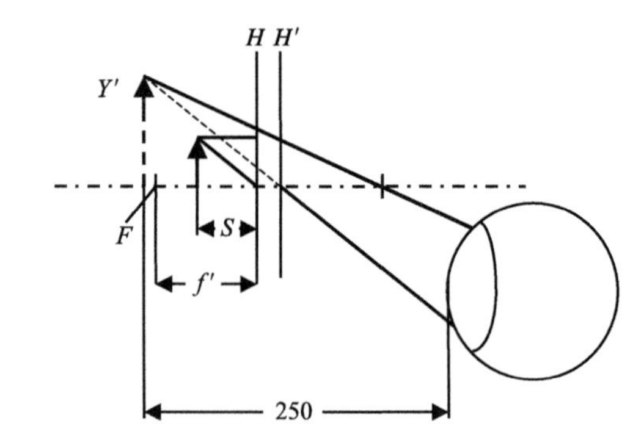

Fig. 1.1.14 A simple magnifier.

$s \approx f$ (of course $s < 0; f < 0$), and for visible magnification of the magnifier we have

$$\Gamma = \frac{250}{f'}. \tag{1.1.15}$$

Since the distance of best vision is much greater than the focal distance of the eye, the rays coming to the eye pupil are almost parallel. In most cases they can be treated just as a parallel beam (or beams).

1.1.2.4. The microscope

Figure 1.1.15a demonstrates the basic layout of a microscope working with the eye and Fig. 1.1.15b shows a microscope working with an electro-optical detector (like a CCD or other video sensor). In both cases the first lens L_1 (called the objective) is a short-focus well-corrected lens creating the first real magnified image of the object (AB) in the plane P. The second lens is the eyepiece (Fig. 1.1.15a) or the relay lens (Fig. 1.1.15b). The eyepiece L_2 functions like a simple magnifier and its visible magnification obeys Eq. (1.1.15). Magnification of the objective, V_1, can be found from Eq. (1.1.5). Usually the object distance S_1 is very close to f_1 and the distance $S'_1 = T$ from the lens L_1 to the plane P is chosen as one of several standardized values accepted by all manufacturers of microscopes (160 mm or 180 mm or 210 mm). Therefore, for the total magnification of the microscope working with eye we get:

$$V_M = V_1 V_2 = \frac{T}{f'_1} \times \frac{250}{f'_2}. \tag{1.1.16}$$

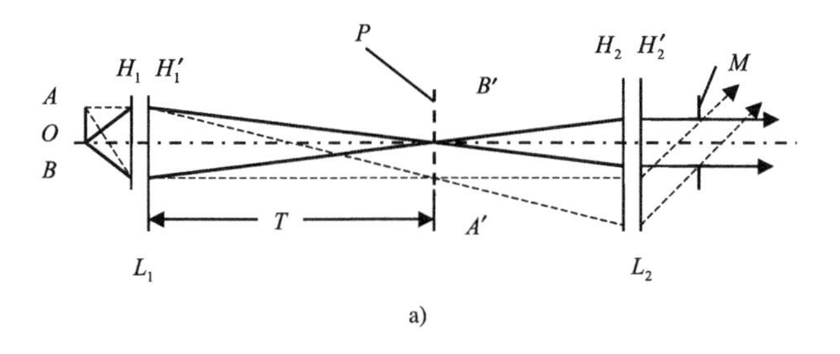

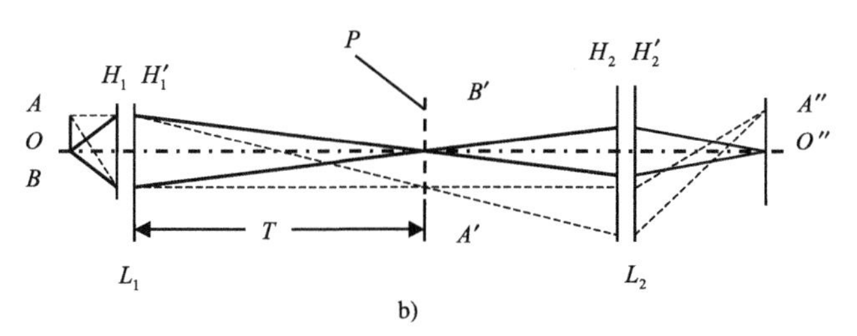

Fig. 1.1.15 Layout of a microscope working with (a) the eye and (b) an area detector.

If instead of an eyepiece a relay lens is exploited its actual linear magnification, V_2, should be taken into account, and then

$$V_M = \frac{T}{f'_1} V_2. \tag{1.1.17}$$

If the microscope is intended for measurements and not only for observation then a glass slab with a special scale (a ruler, a crosshair, etc.) called a reticle is introduced in the plane P. In such a case the eye observes the image overlapped with the scale.

In Fig. 1.1.15 the rays originating from two points of the object are drawn – from the center of the object (point O) and from the side (point A). As can be seen from Fig. 1.1.15a, each point gives a parallel beam after the eyepiece: one is parallel to the optical axis and the other is tilted to OX. Intersection of the beams occurs in the plane M (exit pupil of the microscope) where the operator's eye should be positioned.

For the convenience of the operator the optical layout in most cases is split after the plane P in two branches, each one having a separate eyepiece. Such an output assembly is called binocular and observation is done by two eyes. It should be understood, however, that binocular itself does not render stereoscopic vision, since both eyes are observing the same image created by a single objective L_1. To achieve a real stereoscopic effect two objectives are required in order to observe the object from two different directions. Each image is transferred through a separate branch (a pair of lenses L_1 and L_2).

The architecture shown in Fig. 1.1.16 is actually the combination of the two layouts presented in Fig. 1.1.15 and its output assembly is called trinocular – it creates images on the area sensor as well as in the image plane of both eyepieces. The beam splitter, BS, turns the optical axis in the direction of the relay lens L_3.

In the last few years microscopes have been designed as infinity color-corrected systems (ICS) which means that the object is located in the front focal plane of the objective, its image is projected to infinity, and an additional lens L_4 (the tube lens) is required in order to create an intermediate image in the plane P. Such a layout is

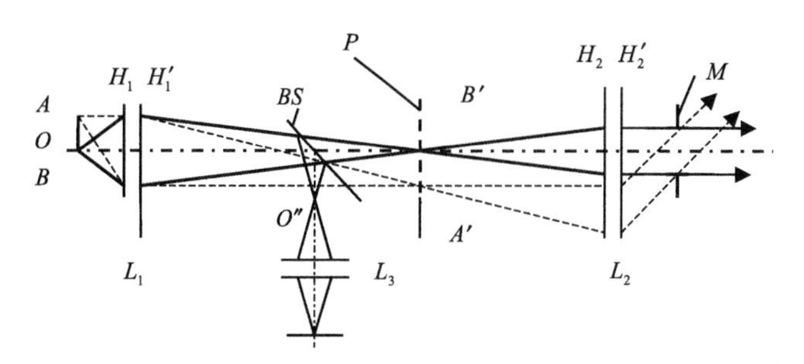

Fig. 1.1.16 Microscope with a trinocular assembly.

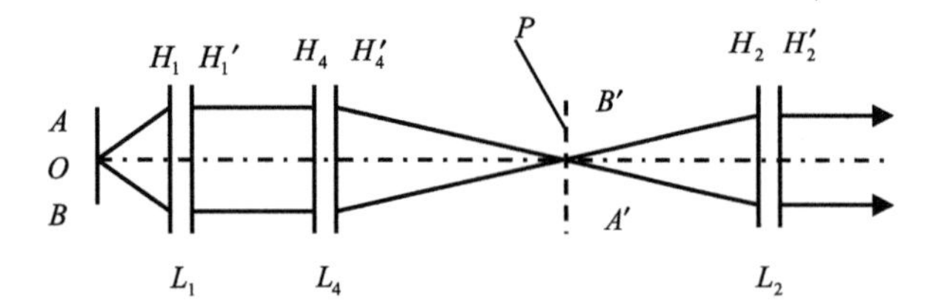

Fig. 1.1.17 Layout of a microscope with ICS optics.

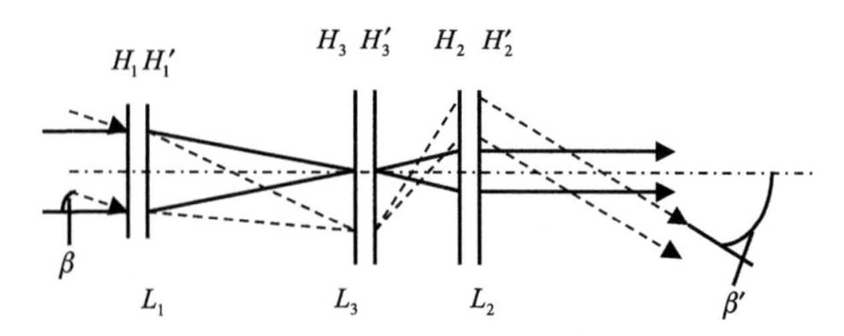

Fig. 1.1.19 Telescope with a field lens.

demonstrated in Fig. 1.1.17. One of the important advantages of ICS optics is that the light beams are parallel between L_1 and L_4 enabling one to introduce here optical filters with no degradation of the optical quality and without relocation of the image plane P.

1.1.2.5. The telescope

Telescopic systems are intended for observation of remote objects. If the distance between the object and the first lens of the system is much greater than the focal length of L_1 all light beams at the entrance of the system can be considered as parallel, whether they are coming from the central point of the object or from the side.

Again, as in the above consideration of a microscope, the central point beam is parallel to the optical axis whereas the side point generates an oblique parallel beam. All incident beams are concentrated by the objective in its back focal plane (passing through the back focus F'_1). The second lens L_2 is positioned in such a way that its front focus F_2 coincides with F'_1. Obviously all beams after lens L_2 become parallel again, but the exit angles of the oblique rays are different from those of the corresponding beams at the entrance (see Fig. 1.1.18) and this causes the angular magnification of the telescope to be (see Eq. (1.1.12))

$$W = \tan(\beta')/\tan(\beta) = f'_1/f'_2. \tag{1.1.18}$$

As follows from Eq. (1.1.18), the longer the focal length of the objective, the greater the magnification. However, along with this the necessary size of the lens L_2 also increases, which might cause a limitation of the visible field of view (the part of the object space visible through the system). To solve this problem an additional lens L_3 (the field lens) is introduced in the system (see Fig. 1.1.19). This lens allows one to vary the vertical location of the oblique beam incident on L_2.

The configurations shown in Figs. 1.1.18 and 1.1.19 are built of positive lenses. In the Galilean architecture the eyepiece L_2 is negative (Fig. 1.1.20). As a result, the total length of the system is shortened. However, the intermediate image is virtual (both focal points F'_1 and F_2 are behind the eyepiece) and there is no way to introduce a measurement scale, if necessary. However, as it turns out, this shortcoming becomes very useful if the Galilean configuration is exploited with high-power lasers (for beam expanding).

Problems

P.1.1.9. If the angular resolution of the eye is 3×10^{-4} rad, what is the average size of the retina cells?

P.1.1.10. A microscope is intended for imaging an object located in the plane P simultaneously in two branches: one for observation by eye and the other for imaging onto a plane area sensor (CCD). The objective of the microscope serving the two branches is of 20 mm focal length and provides linear magnification $V = -10$ to the image plane of the eyepiece where a reticle M of 19 mm diameter is positioned (Fig. 1.1.21). The CCD sensor is 4.8 mm (vertical) $\times$ 5.6 mm (horizontal) in size. In front of the CCD at a distance of 20 mm an additional relay lens L_3 is introduced in order to reach the best compatibility of the field of view in both branches. Assuming the eyepiece L_2 to be of 25 mm focal length and neglecting the thickness of the lenses, find:

(a) the working distance (location of the object plane P with regard to the objective);

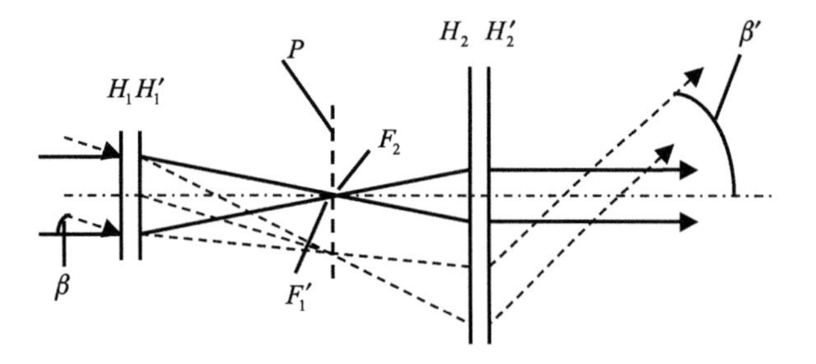

Fig. 1.1.18 Basic layout of a telescope.

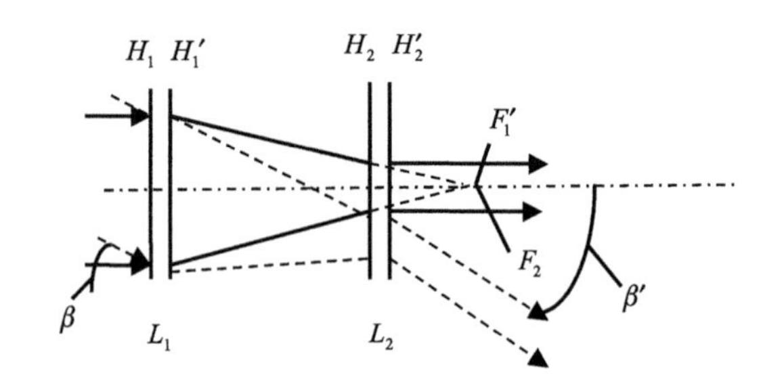

Fig. 1.1.20 Galilean telescope.

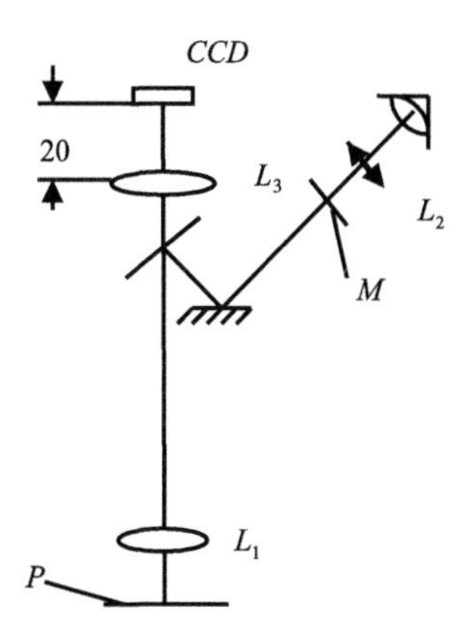

Fig. 1.1.21 Problem P.1.1.10 – Two-branch microscope.

(b) the total magnification in the branch to the eye;

(c) the optical power of lens L_3. [Note: Find two solutions and choose the one which provides the shortest distance between P and the CCD.]

P.1.1.11. *Dual magnification system with negative relay lens.* Such a system is widely used in the microelectronics industry where automatic processing of wafers is a main concern. The object (usually a wafer) is located in the plane P and imaged onto a CCD either at low magnification $V_1 = -3$ (through the right branch of the arrangement, exploited for initial alignment) or at high magnification $V_2 = 2 \times V_1$ (fine alignment through the left branch where lens L_2 and retroreflector R are introduced). While switching the system between two alignment procedures no optical element should be moved, except the aperture D (Fig. 1.1.22). The retroreflector R allows one to vary the high magnification of the system with minimum effort–just replacement of R and L_2, with no other changes in the arrangement. Thus, lens L_2 serves as a negative relay lens of the system. Neglecting the thickness of the lenses and taking all necessary distances from Fig. 1.1.22, find:

(a) the focal length of L_2 and its position with regard to the CCD and the other elements of the arrangement;

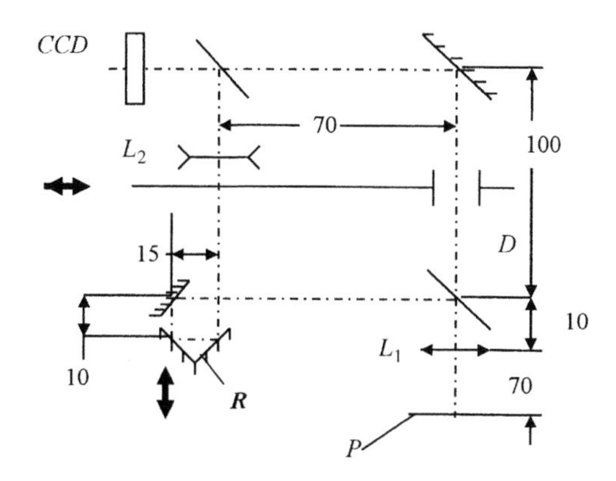

Fig. 1.1.22 A dual magnification system.

(b) the relocation of the retroreflector and the relay lens L_2 from their initial positions required to increase the magnification in the left branch by 10%.

1.1.3. Diaphragms in optical systems. Calculation of aperture angle and field of view. Vignetting

The size of each optical element of a system should be considered properly, since it influences: (i) the quantity of radiant energy passing through the system; (ii) the quality of images; and (iii) the cost of the system. Among all the geometrical parameters the working diameter is of primary importance (remember that we assume that the system is rotationally symmetric) – it acts as the transparent part of the element.

Sometimes an additional diaphragm (a physical element called a stop which has a final size aperture and negligible thickness) is introduced in the system. An aperture stop is a diaphragm which actually limits the size of light bundles passing through the system and consequently it is responsible for the amount of energy collected at each point of the image. The aperture stop is illustrated in Fig. 1.1.23. Assume that the system consists of a number of elements (of which the first and the last curved surfaces are shown in the figure) and also includes the stop cd. The boundaries of each optical surface are also considered as diaphragms. First we "transfer" all the diaphragms into the object space (e.g., we find the size and location of the image of each diaphragm through the rest of the optical elements to the left of it, as if the light beams are propagated from right to left). Such an image of the stop cd is c'd'; the image of the first diaphragm ab is ab itself, since there is no element left of it; the third diaphragm shown in the figure is the image of some other optical surface, etc. Then we connect the ray from the central point O of the object to the side of each image and find the angle of each ray with the optical axis. The smallest angle (in our example it is the angle of the ray Oc') is called the aperture angle, α_{ap},

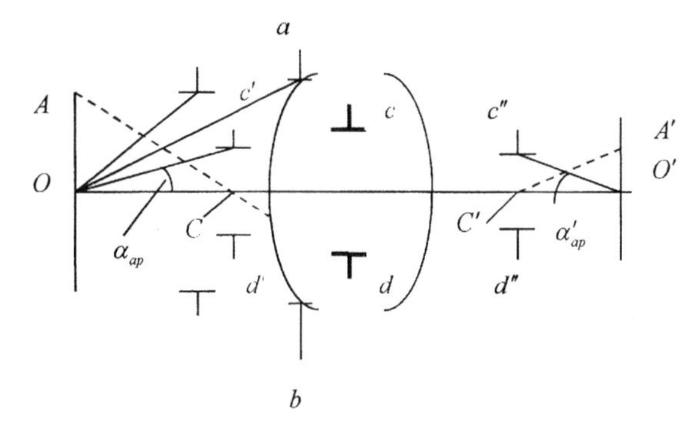

Fig. 1.1.23 Aperture stop and entrance and exit pupils.

and the corresponding physical diaphragm is called the aperture stop (cd in the case of Fig. 1.1.23). Its image in the object space is called the entrance pupil and its image to the image space is called the exit pupil (c′d′ and c″d″, respectively). Obviously, the aperture angle defines the maximum cone of light rays emerging from point O and passing through the system with no obstacle up to point O′ in the image plane. The corresponding angle α'_{ap} is the aperture angle in the image space. Drawing the rays that connect any other point of the object with the entrance pupil we find the corresponding cone of light participating in imaging of that point. The ray connecting the oblique point A with the center C of the entrance pupil is called the chief ray (shown by the dotted line in Fig. 1.1.23). Its position in the image space is C′A′.

Now we consider the entrance pupil, ab, together with any other diaphragm (or its image, gh) in the object space (Fig. 1.1.24). It is understood that the conical bundle originating from point O of the object is not affected by gh at all. The same is true for any other point of the object plane between O and A where the last one is found with the ray passing through the sides a and g of both diaphragms. For remote points above A (point B, for example), the light cone filling the entrance pupil is cut partially by the diaphragm gh (the dotted line originating in B cannot be transferred). This means that the active cone of light passing through the system is reduced gradually until we achieve finally the point C from which no ray can pass the system. The rays emerging from any point above C cannot achieve the image plane at all. Therefore, image formation can be performed only for a part of the object plane (the circle of radius OC). This part of the object plane is called the field of view and the diaphragm gh is called the field aperture. If gh is the image of a real physical diaphragm GH located somewhere in the system then GH is called the field stop.

Reduction of the light cones while moving out from the optical axis causes a decrease of the image brightness in the corresponding parts of the image plane. Even if the object plane is equally illuminated we get a reduction of the brightness in the image plane, as is illustrated by the graph of intensity, $I(r)$, in Fig. 1.1.24.

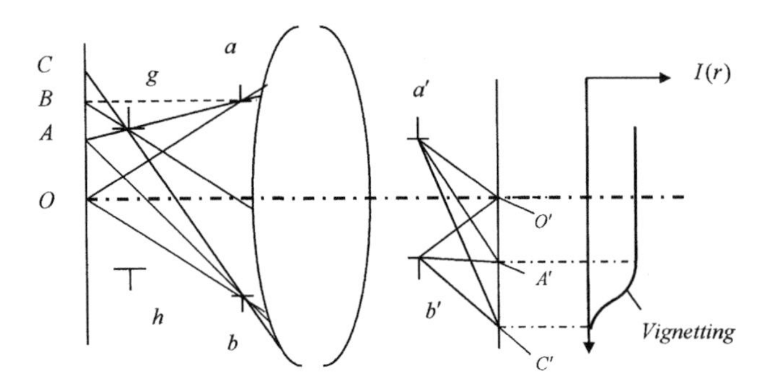

Fig. 1.1.24 Field of view and vignetting.

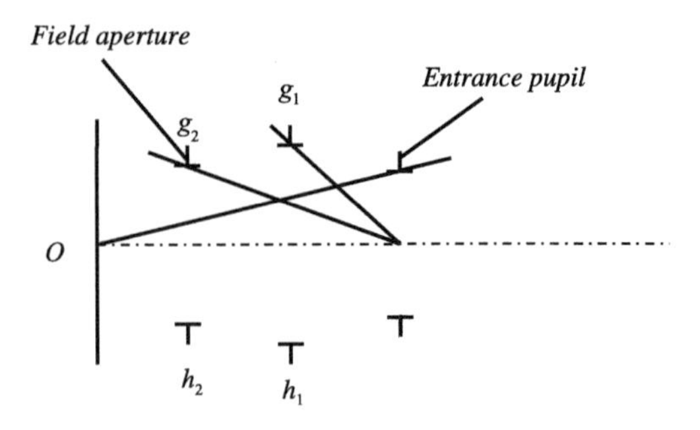

Fig. 1.1.25 Finding the field aperture.

This phenomenon is known as vignetting, and it should be carefully investigated if a new optical system is designed.

To find the field aperture it is necessary to image all physical diaphragms of the system into the object space, to calculate the sizes and location of each image, and then to draw a ray connecting the center of the entrance pupil with the side point of each image and to calculate the corresponding angle with the optical axis. The minimum angle defines the field aperture (and consequently the field stop). The procedure described is illustrated by Fig. 1.1.25 where g_2h_2 serves as the field aperture. It is useful to take into account the fact that to avoid vignetting it is necessary to position the field stop at the plane of the intermediate image of the system.

Problems

P.1.1.12. The system of two thin lenses L_1 (focal length 100 mm, diameter 20 mm) and L_2 (focal length 50 mm, diameter 20 mm) shown in Fig. 1.1.26 forms an image of the object plane P on a screen M at magnification $V = 3$. The distance between P and L_1 is 200 mm.

(a) How can the field stop ab of the system be positioned in order to get imaging with no vignetting?

(b) What should be the size of the field stop if the field of view is 10 mm?

(c) Find the location of all elements of the system and calculate the aperture angle and position of the entrance pupil.

P.1.1.13. In the system of Problem P.1.1.10, find the minimum size of lens L_3 which enables one to get images on the CCD with no vignetting.

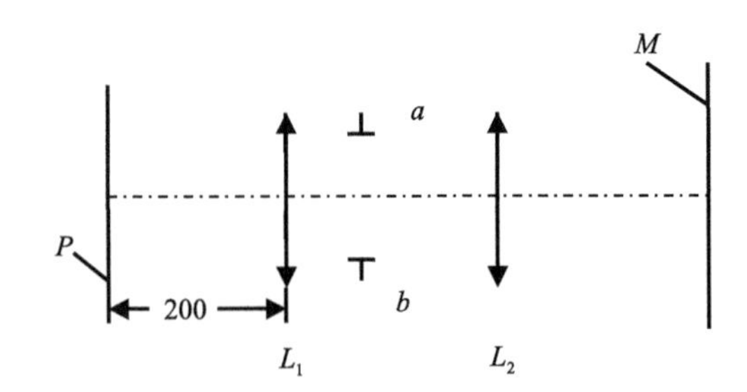

Fig. 1.1.26 Problem P.1.1.12 – Imaging with two lenses.

1.1.4. Prisms in optical systems

Prisms serve three main purposes in optical systems: (i) to fold the optical axis; (ii) to invert the image; and (iii) to disperse light of different wavelengths. Here we will consider the first two purposes. It is quite understandable that both are achieved due to reflection of rays on one or several faces of the prism. So, it is worth keeping in mind how the system of plane reflectors (plane mirrors) can be treated (e.g., see Problem P.1.1.14).

A great variety of prisms are commonly used in numerous optical architectures. Only a few simple cases are described below.

Right-angle prism. This is intended for changing the direction of the optical axis through 90°. The cross-section of this prism is shown in Fig. 1.1.27a. The rays coming from the object (the arrow 1–2) strike the input face AB at 90° and after reflection from the hypotenuse side emerge along the normal to the face BC. It can be seen that beyond the prism the object is inverted. The shortcoming of this prism is revealed if the incoming light is not normal to the prism face. In this case the angle between incoming and outgoing rays differs from 90°. Another issue is concerned with the total reflection of rays on face AC: it might happen that for some tilted rays total reflection does not occur. In such a case a reflecting coating on AC is required.

Penta-prism. This prism has effectively four faces with an angle of 90° between AB and BC and 45° between the two other sides (Fig. 1.1.27b). The shortcoming of the right-angle prism does not occur here, i.e., the outgoing beam is always at 90° to the input beam, independent of the angle of incidence. Also, the object is not inverted. This results from a double reflection in the prism and is evidence of the common rule for any prism or system with reflectors; namely, the image is not inverted if the number of reflections is even.

Dove prism. The angles A and D are of 45° and the input and output beams are usually parallel to the basis face AD (Fig. 1.1.27c). While traveling through the prism the beams are inverted. Another feature of this prism is its ability to rotate an image: when the prism is inserted in an imaging system rotated around the input beam with angular speed ω the image in the system will be rotated at a speed 2ω.

If it is necessary to invert beams around two axes a combination of prisms, like the Amici prism shown in Fig. 1.1.28, can be exploited. This prism is actually a right-angle prism with an additional "roof" (for this reason it is also called the roof-prism). As a result the beams are inverted in both directions: upside-down and left-right.

In general, any prism inserted in an imaging system makes the optical path longer. This effect should be taken into account if a system designed for an unbent configuration has to be bent to a more compact size using prisms and mirrors. With regard to its influence on image quality and optical aberrations the prism acts as a block of glass with parallel faces. As was demonstrated earlier (see Problem P.1.1.6 where the propagation of a divergent–convergent beam through a glass slab of thickness t was considered) the block of glass causes a lengthening of the optical path by $(n - 1)t/n$ compared to the ray tracing in air. Therefore instead of tracing the rays through the slab and calculating the refraction at the entrance and exit surfaces one can replace a real plate by a virtual "air slab" of reduced thickness, t/n, and perform ray tracing for air only. To apply this approach to prisms we have to find the slab equivalent to the prism with regard to the ray path inside the glass. This can be done by the following procedure based on unfolded diagrams (see Fig. 1.1.29). We start moving along the incident ray until the first reflection occurs. Then we build the mirror image of the prism and the rays and proceed moving further along the initial direction until the second reflected surface is met. Then again we build the mirror image of the configuration, including the ray path, and proceed further until the initial ray leaves the last (exit) face of the prism. Details of the procedure can be seen in Problem P.1.1.15.

Creating unfolded diagrams is aimed at calculating the thickness, t_e, of the equivalent glass block. For the cases depicted in Fig. 1.1.29:

(a) right-angle prism with an entrance face of size a: $t_e = a$;

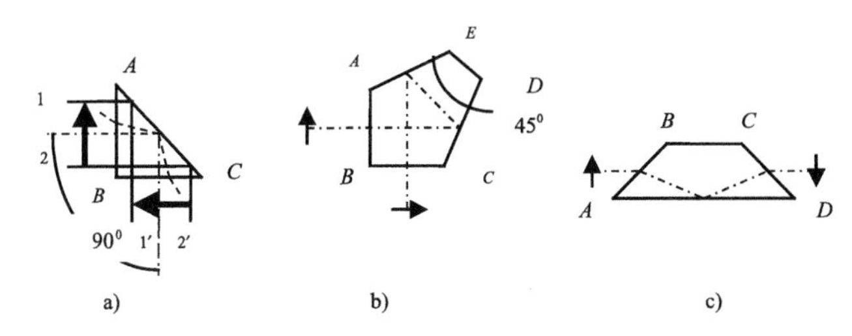

Fig. 1.1.27 Layout of different prisms: (a) right-angle prism; (b) penta-prism; (c) Dove prism.

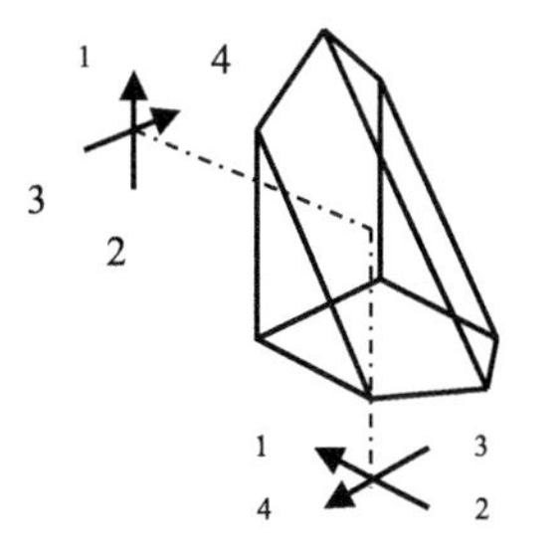

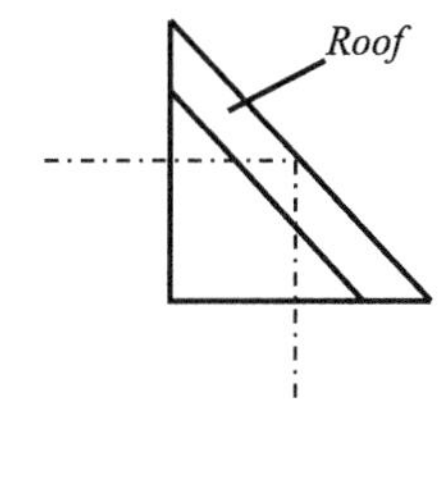

Fig. 1.1.28 Amici prism.

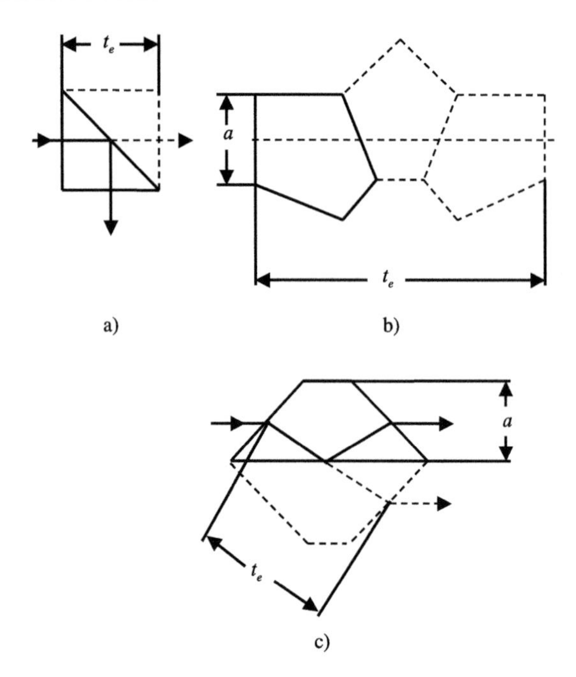

Fig. 1.1.29 Unfolded diagram for (a) the right-angle prism, (b) the penta-prism, and (c) the Dove prism.

(b) penta-prism with the same size a of the entrance face: $t_e = a(2 + \sqrt{2})$;

(c) Dove prism of height a and 45° angles between faces: $t_e = 3.035a$.

Once t_e is known, the apparent thickness in air is calculated from t_e/n.

Problems

P.1.1.14. *Imaging in systems of plane mirrors.* An object AB is positioned as shown in Fig. 1.1.30 in front of a mirror corner of 45°. Find the location of the image beyond the mirrors.

P.1.1.15. Find the reduced (apparent) thickness of a 45° rhomboidal prism of 2 cm face length. The prism is made of BK-7 glass ($n = 1.5163$).

P.1.1.16. Alens L of 30 mm focal length transfers the image of an object AB positioned 40 mm in front of L to a screen P. A penta-prism with 10 mm face size is inserted 20 mm beyond the lens. Find the location of the screen P relative to the prism if it is made of BK-7 glass ($n = 1.5163$).

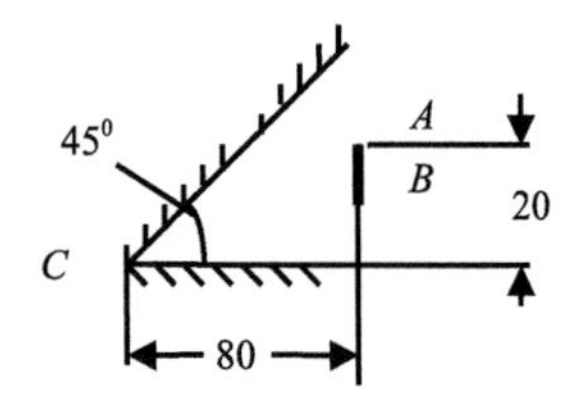

Fig. 1.1.30 Problem P.1.1.14 – Imaging in a mirror corner.

P.1.1.17. *Dispersive prism at minimum deviation.* Find the minimum deviation angle of a prism with vertex angle $\beta = 60°$. The prism is made of SF-5 glass with refractive index $n = 1.6727$.

1.1.5. Solutions to problems

P.1.1.1. We are looking for a solution in the paraxial range and assume the lens is of negligible thickness. To find the image of point A we use two rays emerging from A: ray 1 parallel to the optical axis and ray 2 passing through the center of the lens (Fig. 1.1.31). Ray 1 after passing through the lens goes through the back focus F′. Ray 2 does not change its direction and continues beyond the lens along the incident line. The intersection of the two rays after the lens creates the image A′ of point A. Once the image A′ is found, the image O′ of point O is obtained as the intersection of the normal from point A′ to the optical axis.

It should be noted that instead of ray 1 or 2 one can use ray 3 (dotted line) going through the front focus F in the object space (in front of the lens) and parallel to OX after the lens. Intersection with the two other rays occurs, of course, at the same point A′. Also note that in our approximation of the paraxial range the homocentric beam also remains homocentric in the image space.

P.1.1.1.2. In both cases, Figs. 1.1.32a and b, we draw the ray (dotted line) parallel to AB and passing through the center of the lens. The ray crosses the back focal plane at point C. Since the ray and AB belong to the same parallel oblique bundle and all rays of such a bundle are collected by the lens in a single point of the back focal plane, this must be point C. Therefore, the ray AB after passing through the lens goes from B through C to point A′ at the intersection with the axis. This point is the image of A. In the case of Fig. 1.1.32b the focus F′ and corresponding back focal plane are located to the left of the lens. Hence, not the ray itself but its continuation passes through point C. The intersection with OX is still the image of the point source A which becomes virtual in this case.

P.1.1.3. First we will derive the ray tracing formula valid for the paraxial approximation. By multiplying both

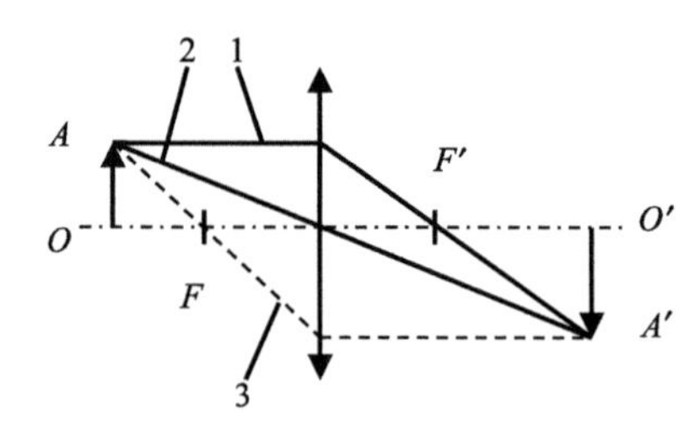

Fig. 1.1.31 Problem P.1.1.1 – Graphical method of finding the image.

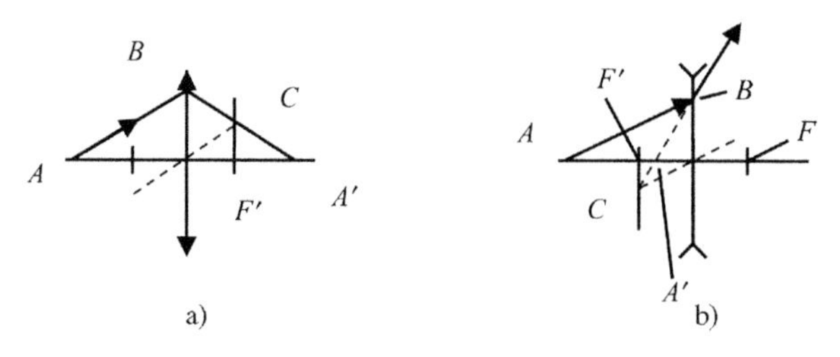

Fig. 1.1.32 Problem P.1.1.2 – Graphical method of finding the image with (a) a positive lens and (b) a negative lens.

sides of Eq. (1.1.4) by h (see Fig. 1.1.33) and denoting $h/S = \tan(u) \approx u$ and $h/S' = \tan(u') \approx u'$ we get

$$u' - u = h\Phi,$$

which yields for a number of lenses ($k = 1,2,\ldots,N$):

$$u_{k+1} = u_k + h_k\Phi_k, \qquad \text{(A)}$$

with the additional geometrical relation

$$h_{k+1} = h_k - u_{k+1}d_{k,k+1}. \qquad \text{(B)}$$

Expressions (A) and (B) enable one to calculate the ray trajectory in a system of thin lenses. To start the calculation we need the values u_1, h_1. Usually these values can be arbitrarily chosen, as they do not affect the final results. Going back to the numerical data of the problem, we choose $u_1 = -0.1$ and then proceed as follows (see Fig. 1.1.34):

$$u_1 = -0.1;\; h_1 = S_1u_1 = (-30)(-0.1) = 3.0$$
$$u_2 = -0.1 + \frac{3.0}{15} = 0.1;$$
$$h_2 = 3.0 - 0.1 \times 40 = -1.0$$
$$u_3 = 0.1 - \frac{1.0}{15} = 0.0333;$$
$$h_3 = -1.0 - 0.333 \times 60 = -3.0$$
$$u_4 = 0.0333 - \frac{3.0}{20} = -0.1167;$$
$$S_3' = \frac{h_3}{u_4} = \frac{-3.0}{-0.1167} = 25.71 \text{ mm}.$$

It can be easily checked that exactly the same result will be obtained if we choose another initial value of u_1 (say, $u_1 = -0.2$). Of course, this results from the linearity of the expressions (A) and (B).

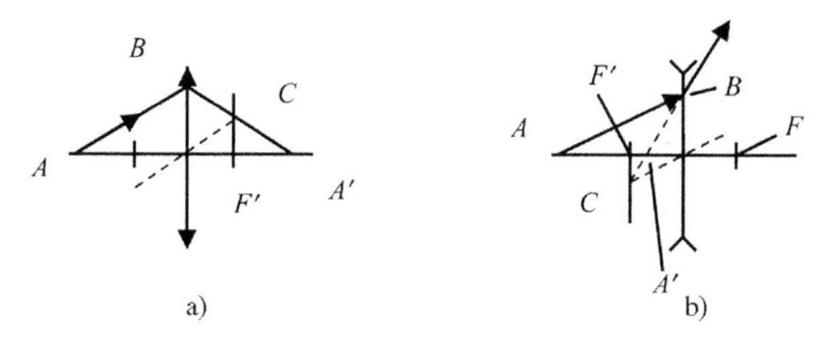

Fig. 1.1.33 Problem P.1.1.3 – Ray tracing through a single lenses.

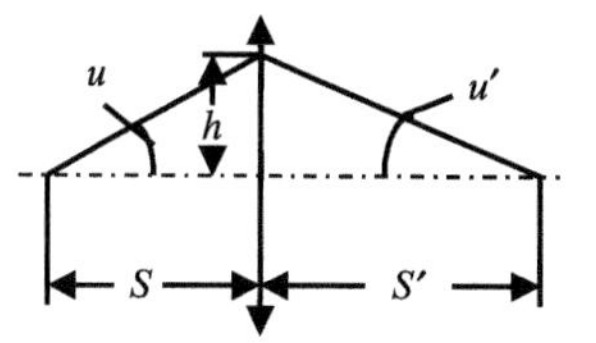

Fig. 1.1.34 Problem P.1.1.3 – Ray tracing through a system of three lenses.

P.1.1.4. Measurement of the optical power of a lens or its focal length is often required in the practice of optical testing. The method described here is particularly useful because it is based only on measurements of linear distance (L) and linear displacement (a) which can be easily and accurately realized.

We start with a derivation of working formulas. Combining Eqs. (1.1.4) and (1.1.5) gives

$$S' = VS;\; \frac{1}{f'} = \frac{1}{VS} - \frac{1}{S} = \frac{1-V}{VS};$$
$$S = f'\frac{1-V}{V};\; S' = f'(1-V).$$

Therefore

$$S' + S = \frac{f'(1-V^2)}{V} = a;\; S' - S = -f'\frac{(1-V)^2}{V} = L$$

(remember that $V < 0$, $S < 0$, and $S' > 0$ in both positions 1 and 2 of the lens). Solving the last equations for V and for f' we get

$$f' = \frac{aV}{1-V^2} = \frac{LV}{(1-V)^2};$$
$$V = -\frac{L+a}{L-a};\; f' = \frac{L^2-a^2}{4L} = \frac{L}{4} - \frac{a^2}{4L}.$$

It is understood that linear magnifications V_1 and V_2 in positions 1 and 2 are reciprocal values ($V_1 = 1/V_2$), and the segments S, S' are just replacing each other while moving from position 1 to 2.

In deriving the above expressions we did not take into account the thickness of the lens, or, more exactly, the distance A between the principal planes. The rigorous relation is $L = S' - S + A$. Actually Δ is unknown and therefore it is the origin of uncertainty in the value of L. Differentiating the above expression for f' with regard to L and denoting $dL = \Delta$ we obtain

$$df = \frac{\Delta}{4}\left(1 + \frac{a^2}{L^2}\right),$$

Now for the numerical data of the problem we have

$$f' = \frac{135}{4} - \frac{45^2}{4 \times 135} = 30.0\ \text{mm}.$$

If the thickness of the lens is about 6 mm then the distance between its principal planes is about 2 mm (approximately one-third of the lens thickness). Hence, for the uncertainty of the focal length we have

$$df = \frac{2}{4}\left[1 + \left(\frac{45}{135}\right)^2\right] = 0.56\ \text{mm}.$$

P.1.1.5. Consider the layout of the thick lens shown in Fig. 1.1.35. We use Eqs. (1.1.7) and (1.1.8) and apply them to two surfaces of the lens. We choose an arbitrary value for h_1 and start with $u_1 = 0$, remembering that in our case $n_1 = n_3 = 1; n_2 = n$. Since we are looking for a solution in the paraxial range, where the heights of all rays are small, one can neglect the segments x_1, x_2 (the latter is not shown in the figure) assuming that the distance between h_1 and h_2 is equal to the lens thickness, t. Then we get

$$u_2 = \frac{1}{n}u_1 + \frac{h_1}{R_1 n}(n-1) = \frac{h_1}{R_1 n}(n-1);$$

$$h_2 = h_1 - u_2 t = h_1 - \frac{h_1 t}{R_1 n}(n-1);$$

$$u_3 = nu_2 + \frac{h_2}{R_2}(1-n) = \frac{h_1}{R_1}(n-1) - \frac{h_1}{R_2}(n-1) + \frac{h_1 t}{R_1 R_2 n}(n-1)^2,$$

which enables one to calculate the focal length and $S_{F'}$ (BFL):

$$\frac{1}{f'} = \frac{u_3}{h_1} = (n-1)\left(\frac{1}{R_1} - \frac{1}{R_2}\right) + \frac{t(n-1)^2}{R_1 R_2 n}$$

$$S_{F'} = \frac{h_2}{u_3} = f'\left[1 - \frac{t(n-1)}{R_1 n}\right].$$

To find the segment SF (FFL) we should repeat the same procedure, but to assume that the ray which is parallel to OX is incident on the surface R_2 of the lens from the right. Then the exit ray intersects the optical axis in the front focus F (left of the lens) and replacing f' by f and R_1 by R_2 in the above expression for $S_{F'}$ we finally get

$$S_F = f\left[1 + \frac{t(n-1)}{R_2 n}\right].$$

Here we should remember that in our problem $f < 0$ and $R_2 < 0$, hence the value of S_F is negative. Using the numerical data of the problem we obtain

$$\begin{aligned}\Phi &= \frac{1}{f'} \\ &= \left[0.5163\left(\frac{1}{50} + \frac{1}{75}\right) - \frac{6(0.5163)^2}{1.5163 \times 50 \times 70}\right] \times 10^3 \\ &= 16.92\ \text{dioptry}\end{aligned}$$

$$\begin{aligned}f' &= \frac{1}{\Phi} = 59.1\,\text{mm};\ S_{F'} = 59.1\left[1 - \frac{6 \times 0.5163}{50 \times 1.5163}\right] \\ &= 56.69\,\text{mm};\ S_F = -59.1\left[1 - \frac{6 \times 0.5163}{75 \times 1.5163}\right] \\ &= -57.49\,\text{mm}.\end{aligned}$$

As we see, the principal planes H and H′ are located 1.61 mm and 2.41 mm, respectively, inside the lens.

P.1.1.6. Consider the ray incident on the slab at a height h_1 along the direction of the angle i (see Fig. 1.1.36). We have $h_2 = h_1 - t \times \tan(r)$ where the refraction angle r is calculated from Eq. (1.1.1). Since the incident angle at point 2 is also r, the refraction angle here (found again from Eq. (1.1.1)) is equal to i, meaning that the exit ray is parallel to the incident one. Then $O'A' = h_2/\tan(i) = h_1/\tan(i) - t\tan(r)/\tan(i)$; $OA = h_1/\tan(i)$; and therefore $AA' = O'A' - (OA - t) = t - t[\tan(r)/\tan(i)]$

$$AA' = t\left[1 - \sqrt{\frac{1 - \sin^2(i)}{n^2 - \sin^2(i)}}\right].$$

As we see, AA' depends on i, which means that each ray of the homocentric incident beam intersects the optical axis after the slab in another point A'. In other words, the homocentricity of the beam is violated. As the measure

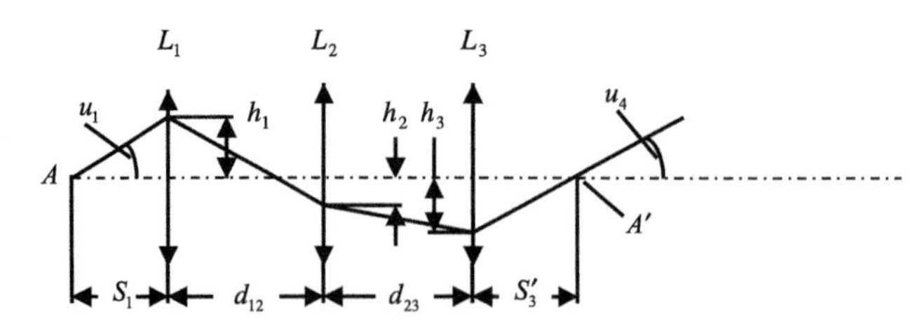

Fig. 1.1.35 Problem P.1.1.5 - Finding the location of the principal planes in a thick lens.

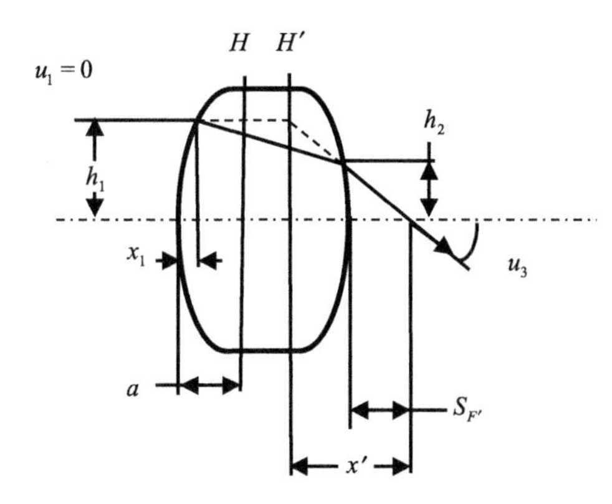

Fig. 1.1.36 Problem P.1.1.6 – Ray tracing through a parallel plate.

of this violation one can choose the value $\delta = (AA')_{i\ max}$. Since i_{max} is related to the given solid angle, ω, as $\omega = 2\pi[1 - \cos(i\ max)]$, we obtain

$$\cos(i_{max}) = 1 - \frac{\omega}{2\pi} = 1 - \frac{1.5}{2\pi} = 0.761;$$

$$\sin(i_{max}) = 0.649;$$

$$\delta = 5\left[1 - \sqrt{\frac{0.761}{2.25 - 0.421}}\right] = 1.775 \text{ mm};$$

$$O'A' = 31.775 \text{ mm}.$$

[Note: The above expression for AA′ is rigorous, it is valid for any angle i. For small angles i (paraxial approximation) we have $\sin i \approx i$; $\sin^2(i)\ 1 \ll n^2$; $AA' \approx t(1 - 1/n)$; and AA′ does not depend on i. If $n = 1.5$ then $AA' = t/3$. This means that in the paraxial approximation the center of the incident beam is just relocated with regard to initial point A by one-third of the glass slab thickness (1.667 mm in our case).]

P.1.1.7. Considering a two-lens system in general, and referring to Fig. 1.1.37 one obtains

$$S_2 = f_1' - d; \quad \frac{1}{S_2'} = \frac{1}{f_2'} + \frac{1}{S_2} = \frac{f_1' - d + f_2'}{f_2'(f_1' - d)};$$

$$S_2' = \frac{1 - \Phi_1 d}{\Phi_1 + \Phi_2 - \Phi_1\Phi_2 d};$$

$$h_2 = h_1\frac{S_2}{f_1'} = h_1(1 - \Phi_1 d);$$

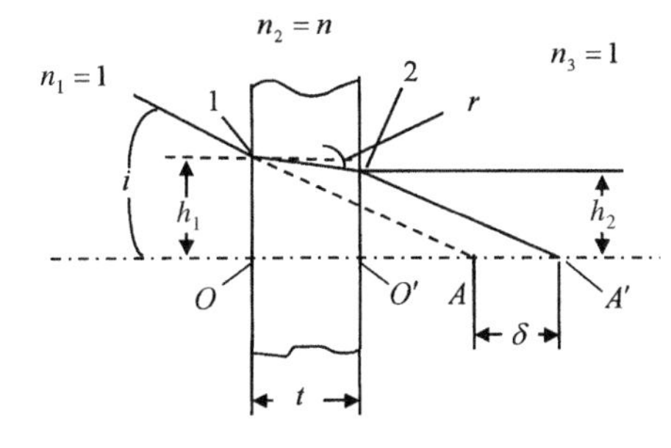

Fig. 1.1.37 Problem P.1.1.7 – Ray tracing through a system of two thin lenses.

$$f_e' = S_2'\frac{h_1}{h_2} = \frac{1 - \Phi_1 d}{\Phi_1 + \Phi_2 - \Phi_1\Phi_2 d} \times \frac{h_1}{h_1(1 - \Phi_1 d)}$$

$$= \frac{1}{\Phi_1 + \Phi_2 - \Phi_1\Phi_2 d} = \frac{1}{\Phi_e}. \qquad \text{(A)}$$

Now, by substituting the problem data in expression (A) we get

$$\Phi_e = \frac{1}{100} + \frac{1}{75} - \frac{30}{100 \times 75}$$

$$= 0.01933 \text{ mm}^{-1} = 19.33 \text{ diopter};$$

$$f_e' = \frac{1}{0.01933} = 51.72 \text{ mm};$$

$$S_2' = \frac{1 - 0.01 \times 30}{0.01933} = 36.2 \text{ mm}.$$

Thus, the two lenses could be considered as a single system with 51.72 mm focal length and the focal point positioned 36.2 mm behind the second component. [Note: Replacing two lenses by a single equivalent lens is useful only if a parallel beam strikes the system. If imaging is performed for an object positioned at a final distance from the first lens then Eq. (A) above becomes useless and calculations should be done according to Eqs. (1.1.4) and (1.1.5), first for the first element and then for the second one.]

P.1.1.8. For a ball lens of radius r the Eqs. (1.1.9) and (1.1.11) are transformed as follows ($d = 2r; r_1 = -r_2$):

$$S_{F'} = f'\left[1 - \frac{2(n-1)}{n}\right] = f'\frac{2-n}{n};$$

$$\frac{1}{f'} = \frac{2(n-1)}{r} - \frac{2(n-1^2)}{rn} = \frac{2(n-1)}{nr} \qquad \text{(A)}$$

and therefore

$$a' = f' - S_{F'} = f'\left(1 - \frac{2-n}{n}\right)$$

$$= \frac{nr}{2(n-1)} \times \frac{2(n-1)}{n} = r. \qquad \text{(B)}$$

Thus, the principal plane H′ is located at the center of the ball. Due to the symmetry of the lens one can state that the front principal plane is located at the same point. From the data of the problem, using the glass data from Appendix 2 ($n_D = 1.67270$), we find

$$S_F = \frac{2 - 1.6727}{2 \times 0.6727}3 = 0.73\,\text{mm};$$

$$f' = \frac{3 \times 1.6727}{2 \times 0.6727} = 3.73.$$

As we see, the focus is distant from the lens surface by 0.73 mm.

P.1.1.9. The angle in air between two chief rays directed to two separate object points still distinguished by the eye is $3 \times 10-^{4}$ rad. Taking into account the "reduced eye" properties, in particular the refractive index of the medium between the eye lens and the retina as $n = 1.336$ and the back focal length as 22.9 mm, we get that the limiting angle in the vitreous is $3 \times 10^{-4}/1.336 = 2.25 \times 10^{-4}$ rad. The corresponding distance between two images on the retina is $2.25 \times 10^{-4} \times 22.9 = 5.15 \times 10^{-3}$ mm and they should fall on two different cells. This means that the retina cell size is about 5 μm.

P.1.1.10. (a) The intermediate image in the branch to the eye is formed in the plane of the reticle M of size 19 mm. As linear magnification of the objective is $V_1 = -10$, it yields

$$S_1' = -10S_1; \quad \frac{1}{-10S_1} - \frac{1}{S_1} = \frac{1}{20}; \quad S_1 = -22 \text{ mm}$$

and this is the working distance of the system. The corresponding field of view is 1.9 mm.

(b) The eyepiece has visual magnification determined from Eq. (1.1.15): $\Gamma = 250/25 = 10$ and therefore the total magnification in the branch to the eye is $\mathbf{V}_{tot} = \mathbf{V}_1\Gamma = 100$.

(c) The optimal imaging in the CCD branch is the one which enables one to see the maximum part of the image created on the circular reticle M. Such a situation shown in Fig. 1.1.38 means that the maximum image size on the CCD is $y' = \sqrt{(4.8^2 + 5.6^2)} = 7.4$ mm and therefore the relay lens L_3 provides an optical magnification $V_3 = 7.4/19 = 0.388$. This can be realized in two possible arrangements. The first is demonstrated in Fig. 1.1.38 and the second in Fig. 1.1.39. In both cases $S'_3 = 20$ mm, but $S_3 = S'_3/V_3 = 20/0.388 = 51.5$ mm in the first case and $S_3 = -51.5$ mm in the second case.

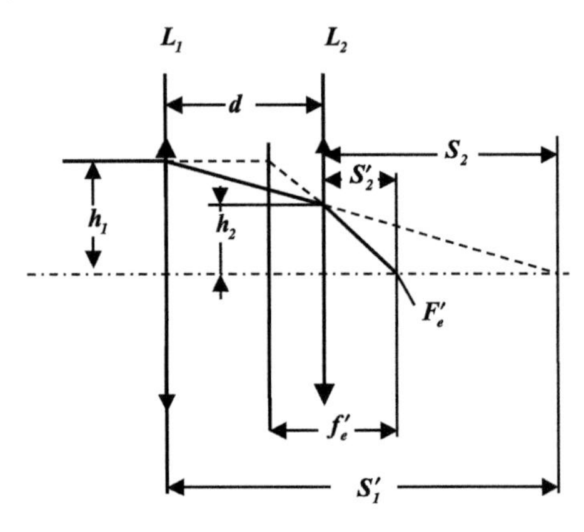

Fig. 1.1.38 Problem P.1.1.10 – Formation of an image onto the CCD plane.

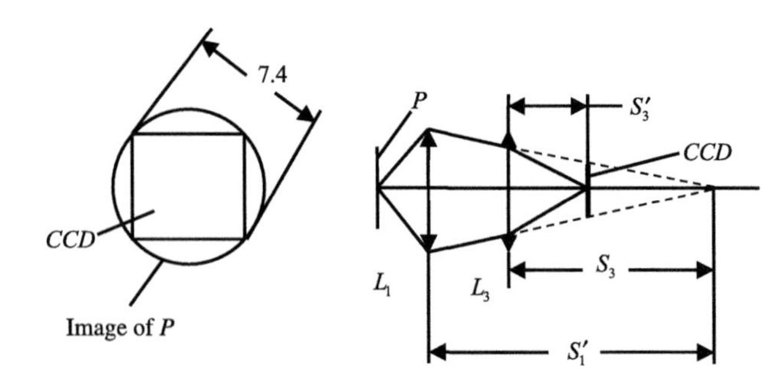

Fig. 1.1.39 Problem P.1.1.10 – Formation of image on CCD, the second case.

Evidently the shortest configuration is that of Fig. 1.1.38. In this case

$$\frac{1}{f_3'} = \frac{1}{20} - \frac{1}{51.5}; \quad f_3' = 32.7 \text{mm},$$

and the distance between P and the CCD is $22 \times 10 - 51.5 = 168.5$ mm. In the second case

$$\frac{1}{f_3'} = \frac{1}{20} - \frac{1}{51.5}; \quad f_3' = 14.4 \text{ mm},$$

and the distance between P and the CCD is $220 + 51.5 = 271.5$ mm.

P.1.1.11. (a) From the numerical data of Fig. 1.1.22 we have $S_1 = -70$; $S'_1 = (-70) \times (-3) = 210$. Therefore, the distance between the CCD and the last beam splitter is $210 - (10 + 100 + 70) = 30$ mm and the focal length of L_1 should be

$$f_2' = \left(\frac{1}{210} + \frac{1}{70}\right)^{-1} = 52.5 \text{ mm}.$$

In the high-magnification branch the image created by lens L_1 (the same size and position as in the low-magnification branch) serves as the virtual object for the second lens, L_2 (negative relay lens). Magnification of L_2 is $V_2 = 2 \times (-3)/V_1 = 2$. Since the distance along the optical axis between L_1 and the CCD is $l = 10 + 70 + 2 \times (10 + 15) + 100 + 30 = 260$ mm and taking into account that $S'_2 - S_2 = l - S'_1 = 50$ mm and $S'_2 = 2 \times S_2$, we get $S_2 = 50$ mm and $S'_2 = 100$ mm and therefore the location of L_2 is 70 mm below the last beam splitter. Its focal length is

$$f_2' = \left(\frac{1}{100} - \frac{1}{50}\right)^{-1} = -100 \text{ mm}.$$

(b) Increasing the high magnification by 10% requires $V_2 = 2.2$ (V_1 remains the same as before). Hence,

$$S_2' = 2.2S_2; \quad \frac{1 - V_2}{V_2 S_2} = \frac{1}{f_2'};$$

which yields $S_2 = 54.54$ mm and $S'_2 = 120$ mm. In other words, the relay lens should be relocated to 90 mm below the last beam splitter. Since in this case $S'_2 - S_2 = 120 - 54.54 = 65.46$ mm it is necessary to add the length 15.46 mm to the optical path of the left branch. This is done by moving down the retroreflector by the segment $\Delta z = 15.46/2 = 7.73$ mm.

P.1.1.12. (a) To get an image with no vignetting it is necessary to position the field stop in the plane of the intermediate image created by the first lens. Referring to Fig. 1.1.40 we get

$$S_1 = -200; \quad S'_1 = \left(\frac{1}{f'_1} + \frac{1}{S_1}\right)^{-1} = 200 \text{ mm};$$
$$V_1 = -1;$$

and therefore the field stop should be of the same size as the field of view, i.e., $d_{ab} = 10$ mm, and positioned 200 mm behind L_1. The total magnification $V = 3 = V_1 \times V_2$ requires $V_2 = -3$, which enables one to find the position of L_2 and M:

$$\frac{1 - V_2}{S_2 V_2} = \frac{1}{f'_2}; \quad S'_2 = 50\frac{1+3}{(-3)} = -66.67 \text{ mm};$$
$$S'_2 = S_2 V_2 = 200 \text{ mm}.$$

(b) The field stop size, as we saw above is 10 mm.

(c) To find the entrance pupil we should build the image of all diaphragms (L_2 and ab in our case) in the object space, e.g., to create their images through L_1 at reverse illumination (as if radiation propagates from right to left). Since ab conjugated with the object plane P, one has to find only the image of L_2 through the first lens. We have:

$$S_{21} = -226.67 \text{ mm}; \quad S'_{21} = \left(\frac{1}{100} - \frac{1}{266.67}\right)^{-1}$$
$$= 160.0 \text{ mm};$$
$$d'_{L_2} = 20 \times \frac{160}{266.67} = 12 \text{ mm}.$$

Calculating the angle of the margin ray coming from the on-axis point of the obje to the side point of the lens L_1 gives $\alpha_1 = 10/200 = 0.05$. The corresponding angle of

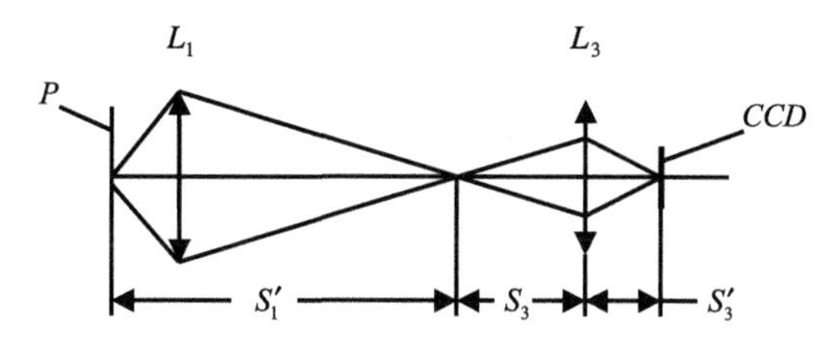

Fig. 1.1.40 Problem P.1.1.12 – Imaging system with the field stop.

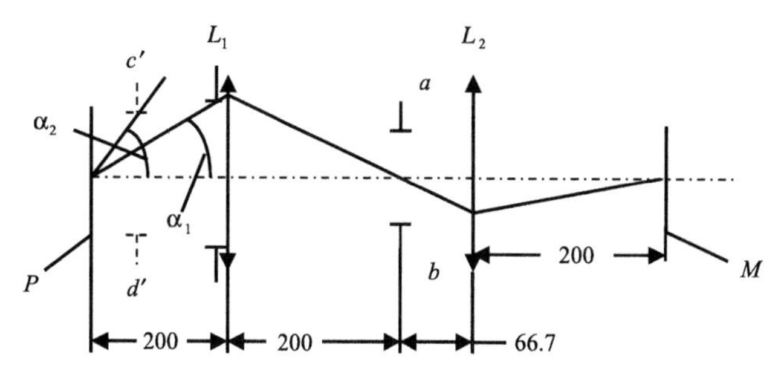

Fig. 1.1.41 Problem P.1.1.13 – The margin ray tracing through lenses L_1 and L_3.

the image of L_2 is $\alpha_2 = 6/(200 - 160) = 0.15 > \alpha_1$ and therefore the entran pupil is the lens L_1 and the aperture angle of the system is $\alpha_{ap} = \alpha_1 = 0.05$.

P.1.1.13. Referring to the solution of Problem P.1.1.10 and Fig. 1.1.41, we first find t size of the lens L_1 using NA = 0.2 and the distance to the object $S_1 = -22$ mm: $D_{L1} = 2 \times 22 \times \tan(\arcsin 0.2) = 9.0$ mm. Then we proceed with the margin ray originating in the off-axis point of object A. This ray comes to the side point A' of the intermediate image (O'A' = 19/2 = 9.5 mm) and it is this ray which determines the active size of lens L3. Geometrical consideration of the figure gives

$$D_{L3} = 2 \times (O'_1N + ND)$$
$$= 2 \times \left[D_{L1}/2 + \frac{168.5}{220}(9.5 - 4.5)\right]$$
$$= 16.66 \text{ mm}.$$

P.1.1.14. To demonstrate the advantage of the unfolded diagram we describe two approaches in solving the problem: first without the diagram and then using unfolding. In the first case we start with imaging through mirror M_1 (see Fig. 1.1.42a) and find the image point A' using the triangle AOiA', where AO1 = (80 − 20) = 60 mm = A'Oi. Obviously the second image point, B', is on the horizontal line passing through A'. Then, referring to A'B' as a new object we find its image in mirror M_2: A″O2 = A'O2 = 80 mm and B″ is again located on the horizontal line passing through A″.

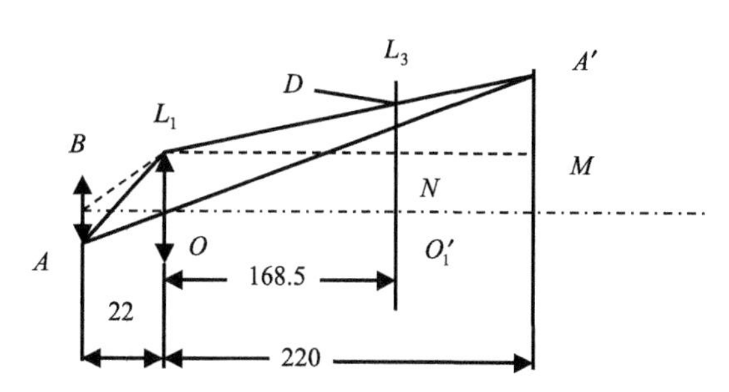

Fig. 1.1.42 Problem P.1.1.14 – Two approaches to finding the image: (a) without unfolded diagram; (b) with unfolding.

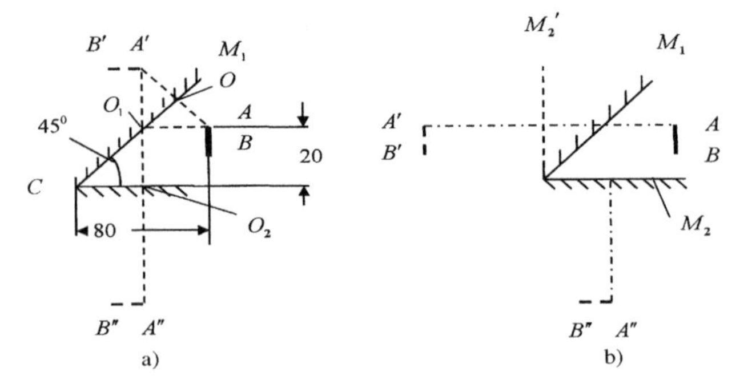

Fig. 1.1.43 Problem P.1.1.15 – Unfolded diagram of a rhomboidal prism.

In the second case (Fig. 1.1.42b) we create the image of the mirror corner in mirror M_1. Then mirror M_2 image, M'_2, is vertical and A′B′ is parallel to AB and distant from M'_2 by 80 mm. Going back to the real mirror M_2 we just put A″B″ beneath M_2 at the same distance 80 mm and 20 mm to the right of the vertex. As we see, the second approach is significantly shorter and easier.

P.1.1.15. We build an unfolded diagram for the prism, as demonstrated in Fig. 1.1.43, and consider the principal ray MNPQ striking the entrance face AB at the height AM = $(a \times \sin 45°)/2$ = 0.707 cm. This is the center of a beam passing through the prism. Obviously MN = AM = PQ = P_1Q_1 = 0.707 cm, $NP_1 = a = 2$ cm, and t_e = MN + NP_1 + P_1Q_1 = 3.414 cm. Hence, the apparent (reduced) thickness is

$$\frac{t_e}{n} = \frac{3.414}{1.5163} = 2.251\,\text{cm}.$$

P.1.1.16. We refer to Fig. 1.1.44 and assume that the lens is working in the paraxial range. Without the prism the distance from the lens to the screen P would be

$$S' = \left(\frac{1}{30} - \frac{1}{40}\right)^{-1} = 120\,\text{mm}.$$

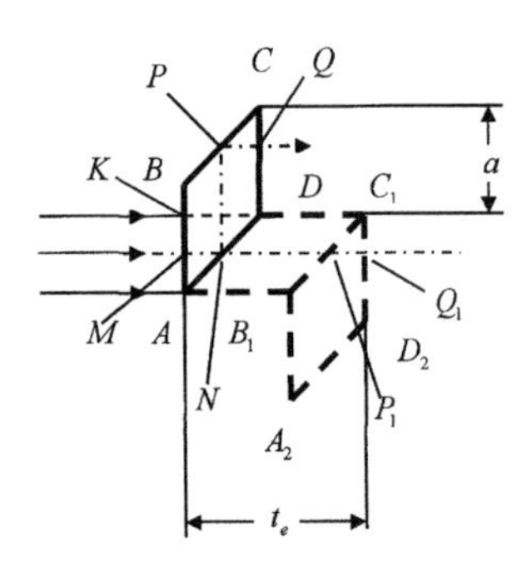

Fig. 1.1.44 Problem P.1.1.16 – Imaging through a penta-prism.

The thickness of the glass block which is equivalent to the prism is $t_e = 3.41a$ = 34.1 mm. The prism makes the ray trajectory longer by the segment $\Delta = t_e(1 - 1/n)$ = 34.1(1 − 1/1.5163) = 11.61 mm. Finally, from the geometry of the figure we get for the distance between the screen P and the exit face of the prism x = 120−20−34.1 + 11.61 =88.39 mm.

P.1.1.17. The prism ABC of refractive index n has a vertex angle β and an input ray strikes the side AB at point O_1 at an incident angle i_1 (see Fig. 1.1.45). The deviation angle φ is defined as the angle between the input direction and the output direction of the ray. Geometrical consideration of the triangles O_1BO_2 and O_1DO_2 yields

$$\varphi = (i_1 - r_1) + (r_2 - i_2);\quad r_1 + i_2 + \gamma = 180° = \beta + \gamma;$$

$$r_1 + i_2 = \beta; \tag{A}$$

$$\varphi = i_1 + r_2 - \beta. \tag{B}$$

Snell's law gives $i_1 = \arcsin(n \sin r_1)$ and $r_2 = \arcsin(n \sin i_2) = \arcsin[n \sin(\beta - r_1)]$. By substituting these expressions in (B) we get

$$\varphi = \text{arc}(n \sin r_1) + \arcsin[n \sin(\beta - r_1)] - \beta. \tag{C}$$

To find the minimum deviation angle we calculate the derivative $d\varphi/dr_1$ and find the angle at which it has a zero value, as usual:

$$\frac{d\varphi}{dr_1} = \frac{n \cos r_1}{\sqrt{1 - n^2 \sin^2 r_1}} - \frac{n \cos(\beta - r_1)}{\sqrt{1 - n^2 \sin^2(\beta - r_1)}} = 0;$$

$$\cos r_1 \sqrt{1 - n^2 \sin^2 \psi} = -\cos \psi \sqrt{1 - n^2 \sin^2 r_1} = 0$$

where the new variable $\psi = \beta - r_1$ is introduced. From the last equation we have

$$\frac{\cos^2 r_1}{\cos^2 \psi} = \frac{1 - n^2 \sin^2 r_1}{1 - n^2 \sin^2 \psi},$$

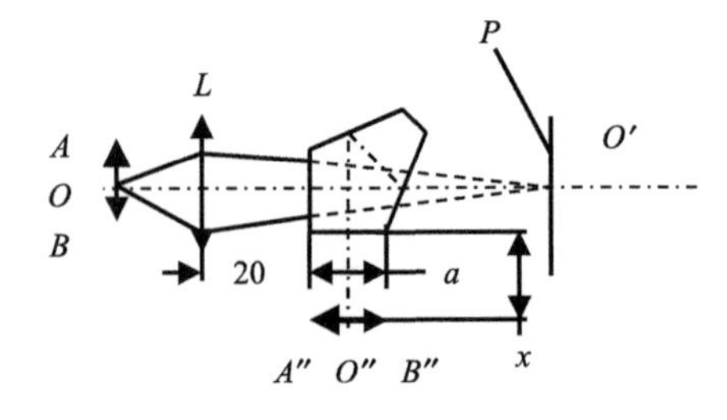

Fig. 1.1.45 Problem P.1.1.17 – Deviation of a ray traveling through a prism ABC.

and denoting $z = \sin^2 r_1$ we proceed as follows:

$$\frac{1-z}{\cos^2\psi} = \frac{1-n^2 z}{1-n^2\sin^2\psi};$$

$$z = \frac{1/\cos^2\psi - 1/(1-n^2\sin^2\psi)}{1/\cos^2\psi - n^2/(1-n^2\sin^2\psi)}$$

$$= \frac{\sin^2\psi \times (1-n^2)}{1-n^2} = \sin^2\psi.$$

The last equation is satisfied if $r_1 = \psi$ and therefore $r_1 = \beta - r_1$; and $r_1 = \beta/2$. With this value we have from (C):

$$\varphi = 2\arcsin\left(n \sin\frac{\beta}{2}\right) - \beta \qquad \text{(D)}$$

and

$$i_1 = \arcsin\left(n \sin\frac{\beta}{2}\right). \qquad \text{(E)}$$

The last two expressions allow one to calculate the angle of minimum deviation of the prism and the incidence angle corresponding to such a deviation. Going back to the problem, we find $\varphi = 2\arcsin(1.6727 \times \sin 30°) - 60° = 53.51°$ and the incidence angle $i_1 = \arcsin(1.6727 \times \sin 30°) = 56.76°$.

Chapter 1.2

1.2

Theory of imaging

Menn

1.2.1 Optical aberrations

1.2.1.1 General consideration. Ray fan and aberration plot. Concept of wave aberrations

We will proceed by considering the concept of imaging as described in Section 1.1.2 of Chapter 1.1 and pay most attention to the real imaging situation experienced in practice. Fig. 1.2.1 demonstrates the basic difference between ideal imaging and real imaging. Let the rays originating in a point source A come to the system, each one at a different angle u. If the medium is homogeneous (has the same refractive index everywhere) the wavefront W in the object space is a sphere. If in the image space *all rays* intersect at a single point A′ then the beam remains homo-centric, with a spherical wavefront W′, and A′ is a stigmatic (ideal) image of A. However, in most situations this does not happen and the rays of different angles u' come to different points on the axis OO′ (or, for tilted beams, to different off-axis locations). As a result, the real wavefront in the image space is not spherical, the homocentricity of the output beam is violated, and instead of a sharp point image there is a blurred spot. Such violation of stigmatic imaging is defined as optical aberrations.

Numerically aberrations are characterized by the deviation of a real image A′ from the ideal image A'_0 obtained in the paraxial range. This deviation can be determined either by the horizontal segment, $\delta s'$, along the optical axis, as in Fig. 1.2.2 (and then it is called the lateral aberration) or it can be related to the vertical segment ρ (then it is called the transverse aberration). The geometrical relation between lateral and transverse aberrations is quite obvious:

$$\rho = \delta s' \times \tan u \approx \delta s' \frac{h}{S'}, \qquad (1.2.1)$$

in which the fact is taken into account that $\delta s' \ll S'$. Since for each ray aberrations depend on the height of the ray

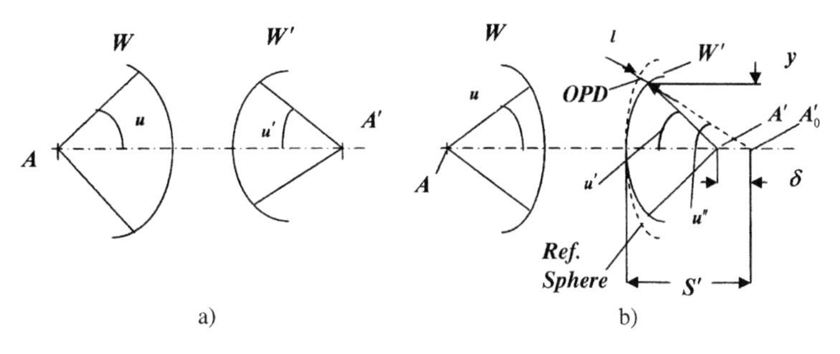

Fig. 1.2.1 (a) Ideal imaging and (b) real imaging.

Practical Optics; ISBN: 9780750684507

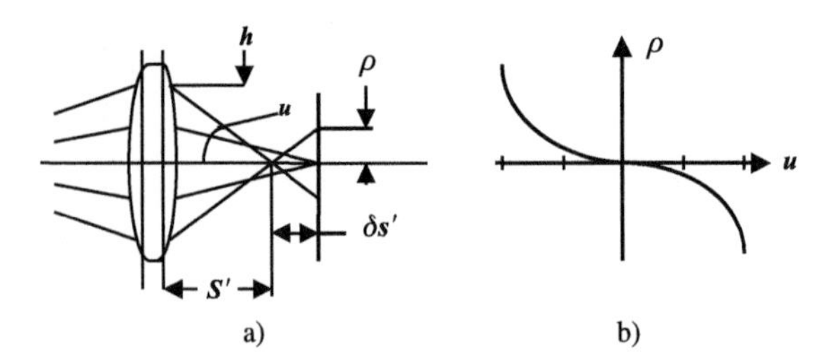

Fig. 1.2.2 (a) Lateral and transverse aberration and (b) the aberration diagram.

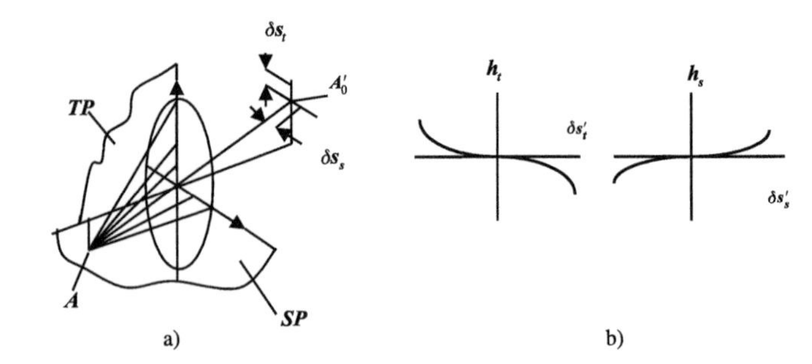

Fig. 1.2.3 (a) The fan of rays in the entrance pupil and (b) the meridional and saggital plots.

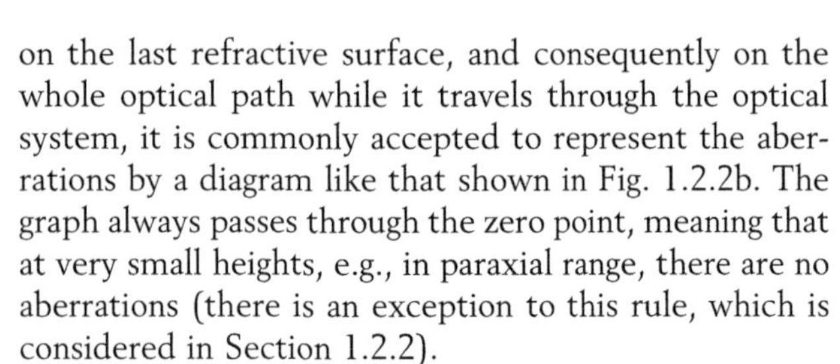

on the last refractive surface, and consequently on the whole optical path while it travels through the optical system, it is commonly accepted to represent the aberrations by a diagram like that shown in Fig. 1.2.2b. The graph always passes through the zero point, meaning that at very small heights, e.g., in paraxial range, there are no aberrations (there is an exception to this rule, which is considered in Section 1.2.2).

There is a great variety of reasons why aberrations happen in optical systems. Some of them are relevant in a specific application whereas some others are not. It was understood at a very early stage of the development of aberration theory that it is worth classifying aberrations in three groups and consider each one separately. These groups are:

(a) chromatic aberrations – chromaticity of location (the only aberration existing also in the paraxial area) and chromaticity of magnification;

(b) monochromatic aberrations of wide beams (spherical aberration and coma);

(c) field aberrations or monochromatic aberrations of tilted beams (astigmatism, field curvature, and distortion).

We will address each group in following sections of this chapter.

To characterize the image quality it is not enough to consider aberrations of several rays coming from an on-axis point. It is necessary to analyze a great number of rays coming from on-axis as well as from off-axis points of the object and to do this for three wavelengths at least if the system is intended for imaging with white light. Usually the ray tracing analysis is carried out for rays propagating in the vertical plane passing through the optical axis (this plane is called the tangential or meridional plane) and for rays propagating in a tilted plane where an off-axis point of the object and horizontal diameter of the entrance pupil are located (this is called the saggital plane). More specifically (see Fig. 1.2.3), a number of points on the vertical and horizontal diameters of the entrance pupil are chosen and the meridional fan of rays (all in the plane TP) and the saggital fan of rays (all in the tilted plane SP) are analyzed aiming at the location of the final destination of each ray in the image plane. Then the meridional plot and the saggital plot, like the two graphs shown in Fig. 1.2.3b, are created followed by the calculation, if necessary, of some other integral parameters of the image (like spot diagrams, energy distribution, vignetting rate, modulation transfer function, etc.).

With regard to the integral characteristics of imaging, one more issue should be considered here. Image blurring can be characterized not only in terms of the geometric parameters of the rays but also in terms of wavefront distortion or wave aberrations (also termed optical path differences, OPDs). Referring to Fig. 1.2.1b, consider the real wavefront W' and the virtual reference sphere (dotted line) of radius S' centered at the point A'_0. The distance l between W' and the reference sphere along the radius passing through A'_0 and tilted to the axis at an angle u'' is called the wave aberration, or OPD. The difference between the angles u' and u'' is small, so that the local wave aberration in terms of lateral aberration can be expressed as

$$l = n \times \delta s' \times (1 - \cos u'), \tag{1.2.2}$$

and the overall (cumulated) wave aberration is defined by the integral

$$l = n \int_0^u \delta s' \times \sin u' du'. \tag{1.2.3}$$

The OPD value can be calculated if aberrations $\delta s'$ are known for all angles from 0 to u. The expression for lateral aberrations in terms of wave aberration is of great importance and allows one to obtain analytical expressions for lateral and transverse ray aberrations in a closed form as far as a third-order approximation is considered (Seidel's formula, discussed in following sections of this chapter).

There exists an important Rayleigh's criterion of acceptable degradation due to aberrations: the image quality is acceptable if the wave aberration l is not greater than 0.25λ.

Problems

P.1.2.1. A lens L of 10 mm diameter and 100 mm focal length working in the paraxial range builds a sharp image at magnification $V = -2$ in the plane P where the observation screen is located. If L is replaced by another lens of the same nominal focus but manufactured with 5% tolerance, what blurring could be expected on the screen?

[Note: Calculate the meridional plot of rays in the plane P.]

P.1.2.2. A lens of 40 mm size designed to form an image at a distance of 125 mm in air was used in a laboratory set-up where the optical axis was turned through 90° by a penta-prism of 30 mm entrance face positioned 35 mm behind the lens. Assuming the prism is made of BK-7 glass ($n = 1.5163$) find the meridional ray plot of the additional aberration introduced by the prism.

1.2.1.2 Chromatic aberrations: principles of achromatic lens design

As we mentioned earlier, chromaticity caused by chromatic aberration of location is the only aberration (except defocusing) experienced even in the paraxial range. The origin of chromaticity is in the dispersion of light inside optical elements (made of glass or crystals).

It is well known that the refractive index of optical glasses varies with wavelength and its spectral behavior can be approximately described by the formula

$$n(\lambda) = A + \frac{B}{(C-\lambda)^2}, \tag{1.2.4}$$

where A, B, and C are constants characterizing a specific material. Usually the refractive index is considered for three main wavelengths, $\lambda_D = 0.589$ µm; $\lambda_F = 0.486$ µm, and $\lambda_C = 0.656$ µm, and the corresponding values n_D, n_F, and n_C are also included in the parameter of dispersion called the Abbe number (or the Abbe value):

$$\nu_D = \frac{n_D - 1}{n_F - n_C}. \tag{1.2.5}$$

Selected data for several optical glasses are presented in Appendix 2.

Since the focal length of a lens is directly related to its refractive index by Eq. (1.2.11), it is quite understandable that if the lens is operated simultaneously at several wavelengths (or with white light illumination) significant chromatic aberration might occur in the system. Usually chromatic aberration is defined as the difference between the focal length at the selected wavelength relative to that of line D:

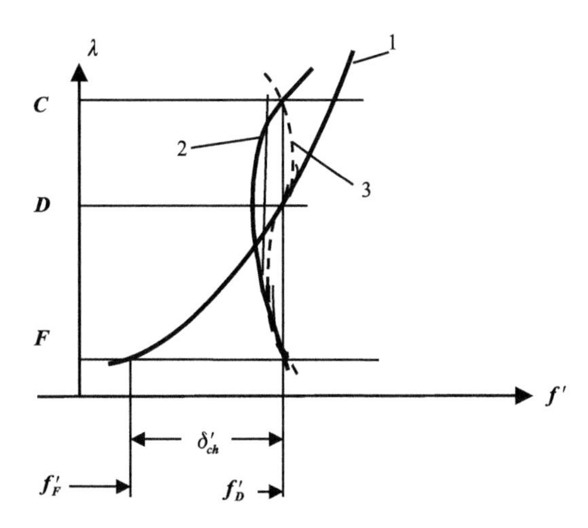

Fig. 1.2.4 Chromaticity of a single lens.

$$\delta_{Ch} = f'_\lambda - f'_D. \tag{1.2.6}$$

In more general cases δ_{Ch} is related to the distances between the lens and the image, $\delta_{Ch} = S'_\lambda - S'_D$, and apparently varies with the magnification of the system.

Chromatic aberration of a single lens is demonstrated in Fig. 1.2.4 (curve 1) and explained by the ray diagrams of Fig. 1.2.5 separately for positive and negative lenses. As can be seen, the aberration plots in these two cases are opposite and this fact is widely exploited for the correction of chromaticity. The lens is divided in two components, one positive and one negative, which are designed according to the rules described below and then brought in contact and cemented in a single element called a doublet lens, or achromat.

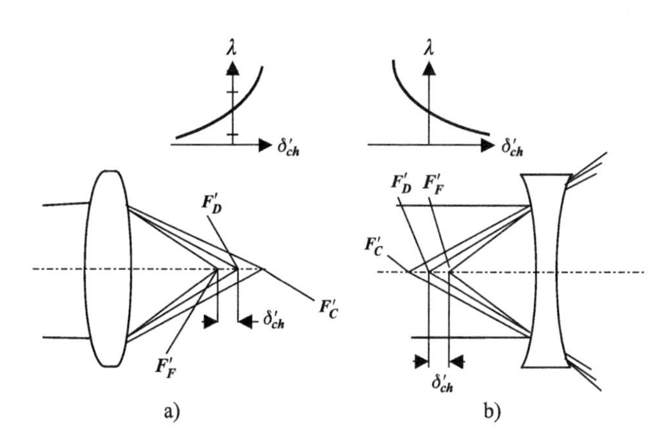

Fig. 1.2.5 Chromatic aberration of (a) positive and (b) negative lenses.

From the variety of available optical glasses we choose two different materials –one for the positive component (Abbe value ν_{D1}) and another for the negative part (with Abbe value ν_{D2}). Neglecting the thickness of the components we have for the total optical power, Φ, of the achromat

$$\Phi = \Phi_1 + \Phi_2. \tag{1.2.7}$$

Each component obeys the single lens equation (1.1.11) which we rewrite in the form

$$\Phi_k = (n-1)c_k, \tag{1.2.8}$$

where $c_k = (1/r_{k1} - 1/r_{k2})$ is the bending parameter independent of wavelength. Considering the variation of the optical power as the wavelength is changed from λ_F to λ_C, we have

$$d\Phi_k = \frac{\Phi_k}{\nu_{Dk}}. \tag{1.2.9}$$

Although $d\Phi_1$ and $d\Phi_2$ both have finite values, we require that the variation of the total optical power be zero: $d\Phi = d\Phi_1 + d\Phi_2 = 0$, which yields the following equation:

$$\frac{\Phi_1}{\nu_{D1}} = -\frac{\Phi_2}{\nu_{D2}}. \tag{1.2.10}$$

Resolving Eq. (1.2.10) together with Eq. (1.2.7) gives the following formula for the optical power of both components of the achromat:

$$\Phi_1 = \frac{\nu_{D1}}{\nu_{D1} - \nu_{D2}}\Phi; \qquad \Phi_2 = -\frac{\nu_{D2}}{\nu_{D1} - \nu_{D2}}\Phi. \tag{1.2.11}$$

To complete the design of the achromat we have to find the bending parameters. Since we have only two conditions (1.2.11) for four independent radii, r_1, r_2, r_3, and r_4, there are two degrees of freedom here. One degree can be reduced if we require that the contacting surfaces of both lenses have the same shape, e.g., $r_2 = r_3$. An additional degree of freedom is one of the two remaining radii. Indeed, we can arbitrarily choose one of them (e.g., $r_4 = \infty$) and then complete the design in the following manner:

$$r_3 = \frac{n_{D2} - 1}{\Phi_2} = r_2; \quad r_1 = \frac{n_{D1} - 1}{\Phi_1}.$$

Or, choose the first surface of the positive lens to be plane and then calculate the rest of the shapes. Several possible forms of doublet lens are shown in Fig. 1.2.6. All of them, however, represent a cemented pair (the adhesive used is of a refractive index very close to that of glass).

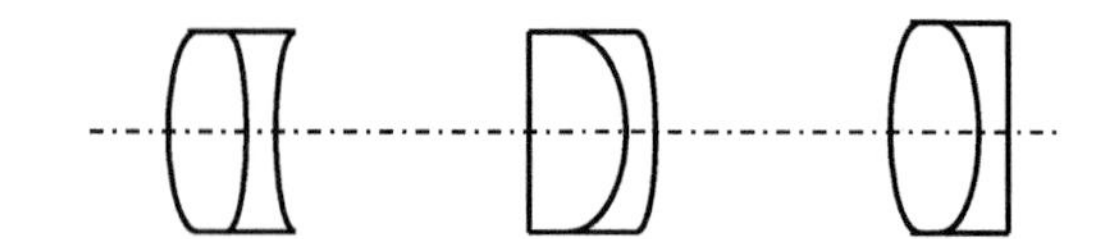

Fig. 1.2.6 Doublet lenses of different shapes.

There also exists the possibility of designing an achromatic lens with an air spacing between the components (e.g., see the detailed explanation in Kingslake, 1979). In any case the achromatic lens provides two focuses to coincide, F_F and F_C. The remaining difference between these two and the focus of the line D is called the secondary spectrum (it is shown by curve 2 in Fig. 1.2.4). In some situations the residual chromatic aberration of the doublet lens is too large and further correction is required. This is realized in triplet lenses or in more complex configurations. The remaining chromatism is called the tertiary spectrum (curve 3 in Fig. 1.2.4). It can be seen that the three focuses coincide in such a case and the residual aberration is very small.

Problems

P.1.2.3. Find the chromatic aberration introduced by the penta-prism in Problem P.1.2.2 and build the aberration plot.

P.1.2.4. Find the doublet lens of 13.33 diopters optical power if the components are made from BK-7 and F-1 glasses and calculate the residual chromatic aberration (the secondary spectrum).

1.2.1.3 Spherical aberration and coma

These two types of aberrations are monochromatic aberrations of a wide beam. Consider first the spherical aberration (see Fig. 1.2.7). Due to the geometry of a spherical shape the rays originating in an on-axis point A and incident on the lens at different distances h from the optical axis are not concentrated in a single point behind the lens, but create images at separate locations (A'_1, A'_2, A'_3, etc.). The lateral spherical aberration $\delta s'_{Sph}(h_i)$ is defined as the distance between the image in the paraxial range (A'_1) and the image corresponding to the height h_i (e.g., the point A'_1). The corresponding transverse spherical aberration $\delta y'_{Sph}$ defines the size of the light spot created in the plane perpendicular to the axis (e.g., on an observation screen). At any position along the axis the spot on the screen has a finite size, but at some location the size is a minimum and this is the point of the best imaging, as far as spherical aberration is concerned.

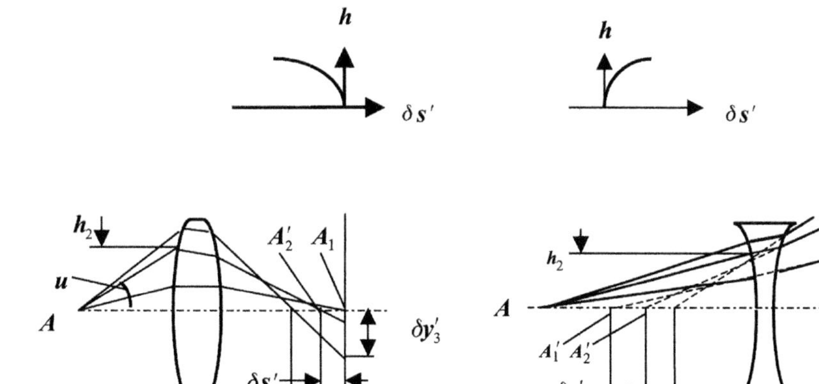

Fig. 1.2.7 Spherical aberration of (a) a single positive lens and (b) a single negative lens.

To describe the spherical aberration analytically it is usually expanded in a power series

$$\delta s'_{\mathrm{Sph}}(h) = a_3 h^2 + a_5 h^4 + \cdots \tag{1.2.12}$$

(it can be easily shown that the terms with coefficients a_0 and a_1 are equal to zero). If only the first term of Eq. (1.2.12) is considered then the solution can be derived in a closed form. Such an approximation is called a third-order aberration and it is commonly known as Seidel's formula. We describe it here as follows (for further details, see Born and Wolf, 1968):

$$\delta s'_{\mathrm{Sph}} = -\frac{1}{2}\frac{h^2 S'^2}{(1-\xi)^2 f'}\left[A + B\frac{1-\xi}{r_1} + (1+2\xi)\left(\frac{1-\xi}{r_1}\right)^2\right], \tag{1.2.13}$$

where

$$A = \frac{3}{S'}C_1 + \frac{1}{f'^2} - \xi(2+\xi)C_2 C_1 + \xi^2(1+2\xi)C_1^2$$

$$B = 2\xi(1+2\xi)C_1 - (2+\xi)C_2$$

$$C_1 = \frac{1}{S'} - \frac{1}{f'}; \quad C_2 = \frac{2}{S'} - \frac{1}{f'}; \quad \xi = \frac{1}{n}.$$

Expression (1.2.13) can be used to estimate the spherical aberration at any position of the object and the image. For the special case of the object in infinity, $S' = f'$ and Eq. (1.2.13) is transformed to

$$\delta s'_{\mathrm{Sph}} = -\frac{1}{2}\frac{h^2 f'}{(1-\xi)^2}\left[\frac{1}{f'2} - \frac{(2+\xi)}{f'}\frac{(1-\xi)}{r_1} + (1+2\xi)\left(\frac{1-\xi}{r_1}\right)^2\right]. \tag{1.2.14}$$

As the aberration value depends explicitly on the shape of the lens (radius r_1), one might minimize aberration by optimizing the shape. $\delta s'_{\mathrm{Sph}}$ achieves its minimum value

$$(\delta s'_{\mathrm{Sph}})_{\min} = -\frac{1}{8}\frac{\xi}{(1-\xi)^2}\frac{(4-\xi)}{(1+2\xi)}\frac{h^2}{f'}, \tag{1.2.15}$$

when its radii obey the relations:

$$r_1 = 2(1-\xi)\frac{1+2\xi}{2+\xi}f'; \quad r_2 = 2\frac{(1-\xi)(1+2\xi)}{2-\xi-4\xi^2}f'. \tag{1.2.16}$$

For instance, assuming $n = 1.5$ and keeping in mind that $h_{\max} = D/2$ we get from Eqs. (1.2.15) and (1.2.16)

$$\delta s'_{\mathrm{Sph}} = -0.268\frac{D^2}{f'}; \quad r_1 = \frac{7}{12}f'; \quad r_2 = -3.5f'. \tag{1.2.17}$$

One should remember that in the aberration blur the radiation energy is not equally distributed. For this reason half the size of the maximum spot caused by aberration and calculated from Eqs. (1.2.13)–(1.2.17) is exploited as an aberration measure. It should also be mentioned again that the above formulas enable one to estimate the spherical aberration of a single lens approximately. To get more rigorous results the ray tracing procedure is inevitably required.

As can be seen from Fig. 1.2.7, the lateral spherical aberrations of a positive lens are negative whereas the aberrations of a negative lens have the opposite sign. This fact allows one to reduce drastically the spherical aberration if the single lens is replaced by a doublet (like the achromat described in Section 1.2.1.2).

Coma is an aberration of a wide tilted beam originating in an off-axis point of the object. This aberration is caused by the fact that the magnification of the system is not constant, but varies with the height of the incident ray: $V = F(h)$.

Figure 1.2.8 demonstrates the formation of coma and explains the parameter, δk, chosen as its numerical measure:

$$\delta k = \frac{1}{2}(y'_1 + y'_2) - y'_C, \qquad (1.2.18)$$

where y'_C is the vertical coordinate of the chief ray of the beam at the image plane P and y'_1 and y'_2 are the vertical coordinates of the upper and the lower rays 1 and 2 on the same plane. The ray bundle starting in the off-axis object point A and coming to the entrance pupil is not symmetrical with regard to the optical axis, so it is not surprising that the spot in the plane P is also not symmetrical. The conditions and methods of coma correction are discussed later in this chapter.

Problems

P.1.2.5. (a) Find the optimal shape of a lens of 30 mm diameter and $f\# = 2.0$ (f-number, $f\#$, defined as the ratio f'/D of a lens focus to its diameter) intended for imaging from infinity if it is made of (i) BK-7 glass and (ii) SF-11 glass, and calculate the maximum transverse aberration in both cases (for imaging in monochromatic light of wavelength D). (b) How will the results be changed if the lens is turned by 180°?

P.1.2.6. *Spherical aberration of a cylinder rod or a sphere: a rigorous ray tracing.*

(a) Calculate the plot of transverse spherical aberration of a cylinder rod of 7 mm diameter made of BK-7 glass working with a point light source (laser diode of 0.59 wavelength) located 2 mm in front of the rod. (b) How will the results of the calculation be affected if the rod is replaced by a lens having the shape of a full sphere of 7 mm diameter (a ball lens)?

P.1.2.7. *Spherical aberration of a plano-convex cylindrical lens: a rigorous ray tracing.* How will the plot of spherical aberration calculated in Problem P.1.2.6 be changed if the cylinder rod is replaced by a plano-convex cylindrical lens of the same radius (3.5 mm) made of BK-7 glass? The plane P remains at the same location as in Problem P.1.2.6. The size of the new lens is shown in Fig. 1.2.9.

1.2.1.4 Aberrations of tilted beams (field aberrations)

This group of aberrations includes astigmatism, curvature of field, and distortion.

Astigmatism

This aberration occurs if a pencil of tilted rays originating in an off-axis point of the object strikes the entrance pupil of the system. Astigmatism is illustrated in Fig. 1.2.10. For a tilted beam (which is initially homocentric) the optical axis is not an axis of symmetry any more and the behavior of the rays in the meridional plane (rays 1 and 2) differs from that of the saggital rays (rays 3 and 4). As a result the lens concentrates the tangential rays and the saggital rays in two different points, A'_t and A'_s. Both are out of the plane P of the paraxial image (point A'_0). Aberration of astigmatism is measured as the distance between the meridional and saggital images originating in the same point of the object (in Fig. 1.2.10a

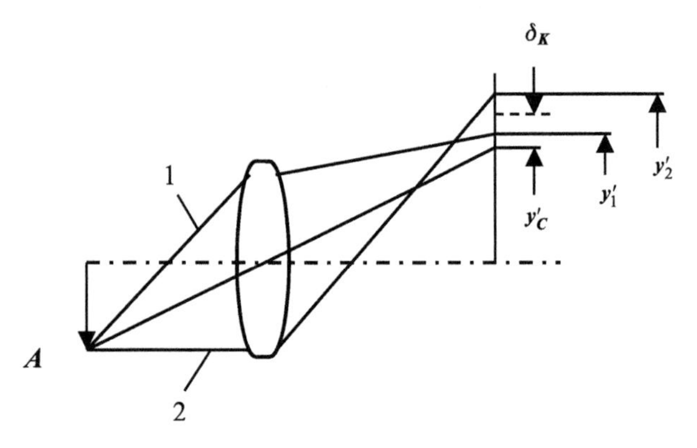

Fig. 1.2.8 Formation of coma.

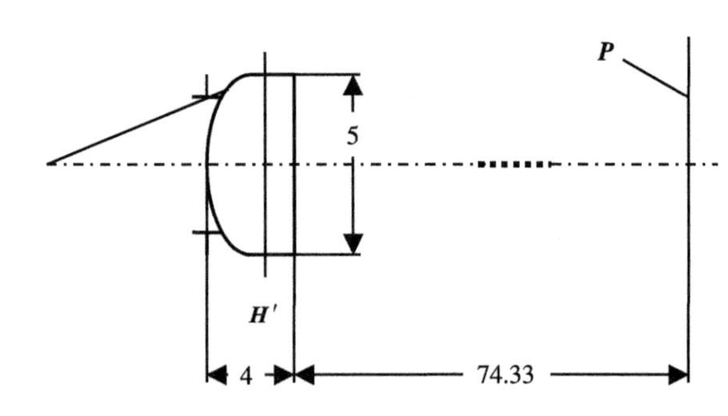

Fig. 1.2.9 Problem P.1.2.7 – Plano-convex cylindrical lens and the image plane.

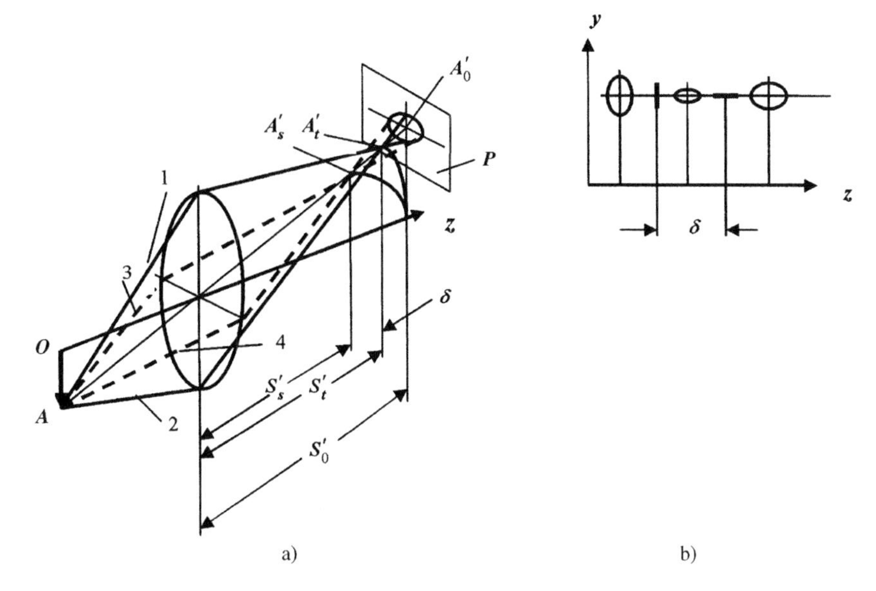

Fig. 1.2.10 Astigmatism of a single lens: (a) imaging by meridional and saggital rays; (b) cross-section of the light spots along the optical axis.

$\delta s'_{\text{Ast}} = S'_t - S'_S$). Obviously the greater the height of point A the larger the difference $S'_t - S'_S$, and for the on-axis point O aberration of astigmatism is approaching zero. The cross-section of the light bundle behind the lens is not homocentric anywhere but varies in a manner demonstrated in Fig. 1.2.10b.

Astigmatic aberration appears not only in elements with optical power (lenses or mirrors), but features also in a parallel plate. In this case the aberration can be described analytically. Referring to Fig. 1.2.11, we consider the tangential and saggital images A'_t and A'_s of a point A having a (virtual) image A'_0 (e.g., the image that would be created in air, without a parallel plate of thickness d). The distance a between the two images is determined by the formula

$$a = \left(1 - \frac{\cos^2 u}{\cos^2 u'}\right)\frac{d}{n \cos u'}, \tag{1.2.19}$$

and the astigmatic aberration becomes

$$\delta s'_{\text{Ast}} = \delta = (n^2 - 1)\frac{\tan^3 u'}{\tan u} d \approx \frac{(n^2 - 1)}{n^3} u^2 d. \tag{1.2.20}$$

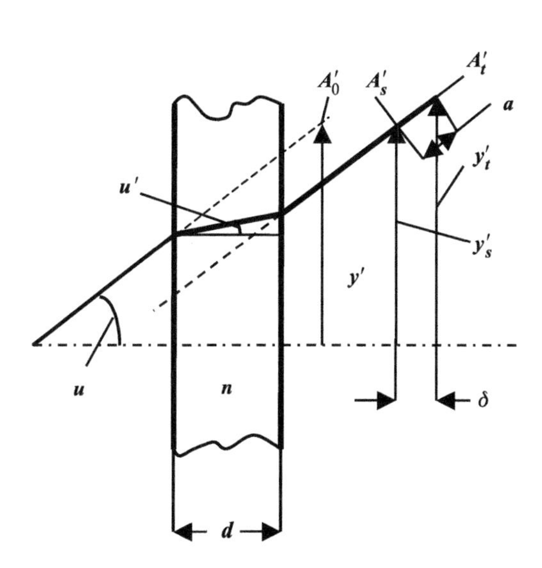

Fig. 1.2.11 Astigmatism in a parallel plate.

Curvature of field

Going back to the astigmatism of a lens-based system as shown in Fig. 1.2.10, one may note the fact that both the tangential and saggital images are not segments of straight lines but rather have noticeable curvature. Furthermore, it is reasonable to assume that the image created simultaneously by tangential as well as by saggital rays is located on a curved surface passing somewhere between the meridional and saggital images, as depicted in Fig. 1.2.12. This is commonly defined as an additional aberration called the curvature of field and is estimated as the radius of curvature, ρ, of the best image. It can be shown that the value of ρ obeys the following expression (Petzval's theorem).

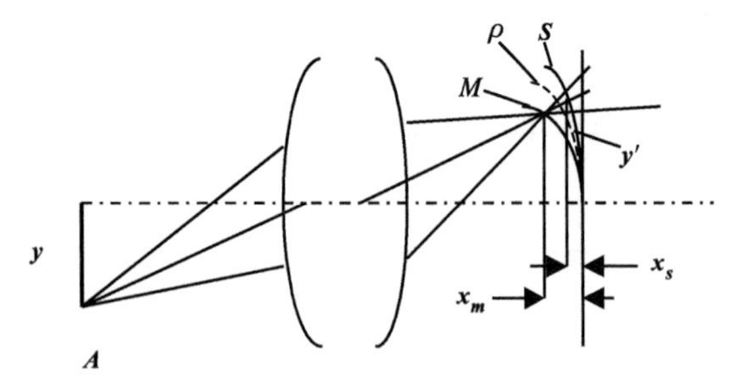

Fig. 1.2.12 Occurrence of curvature of field.

$$\frac{1}{\rho} = -n' \sum_i \frac{1}{r_i} \left(\frac{1}{n} - \frac{1}{n_{i-1}} \right), \qquad (1.2.21)$$

where n' is the refractive index in the image space and the summation is carried out over all refraction surfaces of the system.

Distortion

It is assumed in paraxial optics that linear magnification between two conjugate planes is defined solely by the location of the planes along the optical axis (in other words, by the distance of the object to the lens). In reality this assumption is violated and linear magnification, V, depends not only on the location of the plane along OZ, but also on the distance of the point of interest from the optical axis (distance in the radial direction). Violation of the above condition causes distortion of images, as illustrated in Fig. 1.2.13. The object shown is a regular square of size a with its center O positioned on the optical axis. Since the radial distances from O to points A and B are different, $a/2$ and $a/\sqrt{2}$, respectively, their images A′ and B′ are determined by different magnifications, V_A and V_B, and the whole image of the square is deformed. Distortion is characterized numerically as follows:

$$\Delta = \frac{y' - y'_0}{y'_0} 100\%, \qquad (1.2.22)$$

where y'_0 and y' are the radial displacement (or height) of the paraxial image and the real image of the same point. The value defined by Eq. (1.2.22) is sometimes called the fractional distortion.

Two kinds of distortion can be experienced in imaging systems, positive and negative. In the first case linear magnification in the image plane is increased with radial distance. In the second case the larger the distance the lower the magnification. Both cases are shown in Fig. 1.2.13b.

Distortion might originate not only in lenses or mirrors, but also in prisms or parallel glass plates. Such a case is shown in Fig. 1.2.14 where the off-axis image, y', created by the system is transferred by the parallel plate of thickness d into the final image y''. Distortion can be expressed in terms of the thickness and refractive index of the plate and the skew angle u and the distance p from the entrance pupil to the image plane:

$$\Delta = -\frac{n^2 - 1}{2n^3} \frac{d}{p} u^2. \qquad (1.2.23)$$

Aberration of distortion might be very critical in some applications, for example in optical systems for mapping.

Problems

P.1.2.8. A lens of 30 mm focal length operates in an angular field of view of ±30° and creates an image at magnification $V = -2$. Behind the lens, at a distance of

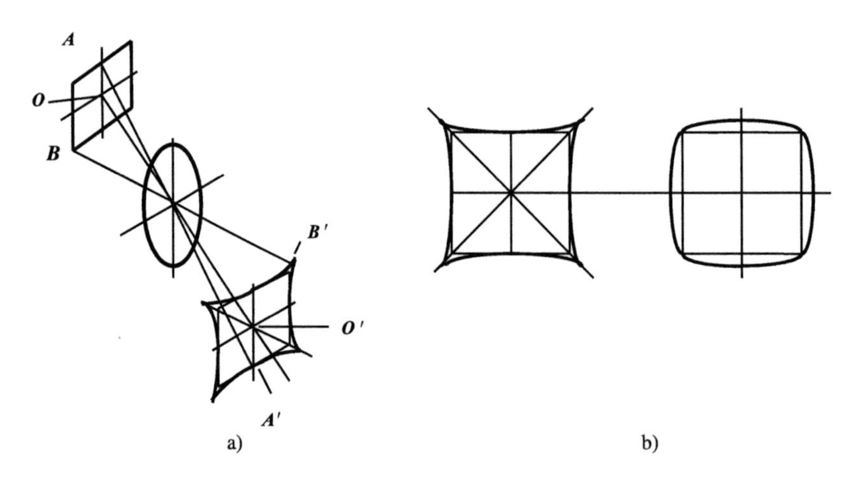

Fig. 1.2.13 (a) Distortion and (b) positive and negative distortion of images.

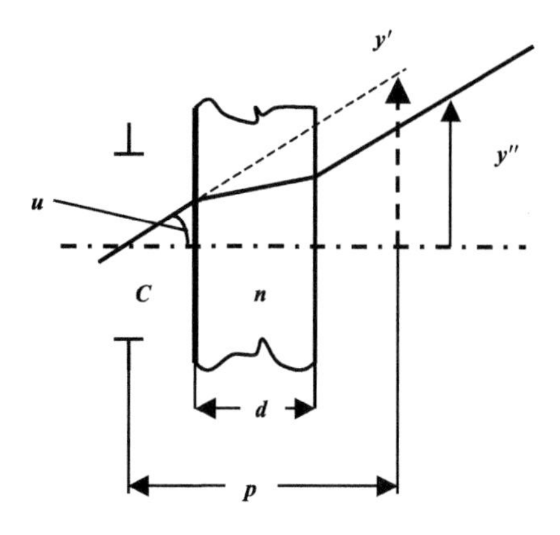

Fig. 1.2.14 Distortion in a parallel plate.

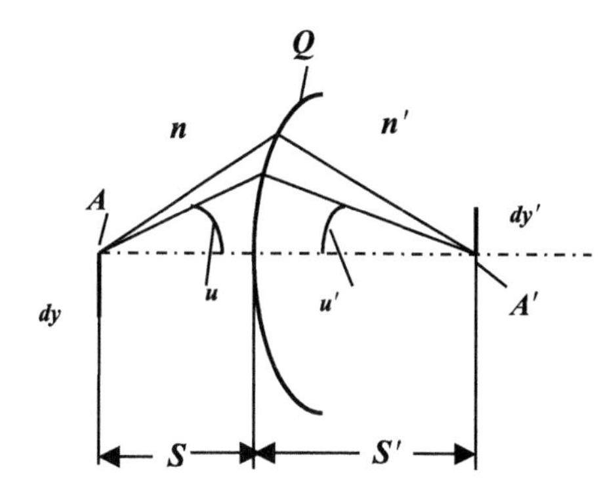

Fig. 1.2.15 The sine condition.

20 mm from it, a right-angle prism of 30 mm × 30 mm size is positioned in order to bend the optical axis by 90°. Find the diagram of astigmatism and distortion across the field of view.

P.1.2.9. *A flattener element in the imaging system*. A biconvex symmetrical lens of 40 mm focal length made of BK-7 glass performs imaging of distant objects to the plane P in a wide field of view. Is it reasonable to expect that the image quality of the off-axis areas will be of the same grade as images close to the optical axis? Find the flattener which is capable of improving image degradation for off-axis points (assume that it is made of SF-11 glass).

1.2.1.5 Sine condition and aplanatic points

Once we realize that imaging in general is accompanied by aberrations, it is quite understandable that finding locations where aberrations are small or even can be avoided completely is of great importance not only from a theoretical point of view but also for practical reasons. It can be shown that such locations do exist for a single surface with curvature, either a reflective or refractive surface, aspherical or spherical. Obviously the latter is more attractive, since manufacturing spherical optics is much easier and cheaper than fabrication of aspherical elements.

Let us consider a refraction surface Q separating media of refractive index n and n', as illustrated in Fig. 1.2.15, and let the conjugate pair A and A′ be the points of stigmatic imaging (imaging with homocentric beams, with no aberrations). This means that any ray emerging from A comes to A′, no matter what the ray angle u, and in terms of aberration it is equivalent to zero spherical aberration. Furthermore, the small object, dy, is imaged by the surface Q into dy′, both the object and the image being perpendicular to the optical axis and starting in the stigmatic points A and A′. If linear magnification $V = dy'/dy$ is independent of the ray angle, u, and remains constant it means that the following relation is valid:

$$V = \frac{n\sin u}{n'\sin u'} = \frac{S'}{S} = \text{const.} \qquad (1.2.24)$$

This relation is known as the sine condition and violation of it is usually called offence against the sine condition (OSC). Obeying the sine condition actually means that the small object dy is imaged with no aberration and the off-axis side of dy' is not blurred (it is a point and not a spot, i.e., there is no coma aberration here).

The pair of conjugate points where the spherical aberration is zero and the sine condition is kept valid are known as aplanatic points of the surface Q. If the object (and the front aplanatic point) is located at infinity the sine condition is transformed into following relation:

$$f' = \frac{h}{\sin u'}, \qquad (1.2.25)$$

for any height h at which the ray strikes the surface Q. Then

$$\text{OSC} = \delta f' = \frac{h}{\sin u'} - f'.$$

Needless to say, aplanatic points of the surface are of great significance, since in the vicinity of these points imaging occurs with no aberration, even for a beam of a wide solid angle.

A spherical surface has at least three pairs of aplanatic points, two of them being trivial, like the point, C, where the surface crosses the optical axis or the center, O, of the sphere curvature (see Fig. 1.2.16a). The third

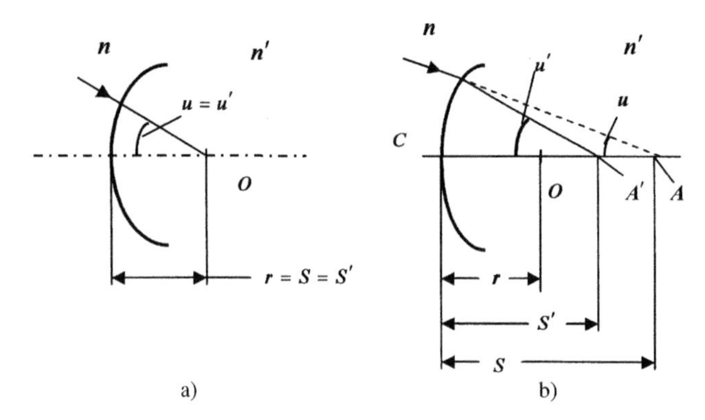

Fig. 1.2.16 Aplanatic points of a spherical refraction surface: (a) in the center of curvature; (b) off-center points.

aplanatic point pair, A, A′ (shown in Fig. 1.2.16b), is defined by the relations

$$S = CA = \frac{n+n'}{n}r; \quad S' = CA' = \frac{n+n'}{n'}r, \tag{1.2.26}$$

where r is the radius of curvature of the surface. Magnification at these points obeys the expression

$$V = \left(\frac{n}{n'}\right)^2, \tag{1.2.27}$$

while magnification of the center point, O, is

$$V_O = \left(\frac{n}{n'}\right). \tag{1.2.28}$$

Combination of two spherical surfaces with two kinds of aplanatic points enables one to create lenses where imaging is performed with no aberration (theoretically). Two such examples, one of a positive lens and another of a negative one, are depicted in Fig. 1.2.17 (see also details in Problem 1.2.11).

Problems

P.1.2.10. *A ball lens.* Find the OSC plot for a sapphire ball lens of 3 mm diameter working with an object at infinity and calculate the maximum diameter of the beam which can be concentrated behind the lens. The refractive index of sapphire is 1.77.

P.1.2.11. How does one design the aplanatic objective of a microscope if the required magnification is $V = -6$ at least and it is known that the gap of 0.7 mm between the object plane and the first lens surface is filled with immersion oil of $n = 1.8$? [Note: The objective should be

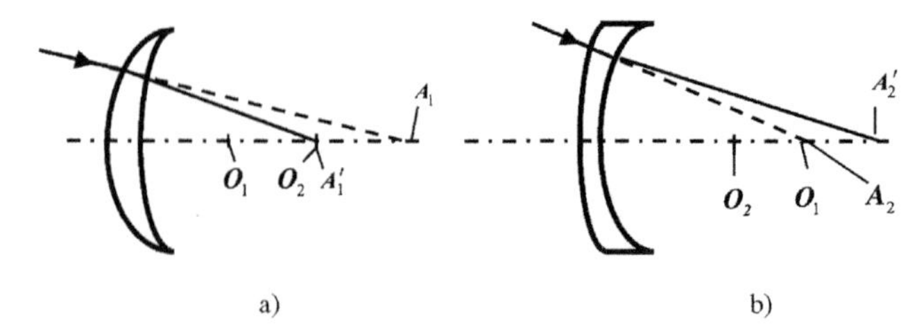

Fig. 1.2.17 Aplanatic points of (a) positive and (b) negative lenses.

constructed from two lenses. The first, which is a hemispherical lens, is 1 mm distant from the second lens of 3 mm thickness. The first component is made of SF-57 glass and the glass for the second lens can be chosen using the data of Appendix 2.]

1.2.1.6 Addition of aberrations

In practical situations when an imaging system comprises several elements (sometimes consisting of ten or more components), estimating the contribution of each element to the total aberration balance is quite useful. There are several rules allowing one to add aberrations of separate elements and to calculate their impact at different locations along the optical axis. One should keep in mind, however, that the main goal is to reveal how aberrations of each element affect the final image.

Addressing the procedure of addition of aberrations we suppose that the i-th element (or a group of elements) performs imaging from its object space to the image space with some linear magnification, V_i, and the image built by the i-th element serves as a virtual object for the next $(i + 1)$-th element. The following rules should be followed:

- the lateral aberrations $\delta s'_{i-1}$ while being transferred from the object space to the image space of the i-th component are multiplied by V_i^2, so that the total lateral aberration at the image space of the i-th element becomes

$$\delta s'_{tot} = \delta s'_{i-1} \times V_i^2 + \delta s'_i; \tag{1.2.29}$$

- the transverse aberrations while being transferred from the object space to the image space of the i-th component are multiplied by V_i, so that the total transverse aberration at the image space of this component becomes

$$\delta s'_{t,tot} = \delta s'_{t,i\text{-}1} \times V_i + \delta s'_{t,i}; \tag{1.2.30}$$

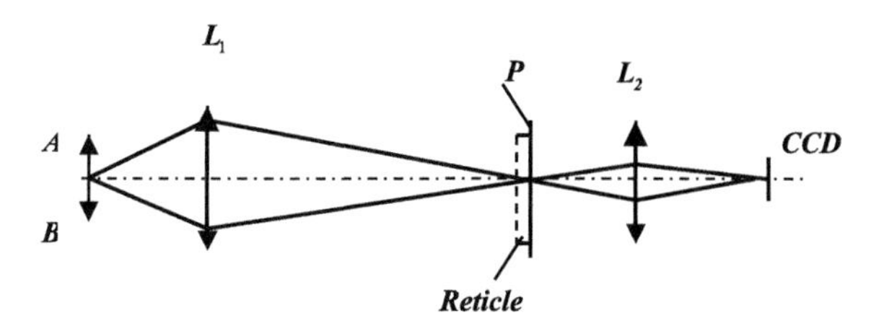

Fig. 1.2.18 Problem P.1.2.12- Two-lens imaging system with reticle.

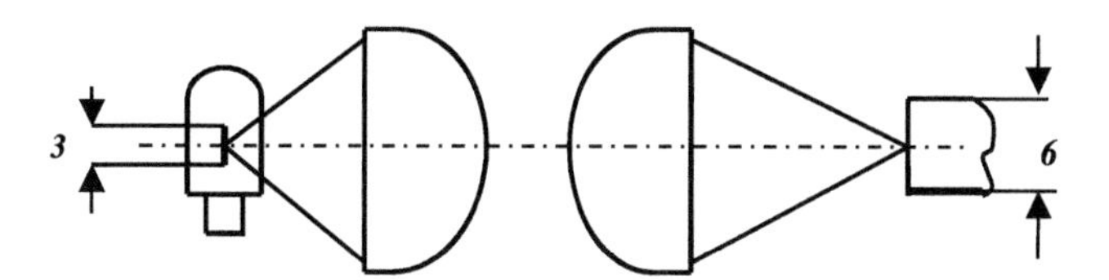

Fig. 1.2.20 Problem P.1.2.14 – Configuration of a two-lens condenser.

- if the i-th element transfers images to infinity it should be treated as if radiation propagates in the opposite direction and the aberrations computed in such a manner should be added to the aberrations of the image space of the $(i-1)$-th element ($\delta_{S_i}^{\leftarrow\prime}$ with its sign, and $\delta_{S_{t,i}}^{\leftarrow\prime}$ with the opposite sign);
- if a parallel beam is created between two subsequent elements they should be considered as a group with a single magnification and aberrations are treated as if they are transferred from the object space of the first element of this group to the image space of the second element of the group;
- addition of aberrations should be done separately for aberrations along the chief ray and for aberrations along the marginal ray.

These simple rules assist in the analysis and synthesis of imaging systems as far as aberrations are concerned.

Problems

P.1.2.12. In a two-lens imaging system (Fig. 1.2.18) initially aligned to get a sharp image of an object of 0.25 mm on a CCD of size 5 mm × 5 mm, a scale reticle R of 2 mm thickness is introduced in the plane P where the intermediate image is formed at magnification $V_1 = -5$. Both lenses are of 8 mm diameter and 15 mm focal length and they are properly corrected (the lens aberrations can be neglected). Find the impact of the reticle on aberrations in the CCD plane and the way to make a correction.

P.1.2.13. A lens of 40 mm in size and $f\# = 1.2$ performs imaging of a distant object to the detector plane P and has the residual aberration shown in the plot of Fig. 1.2.19a. The optical axis should be turned through 90° and two possible configurations are compared: one with a plane mirror and the other with a penta-prism (see Figs. 1.2.19b and 1.2.19c, respectively). What is the advantage of the second layout and what is the optimal size of the prism?

P.1.2.14. *A two-lens condenser.* An illumination system (Fig. 1.2.20) aiming to concentrate radiation from a halogen lamp with 3 mm filament into an optical fiber bundle of 6 mm in size consists of two lenses: L_1 of 30 mm diameter and 60 mm focal length and L_2 of the same diameter and 120 mm focal length, both made of BK-7 glass. Find the optimal shape of the condenser lenses and estimate the spherical aberration at the bundle entrance.

1.2.2 Diffraction effects and resolution

1.2.2.1 General considerations

Diffraction effects result from the wave nature of radiation participating in imaging. In general diffraction is

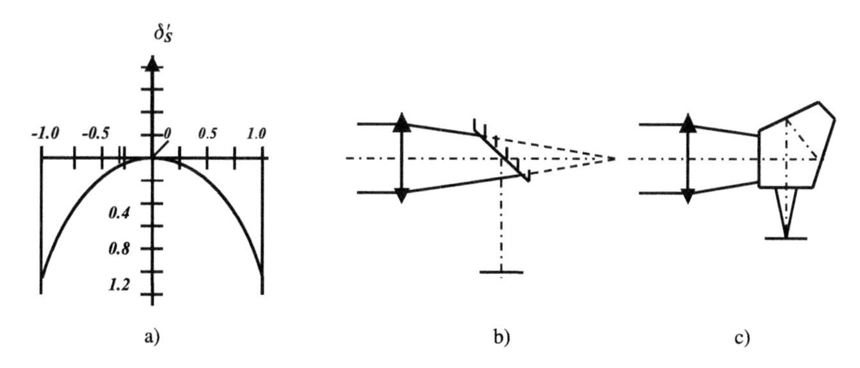

Fig. 1.2.19 Problem P.1.2.13 – (a) Residual lateral aberration of a lens and layouts with bending by (b) a mirror and (c) a penta-prism.

caused by the secondary waves generated in the substance of an obstacle on which electromagnetic waves impinge while traveling in space. An obstacle can be a body of any shape, either transparent or opaque. Interference of the secondary waves changes the spatial distribution of the propagated radiation in such a way that light energy appears not only in the direction of the initial propagation but also to the side of it. Because of this, for example, an ideal lens with no aberration is not capable of concentrating light in a single point of the image plane and some energy is always revealed in a small but finite vicinity of the image. Thus, diffraction is a basic limitation in imaging optics which cannot be avoided. Other effects, like aberrations considered in the previous section, which also "spoil" the image quality appear together with diffraction and cannot neutralize it in any way. If all other effects become negligible diffraction remains a single factor affecting the system performance. In such a case the optical system is termed diffraction limited.

Diffraction occurs at any stop through which light passes. It could be a real aperture, or the mounting of a lens, prism, or mirror, or just the boundaries of an optical element of the system. We shall consider a simple case of propagation of monochromatic light of wavelength λ through a circular non-transparent stop of radius a followed by a lens (see Fig. 1.2.21). It can be shown that the intensity distribution of light in the spot created in the image plane P due to diffraction is governed by the following function (Airy's function):

$$I(r) = I_0\left[\frac{2J(x)}{x}\right]^2; \quad \text{where } x = \frac{2\pi}{\lambda}n'r'\sin u'_{\max}, \tag{1.2.31}$$

n' is the refractive index in the image space, r' is the radial coordinate in the plane P, u'_{max} is the maximum angle of the direction from the stop boundary to the center of the spot, and $J_1(x)$ is the Bessel function of the first order.

Expression (1.2.31) is an oscillating function with a strong central maximum followed by dark and light rings of decreasing intensity. It is commonly accepted that most of the energy of the spot is concentrated in the central maximum limited by the first dark ring which corresponds to the value $x_{\min}^{(1)} = 3.8317$ in Eq. (1.2.31). Hence, the relevant size of the spot in the plane P obeys the relation

$$\delta_{\text{dif}} = \frac{1.22\lambda}{n'\sin u'_{\max}}. \tag{1.2.32}$$

In the case when P is the focal plane of a lens of diameter $D = 2a$, Eq. (1.2.32) is transformed into the well-known expression

$$\delta_{\text{dif}} = \frac{2.44\lambda}{D}f' \qquad (n' = 1).$$

The diffraction spot has a direct impact on limiting resolution which is one of the basic features of any imaging system. Consider two very close images in the plane P, each one generating a diffraction spot. If the distance between the two images is large enough the spots are well separated and an observer looking on the image plane P is capable of perceiving them easily. The smaller the distance, the closer the spots, and at some stage they become overlapped. The question is, what is the minimum distance at which two partially overlapping spots are still recognized as two separate objects? Such a minimal distance is called the limiting resolution and it is defined, according to the Rayleigh criteria, as the situation when the minimum of one spot coincides with the maximum of the second. Fig. 1.2.22 demonstrates the situation when two images, one centered at point A′ and the other centered at B′, are still resolvable. The dotted line in Fig. 1.2.22b shows the distribution of energy after summation of both spots.

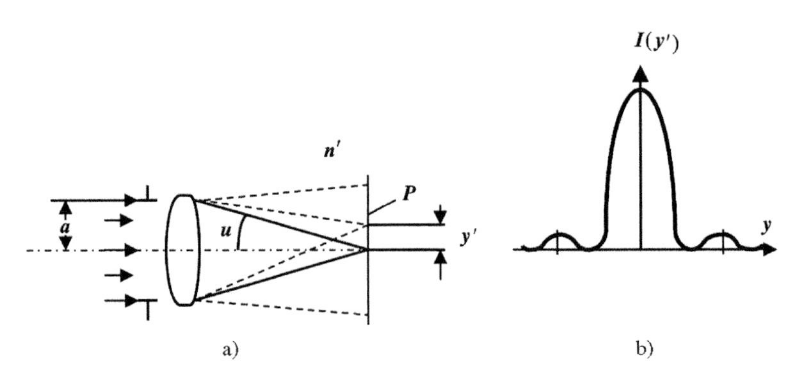

Fig. 1.2.21 Diffraction on (a) a circular stop and (b) the intensity distribution in the diffraction spot.

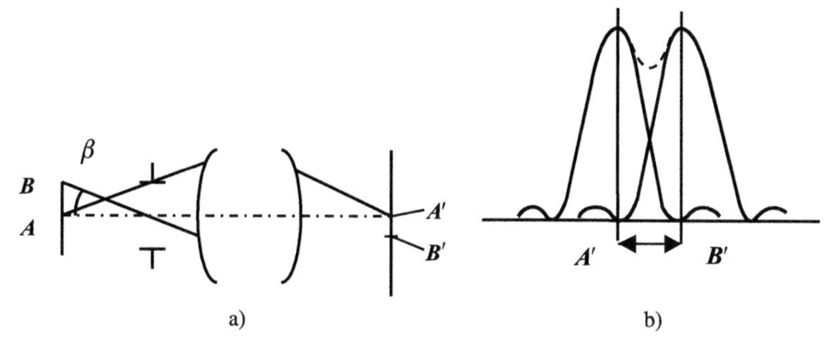

Fig. 1.2.22 (a) System resolution and (b) two spots according to the Rayleigh criteria.

The "valley" between the two maxima is about 70% of the maximum intensity (i.e., about 30% reduction of energy).

What is usually important in practical applications is the distance in the object plane between two points A and B corresponding to limiting resolution in the image plane. Referring to Fig. 1.2.22a, suppose an entrance pupil of size D_p is located at a distance p from the object plane. Taking into account that the product $n \times \sin u \times r$ is the system invariant (it remains constant while transferring through each refraction surface) and using Eq. (1.2.32), one can transform the distance δ_{dif} into the corresponding distance AB and resolvable angle β in the object plane:

$$\mathrm{AB} = \frac{1.22\lambda}{nD_p}p; \quad \tan\beta = \frac{\mathrm{AB}}{p} = \frac{1.22\lambda}{nD_p}, \tag{1.2.33}$$

or, using the expression in angular seconds, $\beta = 120''/D_p$ for $\lambda = 0.5$ μm and $n = 1$.

If aberrations of the system are significant then the diffraction spot should be considered together with the aberration spot. The common practice is to use the square root rule for getting the total spot as follows:

$$\delta_{sum} = \sqrt{\delta_{dif}^2 + \delta s_{ab}^2}. \tag{1.2.34}$$

Apparently the resolution limit is affected by Eq. (1.2.34). A simple way to define resolution with the spot enlarged by aberrations is demonstrated in Fig. 1.2.23. To find the resolution in the object plane in this case one should divide the value δ_{sum} calculated from Eq. (1.2.34) by the magnification of imaging.

1.2.2.2 Diffraction theory of imaging in a microscope

Considering microscopic imaging in terms of diffraction allows one to understand the basic limitations existing in this kind of instrument and to determine relations governing the maximum achievable resolution.

In Abbe's theory of the microscope the object is referred to as a transparent diffraction grating (see Chapter 5) of a spatial period d. This approach is based on the assumption that a real object described by an arbitrary intensity distribution function which can be expanded in a Fourier series of separate harmonics is considered as a collection of sine periodic spatial waves, each one acting as a diffraction grating. Being illuminated by a parallel beam, the grating generates several fans of beams corresponding to different diffraction orders. The zero-order beam is concentrated by the microscope objective at the back focal point whereas the other diffraction orders are collected at other points of the same back focal plane (F_{i+1}; F_i;...). The aperture stop located in the back focal plane comprises all focused centers of diffracted beams (see Fig. 1.2.24). Light of each order proceeds further as

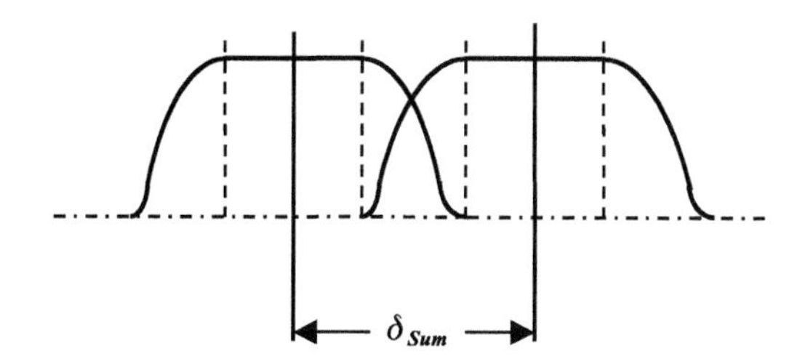

Fig. 1.2.23 Resolution limit caused by aberrations and diffraction.

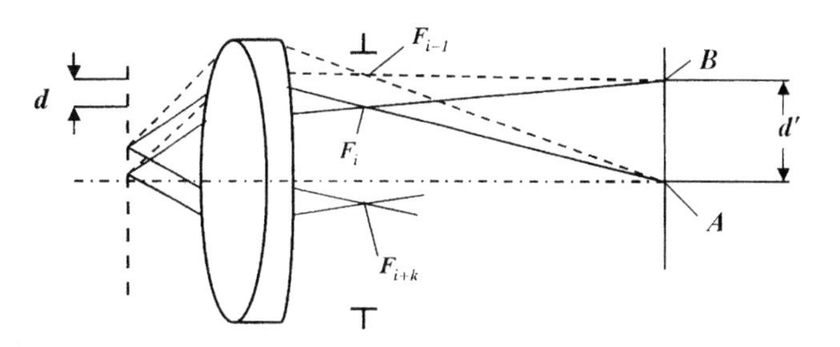

Fig. 1.2.24 Diffraction in microscope imaging.

a divergent fan to the plane of the field stop positioned in the focal plane of the eyepiece. Here the fans overlap and interfere. The resulting fringe pattern with a constant spacing, d', constitutes an image of the initial grating of the object plane and both are related through the system magnification: $d'/d = V$.

To create the fringe pattern at least two divergent beams and therefore two diffraction orders must be present simultaneously in the aperture stop. The location of the diffraction maxima, F_i, in the focal plane of the objective is dictated by the diffraction grating equation (Eq. (5.18); see Chapter 5): $\sin u_{max}^{(i)} = i\lambda/d$ for $i = 0, \pm 1, \pm 2, \ldots$ The aperture stop size, D_{as}, is related to the numerical aperture of the system in the object space: $n \sin u_{max} = D_{as}/(2f')$. These last two expressions allow one to find the minimum diffraction spacing, d, which can be imaged by the microscope.

Two possible methods of illumination should be considered separately: direct illumination when the zero-order diffraction is focused in a point on the optical axis and oblique illumination when the focus of the zero-order diffraction is located in an off-axis point. Fig. 1.2.25 illustrates both situations. The limiting condition for direct illumination requires that the zero-order maximum as well as the 1st and the $(-1)^{st}$ order maxima are inside the aperture stop whereas the corresponding limit for oblique illumination can be realized if the zero-order and only one of the first-order diffraction maxima are inside the circle of diameter D_{as}. As can be seen, the limiting resolution is related to the numerical aperture (NA) of the system as

$$d = \frac{\lambda}{n \sin u_{max}} = \frac{\lambda}{NA}; \quad d = \frac{\lambda}{2n \sin u_{max}} = \frac{\lambda}{2NA}, \tag{1.2.35}$$

for direct (on-axis) and oblique illumination, respectively. In the real practice of microscopy illumination is supplied by a wide-angle condenser coming at both direct and oblique directions. It can be shown that in such a case the limiting resolution of the microscope is determined as

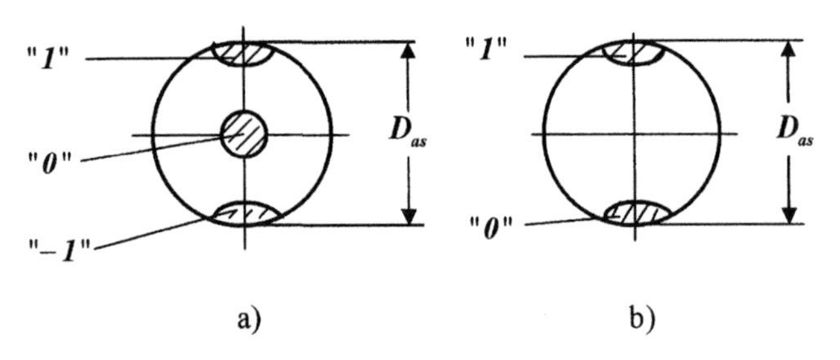

Fig. 1.2.25 Location of diffraction maxima in an aperture stop: (a) on-axis illumination; (b) oblique illumination.

$$d = \frac{\lambda}{NA + NA_C}, \tag{1.2.36}$$

where NA_C is the numerical aperture of the condenser.

Problems

P.1.2.15. Find the minimum required active diameter of the well-corrected imaging optics for visible wavelengths operating at a working distance of 30 mm and providing a resolution of 0.5 µm.

P.1.2.16. A microscope for the visible range is supplied with three objectives: 10 × 0.25 NA, 40 × 0.65 NA, and 100 × 1.2 NA, and a condenser of 0.96 NA. Find the maximum resolution in all three possible configurations.

P.1.2.17. A microscope objective of magnification × 10 has a focal length of 16 mm and is operated with an aperture stop of 5 mm diameter. At which angle of oblique illumination should one expect the resolution to be twice that of normal (on-axis) illumination? Which resolution (in the visible) will be available in this case and how will the resolution be changed if the illumination angle is held at 5°?

1.2.3 Image evaluation

Evaluation of images is carried out (i) at the design stage when it is checked whether the configuration designed is capable of delivering the system performance requirements; and (ii) at the end of manufacturing when a real system with all the tolerances of component fabrication and assembling is aligned and prepared for final testing. Image evaluation at the design stage is performed theoretically, by analyzing aberrations of the system and also by calculating some integral parameters enabling one to estimate the expected image quality. Image evaluation at the manufacturing stage is done with special hardware allowing one to measure resolution, contrast, and other parameters related to the system performance, usually determined in a procedure specific for each tested architecture.

Theoretical evaluation of image quality is usually based on ray tracing of a great number of rays, originating in on-axis and off-axis points of the object and, if necessary, related to several representative wavelengths (mostly, the lines C, D, and F) of the illuminating radiation. Obviously computing is carried out with special software allowing one to calculate and display the location of the rays in the image plane (a spot diagram), energy distribution in a spot, frequency response of the system (modulation transfer function, see below), position of the best focus, and other useful parameters. Diffraction effects are also taken into account while

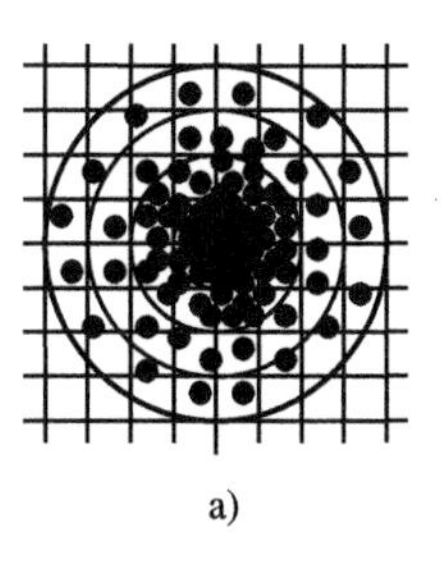

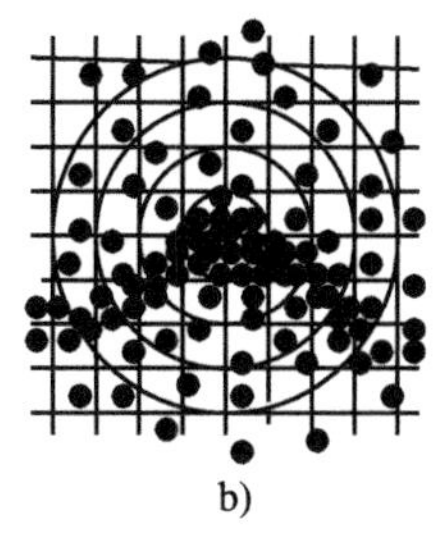

Fig. 1.2.26 Spot diagram for (a) an on-axis point and (b) a field (off-axis) point.

computing the relevant parameters and convolution between geometrical optics results (ray tracing) and the diffraction pattern at each and every image point is accurately calculated.

Examples of spot diagrams for on-axis and off-axis points are depicted in Fig. 1.2.26. Each diagram is calculated by tracing the rays striking the entrance pupil as a uniformly distributed fan and indicating the points of intersection of the rays with the image plane. Apparently for a perfect lens the spot diagram is transformed in a single point located in the paraxial image of a corresponding object point. The on-axis spot is usually symmetrical whereas the off-axis spot might be strongly asymmetric (as shown in Fig. 1.2.26b) which is an indication of strong field aberrations.

The size of the spot, δ, can be used as a simple evaluation parameter. In some cases this value can also be estimated analytically, using expressions for the third-order aberrations (e.g., Eqs. (1.2.14) and (1.2.15)) or for the diffraction spot (Eq. (1.2.32)).

Once the spot diagram is found it is possible to count the number of rays intersecting the image plane inside a circle of a chosen size. With the assumption that each ray bears the same amount of energy such a procedure (performed several times, for circles of different diameter, d) gives another parameter called the "encircled energy distribution" (see Fig. 1.2.27). The circle diameters can be taken in absolute units or in relative units, in terms of the unit $z = d/\delta_{\mathrm{dif}}$, where δ_{dif} is the spot of a diffraction-limited system, as per Eq. (1.2.32).

The distributions shown in Fig. 1.2.27c illustrate the action of a perfect lens (curve 1) compared to a real lens with aberrations (curve 2). As can be seen, in the first case there are some oscillations on the graph which are evidently related to the interference rings of Airy's function. Curve 1 can be found analytically. Denoting the relative energy inside the circle $d(E(d)/E_{\mathrm{tot}})$ as $L(d)$, one finds the following expression (for details, see Born and Wolf, 1968):

$$L(d) = 1 - J_0^2(z) - J_1^2(z); \quad z = \frac{\pi D_p d}{2\lambda p}, \qquad (1.2.37)$$

where D_p is the exit pupil size located at a distance p from the image plane. The first minimum occurs at $z = 3.8317$ and the corresponding encircled energy is about 84%. Aberrations influence significantly the energy distribution and therefore this distribution can be used as a tool for image quality evaluation.

The most common way to evaluate images is based on the modulation transfer function (MTF). To explain this approach we consider the basic relation between an object $T(x, y)$ and its image $I(x', y')$ created by an optical system. The system is characterized by the point spread function (PSF) $S(x, x', y, y')$ which is actually the pattern created in the image space resulting from a single point object. Then the image of an arbitrary object T can be represented as follows:

$$I(x', y') = \iint S(x - x', y - y')T(x, y)dxdy, \qquad (1.2.38)$$

which is the convolution between the object and the PSF. By performing the Fourier transform of Eq. (1.2.38) we get

$$R(k_x, k_y) = Q(k_x, k_y) \times H(k_x, k_y), \qquad (1.2.39)$$

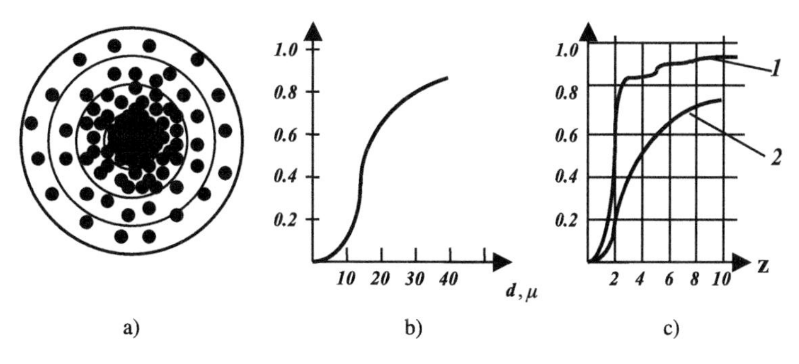

Fig. 1.2.27 Encircled energy distribution: (a) schematic of rays inside different circles; (b) relative energy distribution vs. circle size in absolute units; (c) relative energy distribution vs. circle size in units of z.

where the term on the left-hand side, called the frequency response in the image space and given by

$$R(kx,k_y) = \frac{1}{2\pi}\iint I(x',y')\exp[i(k_x x' + k_y y')]dx'dy', \tag{1.2.40}$$

is related through the system optical transfer function (OTF),

$$Q(k_x,k_y) = \frac{1}{2\pi}\iint S(x,y)\exp[i(k_x x + k_y y)]dxdy, \tag{1.2.41}$$

to the harmonics of the object, $H(k_x, k_y)$. In the above expressions $k = 2\pi/v$, where v is the spatial frequency in cycles/mm.

Since the OTF is generally a function in a complex space characterized by its amplitude and phase, it is valuable to consider its modulus (a real function) called the MTF. $\mathrm{MTF}(v) = |Q(v)|$. The MTF does not depend on the object, but only on the system properties and this is the reason why it is widely used for image quality evaluation. The MTF also can be interpreted in terms of modulation, which is a feature related to the intensity of light and can be easily measured in practice. Let the light intensity in the object vary from I_{max} to I_{min}. Then the contrast revealed in the object plane can be characterized by the modulation M_o:

$$M_o = \frac{I_{max} - I_{min}}{I_{max} + I_{min}}, \tag{1.2.42}$$

and the contrast in the image plane is described by the corresponding modulation M_i. The ratio between the two modulations is governed by the MTF:

$$M_i(v)/M_o(v) = \mathrm{MTF}(v), \tag{1.2.43}$$

if all three values are determined for the same spatial frequency v.

Another attractive feature of the MTF is that the MTF of an imaging system is just the product of the MTFs of the separate components constituting the system. This results from linearity and other features of the Fourier transform. Thus, adding or replacing an element can be easily analyzed with regard to the new image quality.

Computation of MTF(v) is a cumbersome and time-consuming procedure which in most cases is performed by special software. However, there are a few cases when it can be expressed explicitly, in analytical form. For example, for a diffraction-limited system, with no aberration, the PSF is Airy's function described by Eq. (1.2.31). Its OTF and MTF can be found analytically as follows (e.g., see Smith, 1984):

$$\mathrm{MTF}(v) = \frac{2}{\pi}(F - \sin F\cos F)(\cos\beta)^n;$$

$$F = \arccos\left(\frac{\lambda v}{2\mathrm{NA}}\right), \tag{1.2.44}$$

where β is half of the full-field angle, NA is the numerical aperture in the image space ($\mathrm{NA} = n' \sin u'$), and the power $n = 1$ or $n = 3$ for radial or tangential directions, respectively. Obviously MTF $= 0$ at $F = 0$, meaning that the spatial frequency

$$v_c = \frac{2\mathrm{NA}}{\lambda}, \tag{1.2.45}$$

is the maximum frequency transferred by the system from the object to the image space (it is called the cut-off frequency).

Figure 1.2.28 illustrates theoretically calculated MTFs. The diffraction-limited system (curve 1) features the highest MTF at any spatial frequency. It can also be seen that the influence of aberrations is more significant at higher frequencies (difference between curve 1 and curves 2 and 3). Curve 3 also shows that in some cases the MTF might have negative values, which means a 180° phase inversion (black zones become white and vice versa).

According to the explanation above, the MTF (and OTF), strictly speaking, refers to the harmonics, or sine waves, in the object space. In reality, however, the same approach is also exploited for "square wave" objects, like a bar code. As is demonstrated in Fig. 1.2.29, the contrast and the modulation M_i in the image plane decrease when the spacing (period) of the object square wave, T(x), decreases.

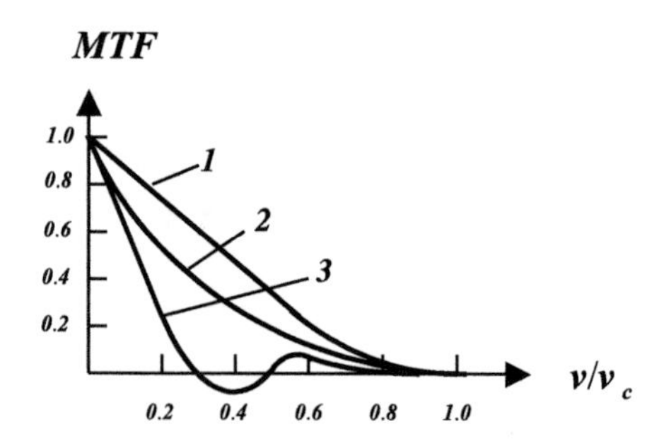

Fig. 1.2.28 Modulation transfer functions for a diffraction-limited system (1) and for systems with aberrations (2 and 3).

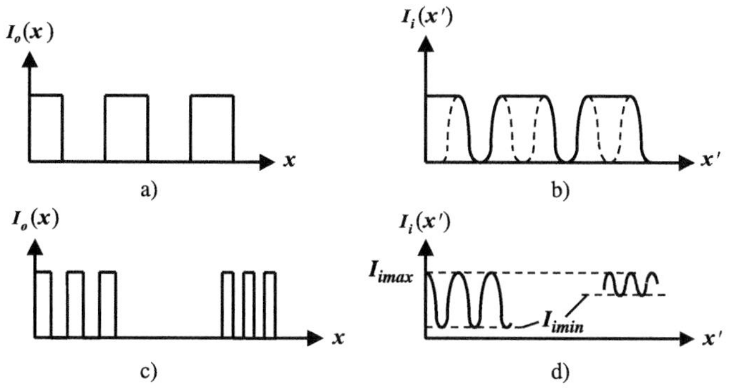

Fig. 1.2.29 (a,c) Input square waves and (b,d) the corresponding output patterns.

Using a target bar code with several groups of well-defined spatial frequencies as the system object and measuring the modulation M_i of the corresponding images at the system output allows one to find the MTF (see Eq. (1.2.43)). This method is commonly exploited in image quality evaluation at the final testing stage. The limiting resolution of the system is defined as the spatial frequency of the group still visible at the image plane with a minimum contrast of 3–5% (which is considered as the limit of the perception capability of a human eye).

Problems

P.1.2.18. What could be concluded about imaging optics if an analysis of the encircled energy revealed 45% of the total energy of the spot corresponding to an on-axis image point is inside the circle diameter which is half the size of the whole spot?

P.1.2.19. Imaging optics operated with monochromatic illumination of 0.6 μm and having NA = 0.25 in the object space is ended by a CCD area sensor. What should be the minimum pitch of the CCD in order to acquire all spatial frequencies transferred by the optics?

P.1.2.20. MTF measurements are carried out with a square-wave target of variable frequencies made of chrome on glass in a bar code pattern. Data are collected for low spatial frequency ($\upsilon_1 = 10$ cycles/mm) and for high spatial frequency ($\upsilon_2 = 200$ cycles/mm) and the respective measured modulations are 70% and 20%. Assuming that the reflectance of chrome is 70% and the reflectance of glass is 4% and also keeping in mind that the contrast of images is slightly degraded by the background light scattered inside the measurement set-up, find the true MTF value at higher spatial frequency.

P.1.2.21. A diffraction-limited optical system operated in the visible range and having NA = 0.15 creates an image on a CCD sensor followed by a video monitor. The MTF of CCD + monitor is 60%. Could we expect to see on the screen the tiny details of an object corresponding to the spatial frequency of 575 cycles/mm?

1.2.4 Two special cases

1.2.4.1 Telecentric imaging system

This kind of architecture is usually exploited in measurement systems where errors caused by the third dimensions (along the optical axis) of an object have to be minimized. To explain this error (sometimes called the parallax error, or perspective error) we refer to Fig. 1.2.30a where simple imaging with a single lens is depicted. Two objects, O_1A and O_2B, having the same height and located at different distances from the lens, after imaging are transformed into images O'_1A' and O'_2B' of different heights. The error $\Delta y'$ might cause problems if the defocusing $\Delta x'$ is small (not revealed by the system observer). The telecentric imaging system shown in Fig. 1.2.30b is free of this error. The system is configured as an afocal lens pair where the back focal plane of the first lens coincides with the front focal plane of the second. What is also important is that the aperture stop ab is located in this plane P. As a result,

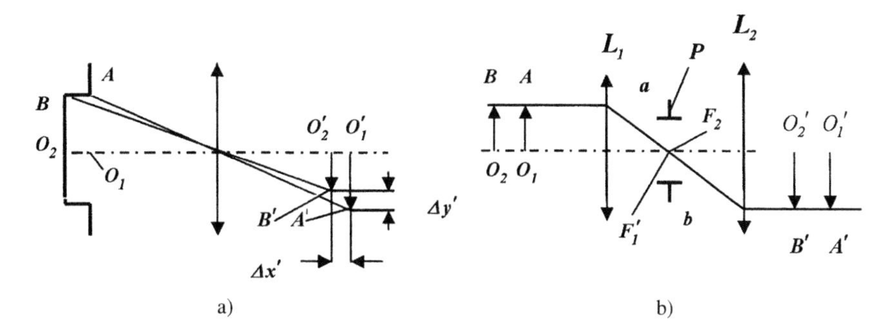

Fig. 1.2.30 Imaging with (a) parallax error and (b) the telecentric configuration.

the entrance pupil and the exit pupil are both located at infinity (one on the object space side and the other on the image space side) and the chief rays originating in points A and B are parallel to the optical axis in both spaces. Hence the images O'_1A' and O'_2B' are of the same size and the parallax error does not occur.

If the system is operated with a video area sensor (like a CCD) it should be positioned in such a way that both images are sharp enough. Aberration of defocusing (like the other aberrations) strongly depends on the active lens size, but a reduction in the lens diameter is accompanied by the increasing impact of diffraction, as discussed in Section 1.2.2, and also a decrease in the image illumination. Therefore, a compromise should be found. In any case, symmetrical configurations are preferred where the shapes of the lenses are equally positioned with regard to the plane P (or one of them is scaled in a symmetrical manner, if magnification/minification is required). Estimation of aberrations can be carried out by the method described in Section 1.2.1.6 and Problem P.1.2.14.

1.2.4.2 Telephoto lens

There are numerous situations where the effective focal length of the objective has to be long while the actual size of the lens should be kept as small as possible. A possible architecture in such a case is a two-lens configuration, one of positive and the other of negative optical power (see Fig. 1.2.31). Usually what is known is the equivalent focal length, f'_e, and the desired length of the configuration, l. The optical power of each component and their locations with regard to the image plane should be found.

Considering the system in terms of first-order optics (paraxial approximation) we have for this two-lens system (see Problem P.1.1.7)

$$\Phi = 1/f'_e = \Phi_1 + \Phi_2 - \Phi_1\Phi_2 d, \qquad (1.2.46)$$

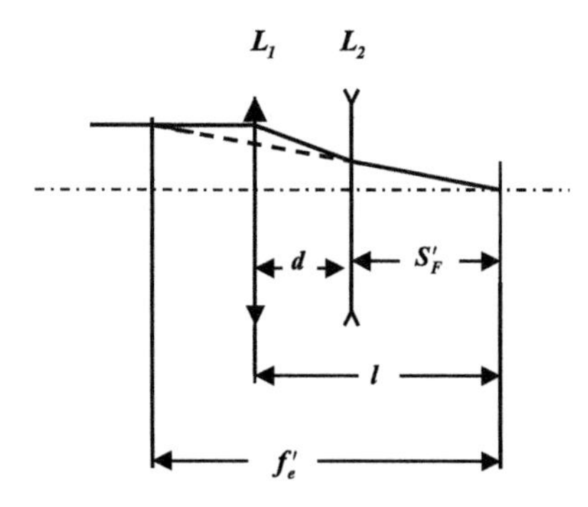

Fig. 1.2.31 Configuration of a telephoto lens.

and taking into account that $d + S'_F = l$ we also get

$$\Phi_1 d = 1 - \Phi(l - d), \qquad (1.2.47)$$

Equations (1.2.46) and (1.2.47) for three unknowns, Φ_1, Φ_2, and d, allow one to introduce an additional condition to optimize the configuration with regard to aberration. This could be either the requirements for a minimum optical power of the second element (which in general might result in lower residual aberrations) or the requirements for a configuration with minimal (better zero) curvature of the image surface. In the first case the best results, as can be shown, are obtained with $d = 0.5l$ and the corresponding focal lengths of the elements are

$$f'_1 = \frac{lf'_e}{2f' - l_e}; \quad f'_2 = -\frac{l^2}{4(f' - l)}. \qquad (1.2.48)$$

In the second case a zero Petzval's sum (see Section 1.2.1.4) is required which is achieved with $\Phi_2 = -\Phi_1$. By introducing this condition in Eqs. (1.2.46) and (1.2.47) we have

$$l = 0.75f'_e; \quad f'_1 = -f'_2 = 0.5f'_e; \quad S'_F = 0.5f'_e. \qquad (1.2.49)$$

The latter approach is widely used in the design of telephoto lenses intended for imaging in large angular fields of view.

Problems

P.1.2.22. How does one design a telecentric imaging system which is operated at magnification $V = -3$ in an angular field of view of $\pm 5°$ and provides a resolution of 2 μm in the visible spectral interval?

[Note: Assume the system is free of aberration.]

P.1.2.23. A telephoto lens forms images with negligible curvature at a distance of 60 mm from the first (front) element. What are the focal lengths and the distance between the lenses?

1.2.5 Solutions to problems

P.1.2.1. Since the lens is working in the paraxial range ($f\# = 10$) we can find the distance to the plane P where an ideal image is formed by the lens with nominal focal length:

$$\frac{1 - V}{S'} = \frac{1}{f'}; \quad S' = 100 \times (1 + 2) = 300 \text{ mm}.$$

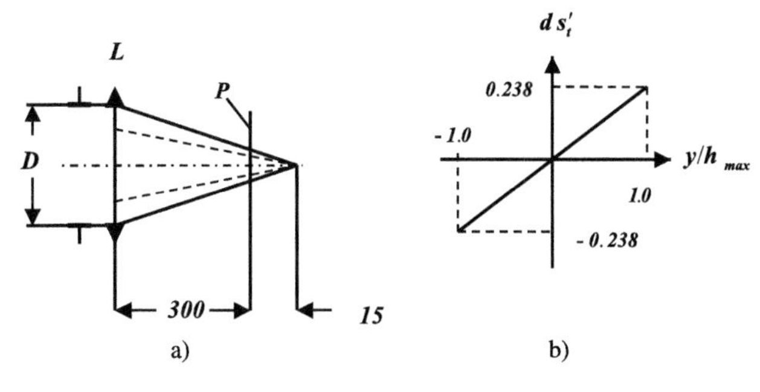

Fig. 1.2.32 Problem P.1.2.1 – (a) Defocusing in the plane P and (b) the meridional aberration plot.

If the lens is manufactured with 5% tolerance the focal length might be as long as 315 mm, and in this case blurring due to defocusing occurs in the plane P, as demonstrated in Fig. 1.2.32a. The maximum lateral aberration of defocusing, $\delta s_l' = 15\text{mm}$, gives the corresponding transverse aberration calculated as in Eq. (1.2.1): $\delta s_t' = \rho_{\max} = \delta s_l' \times \tan u_{\max} = 15 \times 5/315 = 0.238$ mm. Obviously this aberration is a linear function of the height, y, of the ray at the entrance pupil. In the plot shown in Fig. 1.2.32b the relative vertical coordinate is exploited, $y/h_{\max}$, which varies in the range from 1.0 to (– 1.0).

P.1.2.2. We start by calculating the thickness of the glass block equivalent to the penta-prism (see Section 1.1.4). We have $t_e = 3.41a = 3.41 \times 30 = 102.43$ mm and the prism makes the optical path of a chief ray longer by $t_e(1 - n)/n = 102.4(1 - 1.5163)/1.5163 = 34.87$ mm. Figure 1.2.33 demonstrates the divergent beam traveling through the slab of thickness t_e and explains the appearance of lateral aberration $\delta s'$ as a function of the incidence angle u. Using the rigorous formula from Problem P.1.1.6, one can find the aberration $\delta s'$ as a difference between lateral segments calculated for the paraxial range and for any final angle u:

$$\delta s' = t_e\left[(1 - 1/n) - \left(1 - \sqrt{\frac{1 - \sin^2 u}{n^2 - \sin^2 u}}\right)\right]$$

$$= \frac{t_e}{n}\left(1 - n\sqrt{\frac{1 - \sin^2 u}{n^2 - \sin^2 u}}\right).$$

For the maximum angle defined as $\tan u_{\max} = 20/125$ we get from the above equation $\delta s' = 0.48$ mm and for half of the maximum height we obtain $\tan u = 10/125$; $\delta s' = 0.11$ mm. The final plot of aberration introduced by the prism is presented in Fig. 1.2.33b.

P.1.2.3. Proceeding with the penta-prism considered in Problem P.1.2.2, we refer to Fig. 1.2.34a and calculate the displacement segment $AA' = L$ separately for three main wavelengths, C, D, and F. Chromatic aberration is determined as $L_c - L_d$ and $L_F - L_D$.

Keeping in mind that for BK-7 glass (see Appendix 2) $n_D = 1.5168$, $n_C = 1.51432$, and $n_F = 1.52238$, and starting with the simplified expression for L valid in the paraxial range we have

$$\delta s'_{Ch} = L_C - L_D = t_e\left[\left(1 - \frac{1}{n_C}\right) - \left(1 - \frac{1}{n_D}\right)\right]$$

$$= 102.43\left(\frac{1}{1.5168} - \frac{1}{1.51432}\right) = -0.111 \text{ mm}.$$

Before proceeding further, we compare the result obtained above with the calculation by the rigorous

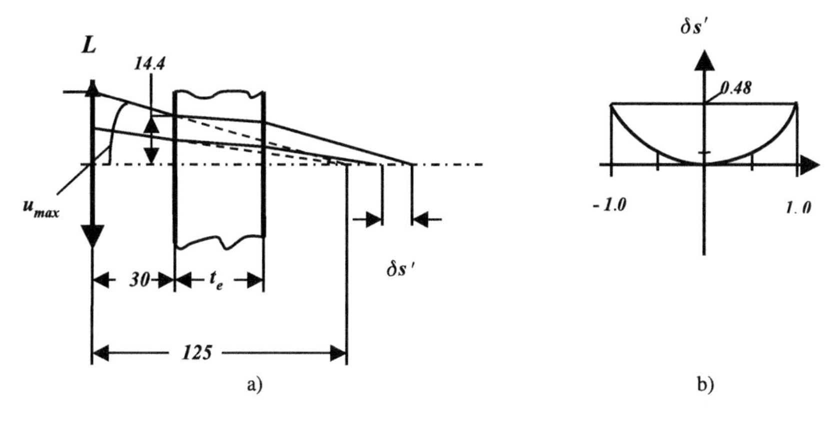

Fig. 1.2.33 Problem P.1.2.2 – (a) Geometry of rays traveling through an unfolded prism and (b) the aberration plot.

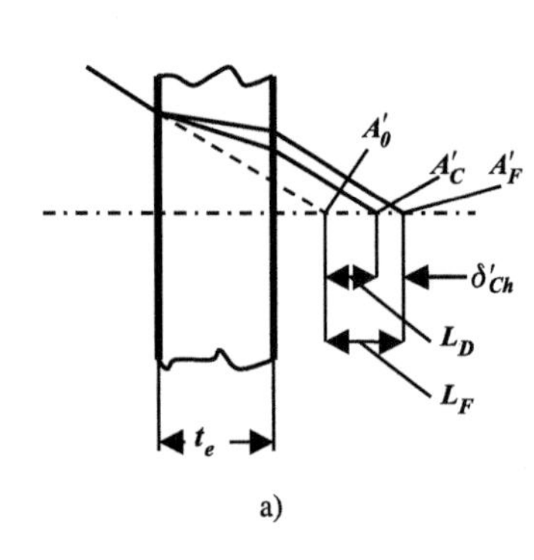

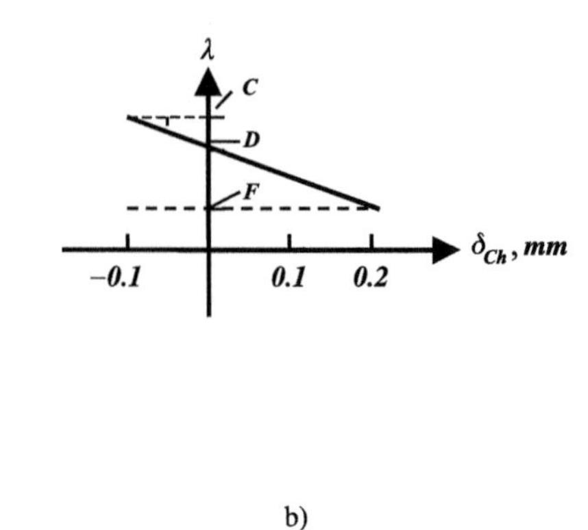

Fig. 1.2.34 Problem P.1.2.3 – Chromatism of a penta-prism: (a) the ray diagram; (b) the aberration plot.

formula for the maximum angle of incidence, $i_{max} = \arctan(10/125) = 9.09°$:

$$\delta'_{Ch} = t_e\left[\left(1 - \sqrt{\frac{1-\sin^2 i_{max}}{n_C^2 - \sin^2 i_{max}}}\right) - \left(1 - \sqrt{\frac{1-\sin^2 i_{max}}{n_D^2 - \sin^2 i_{max}}}\right)\right] = -0.112 \text{ mm}.$$

As we see, the difference between exact solution and the paraxial approximation is very small in our case, so that we proceed with the simplified formula and calculate

$$\delta s'^{(2)}_{Ch} = L_F - L_D = t_e\left[\left(1 - \tfrac{1}{n_F}\right) - \left(1 - \tfrac{1}{n_D}\right)\right]$$

$$= 102.43\left(\frac{1}{1.5168} - \frac{1}{1.52238}\right) = 0.248 \text{ mm}.$$

The plot of the chromatic aberration of the prism is shown in Fig. 1.2.34b.

P.1.2.4. Using the glass data from Appendix 2, we get for the components of the doublet lens (achromat) the following (see Eq. (1.2.11)):

$$\Phi_1 = \Phi\frac{\nu_{D1}}{\nu_{D1} - \nu_{D2}} = 0.01333\frac{64.12}{64.12 - 33.686}$$

$$= 0.028084$$

$$\Phi_2 = -\Phi\frac{\nu_{D2}}{\nu_{D1} - \nu_{D2}} = 0.01333\frac{33.686}{64.12 - 33.686}$$

$$= -0.014751.$$

Assuming the first surface is plane ($r_1 = \infty$), we can find the second radius from Eq. (1.1.11) for a thin lens: $r_2 = -(0.5168/0.028084) - 18.40$ mm and this is also the first radius (r_3) of the second element. Hence the last radius can be calculated as follows:

$$\frac{1}{r_4} = \frac{1}{r_3} - \frac{\Phi_2}{n_{D2} - 1} = -\frac{1}{18.4} + \frac{0.014751}{0.62588}$$

$$= -0.0307794; \quad r_4 = -32.489 \text{ mm}.$$

To find the residual aberration we calculate the focal length of the doublet in all three main wavelengths using Eq. (1.2.8). We have for the positive component

$$\Phi_{1C} = \frac{0.51432}{18.40} = 0.027952; \quad \Phi_{1D} = 0.028084;$$

$$\Phi_{1F} = \frac{0.52238}{18.40} = 0.02839,$$

and for the negative component:

$$\Phi_{2C} = -0.62074\left(\frac{1}{18.40} - \frac{1}{32.49}\right) = -0.14630;$$

$$\Phi_{2D} = -0.014751;$$

$$\Phi_{2F} = -0.63932 \times 0.023568 = -0.015067.$$

Then

$$\Phi_C = \Phi_{1C} + \Phi_{2C} = 0.027952 - 0.014630$$
$$= 0.013322; \quad f'_C = 75.06 \text{ mm}$$
$$\Phi_F = \Phi_{1F} + \Phi_{2F} = 0.02839 - 0.015067$$
$$= 0.013323; \quad f'_F = 75.06 \text{ mm}$$

and the residual aberration (secondary spectrum) is $\delta s'_{Ch} = f'_C - f'_D = 0.06$ mm (see the plot depicted in Fig. 1.2.35b).

P.1.2.5. From the problem data it follows that the focal length of the lens is $f' = f\# \times D = 60$ mm

(a) Starting with the lens made of BK-7 glass and data from Appendix 2 we have $\xi = 1/n_D = 1/1.5168 = 0.6593$. By substituting this value in Eq. (1.2.16) we get the radii of the lens of the optimal shape:

$$r_1 = 2(1 - 0.6593)\frac{2.319}{2.6593}60 = 35.65 \text{ mm};$$

$$r_2 = \frac{2 \times 0.3407 \times 2.319}{2 - 0.6593 - 4 \times 0.6593^2} = -238.2 \text{ mm},$$

and the lateral spherical aberration at the maximum height $h = D/2 = 15$ mm, as per Eq. (1.2.15), is

$$\delta s'_{Sph} - \frac{1}{8} \times \frac{0.6593}{0.3407^2} \times \frac{3.3405}{2.319} \times \frac{15^2}{60} = 3.84 \text{ mm}.$$

This yields the transverse spherical aberration as $\delta s'_t = \delta s'_{Sph} \times \frac{15}{60} = 0.96$ mm.

If the lens is made of SF-11 glass the optimal shape is different. Doing just as above, but with $n_D = 1.78472$, we obtain $r_1 = 2 \times 0.43969 \times 2.1206 \times (60/2.5603) = 43.7$ mm and the corresponding value for the second radii becomes

$$r_2 = 2 \times 0.43969 \times 2.1206 \times \frac{60}{2 - 0.5603 - 4 \times 0.5603^2} = 608.2\text{mm}.$$

As we see, the second radius in this case is also positive so that the optimal shape is a meniscus with a very large second radius. The value of the lateral spherical aberration, if the lens stands optimally, is

$$\delta s'_{Sph} = -\frac{1}{8} \times \frac{0.5603}{0.43969^2} \times \frac{3.4397}{2.1206} \times \frac{15^2}{60} = 2.20\text{mm},$$

which gives the transverse aberration as $\delta s'_t = 0.55$ mm.

(b) If the object is at infinity, but the lens does not stand optimally (e.g., the second radius, r_2, is directed to the object, meaning that the lens is turned by 180°), the calculations should be based on the more general formula (Eq. (1.2.14)) which gives for the BK-7 lens

$$\delta s'_{Sph} = -\frac{1}{2} \times \frac{225}{0.3407^2}\left(\frac{1}{60} - \frac{2.6593 \times 0.3407}{238.2} + 2.319 \times 60 \times \frac{0.3407^2}{238.2^2}\right) = -12.75\text{mm},$$

and $\delta s'_t = -3.19$ mm. For the SF-11 lens $\delta s'_{sph} = -10.77$ mm and $\delta s'_t = -2.69$ mm. As we see, optimization of the lens position causes an about three times reduction of the spherical aberration for the BK-7 lens and an about five time reduction for the SF-11 lens.

P.1.2.6. (a) Let us find first the paraxial parameter of a cylinder lens (a rod). Since $r_1 = -r_2 = 7/2 = 3.5$ mm and $d = 7$ mm, we obtain from the lens formula (Eq. (1.2.11))

$$\frac{1}{f'} = (1.5168 - 1)\frac{2}{3.5} - \frac{7 \times (1.5168 - 1)^2}{3.5^2 \times 1.5168} = 0.1947;$$
$$f' = 5.136\text{mm},$$

and using this value in Eq. (1.2.10) we find the location of the principal planes:

$$a' = f' - S'_F = \frac{5.136 \times 7 \times 0.5168}{3.5 \times 1.5168} = 3.5 \text{ mm}.$$

Therefore, both principal planes are located in the center of the lens and the distance to the object is $S = -(3.5 + 2) = -5.5$ mm. Then the distance to the paraxial image is $S' = (1/S + 1/f')^{-1} = (1/5.136 - 1/5.5)^{-1} = 77.83$ mm, i.e., the plane P passing through the paraxial image is located at a distance $l'_0 = 74.33$ mm from the lens. It is the plane P where we should calculate the transverse spherical aberrations.

Before proceeding further, we will consider the general case of ray tracing through a full cylinder lens. Referring to Fig. 1.2.36, we find the segment Δ_1 from two triangles, AA_1B and OA_1B: $\Delta_1 = \rho - \sqrt{\rho^2 - y_1^2} - (y_1/\tan u) \quad l,$

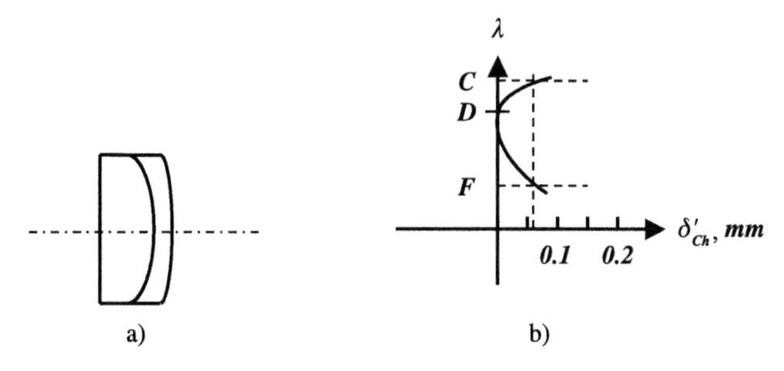

Fig. 1.2.35 Problem P.1.2.4 – (a) The doublet lens and (b) its secondary spectrum.

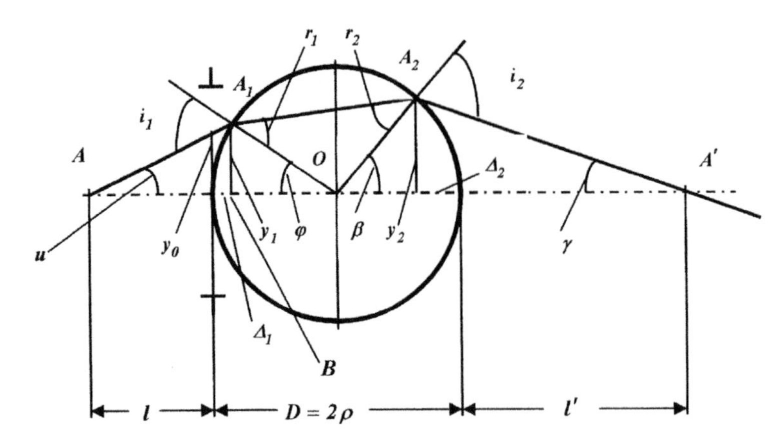

Fig. 1.2.36 Problem P.1.2.6 – Ray tracing through a cylindrical rod.

where $\rho = D/2$ is the radius of the lens. By dividing both sides by ρ and denoting

$$a = 1 + l/\rho; \quad b = 1/\tan u; \quad \sin\varphi = z, \tag{A}$$

we have the following equation with regard to z:

$$1 - z^2 = a^2 - 2abz + b^2z^2. \tag{B}$$

Solving this one can find the angle φ as follows:

$$\sin\varphi = z = \frac{ab - \sqrt{a^2b^2 - (a^2-1)(b^2+1)}}{b^2+1}. \tag{C}$$

This yields further $i_1 = u + \varphi$; $r_1 = \arcsin(\sin i_1/n)$ and from the triangle A_1OA_2, taking into account that $r_1 = r_2$; and $i_1 = i_2$, we have $180° = \varphi + \beta + (180° - 2r_1)$ which gives

$$\beta = 2r_1 - \varphi; \quad \gamma = i_1 - \beta. \tag{D}$$

As $y_2 = \rho\sin\beta$ and $\Delta_2 = \rho - \rho\cos\beta$ we finally obtain:

$$l' = \frac{y_2}{\tan\gamma} - \Delta_2 = \rho\left(\frac{\sin\beta}{\tan\gamma} - 1 + \cos\beta\right). \tag{E}$$

The lateral spherical aberration for each ray angle u is found as the difference between l'_0 and the value l' from Eq. (E). The corresponding transverse spherical aberration is determined as

$$\delta s'_t = \tan\gamma \times (l' - l'_0). \tag{F}$$

To build the aberration plot we should repeat the procedure as per Eqs. (A)–(F) for several angles u. The whole range of u can be found as the following: $\sin u_{max} = \rho/(l+\rho)$ and according to the data of the problem we find $u_{max} = 39.52°$. If the entrance pupil is located at the distance $l = 2$ mm from the light source (the object) the maximum coordinate in the pupil plane is $y_{max} = 1 \times \tan 39.52 = 1.65$ mm and the relative coordinate of a point in the entrance pupil is calculated as $\tilde{y} = y/1.65$.

We start with $u_1 = 35°$. This gives $y_{01} = 2\tan 35° = 1.4009$; $\tilde{y}_1 = 0.849$. Then from Eqs. (A)–(E) we find step-by-step: $a = 1.5714$; $b = 1.428$; $z = \sin\varphi = 0.490$; $\varphi = 29.33°$; $i_1 = u_1 + \varphi = 64.33°$; $\sin r_1 = 0.5942$; $r_1 = 36.46°$; $\beta = 43.58°$; $\gamma = 20.75°$; $l' = 5.403$ mm. This leads to $\delta s'_{Sph} = 5.403 - 74.33 = -68.93$ mm; $\delta s'_t = -\tan 20.75° \times 68.93 = -26.11$ mm.

Choosing $u_2 = 25°$ we get $y_{02} = 2\tan 25° = 0.9326$; $\tilde{y}_2 = 0.565$. Then $a = 1.5714$; $b = 2.1445$; $z = \sin\varphi = 0.2862$; $\varphi = 16.63°$; $i_1 = u_1 + \varphi = 41.63°$; $\sin r_1 = 0.4380$; $r_1 = 25.97°$; $\beta = 35.32°$; $\gamma = 6.31°$; $l' = 17.65$ mm. This leads to $\delta s'_{Sph} = 17.65 - 74.33 = -56.68$ mm; $\delta s'_t = -\tan 6.31° \times 56.68 = -6.26$ mm.

Choosing $u_3 = 10°$ we get $y_{03} = 2\tan 10° = 0.3527$; $\tilde{y}_1 = 0.214$. Then $a = 1.5714$; $b = 5.671$; $z = \sin\varphi = 0.1017$; $\varphi = 5.835°$; $i_1 = u_1 + \varphi = 15.835°$; $\sin r_1 = 0.1798$; $r_1 = 10.36°$; $\beta = 14.893°$; $\gamma = 0.942°$; $l' = 54.59$ mm. This leads to $\delta s'_{Sph} = 54.59 - 74.33 = -19.73$ mm; $\delta s'_t = -\tan 0.942° \times 19.73 = -0.324$ mm.

Finally the aberration plot is as shown in Fig. 1.2.37 (we take into account that the rays incident on the lower half of the entrance pupil are going up at the exit, so that the transverse aberrations here are positive and the diagram is anti-symmetric).

(b) Let the lens be a full sphere (a ball lens) and the object is an on-axis point. Tracing any ray coming to the lens, we consider the plane of incidence which includes the ray and the optical axis. Such a plane is a meridional plane and it is similar to the one shown in Fig. 1.2.36 for the cylindrical rod. Therefore Eqs. (A)–(F) remain valid

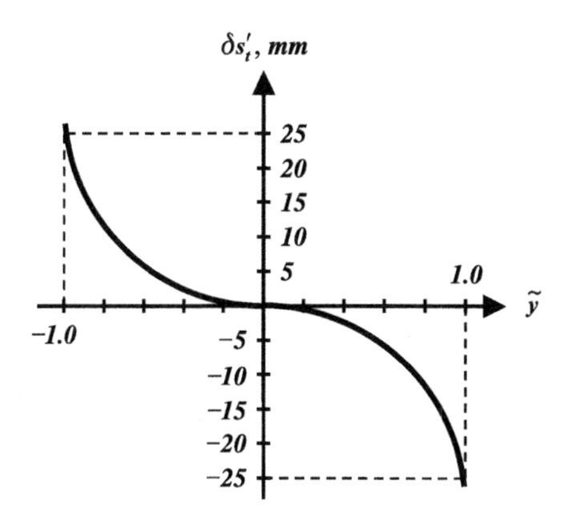

Fig. 1.2.37 Problem P.1.2.6 – Transverse spherical aberration plot.

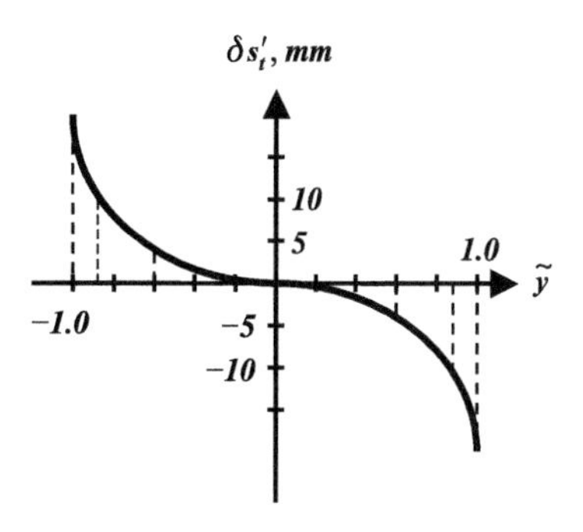

Fig. 1.2.38 Problem P.1.2.7 –Transverse spherical aberration plot (plano-convex cylindrical lens).

and the results will be identical to those obtained in (a) above.

P.1.2.7. As was done in Problem P.1.2.6, we find the parameters of paraxial imaging. Referring to Fig. 1.2.9, we find the focal length and location of the principal planes of a new lens from Eqs. (1.1.9)–(1.1.11), keeping in mind that $r_1 = 3.5$ mm and $r_2 = \infty$:

$$\frac{1}{f'} = \frac{0.5168}{3.5}; \quad f' = 6.772 \text{ mm};$$

$$a' = f'\frac{d(n-1)^2}{r_1 n} = 6.772\frac{4 \times 0.5168^2}{3.5 \times 1.5168} = 2.637 \text{ mm}.$$

To get the paraxial image in the plane P it is necessary to locate the object A at the distance

$$S = -\left(\frac{1}{S'} - \frac{1}{f'}\right)^{-1}$$

$$= -\left(\frac{1}{74.33 + 2.637} - \frac{1}{6.772}\right)^{-1} = -7.43 \text{ mm},$$

from the front surface of the lens. We choose also the entrance pupil to be 7.43 mm distant from A and assume its size $D_p = 4.4$ mm (to ensure that any ray incident on the entrance pupil will pass the lens and proceed to the plane P). We calculate the aberration plot by performing rigorous ray tracing (see Fig. 1.2.38). Consideration of the first (bended) surface gives exactly the same expressions (A)–(C) as in Problem P.1.2.6 and the same formulas for the angles i_1 and r_1:

$$i_1 = u + \varphi; \quad r_1 = \arcsin\left(\frac{\sin i_1}{n}\right).$$

The rest is different. That is,

$$r_2 = \varphi - r_1; \quad y_2 = \rho \sin\varphi - [t - \rho(1 - \cos\varphi)]\tan(r_2);$$
$$\gamma = \arcsin(n \sin r_2); \quad l' = \frac{y_2}{\tan\gamma}. \tag{G}$$

Once the value of l' is found the transverse aberration is calculated from Eq. (F) of Problem P.1.2.6.

We apply the above approach choosing first $u_1 = 15°$. We get step-by-step: $y_{01} = 7.43 \times \tan 15° = 1.99$; $\tilde{y}_1 = 1.99/2.2 = 0.88$. Then $a = 3.123$; $b = 3.732$; $z = \sin\varphi = 0.6284$; $\varphi = 38.43°$; $i_1 = u_1 + \varphi = 53.43°$; sin $r_1 = 0.4143$; $r_1 = 31.97°$; $r_2 = 38.43 - 31.97 = 6.46°$; $\gamma = 9.824°$; $y_2 = 1.833$; $l' = 10.58$. Therefore, $\delta s'_{\text{Sph}} = 10.58 - 74.33 = -63.74$ mm; $\delta s'_{\text{t}} = -11.03$ mm.

Choosing then $u_2 = 10°$ we have $y_{01} = 7.43 \times \tan 10° = 1.31$; $\tilde{y}_1 = 1.31/2.2 = 0.596$. Then $a = 3.123$; $b = 5.671$; $z = \sin\varphi = 0.388$; $\varphi = 22.83°$; $i_1 = u_1 + \varphi = 32.48°$; sin $r_1 = 0.3574$; $r_1 = 20.94°$; $r_2 = 22.83 - 20.94 = 1.89°$; $\gamma = 2.86°$; $y_2 = 1.235$; $l' = 24.70$; and $\delta s'_{\text{Sph}} = -49.63$ mm; $\delta s'_{\text{t}} = -2.48$ mm.

For the angle $u_2 = 5°$ we obtain $y_{01} = 7.43 \times \tan 5° = 0.65$; $\tilde{y}_1 = 0.65/2.2 = 0.295$. Then $a = 3.123$; $b = 11.43$; $z = \sin\varphi = 0.1873$; $\varphi = 10.795°$; $i_1 = u_1 + \varphi = 15.795°$; sin $r_1 = 0.180$; $r_1 = 10.338°$; $r_2 = 10.795 - 10.338 = 0.457°$; $\gamma = 0.694°$; $y_2 = 0.626$; $l' = 51.68$; and finally $\delta s'_{\text{Sph}} = -22.65$ mm; $\delta s'_{\text{t}} = -0.274$ mm.

The resulting aberration plot is presented in Fig. 1.2.38. Comparing this diagram with that of Fig. 1.2.37 makes very clear that the full rod has a much greater aberration than the plano-convex lens.

P.1.2.8. Assuming the lens performs ideal imaging we find the location of the object and the image from paraxial relations:

$$S = \frac{1 - V}{V}f' = -\frac{3}{2}30 = -45 \text{ mm}; \quad S' = 90 \text{ mm}.$$

The right-angle prism located 20 mm behind the lens can be unfolded as a parallel glass slab of thickness $t_e = 30$ mm (the dotted line in Fig. 1.2.39a). So, we can estimate the astigmatism introduced by the prism using Eq. (1.2.20) for the plate with $d = t_e$. Doing the calculation for several incidence angles $u < 30°$ we find

$$u = 10° : u' = \arcsin(\sin 10°/1.5168) = 6.574°;$$

$$\delta'_{Ast} = (1.5168^2 - 1)\frac{\tan^3 6.574}{\tan 10°}30 = 0.338 \text{ mm}$$

$$u = 20° : u' = \arcsin(\sin 20°/1.5168) 13.03°;$$

$$\delta'_{Ast} = (1.5168^2 - 1)\frac{\tan^3 13.03°}{\tan 20°}30 = 1.32 \text{ mm}$$

$$u = 30° : u' = \arcsin(\sin 30°/1.5168) = 19.25°;$$

$$\delta'_{Ast} = (1.5168^2 - 1)\frac{\tan^3 19.25°}{\tan 30°}30 = 2.88 \text{ mm.}$$

The astigmatism plot is shown in Fig. 1.2.39b.

As to distortion, we have to find first the location of the exit pupil. Since imaging is done by the lens only and the prism is big enough and does not affect the angular field of view, we can assume that the lens itself is the aperture diaphragm and the exit pupil. Then the distance p appearing in Eq. (1.2.23) is equal to 90 mm. Calculation for several incident angles, u, using Eq. (1.2.23), yields for the angle $u = 10°$

$$\Delta = \frac{1.5168^2 - 1}{1.5168^3} \times \left(\frac{10}{180}\pi\right)^2 = 0.189\%$$

for the angle $u = 20°$, $\Delta = 0.76\%$
for the angle $u = 30°$, $\Delta = 1.71\%$.

The distortion plot is depicted in Fig. 1.2.39c.

P.1.2.9. We should address here the curvature of field, and, more specifically, check Petzval's condition, as in Eq. (1.2.21). To find the radii of the lens we use the lens formula in the paraxial range, remembering that the lens is symmetrical ($r_1 = -r_2$):

$$r_1 = 2f'(n-1) = 2 \times 40 \times 0.5168 = 41.34 \text{ mm}$$
$$= -r_2.$$

Now Petzval's sum is as follows:

$$\frac{1}{\rho} = -\frac{1}{41.34}\left(\frac{1}{1.5168} - 1\right) + \frac{1}{41.34}\left(1 - \frac{1}{1.5168}\right)$$
$$= 0.01648,$$

and the curvature of Petzval's surface is $\rho = 60.67$ mm, i.e., the image will suffer significant degradation since off-axis areas have sharp images not in the plane P but rather on the surface of curvature ρ.

The flattener is a negative lens, usually plano-concave, positioned with its plane surface just in the image plane (in the plane P in our case) in a manner demonstrated in Fig. 1.2.40. Because the flattener practically coincides with the image it does not affect the image magnification. On the other hand, its contribution to Petzval's sum might improve the field curvature. Denoting the first radius of the flattener as r_3, we rewrite Eq. (1.2.21) in the following manner:

$$\frac{1}{\rho} = -\frac{2}{r_1}\left(\frac{1}{n} - 1\right) + \frac{1}{r_3}\left(\frac{1}{n_{Fl}} - 1\right),$$

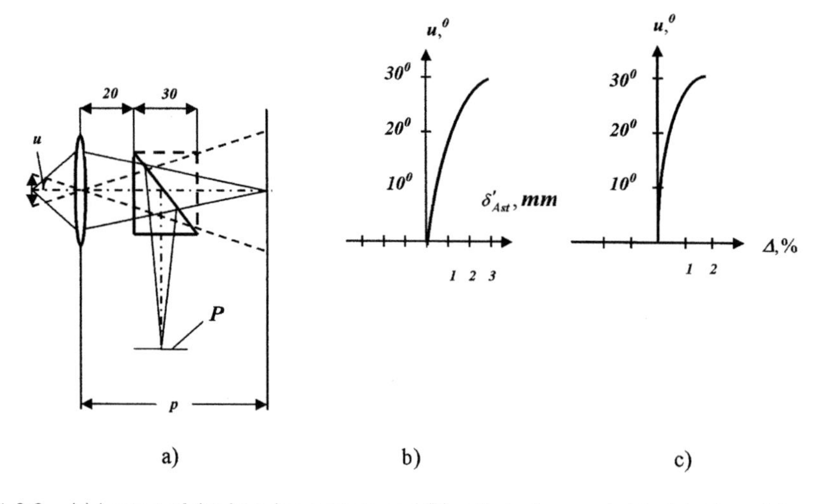

Fig. 1.2.39 Problem P.1.2.8 – (a) Layout of the imaging system and (b) astigmatism and, (c) distortion of the right-angle prism.

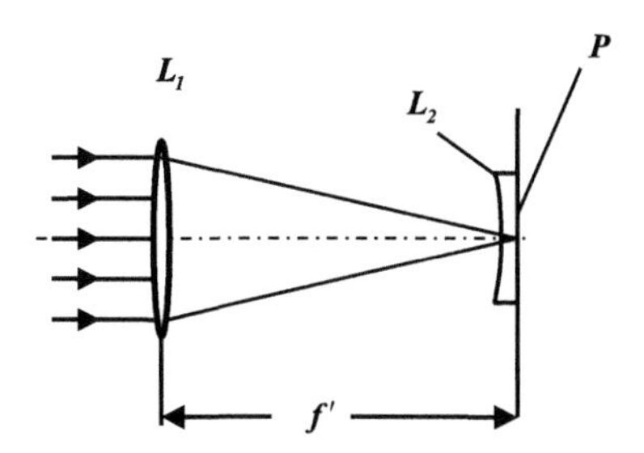

Fig. 1.2.40 Problem P.1.2.9 – The flattener and the image plane.

where n_{Fl} is the refractive index of the flattener glass. Optimization of r_3 might cause zero curvature ($\rho \to \infty$) of Petzval's sum (i.e., the image plane becomes flat). This occurs if r_3 is as follows:

$$r_3 = \frac{1/n_{Fl} - 1}{2(1 - 1/n)} r = \frac{1/1.78477 - 1}{2(1 - 1/1.5168)} 41.34$$
$$= -26.67 \text{ mm}.$$

P.1.2.10. To calculate the OSC we use the formulas for rigorous ray tracing derived in Problem P.1.2.6. For the case relevant here when the incident rays are coming from infinity and therefore $u = 0$ we rewrite the expressions of Problem P.1.2.6 as follows:

$$i_1 = \varphi = \arcsin(h/\rho); \quad r_1 = \arcsin(\sin\varphi/n);$$
$$\beta = 2r_1 - \varphi; \quad \gamma = 2(\varphi - r_1)$$

where h is the height of the ray striking the lens and ρ is the ball radius. The OSC for any chosen value h is calculated from Eq. (1.2.25) by substituting the angle u' with y found from Eq. (A) and the focal length,f' , as per Eq. (1.1.11) in a complete form:

$$\frac{1}{f'} = \frac{2(n-1)}{\rho} - \frac{2\rho(n-1)^2}{\rho^2 n} =$$
$$\frac{2 \times 0.77}{1.5} - \frac{2(0.77)^2}{1.5 \times 1.77} = 0.580;$$
$$f' = 1.724 \text{ mm}.$$

For $h = 0.25$ mm: $\sin\varphi = 0.25/1.5 = 0.1667$; $\varphi = 9.59°$; $r_1 = 5.40°$; $\gamma = 8.38°$; and

$$\text{OSC} = \frac{0.25}{\sin 8.38°} - 1.724 = -8.59 \times 10^{-3} \text{ mm}.$$

For $h = 0.5$ mm: $\sin\varphi = 0.5/1.5 = 0.333$; $\varphi = 19.47°$; $r_1 = 10.854°$; $\gamma = 17.232°$; and

$$\text{OSC} = \frac{0.5}{\sin 17.232°} - 1.724 = -0.0362 \text{ mm}.$$

For $h = 0.75$ mm we get in a similar manner OSC $= -0.0822$ mm and for $h = 1.0$ mm OSC $= -0.147$ mm. The plot of OSC is presented in Fig. 1.2.41a. The horizontal coordinate on the diagram, $\tilde{y} = h/h_{max}$, is defined as the ratio of the real height to the maximum possible height dictated by the refractive index n. To determine h_{max} we refer to the limiting situation shown in Fig. 1.2.41b. Considering geometry of the ray we obtain

$$\varphi = 2r_1; \quad \frac{\sin\varphi}{\sin r_1} = \frac{\sin\varphi}{\sin\varphi/2} = 2\cos\frac{\varphi}{2} = n,$$

and further $\varphi = 2\arccos(n/2) = 2\arccos(1.\ 77/2) = 55.5°$; $h_{max} = 1.5 \sin 55.5° = 1.236$ mm. Therefore, the maximum diameter of the beam which is concentrated by the lens somewhere behind the ball is 2.472 mm. The rest of the rays striking the ball cross the optical axis inside the lens and cannot be exploited for imaging or any other application related to energy concentration.

P.1.2.11. The design of the objective will be based on aplanatic points of spherical surfaces. Choosing the concept depicted in Fig. 1.2.42, we start with the relations for the first component. Since the immersion oil has practically the same refractive index as that of the first lens ($n_{D1} = 1\ .\ 84666$), the ray originating in object point A travels to point B with no change in direction and $S_2 = (-t_0 + r_2)$. On the other hand, if point A is the

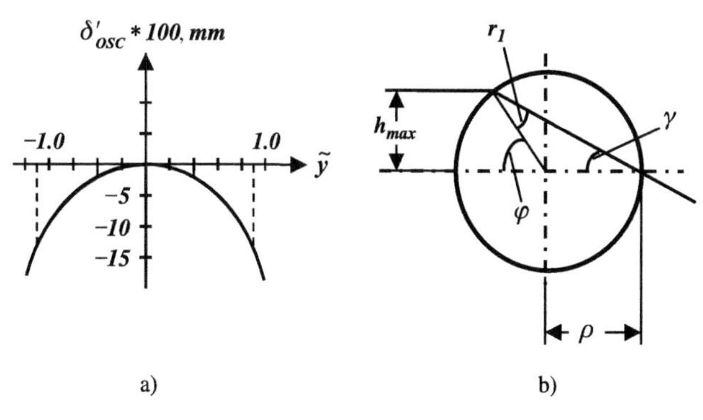

Fig. 1.2.41 Problem P.1.2.10 – (a) OSC of a sapphire ball lens and (b) definition of the maximum ray height.

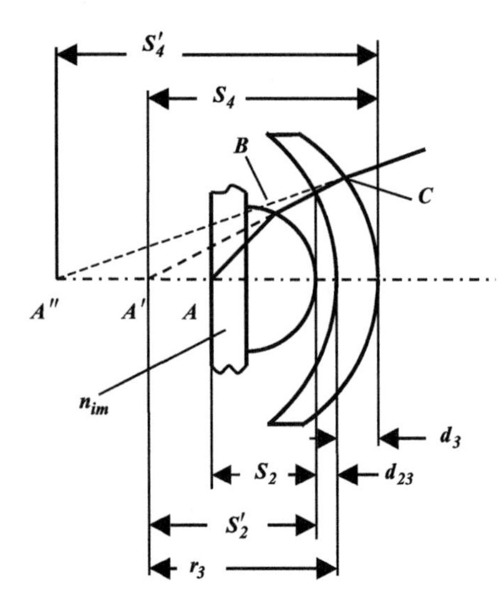

Fig. 1.2.42 Problem P.1.2.11 – Aplanatic objective consisting of two components.

aplanatic point of the second (spherical) surface one may use Eq. (1.2.26) and to get

$$S_2 = \frac{n'_2 + n_2}{n_2} r_2 = r_2 - t_0; \qquad \text{(A)}$$

which gives

$$r_2 = -t_0 \frac{n_2}{n'_2} = -0.7 \times \frac{1.84666}{1} = -1.29 \text{ mm}.$$

Then $S_2 = -1.29 - 0.7 = -1.99$ mm and Eq. (1.2.26) yields

$$S'_2 = S_2 \frac{n_2}{n'_2} = -1.99 \times 1.84666 = -3.67 \text{ mm}.$$

Magnification of the first component is governed by Eq. (1.2.27): $V_1 = 1.84666^2 = 3.41$. Then the magnification of the second lens should be $V_{t2} = 6/3.41 = 1.76$. According to the schematics of Fig. 1.2.42, point A′ which is the image of A created by the first lens serves as a center of curvature of the third (spherical) surface (e.g., the first surface of the second component). This point is also aplanatic, but there is no bending of rays here and the magnification is determined from Eq. (1.2.28): $V_3 = 1/n_{D2}$. The ray travels further to point C of the fourth surface for which A′ is the aplanatic point and the image after refraction is point A″. Here again the magnification is determined by Eq. (1.2.27): $V_4 = (n_{D2})^2$. Thus the entire magnification of the second element becomes $V_{t2} = V_3 \times V_4 = (n_{D2})^2/n_{D2} = n_{D2}$. As we see above, this value should be 1.76. The closest glass from the data of Appendix 2 is SF-11 with $n_D = 1.78472$. We choose this for the second element of the objective.

Going back to the radii of the second lens, for the third surface we have $r_3 = S'_2 - d_2 = -3.67 - 1 = -4.67$ mm. Keeping in mind that $S_4 = r_3 - d_3 = -4.67 - 3 = -7.67$ mm we get, again using Eq. (1.2.26)

$$r_4 = -7.67 \frac{1.78472}{1.78472 + 1} = -4.912 \text{ mm},$$

and $S'_4 = 1.78472 \times S_4 = -13.69$ mm. Thus, the objective is a compound of two elements performing imaging around the aplanatic points only. The total magnification is $V_{tot} = V_1 \times V_{t2} = 3.41 \times 1.78472 = 6.085$, i.e., 1.4% deviation from the required value (such a tolerance is usually acceptable).

It should be mentioned that the design presented here is for demonstration and teaching purposes only. In reality many more ray tracing operations followed by image quality analysis are required.

P.1.2.12. We start with the positioning of the system components and use the paraxial formulas for thin lenses. As total magnification $V = 5/0.25 = 20$ and $V_1 = -5$, we get

$$S'_1 = f'(1 - V_1) = 15 \times 6 = 90 \text{ mm};$$

$$V_2 = \frac{V}{V_1} = \frac{20}{(-5)} = -4;$$

$$S_2 = 15\frac{1+4}{-4} = -18.75 \text{ mm};$$

$$S'_2 = S_2 \times V_2 = 75 \text{ mm}.$$

The reticle of thickness $d = 2$ mm makes the optical path longer by $d(1 - 1/n) = 0.67$ mm (see Problem P.1.1.6). This causes the defocusing aberration in the plane P, as depicted in Fig. 1.2.43. Assuming the active size of the

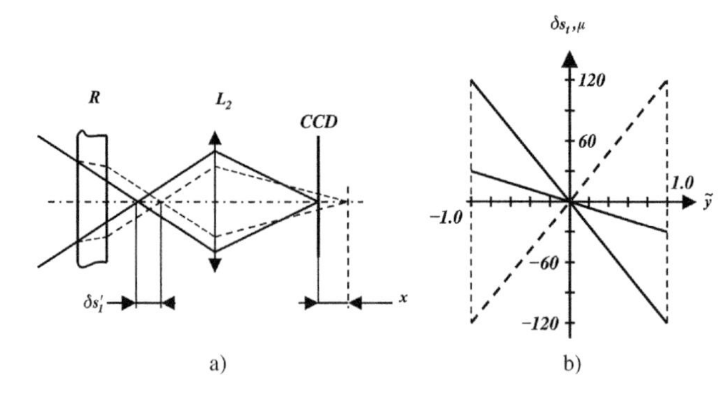

Fig. 1.2.43 Problem P.1.2.12 – (a) Defocusing caused by reticle and (b) the aberration plot in the CCD plane.

first lens is dictated by its diameter, we have for the coordinate of the plot in the plane P

$$\delta s'_{1,t} = -\frac{y_1}{S'_1}\delta_R = -\frac{0.67}{90}y_1;$$

$$\delta s'_{1,t\max} = -\frac{0.67}{90}4 = -29.6\ \mu\text{m}.$$

Transferring this aberration to the CCD plane, we obtain the corresponding straight line on the plot, with the maximum deviation $\delta s'_{2,t\ \max} = \delta s'_{1,t\ \max}V = 29.6\times(-4) = 118.4\ \mu$m. This means there is a noticeable defocusing on the CCD. Correction can be done by displacement of the CCD to a new position where the aberration has the same value, but with opposite sign (shown by the dotted line in Fig. 1.2.43b). To calculate the displacement required to get the defocusing of $(-118.6)\ \mu$m we first find the active size of the second lens: $h_2 = h_1S_2/(-S'_1) = 0.83$ mm, which gives the necessary displacement as follows:

$$\delta s'_2 = x = 0.1184 \times 75/0.83 = 10.84\ \text{mm}.$$

Instead of referring to the transverse aberration we could consider the lateral aberration. In such a case we get $x = 0.67 \times 4^2 = 10.8$ mm which is actually the same result.

Of course, the case considered in the problem is quite trivial, but it demonstrates the principle of aberration transfer and summation.

P.1.2.13. Considering the on-axis point we find the angular range of rays creating the image as $\pm\beta = \pm\arctan(D/2S') = \pm\arctan(1/2f) = \pm 23°$ (it is assumed here that the difference between S' and focal length is small while imaging distant objects). As is evident from the aberration plot, the residual (uncorrected) aberrations of lens L are significant and cannot be neglected. Correction by additional elements of the system is desirable. This can be realized in a layout with a penta-prism and cannot be done if a mirror is used for bending. The prism introduces additional spherical aberration described by Eq. (A) of Problem P.1.2.2. By substituting in that expression different values of u, from 0° to 23°, we obtain the plot shown in Fig. 1.2.44a by the solid line. Comparing this to the residual aberration of the lens, shown by the dotted line, we find that they have opposite sign and therefore noticeable correction can be done if the size of the prism is properly chosen. We will do a full correction for the maximum angle $u = 23°$ where aberration of the lens is as high as 1.1 mm. Since the penta-prism is equivalent to a parallel glass slab of thickness $t_e = 3.414$a, where a is the entrance face size (see Section 1.1.4 for unfolded diagram of the prisms), we get

$$\frac{3.414a}{n}\left(1 - \frac{n\cos u}{\sqrt{n^2 - \sin^2 u_{\max}}}\right) = 3.414a \times 0.031$$

$$= 1.1\ \text{mm}; \quad a = 10\ \text{mm}.$$

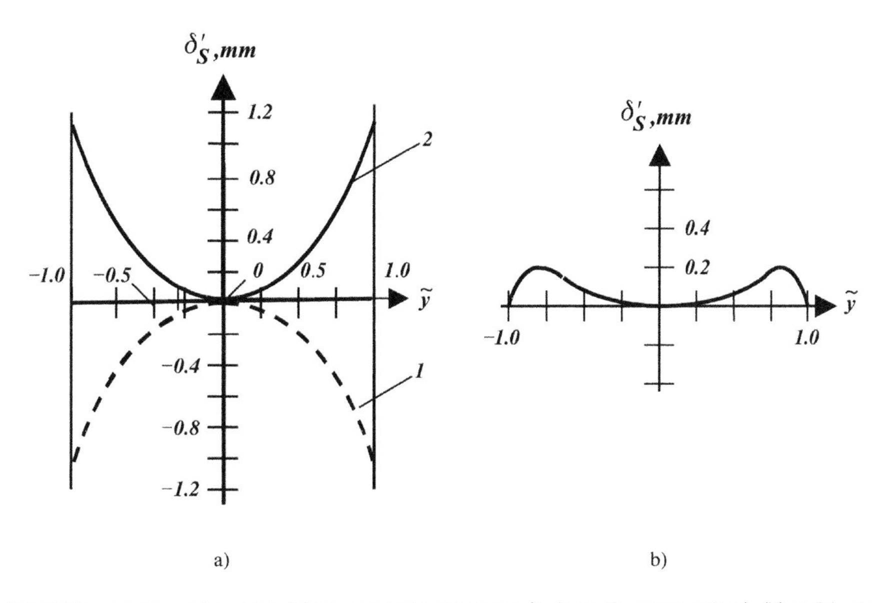

Fig. 1.2.44 Problem P.1.2.13 – an aberration plot: (a) for separate elements (1, lens; 2, penta-prism); (b) residual values after correction.

Calculating the aberration of the prism for different angles and subtracting the results from the plot of the lens aberration we obtain the plot of the final residual aberrations of the system after correction (Fig. 1.2.44b).

P.1.2.14. In general it can be stated that a symmetrical configuration with a parallel trace between two components yields the best results with regard to residual aberrations. In our case each lens is optimized for imaging from infinity to its focal plane. Then, according to the rules described in Section 1.2.1.6 we should find the aberrations of lens L_1 in reverse operation mode (light propagating from the right to the left) and transfer them, keeping in mind the linear magnification of the system, to the focal plane of lens L_2 where they are added to the aberrations of the second lens. Obviously the smaller the aberration of separate elements the lower the total sum of aberrations in the bundle entrance.

Optimization of the first lens can be done using Eqs. (1.2.15) and (1.2.16). For $n_D = 1.5168$ we get the radii of the lens as follows (r_1 is going towards the parallel beam, i.e., inside the condenser; $h = D/2 = 15\,\text{mm}$):

$$\xi = 1/1.5168 = 0.6593;$$

$$r_1 = -2 \times (1 - 0.6593)\frac{2.319}{2.6593}60 = -35.65\ \text{mm};$$

$$r_2 = 2\frac{0.3407 \times 2.319}{2 - 0.6593 - 1.739} = 238.2\ \text{mm}.$$

Similar calculations for the second lens give

$$r_3 = -r_1\frac{f'_2}{f'_1} = 71.3\ \text{mm};$$

$$r_4 = -r_2\frac{f'_2}{f'_1} = -238.2 \times 2 = -476.4\ \text{mm}.$$

Now, using Eq. (1.2.15) we find the transverse aberrations of separate components:

$$\delta s'_{t1} = -\frac{1}{8}\frac{0.6593 \times 3.3407}{0.1161 \times 2.319}\frac{h^3}{f_1'2} = -0.959\ \text{mm};$$

$$\delta s'_{t2} = -\frac{1}{8}\frac{0.6593 \times 3.3407}{0.1161 \times 2.319}\frac{h^3}{f_2'2} = -0.24\ \text{mm}$$

and the total transverse aberration in the plane of the bundle entrance:

$$\delta s'_{\text{tot},t} = -\delta s'_{t1}V + \delta s'_{t2} = 0.959 \times 2 - 0.24 = 1.68\ \text{mm}.$$

It is interesting to mention that if two components were identical ($f'_1 = f''_2$) then $\delta s'_{t1} = \delta s'_{t2}$; $|V| = 1$; and the total transverse aberrations of the condenser would approach zero.

P.1.2.15. Expression (1.2.33) defines the minimum resolvable spot as follows:

$$\delta_{\text{dif}} = AB = \frac{1.22\lambda p}{D_p} = 0.5\ \mu\text{m}.$$

Assuming the entrance pupil is located at the mounting of the imaging lens we have $p = 30\,\text{mm}$. Then, taking also $\lambda = 0.5\ \mu\text{m}$, we get the necessary size of the lens (the entrance pupil) as

$$D_p = \frac{1.22 \times 0.5 \times 30}{0.5} = 36.6\ \text{mm}.$$

P.1.2.16. We use Eq. (1.2.36), keeping in mind that the illumination angle cannot be greater than that defined by the numerical aperture of the objective. Therefore for the first two objectives only a fraction of NAC can be exploited and the resolution is

$$R_1 = \frac{\lambda}{2\text{NA}} = \frac{0.5}{2 \times 0.25} = 1\ \mu\text{m};$$

$$R_2 = \frac{\lambda}{2\text{NA}} = \frac{0.5}{2 \times 0.65} = 0.38\ \mu\text{m}.$$

The NA of the last objective is greater than that of the condenser and in this case from Eq. (1.2.36) we get

$$R_3 = \frac{\lambda}{\text{NA} + \text{NA}_C} = \frac{0.5}{1.2 + 0.96} = 0.23\ \mu\text{m}.$$

P.1.2.17. The key equation here is

$$\sin\beta - \sin\beta_0 = \frac{m\lambda}{d}, \qquad \text{(A)}$$

where β_0 is the illumination angle and β is the angle of the diffraction maximum of order m. If $m = -1$ and $\beta = -\beta_0$ we have the situation depicted in Fig. 1.2.25b and the resolution is improved by a factor of 2. This occurs if the following condition is obeyed:

$$\tan \beta_0 = \frac{D/2}{f'} = \frac{2.5}{16} = 0.156; \quad \beta_0 = 8.9^\circ.$$

Expression (1.2.35) renders resolution in this case: $d = 0.5/2 \ \sin 8.9^\circ = 1.6\,\mu\text{m}$. If the illumination angle is smaller than 8.9° we still can improve the resolution, as follows from the geometry shown in Fig. 1.2.45. In this case the position of the zero order is determined by $y_0 = f' \tan \beta_0 = 16 \times \tan 5^\circ = 1.4\,\text{mm}$ whereas the $(1)^{\text{st}}$ order comes at the side of the aperture stop:

$$\tan \beta = 2.5/16 = 0.156; \quad \beta = 8.9^\circ.$$

Then, from Eq. (A) one obtains the resolution as follows:

$$d = \frac{0.5}{\sin 5^\circ + \sin 8.9^\circ} = 2.07\ \mu\text{m}.$$

P.1.2.18. For a diffraction-limited system the parameter z in Eq. (1.2.37) becomes

$$z = \frac{\pi D d}{2\lambda p} = 3.8317 \frac{d}{d_{\text{dif}}},$$

where d_{dif}, determined from Eq. (1.2.32), is almost the full spot size. For a circle of $d = 0.5 d_{\text{dif}}$ we get $z = 1.91$ and Eq. (1.2.37) yields $L = 60\%$. This means that if our system is limited by diffraction only up to 60% of the full energy of the spot will be collected inside 50% of its diameter. Since this differs from what was actually found (45%), one can draw the conclusion that the system has noticeable aberrations.

P.1.2.19. The maximum spatial frequency transferred by the system (the cut-off frequency) is determined from Eq. (1.2.45):

$$\nu_C = \frac{2\text{NA}}{\lambda} = \frac{2. \times 0.25}{0.6 \times 10^{-3}} = 833\ \text{cycles/mm},$$

which corresponds to a period of 1.2 μm in the object space or 12.0 μm in the CCD plane (after × 10 magnification). According to the Nyquist theorem the sampling frequency has to be two times higher than the tested frequency, i.e., two pixels of the CCD are necessary for one period of 12 μm. Therefore the CCD elements (pixels) should be arranged with a 6 μm pitch (center-to-center distance).

P.1.2.20. We start with the calculation of modulation, M_o, in the object space. Let the illumination intensity be I_0 and it be spread uniformly on the target. The square wave segments of the target coated with chrome reflect $0.7 I_0$ whereas the target segments with no coating reflect $0.04 I_0$. This gives the following value for the modulation of the object:

$$M_0 = \frac{0.7I_0 - 0.04I_0}{0.7I_0 + 0.04I_0} = 0.89.$$

We denote the background scattered light as I_S and assume that it is the same at all locations (uniformly spread in the system). Therefore the true modulation in the image plane, M_i, and the apparent modulation registered with the scattered light, M'_i, are related as follows:

$$\frac{1}{M'_i} = \frac{I_{\max} + I_{\min} + 2I_S}{I_{\max} - I_{\min}} = \frac{1}{M_i} + \frac{1}{M_S} = \frac{1}{\text{MTF}' \times M_0}, \tag{A}$$

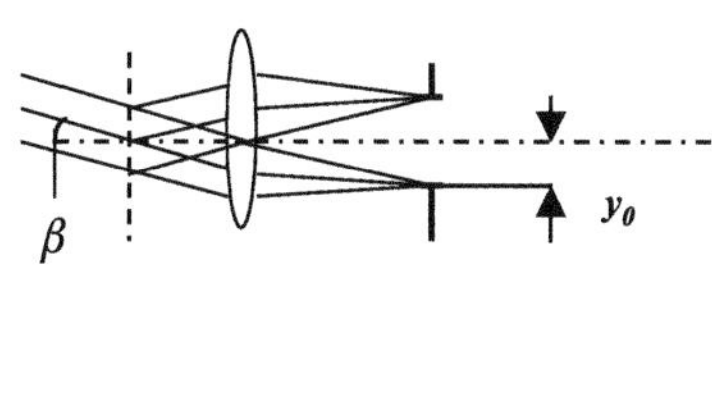

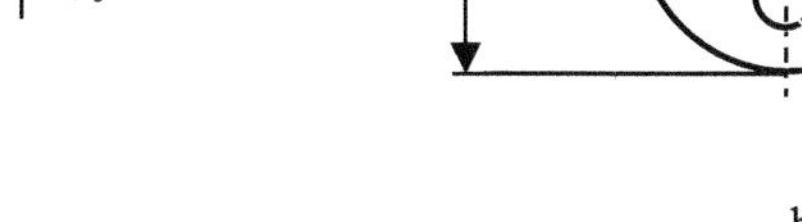

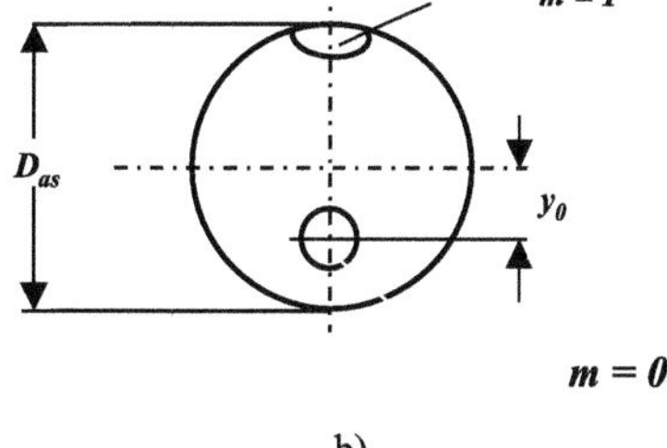

Fig. 1.2.45 Problem P.1.2.17 – (a) Oblique illumination in a microscope objective and (b) location of diffraction maxima in the aperture stop.

where MTF′ is the measured MTF value when the scattering is present. The influence of the scattered light, $1/M_S$, can be found from Eq. (A) and we will find it separately for low frequencies and for high frequencies. Since scattering does not depend on the frequency chosen for measurement we can write:

$$\frac{1}{M_S} = \frac{1}{M'_{iH}} - \frac{1}{M_{iH}} = \frac{1}{M'_{iL}} - \frac{1}{M_{iL}};$$

$$\frac{1}{M_{iH}} = \frac{1}{M'_{iH}} - \frac{1}{M'_{iL}} + \frac{1}{M_{iL}} = \frac{1}{M'_{iH}} - \frac{1}{M'_{iL}} + \frac{1}{M_0}. \qquad \text{(B)}$$

where it is taken into account that at very low spatial frequency MTF is close to 100% (if no scattering is present in the system): $M_{iL} = M_0$. Using in Eq. (B) the definition of MTF as per Eq. (1.2.43), we get for the true MTF at high frequency

$$\text{MTF}_H = \frac{M_{iH}}{M_0} = \left(\frac{1}{\text{MTF}'_H} - \frac{1}{\text{MTF}'_L} + 1\right)^{-1}, \qquad \text{(C)}$$

and by substituting the problem data in Eq. (C) we have

$$\text{MTF}_H = \left(\frac{1}{0.2} - \frac{1}{0.7} + 1\right)^{-1} = 0.22.$$

Obviously, if there is no scattering MTF'_L approaches unity and $\text{MTF}_H = \text{MTF}'_H$.

P.1.2.21. We accept that the minimum overall MTF of the system, optics + CCD + monitor, should be 5% at least for a spatial frequency which can be observed. Since the total MTF is the product of the MTFs of separate elements, we obtain the minimum requirements for the MTF of the imaging optics: $\text{MTF}_o = \text{MTF}_{tot}/\text{MTF}_{CCD} = 0.05/0.6 = 0.083$. As our system is diffraction limited its MTF obeys Eq. (1.2.44) with cut-off frequency (Eq. (1.2.45)) given by

$$v_C = \frac{2\text{NA}}{\lambda} = \frac{2 \times 0.15}{0.5 \times 10^{-3}} = 600 \text{ cycles/mm},$$

where MTF = 0. Although the graph of MTF(v) described by Eq. (1.2.44) is a curve, its deviation from a straight line is not very noticeable. Then, approximating the graph by a straight line we find the frequency which corresponds to MTF = 0.083. This frequency is $v = 600(1 - 0.083) = 550$ cycles/mm. Therefore, it is impossible to see on the monitor the details originating in a spatial frequency of 575 cycles/mm in the object plane.

P.1.2.22. We choose the architecture of the telecentric configuration as that of Fig. 1.2.30b and restrict ourselves to the paraxial range. Then we have

$$f'_1 + f'_2 = 100 \text{ mm}; \quad f'_2 = f'_1 \times |V| = 3f'_1;$$

$$f'_1 = 25 \text{ mm}; \quad f'_2 = 75 \text{ mm}.$$

To find the size of the aperture stop we should calculate the required NA in the object space. Suppose the system is free of aberration. Then, using the first relation of Eq. (1.2.35) and keeping in mind that the required resolution should be equal to 2 μm, we obtain

$$d = \frac{\lambda}{\text{NA}} = 2\ \mu\text{m}; \quad \text{NA} = 0.5/2 = 0.25$$
$$= \sin u_{max}; \quad u_{max} = 14.5°.$$

This yields the size of the aperture stop as follows (see Fig. 1.2.46): $D_{ab} = 2f'_1 \tan u_{max} = 2 \times 0.258 \times 25 = 12.9$ mm. By considering further the angular field of view and taking into account that the marginal chief ray should pass through the center of the aperture stop, we get $y = f'_1 \tan\beta = 25 \times \tan 5° = 2.2$ mm and the size of the first lens becomes $D_1 = 2h_1 = 2(f'_1 \tan\beta + D_{ab}/2) = 2(25 \tan 5° + 12.9/2) = 17.3$ mm. The size of the second lens is calculated in a similar way: $D_2 = 2h_2 = 2(f'_2 \tan\beta + D_{ab}/2) = 26$ mm. The object is positioned 25 mm in front of lens L_1 and the image is created at a distance of 75 mm behind lens L_2.

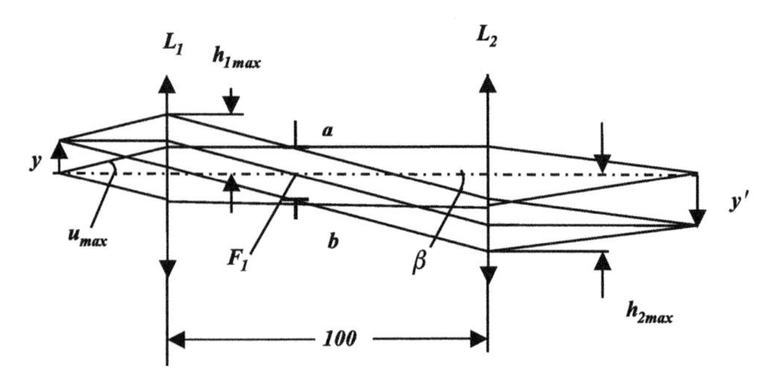

Fig. 1.2.46 Problem P.1.2.22 – A telecentric system.

P.1.2.23. Referring to the system depicted in Fig. 1.2.31, we use the second approach and Eq. (1.2.49) which gives for the configuration with a distance $l = 60$ mm between the first lens and the image

$$f'_e = 60/0.75 = 80\text{ mm};\quad f'_1 = -f'_2 = 40\text{ mm};$$
$$S'_F = 40\text{ mm}$$

and therefore the distance between the lenses is 20 mm. Obviously Petzval's sum is equal to zero and therefore the image curvature is negligible.

Section Two

Optical waveguides

Chapter 2.1

2.1

Wave theory of optical waveguides

Okamoto

The basic concepts and equations of electromagnetic wave theory required for the comprehension of lightwave propagation in optical waveguides are presented. The light confinement and formation of modes in the waveguide are qualitatively explained, taking the case of a slab waveguide. Maxwell's equations, boundary conditions, and the complex Poynting vector are described as they form the basis for the following chapters.

2.1.1 Waveguide structure

Optical fibers and optical waveguides consist of a core, in which light is confined, and a cladding, or substrate surrounding the core, as shown in Fig. 2.1.1. The refractive index of the core n_1 is higher than that of the cladding n_0. Therefore the light beam that is coupled to the end face of the waveguide is confined in the core by total internal reflection. The condition for total internal reflection at the core–cladding interface is given by $n_1 \sin(\pi/2 - \phi) \geq n_0$. Since the angle ϕ is related with the incident angle θ by $\sin\theta = n_1 \sin\phi \leq \sqrt{n_1^2 - n_0^2}$, we obtain the critical condition for the total internal reflection as

$$\theta \leq \sin^{-1}\sqrt{n_1^2 - n_0^2} \equiv \theta_{max}. \tag{2.1.1}$$

The refractive-index difference between core and cladding is of the order of $n_1 - n_0 = 0.01$. Then θ_{max} in Eq. (2.1.1) can be approximated by

$$\theta_{max} \cong \sqrt{n_1^2 - n_0^2}. \tag{2.1.2}$$

θ_{max} denotes the maximum light acceptance angle of the waveguide and is known as the *numerical aperture (NA)*.

The relative refractive-index difference between n_1 and n_0 is defined as

$$\Delta = \frac{n_1^2 - n_0^2}{2n_1^2} \cong \frac{n_1 - n_0}{n_1}. \tag{2.1.3}$$

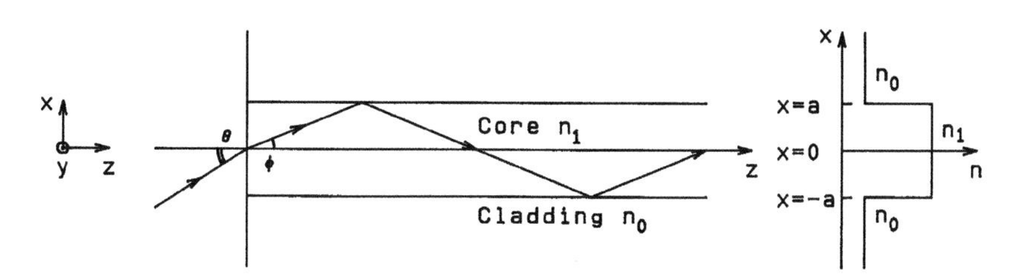

Fig. 2.1.1 Basic structure and refractive-index profile of the optical waveguide.

Fundamentals of Optical Waveguides; ISBN: 9780125250962

Δ is commonly expressed as a percentage. The numerical aperture NA is related to the relative refractive-index difference Δ by

$$\mathrm{NA} = \theta_{max} \cong n_1\sqrt{2\Delta}. \tag{2.1.4}$$

The maximum angle for the propagating light within the core is given by $\phi_{max} \cong \theta_{max}/n_1 \cong \sqrt{2\Delta}$. For typical optical waveguides, NA $= 0.21$ and $\theta_{max} = 12°(\phi_{max} = 8.1°)$ when $n_1 = 1.47$, $\Delta = 1\%$ (for $n_0 = 1.455$).

2.1.2 Formation of guided modes

We have accounted for the mechanism of mode confinement and have indicated that the angle ϕ must not exceed the critical angle. Even though the angle ϕ is smaller than the critical angle, light rays with arbitrary angles are not able to propagate in the waveguide. Each mode is associated with light rays at a discrete angle of propagation, as given by electromagnetic wave analysis. Here we describe the formation of modes with the ray picture in the slab waveguide [1], as shown in Fig. 2.1.2. Let us consider a plane wave propagating along the z-direction with inclination angle ϕ. The phase fronts of the plane waves are perpendicular to the light rays. The wavelength and the wavenumber of light in the core are λ/n_1 and kn_1 ($k = 2\pi/\lambda$) respectively, where λ is the wavelength of light in vacuum. The propagation constants along z and x (lateral direction) are expressed by

$$\beta = kn_1 \cos\phi, \tag{2.1.5}$$

$$\kappa = kn_1 \sin\phi. \tag{2.1.6}$$

Before describing the formation of modes in detail, we must explain the phase shift of a light ray that suffers total reflection. The reflection coefficient of the totally reflected light, which is polarized perpendicular to the incident plane (plane formed by the incident and reflected rays), as shown in Fig. 2.1.3, is given by [2]

$$r = \frac{A_r}{A_i} = \frac{n_1 \sin\phi + j\sqrt{n_1^2\cos^2\phi - n_0^2}}{n_1 \sin\phi - j\sqrt{n_1^2\cos^2\phi - n_0^2}}. \tag{2.1.7}$$

When we express the complex reflection coefficient r as $r = \exp(-j\Phi)$, the amount of phase shift Φ is obtained as

$$\Phi = -2\tan^{-1}\frac{\sqrt{n_1^2\cos^2\phi - n_0^2}}{n_1 \sin\phi} = -2\tan^{-1}\sqrt{\frac{2\Delta}{\sin^2\phi} - 1}. \tag{2.1.8}$$

where Eq. (2.1.3) has been used. The foregoing phase shift for the totally reflected light is called the Goos–Hänchen shift [1, 3].

Let us consider the phase difference between the two light rays belonging to the same plane wave in Fig. 2.1.2. Light ray PQ, which propagates from point P to Q, does not suffer the influence of reflection. On the other hand, light ray RS, propagating from point R to S, is reflected two times (at the upper and lower core-cladding interfaces). Since points P and R or points Q and S are on the same phase front, optical paths PQ and RS (including the Goos–Hänchen shifts caused by the two total reflections) should be equal, or their difference should be an integral multiple of 2π. Since the distance between points Q and R is $2a/\tan\phi - 2a\tan\phi$ the distance between points P and Q is expressed by

$$\ell_1 = \left(\frac{2a}{\tan\phi} - 2a\tan\phi\right)\cos\phi = 2a\left(\frac{1}{\sin\phi} - 2\sin\phi\right). \tag{2.1.9}$$

Also, the distance between points R and S is given by

$$\ell_2 = \frac{2a}{\sin\phi}. \tag{2.1.10}$$

The phase-matching condition for the optical paths PQ and RS then becomes

$$(kn_1\ell_2 + 2\Phi) - kn_1\ell_1 = 2m\pi, \tag{2.1.11}$$

where m is an integer. Substituting Eqs. (2.1.8)–(2.1.10) into Eq. (2.1.11) we obtain the condition for the propagation angle ϕ as

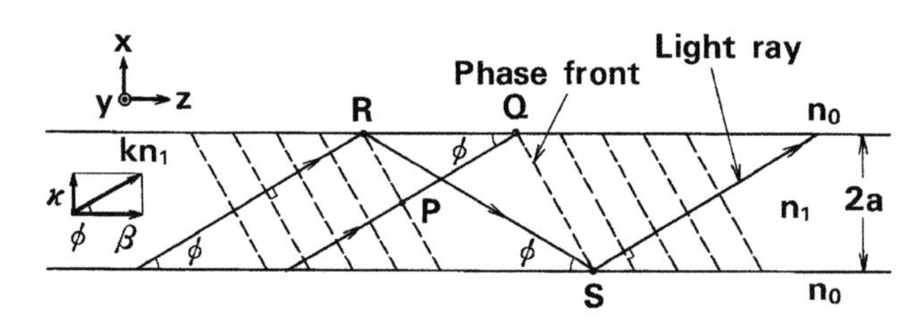

Fig. 2.1.2 Light rays and their phase fronts in the waveguide.

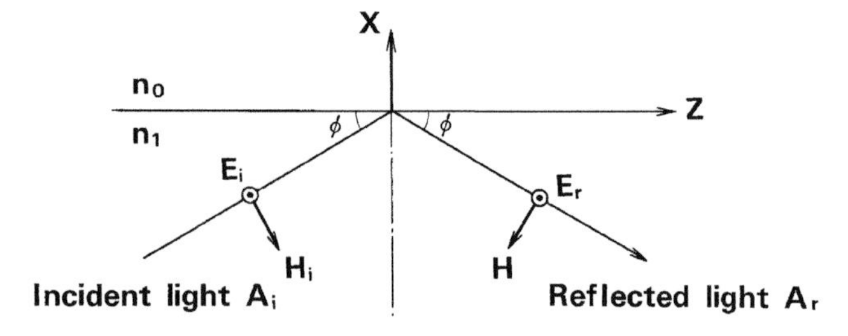

Fig. 2.1.3 Total reflection of a plane wave at a dielectric interface.

$$\tan\left(kn_1 a \sin\phi - \frac{m\pi}{2}\right) = \sqrt{\frac{2\Delta}{\sin^2\phi} - 1}. \qquad (2.1.12)$$

Equation (2.1.12) shows that the propagation angle of a light ray is discrete and is determined by the waveguide structure (core radius a, refractive index n_1 refractive-index difference Δ) and the wavelength λ of the light source (wavenumber is $k = 2\pi/\lambda$) [4]. The optical field distribution that satisfies the phase-matching condition of Eq. (2.1.12) is called the *mode.* The allowed value of propagation constant β [Eq. (2.1.5)] is also discrete and is denoted as an eigenvalue. The mode that has the minimum angle ϕ in Eq. (2.1.12) ($m = 0$) is the fundamental mode; the other modes, having larger angles, are higher-order modes ($m \geq 1$).

Figure 2.1.4 schematically shows the formation of modes (standing waves) for (a) the fundamental mode and (b) a higher-order mode, respectively, through the interference of light waves. In the figure the solid line represents a positive phase front and a dotted line represents a negative phase front, respectively. The electric field amplitude becomes the maximum (minimum) at the point where two positive (negative) phase fronts interfere. In contrast, the electric field amplitude becomes almost zero near the core–cladding interface, since positive and negative phase fronts cancel out each other. Therefore the field distribution along the x-(transverse) direction becomes a standing wave and varies periodically along the z direction with the period $\lambda_p = (\lambda/n_1)/\cos\phi = 2\pi/\beta$.

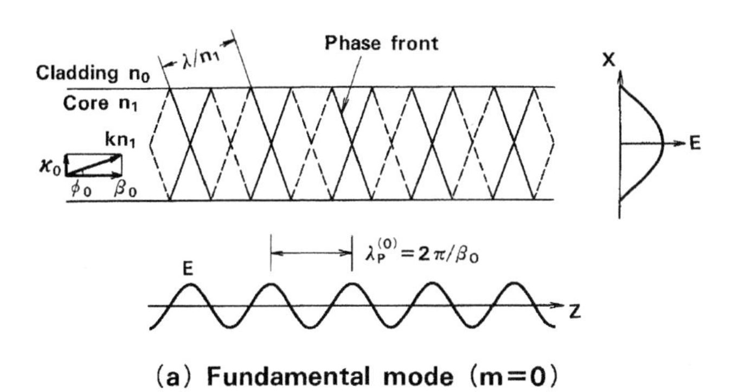

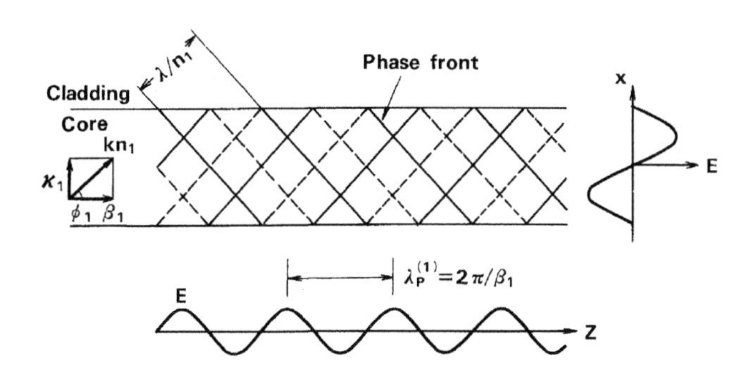

Fig. 2.1.4 Formation of modes: (a) Fundamental mode, (b) higher-order mode.

Since $n_1 \sin\phi = \sin\theta \leq \sqrt{n_1^2 - n_0^2}$ from Fig. 2.1.1, Eqs. (2.1.1) and (2.1.3) give the propagation angle as sin $\sin\phi \leq \sqrt{2\Delta}$. When we introduce the parameter

$$\xi = \frac{\sin\phi}{\sqrt{2\Delta}}, \qquad (2.1.13)$$

which is normalized to 1, the phase-matching Eq. (2.1.12) can be rewritten as

$$kn_1 a\sqrt{2\Delta} = \frac{\cos^{-1}\xi + m\pi/2}{\xi}. \qquad (2.1.14)$$

The term on the left-hand side of Eq. (2.1.14) is known as the *normalized frequency,* and it is expressed by

$$v = kn_1 a\sqrt{2\Delta}. \qquad (2.1.15)$$

When we use the normalized frequency v, the propagation characteristics of the waveguides can be treated generally (independent of each waveguide structure). The relationship between normalized frequency v and ξ (propagation constant β), Eq. (2.1.14), is called the *dispersion equation.* Figure 2.1.5 shows the dispersion curves of a slab waveguide. The crossing point between $\eta = (\cos^{-1}\xi + m\pi/2)/\xi$ and $\eta = v$ gives ξ_m for each mode number m, and the propagation constant β_m is obtained from Eqs. (2.1.5) and (2.1.13).

It is known from Fig. 2.1.5 that only the fundamental mode with $m = 0$ can exist when $v < v_c = \pi/2$. v_c determines the single-mode condition of the slab waveguide—in other words, the condition in which higher-order modes are cut off. Therefore it is called the cutoff v-value. When we rewrite the cutoff condition in terms of the wavelength we obtain

$$\lambda_c = \frac{2\pi}{v_c} a n_1 \sqrt{2\Delta}. \qquad (2.1.16)$$

λ_c is called the *cutoff* (free-space) *wavelength.* The waveguide operates in a single mode for wavelengths longer than λ_c. For example, $\lambda_c = 0.8$ μm when the core width $2a = 3.54$ μm for the slab waveguide of $n_1 = 1.46$, $\Delta = 0.3\%(n_0 = 1.455)$.

2.1.3 Maxwell's equations

Maxwell's equations in a homogeneous and lossless dielectric medium are written in terms of the electric field **e** and magnetic field **h** as [5]

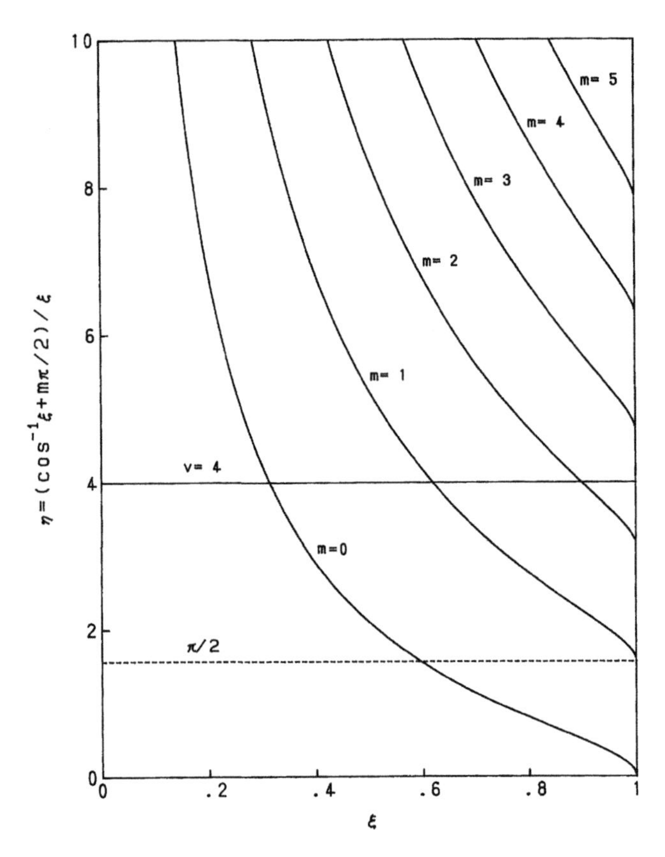

Fig. 2.1.5 Dispersion curves of a slab waveguide.

$$\nabla \times \mathbf{e} = -\mu \frac{\partial \mathbf{h}}{\partial t}, \tag{2.1.17}$$

$$\nabla \times \mathbf{h} = \varepsilon \frac{\partial \mathbf{e}}{\partial t}, \tag{2.1.18}$$

where ε and μ denote the permittivity and permeability of the medium, respectively, ε and μ are related to their respective values in a vacuum of $\varepsilon_0 = 8.854 \times 10^{-12}$[F/m] and $\mu_0 = 4\pi \times 10^{-7}$[H/m] by

$$\varepsilon = \varepsilon_0 n^2, \tag{2.1.19a}$$

$$\mu = \mu_0, \tag{2.1.19b}$$

where n is the refractive index. The wavenumber of light in the medium is then expressed as [5]

$$\Gamma = \omega\sqrt{\varepsilon\mu} = \omega n\sqrt{\varepsilon_0\mu_0} = kn. \tag{2.1.20}$$

In Eq. (2.1.20), ω is an angular frequency of the sinusoidally varying electromagnetic fields with respect to time; k is the wavenumber in a vacuum, which is related to the angular frequency ω by

$$k = \omega\sqrt{\varepsilon_0\mu_0} = \frac{\omega}{c}. \tag{2.1.21}$$

In Eq. (2.1.21), c is the light velocity in a vacuum, given by

$$c = \frac{1}{\sqrt{\varepsilon_0\mu_0}} = 2.998 \times 10^8 [\text{m/s}]. \tag{2.1.22}$$

The fact that the units for light velocity c are m/s is confirmed from the units of the permittivity ε_0 [F/m] and permeability μ_0 [H/m] as

$$\frac{1}{\sqrt{[\text{F/m}][\text{H/m}]}} = \frac{\text{m}}{\sqrt{\text{F}\cdot\text{H}}} = \frac{\text{m}}{\sqrt{[\text{A}\cdot\text{s/V}][\text{V}\cdot\text{s/A}]}} = \frac{\text{m}}{\text{s}}.$$

When the frequency of the electromagnetic wave is f[Hz], it propagates c/f[m] in one period of sinusoidal variation. Then the wavelength of electromagnetic wave is obtained by

$$\lambda = \frac{c}{f} = \frac{\omega/k}{f} = \frac{2\pi}{k}, \tag{2.1.23}$$

where $\omega = 2\pi f$.

When the electromagnetic fields **e** and **h** are sinusoidal functions of time, they are usually represented by complex amplitudes, i.e., the so-called phasors. As an example consider the electric field vector

$$\mathbf{e}(t) = |\mathbf{E}|\cos(\omega t + \phi), \tag{2.1.24}$$

where $|\mathbf{E}|$ is the amplitude and ϕ is the phase. Defining the complex amplitude of $\mathbf{e}(t)$ by

$$\mathbf{E} = |\mathbf{E}|e^{j\phi}, \tag{2.1.25}$$

Eq. (2.1.24) can be written as

$$\mathbf{e}(t) = \text{Re}\{\mathbf{E}e^{j\omega t}\}. \tag{2.1.26}$$

We will often represent $\mathbf{e}(t)$ by

$$\mathbf{e}(t) = \mathbf{E}e^{j\omega t} \tag{2.1.27}$$

instead of by Eq. (2.1.24) or (2.1.26). This expression is not strictly correct, so when we use this phasor expression we should keep in mind that what is meant by Eq. (2.1.27) is the real part of $\mathbf{E}e^{j\omega t}$. In most mathematical manipulations, such as addition, subtraction, differentiation and integration, the replacement of Eq. (2.1.26) by the complex form (2.1.27) poses no problems. However, we should be careful in the manipulations that involve the product of sinusoidal functions. In these cases we must use the real form of the function (2.1.24) or complex conjugates [see Eqs. (2.1.42)].

When we consider an electromagnetic wave having angular frequency ω and propagating in the z direction with propagation constant β, the electric and magnetic fields can be expressed as

$$\mathbf{e} = \mathbf{E}(\mathbf{r})e^{j(\omega t-\beta z)}, \quad (2.1.28)$$

$$\mathbf{h} = \mathbf{H}(\mathbf{r})e^{j(\omega t-\beta z)}, \quad (2.1.29)$$

where **r** denotes the position in the plane transverse to the z-axis. Substituting Eqs. (2.1.28) and (2.1.29) into Eqs. (2.1.17) and (2.1.18), the following set of equations are obtained in Cartesian coordinates:

$$\begin{cases} \dfrac{\partial E_z}{\partial y} + j\beta E_y = -j\omega\mu_0 H_x \\ -j\beta E_x - \dfrac{\partial E_z}{\partial x} = -j\omega\mu_0 H_y \\ \dfrac{\partial E_y}{\partial x} - \dfrac{\partial E_x}{\partial y} = -j\omega\mu_0 H_z \\ \dfrac{\partial H_z}{\partial y} + j\beta H_y = j\omega\varepsilon_0 n^2 E_x \\ -j\beta H_x - \dfrac{\partial H_z}{\partial x} = j\omega\varepsilon_0 n^2 E_y \\ \dfrac{\partial H_y}{\partial x} - \dfrac{\partial H_x}{\partial y} = j\omega\varepsilon_0 n^2 E_z. \end{cases} \quad (2.1.30)$$

The foregoing equations are the bases for the analysis of slab and rectangular waveguides.

For the analysis of wave propagation in optical fibers, which are axially symmetric, Maxwell's equations are written in terms of cylindrical coordinates:

$$\begin{cases} \dfrac{1}{r}\dfrac{\partial E_z}{\partial \theta} + j\beta E_\theta = -j\omega\mu_0 H_r \\ -j\beta E_r - \dfrac{\partial E_z}{\partial r} = -j\omega\mu_0 H_\theta \\ \dfrac{1}{r}\dfrac{\partial}{\partial r}(rE_\theta) - \dfrac{1}{r}\dfrac{\partial E_r}{\partial \theta} = -j\omega\mu_0 H_z \\ \dfrac{1}{r}\dfrac{\partial H_z}{\partial \theta} + j\beta H_\theta = j\omega\varepsilon_0 n^2 E_r \\ -j\beta H_r - \dfrac{\partial H_z}{\partial r} = j\omega\varepsilon_0 n^2 E_\theta \\ \dfrac{1}{r}\dfrac{\partial}{\partial r}(rH_\theta) - \dfrac{1}{r}\dfrac{\partial H_r}{\partial \theta} = j\omega\varepsilon_0 n^2 E_z. \end{cases} \quad (2.1.31)$$

Maxwell's Eqs. (2.1.30) or (2.1.31) do not determine the electromagnetic field completely. Out of the infinite possibilities of solutions of Maxwell's equations, we must select those that also satisfy the boundary conditions of the respective problem. The most common type of boundary condition occurs when there are discontinuities in the dielectric constant (refractive index), as shown in Fig. 2.1.1.

At the boundary the tangential components of the electric field and magnetic field should satisfy the conditions

$$E_t^{(1)} = E_t^{(2)} \quad (2.1.32)$$

$$H_t^{(1)} = H_t^{(2)}, \quad (2.1.33)$$

where the subscript t denotes the tangential components to the boundary and the superscripts (1) and (2) indicate the medium, respectively. Equations (2.1.32) and (2.1.33) mean that the tangential components of the electromagnetic fields must be continuous at the boundary. There are also natural boundary conditions that require the electromagnetic fields to be zero at infinity.

2.1.4 Propagating power

Consider Gauss's theorem (see Section 10.1) for vector **A** in an arbitrary volume V

$$\iiint_V \nabla\cdot\mathbf{A}\, dv = \iint_S \mathbf{A}\cdot\mathbf{n}ds, \quad (2.1.34)$$

where **n** is the outward-directed unit vector normal to the surface S enclosing V and dv and ds are the differential volume and surface elements, respectively. When we set $\mathbf{A} = \mathbf{e}\times\mathbf{h}$ in Eq. (2.1.34) and use the vector identity

$$\nabla\cdot(\mathbf{e}\times\mathbf{h}) = \mathbf{h}\cdot\nabla\times\mathbf{e} - \mathbf{e}\cdot\nabla\times\mathbf{h}, \quad (2.1.35)$$

we obtain the following equation for electromagnetic fields:

$$\iiint_v (\mathbf{h}\cdot\nabla\times\mathbf{e} - \mathbf{e}\cdot\nabla\times\mathbf{h})\, dv = \iint_S (\mathbf{e}\times\mathbf{h})\cdot\mathrm{n}\, ds. \quad (2.1.36)$$

Substituting Eqs. (2.1.17) and (2.1.18) into Eq. (2.1.36) results in

$$\iiint_v \left(\varepsilon\mathbf{e}\cdot\frac{\partial\mathbf{e}}{\partial t} + \mu\mathbf{h}\cdot\frac{\partial\mathbf{h}}{\partial t}\right) dv = -\iint_S (\mathbf{e}\times\mathbf{h})\cdot\mathbf{n}\, ds. \quad (2.1.37)$$

The first term in Eq. (2.1.37)

$$\varepsilon\mathbf{e}\cdot\frac{\partial\mathbf{e}}{\partial t} = \frac{\partial}{\partial t}\left(\frac{\varepsilon}{2}\mathbf{e}\cdot\mathbf{e}\right) = \frac{\partial W_e}{\partial t}, \quad (2.1.38)$$

represents the rate of increase of the electric stored energy W_e and the second term

$$\mu\mathbf{h}\cdot\frac{\partial\mathbf{h}}{\partial t} = \frac{\partial}{\partial t}\left(\frac{\mu}{2}\mathbf{h}\cdot\mathbf{h}\right) \equiv \frac{\partial W_h}{\partial t}, \quad (2.1.39)$$

represents the rate of increase of the magnetic stored energy W_h, respectively. Therefore, the left-hand side of Eq. (2.1.37) gives the rate of increase of the electromagnetic stored energy in the whole volume V; in other words, it represents the total power flow into the volume bounded by S. When we replace the outward-directed unit vector $\mathbf{n}$ by the inward-directed unit vector $\mathbf{u}_z(= -\mathbf{n})$, the total power flowing into the volume through surface S is expressed by

$$P = \iint_S -(\mathbf{e} \times \mathbf{h}) \cdot \mathbf{n}\, ds = \iint_S -(\mathbf{e} \times \mathbf{h}) \cdot \mathbf{u}_z\, ds. \tag{2.1.40}$$

Equation (2.1.40) means that $\mathbf{e} \times \mathbf{h}$ is the vector representing the power flow, and its normal component to the surface $(\mathbf{e} \times \mathbf{h}) \cdot \mathbf{u}_z$ gives the amount of power flowing through unit surface area. Therefore, vector $\mathbf{e} \times \mathbf{h}$ represents the power-flow density, and

$$\mathbf{S} = \mathbf{e} \times \mathbf{h} [\mathrm{W/m^2}] \tag{2.1.41}$$

is called the *Poynting vector*. In this equation, $\mathbf{e}$ and $\mathbf{h}$ denote instantaneous fields as functions of time t. Let us obtain the average power-flow density in an alternating field. The complex electric and magnetic fields can be expressed by

$$\mathbf{e}(t) = \mathrm{Re}\{\mathbf{E}e^{j\omega t}\} = \frac{1}{2}\{\mathbf{E}e^{j\omega t} + \mathbf{E}^* e^{-j\omega t}\}, \tag{2.1.42a}$$

$$\mathbf{h}(t) = \mathrm{Re}\{\mathbf{H}e^{j\omega t}\} = \frac{1}{2}\{\mathbf{H}e^{j\omega t} + \mathbf{H}^* e^{-j\omega t}\}, \tag{2.1.42b}$$

where * denotes the complex conjugate. The time average of the normal component of the Poynting vector is then obtained as

$$\begin{aligned}\langle \mathbf{S} \cdot \mathbf{u}_z \rangle &= \langle (\mathbf{e} \times \mathbf{h}) \cdot \mathbf{u}_z \rangle \\ &= \frac{1}{4}\langle [\mathbf{E}e^{j\omega t} + \mathbf{E}^* e^{-j\omega t} \times (\mathbf{H}e^{j\omega t} + \mathbf{H}^* e^{-j\omega t})] \cdot \mathbf{u}_z \rangle \\ &= \frac{1}{4}(\mathbf{E} \times \mathbf{H}^* + \mathbf{E}^* \times \mathbf{H}) \cdot \mathbf{u}_z \\ &= \frac{1}{2}\mathrm{Re}\{(\mathbf{E} \times \mathbf{H}^*) \cdot \mathbf{u}_z\},\end{aligned} \tag{2.1.43}$$

where $\langle\ \rangle$ denotes a time average. Then the time average of the power flow is given by

$$P = \iint_S \frac{1}{2}\mathrm{Re}\{(\mathbf{E} \times \mathbf{H}^*) \cdot \mathbf{u}_z\}\, ds. \tag{2.1.44}$$

Since $\mathbf{E} \times \mathbf{H}^*$ often becomes real in the analysis of optical waveguides, the time average propagation power in Eq. (2.1.44) is expressed by

$$P = \iint_S \frac{1}{2}(\mathbf{E} \times \mathbf{H}^*) \cdot \mathbf{u}_z ds. \tag{2.1.45}$$

References

[1] Marcuse, D. 1974. *Theory of Dielectric Optical Waveguides*. New York: Academic Press.

[2] Born, M. and E. Wolf. 1970. *Principles of Optics*. Oxford: Pergamon Press.

[3] Tamir, T. 1975. *Integrated Optics*. Berlin: Springer-Verlag.

[4] Marcuse, D. 1972. *Light Transmission Optics*. New York: Van Nostrand Rein-hold.

[5] Stratton, J. A. 1941. *Electromagnetic Theory*. New York: McGraw-Hill.

Chapter 2.2

Planar optical waveguides

Okamoto

Planar optical waveguides are the key devices to construct integrated optical circuits and semiconductor lasers. Generally, rectangular waveguides consist of a square or rectangular core surrounded by a cladding with lower refractive index than that of the core. Three-dimensional analysis is necessary to investigate the transmission characteristics of rectangular waveguides. However, rigorous three-dimensional analysis usually requires numerical calculations and does not always give a clear insight into the problem. Therefore, this chapter first describes two-dimensional slab waveguides to acquire a fundamental understanding of optical waveguides. Then several analytical approximations are presented to analyze the three-dimensional rectangular waveguides. Although these are approximate methods, the essential lightwave transmission mechanism in rectangular waveguides can be fully investigated. The rigorous treatment of three-dimensional rectangular waveguides by the finite element method will be presented in Chapter 6.

2.2.1 Slab waveguides

2.2.1.1 Derivation of basic equations

In this section, the wave analysis is described for the slab waveguide (Fig. 2.2.1) whose propagation characteristics have been explained [1–3]. Taking into account the fact that we treat dielectric optical waveguides, we set permittivity and permeability as $\varepsilon = \varepsilon_0 n^2$ and $\mu = \mu_0$ in the Maxwell's Eq. (1.17) and (1.18) as

$$\nabla \times \tilde{\mathbf{E}} = -\mu_0 \frac{\partial \tilde{\mathbf{H}}}{\partial t}, \tag{2.2.1a}$$

$$\nabla \times \tilde{\mathbf{H}} = \varepsilon_0 n^2 \frac{\partial \tilde{\mathbf{E}}}{\partial t}, \tag{2.2.1b}$$

where n is the refractive index. We are interested in plane-wave propagation in the form of

$$\tilde{\mathbf{E}} = \mathbf{E}(x,y)e^{j(\omega t-\beta z)}, \tag{2.2.2a}$$

$$\tilde{\mathbf{H}} = \mathbf{H}(x,y)e^{j(\omega t-\beta z)}, \tag{2.2.2b}$$

Substituting Eqs. (2.2.2a) and (2.2.2b) into Eqs. (2.2.1a) and (2.2.1b), we obtain the following set of equations for the electromagnetic field components:

$$\begin{cases} \frac{\partial E_z}{\partial y} + j\beta E_y = -j\omega\mu_0 H_x \\ -j\beta E_x - \frac{\partial E_z}{\partial x} = -j\omega\mu_0 H_y \\ \frac{\partial E_y}{\partial x} - \frac{\partial E_x}{\partial y} = -j\omega\mu_0 H_z \end{cases} \tag{2.2.3}$$

$$\begin{cases} \frac{\partial E_z}{\partial y} + j\beta \mathrm{H}_y = j\omega\varepsilon_0 n^2 E_x \\ -j\beta \mathrm{H}_x - \frac{\partial H_z}{\partial x} = j\omega\varepsilon_0 n^2 E_y \\ \frac{\partial H_y}{\partial x} - \frac{\partial H_x}{\partial y} = j\omega\varepsilon_0 n^2 E_z. \end{cases} \tag{2.2.4}$$

In the slab waveguide, as shown in Fig. 2.2.1, electromagnetic fields **E** and **H** do not have y axis dependency.

Fundamentals of Optical Waveguides; ISBN: 9780125250962

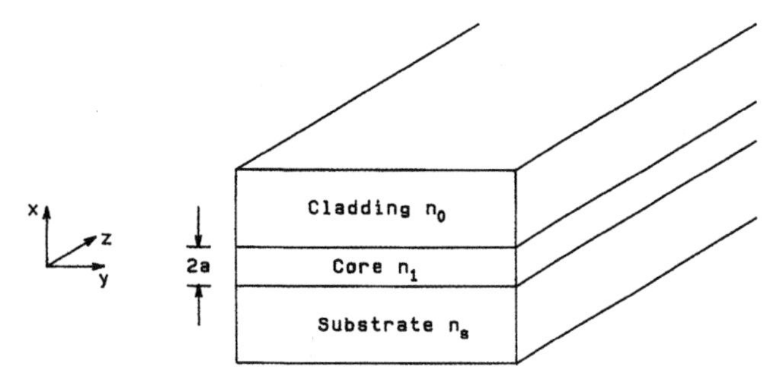

Fig. 2.2.1 Slab optical waveguide.

Therefore, we set $\partial E/\partial y = 0$ and $\partial H/\partial y = 0$. Putting these relations into Eqs. (2.2.3) and (2.2.4), two independent electromagnetic modes are obtained, which are denoted as TE mode and TM mode, respectively. TE mode satisfies the following wave equation:

$$\frac{d^2E_y}{dx^2} + (k^2n^2 - \beta^2)E_y = 0, \tag{2.2.5a}$$

where

$$H_x = -\frac{\beta}{\omega\mu_0}E_y, \tag{2.2.5b}$$

$$H_z = \frac{j}{\omega\mu_0}\frac{dE_y}{dx}, \tag{2.2.5c}$$

and

$$E_x = E_z = H_y = 0. \tag{2.2.5d}$$

Also the tangential components E_y and H_z should be continuous at the boundaries of two different media. As shown in Eq. (2.2.5d) the electric field component along the z-axis is zero ($E_z = 0$). Since the electric field lies in the plane that is perpendicular to the z-axis, this electromagnetic field distribution is called transverse electric (TE) mode.

The TM mode satisfies the following wave equation:

$$\frac{d}{dx}\left(\frac{1}{n^2}\frac{dH_y}{dx}\right) + \left(k^2 - \frac{\beta^2}{n^2}\right)H_y = 0, \tag{2.2.6a}$$

where

$$E_x = \frac{\beta}{\omega\varepsilon_0 n^2}H_y, \tag{2.2.6b}$$

$$E_z = -\frac{j}{\omega\varepsilon_0 n^2}\frac{dH_y}{dx}, \tag{2.2.6c}$$

$$E_y = H_x = H_z = 0. \tag{2.2.6d}$$

As shown in Eq. (2.2.6d) the magnetic field component along the z-axis is zero ($H_z = 0$). Since the magnetic field lies in the plane that is perpendicular to the z-axis, this electromagnetic field distribution is called transverse magnetic (TM) mode.

2.2.1.2 Dispersion equations for TE and TM modes

Propagation constants and electromagnetic fields for TE and TM modes can be obtained by solving Eq. (2.2.5) or (2.2.6). Here the derivation method to calculate the dispersion equation (also called the eigenvalue equation) and the electromagnetic field distributions is given. We consider the slab waveguide with uniform refractive-index profile in the core, as shown in Fig. 2.2.2. Considering the fact that the guided electromagnetic fields are confined in the core and exponentially decay in the cladding, the electric field distribution is expressed as

$$E_y = \begin{cases} A\cos(\kappa a - \phi)e^{-\sigma(x-a)} & (x>a) \\ A\cos(\kappa x - \phi) & (-a \le x \le a) \\ A\cos(\kappa a + \phi)e^{\xi(x+a)} & (x < -a), \end{cases} \tag{2.2.7}$$

where κ, σ, and ξ are wavenumbers along the x-axis in the core and cladding regions and are given by

$$\begin{cases} \kappa = \sqrt{k^2n_1^2 - \beta^2} \\ \sigma = \sqrt{\beta^2 - k^2n_0^2} \\ \xi = \sqrt{\beta^2 - k^2n_s^2}. \end{cases} \tag{2.2.8}$$

The electric field component E_y in Eq. (2.2.7) is continuous at the boundaries of core–cladding interfaces ($x = \pm a$). There is another boundary condition, that the magnetic field component H_z should be continuous at the boundaries. H_z is given by Eq. (2.2.5c). Neglecting the terms independent of x, the boundary condition for H_z is treated by the continuity condition of dE_y/dx as

$$\frac{dE_y}{dx} = \begin{cases} -\sigma A\cos(\kappa a - \phi)e^{-\sigma(x-a)} & (x>a) \\ -\kappa A\sin(\kappa x - \phi) & (-a \le x \le a) \\ \xi A\cos(\kappa a + \phi)e^{\xi(x+a)} & (x < -a). \end{cases} \tag{2.2.9}$$

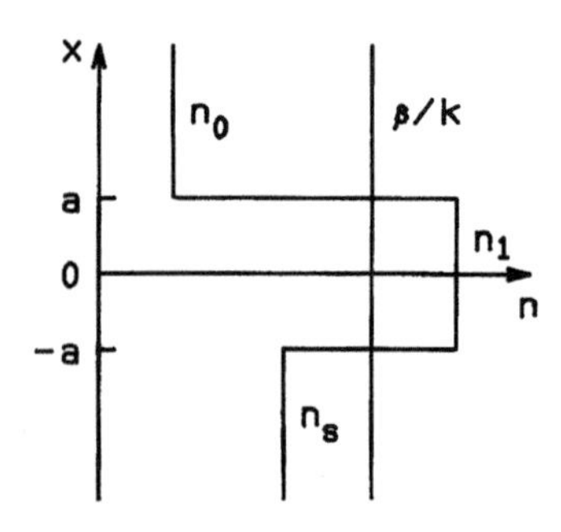

Fig. 2.2.2 Refractive-index profile of slab waveguide.

From the conditions that dE_y/dx are continuous at $x = \pm a$, the following equations are obtained:

$$\begin{cases} \kappa A \sin(\kappa a + \phi) = \xi A \cos(\kappa a + \phi) \\ \sigma A \cos(\kappa a - \phi) = \kappa A \sin(\kappa a - \phi). \end{cases}$$

Eliminating the constant A, we have

$$\tan(u + \phi) = \frac{w}{u}, \quad (2.2.10a)$$

$$\tan(u - \phi) = \frac{w'}{u}, \quad (2.2.10b)$$

where

$$\begin{cases} u = \kappa a \\ w = \xi a \\ w' = \sigma a. \end{cases} \quad (2.2.11)$$

From Eqs. (2.2.10) we obtain the eigenvalue equations as

$$u = \frac{m\pi}{2} + \frac{1}{2}\tan^{-1}\left(\frac{w}{u}\right) + \frac{1}{2}\tan^{-1}\left(\frac{w'}{u}\right) \quad (m = 0, 1, 2, \ldots) \quad (2.2.12)$$

$$\phi = \frac{m\pi}{2} + \frac{1}{2}\tan^{-1}\left(\frac{w}{u}\right) - \frac{1}{2}\tan^{-1}\left(\frac{w'}{u}\right). \quad (2.2.13)$$

The normalized transverse wavenumbers u, w and w' are not independent. Using Eqs. (2.2.8) and (2.2.11) it is known that they are related by the following equations:

$$u^2 + w^2 = k^2 a^2 \left(n_1^2 - n_s^2\right) \equiv v^2, \quad (2.2.14)$$

$$w' = \sqrt{\gamma v^2 + w^2}, \quad (2.2.15a)$$

$$\gamma = \frac{n_s^2 - n_0^2}{n_1^2 - n_s^2} \quad (2.2.15b)$$

where v is the normalized frequency, defined as Eq. (2.2.15) and γ is a measure of the asymmetry of the cladding refractive indices. Once the wavelength of the light signal and the geometrical parameters of the waveguide are determined, the normalized frequency v and γ are determined. Therefore u, w, w' and ϕ are given by solving the eigenvalue equations Eqs. (2.2.12) and (2.2.13) under the constraints of Eqs. (2.2.14)–(2.2.15). In the asymmetrical waveguide ($n_s > n_0$) as shown in Fig. 2.2.2, the higher refractive index n_s is used as the cladding refractive index, which is adopted for the definition of the normalized frequency v. It is preferable to use the higher refractive index n_s because the cutoff conditions are determined when the normalized propagation constant β/k coincides with the higher cladding refractive index. Equations (2.2.12), (2.2.14) and (2.2.15) are the dispersion equations or eigenvalue equations for the TE_m modes. When the wavelength of the light signal and the geometrical parameters of the waveguide are determined—in other words, when the normalized frequency v and asymmetrical parameter γ are determined—the propagation constant β can be determined from these equations. As is known from Fig. 2.2.2 or Eqs. (2.2.7) and (2.2.8), the transverse wavenumber κ should be a real number for the main part of the optical field to be confined in the core region. Then the following condition should be satisfied:

$$n_s \le \frac{\beta}{k} \le n_1, \quad (2.2.16)$$

β/k is a dimensionless value and is a refractive index itself for the plane wave. Therefore it is called the *effective index* and is usually expressed as

$$n_e = \frac{\beta}{k}. \quad (2.2.17)$$

When $n_e < n_s$, the electromagnetic field in the cladding becomes oscillatory along the transverse direction; that is, the field is dissipated as the radiation mode. Since the condition $\beta = kn_s$ represents the critical condition under which the field is cut off and becomes the nonguided mode (radiation mode), it is called the *cutoff* condition. Here we introduce a new parameter, which is defined by

$$b = \frac{n_e^2 - n_s^2}{n_1^2 - n_s^2}. \quad (2.2.18)$$

Then the conditions for the guided modes are expressed, from Eqs. (2.2.16) and (2.2.17), by

$$0 \le b \le 1, \quad (2.2.19)$$

and the cutoff condition is expressed as

$$b = 0. \quad (2.2.20)$$

b is called the *normalized* propagation constant. Rewriting the dispersion Eq. (2.2.12) by using the normalized frequency v and the normalized propagation constant b, we obtain

$$2v\sqrt{1-b} = m\pi + \tan^{-1}\sqrt{\frac{b}{1-b}} + \tan^{-1}\sqrt{\frac{b+\gamma}{1-b}}. \quad (2.2.21)$$

Also Eq. (2.2.8) is rewritten as

$$\begin{cases} u = v\sqrt{1-b} \\ w = v\sqrt{b} \\ w' = v\sqrt{b+\gamma}. \end{cases} \quad (2.2.22)$$

For the symmetrical waveguides with $n_0 = n_s$, we have $\gamma = 0$ and the dispersion Eqs. (2.2.12) and (2.2.13) are reduced to

$$u = \frac{m\pi}{2} + \tan^{-1}\left(\frac{w}{u}\right), \tag{2.2.23a}$$

$$\phi = \frac{m\pi}{2}. \tag{2.2.23b}$$

Equation (2.2.23a) is also expressed by

$$w = u\tan\left(u - \frac{m\pi}{2}\right), \tag{2.2.24}$$

or

$$v\sqrt{1-b} = \frac{m\pi}{2} + \tan^{-1}\sqrt{\frac{b}{1-b}}. \tag{2.2.25}$$

If we notice that the transverse wavenumber $kn_1 a \sin\phi$ in Eq. (1.12) can be expressed by using the present parameters as $u = \kappa a = kn_1 a \sin\phi$, then Eq. (1.12) coincides completely with Eq. (2.2.24).

2.2.1.3 Computation of propagation constant

First the graphical method to obtain qualitatively the propagation constant of the symmetrical slab waveguide is shown, and then the quantitative numerical method to calculate accurately the propagation constant is described. The relationship between u and w for the symmetrical slab waveguide, which is shown in Eq. (2.2.24), is plotted in Fig. 2.2.3. Transverse wavenumbers u and w should satisfy Eq. (2.2.14) for a given normalized frequency v. This relation is also plotted in Fig. 2.2.3 for the case of $v = 4$ as the semicircle with the radius of 4. The solutions of the dispersion equation are then given as the crossing points in Fig. 2.2.3. For example, the transverse wavenumbers u and w for the fundamental mode are given by the crossing point of the curve tangential with $m = 0$ and the semicircle. The propagation constant (or eigenvalue) β is then obtained by using Eqs. (2.2.8) and (2.2.11). In Fig. 2.2.3, there is only one crossing point for the case of $v < \pi/2$. This means that the propagation mode is the only one when the waveguide structure and the wavelength of light satisfy the inequality $v < \pi/2$. The value of $v_c = \pi/2$ then gives the critical point at which the higher-order modes are cut off in the symmetrical slab waveguide. v_c is called the *cutoff normalized frequency*, which is obtained from the cutoff condition for the $m = 1$ mode,

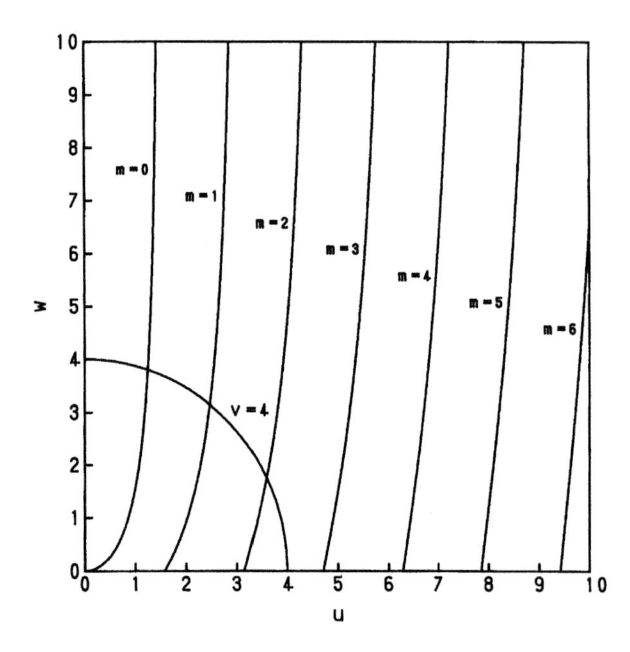

Fig. 2.2.3 *u-w* relationship in slab waveguide.

$$\begin{cases} b = w = 0 & \text{(a)} \\ u = v = \frac{\pi}{2}, & \text{(b)} \end{cases} \tag{2.2.26}$$

where Eqs. (2.2.20) and (2.2.22) have been used. Generally, the cutoff v-value for the TE mode is given by Eq. (2.2.21) as

$$v_{c,\mathrm{TE}} = \frac{m\pi}{2} + \frac{1}{2}\tan^{-1}\sqrt{\gamma}, \tag{2.2.27}$$

and that for the TM mode is given by Eq. (2.2.38) (explained in Section 2.2.1.5) as

$$v_{c,\mathrm{TM}} = \frac{m\pi}{2} + \frac{1}{2}\tan^{-1}\left(\frac{n_1^2}{n_0^2}\sqrt{\gamma}\right). \tag{2.2.28}$$

A qualitative value can be obtained by this graphical solution for the dispersion equation. However, in order to obtain an accurate solution of the dispersion equation, we should rely on the numerical method. Here, we show the numerical treatment for the symmetrical slab waveguide so as to compare with the previous graphical method. We first rewrite the dispersion Eq. (2.2.25) in the following form:

$$f(v,m,b) = v\sqrt{1-b} = \frac{m\pi}{2} - \tan^{-1}\sqrt{\frac{b}{1-b}} = 0. \tag{2.2.29}$$

Figure 2.2.4 shows the plot of $f(v, m, b)$ for $v = 4$. The b-value at which $f = 0$ gives the normalized propagation constant b for the given v-value. The solution of Eq. (2.2.29) is obtained by the Newton–Raphson method or the bisection method or the like. Here, the subroutine program of the most simple bisection method is shown in Fig. 2.2.5.

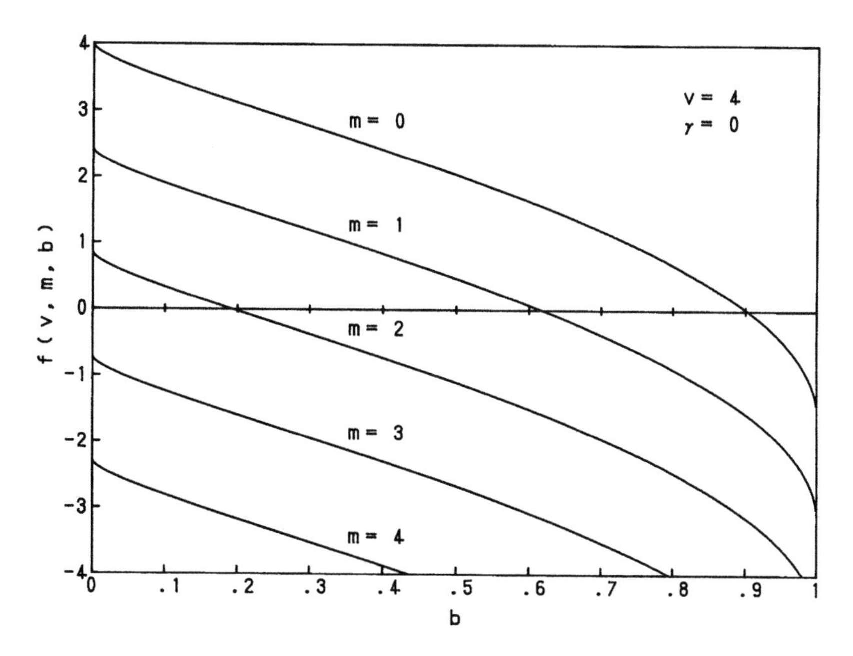

Fig. 2.2.4 Plot of $f(v, m, b)$ for the calculation of the eigenvalue.

The normalized propagation constant b is calculated for each normalized frequency v. Figure 2.2.6 shows the v-b relationship, which is called *dispersion curve*, for TE mode. The mode number is expressed by the subscript m, such as TE_m or TM_m mode. The parameter in Fig. 2.2.6 is the measure of asymmetry γ. It is known that there is no cutoff for the lowest TE_0 mode in the symmetrical waveguide ($\gamma = 0$). On the other hand, cutoff exists for the TE_0 mode in the asymmetrical waveguide ($\gamma \neq 0$).

2.2.1.4 Electric field distribution

Once the eigenvalue of the waveguide is obtained, the electric field distribution given by Eq. (2.2.7) is determined except for the arbitrary constant A. Constant

```
C       ********************** EIGEN **********************
C       *                                                 *
C       *           Eigen Value of Single Waveguide       *
C       *                                                 *
C       ***************************************************
        SUBROUTINE EIGEN(V,M,B)
C
        IMPLICIT INTEGER(I-N),REAL(A-H,O-Z)
        PAI=3.141592653589793
C
        EPS=1.0E-6
        B0=1.0-EPS
        DIVSN=0.01
        B1=B0
        B2=B1-DIVSN
        F1=V*SQRT(1.0-B1)-FLOAT(M)*PAI/2.0-ATAN(SQRT(B1/(1.0-B1)))
C
     10 F2=V*SQRT(1.0-B2)-FLOAT(M)*PAI/2.0-ATAN(SQRT(B2/(1.0-B2)))
         IF(F1*F2.LE.0.0) GO TO 20
        B1=B2
        B2=B1-DIVSN
        F1=F2
         GO TO 10
     20  IF(DIVSN.LE.EPS) GO TO 30
        B2=(B1+B2)/2.0
        DIVSN=DIVSN/2.0
         GO TO 10
C
     30 B=(B1+B2)/2.0
        RETURN
        END
```

Fig. 2.2.5 Subroutine program of the bisection method to calculate the eigenvalue.

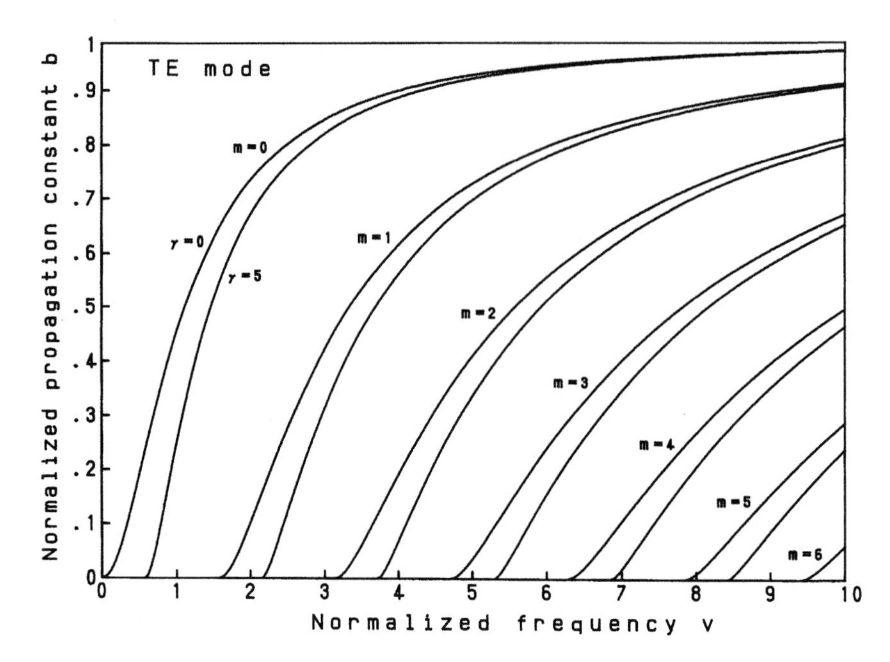

Fig. 2.2.6 Dispersion curves for the TE modes in the slab waveguide.

A is determined when we specify the optical power P carried by the waveguide. Power P is expressed, by using Eq. (1.45) as

$$P = \int_0^1 dy \int_{-\infty}^{\infty} \frac{1}{2}(\mathbf{E} \times \mathbf{H}^*) \cdot u_z \, dx = \int_{-\infty}^{\infty} \frac{1}{2}(E_x H_y^* - E_y H_x^*) dx. \tag{2.2.30}$$

For the TE mode we can rewrite Eq. (2.2.30), by using Eqs. (2.2.5) as

$$P = \frac{\beta}{2\omega\mu_0} \int_{-\infty}^{\infty} |E_y|^2 dx. \tag{2.2.31}$$

Substituting Eq. (2.2.7) into (2.2.31) we obtain power fraction in core, substrate, and cladding regions, respectively, as

$$P_{\text{core}} = \frac{\beta a A^2}{2\omega\mu_0}\left\{1 + \frac{\sin^2(u+\phi)}{2w} + \frac{\sin^2(u-\phi)}{2w\prime}\right\} \quad (-a \le x \le a) \tag{2.2.32a}$$

$$P_{\text{sub}} = \frac{\beta a A^2}{2\omega\mu_0} \frac{\cos^2(u+\phi)}{2w} \quad (x \le -a) \tag{2.2.32b}$$

$$P_{\text{clad}} = \frac{\beta a A^2}{2\omega\mu_0} \frac{\cos^2(u-\phi)}{2w\prime} \quad (x>a). \tag{2.2.32c}$$

For the calculation of Eq. (2.2.32a) we use Eq. (2.2.10). The total power P is then given by

$$P = P_{\text{core}} + P_{\text{sub}} + P_{\text{clad}} = \frac{\beta a A^2}{2\omega\mu_0}\left\{1 + \frac{1}{2w} + \frac{1}{2w\prime}\right\}. \tag{2.2.33}$$

Here the constant A is determined by

$$A = \sqrt{\frac{2\omega\mu_0 P}{\beta a(1 + 1/2w + 1/2w\prime)}}. \tag{2.2.34}$$

Figure 2.2.7 shows the electric field distributions of the TE mode for $v = 4$ in the waveguide with $n_1 = 3.38$, $n_s = 3.17$, $n_0 = 1.0 (\gamma = 6.6)$. The power confinement factor in the core is important to calculate the threshold current density J_{th} of semiconductor lasers [4]. The confinement factor is calculated by using Eqs. (2.2.32) and (2.2.33) as

$$\Gamma = \frac{P_{core}}{P} = \frac{1 + \dfrac{\sin^2(u+\phi)}{2w} + \dfrac{\sin^2(u-\phi)}{2w\prime}}{1 + \dfrac{1}{2w} + \dfrac{1}{2w\prime}}. \tag{2.2.35}$$

The power-confinement factor Γ and the ratio to the core width $2a/\Gamma$ for the fundamental mode are shown in Fig. 2.2.8. The vertical lines in the figure express the single mode core width.

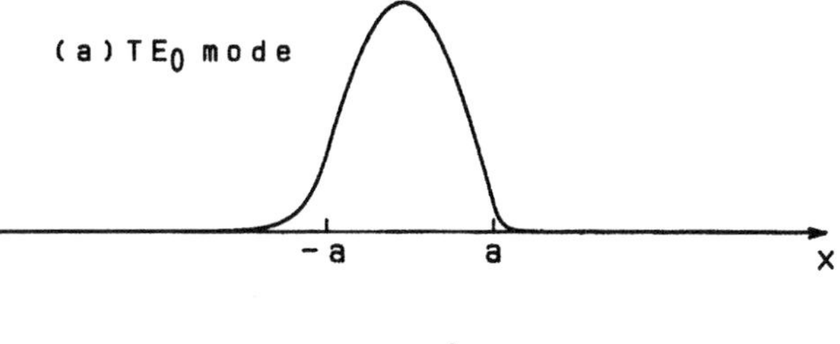

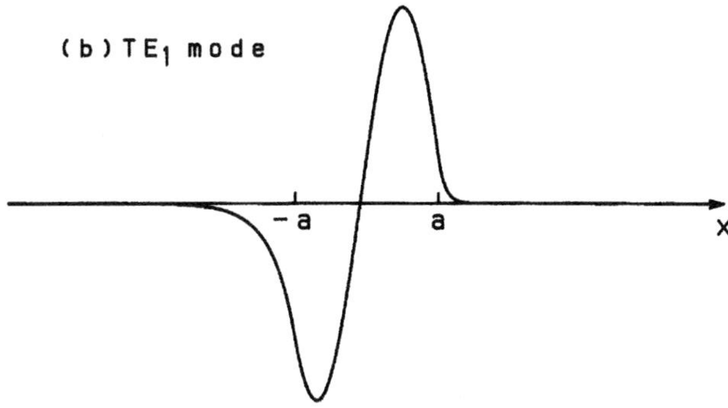

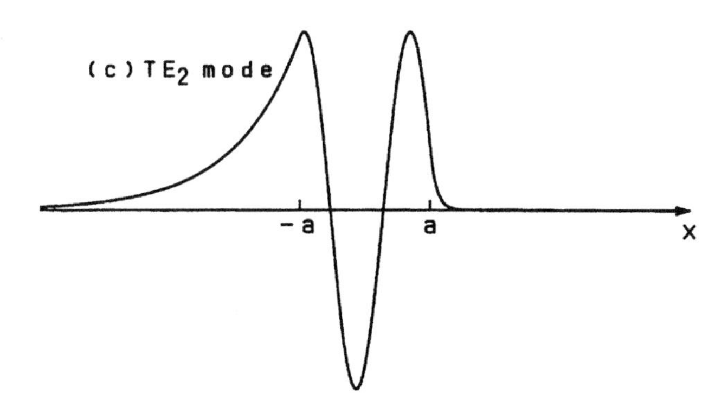

Fig. 2.2.7 Electric field distributions in the slab waveguide.

2.2.1.5 Dispersion equation for TM mode

Based on Eq. (2.2.6), the dispersion equation for the TM mode is obtained in a similar manner to that of the TE mode. We first express the magnetic field distribution H_y as

$$H_y = \begin{cases} A\cos(\kappa a - \phi)e^{-\sigma(x-a)} & (x>a) \\ A\cos(\kappa x - \phi) & (-a \le x \le a) \\ A\cos(\kappa a + \phi)e^{\xi(x+a)} & (x < -a). \end{cases} \tag{2.2.36}$$

Applying the boundary conditions that H_y and E_z should be continuous $x = \pm a$, the following dispersion equation is obtained:

$$u = \frac{m\pi}{2} + \frac{1}{2}\tan^{-1}\left(\frac{n_1^2 w}{n_s^2 u}\right) + \frac{1}{2}\tan^{-1}\left(\frac{n_1^2 w\prime}{n_0^2 u}\right). \tag{2.2.37}$$

Rewriting the above equation by using the normalized frequency v and the normalized propagation constant b, it reduces to

$$2v\sqrt{1-b} = m\pi + \tan^{-1}\left(\frac{n_1^2}{n_s^2}\sqrt{\frac{b}{1-b}}\right) + \tan^{-1}\left(\frac{n_1^2}{n_0^2}\sqrt{\frac{b+\gamma}{1-b}}\right). \tag{2.2.38}$$

The dispersion curve of the TM modes in the waveguide with $n_1 = 3.38$, $n_s = n_0 = 3.17(\gamma = 0)$ are shown in Fig. 2.2.9 and compared with those for the TE modes. It is known that the normalized propagation constant b for

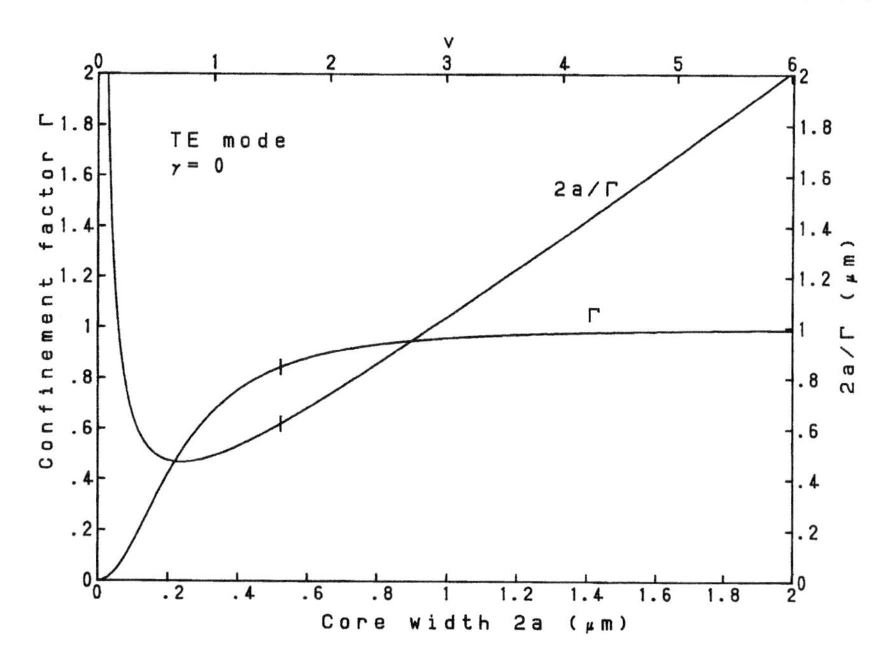

Fig. 2.2.8 Power-confinement factor of the symmetrical slab waveguide.

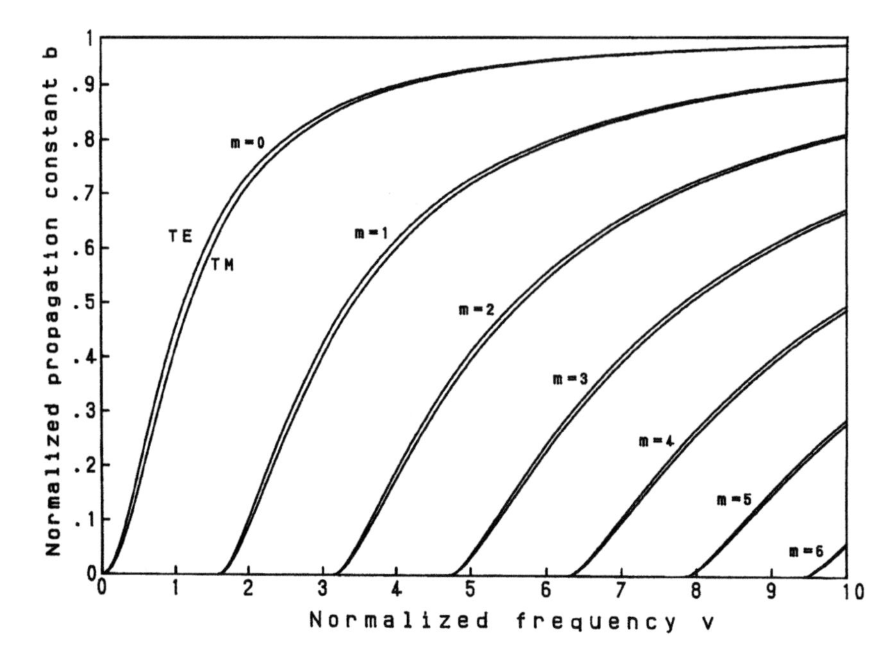

Fig. 2.2.9 Dispersion curves of TE and TM modes in the slab waveguide.

the TM mode is smaller than that for the TE mode with respect to the same v. That means the TE mode is slightly better confined in the core than the TM mode. The power carried by the TM mode is obtained, from Eqs. (2.2.6) and (2.2.30) by

$$P = \frac{\beta}{2\omega\varepsilon_0}\int_{-\infty}^{\infty}\frac{1}{n^2}|H_y|^2 dx. \tag{2.2.39}$$

2.2.2 Rectangular waveguides

2.2.2.1 Basic equations

In this section the analytical method, which was proposed by Marcatili [5], to deal with the three-dimensional optical waveguide, as shown in Fig. 2.2.10, is described. The important assumption of this method is that the electromagnetic field in the shaded area in Fig. 2.2.10 can be neglected since the electromagnetic field of the well-guided mode decays quite rapidly in the cladding region. Then we do not impose the boundary conditions for the electromagnetic field in the shaded area.

We first consider the electromagnetic mode in which E_x and H_y are predominant. According to Marcatili's treatment, we set $H_x = 0$ in Eqs. (2.2.3) and (2.2.4). Then the wave equation and electromagnetic field representation are obtained as

$$\frac{\partial^2 H_y}{\partial x^2} + \frac{\partial^2 H_y}{\partial y^2} + (k^2n^2 - \beta^2)H_y = 0, \tag{2.2.40}$$

$$\begin{cases} H_x = 0 \\ E_x = \dfrac{\omega\mu_0}{\beta}H_y + \dfrac{1}{\omega\varepsilon_0 n^2\beta}\dfrac{\partial^2 H_y}{\partial x^2} \\ E_y = \dfrac{1}{\omega\varepsilon_0 n^2\beta}\dfrac{\partial^2 H_y}{\partial x \partial y} \\ E_z = \dfrac{-j}{\omega\varepsilon_0 n^2}\dfrac{\partial H_y}{\partial x} \\ H_z = \dfrac{-j}{\beta}\dfrac{\partial H_y}{\partial y}. \end{cases} \tag{2.2.41}$$

On the other hand, we set $H_y = 0$ in Eqs. (2.2.3) and (2.2.4) to consider the electromagnetic field in

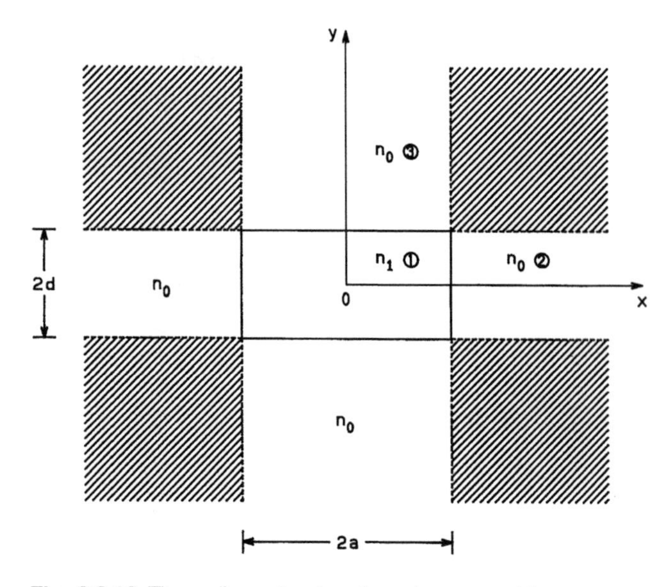

Fig. 2.2.10 Three-dimensional rectangular waveguide.

which E_y and H_x are predominant. The wave equation and electromagnetic field representation are given by

$$\frac{\partial^2 H_x}{\partial x^2}+\frac{\partial^2 H_x}{\partial y^2}+(k^2n^2-\beta^2)H_x=0, \quad (2.2.42)$$

$$\begin{cases} H_y = 0 \\ E_x = -\dfrac{1}{\omega\varepsilon_0 n^2\beta}\dfrac{\partial^2 H_x}{\partial x\partial y} \\ E_y = -\dfrac{\omega\mu_0}{\beta}H_x - \dfrac{1}{\omega\varepsilon_0 n^2\beta}\dfrac{\partial^2 H_x}{\partial y^2} \\ E_z = \dfrac{j}{\omega\varepsilon_0 n^2}\dfrac{\partial H_x}{\partial y} \\ H_z = \dfrac{-j}{\beta}\dfrac{\partial H_x}{\partial x}. \end{cases} \quad (2.2.43)$$

The modes in Eqs. (2.2.40) and (2.2.41) are described as E^x_{pq} (p and q are integers) since E_x and H_y are the dominant electromagnetic fields. On the other hand, the modes in Eqs. (2.2.42) and (2.2.43) are called E^y_{pq} since E_y and H_x are the dominant electromagnetic fields. In the following section, the solution method of the dispersion equation for the E^x_{pq} mode is described in detail, and only the results are shown for the E^y_{pq} mode.

2.2.2.2 Dispersion equations for E^x_{pq} and E^y_{pq} modes

Since the rectangular waveguide shown in Fig. 2.2.10 is symmetrical with respect to the x- and y-axes, we analyze only regions ①–②. We first express the solution fields, which satisfy the wave equation (2.2.40), as

$$H_y = \begin{cases} A\cos(k_x x-\phi)\cos(k_y y-\psi) & \text{region ①} \\ A\cos(k_x a-\phi)e^{-\gamma_x(x-a)}\cos(k_y y-\psi) & \text{region ②} \\ A\cos(k_x x-\phi)e^{-\gamma_y(y-d)}\cos(k_y d-\psi) & \text{region ③} \end{cases} \quad (2.2.44)$$

where the transverse wavenumbers k_x, k_y, γ_x, and γ_y and the optical phases ϕ and ψ are given by

$$\begin{cases} -k_x^2-k_y^2+k^2n_1^2-\beta^2=0 & \text{region ①} \\ \gamma_x^2-k_y^2+k^2n_0^2-\beta^2=0 & \text{region ②} \\ -k_x^2+\gamma_y^2+k^2n_0^2-\beta^2=0 & \text{region ③} \end{cases} \quad (2.2.45)$$

and

$$\begin{cases} \phi=(p-1)\dfrac{\pi}{2} \quad (p=1,2,\ldots) \\ \psi=(q-1)\dfrac{\pi}{2} \quad (q=1,2,\ldots). \end{cases} \quad (2.2.46)$$

We should note here that the integers p and q start from 1 because we follow the mode definition by Marcatili. To the contrary the mode number m in Eq. (2.2.12) for the slab waveguides starts from zero. By the conventional mode definition, the lowest mode in the slab waveguide is the $TE_{m=0}$ mode (Fig. 2.2.7(a)) which has one electric field peak. On the other hand, the lowest mode in the rectangular waveguides is $E^x_{p=1,q=1}$ or $E^y_{p=1,q=1}$ mode (Fig. 2.2.11) which has only one electric field peak along both x- and y-axis directions. Therefore in the mode definition by Marcatili, integers p and q represent the number of local electric field peaks along the x- and y-axis directions.

When we apply the boundary conditions that the electric field $E_z \propto (1/n^2)\partial H_y/\partial x$ should be continuous at $x=a$ and the magnetic field $H_z \propto \partial H_y/\partial x$ should be continuous at $y=d$, we obtain the following dispersion equations:

$$k_x a=(p-1)\frac{\pi}{2}+\tan^{-1}\left(\frac{n_1^2\gamma_x}{n_0^2k_x}\right), \quad (2.2.47a)$$

$$k_y d=(q-1)\frac{\pi}{2}+\tan^{-1}\left(\frac{\gamma_y}{k_y}\right). \quad (2.2.47b)$$

Transversal wavenumbers k_x, k_y, γ_x, and γ_v are related, by Eq. (2.2.45) as

$$\gamma_x^2=k^2(n_1^2-n_0^2)-k_x^2, \quad (2.2.48)$$

$$\gamma_y^2=k^2(n_1^2-n_0^2)-k_y^2. \quad (2.2.49)$$

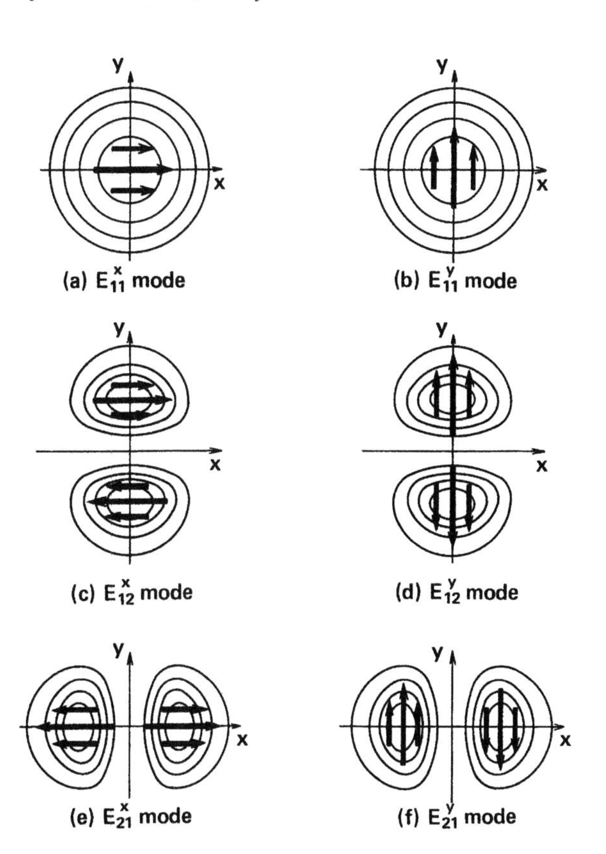

Fig. 2.2.11 Mode definitions and electric field distributions in Marcatili's method.

k_x is obtained from Eqs. (2.2.47a) and (2.2.48), and k_y is determined from Eqs. (2.2.47b) and (2.2.49), respectively. The propagation constant β is then obtained from

$$\beta^2 = k^2 n_1^2 - (k_x^2 + k_y^2). \quad (2.2.50)$$

In order to calculate the dispersion equation for the E_{pq}^y mode, we express the magnetic field H_x as

$$H_x = \begin{cases} A\cos(k_x x - \phi)\cos(k_y y - \psi) & \text{region ①} \\ A\cos(k_x a - \phi)e^{-\gamma_x(x-a)}\cos(k_y y - \psi) & \text{region ②} \\ A\cos(k_x x - \phi)e^{-\gamma_y(y-d)}\cos(k_y d - \psi) & \text{region ③} \end{cases} \quad (2.2.51)$$

Applying the boundary conditions that the magnetic field $H_z \propto \partial H_x/\partial x$ should be continuous at $x = a$ and the electric field $E_z \propto (1/n^2)\partial H_x/\partial y$ should be continuous at $y = d$, we obtain the following dispersion equations

$$k_x a = (p-1)\frac{\pi}{2} + \tan^{-1}\left(\frac{\gamma_x}{k_x}\right), \quad (2.2.52)$$

$$k_y d = (q-1)\frac{\pi}{2} + \tan^{-1}\left(\frac{n_1^2\gamma_y}{n_0^2 k_y}\right). \quad (2.2.53)$$

2.2.2.3 Kumar's method

In Marcatili's method, electromagnetic fields and the boundary conditions in the shaded area in Fig. 2.2.10 are not strictly satisfied. In other words, the hybrid modes in the rectangular waveguides are approximately analyzed by separating into two independent slab waveguides as shown in Fig. 2.2.12. It is well understood when we compare the dispersion Eqs. (2.2.47a) and (2.2.47b) for the E_{pq}^x mode with the slab dispersion Eqs. (2.2.12) and (2.2.37). Equation (2.2.47a) corresponds to the TM mode dispersion equation of the symmetric slab waveguide [Fig. 2.2.12(b)], and Eq. (2.2.47b) corresponds to the TE mode dispersion equation [Fig. 2.2.12(c)], respectively.

Kumar et al. proposed an improvement of accuracy for the Marcatili's method by taking into account the contribution of the fields in the shaded area in Fig. 2.2.10 [6]. We call this method this *Kumar's method* and describe an example of it by analyzing the E_{pq}^x mode in rectangular waveguides.

In Kumar's method, the refractive-index distribution of the rectangular waveguide is expressed by

$$n^2(x,y) = N_x^2(x) + N_y^2(y) + O(n_1^2 - n_0^2), \quad (2.2.54)$$

where

$$N_x^2(x) = \begin{cases} n_1^2/2 & |x| \le a \\ n_0^2 - n_1^2/2 & |x| > a \end{cases} \quad (2.2.55a)$$

$$N_y^2(y) = \begin{cases} n_1^2/2 & |y| \le d \\ n_0^2 - n_1^2/2 & |y| > d. \end{cases} \quad (2.2.55b)$$

The refractive-index distribution that is expressed by Eqs. (2.2.54) and (2.2.55) is shown in Fig. 2.2.13. Generally, the refractive-index difference between core and cladding is quite small ($n_x \approx n_0$) and then we have $O(n_1^2 - n_0^2) \approx 0$ in Eq. (2.2.54). Also the refractive index in the shaded area is approximated as

$$\sqrt{2n_0^2 - n_1^2} \approx n_0. \quad (2.2.56)$$

Therefore it is known that the refractive index expressed by Eqs. (2.2.54) and (2.2.55) approximates quite well the actual refractive-index distribution of the rectangular waveguide. Although the approximation is good, there still remains the small difference in the refractive-index expression of Eq. (2.2.54) for the shaded area from the actual value. In the present method, the correction is made by using the perturbation method as shown in the following.

We first express the solution of the wave Eq. (2.2.40) for the E_{pq}^x mode, by using the separation of variables, as

$$H_y(x,y) = X(x)Y(y). \quad (2.2.57)$$

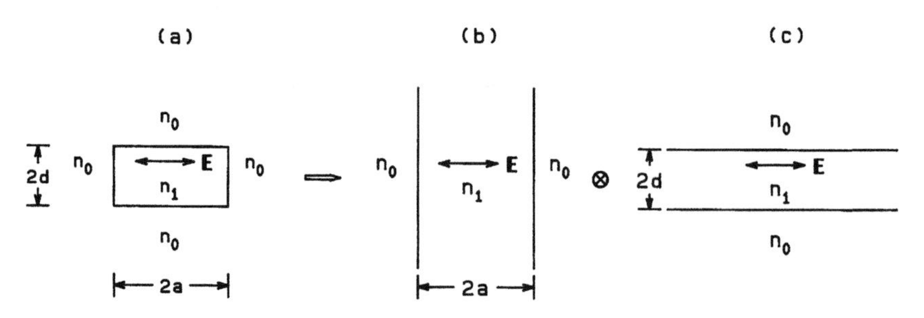

Fig. 2.2.12 Rectangular waveguide and its equivalent, two independent slab waveguides, in Marcatili's method.

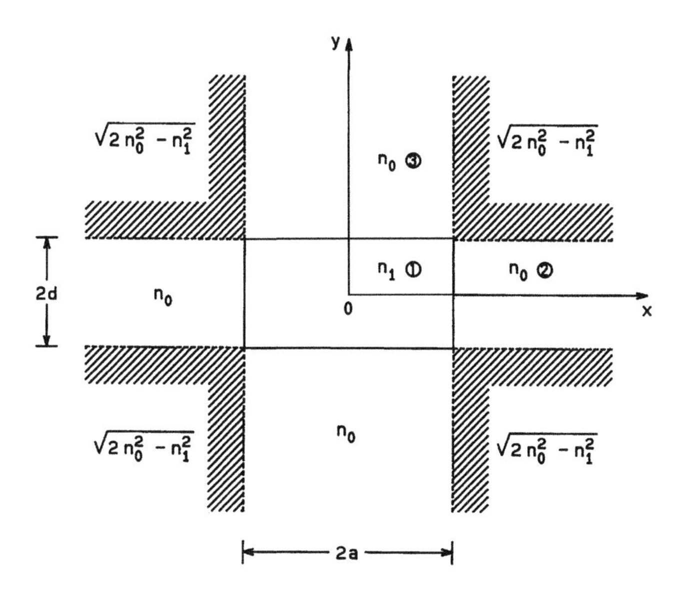

Fig. 2.2.13 Refractive-index profile in Kumar's method.

Substituting Eqs. (2.2.54) and (2.2.57) into (2.2.40), the wave equation reduces to

$$\frac{d^2X}{dx^2}Y+X\frac{d^2Y}{dy^2}+\left[k^2(N_x^2+N_y^2)-\beta^2\right]XY=0, \tag{2.2.58}$$

where small quantities of the order of $O(n_1^2-n_0^2)$ have been neglected. Dividing Eq. (2.2.58) by XY, it can be separated into two terms: one is dependent on variable x and the other is dependent on variable y, respectively, as

$$\frac{1}{X}\frac{d^2X}{dx^2}+k^2N_x^2(x)+\frac{1}{Y}\frac{d^2Y}{dy^2}+k^2N_y^2(x)=\beta^2. \tag{2.2.59}$$

The necessary conditions for Eq. (2.2.59) to be satisfied for arbitrary values of x and y are

$$\frac{1}{X}\frac{d^2X}{dx^2}+k^2N_x^2(x)=\beta_x^2, \tag{2.2.60a}$$

$$\frac{1}{Y}\frac{d^2Y}{dy^2}+k^2N_y^2(y)=\beta_y^2, \tag{2.2.60b}$$

where β_x and β_y are constants that are independent from x and y. We then have two independent wave equations as

$$\frac{d^2X}{dx^2}+\left[k^2N_x^2(x)-\beta_x^2\right]X(x)=0, \tag{2.2.61a}$$

$$\frac{d^2Y}{dy^2}+\left[k^2N_y^2(y)-\beta_y^2\right]Y(y)=0. \tag{2.2.61b}$$

From Eqs. (2.2.59) and (2.2.60) we can derive an equation to determine the propagation constant:

$$\beta^2=\beta_x^2+\beta_y^2. \tag{2.2.62}$$

The solution fields of Eqs. (2.2.61a) and (2.2.61b) are expressed in a similar manner as previously in Section 2.2.2.2 by

$$X(x)=\begin{cases}A\cos(k_xx-\phi) & (0\le x\le a)\\ A\cos(k_xa-\phi)e^{-\gamma_x(x-a)} & (x>a)\end{cases} \tag{2.2.63}$$

$$Y(y)\begin{cases}B\cos(k_yy-\psi) & (0\le y\le d)\\ B\cos(k_yd-\psi)e^{-\gamma_y(y-d)} & (y>d),\end{cases} \tag{2.2.64}$$

where only the first quadrant is considered due to the symmetry of the waveguide, and transverse wavenumbers k_x, γ_x, k_y, and γ_y are related with β_x and β_y by

$$\gamma_x^2=k^2(n_1^2-n_0^2)-k_x^2, \tag{2.2.65a}$$

$$\gamma_y^2=k^2(n_1^2-n_0^2)-k_y^2, \tag{2.2.65b}$$

$$\beta_x^2=\frac{k^2n_1^2}{2}-k_x^2, \tag{2.2.66a}$$

$$\beta_y^2=\frac{k^2n_1^2}{2}-k_y^2, \tag{2.2.66b}$$

and optical phases are expressed by

$$\begin{cases}\phi=(p-1)\frac{\pi}{2} & (p=1,2,\ldots)\\ \psi=(q-1)\frac{\pi}{2} & (1=1,2,\ldots).\end{cases} \tag{2.2.67}$$

When we apply the boundary conditions that the electric field

$$E_z\propto(1/n^2)\partial H_y/\partial x=(Y/n^2)dX/dx$$

should be continuous at $x=a$ and the magnetic field $H_z\propto\partial H_y/\partial y=XdY/dy$ should be continuous at $y=d$, we obtain the following dispersion equations:

$$k_xa=(p-1)\frac{\pi}{2}+\tan^{-1}\left(\frac{n_1^2\gamma_x}{n_0^2k_x}\right), \tag{2.2.68}$$

$$k_yd=(q-1)\frac{\pi}{2}+\tan^{-1}\left(\frac{\gamma_y}{k_y}\right). \tag{2.2.69}$$

The propagation constant is obtained from Eqs. (2.2.62) and (2.2.66a), by

$$\beta^2=k^2n_1^2-(k_x^2+k_y^2). \tag{2.2.70}$$

Eqs. (2.2.68)–(2.2.70) for the dispersion and the propagation constant are known to be the same as the Eqs. (2.2.47a), (2.2.47b) and (2.2.50) by the Marcatili's method. But in the present Kumar's method, an improvement of the accuracy of the propagation constant can be obtained by the perturbation method with respect to the shaded area in Fig. 2.2.13. We rewrite the refractive-index distribution for the rectangular waveguide as

$$n^2(x,y) = N_x^2(x) + N_y^2(y) + \delta \cdot \eta(x,y), \qquad (2.2.71)$$

where δ is a small quantity and $\delta \cdot \eta(x,y)$ denotes the perturbation term which is expressed by

$$\delta \cdot \eta(x,y) = \begin{cases} (n_1^2 - n_0^2) & |x|>a \quad and \quad |y|>d \\ 0 & |x| \le a \quad or \quad |y| \le d. \end{cases} \qquad (2.2.72)$$

Generally the wave equation is expressed by

$$\nabla^2 f + (k^2 n^2 - \beta^2) f = 0, \qquad (2.2.73)$$

where $\nabla^2 = \partial^2/\partial x^2 + \partial^2/\partial y^2$. The solution field f and the eigenvalue β^2 of the above equation are expressed in the first-order perturbation form as

$$f = f_0 + \delta \cdot f_1 \qquad (2.2.74)$$

$$\beta^2 = \beta_0^2 + \delta \cdot \beta_1^2. \qquad (2.2.75)$$

Substituting Eqs. (2.2.74) and (2.2.75) into (2.2.73) and comparing the terms for each order of δ, the following equations are obtained:

$$\nabla^2 f_0 + [k^2(N_x^2 + N_y^2) - \beta_0^2] f_0 = 0, \qquad (2.2.76)$$

$$\nabla^2 f_1 + [k^2(N_x^2 + N_y^2) - \beta_0^2] f_1 + k^2 \eta f_0 - \beta_1^2 f_0 = 0. \qquad (2.2.77)$$

Here we consider the integral

$$\iint [\text{Eq.}(2.76)^* \cdot f_1 - \text{Eq.}(2.77) \cdot f_0^*] dx\, dy$$

in the region D. Then we have

$$\beta_1^2 \iint_D |f_0|^2 dx\, dy = \iint_D [f_0^* \nabla^2 f_1 - f_1 \nabla^2 f_0^*] dx\, dy + k^2 \iint_D \eta |f_0|^2 dx\, dy. \qquad (2.2.78)$$

The first term in the right-hand side of the above equation is rewritten, by using Green's theorem (refer to Chapter 10), as

$$\iint_D [f_0^* \nabla^2 f_1 - f_1 \nabla^2 f_0^*] dx\, dy = \oint \left[f_0^* \frac{\partial f_1}{\partial n} - f_1 \frac{\partial f_0^*}{\partial n} \right] d\ell, \qquad (2.2.79)$$

where $\partial/\partial n$ represents the differentiation along the outside normal direction on the periphery of the integration region, and $\oint d\ell$ represents the line integral along the periphery. The line integral of Eq. (2.2.79) becomes zero when region D is enlarged to infinity. Therefore we have

$$\beta_1^2 = \frac{k^2 \int_{-\infty}^{\infty} \int_{-\infty}^{\infty} \eta(x,y) |f_0|^2 dx\, dy}{\int_{-\infty}^{\infty} \int_{-\infty}^{\infty} |f_0|^2 dx\, dy}. \qquad (2.2.80)$$

Equation (2.2.73) corresponds to (2.2.40) and Eq. (2.2.76) corresponds to (2.2.58). Then β_0 is the eigenvalue for the dispersion Eqs. (2.2.68)–(2.2.70) and $f_o(x, y)$ is the field distribution given by Eqs. (2.2.57), (2.2.63) and (2.2.64). The eigenvalue which is given by the first-order perturbation is therefore expressed from Eqs. (2.2.72), (2.2.75) and (2.2.80) by

$$\begin{aligned} \beta^2 &= \beta_0^2 + \frac{k^2 \int_{-\infty}^{\infty} \int_{-\infty}^{\infty} \delta \cdot \eta(x,y) |X(x)Y(y)|^2 dx\, dy}{\int_{-\infty}^{\infty} \int_{-\infty}^{\infty} |X(x)Y(y)|^2 dx\, dy} \\ &= (k^2 n_1^2 - k_x^2 - k_y^2) \\ &\quad + \frac{k^2 (n_1^2 - n_0^2) \int_a^{\infty} |X(x)|^2 dx \int_d^{\infty} |Y(y)|^2 dy}{\int_0^{\infty} |X(x)|^2 dx \int_0^{\infty} |Y(y)|^2 dy} \\ &= (k^2 n_1^2 - k_x^2 - k_y^2) \\ &\quad + \frac{k^2 (n_1^2 - n_0^2) \cos^2(k_x a - \phi) \cos^2(k_y d - \psi)}{(1 + \gamma_x a)(1 + \gamma_y d)}. \end{aligned} \qquad (2.2.81)$$

In the second term of Eq. (2.2.81), we approximated $n_0^2/n_1^2 \approx 1$. The normalized propagation constant is obtained from Eqs. (2.2.17), (2.2.18) and (2.2.81) as

$$b = 1 - \frac{k_x^2 + k_y^2}{k^2(n_1^2 - n_0^2)} + \frac{\cos^2(k_x a - \phi) \cos^2(k_y d - \psi)}{(1 + \gamma_x a)(1 + \gamma_y d)}. \qquad (2.2.82)$$

Figure 2.2.14 shows the dispersion curves for the rectangular waveguides with core aspect ratios $a/d = 1$ and $a/d = 2$, which are calculated by using different analysis methods for the scalar wave equations [6]. Analysis by the point matching method [7] gives the most accurate value among four of the analyses and is used as the standard for the comparison of accuracy. The effective index method will be described in the following section. It is known from Fig. 2.2.14 that Kumar's method gives the more accurate results than Marcatili's method. The accuracy of the effective index method is almost the same as that of Marcatili's method; but the effective index method gives the larger estimation than the accurate solution, whereas, Marcatili's method gives the lower estimation than the accurate solution, respectively. For practicality, however, the effective index method is

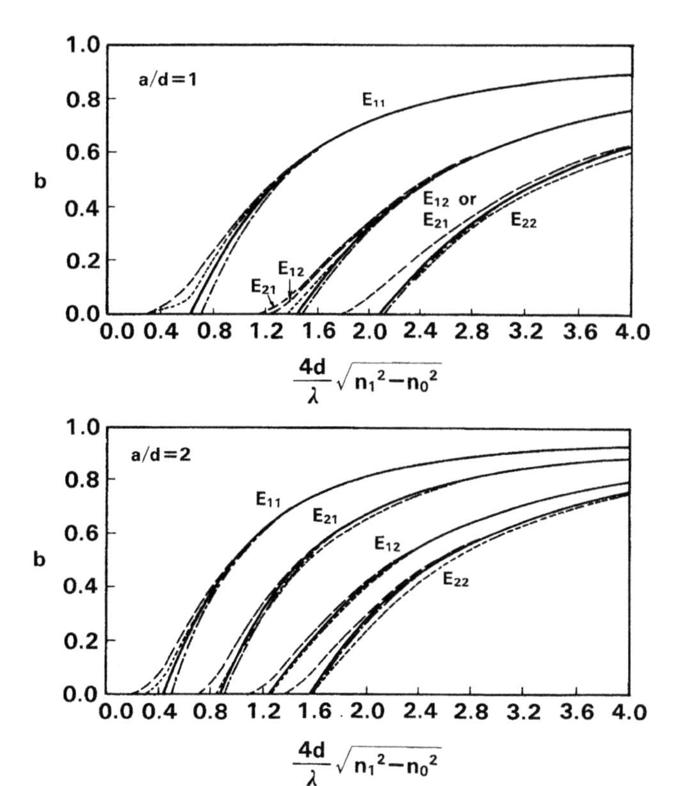

Fig. 2.2.14 Comparison of the dispersion curves calculated with different analytical methods.—— Marcatili's method [6];———Kumar's method, ... point matching method [7];——— effective index method.(After Ref. [6]).

a very important method to analyze, for example, ridge waveguides which require a numerical method such as the finite element method.

2.2.2.4 Effective index method

The ridge waveguide, such as shown in Fig. 2.2.15, is difficult to analyze by Marcatili's method or Kumar's method since the waveguide structure is too complicated to deal with by the division of waveguides. In order to analyze ridge waveguides, we should use numerical methods, such as the finite element method and finite difference method. The effective index method [8]–[9] is an analytical method applicable to complicated waveguides such as ridge waveguides and diffused waveguides in $LiNbO_3$. In the following, the effective index method of analysis is described, taking as an example the E^x_{pq} mode in the ridge waveguide.

The wave equation for the E^x_{pq} mode is given by Eq. (2.2.40) as

$$\frac{\partial^2 H_y}{\partial x^2} + \frac{\partial^2 H_y}{\partial y^2} + [k^2 n^2(x,y) - \beta^2]H_y = 0. \tag{2.2.83}$$

The basic assumption of the effective index method is that the electromagnetic field can be expressed, with the separation of variables, as

$$H_y(x,y) = X(x)Y(y). \tag{2.2.84}$$

Therefore if the assumption of separation of variables are not accurate, due to the waveguide structure or the wavelength of light, the accuracy of the method itself becomes very poor. Substituting Eq. (2.2.84) into Eq. (2.2.83) and dividing it by XY, we obtain

$$\frac{1}{X}\frac{d^2X}{dx^2} + \frac{1}{Y}\frac{d^2Y}{dy^2} + [k^2 n^2(x,y) - \beta^2] = 0. \tag{2.2.85}$$

Here we add to, and subtract from, Eq. (2.2.85) the y-independent value of $k^2 n^2_{\mathrm{eff}}(x)$ and separate the equation into two independent equations:

$$\frac{1}{Y}\frac{d^2Y}{dy^2} + [k^2 n^2(x,y) - k^2 n^2_{\mathrm{eff}}(x)] = 0, \tag{2.2.86a}$$

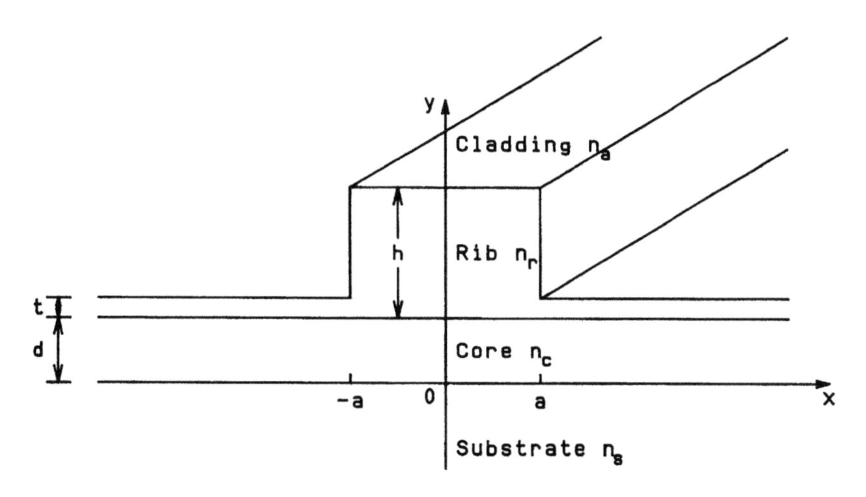

Fig. 2.2.15 Ridge waveguide.

$$\frac{1}{X}\frac{d^2Y}{dx^2} + [k^2 n_{\text{eff}}^2(x) - \beta^2] = 0. \quad (2.2.86b)$$

$n_{\text{eff}}(x)$ is called the *effective index distribution*. We first solve Eq. (2.2.86) and determine the effective index distribution $n_{\text{eff}}(x)$. The variation of the actual refractive-index profile $n(x, y)$ is depicted in Fig. 2.2.16 where $n_r = n_s$ and s is the height of the rib, which takes the following values, depending on the position x:

$$s = \begin{cases} h & 0 \le |x| \le a \\ t & |x| > a. \end{cases} \quad (2.2.87)$$

From the boundary condition that $H_z \propto \partial H_y/\partial y$ should be continuous at $y = 0$, d and $d + s$, we have the continuity condition for dY/dy at the foregoing boundaries. The dispersion equations for the four-layer slab waveguide shown in Fig. 2.2.16 is given by

$$\sin(\kappa d - 2\phi) = \sin(\kappa d)e^{-2(\sigma s+\psi)}, \quad (2.2.88)$$

where the parameters are

$$\phi = \tan^{-1}\left(\frac{\sigma}{\kappa}\right), \quad (2.2.89a)$$

$$\psi = \tanh^{-1}\left(\frac{\sigma}{\gamma}\right), \quad (2.2.89b)$$

$$\kappa = k\sqrt{n_c^2 - n_{\text{eff}}^2}, \quad (2.2.89c)$$

$$\sigma = k\sqrt{n_{\text{eff}}^2 - n_s^2}, \quad (2.2.89d)$$

$$\gamma = k\sqrt{n_{\text{eff}}^2 - n_a^2}. \quad (2.2.89e)$$

The solution of Eq. (2.2.88) with $s = h(0 \le |x| \le a)$ gives the effective index $n_{\text{eff}}(h)$ for $0 \le |x| \le a$, and the solution of Eq. (2.2.88) with $s = t(|x| > a)$ gives the effective index $n_{\text{eff}}(t)$ for $|x| > a$ respectively. Then the effective index distribution $n_{\text{eff}}(x)$ is obtained (see Fig. 2.2.17). The solution of the wave equation (2.2.86b) is calculated by solving the three-layer symmetrical slab waveguide. The boundary condition is that $E_z \propto (1/n^2)\partial H_y/\partial x$ should be continuous at $x = \pm a$. Therefore $(1/n^2)X$ should be continuous at $x = \pm a$. Under the above boundary condition, the dispersion equation is obtained as

$$u \tan(u) = \frac{n_{\text{eff}}^2(h)}{n_{\text{eff}}^2(t)} w, \quad (2.2.90)$$

where

$$u = ka\sqrt{n_{\text{eff}}^2(h) - \left(\frac{\beta}{k}\right)^2}, \quad (2.2.91a)$$

$$w = ka\sqrt{\left(\frac{\beta}{k}\right)^2 - n_{\text{eff}}^2(t)}. \quad (2.2.91b)$$

The dispersion equations for the E_{pq}^y mode are obtained in the similar manner:

$$\sin(\kappa d - 2\phi) = \sin(\kappa d)e^{-2(\sigma s+\psi)}, \quad (2.2.92)$$

$$u \tan(u) = w, \quad (2.2.93)$$

where

$$\phi = \tan^{-1}\left(\frac{\sigma n_c^2}{\kappa n_s^2}\right), \quad (2.2.94a)$$

$$\psi = \tanh^{-1}\left(\frac{\sigma n_a^2}{\gamma n_s^2}\right). \quad (2.2.94b)$$

2.2.3 Radiation field from waveguide

The radiation field from an optical waveguide into free space propagates divergently. The radiation field is different from the field in the waveguide. Therefore it is important to know the profile of the radiation field for

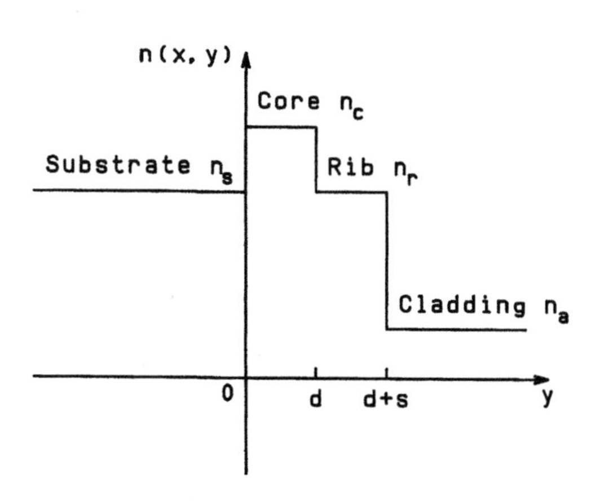

Fig. 2.2.16 Variation of the actual refractive-index profile $n(x, y)$.

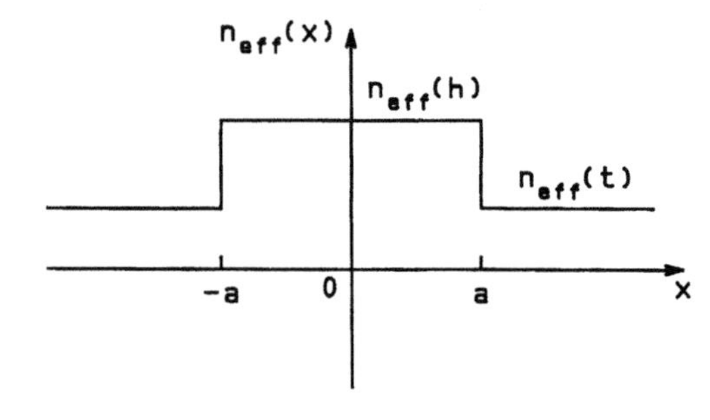

Fig. 2.2.17 Effective index distribution $n_{\text{eff}}(x)$.

efficiently coupling the light between two waveguides or between a waveguide and an optical fiber. In this section, we describe the derivation of the radiation field pattern from the rectangular waveguide.

2.2.3.1 Fresnel and Fraunhofer regions

We consider the coordinate system shown in Fig. 2.2.18, where the endface of the waveguide is located at $z = 0$ and the electromagnetic field is radiated into the free space with refractive index n. The electric field at the endface of the waveguide is denoted by $g(x_0, y_0, 0)$ and the electric field distribution on the observation plane at distance z is expressed as $f(x, y, z)$. By the Fresnel–Kirchhoff diffraction formula [10] (see Chapter 10), the radiation pattern $f(x, y, z)$ is related to the endface field $g(x_0, y_0, 0)$ as

$$f(x,y,z) = \frac{jkn}{2\pi}\int_{-\infty}^{\infty}\int_{-\infty}^{\infty} g(x_0,y_0,0)\frac{1}{r}e^{-jknr}dx_0\,dy_0, \tag{2.2.95}$$

where k is the free-space wavenumber $k = 2\pi/\lambda$ and the distance r between Q and P is given by

$$r = [(x-x_0)^2 + (y-y_0) + z^2]^{1/2}. \tag{2.2.96}$$

When the distance of the observation plane z is very large compared with $|x-x_0|$ and $|y-y_0|$, Eq. (2.2.96) is approximated by

$$\begin{aligned} r &= z\left[1 + \frac{(x-x_0)^2 + (y-y_0)^2}{z^2}\right]^{1/2} \\ &= z + \frac{(x-x_0)^2 + (y-y_0)}{2z} + \ldots \\ &= z + \frac{x^2+y^2}{2z} - \frac{xx_0+yy_0}{z} + \frac{x_0^2+y_0^2}{2z} + \ldots. \end{aligned} \tag{2.2.97}$$

The number of expansion terms to approximate r accurately depends on the distance z between the endface of the waveguide and the observation plane. Generally, the electromagnetic field in the waveguide is confined in a small area of the order of 10 μm. Therefore if z is larger than, for example, 1 mm, any term higher than the fourth term in the right-hand side of Eq. (2.2.97) can be neglected. The radiation field where this condition is satisfied is called the far-field region or Fraunhofer region. On the other hand, when z is not so large, we should take into account up to the fourth term in Eq. (2.2.97). The radiation field in which this condition holds is called the near-field region or Fresnel region. However, we should note that even the Fresnel approximation is not satisfied in the region close to the waveguide endface. The fourth term in the extreme right of Eq. (2.2.97) determines which approximation we should adopt. Generally the contribution to knr by the fourth term, $kn(x_0^2 + y_0^2/2z)$ determines whether the Fresnel or Fraunhofer approximation should be used. The measure for the judgment is $kn(x_0^2+y_0^2)/2z = \pi/2$. If, for example, the optical field is confined in the rectangular region with square core area D^2, then $kn(x_0^2 + y_0^2)/2z = \pi/2$ at $z = nD^2/\lambda$ and we have the criteria:

$$\begin{cases} z > n\dfrac{D^2}{\lambda} & \text{Fraunhofer region} \\ z < n\dfrac{D^2}{\lambda} & \text{Fresnel region.} \end{cases} \tag{2.2.98}$$

When we apply the Fraunhofer approximation to r, Eq. (2.2.95) reduces to

$$\begin{aligned} f(x,y,z) = {} & \frac{jkn}{2\pi z}\exp\left\{-jkn\left[z + \frac{x^2+y^2}{2z}\right]\right\} \\ & \times \int_{-\infty}^{\infty}\int_{-\infty}^{\infty} g(x_0,y_0,0) \\ & \exp\left\{jkn\frac{(xx_0+yy_0)}{z}\right\}dx_0dy_0. \end{aligned} \tag{2.2.99}$$

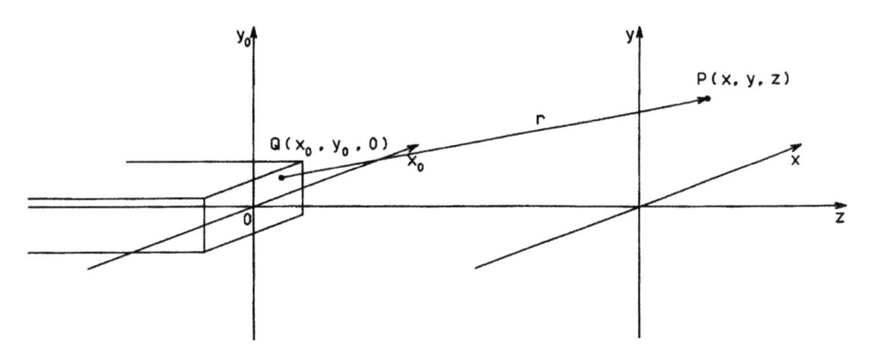

Fig. 2.2.18 Coordinate system for the waveguide endface ($z = 0$) and observation plane ($z = z$).

It is known from the above equation that the Fraunhofer pattern $f(x, y, z)$ is a spatial Fourier transformation of the field profile at the waveguide endface $g(x_0, y_0, 0)$.

2.2.3.2 Radiation pattern of Gaussian Beam

It has been described that the radiation pattern from the waveguide is expressed by Eq. (2.2.95) or (2.2.99). The accurate electromagnetic field distribution in the rectangular waveguide $g(x_0, y_0, 0)$ is determined numerically by, for example, the finite element method, as described in Chapter 6. An analytical method such as Marcatili's method does not give the accurate field distribution, especially for the cladding region. Even though the accuracy of the eigenvalue is improved by Kumar's method, the field distribution is not accurate since Eq. (2.2.77) is difficult to solve to obtain the perturbation field f_1. Therefore it is not easy to calculate the radiation pattern from the rectangular waveguide analytically. Here we approximate the electric field distribution in the rectangular waveguide by a Gaussian profile to obtain the radiation pattern analytically. The Gaussian electric field profile in the waveguide is expressed by

$$g(x_0, y_0, 0) = A\exp\left\{-\left[\frac{x_0^2}{w_1^2}+\frac{y_0^2}{w_2^2}\right]\right\}, \qquad (2.2.100)$$

where w_1 and w_2 are the spot size of the field (the position at which electric field $|g|$ becomes $1/e$ to the peak value) along the x_0- and y_0-axis directions, respectively, and A is a constant. Substituting Eq. (2.2.97) and (2.2.100) into (2.2.95) we obtain

$$\begin{aligned} f(x,y,z) &= \frac{jkn}{2\pi z}\int_{-\infty}^{\infty}\int_{-\infty}^{\infty} g(x_0,y_0,0) \\ &\times \exp\left\{-jkn\left[z+\frac{(x-x_0)^2+(y-y_0)^2}{2z}\right]\right\}dx_0dy_0 \\ &= \frac{jkn}{2\pi z}Ae^{-jknz}\int_{-\infty}^{\infty}\exp\left\{-\frac{x_0^2}{w_1^2}-j\frac{kn}{2n}(x-x_0)^2\right\}dx_0 \\ &\times\int_{-\infty}^{\infty}\exp\left\{-\frac{y_0^2}{w_2^2}-j\frac{kn}{2z}(y-y_0)^2\right\}dy_0, \end{aligned} \qquad (2.2.101)$$

where the Fresnel approximation to r has been used. Since the integral in Eq. (2.2.101) for x_0 and y_0 has the same form, detailed calculation only for x_0 will be described. When we define the parameter p as

$$p = \frac{1}{w_1^2}+j\frac{\pi n}{\lambda z}, \qquad (2.2.102)$$

the integral with respect to x_0 in Eq. (2.2.101) becomes

$$\begin{aligned} &\int_{-\infty}^{\infty}\exp\left\{-\frac{x_0^2}{w_1^2}-j\frac{kn}{2z}(x-x_0)^2\right\}dx_0 \\ &= \exp\left\{-j\frac{\pi n}{\lambda n}x^2-\frac{\pi^2n^2x^2}{p\lambda^2z^2}\right\} \\ &\int_{-\infty}^{\infty}\exp\left\{-p\left(x_0-j\frac{\pi nx}{p\lambda z}\right)^2\right\}dx_0 \\ &= \sqrt{\frac{\pi}{p}}\exp\left\{-j\frac{\pi n}{\lambda n}x^2-\frac{\pi^2n^2x^2}{p\lambda^2z^2}\right\} \\ &= \sqrt{\frac{\pi}{p}}\exp\left\{-j\frac{\pi nx^2}{\lambda}\frac{\left(z-j\frac{\pi nw_1^2}{\lambda}\right)}{\left(z^2+\frac{\pi^2n^2w_1^2}{\lambda^2}\right)}\right\}. \end{aligned} \qquad (2.2.103)$$

We further introduce new variables, whose physical meanings are explained later, as

$$W_1(z) = w_1\sqrt{1+\left(\frac{\lambda z}{\pi nw_1^2}\right)^2}, \qquad (2.2.104a)$$

$$R_1(z) = z\left[1+\left(\frac{\pi nw_1^2}{\lambda z}\right)^2\right], \qquad (2.2.104b)$$

$$\Theta_1(z) = \tan^{-1}\left(\frac{\lambda z}{\pi nw_1^2}\right). \qquad (2.2.104c)$$

Parameter p of Eq. (2.2.102) can be rewritten by using (2.2.104) as

$$p = j\frac{\pi nW_1(z)}{\lambda zw_1}e^{-j\Theta_1}. \qquad (2.2.105)$$

Substituting Eq. (2.2.105) into (2.2.103), Eq. (2.2.103) is finally expressed as

$$\begin{aligned} &\int_{\infty}^{\infty}\exp\left\{-\frac{x_0^2}{w_1^2}-j\frac{kn}{2z}(x-x_0)^2\right\}dx_0 \\ &= \sqrt{\frac{\lambda w_1z}{jnW_1(z)}}\exp\left\{-\left[\frac{1}{W_1^2(z)}+\frac{jkn}{2R_1(z)}\right]x^2+j\frac{\Theta_1(z)}{2}\right\}. \end{aligned} \qquad (2.2.106)$$

Similarly, the integral with respect y_0 in Eq. (2.2.101) is given by

$$\begin{aligned} &\int_{-\infty}^{\infty}\exp\left\{-\frac{y_0^2}{w_2^2}-j\frac{kn}{2z}(y-y_0)^2\right\}dy_0 \\ &= \sqrt{\frac{\lambda w_2z}{jnW_2(z)}}\exp\left\{-\left[\frac{1}{W_2^2(z)}+\frac{jkn}{2R_2(z)}\right]y^2+j\frac{\Theta_2(z)}{2}\right\}, \end{aligned} \qquad (2.2.107)$$

where the parameters W_2, R_2 and Θ_2 are defined by

$$W_2(z) = w_2\sqrt{1+\left(\frac{\lambda z}{\pi n w_2^2}\right)^2}, \quad (2.2.108a)$$

$$R_2(z) = z\left[1+\left(\frac{\pi n w_2^2}{\lambda z}\right)^2\right], \quad (2.2.108b)$$

$$\Theta_2(z) = \tan^{-1}\left(\frac{\lambda z}{\pi n w_2^2}\right), \quad (2.2.108c)$$

Substituting Eqs. (2.2.106) and (2.2.107) into (2.2.101), the radiation pattern from the rectangular waveguide $f(x, y, z)$ is expressed by

$$f(x,y,z) = \sqrt{\frac{w_1 w_2}{W_1 W_2}}A \exp\left\{-\left[\frac{x^2}{W_1^2}+\frac{y^2}{W_2^2}\right] - jkn\left[\frac{x^2}{2R_1}+\frac{y^2}{2R_2}+z\right]+j\frac{(\Theta_1+\Theta_2)}{2}\right\}. \quad (2.2.109)$$

It is known from the above equation that $W_1(z)$ and $W_2(z)$ represent the spot sizes of the radiation field, and $R_1(z)$ and $R_2(z)$ represent the radii of curvature of the wavefronts, respectively. If the observation point P is sufficiently far from the endface of the waveguide, and the following conditions $z \gg \pi n w_1^2/\lambda$ and $\pi n w_2^2/\lambda$ are satisfied in the Fraunhofer region, Eqs. (2.2.104) and (2.2.108) are approximated as

$$\begin{cases} W_1(z) \cong \dfrac{\lambda z}{\pi n w_1} \\ W_2(z) \cong \dfrac{\lambda z}{\pi n w_2} \\ R_1(z) \cong R_2(z) \cong z. \end{cases} \quad (2.2.110)$$

In this Fraunhofer region, the divergence angles θ_1 (Fig. 2.2.19) and θ_2 of the radiation field along the x- and y-axis directions are expressed by

$$\theta_1 = \tan^{-1}\left(\frac{W_1(z)}{z}\right) = \tan^{-1}\left(\frac{\lambda}{\pi n w_1}\right), \quad (2.2.111a)$$

$$\theta_2 = \tan^{-1}\left(\frac{W_2(z)}{z}\right) = \tan^{-1}\left(\frac{\lambda}{\pi n w_2}\right). \quad (2.2.111b)$$

Let us calculate the divergence angles of the radiation field from a semiconductor laser diode operating at $\lambda = 1.55$ µm and having the active-layer (core) refractive index $n_1 = 3.5$, cladding index $n_0 = 3.17$, and core width and thickness of $2a = 1.5$ µm and $2d = 0.15$ µm, respectively. The electric field distribution of the waveguide is calculated by using the finite element method waveguide analysis, which will be described in Chapter 6, and is Gaussian fitted to obtain the spot sizes w_1 and w_2

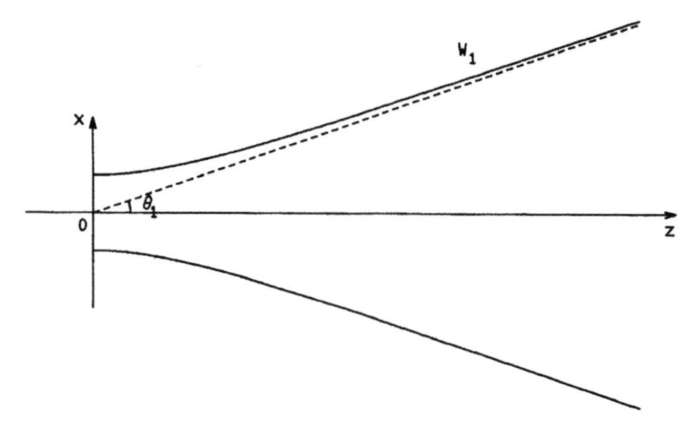

Fig. 2.2.19 Variation of the spot size along x-axis direction, $W_1(z)$.

along x- and y-axis directions. Gaussian-fitted spot sizes w_1 and w_2 are

$$\begin{cases} w_1 = 0.88 \text{ µm} \\ w_2 = 0.35 \text{ µm}. \end{cases} \quad (2.2.112)$$

The divergence angles θ_1 and θ_2 are then obtained, by Eqs. (2.2.111b) and (2.2.112) as

$$\begin{cases} \theta_1 = 0.51(\text{rad.}) = 29.4(\text{degree}) \\ \theta_2 = 0.95(\text{rad.}) = 54.4(\text{degree}). \end{cases} \quad (2.2.113)$$

It is known from the above result that the radiation field from the semiconductor laser diode has an elliptic shape and the divergence angle along the thin active-layer (y-axis direction) is much larger than that along the wide active-layer direction (x-axis direction).

2.2.4 Multimode interference (MMI) device

Multimode interference devices, based on self-imaging effect [11,12], are very important integrated optical components which can perform many different splitting

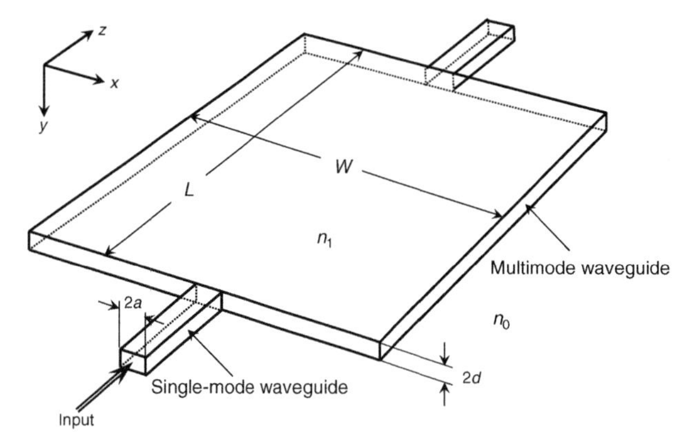

Fig. 2.2.20 Schematic configuration of multimode interference (MMI) waveguide.

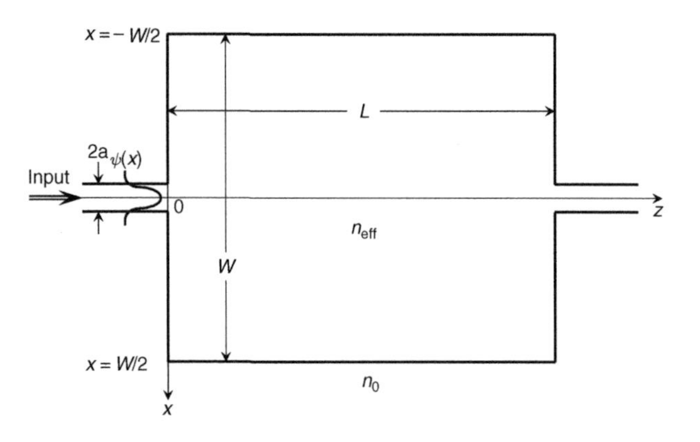

Fig. 2.2.21 Two-dimensional representation of an MMI waveguide.

and combining functions [13–16]. Figure 2.2.20 shows a schematic configuration of MMI waveguide. The key structure of an MMI device is a waveguide designed to support a large number of modes. The width, thickness and length of the multimode region are W, $2d$ and L, respectively. Single-mode waveguide with core width $2a$ and thickness $2d$ is connected to the multi-mode waveguide. Refractive indices of the core of single-mode and multimode waveguides are equal to n_1 and the refractive index of the cladding is n_0. Three-dimensional waveguide structure can be reduced to a two-dimensional problem by using an effective index method. The effective index n_{eff} of the core is calculated by solving the eigen-mode equation along y-axis. Figure 2.2.21 shows a two-dimensional configuration of an MMI waveguide of core effective index n_{eff} and cladding refractive index n_0. Electric field in the multimode waveguide is calculated by using Eqs. (2.2.7)–(2.2.15). Here waveguide parameters n_1, a and n_s in Eqs. (2.2.7)–(2.2.15) are replaced by n_{eff}, $W/2$ and n_0, respectively. Then electric field profile for TE_m mode in the multimode waveguide is expressed by

$$E_y^m(x,y) = \begin{cases} A_m\cos\left(u_m + \dfrac{m\pi}{2}\right)\exp\left[\dfrac{2w_m}{W}\left(x + \dfrac{W}{2}\right) - j\beta_m z\right] & \left(x < -\dfrac{W}{2}\right) \\ A_m\cos\left(\dfrac{2u_m}{W}x - \dfrac{m\pi}{2}\right)\exp(-j\beta_m z) & \left(|x| \le \dfrac{W}{2}\right) \\ A_m\cos\left(u_m - \dfrac{m\pi}{2}\right)\exp\left[-\dfrac{2w_m}{W}\left(x - \dfrac{W}{2}\right) - j\beta_m z\right] & \left(x > \dfrac{W}{2}\right), \end{cases} \quad (2.2.114)$$

where u_m and w_m denote the transverse wavenumbers of the m-th mode in the core and cladding and A_m is constant. Transverse wavenumbers are obtained from the eigenvalue equation as

$$w_m = u_m \tan\left(u_m - \frac{m\pi}{2}\right) \quad (2.2.115)$$

and

$$u_m^2 + w_m^2 = k^2\left(\frac{W}{2}\right)^2 (n_{eff}^2 - n_0^2) \equiv v^2 \quad (2.2.116)$$

When the core width W of MMI region is large, the normalized frequency v also becomes large. It is known from Fig. 2.2.3 that when v becomes large, u_m approaches $u_m \cong (m + 1)\,\pi/2$. Then the propagation constant β_m is approximately expressed as

$$\beta_m = \sqrt{k^2 n_{eff}^2 - (2u_m/W)^2} \cong k n_{eff} - \frac{(m+1)^2\lambda}{4 n_{eff} W^2}\pi. \quad (2.2.117)$$

The total electric field in the MMI region is obtained by

$$\begin{aligned}\Psi(x,z) &= \sum_{m=0}^{M} E_y^m(x,z) \\ &= e^{-jkn_{eff}z} \sum_{m=0}^{M} A_m \cos\left[\frac{(m+1)\pi}{W}x - \frac{m\pi}{2}\right] \\ &\quad \times \exp\left[j\frac{(m+1)^2\pi\lambda}{4 n_{eff} W^2} z\right],\end{aligned} \quad (2.2.118)$$

where M denotes the maximum mode number. At $z = 0$, Ψ $(x, 0)$ coincides with the electric field of the input waveguide $\psi(x)$. Then the electric field amplitude A_m is obtained from Eq. (2.2.118) as

$$A_m = \frac{2}{W}\int_{-W/2}^{W/2} \psi(x)\cos\left[\frac{(m+1)\pi}{W}x - \frac{m\pi}{2}\right] dx. \quad (2.2.119)$$

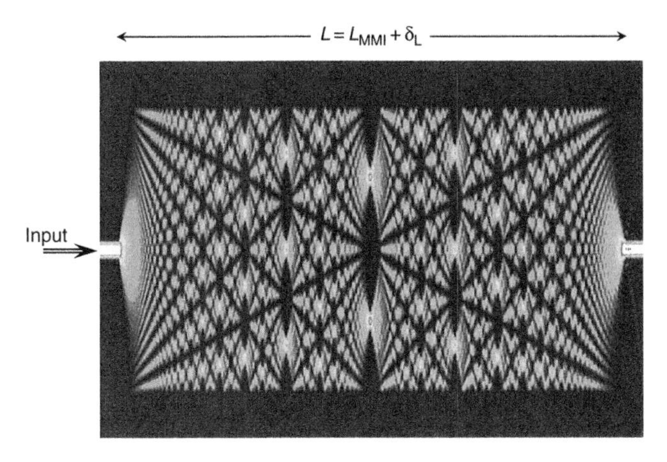

Fig. 2.2.22 Image formation for light input at the center of MMI waveguide.

For simplicity, we consider the case in which the input single-mode waveguide is connected to the center of the multimode waveguide. In this condition, modes in MMI region become only symmetrical modes; that is, m becomes an even number $m = 2\ell(\ell$: integer). At the point $z = n_{\rm eff}W^2/\lambda$, the phase term in Eq. (2.2.118) reduces to

$$\exp\left[j\frac{(m+1)^2\pi\lambda}{4n_{\rm eff}W^2}z\right] = \exp\left[j\ell(\ell+1)\pi + j\frac{\pi}{4}\right] = \exp\left(j\frac{\pi}{4}\right). \quad (2.2.120)$$

We now define the characteristic length $L_{\rm MMI}$ as

$$L_{\rm MMI} = \frac{n_{\rm eff}W^2}{\lambda}. \quad (2.2.121)$$

The electric field profile at the MMI length $L_{\rm MMI}$ is then obtained from Eq. (2.2.118) as

$$\Psi(x, L_{\rm MMI}) = e^{-jkn_{\rm eff}\,L_{\rm MMI}+j\pi/4} \times \sum_{m=0}^{M} A_m \cos\left[\frac{(m+1)\pi}{W}x - \frac{m\pi}{2}\right]. \quad (2.2.122)$$

Since Ψ $(x, 0)$ in Eq. (2.2.118) is the electric field of the input waveguide ψ (x), the above equation is rewritten as

$$\Psi(x, L_{\rm MMI}) = \psi(x)e^{-jkn_{\rm eff}L_{\rm MMI+j\pi/4}}. \quad (2.2.123)$$

It is confirmed that the input electric field ψ (x) is reproduced at the specific length $L_{\rm MMI}$ with slight phase retardation. Self-imaging characteristics in MMI waveguide are confirmed by the Beam Propagation Method (BPM) simulation. Figure 2.2.22 shows the image formation for light input at the center of MMI waveguide. Refractive-index difference of the waveguide is $\Delta = 0.75\%$ and the wavelength of light is $\lambda = 1.55$ μm. Waveguide parameters of the single-mode input waveguide are $2a = 7$ μm and $2d = 6$ μm and those of MMI are $W = 150$ μm and $L = 25.99$ mm. Specific length $L_{\rm MMI}$ that is given by Eq. (2.2.121) is 25.89 mm. Since there is a Goos-Hanshen effect, light field slightly penetrates into cladding region. Therefore, slight correction is necessary for either width W or length L of the MMI region. In Fig. 2.2.22, correction length of $\delta_L = 100$ μm is added to the analytical self-imaging length $L_{\rm MMI}$. It is known from Fig. 2.2.22 that N images are formed at $z = L_{\rm MMI}/N$, for any integer N. Figure 2.2.23 shows light-splitting characteristics of MMI waveguide with a length of $L_{\rm MMI}/8+\delta_L'$, where $\delta_L' = 105\mu$m. Output waveguides

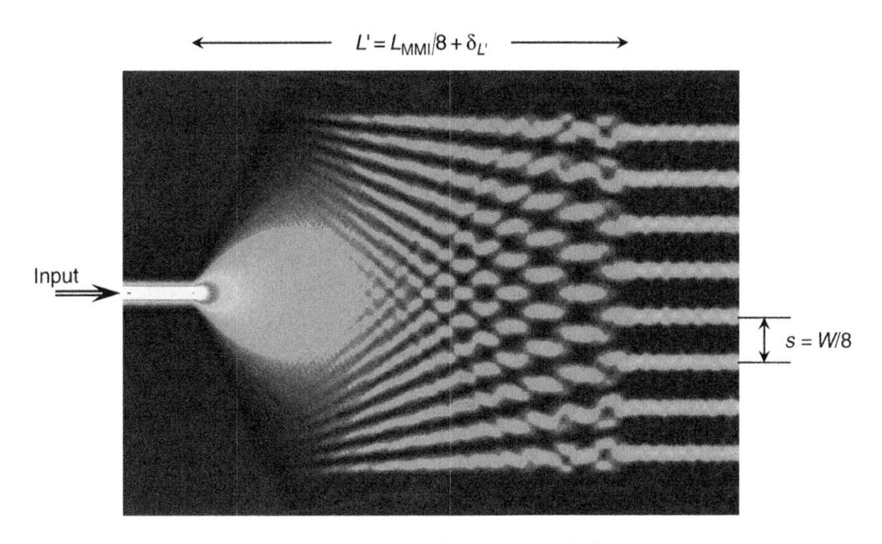

Fig. 2.2.23 Light-splitting characteristics of MMI waveguide with a length of LMMI/8.

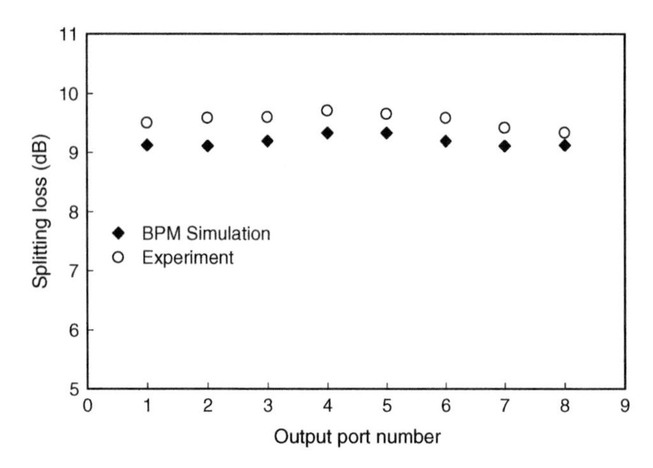

Fig. 2.2.24 Theoretical and experimental splitting characteristics of 1 × 8 MMI splitter.

are located at $x_i = (4.5 - i)\ W/8$ for i-th (i = 1,2..., 8) waveguide. Figure 2.2.24 shows theoretical and experimental splitting characteristics of 1 × 8 MMI splitter at $\lambda = 1.55$ μm. MMI splitter is made of silica waveguide with $2a = 7$ μm, $2d = 6$ μm and $\Delta = 0.75\%$. Splitting loss of 1/8 corresponds to 9dB. Therefore, it is known that the excess loss of each output port is less than 1 dB.

When an input waveguide is placed at the proper position from the center of MMI waveguide, two-fold images are formed with equal amplitude at the distance of [17]

$$L_{3\ \mathrm{dB}} = \frac{2}{3} \cdot \frac{n_{\mathrm{eff}} W^2}{\lambda}, \tag{2.2.124a}$$

$$W = (N+1)s \quad (N = 2), \tag{2.2.124b}$$

where s denotes the separation of output waveguides. Figure 2.2.25 shows a 3-dB coupler based on the 2 × 2 MMI splitter. BPM simulation is required in order to accurately determine the 3-dB coupler configuration. Light propagation characteristics in the MMI 3-dB coupler are shown in Fig. 2.2.26. The length of 3-dB coupler is determined to be $L = L_{3\,\mathrm{dB}} + \delta_{3\,\mathrm{dB}}$ ($\delta_{3\mathrm{dB}} = 200$ μm) by the BPM calculation. Figure 2.2.27 shows spectral splitting ratios of MMI 3-dB coupler and codirectional coupler (see Section 4.2) and spectral insertion loss of MMI 3-dB coupler. It is known from the figure that the coupling ratio of a MMI 3-dB coupler is almost insensitive to wavelength as compared to a codirectional coupler. This is a great advantage of a MMI 3-dB coupler over a codirectional coupler. However, we should take notice of two facts. First, insertion loss of MMI is not zero even in theoretical simulation. Theoretical insertion loss of the MMI 3-dB coupler is about 0.22 dB at the minimum. Second, the insertion loss of the MMI coupler rapidly increases as wavelength departs from the optimal wavelength.

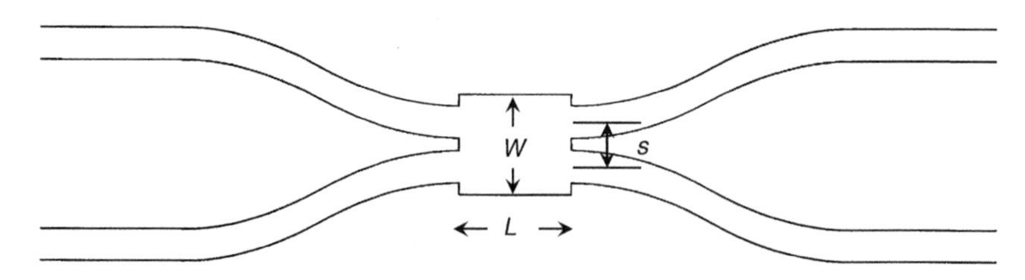

Fig. 2.2.25 3-dB coupler based on the 2 × 2 MMI splitter.

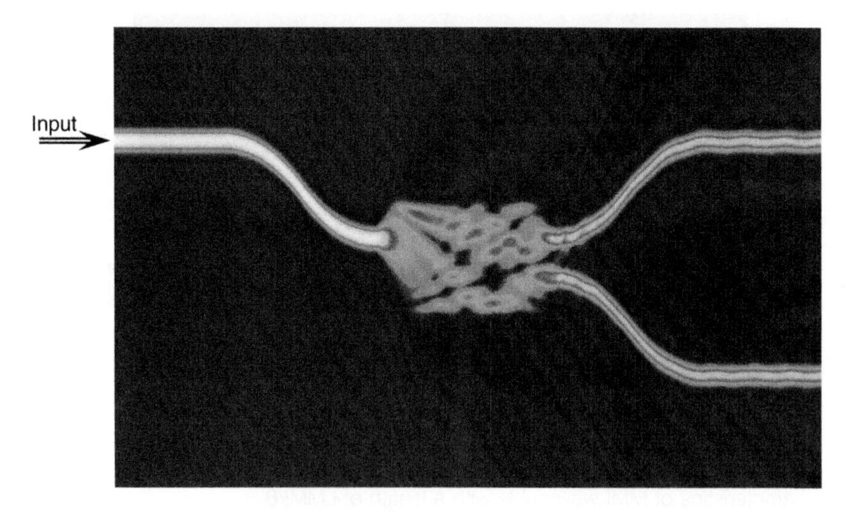

Fig. 2.2.26 Light propagation characteristics of MMI 3-dB coupler.

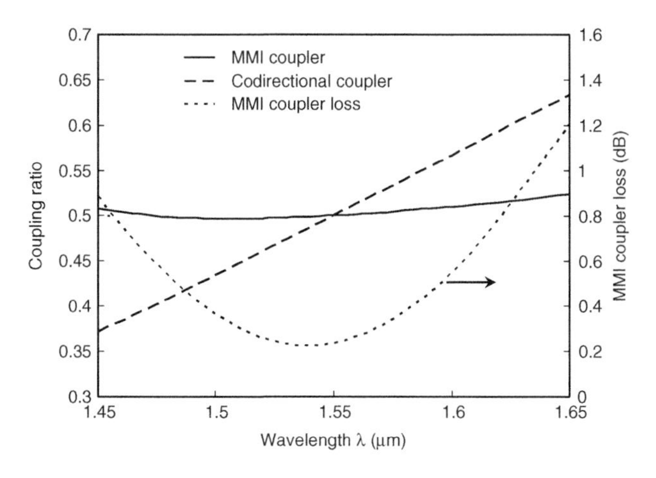

Fig. 2.2.27 Spectral splitting ratios of MMI 3-dB coupler and codirectional coupler and spectral insertion loss of MMI 3-dB coupler.

When phase of the input light is properly adjusted a $1 \times N$ MMI splitter functions as a $N \times 1$ combiner. Figure 2.2.28 shows light propagation characteristics in a MMI 16×1 combiner. Complex electric field g_i in the i-th input waveguide is given by $g_i = a_i \exp(j\theta_i)$, where

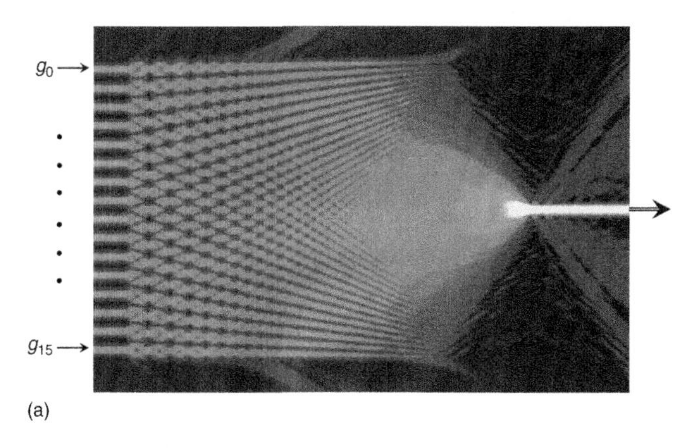

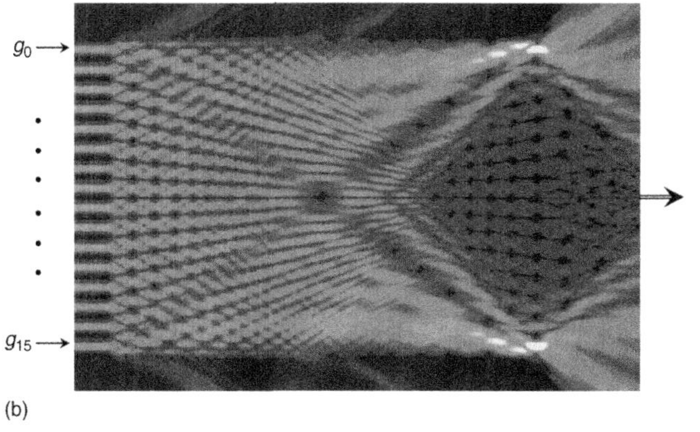

Fig. 2.2.28 Light propagation characteristics in the MMI 16×1 combiner: (a) Amplitudes of g_i's are equal and their phases are properly adjusted; and (b) amplitudes of g_i's are equal and their phases are completely out of phase.

a_i and θ_i denote amplitude and phase of g_i. In the BPM simulation shown in Fig. 2.2.28a, amplitudes of g_i's are set to be equal and their phases are properly adjusted. Every light beam from 16 input ports constructively interfere to form a single output image at the distance of $L = L_{MMI}/16 + \delta_L$, where $\delta_L = 140$ μm. It is shown by a number of numerical simulations that the MMI $N \times 1$ combiner has a unique property in which we can obtain a coherent summation of the input electric field as

$$T = \left|\frac{1}{N}\sum_{i=0}^{N-1} g_i\right|^2 = \left|\frac{1}{N}\sum_{i=0}^{N-1} a_i e^{j\theta_i}\right|^2. \qquad (2.2.125)$$

Numerical simulation was confirmed by the experiment using a 16-tap ($N = 16$) coherent transversal filter [18]. The device was fabricated using silica-based planar lightwave circuits. The core size and refractive-index difference of the waveguides are 7 μm × 7 μm and 0.75%, respectively. Figure 2.2.29 compares the experimental normalized output power with the theoretical one given by Eq. (2.2.125). Complex electric fields of even and odd ports were set to be $g_i = 1$ ($i = 0,2,\ldots, 14$) and $g_i = \exp(j\alpha)$ ($i = 1, 3,\ldots, 15$), respectively. Phase shift a was introduced to the waveguide by the thermo-optic effect. The output power of the MMI $N \times 1$ combiner is given from Eq. (2.2.125) by

$$T(\alpha) = \left|\frac{1}{N}(1+e^{j\alpha})\frac{N}{2}\right|^2 = \left|\cos\left(\frac{\alpha}{2}\right)\right|^2. \qquad (2.2.126)$$

Solid line in Fig. 2.2.29 shows theoretical curve $|\cos(\alpha/2)|^2$ and circles are experimental values for $\alpha = 0$, $\pi/3$, $\pi/2$, $2\pi/3$ and π, respectively. It is confirmed that the collective summation of complex electric fields is obtained by using the MMI combiner.

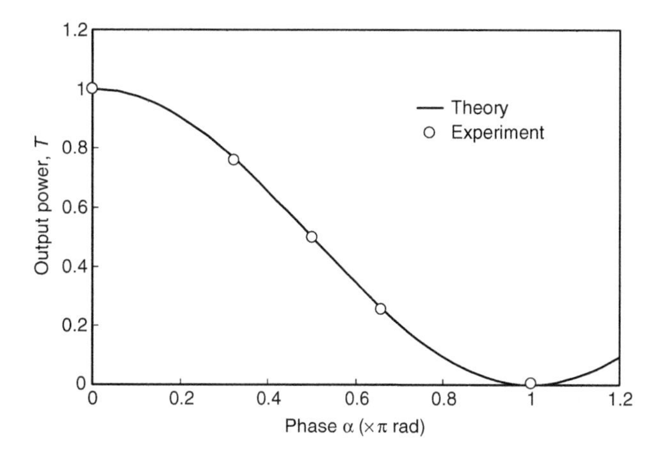

Fig. 2.2.29 Normalized output power T of the 16×1 MMI combiner.

References

[1] Marcuse, D. 1982. *Light Transmission Optics*. New York: Van Nostrand Reinhold.

[2] Marcuse, D. 1974. *Theory of Optical Waveguides*. New York: Academic Press.

[3] Unger, H. G. 1977. *Planar Optical Waveguides and Fibers*. Oxford: Clarendon Press.

[4] Kressel, H. and J. K. Butler. 1977. *Semiconductor Lasers and Heterojunction LEDs*. New York: Academic Press.

[5] Marcatili, E. A. J. 1969. Dielectric rectangular waveguide and directional coupler for integrated optics. *Bell Syst. Tech. J.* 48:2071–2102.

[6] Kumar, A., K. Thyagarajan and A. K. Ghatak. 1983. Analysis of rectangular-core dielectric waveguides—An accurate perturbation approach. *Opt. Lett.* 8: 63-65.

[7] Goell, J. E. 1969. A circular harmonic computer analysis of rectangular dielectric waveguides. *Bell Syst. Tech. J.* 48:2133–2160.

[8] Knox, R. M. and P. P. Toulios. 1970. Integrated circuits for the millimeter through optical frequency range. *Symposium on Submillimeter Waves*, Polytechnic Institute of Brooklyn, pp. 497–516.

[9] Tamir, T. 1975. *Integrated Optics*. Chap. 2, Berlin: Springer-Verlag.

[10] Born, M. and E. Wolf. 1970. *Principles of Optics*. Oxford: Pergamon Press.

[11] Bryngdahl, O. 1973. Image formation using self-imaging techniques. *J. Opt. Soc. Amer.* 63:416–419.

[12] Ulrich, R. 1975. Image formation by phase coincidences in optical waveguides. *Opt. Commun.* 13: 259–264.

[13] Niemeier, T. and R. Ulrich. 1986. Quadrature outputs from fiber interferometer with 4 × 4 coupler. *Opt. Lett.* 11:677–679.

[14] Veerman, F. B., P. J. Schalkwijk, E. C. M. Pennings, M. K. Smit and B. H. Verbeek. 1992. An optical passive 3-dB TMI-coupler with reduced fabrication tolerance sensitivity. *IEEE J. Lightwave Tech.* 10: 306–311.

[15] Bachmann, M., P. A. Besse and H. Melchior. 1994. General self-imaging properties in N × N multimode interference couplers including phase relations. *Appl. Opt.* 33:3905–3911.

[16] Heaton, J. M. and R. M. Jenkins. 1999. General matrix theory of self-imaging in multimode interference (MMI) couplers. *IEEE Photon. Tech. Lett.* 11:212–214.

[17] Soldano, L. B. and E. C. M. Pennings. 1995. Optical multi-mode interference devices based on self-imaging: Principles and applications. *IEEE J. Lightwave Tech.* 13:615–627.

[18] Okamoto, K., H. Yamada and T. Goh. 1999. Fabrication of coherent optical transversal filter consisting of MMI splitter/combiner and thermo-optic amplitude and phase controllers. *Electron. Lett.* 35:1331–1332.

Section Three

Optical fibers

Chapter 3.1

3.1

Optical fibers for broadband communication

Pal

3.1.1 Introduction

Development of optical fiber technology is considered to be a major driver behind the information technology revolution and the tremendous progress in global telecommunications that has been witnessed in recent years. Fiber optics, from the point of view of telecommunication, is now almost taken for granted in view of its wide-ranging application as the most suitable singular transmission medium for voice, video, and data signals. Indeed, optical fibers have now penetrated virtually all segments of telecommunication networks, whether transoceanic, transcontinental, intercity, metro, access, campus, or on-premise. The first fiber optic telecom link went public in 1977. Since that time, growth in the lightwave communication industry until about 2000 has been indeed mind boggling. According to a Lucent technology report [1], in the late 1990s optical fibers were deployed at approximately 4800 km/hr, implying a total fiber length of almost three times around the globe each day until it slowed down when the information technology bubble burst!

The Internet revolution and deregulation of the telecommunication sector from government controls, which took place almost globally in the recent past, have substantially contributed to this unprecedented growth within such a short time, which has been rarely seen in any other technology. Initial research and development (R&D) in this field had centered on achieving optical *transparency* in terms of exploitation of the *low-loss* and *low-dispersion* transmission wavelength windows of high-silica optical fibers. Though the low-loss fiber with a loss under 20 dB/km that was reported for the first time was a single-mode fiber (SMF) at the He–Ne laser wavelength [2], the earliest fiber optic lightwave systems exploited the first low-loss wavelength window centered on 820 nm with graded index multi-mode fibers forming the transmission media. However, primarily due to the unpredictable nature of the bandwidth of jointed multimode fiber links, since the early 1980s the system focus shifted to SMFs by exploiting the zero material dispersion characteristic of silica fibers, which occurs at a wavelength of 1280 nm [3] in close proximity to its second low-loss wavelength window centered at 1310 nm [4].

The next revolution in lightwave communication took place when *broadband* optical fiber *amplifiers* in the form of erbium-doped fiber amplifiers (EDFA) were developed in 1987 [5], whose operating wavelengths fortuitously coincided with the lowest-loss transmission wavelength window of silica fibers centered at 1550 nm [6] and heralded the emergence of the era of *dense wavelength division multiplexing* (DWDM) technology in the mid-1990s [7]. By definition, DWDM technology implies simultaneous optical transmission through one SMF of at least four wavelengths within the gain bandwidth of an EDFA (Fig. 3.1.1). Recent development of the so-called AllWave™ and SMF-28e™ fibers devoid of the characteristic OH^- loss peak (centered at 1380 nm) extended the low-loss wavelength window in high-silica fibers from 1280 nm (235 THz) to 1650 nm (182 THz), thereby offering, in principle, an enormously broad 53 THz of optical transmission bandwidth to be potentially tapped through the DWDM technique! These fibers are usually referred to as enhanced SMF (G.652.C) and are characterized with an additional low-loss window in the E-band (1360–1460 nm), which is about 30% more than the two low-loss windows centered

Guided Wave Optical Components and Devices; ISBN: 9780120884810

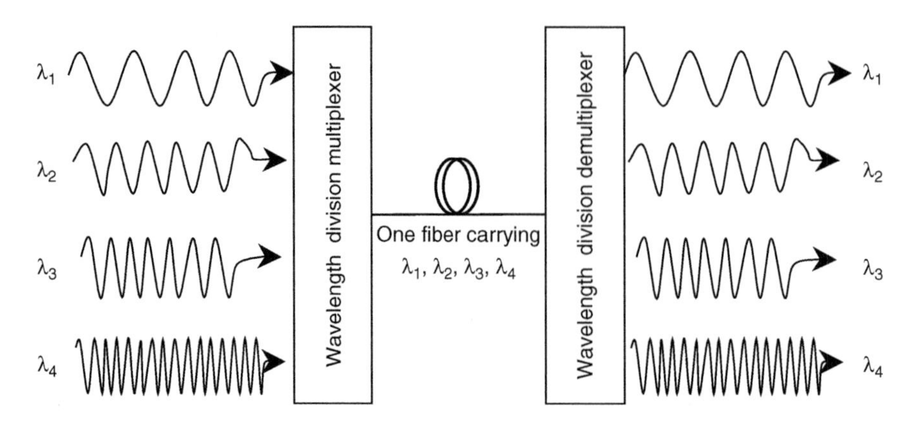

Fig. 3.1.1 Schematic representing DWDM optical transmission with a minimum of four wavelengths in the EDFA band as a sample.

about 1310 and 1550 nm in legacy SMFs. The emergence of DWDM technology has also driven development of various specialty fibers and all-fiber components for seamless growth of the lightwave communication technology. These fibers were required to address new features like nonlinearity-induced potential impairments in optical transmission due to large optical throughput, broadband dispersion compensation, bend-loss sensitivity to variation in signal wavelengths, and so on.

In this chapter we attempt to present evolutionary trends in the design of single-mode optical transmission fibers, in particular for lightwave communication seen in the last 30 years or so. Multimode fibers in the form of plastic fibers, which hold promise for on-premise and other applications, are discussed in Chapter 2 authored by P. L. Chu.

3.1.2 Optical transparency

3.1.2.1 Loss spectrum

SMFs constitute an integral component of any DWDM link meant to transport high volumes of signals and data. Its characteristics greatly influence the choice of auxiliary components that go into a network and also the overall cost and eventual performance of the communication system. *Loss* and *dispersion spectra* are the two most important propagation characteristics of a single-mode optical fiber. Fig. 3.1.2 gives an example of the loss spectrum of a state-of-the-art, commercially available, conventional G.652 type of SMF. Except for a portion of the loss spectrum around 1380 nm at which a peak appears due to absorption by minute traces (in parts per billion) of OH^- present in the fiber, the rest of the spectrum in a G.652 fiber more or less varies, with wavelength as $A\lambda^{-4}$, meaning that signal loss in a state-of-the-art SMF is essentially caused by Rayleigh scattering. Rayleigh scattering loss coefficient A in a fiber may be approximately modeled through the relation [8]

$$A = A_1 + A_2\Delta^n \tag{3.1.1}$$

where $A_1 = 0.7$ dB-μm^4/km in fused silica and $A_2 \approx 0.4$–1 dB-μm^4/km, whereas Δ is the relative core-cladding index difference expressed in % and exponent $n = 0.7$–1, whose value depends on the dopant. With GeO_2 as the dopant and $\Delta \approx 0.37\%$, estimated Rayleigh scattering loss in a high-silica fiber is about 0.18–0.2 dB/km at 1550 nm. Superimposed on this curve over the wavelength range of 1360–1460 nm (often referred to as the E-band) is a dotted curve, which is devoid of any peak but otherwise overlaps with rest of the loss spectrum;

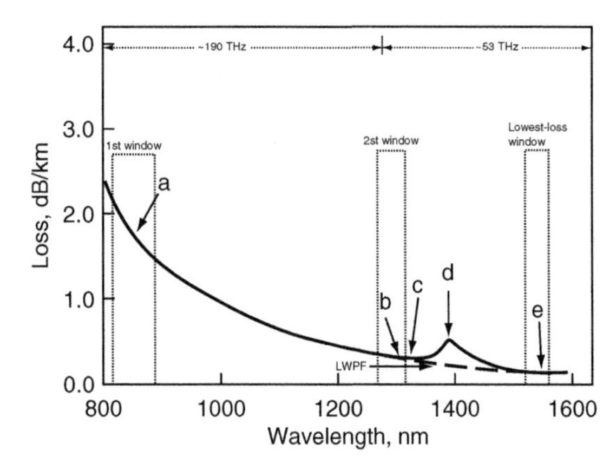

Fig. 3.1.2 A sample loss spectrum (full curve) of a state-of-the-art G.652 type single-mode fiber, e.g. SMF-28 (adapted from Corning product catalog © Corning Inc.): (a) 1.81 dB/km at 850 nm, (b) 0.35 dB/km at 1300 nm, (c) 0.34 dB/km at 1310 nm, (d) 0.55 dB/km at 1380 nm, (e) 0.19 dB/km at 1550 nm. The dashed portion of the curve corresponds to that of an LWPF due to reduction of the OH^- peak in enhanced SMF; available theoretical transmission bandwidths at different low loss spectral windows are also shown.

this modified spectrum corresponds to that of a low water peak fiber (LWPF), classified by the International Telecommunications Union (ITU) as G.652.C fibers, examples of which are AllWave™/SMF-28e™ fibers. LWPFs, also referred to as enhanced SMFs, opened up an additional 30% transmission bandwidth over and above standard SMFs of the G.652 type.

In real-world systems, however, there are other sources of loss, which are required to be budgeted on a case-to-case basis and which could add up to more than the inherent loss. Examples of these are cabling-induced losses due to microbending, bend-induced losses along a fiber route, and losses due to splices and connectors, including those involving insertion of various components in a fiber link. Several of these also depend to a certain extent on the refractive index profile of the fiber in question. Detailed studies have indicated that these extraneous sources of loss could be addressed by optimizing mode field radius (W_P, known as Petermann spot size) and effective cutoff wavelength (λ_{ce}) [9]. The parameter *WP* effectively determines transverse offset-induced loss at a fiber splice and sensitivity to microbend-induced losses, and λ_{ce} essentially determines sensitivity to bend-induced loss in transmission. Both of these are well-known important characteristic parameters of a SMF [10]. For operating at 1310 nm, optimum values of these parameters turned out to be $4.5 < W_P(\mu m) < 5.5$ and $1100 < \lambda_{ce}\ (nm) < 1280$ [11]. An indirect way to test that λ_{ce} indeed falls within this range is to determine whether the measured excess loss of 100 turns of the fiber loosely wound around a cylindrical mandrel of diameter 7.5 cm falls below 0.1 dB at 1310 nm and below 1.0 dB at 1550 nm [8].

In the early 1980s it was observed that many of the installed (multimode) fibers exhibited an increase in transmission loss at $\lambda > 1200$ nm after aging as little as 3 years. Detailed studies indicated that this phenomenon was attributable to formation of certain chemical bonds between silicon and various dopants with hydroxyl ions in the form of Si—OH, Ge—OH, and P—OH because of trapping of hydrogen at the defect centers formed in the Si—O network due to incorporation of various dopants. Of the two most often used dopants, namely GeO_2 and P_2O_5, the latter was found to be more troublesome because defect densities formed with it were much higher in number because of a difference in valency between Si and P [11]. The phenomenon of hydrogen-induced loss increase with aging of a fiber could be effectively eliminated (at least over the life of a system assumed to be about 25 years) by avoiding phosphorus as a refractive index modifier of silica and through complete polymerization of the fiber coating materials, for which ultraviolet-curable epoxy acrylates were found to be the best choice; in fact, these are now universally chosen as the optimum fiber coating material before a fiber is sent to a cabling plant.

3.1.2.2 Dispersion spectrum

Chromatic dispersion is another important transmission characteristic (along with loss) of an SMF, and its magnitude is a measure of its information transmission capacity. In a fiber optical network, signals are normally transmitted in the form of temporal optical pulses (typically in a high-speed fiber optical link, individual pulse width is about 30 ps [12]). For example, under the most extensively used pulse code modulation technique for digitized signal transmission, a telephone voice analog signal is at first converted into 64,000 electrical pulses. Likewise, several other telephone signals are converted to 64,000 electrical pulses, and all of these are electronically time division multiplexed in the electric domain at the telephone exchange before being fed to the electronic circuit driving a laser diode to produce equivalent optical pulses. Thereby, these optical pulses replicate the time division multiplexed signals in the optical domain, which are then launched into an SMF for transmission to a distant receiver. The above-mentioned process steps are schematically depicted in Fig. 3.1.3. Due to the phenomenon of *chromatic dispersion*, these signal pulses in general broaden in time with propagation through the fiber. Chromatic dispersion arises because of the dispersive nature of an optical fiber due to which the group velocity of a propagating signal pulse becomes a function of frequency (usually referred to as *group-velocity dispersion* [GVD] in the literature), and this phenomenon of GVD induces frequency chirp to a propagating pulse, meaning thereby that the *leading edge* of the propagating pulse differs in frequency from the *trailing edge* of the pulse. The resultant frequency chirp (i.e., $\omega(t)$) leads to inter-symbol interference, in the presence of which the receiver fails to resolve the digital signals as individual pulses when the pulses are transmitted too close to each other [13, 14]. Thus these pulses, though started as individually distinguishable pulses at the transmitter, may become indistinguishable at the receiver depending on the amount of chromatic dispersion-induced broadening introduced by the fiber (Fig. 3.1.4). In fact, the phenomenon of GVD limits the number of pulses that can be sent through the fiber per unit time. For self-consistency of our discussion on pulse dispersion, in the following we outline basic principles that underlay pulse dispersion in an SMF [13,14].

The *phase velocity* of a plane wave propagating in a medium of refractive index n having propagation constant k, which determines the velocity of propagation of its phase front, is given by

$$v_p = \frac{\omega}{k} = \frac{c}{n(\omega)}. \tag{3.1.2}$$

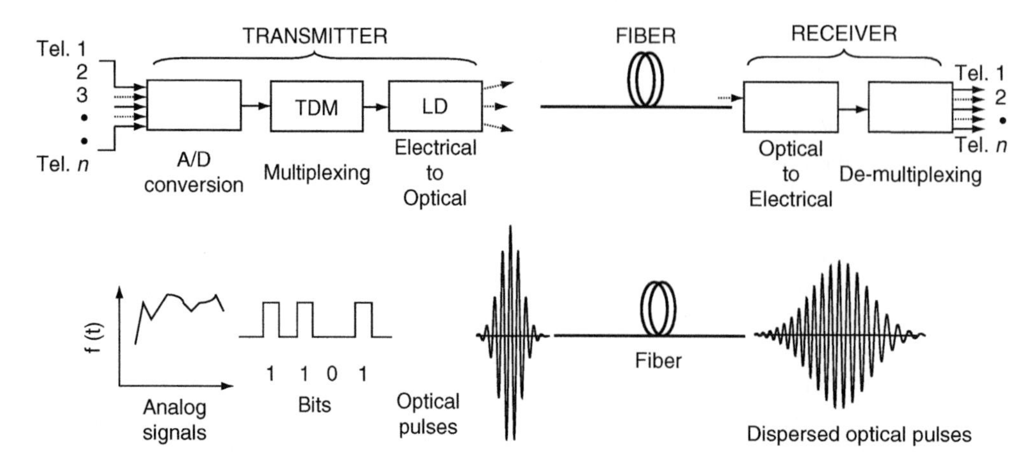

Fig. 3.1.3 Block diagram depicting major components of an optical communication system; lower part of the figure depicts conversion of analog to digital (A/D) form (in electrical domain) and subsequent transmission of same in the form of equivalent optical pulses through the fiber where these get chirped and broadened due to chromatic dispersion. LD, laser diode; TDM, time division multiplexing.

On the other hand, in case of a propagating optical pulse in a *dispersive medium* (for which k is not a linear function of ω), the characteristic velocity at which pulse energy propagates through the medium is decided by its *group velocity* v_g given by

$$\frac{1}{v_g} = \frac{1}{c}\left[n(\omega) + \omega\frac{dn}{d\omega}\right] = \frac{1}{c}\left[n(\lambda_0) - \lambda_0\frac{dn}{d\lambda_0}\right]. \quad (3.1.3)$$

As an optical pulse propagates through a SMF, it propagates via the LP_{01} guided mode of the fiber, and in that case the group velocity of the pulse is decided by the following equation in place of Eq. (3.1.3)

$$\frac{1}{v_g} = \frac{1}{c}\left[n_{eff}(\lambda_0) - \lambda_0\frac{dn_{eff}}{d\lambda_0}\right] = \frac{N_g}{c} \quad (3.1.4)$$

where $n_{eff}(\omega)(= \beta/k_0$; β being propagation constant of the guided mode, and k_0 is k in free space) and N_g respectively represent *effective index* and *group index* of the LP_{01} mode. Likewise, for determining phase velocity of the propagating pulse in an optical fiber, $n(\omega)$ in Eq. (3.1.2) is required to be replaced by $n_{eff}(\omega)$. We now consider a *Fourier transform limited* Gaussian-shaped pulse, which is launched into an SMF such that at $z = 0$

$$f(x, y, z = 0, t) = E_0(x, y)e^{-t^2/2\tau_0^2}\exp(i\omega_c t). \quad (3.1.5)$$

The quantity $E_0(x,y)$ represents the LP_{01} modal field of the fiber, ω_c represents the frequency of the optical carrier, which is being modulated by the pulse, and τ_0 corresponds to the characteristic width of the Gaussian pulse, and $\omega_c t$ represents the phase of the pulse. By obtaining Fourier transform of this pulse in the frequency domain and incorporating Taylor's series expansion (retaining only terms up to second order) of the phase constant β around $\omega = \omega_c$ (in view of the fact that Fourier transform of such a Fourier transform limited pulse is sharply peaked around a narrow frequency domain $\Delta\omega$ around $\omega = \omega_c$), we would get

$$f(x, y, z = L, t) = E_0(x, y)\exp\left[i\left(\omega_c t - \beta|_{\omega=\omega_c}\right)\right] \cdot \Psi(z = L, t) \quad (3.1.6)$$

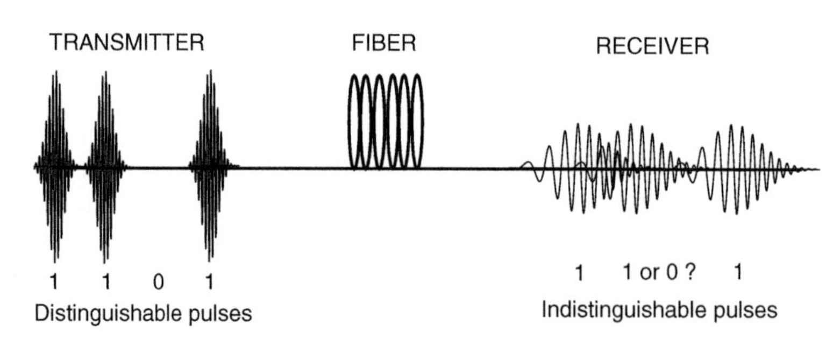

Fig. 3.1.4 Dispersion-induced broadening of optical pulses and the resultant intersymbol interference at the receiver.

where

$$\Psi(z = L,\ t) = \frac{1}{2\pi}\int_{-\infty}^{+\infty}\left\{F(z = 0,\ \Omega) \times \exp\left[-i\left(\frac{\Omega}{v_g} + \frac{1}{2}\beta''\Omega^2\right)L\right]\right\}e^{i\Omega t}d\Omega. \tag{3.1.7}$$

represents the envelope of the dispersed pulse at a fiber length L, and $F(\Omega)$ is Fourier transform of the pulse in the frequency domain with $\Omega = \omega - \omega_c$; the ratio L/v_g simply represents the delay, that is, the time taken by the pulse to propagate through a fiber length L. In Eq. (3.1.7), the quantity $\beta'' = (d^2\beta/d\omega^2)|_{\omega=\omega_c}$ is known as the GVD parameter. We may rewrite the Fourier transform [cf. Eq. (3.1.7)] in a new time frame, which includes effect of delay due to propagation through $z = L$, that is, we replace t by $(t - \frac{L}{v_g})$ and obtain

$$\text{F.T. of}\left\{\Psi\left(z = L,\ t - \frac{L}{v_g}\right)\right\} = F(z = 0,\ \Omega)\exp\left(-i\tfrac{1}{2}\beta''\Omega^2 L\right). \tag{3.1.8}$$

Thus, the right-hand side of Eq. (3.1.8) may be taken as the transfer function of an SMF of length L. By carrying out the integration in Eq. (3.1.7), we would obtain from Eq. (3.1.6)

$$f(x,y,L,t) = \frac{E_0(x,y)}{(\tau(L)/\tau_0)^{1/2}}\exp\left[-\frac{(t-\frac{L}{v_g})^2}{2\tau^2(L)}\right]\exp[i(\Phi(L,t) - \beta_c L)] \tag{3.1.9}$$

where

$$\Phi(L,t) = \omega_c t + \kappa(t - L/v_g)^2 - \frac{1}{2}\tan^{-1}(\bar{\alpha}) \tag{3.1.10}$$

$$\kappa = \frac{1}{2\tau_0^2}\frac{\bar{\alpha}}{(1+\bar{\alpha}^2)} \tag{3.1.11}$$

$$\bar{\alpha} = \frac{\beta'' L}{\tau_0^2} \tag{3.1.12}$$

and

$$\tau^2(L) = \tau_0^2(1+\bar{\alpha}^2). \tag{3.1.13}$$

Thus, the Gaussian temporal pulse remains Gaussian in shape with propagation but its characteristic width increases to $\tau(L)$ (given by Eq. (3.1.13)) as it propagates through a fiber of length L. After propagating through a distance of $L_{DISP}(\tau_0^2/\beta'')$, the pulse assumes a width of $\sqrt{2}\tau_0$. This quantity L_{DISP} is a characteristic (dispersion-related) parameter of an SMF and is referred to as the *dispersion length*, which is often used to describe dispersive effects of a medium. The smaller the dispersion, the longer the L_{DISP}. Further, it can be seen from Eq. (3.1.13) through differentiation that for a given length of a fiber there is an optimum input pulse width $\tau_0^{opt} = \sqrt{\beta'' L}$, for which the output pulse width is minimum given by $\tau_{min} = \sqrt{2\beta'' L}$. The quantity

$$\Delta_\tau = \left[\tau^2(L) - \tau_0^2\right]^{1/2} = \frac{L}{\tau_0}\beta'' \tag{3.1.14}$$

may be defined as the *chromatic dispersion–induced* increase in the width of a Fourier transform limited temporal pulse due to propagation through the fiber. The broadening of a Gaussian temporal pulse with propagation is shown in Fig. 3.1.5.

The GVD parameter β'' may be alternatively expressed as

$$\beta'' = \left.\frac{d^2\beta}{d\omega^2}\right|\omega = \omega_c = \frac{\lambda_0^3}{2\pi c^2}\frac{d^2 n_{eff}}{d\lambda_0^2} = \frac{\lambda_0^2}{2\pi c}D \tag{3.1.15}$$

where D, known as the dispersion coefficient, is given by

$$D = \frac{1}{L}\frac{d\tau}{d\lambda_0} = -\frac{\lambda_0}{c}\frac{d^2 n_{eff}}{d\lambda_0^2} \tag{3.1.16}$$

expressed in units of ps/nm·km. This parameter D is very extensively used in the literature to describe pulse dispersion in an optical fiber. The mode effective index n_{eff} is often expressed through [14]

$$n_{eff} \approx n_{cl}(1 + b\Delta) \tag{3.1.17}$$

where $0 < b < 1$ represents the normalized propagation constant of the LP_{01} mode

$$b = \frac{n_{eff}^2 - n_{cl}^2}{n_c^2 - n_{cl}^2} \tag{3.1.18}$$

and $\Delta \approx ((n_c - n_{cl})/n_{cl})$ stands for relative core-cladding index difference in an SMF. Thus, n_{eff} is composed of two terms: The first is purely material related (n_{cl}) and

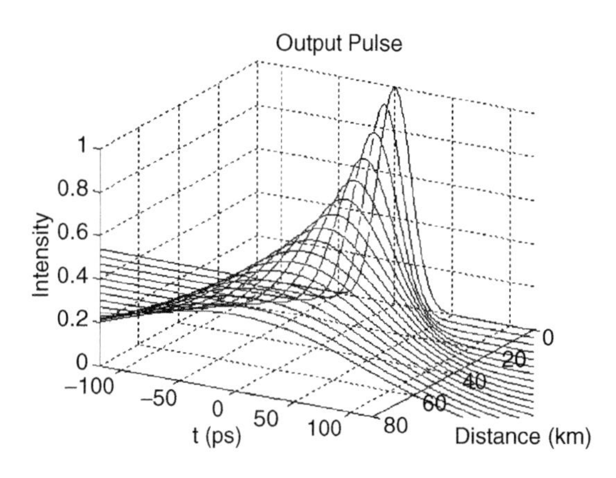

Fig. 3.1.5 Broadening of a Gaussian temporal pulse with propagation through a single-mode fiber.

the second arises due to the waveguide effect (*b*), and hence total dispersion in an SMF could be attributed to two types of dispersion, namely material dispersion and waveguide dispersion. Indeed, it can be shown that the total dispersion coefficient (D_T) is given by the following algebraic sum to a very good accuracy [14,15]:

$$D_T \approx D_M + D_{WG} \tag{3.1.19}$$

where D_M and D_{WG} correspond to material and waveguide contributions to D, respectively. The material contribution can be obtained from Eq. (3.1.16) by replacing n_{eff} with n_{cl} and the waveguide contribution is

$$D_{WG} = -\frac{n_2\Delta}{c\lambda_0}\left(V\frac{d^2(bV)}{dV^2}\right) \tag{3.1.20}$$

where V is the well-known normalized waveguide parameter defined through

$$V = \frac{2\pi}{\lambda_0} a n_c \sqrt{2\Delta}. \tag{3.1.21}$$

In Eq. (3.1.21), a represents the core radius of the fiber. A plot of a typical dispersion spectrum of an SMF (i.e., D vs. λ) along with its components D_M and D_{WG} are shown in Fig. 3.1.6. It is apparent from Fig. 3.1.6 that D_{WG} is negative and D_M changes from negative to positive (going through zero at a wavelength of ~ 1280 nm [16]) as the wavelength increases; as a result, the two components cancel each other at a wavelength of about 1300 nm, which is referred to as the *zero dispersion wavelength* (λ_{ZD}), a very important design parameter of SMFs. Realization of this fact led system operators to choose the operating wavelength of first generation SMFs as 1310 nm. These fibers optimized for transmission at 1310 nm are now referred to as G.652 fibers per ITU standards; millions of kilometers of these fibers are laid

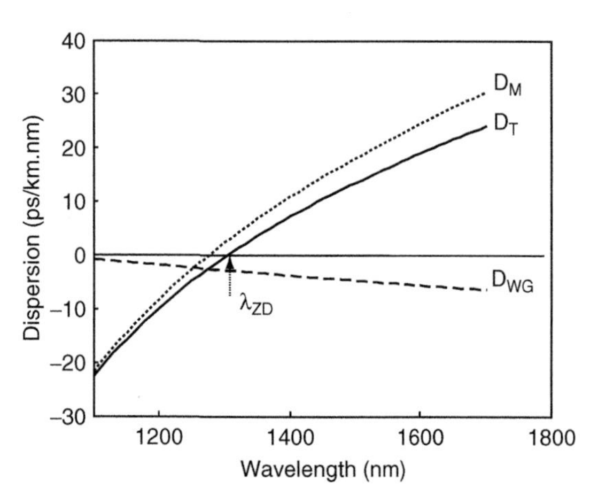

Fig. 3.1.6 Material (D_M), waveguide (D_{WG}), and total (D_T) dispersion coefficients of a high-silica matched clad SMF.

underground all over the world. Though it appears that if operated at λ_{ZD} one might get infinite transmission bandwidth, in reality zero dispersion is only an approximation (albeit a very good approximation) because it signifies that only the second-order dispersive effects would be absent. In fact, as per ITU recommendations, for G.652 fibers to qualify for deployment as telecommunication media provided at the 1310-nm wavelength the D_T is < 3.5 ps/nm·km. At a wavelength around λ_{ZD}, higher order dispersions, namely third-order dispersions characterized by $d^3\beta/d\omega^3$, would determine the net dispersion of a pulse. Thus, in the absence of second-order dispersions, pulse dispersion is quantitatively determined by the dispersion slope S_0 at $\lambda = \lambda_{ZD}$ through

$$\Delta_\tau = \frac{L(\Delta\lambda_0)^2}{2} S_0 \tag{3.1.22}$$

where $S_0 = (dD/d\lambda_0)|_{\lambda_0=\lambda_{ZD}}$, which is measured in units of ps/nm^2·km; S_0 in G. 652 fibers at 1310 nm is ≤ 0.09 ps/nm·km. If third-order dispersion is the sole determining factor of pulse dispersion, the output pulse does not remain symmetric and acquires some oscillation at the tails [17]. A knowledge of D and S_0 enables the determination of the dispersion (D) at any arbitrary wavelength within a transmission window, for example, the EDFA band in which D in G.652 fibers varies approximately linearly with λ. This feature often finds applications in component designs, and $D(\lambda)$ is usually explicitly stated in commercial fiber data sheets as

$$D(\lambda_0) = \frac{S_0}{4}\left[\lambda_0 - \frac{\lambda_{ZD}^4}{\lambda_0^3}\right]. \tag{3.1.23}$$

The genesis of this relation lies in the following three-term polynomial equation often used as a fit to measured data for delay (τ) versus λ in an SMF:

$$\tau(\lambda_0) = \frac{A}{\lambda_0^2} + B + C\lambda_0^2 \tag{3.1.24}$$

where coefficients A, B, and C are determined through a least-square fit to the data for measured $\tau(\lambda_0)$. A typical plot for τ versus λ for a G.652 fiber is shown in Fig. 3.1.7. Equation (3.1.24) yields the following for $D(\lambda_0)$:

$$D(\lambda_0) = \frac{d\tau}{d\lambda_0} = 2C\lambda_0 - \frac{2A}{\lambda_0^2}. \tag{3.1.25}$$

Setting $D(\lambda_0 = \lambda_{ZD})$ to zero yields $\lambda_{ZD} = (A/C)^{1/4}$. Thus,

$$S_0 = 8C. \tag{3.1.26}$$

In addition to pulse broadening, because the energy in the pulse gets reduced within its time slot, the corresponding

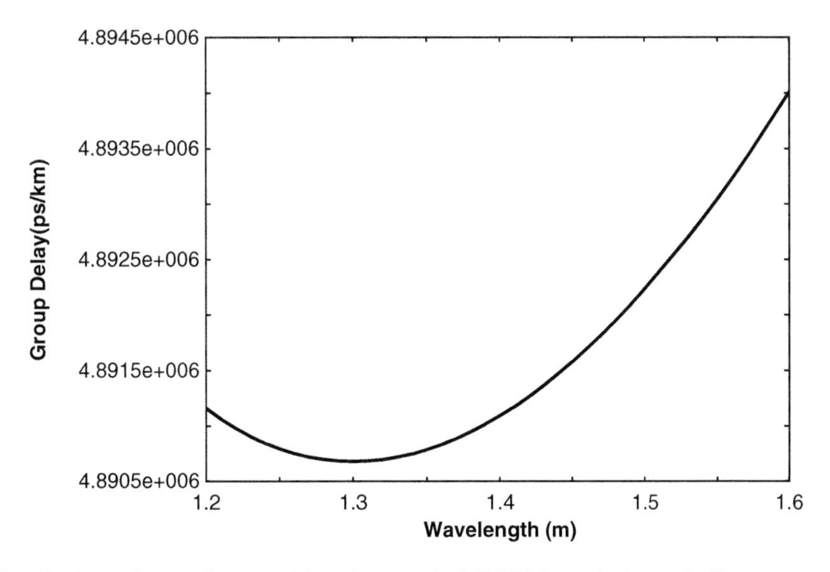

Fig. 3.1.7 Typical wavelength dependence of group delay of a standard G.652 type single-mode fiber.

signal-to-noise ratio decreases, which could be compensated by increasing the power in the input pulses. This additional power requirement is termed the *dispersion power penalty.* For estimating the dispersion power penalty, a more general definition of pulse width in terms of root mean square (rms) width becomes more useful, especially for pulse shapes, which may not be of a well-defined shape like a Gaussian pulse. The rms width of a temporal pulse is defined through

$$\tau_{rms} = \sqrt{\langle t^2 \rangle - \langle t \rangle^2} \tag{3.1.27}$$

where

$$\langle t^n \rangle = \frac{\int_{-\infty}^{\infty} t^n |f(z,t)|^2 dt}{\int_{-\infty}^{\infty} |f(z,t)|^2 dt}. \tag{3.1.28}$$

For a Gaussian output pulse, its $\tau_{\rm rms}$ is related to the input Gaussian pulse width τ_0 through $\tau_{\rm rms} = \tau_0/\sqrt{2}$. If a system is operating at a bit rate of B bits per second, the bit period is $1/B$ s. To keep the interference between neighboring bits at the output below a specified level, $\tau_{\rm rms}$ of the dispersed pulse must be kept below a certain fraction ε of the bit period B. Accordingly, for maximum allowed rms pulse width (assuming it is Gaussian in shape) of the output pulse, $\tau_{\rm rms}$ of the dispersed pulse should satisfy

$$\tau_{\rm rms} = \frac{\tau_0}{\sqrt{2}} < \frac{\varepsilon}{B} \tag{3.1.29}$$

which implies

$$B_{\tau_0} < \sqrt{2}\varepsilon. \tag{3.1.30}$$

On substitution of the optimum value for output pulse width, that is, $\tau_{\rm min} = \sqrt{2\beta'' L}$ in Eq. (3.1.30) and making use of Eq. (3.1.15), we get

$$B^2 DL < \frac{2\pi c}{\lambda_0^2} \varepsilon^2. \tag{3.1.31}$$

For a 2-dB power penalty, it is known that $\varepsilon = 0.491$, whereas for a 1-dB dispersion power penalty it is 0.306; these values for ε were specified by Bellcore standards (Document no. TR-NWT-000253). For a 1-dB dispersion power penalty at the wavelength of 1550 nm, we can write the inequality [Eq. (3.1.31)] approximately as

$$B^2 DL \leq 10^5 \text{ Gb}^2 \cdot \text{ps/nm} \tag{3.1.32}$$

where B is measured in Gbits, D in ps/nm·km, and L in km. Based on Eq. (3.1.32), Table 3.1.1 lists the maximum allowed dispersion for different standard bit rates assuming a dispersion power penalty of 1 dB.

By the mid-1980s, proliferation of optical-trunk networks and a steady increase in bit transmission rates gave

Table 3.1.1 Maximum allowed dispersions for different standard bit rates and for 1-dB dispersion power penalty.

Data rate (*B*)	Maximum allowed dispersion (*D.L.*)
2.5 Gb/s (OC-48)	~16,000 ps/nm
10 Gb/s (OC-192)	~l,000 ps/nm
40 Gb/s (OC-768)	~60 ps/nm

OC: optical channels.

rise to the requirement for a standard transmission format of the digital optical signals so that they could be shared without any need for an interface (this is the so-called mid-fiber meet) between the American, European, and Japanese telephone networks, each of which until then were following a different digital hierarchy [18]. This lack of standards led to the development of synchronous digital hierarchy (SDH), which has now become the global industry standard (in the United States this is known as SONET, for Synchronous Optical NETworks) for transmission of digital *optical channels* (OCs) [19]. An SDH network has a master clock that controls the timing of events all through the network. The base rate (OC-1) in SDH is taken as 51.84 Mb/s; thus, OC-48 rate corresponds to a transmission rate of 2.488 (≈2.5 Gb/s), which became the standard signal transmission rate at the 1310-nm wavelength window over standard G.652 SMFs. Although in principle a very high bit transmission rate was achievable at 1310 nm, maximum link length/repeater spacing was limited by transmission loss (~0.34 dB/km at this wavelength) to ~ 40 km in these systems. Thus, it became evident that it would be an advantage to shift the operating wavelength to the 1550-nm window, where the loss is lower (cf. Fig. 3.1.2), to overcome transmission loss-induced distance limitation of the 1310-nm window.

3.1.2.3 Dispersion shifted fibers

As early as 1979 the Nippon Telephone and Telegraph (NTT) team from Japan had succeeded in achieving a single-mode silica fiber that exhibited a loss of 0.2 dB/km at 1550 nm, which was close to the fundamental limit determined by Rayleigh scattering alone [6]. This figure for fiber loss is 40% less than its value at 1310 nm! Thus, it was only natural for systems engineers to exploit this lowest-loss window by shifting the operating wavelength from 1310 to 1550 nm. However, commercial introduction of next generation systems based on the exploitation of the 1550-nm lowest-loss window had to wait almost another 10 years. This was mainly because conventional G.652 type fibers optimized for transmission at the 1310-nm window exhibited excessive chromatic dispersion of approximately +17 ps/nm·km at 1550 nm (Fig. 3.1.6). For these fibers, which are characterized by a nominal step refractive index profile, typically, the chosen Δ and core diameter ($2a$) were ≤ 0.36% and ≥ 8.2 μm, respectively, which led to a resultant mode field diameter of 9.2 ± 0.4 μm at 1310 nm (10.4 ± 0.8 μm at 1550 nm) and a cutoff wavelength (λ_c) ≤ 1260 nm. Due to a D of approximately +17 ps/nm·km, repeater spacing at 1550 nm with such fiber-based links were limited by chromatic dispersion. By the mid-1980s, it was realized that repeater spacing of 1550-nm-based systems could be pushed to a much longer distance if the fiber designs could be tailored to shift λ_{ZD} to coincide with this wavelength so as to realize dispersion shifted fibers (DSFs), which are given the generic classification as G.653 fibers by the ITU. Comparative plots of dispersion spectra for G.652 and G.653 types of SMFs are shown in Fig. 3.1.8.

On the laser transmitter front, the most common types of laser diodes available for operation at 1550 nm were *Fabry-Perot* (FP) and *distributed feedback* (DFB) lasers, which were based on InGaAsP semiconductors. FP lasers are characterized by spectral widths (full width at half maximum = 2–5 nm), which are broader than the more expensive DFB lasers (continuous wave (cw) spectral width ≪ 1 nm). However, because of chirping and mode hopping, DFB lasers exhibit broader dynamic spectral widths (~ 1–2 nm) at high modulation rates (≥ 2 Gb/s). In view of these factors, FP lasers in conjunction with DSFs seemed to be the best *near-term option* for optimum exploitation of the 1550-nm wavelength window [20]. Typically, D_T is ≤ 2.7 ps/nm·km with $S_0 \leq 0.058$ ps/nm^2·km at the 1550-nm wavelength in a DSF while maintaining inherent scattering loss low (~ 0.21 dB/km at 1550 nm). One serious problem with these fibers is their sensitivity to microbend and bend-induced losses at a $\lambda_0 > 1550$ nm due to their low effective cutoff wavelength of about 1100 nm [11]. A theoretical design was suggested in 1992 to overcome this issue of bend-loss sensitivity by which the theoretical cutoff wavelength was moved up close to about 1550 nm through a spot size optimization scheme [21]. These new generation systems operating at 2.5 Gb/s became commercial in 1990, with the potential to work at bit rates in excess of 10 Gb/s with careful design of sources and receivers and use of DSFs. DSFs were extensively deployed in the backbone trunk

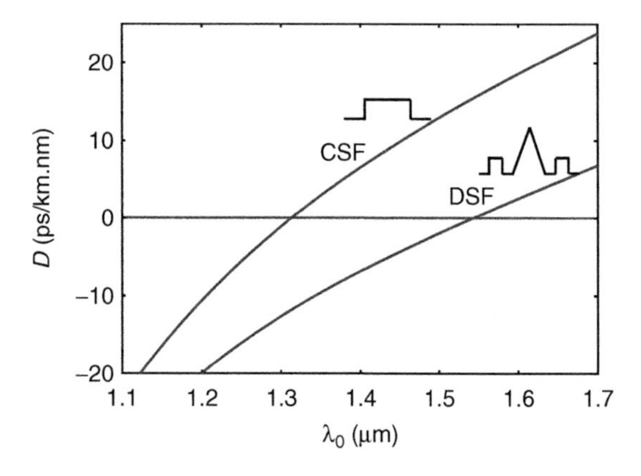

Fig. 3.1.8 Dispersion spectra of a conventional single-mode fiber (CSF) of the type G.652 and a G.653 type DSF along with typical refractive index profiles as their labels. For CSF, λ_{ZD} falls at 1310 nm, and for DSF it occurs at 1550 nm.

network in Japan in the early 1990s [11]. This trend persisted for a while before EDFAs emerged on the scene and led to a dramatic change in technology trends.

3.1.3 Emergence of fiber amplifiers and DWDM systems

3.1.3.1 EDFAs

In the late 1980s typical state-of-the-art, repeater-less, transmission distances were about 40–50 km at the 560 Mb/s transmission rate. Because maximum launched optical power was below 100 μW, it was difficult to improve system lengths beyond this specification, and the use of electronic repeaters became inevitable. At a repeater, the so-called 3R-regeneration functions (*reamplification, retiming,* and *reshaping*) are performed in the electric domain on the incoming attenuated as well as distorted (due to dispersion) signals after detection by a photode-tector and before the revamped signal is fed to a laser diode drive circuit, from which these cleaned optical pulses are launched to the next section of the fiber link. However, these complex functions are expensive and unit replacement is required when network capacity is to be upgraded to higher bit transmission rates because electronic components are bit rate sensitive. Because these units are required to convert photons to electrons and back to photons, often at modulation rates approaching the limits of the then available electronic switching technology, a bottleneck was encountered in the late 1980s. *What was needed was an optical amplifier to bypass this electronic bottleneck.* In 1986, the research group at Southampton University in England reported success in incorporating rare earth trivalent erbium ions into host silica glass during fiber fabrication [22]. Erbium is a well-known lasing species characterized by strong fluorescence at 1550 nm. Subsequently, the same group demonstrated that excellent noise and gain performance is feasible in a large part of the 1550-nm window with erbium-doped standard silica fibers [5].

The concept of optical amplification in fiber is almost as old as the laser itself. Today, EDFAs seem like an outstanding breakthrough, but they are really an old idea. In 1964, Koester and Snitzer [23] demonstrated a gain of 40 dB at 1.06 μm in a 1-meter-long Nd-doped fiber side pumped with flash lamps. The motivation at that time was to find optical sources for communication, but the impressive development of semiconductor lasers that took place in subsequent years pushed fiber lasers to the background. The operation of an EDFA is very straightforward [24]. The electrons in the 4*f* shell of the erbium ions are excited to higher energy states by absorption of energy from a pump. Absorption bands most suitable as pumps for obtaining amplification of 1550-nm signals are the 980- and 1480-nm wavelengths (Fig. 3.1.9a); Fig. 3.1.9b shows a schematic of an EDFA. When pumped at either of these wavelengths, an erbium-doped fiber was found to amplify signals over a band of almost 30–35 nm at the 1550-nm wavelength region (see Fig. 8.9 in Chapter 8).

Typical pump powers required for operating an EDFA as an amplifier range from 20 to 100 mW. Absorption of pump energy by the erbium ions leads to *population inversion* of these ions from the ground state ($4I_{15/2}$) to either $4I_{11/2}$ (980 nm) or $4I_{13/2}$ (1480 nm) excited states: the $4I_{13/2}$ level effectively acts as a storage of pump power from which the incoming weak signals may stimulate emission and experience amplification [13]. Stimulated events are extremely fast, and hence the amplified signal slavishly follows the amplitude modulation of the input signal. EDFAs are accordingly *bit rate transparent.* EDFAs are new tools that system planners now almost routinely use for designing networks. EDFAs can be incorporated in a fiber link with an insertion loss of ~0.1 dB, and almost full population inversion is achievable at the 980-nm pump band. Practical EDFAs with an output power of around 100 mW (20 dBm), a 30-dB small signal gain, and a noise figure of <5 dB are now available commercially. In a 10-Gb/s transmission experiment, a record receiver sensitivity of 102 photons per bit was attained in a two-stage composite EDFA having a noise figure of 3.1 dB [25].

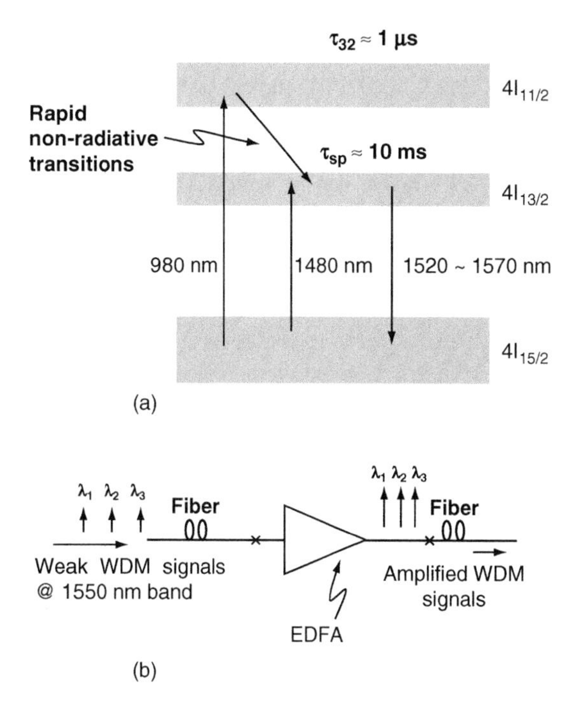

Fig. 3.1.9 (a) Energy level diagram of an Er^{+3} ion. (b) Schematic layout of an EDFA (for details see Chapter 8).

3.1.3.2 DWDM

A very important and attractive feature of EDFAs is that they exhibit a fairly smooth gain versus wavelength curve (especially when the fiber is doped with Al) almost 30–35 nm wide (see Chapter 8). Thus, multichannel operation via WDM within this gain spectrum became feasible, each wavelength channel being simultaneously amplified by the same EDFA. A relatively long lifetime of the excited state ($\sim$ 10 ms) leads to slow gain dynamics and therefore minimal cross-talk between WDM channels. In view of this system, designers were blessed with a great degree of freedom to boost the capacity of a system as and when required due to the flexibility to make the network responsive to future demands simply by introducing additional wavelengths through the same fiber, each multiplexed at, for example, 2.5 or 10 Gb/s or even higher. With the development of L-band EDFA characterized by a gain spectrum that extends from 1570 to 1620 nm [26], the potential for large-scale increase in capacity of already installed links through DWDM became enormous. To avoid haphazard growth of multiple wavelength transmitting fiber links, the ITU introduced certain wavelength standards that are now referred to as ITU wavelength grids for DWDM operation. As per ITU standards, the reference wavelength is chosen to be 1552.52 nm corresponding to the Krypton line, which is equivalent to 193.1 THz (f_0) in the frequency domain; the chosen channel spacing away from f_0 in terms of frequency (Δf) is supposed to follow the relation $\Delta f = 0.1I$THz, with I = positive/negative integers [27].

Recommended channel spacings are 200 GHz ($\equiv$1.6 nm), 100 GHz ($\equiv$0.8 nm), and 50 GHz ($\equiv$0.4 nm); the quantities within parentheses have been calculated by assuming a central wavelength as 1550 nm.

Today's DFB lasers can be tuned to exact ITU wavelength grids. All the terabit transmission experiments that were reported in recent years took the route of DWDM.

The possibility of introducing broadband services in the 1990s led to a boom in the communication industry, blurring the distinction between voice, video, and data service providers, and the demand for more and more bandwidth began to strain the capacity of installed communications links [28]. DSF in combination with EDFA appeared to be the ideal solution to meet the demand for high data rate and long-haul applications. However, it was soon realized that large optical power throughput that is encountered by a fiber in a DWDM system due to simultaneous amplification of multiple wavelength channels poses problems because of the onset of nonlinear propagation effects, which may completely offset the attractiveness of multichannel transmission with DSF. It turned out that due to negligible temporal dispersion that is characteristic of a DSF at the 1550-nm band, these are highly susceptible to nonlinear optical effects, which may induce severe degradation to the propagating signals [29] (see Chapter 6).

3.1.3.3 Fibers for DWDM transmission

By late 1990s the demand for bandwidth had been steadily increasing, and it became evident that new fibers were required to handle transmission of large number of wavelengths, each of which was expected to carry a large amount of data especially for undersea applications and long-haul terrestrial links. A very useful figure of merit (FOM) for a DWDM link is known as *spectral efficiency,* which is defined as the *ratio* of bit rate to channel spacing [11]. Because bit rate cannot be increased arbitrarily due to the constraints of availability of electronic components relevant to bit rates > 40 Gb/s, a decrease in channel spacing appeared to be the best near-term option to tap the huge gain bandwidth of an EDFA for DWDM applications. However, it was soon realized that with a decrease in channel spacing, the fiber became more strongly sensitive to detrimental nonlinear effects such as four-wave mixing (FWM), which could be relaxed by allowing the propagating signals to experience a finite dispersion in the fiber (Fig. 3.1.10). If the number of wavelength channels in a DWDM stream is N, the FWM effect, if present, leads to the generation of $N^2(N-1)$ sidebands! These sidebands would naturally draw power from the propagating signals and hence could result in serious cross-talk [29]. Therefore, for DWDM applications, fiber designers came up with new designs for the signal fiber for low-loss and dedicated DWDM signal transmission at the 1550-nm band, which were generically named non-zero (NZ) DSF. These fibers were designed to meet the requirement of low sensitivity to nonlinear effects

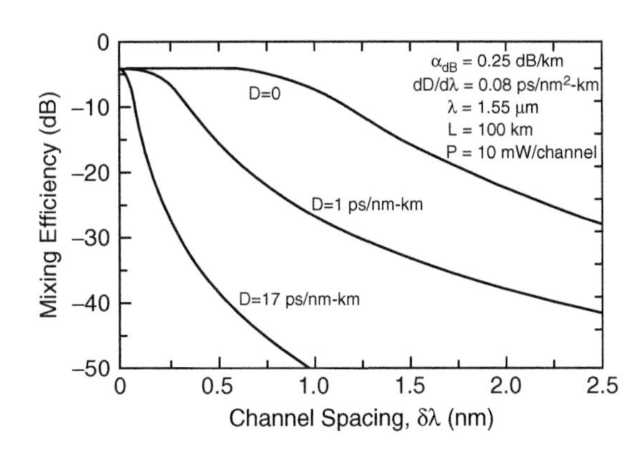

Fig. 3.1.10 Nonlinearity-induced FWM efficiency in a DWDM system as a function of channel separation in fibers having different values for *D*, which are shown as labeling parameters. Other important fiber parameters are shown in the inset. (Adapted from [29]; © IEEE 1995.)

especially to counter FWM, which could be substantially suppressed if each wavelength channel is allowed to experience a finite amount of dispersion during propagation [30] (see Chapter 6). This requirement is precisely counter to the dispersion characteristics of a DSF, hence the name of NZ-DSF [31–33].

ITU has christened such fibers as G.655 fibers, which are generally characterized with refractive index profiles, which are relatively more complex than a simple matched clad step index design of G.652 fibers; a schematic of a specimen is shown in Fig. 3.1.11. The need to deploy such advanced fibers became so imminent that the ITU-T Committee had to evolve new standards for G.655 fibers (ITU-T G.655). As per ITU recommendations, G.655 fibers should exhibit finite dispersion $2 \leq D$ ps/nm·km ≤ 6 in the 1550-nm band to detune the phase matching condition required for detrimental nonlinear propagation effects like FWM and cross-phase modulation to take place during multichannel transmission. ITU arrived at the above range for D by assuming a channel spacing of 100 GHz or more. For smaller channel spacing like 50 or 25 GHz, it turned out that the above range for D is insufficient to suppress potential nonlinear effects unless (1) the power per channel is reduced substantially and (2) the number of amplifiers is limited. Unfortunately, these steps would amount to a decrease in the repeater spacing, which would be counterproductive because this would defeat the very purpose for which G.655 fibers were proposed [34].

To overcome this disadvantage of an NZ-DSF, a more advanced version of NZ-DSF as a transmission fiber was proposed for super-DWDM, which was christened by ITU as advanced NZ-DSF/G.655b fibers. As per the ITU recommendation, the upper limit on D for an advanced NZ-DSF should be 10 ps/nm·km at the longer wavelength edge of the C-band, that is, at 1565 nm. For an advanced NZ-DSF, its D falls in the range $6.8 \leq D$ ps/nm km ≤ 8.9 (C-band) and $9.1 \leq D$ ps/nm·km ≤ 12.0 (L-band) [35]. In view of its reduced sensitivity to detrimental nonlinear effects, it can accommodate transmission of as many as 160 channels of 10 Gb/s at 25 GHz channel spacing within the C-band alone and also offers ease in deployment of 40-Gb/s systems. Another attractive feature of an advanced NZ-DSF is that it can be used to transmit signals in the S-band as well, for which gain-shifted thulium-doped fiber amplifiers and Raman fiber amplifiers have emerged as attractive options. Figure 3.1.12 shows typical dispersion spectra for G.652, G.653, and G.655 types of transmission fibers. There are variations in G.655 fibers, for example, large effective area fibers (LEAF™) [32], reduced slope (TrueWave RS™) [33], and Tera-Light™ [36], each of which are proprietary products of well-known fiber manufacturing giants. Table 3.1.2 depicts typical characteristics of some of these fibers at 1550 nm along with those of G.652 fibers.

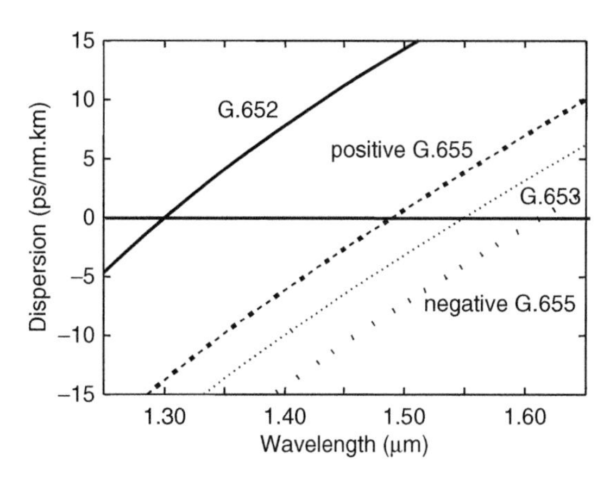

Fig. 3.1.12 Typical dispersion spectra of ITU standard G.652, G.653, and G.655 (both positive and negative versions) fibers.

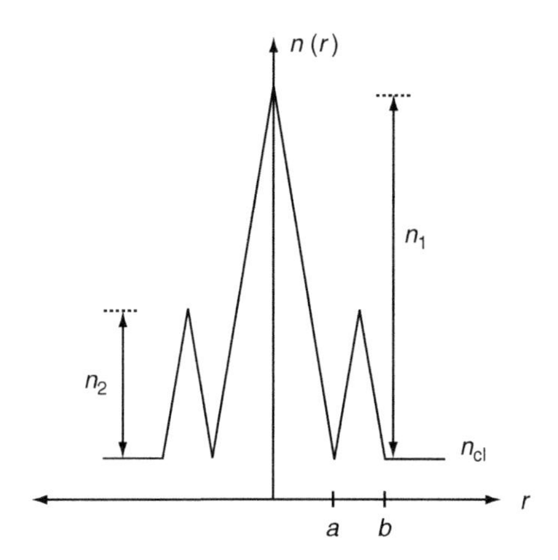

Fig. 3.1.11 A sample schematic refractive index profile of an NZ-DSF.

Table 3.1.2 Characteristics of fibers.

ITU standard fiber type	Dispersion coefficient *D* (ps/nm km) at 1550 nm	Dispersion slope *S* (ps/nm² km)	Mode effective area A_{eff}(μm²) at 1550 nm
G.652	~17	~0.058	~80
G.655			
LEAF™	~4.2	~0.085	~65–80
TrueWave RS™	~4.5	~0.045	~55
TeraLight™	~8	~0.058	—

A_{eff}: mode effective area [see Eq. (3.1.15) in Chapter 6].

3.1.3.4 Dispersion compensating fibers

In G.655 fibers for DWDM transmission, because a finite (albeit low) D is deliberately kept at the design stage itself to counter potentially detrimental nonlinear propagation effects, one would accumulate dispersion between EDFA sites! Assuming a D of 2 ps/nm km, though a fiber length of about 500 km could be acceptable at 10 Gb/s before requiring correction for dispersion, at 40 Gb/s a corresponding unrepeated span would hardly extend to 50 km (see Eq. (3.1.32)). The problem is more severe in G.652 fibers, when at 2.5 Gb/s a link length of about 1000 km would be feasible at the 1550-nm window; if the bit rate is increased to 10 Gb/s, a tolerable D in this case over 1000 km would be hardly about 1 ps/nm·km! Thus, repeater spacing of 1550-nm links based on either of these fiber types would be chromatic dispersion limited for a given data rate.

Around the mid-1990s before the emergence of G.655 fibers, it was believed that for network upgrades it would be prudent to exploit the EDFA technology to "mine" the available bandwidth of the already embedded millions of kilometers of G.652 fibers all over the world by switching over transmission through these at the 1550-nm window. This route appeared to be economically very attractive for attaining network upgrade(s) because it would not require any new fibers to be installed—one could make use of the already installed huge base of G.652 fibers and simply replace the 1310-nm transmitter with a corresponding transmitter operating at 1550 nm. Though transmission of 1550-nm signals through G.652 fibers would imply an accumulation of unacceptably high chromatic dispersion at EDFA sites (ideally spaced approximately every 80–120 km) due to the fiber's D being +17 ps/nm·km, such a large D could substantially reduce the possibility of potentially harmful nonlinear effects. Hence, to reap the benefit of the availability of EDFAs, system upgrade through the DWDM technique in the 1550-nm band of the already installed G.652 fiber links would necessarily require some dispersion compensation scheme. Realization of this immediately triggered a great deal of R&D efforts to develop some dispersion compensating modules, which could be integrated to an SMF optic link so that net dispersion of the link could be maintained/managed within desirable limits. Three major, state-of-the-art, fiber-based, optical technologies available as options for dispersion management are *dispersion compensating fibers* (DCFs) [37–42], *chirped fiber Bragg gratings* [43–46] (see Chapter 15), and *high-order-mode fibers* [35, 47–49] (see Chapter 19).

To understand the logic behind dispersion compensation techniques, we notice from Eqs. (3.1.9) and (3.1.10) that the instantaneous frequency of the output pulse is given by

$$\omega(t) = \frac{d\Phi}{dt} = \omega_c + 2\kappa\left(t - \frac{L}{v_g}\right). \qquad (3.1.33)$$

The center of the pulse corresponds to $t = L/v_g$. Accordingly, the leading and trailing edges of the pulse correspond to $t < L/v_g$ and $t > L/v_g$, respectively. In the normal dispersion regime where $d^2\beta/d\lambda_0^2$ is positive (implying that the parameter κ defined in Eq. (3.1.11) is positive), the leading edge of the pulse is down-shifted, that is, *red-shifted*, in frequency whereas the trailing edge is up-shifted, that is, *blue-shifted*, in frequency with respect to the center frequency ω_c. Thus, red spectral components (i.e., longer wavelength components) of the signal pulse would travel faster than their blue spectral components (thereby meaning shorter wavelength components), that is, the group delay would increase with a decrease in wavelength. The converse would be true if the signal pulse wavelength corresponds to the anomalous dispersion region where $d^2\beta/d\lambda_0^2$ is negative. Hence, as the pulse broadens with propagation due to this variation in its group velocity with wavelength in a dispersive medium like SMF, it also gets chirped, meaning thereby that its instantaneous frequency within the duration of the pulse itself changes with time. Chirping of a temporal pulse with propagation both in the normal and anomalous dispersion regimes in a fiber is schematically shown in Fig. 3.1.13. If we consider propagation of signal pulses through a G.652 fiber at the 1550-nm wavelength band at which its D is positive (i.e., its group delay increases with increase in wavelength), it would exhibit *anomalous* dispersion. If this broadened temporal pulse were transmitted through a DCF, which exhibits *normal* dispersion (i.e., its dispersion coefficient D is negative) at this wavelength band, then the broadened pulse would

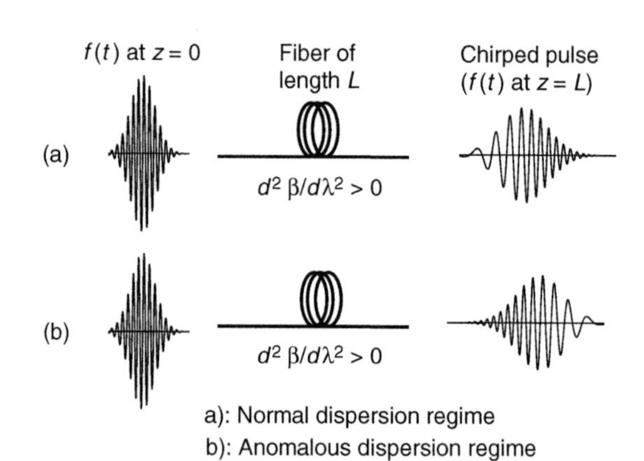

Fig. 3.1.13 Propagation of a Gaussian pulse at a wavelength (a) in the normal dispersion regime and (b) in the anomalous dispersion regime and resulting chirping of the propagated pulse.

get compressed with propagation through the DCF. This could be understood with the help of the transfer function of a fiber given by Eq. (3.1.8) and studying evolution of the pulse as it propagates through different segments of the fiber link shown in Fig. 3.1.14. At stage (1), let the F.T. of the input pulse $f_1(t)$ be $F_1(\Omega)$, which transforms in view of Eq. (3.1.8) at subsequent stages to [12]

$$\begin{aligned} stage(2): \quad F_2(\Omega) &= F_1(\Omega)\exp(-i\beta_T''L_T\Omega^2) \\ stage(3): \quad F_3(\Omega) &= F_2(\Omega)\exp(-i\beta_D''L_D\Omega^2) \\ &= F_1(\Omega)\exp[-i(\beta_T''L_T + \beta_D''L_D)\Omega^2] \end{aligned}$$

where β_T'' and β_D'' represent GVD parameters of the transmission fiber and the DCF, respectively, and $L_{T,D}$ refer to corresponding fiber lengths traversed by the signal pulse. It is apparent from the above that if the following condition is satisfied,

$$\beta_T''L_T + \beta_D''L_D = 0 \tag{3.1.34}$$

the dispersed pulse would recover its original shape (i.e., evolve back to $f_1(t)$) at stage (3). This is more explicitly illustrated in Fig. 3.1.15. Thus, GVD parameters (and hence coefficients $D_{T,D}$) of the transmission fiber and the DCF should be opposite in sign. Consequently, if a G.652 fiber as the transmission fiber is operated at the EDFA band, corresponding DCF must exhibit negative dispersion at this wavelength band [50]. Fig. 3.1.16 illustrates as an example the concept of dispersion compensation by a DCF for the dispersion suffered by 10 Gb/s externally modulated signal pulses over a 600-km span of a G.652 fiber; the solid line represents delay versus length in the transmission fiber and dashed line corresponds to delay versus length for the DCF [51]. Fig. 3.1.16 shows three possible routes for introducing the DCF: (1) at the beginning of the span (dashed line), (2) at the end of the span (solid line), or (3) every 100 km (dotted line), say, at EDFA sites. Further, the larger the magnitude of D_D, the smaller the length of the required DCF. This is achievable if D_{WG} of the DCF far exceeds its D_M in absolute magnitude. Large negative D_{WG} is achievable through an appropriate choice of the refractive index profile of the fiber so that at the wavelengths of interest a large fraction of its modal power rapidly spreads into the cladding region for a small change in the propagating wavelength. In other words, the rate of increase in mode size is large for a small increase in λ. Accordingly, the modal power distribution is strongly influenced by the cladding, where the mode travels effectively faster due to lower refractive index [51].

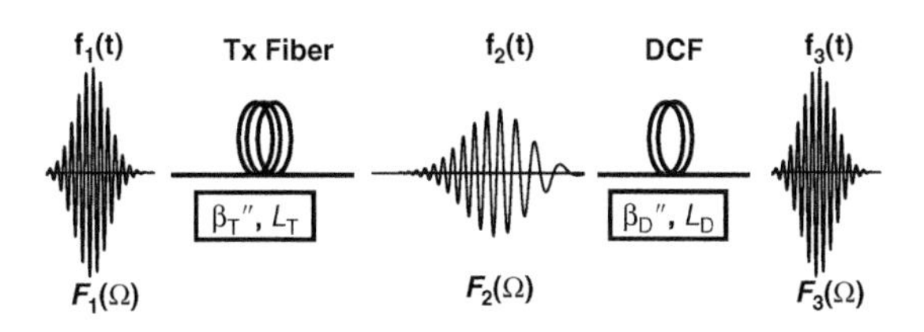

Fig. 3.1.14 Schematic illustration of dispersion-induced pulse broadening with propagation in terms of fiber transfer function through Fourier transform pairs and compression of the dispersed pulse through propagation in a DCF.

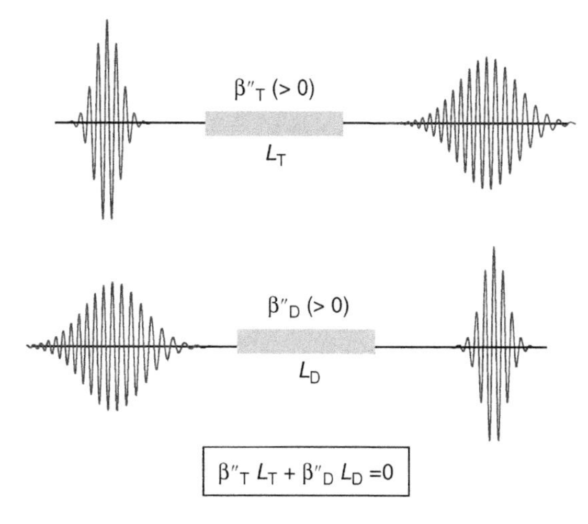

Fig. 3.1.15 Illustration of the role of a DCF to neutralize dispersion-induced broadening of a pulse due to propagation through a fiber having GVD parameter > 0 at the operating wavelength.

The *first-generation* DCFs relied on narrow core and high core cladding refractive index contrast (Δ) fibers to fulfill this task; typically, in these the fiber refractive index profiles were similar to those of the matched clad SMFs with the difference that Δ of the DCF were much larger ($\geq$2%) (Fig. 3.1.17). These DCFs were targeted to compensate for dispersion in G.652 fibers at a single wavelength and were characterized with a D of

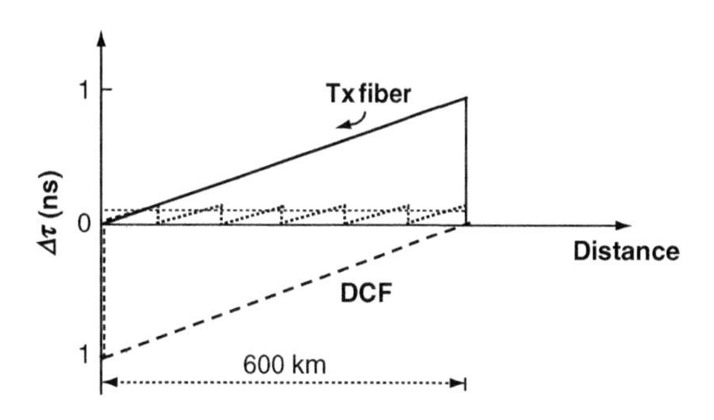

Fig. 3.1.16 Schematic illustrating the concept of dispersion compensation of 10-Gb/s pulses transmitted through 600 km of a G.652 fiber. The full line represents variation of dispersion with length for a G.652 fiber; dashed lines represent corresponding variations for a compatible DCF when placed at the beginning or at the end (full line) or at every 100 km (dotted line) along the link. (Adapted from [51]; © IEEE.)

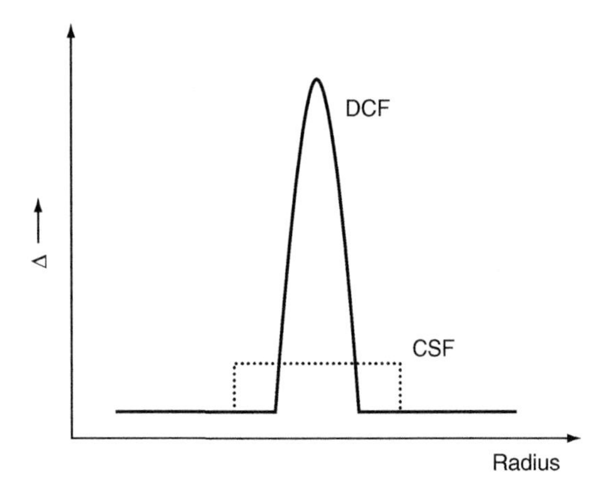

Fig. 3.1.17 Typical refractive index profile of a first-generation DCF relative to that of a conventional single-mode fiber (CSF).

approximately −50 to −100 ps/nm·km and a positive dispersion slope. A schematic for D_D versus λ for a first-generation DCF relative to D_T for the transmission fiber is shown in Fig. 3.1.18. Because of high Δ and poor mode confinement as explained above, a DCF necessarily involved large insertion loss (typically attenuation in DCFs could vary in the range of 0.5–0.7 dB/km [11]) and sensitivity to bend-induced loss. To simultaneously achieve a high negative dispersion coefficient and low attenuation coefficient, α_D, DCF designers have ascribed an FOM to a DCF, which is defined through

$$\mathrm{FOM} = \frac{-D_D}{\alpha_D} \tag{3.1.35}$$

expressed in ps/nm·DB. Total attenuation and dispersion in dispersion-compensated links is given by

$$\alpha = \alpha_T L_T + \alpha_D L_D \tag{3.1.36}$$

$$D = D_T L_T + D_D L_D. \tag{3.1.37}$$

It could be shown from Eqs. (3.1.35) to (3.1.37) that for $D = 0$ [42]

$$\alpha = \left(\alpha_T + \frac{D_T}{\mathrm{FOM}}\right) L_T \tag{3.1.38}$$

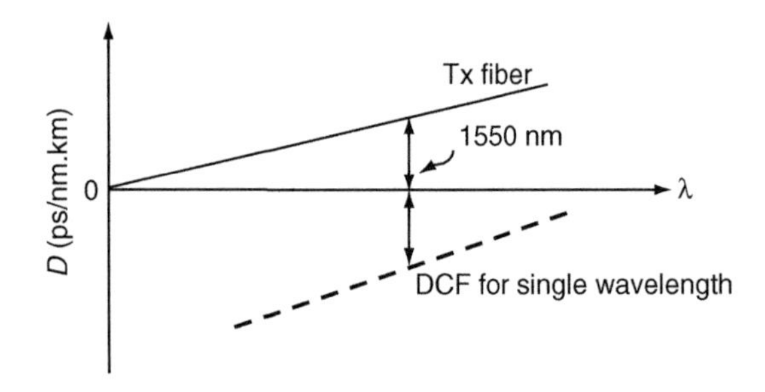

Fig. 3.1.18 Dispersion spectra for a single wavelength DCF relative to that of a CSF at the 1550-nm band.

which shows that any increase in total attenuation in dispersion-compensated links would be solely through FOM of the DCF; thus, the larger the FOM, the smaller the incremental attenuation in the link due to insertion of the DCF. Dispersion slopes for the first-generation DCF were of the same sign as the transmission fiber (cf. Fig. 3.1.18) and hence perfect dispersion compensation was realizable only at a single wavelength in this class of DCFs.

Because DWDM links involve multichannel transmission, ideally one would require a broadband DCF so that dispersion could be compensated for at all the wavelength channels simultaneously. The key to realize a broadband DCF lies in designing a DCF in which not only that D versus λ is negative at all those wavelengths in a DWDM stream, but also that the dispersion slope is negative. The broadband dispersion compensation ability of a DCF is quantifiable through a parameter known as relative dispersion slope (RDS) expressed in units of nm^{-1}, which is defined through

$$\mathrm{RDS} = \frac{S_D}{D_D}. \tag{3.1.39}$$

A related parameter, referred to as κ (in units of nm), also referred to in the literature, is simply the inverse of RDS [52]. Values of RDS for LEAF™, True Wave RS™, and Teralight™ are 0.026, 0.01, and 0.0073, respectively. Thus, if a DCF is so designed that its RDS (or κ) matches that of the transmission fiber, then that DCF would ensure perfect compensation for all wavelengths. Such DCFs are known as dispersion slope compensating fibers (DSCFs). A schematic for the dispersion spectrum of such a DSCF along with single-wavelength DCF is shown in Fig. 3.1.19. Any differences in the value of RDS (or κ) between the transmission fiber and the DCF would result in under- or over-compensation of dispersion, leading to increased *bit error rates* (BERs) at those channels. In practice, a fiber designer targets to match values of the RDS for the Tx fiber and the DCF at the

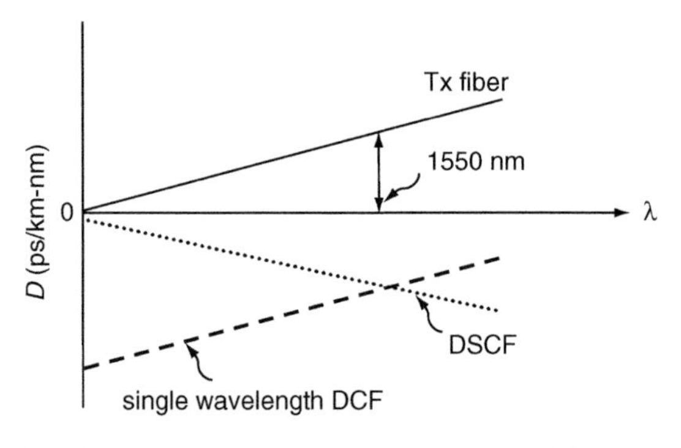

Fig. 3.1.19 Dispersion spectra for a broadband DSCF relative to that of a conventional single-mode fiber (Tx) and a single wavelength DCF at the 1550 nm band.

Table 3.1.3 Tolerable mismatch (ΔS) in S at different bit rates.

Bit rate	$\Delta S/S$ (%)
10 Gb/s	~20
40 Gb/s	~1

After [11]

median wavelength of a particular amplification band, that is, C- or L-band. In principle, this is sufficient to achieve dispersion slope compensation across that particular amplifier band because the dispersion spectrum of the transmission fiber is approximately linear within a specific amplification band. However, other propagation issues, such as bend loss sensitivity and countenance of nonlinear optical effects through large-mode effective area, often demand a compromise between 100% dispersion slope compensation and the largest achievable mode effective area. In such situations, the dispersion slope S may not precisely match those wavelength channels, which fall at the edges of a particular amplification band. Table 3.1.3 shows the impact of this mismatch (ΔS) in S at different bit rates [11].

A typical value of RDS for G.652 type of SMF at 1550 nm is about 0.00335 nm^{-1} (or κ as 298 nm), whereas in LEAF™, TrueWave™, and Teralight™ fibers, its values are about 0.026 nm^{-1}, 0.01 nm^{-1} and 0.0073 nm^{-1}, respectively. The insertion loss of an approximately matching commercial DCF for an 80-km span is ~ 6–7 dB, which includes the loss due to splices. Ideally, a DCF design should be so optimized that it has a low insertion loss, a low sensitivity to bend loss, a large negative D with an appropriate negative slope for broadband compensation, a large mode effective area (A_{eff}) for reduced sensitivity to nonlinear effects, and a low polarization mode dispersion (PMD).

A few proprietary designs of DCF index profiles are shown in Fig. 3.1.20. It is apparent from these index profiles that multiple claddings were introduced to achieve better control on the mode expansion–induced changes in the guided mode's effective index. One recent design (Fig. 3.1.21), which yielded the record for largest negative D (−1800 ps/nm·km at 1558 nm), was based on a coaxial dual-core refractive index profile [53];

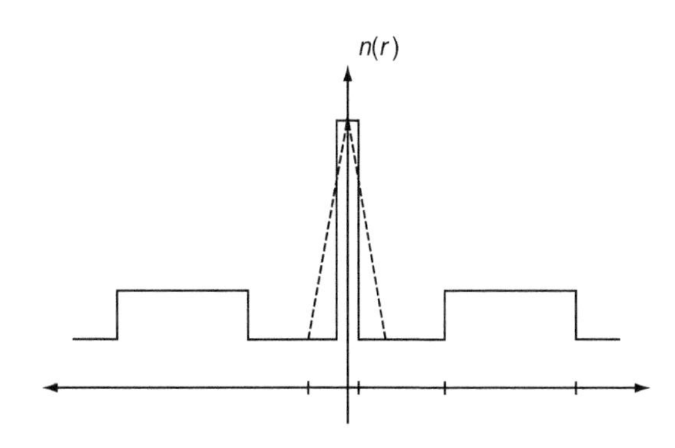

Fig. 3.1.21 Refractive index profile of the coaxial dual-core DCF.

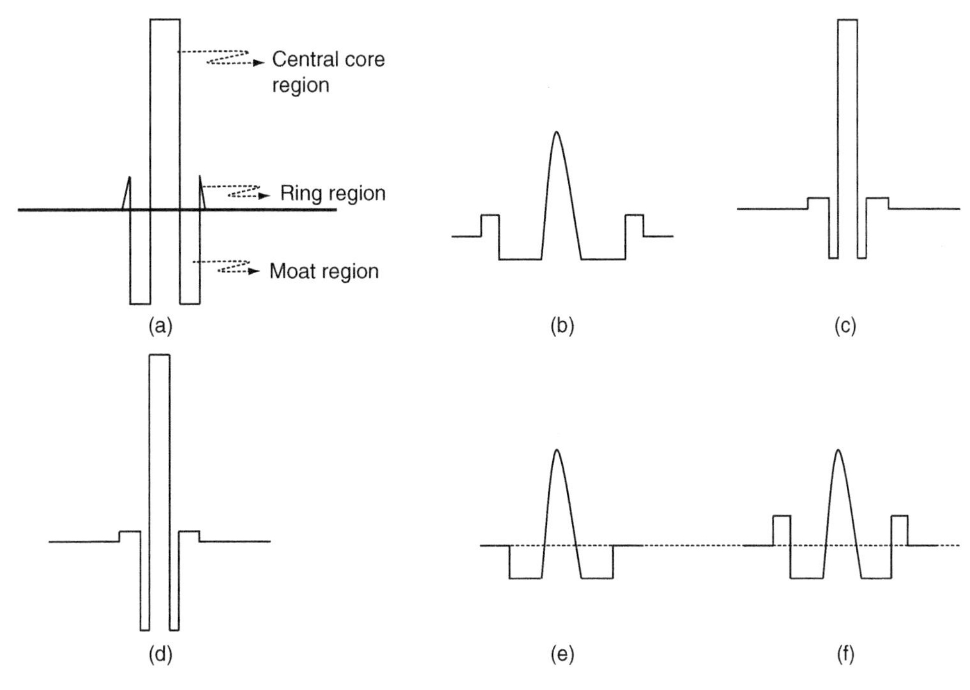

Fig. 3.1.20 Schematic refractive index profiles of DCFs of (a) Corning Inc.; (b), (c), and (d) Lucent Technology; (e) and (f) Furukawa as available in the literature.

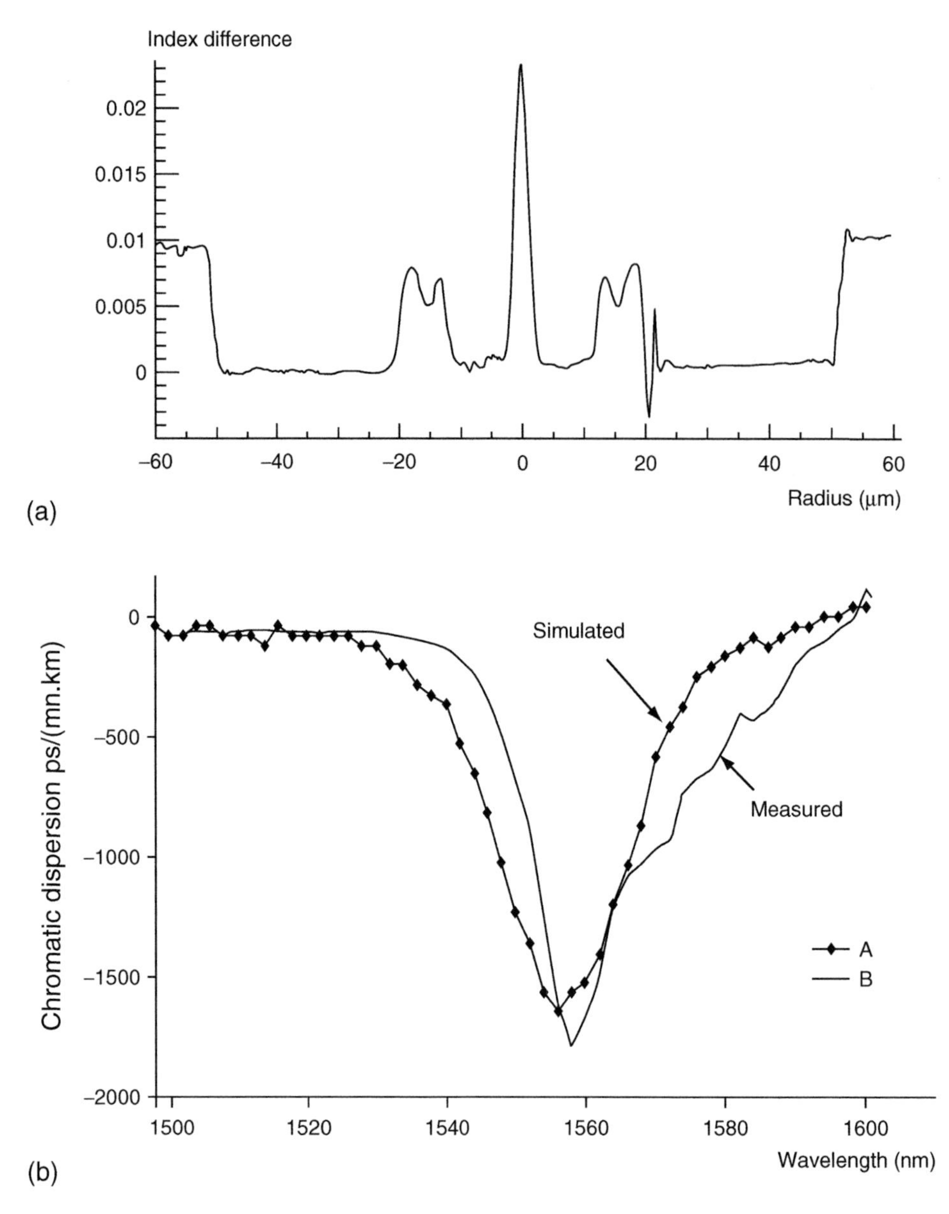

Fig. 3.1.22 (a) Refractive index profile of the fabricated dual-core DCF preform. (b) Measured dispersion spectrum of the fabricated DCF (Adapted from [54]; © Copyright IEE).

Fig. 3.1.22 [54] shows the fabricated DCF's actual index profile and its measured chromatic dispersion spectrum. The two cores are interconnected through a matched cladding, in contrast to most of the other designs in which there is usually a depressed index clad or a moat region in the cladding. In the fabricated preform, the central core was more of a triangular shape (Fig. 3.1.21, dashed line) than a step distribution. The two cores essentially function like a directional coupler. Because these two concentric fibers are not significantly identical, through adjustments of index profile parameters their mode effective indices could be made equal at some desired wavelength (called phase matching wavelength, λ_p), in which case the effective indices as well as modal field distributions of the normal modes of this dualcore fiber would exhibit rapid variations with λ around λp [53].

Typical modal power distributions for three different wavelengths around the phase matching wavelength (λ_p) are shown in Fig. 3.1.23. These rapid variations in modal distributions with wavelengths could be exploited not only to tailor dispersion characteristics of the DCF, for example, to achieve very large negative dispersion or broadband dispersion compensation characteristics, but also to achieve inherent gain flattened EDFAs (see Chapter 8) or broadband Raman fiber amplifiers [55]. Because the index profile involves multitudes of profile parameters, one could optimize the design through an appropriate choice of these parameters to achieve control

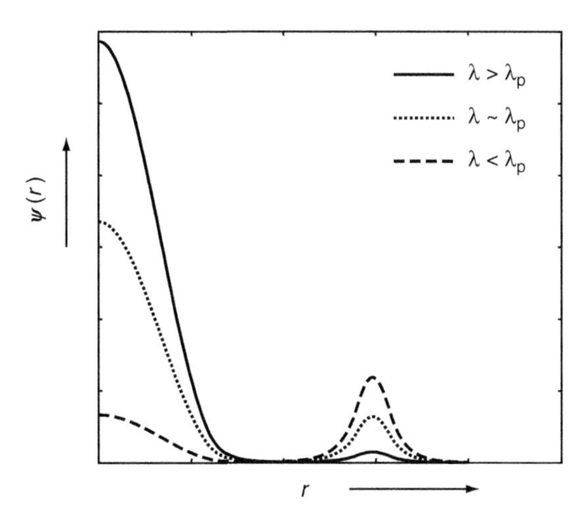

Fig. 3.1.23 Typical modal power distributions for three different operating wavelengths relative to phase matching wavelength in a coaxial dual-core DCF.

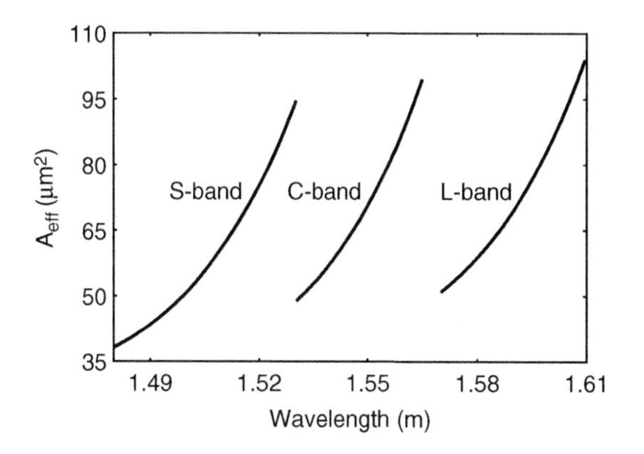

Fig. 3.1.24 Wavelength dependence of mode effective area of the various coaxial dual-core DSCFs designed (see Table 3.1.4) for S-, C-, and L-band fiber amplifiers [57].

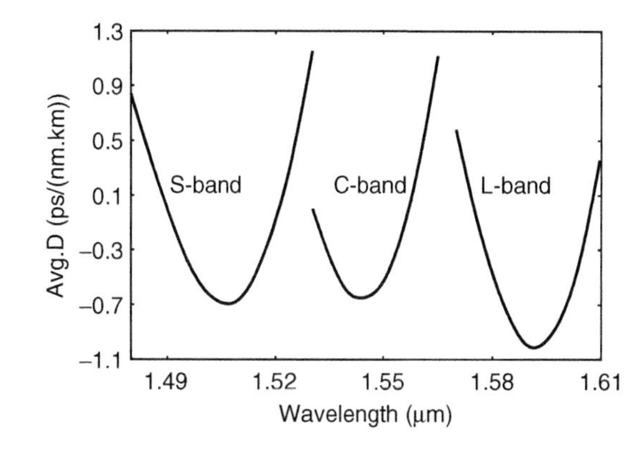

Fig. 3.1.25 Net residual dispersion spectra at different amplifier bands of a dispersion compensated link consisting of 100 km of G.652 fiber jointed with about 10 km of the designed coaxial dual-core DSCFs [57].

of its propagation characteristics. Further research in this direction indeed has led to designs of coaxial dual-core DSCFs for broadband dispersion compensation in G.652 as well as G.655 fibers within various amplifier bands such as the S-, C- and L-bands [56–59]. Mode effective areas of various DCF designs based on Fig. 3.1.20 typically range between 15 and 25 μm^2, which make these susceptible to nonlinear effects unless care is taken to reduce launched power ($\leq$1 dBm per channel) into the so designed DCFs. In contrast, the designs based on dual-core DSCFs could be designed to attain A_{eff}, which are comparable with that of the G.652 fiber (~70–80 μm^2), as shown in Fig. 3.1.24. The net residual dispersion spectra of a 100-km-long G.652 fiber link along with so designed DSCFs (in a ratio of approximately 10:1) at each of the amplifier bands are shown in Fig. 3.1.25. It can be seen that residual average dispersion is well within $\pm$1 ps/nm·km within all three bands. Other estimated performance parameters of the designed DSCFs are shown in Table 3.1.4. Results for residual dispersion for standard G.655 fibers jointed with the designed dual-core DSCFs for various amplifier bands are shown in Fig. 3.1.26 [58]. Table 3.1.5 displays performance parameters similar to those shown in Table 3.1.4 of these coaxial dual-core design-based DSCFs for standard G.655 fibers.

3.1.3.5 Reverse/Inverse dispersion fibers

The above-mentioned DCFs are normally introduced in a link in the form of a spool as a standalone module at

Table 3.1.4 Important performance parameters of the designed DSCFs for G.652 fibers [57].

Amplifier band (central λ) (nm)	*D* (ps/nm km)	RDS (nm^{-1})	A_{eff} (μm^2)	Mode field diameter (μm)	Estimated FOM (ps/nm dB)	Bend loss* (dB)
S (1500)	−182	0.0056	51	7.15	771	0.0221
C (1550)	−191	0.0027	70	7.97	941	0.095
L (1590)	−162	0.0034	70	8.12	837	0.0014

* For a single-turn bend of diameter 32 mm.

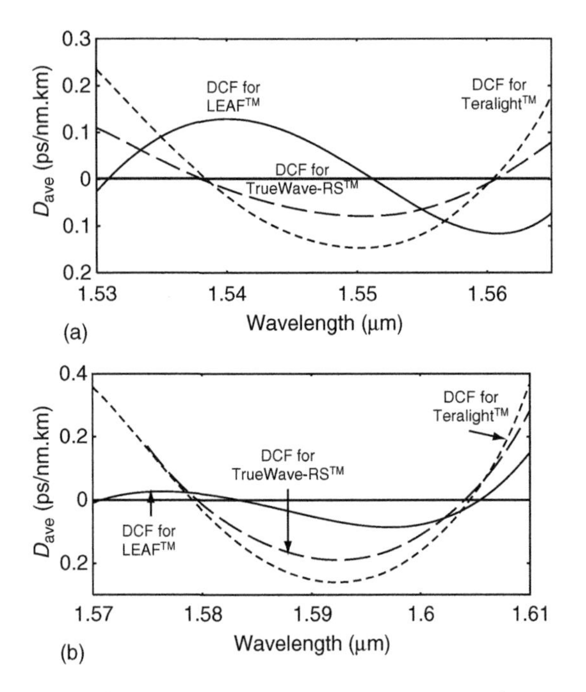

Fig. 3.1.26 (a) Net residual dispersion spectra at the C-band of standard NZ-DSF fiber links jointed with designed dual-core DSCFs. (b) Net residual dispersion spectra at the L-band of standard NZ-DSF fiber links jointed with designed dual-core DSCFs.

an EDFA site. An alternative scheme involves the use of the compensating fiber within the cable itself after having been jointed with the transmission fiber as part of the overall fiber link. Such DCFs are referred as *reverse/inverse* dispersion fibers (RDF/IDF) [60–62]. Reverse dispersion fibers are used with the transmission fiber almost in the ratio of 1:1, meaning their characteristic dispersion parameters D and S are almost the same as that of the transmission fiber except for the signs. Thus, for the reverse dispersion fiber used in [60], D and S were -16 ps/nm·km and -0.050 ps/nm^2·km, respectively. Dispersion ramped fiber links, which involve use of RDFs, alternate fiber segments with the link and consist of positive and negative dispersion fibers, so that the overall dispersion is low, and D versus length curve mimicks a sawtooth curve. One major attribute of these fibers was that loss in such fibers was low (in contrast to high negative D DCFs), ~ 0.25 dB/km. Terabit transmission experiments through a repeater span of over 125 km were reported in such dispersion managed fiber links involving RDFs in conjunction with G.652 fibers over a bandwidth of 50 nm [63]. Alternate fiber segments in these links consist of positive and negative D (in 1:1 ratio) fibers such that overall dispersion is low and D versus length curve mimicks a sawtooth curve. RDFs were also used with NZ-DSFs (in a ratio of 1:2) [64].

3.1.4 Fibers for metro networks

In recent years, metro optical networks have attracted a great deal of attention from lightwave communication engineers due to potentials for high growth. A metro network provides generalized telecommunication services transporting any kind of signal from point to point in a metro. First-generation metropolitan optical networks based on SDH/SONETrelied on rings laced with nodes at which information is electronically exchanged. *Access* that aggregates a wide variety of traffic from business and residential end users is required to port this traffic directly to the transport product for distribution throughout an optical network. In *transport*, DWDM is the key enabling technology to expand the capacity of existing and new fiber cables without optical-to-electrical-to-optical

Table 3.1.5 Important performance parameters of the designed DSCFs for 6.655 fibers (After [58]; copyright Optical Society of America).

Designed dual-core DSCF for	D (ps/nm km)	RDS (nm^{-1})	A_{eff} (μm^2)	Mode field diameter (μm)	FOM (ps/nm dB)	Bend loss* (dB)
LEAF™ (C-band)	−264	0.033	41.8	3.34	1248	2.54 × 1.0e–02
LEAF™ (L-band)	−172	0.017	43.63	3.47	885	4.28 × 1.0e–03
True Wave RS™ (C-band)	−173	0.0099	49.92	3.68	760	2.35 × 1.0e–06
True Wave RS™ (L-band)	−173	0.0075	61.4	4.02	851	1.89 × 1.0e–06
TeraLight™ (S-band)	−201	0.01	45.97	3.42	844	5.2 × 1.0e–02
TeraLight™ (C-band)	−187	0.0084	49.99	3.55	873	6.51 × 1.0e–03
TeraLight™ (L-band)	−150	0.006	57.77	3.92	771	1.4 × 1.0e–02

Values correspond to central λ in each band as in Table 1.4.
* For a single-turn bend of diameter 32 mm.

conversions. Metro networks with distances of less than 200 km, which bridges the gap between local/access networks and long distance telecommunication networks, were originally designed to essentially link local switching offices, from which telephone lines branch out to individual customers in the access portion of the network. In larger cities, it may consist of two layers, *metro access or edge loops* (20–50 km), that connect groups of switching offices, and these loops, in turn, link to a *metro core* or *backbone* (50–200 km) network (having several nodes) that serves the whole metropolitan area [65].

However, with the growth in Internet usage and increasing demand in data services, a need arose to migrate from conventional network infrastructure to an *intelligent* and *data-centric* metro network. Accordingly, the network trend in the metro sector has been to move toward *transparent* rings, in which wavelength channels are routed past or dropped off at the nodes [66]. The futuristic transparent metro optical networks must have flexibility to route/drop off signals at any node in the network so that the signal(s) may travel along the entire ring in the network. New services are now fed optically at the network edge, for example, *ESCON* (enterprise system connection, which is a 200-Mb/s protocol meant for providing connectivity of a main frame to other mainframe storage devices), *Gigabit Ethernet* (which is a simple protocol that uses inexpensive electronics and interfaces for sharing a local area network), *FDDI* (a local area network standard covering transmission at 100 Mb/s, which can support up to 500 nodes on a dual ring network with an internode spacing up to 2 km), and digital video. Gigabit Ethernet is fast evolving as a universal protocol for optical packet switching. For example, most of the data on the Internet start as Ethernet packets generated by system servers. Thus, in addition to voice, video, and data, a metro network should be able to support various protocols like Ethernet, fast Ethernet, and 10-Gb/s Ethernet.

A modern metro network is a *distribution system* and not just an information transport pipeline connecting a pair of points. It may be described as an updated version of a regional network designed to function as a high capacity system, which provides connectivity between many sites. For example, telephone switching offices, Internet service provider servers, corporate, and university campuses are often structured as loops with add/drops at nodes of similar importance.

To offer service providers the flexibility to increase network capacity in response to customer demand, a metro network is required to address unique features like low first cost, high degree of scalability to efficiently accommodate unpredicted traffic growth, flexibility to add/drop individual signals at any central office in the network, interoperability to carry a variety of signals, and provide connectivity to variety of equipment such as cell phones, SONET/SDH, legacy equipment, asynchronous transfer mode (ATM), and Internet protocol (IP) [65]. Legacy metro networks relied heavily on low-cost G.652 fibers having near-zero dispersion at 1310 nm. With G.652 and G.652.C fibers, dispersion limited distances at 10 Gb/s transmission rate would be about 70 km beyond which a conventional DCF with a negative D is required to achieve signal reach in longer metro rings [66]. However, this would add to overall cost, more so because EDFA(s) may be required to offset the extra insertion loss introduced by a DCF.

Directly modulated, distributed feedback (DBF)-lasers in contrast to externally modulated DFB lasers, are an economic advantage in a metro environment. However, directly modulated lasers are usually accompanied with a positive chirp, which could introduce severe pulse dispersion in the EDFA band if the transmission fiber is characterized with a positive D. The positive chirp-induced pulse broadening can be countered with a transmission fiber if it is so designed that it exhibits normal dispersion (i.e., negative D) at the EDFA band. This is precisely the design philosophy followed by certain fiber manufacturers for deployment in a metro network such as MetroCor™ fiber [67]. As a typical example, a schematic of the dispersion spectrum of one such negative dispersion metro fiber relative to that of a standard SMF is shown in Fig. 3.1.27a. Before the introduction of metro-specific fibers, capacity upgrades were based on lighting of new fibers. In a metro network design, technological adaptability is extremely important to outweigh obsolescence because outside plant fixed costs such as trenching and ducts are almost twice that of long-haul networks [67]. DWDM in a metro environment is attractive in this regard for improved speed in provisioning due to possibility of allocating dynamic bandwidth to customers on demand and for better cost efficiency in futuristic transparent networks running up to 200 km or more. For such distances, a DCF in the form of a standard SMF with positive D could be used to compensate for dispersion in a MetroCor™ kind of fiber. However, due to a relatively low magnitude of D in an SSMF, long lengths of standard SMF would be required, which might require careful balancing of overall loss budget and system cost.

An alternative type of metro fiber has also been proposed and realized, which exhibits positive $D \sim 8$ ps/nm·km at 1550 nm as shown in Fig. 1.27b [66]. The advantage cited by the manufacturer of such positive dispersion metro fibers is that dispersion compensation could be achieved with efficient already available conventional DCFs of much shorter length(s) as compared with standard SMFs required for negative dispersion equivalent metro fibers. Nevertheless a careful cost analysis for the network as a whole on a case-by-case basis would decide whether to deploy positive or negative NZ-DSFs in a metro network.

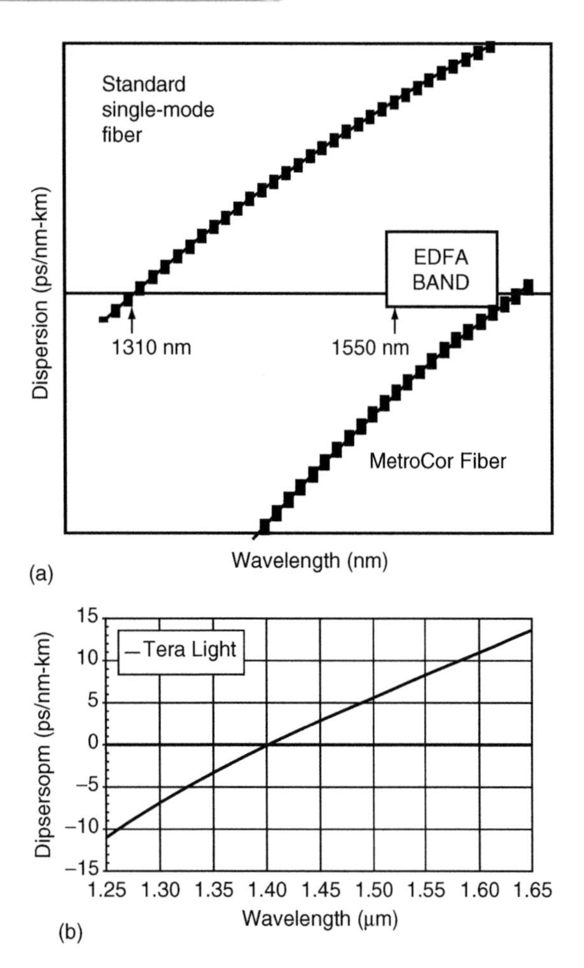

Fig. 3.1.27 (a) Dispersion spectrum of a negative dispersion MetroCor™ fiber relative to a standard G.652 single-mode fiber. (Adapted from [67]; © Corning Inc.) (b) Dispersion spectrum of a positive dispersion NZ-DSF TeraLight™ brand metro fiber. (Adapted from [66]; © Alcatel Telecommunications Review 2002) (© Alcatel Telecommunications Review)

3.1.5 Coarse wavelength division multiplexing

With the availability of LWPF SMFs (Fig. 3.1.2) christened as G.652C fibers by ITU, examples of which are AllWave™ and SMF-28e™, a new market has emerged for configuring short distance communication (up to ~ 20 km, e.g., for campus backbone) networks with these fibers. Because of the low-loss wavelength window being wider by 30% (in the E-band) offered by these fibers in comparison with G.652 legacy fibers, one could use these to transmit widely spaced wavelengths as carriers with low-cost, uncooled, FP lasers as the light sources. Recently introduced ITU standards allow channel spacing as wide as 20 nm under a coarse wavelength division multiplexing scheme due to which laser cooling becomes unnecessary because environmental temperature–induced wavelength wander would not be an issue. This would bring down system complexity and overall network cost because demands on wavelength tolerance of other ancillary components like filters would also be much less stringent as compared with those required for DWDM systems.

3.1.6 Combating PMD in a fiber

With growth in transmission capacity to 10 Gb/s and beyond, systems engineers began to encounter a problem in about 20–30% of legacy fibers, which were deployed before 1990, due to what is known as polarization mode dispersion (PMD). PMD arises because of a certain in-built geometric birefringence, as a result of slight noncircularity/asymmetry in fiber cross-section, which could be attributed to the process of fiber manufacture. Because of asymmetry, however small it may be, and the resultant birefringence, an SMF in general supports two orthogonal polarization modes. The birefringence may also be independently caused by mechanical stress due to diurnal or seasonal heating and cooling of a deployed fiber. It could also arise due to, for example, vibrations from a moving train, if deployed in the neighborhood of railway tracks (sometimes laid to overcome the problem of right of way) [68]. Even if deployed as an aerial cable, it can be subjected to stress due to swaying caused by wind.

Because of birefringence, whatever the cause, the input pulse power gets distributed among the two *eigen polarization* modes of the fiber as shown in Fig. 3.1.28. The eigen modes could be understood by considering the example of an elliptic core fiber of dimension 2a for the major axis (say, along x) $\times$ $2b$ for the minor axis (say, along y). A light polarized parallel to the major axis of the fiber would preferentially launch light into the natural mode of the fiber (of, say, effective index n^x_{eff}), which is polarized along the major axis because light guidance would be strongly dictated by the V-number, which would involve a as the core dimension. Likewise, a

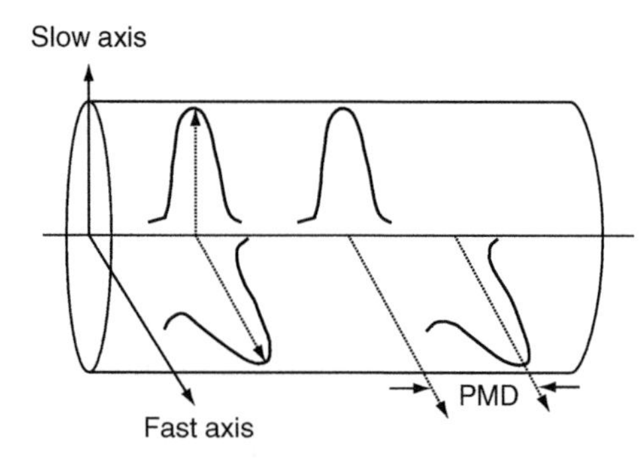

Fig. 3.1.28 Schematic demonstrating the build-up of PMD with propagation in a single-mode fiber. (After [70].)

y-polarized incident light would preferentially excite a y-polarized natural mode (of, say, effective index n^{y}_{eff}) of the fiber. As a result, the pulse energy via one of these modes would travel at a speed slower or faster than the other, giving rise to a differential group delay (DGD), which may be sufficient to cause enough pulse spread to make one pulse overlap with the other and cause intersymbol interference. This group delay difference is formally known as the PMD, and the PMD coefficient (D_{PMD}) is defined for a short length of the fiber as $\Delta N_g/c$, where ΔN_g represents a difference in the group indices (cf. Eq. (3.1.4)) of the two eigen polarization modes of the fiber and c is the velocity of light in a vacuum [69]. It is expressed in ps/km; the product $D_{PMD} \times$ fiber length yields short-fiber length PMD in ps. It can create problems at high bit rates, such as for 10 and 40 Gb/s systems [14]. It would be an alarming effect if the short-length PMDs were extrapolated linearly to long fiber lengths!

However fortuitously, usually cross-coupling of power between the eigen modes takes place along the fiber length, which eventually populates both the polarization states equally [69]. The instantaneous value of PMD varies randomly with time, temperature, and wavelength. Accordingly, it is described in the form of an average of the Maxwellian density function whose mean value increases with square root of fiber length [70]. The maximum instantaneous DGD is approximately 3.2 times its average value [68]. The PMD of a fiber along with other optical components in a network should be less than $140/B$; B is measured in Gb/s [71]. At 40 and 10 Gb/s, the tolerable total PMD across a link has been estimated to be 3.5 and 14 ps, respectively. For a given bit rate and D_{PMD}, and assuming all mode mixing between the eigen modes, the possible link length L (in km) is given by [72]

$$L = \left[\frac{10^3 f}{B(\mathrm{Gb/s}) \times D_{PMD}(\mathrm{ps}/\sqrt{\mathrm{km}})}\right]^2. \quad (3.1.40)$$

The parameter f representing *allowed bit period fraction* is given by [72]

$$f = \mathrm{PMD(ps)} \times B(\mathrm{Gb/s}) \quad (3.1.41)$$

with PMD as the PMD_{total}, that is, it includes PMD of all the components that exhibit PMD and that populate the link between the transmitter and the receiver

$$\mathrm{PMD}_{total} = (\mathrm{PMD}_1^2 + \mathrm{PMD}_2^2 + \mathrm{PMD}_3^2 + \ldots)^{1/2} \quad (3.1.42)$$

where $\mathrm{PMD} = D_{PMD} \times \sqrt{L}$. Equation (3.1.40) is a generalization of a result that assumed $f = 0.1$ [69,72]. It is generally recommended that PMD on a cabled fiber for high-capacity systems should be less than 0.5 ps/$\sqrt{\mathrm{km}}$ [71]. In some of the legacy fibers installed before the early 1990s, D_{PMD} was 1 – 2 ps/$\sqrt{\mathrm{km}}$. Such fibers would introduce intolerable DGD even over distances below 500 km at 10 Gb/s. However, since the early 1990s through certain innovations introduced during the fiber drawing step, for example, controlled spinning [73,74], state-of-the-art transmission fibers may exhibit PMD below 0.1 ps/$\sqrt{\mathrm{km}}$, which is sufficient to maintain DGD below 3.5 ps at 40 Gb/s.

3.1.7 Conclusion

In this chapter, evolving R&D trends in SMF designs since the early 1980s as well as certain specialty fibers like NZ-DSFs, DCFs, LWPFs, and fibers for metro networks were described. Issues such as PMD, which emerged in more recent years, were also discussed. It should serve as a state-of-the-art review of SMFs as well as an introduction to other chapters on specialty fibers that follow in this book.

Acknowledgments

I gratefully acknowledge academic interactions with several of my colleagues in our Fiber Optics group and with my graduate students over the last two decades. In particular, I acknowledge Professors Ajoy Ghatak and K. Thyagarajan and graduate student Ms. Kamna Pande for their constructive criticism and help in drawing some of the figures on dispersion calculation. My colleague Dr. Ravi Varshney is thanked for his help in drawing the figure on group delay versus wavelength. I also wish to thank Dr. R. S. Vodhanel of Corning Inc.'s Research Department, Dr. P. Sansonetti of Drake/Alcatel, and Dr. D. Bayert of Alcatel's Research and Innovations Department (Marcoussis) for their help in obtaining published materials related to metro network-centric NZ-DSFs.

References

1. A.M. Glass, D.J. DiGiovanni, T.A. Straser, A.J. Stenz, R.E. Slusher, A.E. White, A. Refik Kortan, and B. Eggleton, Advances in Fiber Optics, Bell Labs Tech. J., **Jan-Mar.**, pp. 163–187 (2000).
2. F.P. Kapron, D.B. Keck, and R.D. Maurer, Radiation losses in glass optical waveguides, *App. Phys. Lett.* **17**, pp. 423–425 (1970).
3. D.N. Payne and W.A. Gambling, Zero material dispersion in optical fibers, *Electron. Lett.* **11**, pp. 176–178 (1975).
4. J.I. Yamada, S. Machida, and T. Kimura, 2 Gb/s optical transmission

experiments at 1.3 μm with 44 km single-mode fiber, *Electron. Lett.* **17,** pp. 479–480 (1981).

5. R.J. Mears, L. Reekie, I.M. Jauncy, and D.N. Payne, Low-noise erbium-doped fiber amplifier operating at 1.54 μm, *Electron. Lett.* **23**, pp. 1026–1027 (1987).
6. T. Miya, Y. Terunume, T. Hosaka, and T. Miyashita, An ultimate low-loss single-mode fiber at 1.55 mm, *Electron. Lett.* **15,** pp. 106–108 (1979).
7. S.K. Kartalopoulos, *Introduction to DWDM Technology,* SPIE Press, Bellingham, Washington and IEEE Press, Piscataway, NJ (2000).
8. F.P. Kapron, Chapter on "Transmission properties of optical fibers" in *Optoelectronic Technology and Lightwave Communication Systems,* Chinlon Lin (Ed.), Van Nostrand, New York (1989).
9. K. Kitayama, Y. Kato, M. Ohashi, Y. Ishida, and N. Uchida, Design considerations for the structural optimization of a single-mode fiber, IEEE J. Lightwave Tech. **LT-1,** pp. 363–369 (1983).
10. B.P. Pal, Chapter on "Transmission characteristics of telecommunication optical fibers" in *Fundamentals of Fiber Optics in Telecommunication and Sensor Systems,* B.P. Pal (Ed.), John Wiley, New York (1992).
11. N. Uchida, Development and future prospects of optical fiber technologies, IEICE Trans. Electron. **E85** C, pp. 868–880 (2002).
12. A. Yariv, *Optical Electronics in Modern Communication,* Oxford University Press, New York (1997).
13. A. Ghatak and K. Thyagarajan, *Introduction to Fiber Optics,* Cambridge University Press, Cambridge (1998).
14. B.P. Pal, Chapter on "Optical Transmission" in *Perspective in Optoelectronics,* S.S. Jha (Ed.), World Scientific, Singapore (1995).
15. D. Marcuse, Interdependence of waveguide and material dispersion, *App. Opt.* **18**, pp. 2930–2932 (1979).
16. D.N. Payne and W.A. Gambling, Zero material dispersion in optical fibers, *Electron. Lett.* **11**, pp. 176–178 (1975).
17. G.P. Agrawal, *Lightwave Technology: Components and Devices,* John Wiley and Sons, New Jersey (2004).
18. J. Powers, *An Introduction to Fiber Optic Systems,* 2nd ed., Irwin, Chicago (1997).
19. R.J. Boehm, Y.C. Ching, C.G. Griffith, and F.A. Saal, Standardized fiber optic transmission system—a synchronous optical network view, *IEEE J. Selected Areas in Commn.* **SAC-4,** pp. 1424–1431 (1986).
20. B.P. Pal, Chapter on "Optical fibers for lightwave communication: design issues," in *Fiber Optics and Applications,* A.K. Ghatak, B. Culshaw, V. Nagarajan and B.D. Khurana (Eds.), VIVA publishers, New Delhi (1995).
21. R. Tewari, B.P. Pal, and U.K. Das, Dispersion-shifted dual-shape-core fibers: optimization based on spot size definitions, *IEEE J. Lightwave Tech.* **10,** pp. 1–5 (1992).
22. R.J. Mears, L. Reekie, S.B. Poole, and D.N. Payne, Low-threshold tunable cw and Q-switched fiber laser operating at 1.55 μm, *Electron. Lett.* **22**, pp. 159–160 (1986).
23. C.J. Koester and E. Snitzer, Amplification in a fiber laser, *App. Opt.* **3**, pp. 1182–1184 (1964).
24. E. Desurvire, *Erbium Doped Fiber Amplifier: Principle And Applications,* Wiley-Interscience, New York (1994).
25. R.I. Laming, A.H. Gnauck, C.R. Giles, M.N. Zervas, and D.N. Payne, "High sensitivity optical preamplifier at 10 Gbit/s employing a low noise composite EDFA with 46 dB gain," in Post-deadline session, Opt. Amp. and their Applications, OSA Topical Meeting, Washington, DC, Post-deadline Paper PD13, (1992).
26. A.K. Srivastava, Y. Sun, J.W. Sulhoff, C. Wolf, M. Zirngibl, R. Monnard, A.R. Chraplyvy, A.A. Abra-mov, R.P. Espindola, T.A. Strasser, J.R. Pedrazzani, A.M. Vengsarkar, J.L. Zyskind, J. Zhou, D.A. Ferrand, P.F. Wysocki, J.B. Judkins, S.W. Granlund, and Y.P. Li, "1 Tb/s Transmission of 100 WDM 10 Gb/s Channels over 400 km of Truewave™ Fiber," Technical Digest *Optical Fiber Communication Conference* 1998, Opt. Soc. Am., Washington, DC, Post-deadline paper PD10 (1998).
27. ITU-T Recommendation G.692: *Optical Interfaces for Multichannel Systems With Optical Amplifiers,* Inter-national Telecommunications Union, Geneva, Switzerland.
28. K.M. Abe, Optical fiber designs evolve, *Lightwave* (February 1998).
29. T. Li, The impact of optical amplifiers on long-distance lightwave telecommunications, *Proc. IEEE,* **81,** pp. 1568–1579 (1995).
30. T. Li and C.R. Giles, "Optical amplifiers in lightwave telecommunications," Chapter 9, in *Perspective in Optoelectronics,* S.S. Jha (Ed.), World Scientific, Singapore (1996).
31. A.R. Chraplyvy, R.W. Tkach, and K.L. Walker, "Optical fiber for wavelength division multiplexing," U.S. Patent 5,327,516 (issued July 5,1994).
32. D.W. Peckham, A.F. Judy, and R.B. Kummer, "Reduced dispersion slope, non-zero dispersion fiber," in *Proceedings of the 24th European Conference on Optical Communication ECOC'98* (Madrid, 1998), pp. 139-140 (1998).
33. Y. Liu, W.B. Mattingly, D.K. Smith, C.E. Lacy, J.A. Cline, and E.M. De Liso, "Design and fabrication of locally dispersion-flattened large effective area fibers," in *Proceedings of the 24th European Conference on Optical Communication ECOC'98* (Madrid, 1998), pp. 37–38 (1998).
34. J. Ryan, Special report: ITU G.655 adopts higher dispersion for DWDM, *Lightwave,* **18**, no. 10 (2001).
35. Y. Danziger and D. Askegard, High-order-mode fiber an innovative approach to chromatic dispersion man agement that enables optical networking in long-haul high-speed transmission systems, *Opt. Networks Mag.* **2**, pp. 40–50 (2001).
36. Y. Frignac and S. Bigo, "Numerical optimization of residual dispersion in dispersion managed systems at 40 Gb/s," in *Proceedings of the Optical Fiber Communications Conference OFC 2000* (Baltimore, MD, 2000), pp. 48–50 (2000).
37. H. Izadpanah, C. Lin, H. Gimlett, H. Johnson, W. Way, and P. Kaiser, "Dispersion compensation for upgrading interoffice networks built with 1310 nm optimized SMFs using an equalizer fiber, EDFAs, and 1310/1550 nm WDM," Tech Digest Optical Fiber Communication Conference, Post-deadline paper PD15, pp. 371–373 (1992).
38. Y. Akasaka, R. Suguzaki, and T. Kamiya, "Dispersion Compensating Technique of 1300 nm Zero-dispersion SM Fiber to get Flat Dispersion at 1550 nm Range," in *Digest of European Conference on Optical Communication,* paper We.B.2.4 (1995).
39. A.J. Antos and D.K. Smith, Design and characterization of dispersion compensating fiber based on LP_{01} mode, *IEEE J. Lightwave Tech.* **LT–12,** pp. 1739–1745 (1994).
40. D.W. Hawtoff, G.E. Berkey, and A.J. Antos, "High Figure of Merit Dispersion Compensating Fiber," in

Technical Digest Optical Fiber Communication Conference (San Jose, CA), Post-deadline paper PD6 (1996).
41. A.M. Vengsarkar, A.E. Miller, and W.A. Reed, "Highly efficient single-mode fiber for broadband dispersion compensation," in *Proceedings of Optical Fiber Communications Conference OFC'93* (San Jose, CA), pp. 56–59 (1993).
42. L. Grüner-Nielsen, S.N. Knudsen, B. Edvold, T. Veng, B. Edvold, T. Magnussen, C.C. Larsen, and H. Daamsgard, Dispersion Compensating Fibers, *Opt. Fib. Tech.* **6**, pp. 164–180 (2000).
43. F. Ouellette, Dispersion cancellation using linearly chirped Bragg grating filters in optical waveguides, *Opt. Lett.* **12,** pp. 847–849 (1987).
44. H.G. Winful, Pulse compression in optical fibers, *Appl. Phys. Lett.* **46**, pp. 527–529 (1985).
45. B.J. Eggleton, T. Stephens, P.A. Krug, G. Dhosi, Z. Brodzeli, and F. Ouellette, "Dispersion compensation over 100 km at 10 Gb/s using a fiber grating in transmission," in *Technical Digest Optical Fiber Communication Conference* (San Jose, CA), Post-deadline paper PD5 (1996).
46. B.P. Pal, "All-fiber guided wave components," in *Electromagnetic Fields in Unconventional Structures and Materials,* O.N. Singh and A. Lakhtakia (Eds.), John Wiley, New York, pp. 359–432 (2000).
47. C.D. Poole, J.M. Weisenfeld, and D.J. Giovanni, Elliptical-core dual-mode fiber dispersion compensator, *IEEE Photon. Tech. Lett.* **5**, pp. 194–197 (1993).
48. S. Ramachandran, B. Mikkelsen, L.C. Cowsar, M.F. Yan, G. Raybon, L. Boivin, M. Fishteyn, W.A. Reed, P. Wisk, D. Brownlow, R.G. Huff, and L. Gruner-Nielsen, All-fiber, grating-based, higher-order-mode dispersion compensator for broadband compensation and 1000-km transmission at 40 Gb/s, *IEEE Photon. Tech. Lett.* **13**, pp. 632–634 (2001).
49. S. Ramachandran, S. Ghalmi, S. Chandrasekhar, I. Ryazansky, M. Yan, F. Dimarcello, W.A. Reed, and P. Wisk, Wavelength-continuous broadband adjustable dispersion compensator using higher order mode fibers and switchable fiber-gratings, *IEEE Photon. Tech. Lett.* **15**, pp. 727–729 (2003).
50. M. Onishi, Y. Koyana, M. Shigematsu, H. Kanamori, and M. Nishimura, Dispersion compensating fiber with a high figure of merit of 250 ps/nm-dB, *Electron. Letts.* **30**, pp. 161–163 (1994).
51. B. Jopson and A. Gnauck, Dispersion compensation for optical fiber systems, *IEEE Comm. Mag.* **June,** pp. 96–102 (1995).
52. V. Srikant, "Broadband dispersion and dispersion slope compensation in high bit rate and ultra long haul systems," in *Technical Digest Optical Fiber Communication Conference,* Anaheim, CA, paper TuH1 (2001).
53. K. Thyagarajan, R.K. Varshney, P. Palai, A.K. Ghatak, and I.C. Goyal, A novel design of a dispersion compensating fiber, *IEEE Photon. Tech. Letts.* **8**, pp. 1510–1512 (1994).
54. J.L. Auguste, R. Jindal, J.M. Blondy, M. Clapeau, J. Marcou, B. Dussardier, G. Monnom, D.B. Ostrowsky, B.P. Pal, and K. Thyagarajan, −1800 ps/km-nm chromatic dispersion at 1.55 mm in a dual concentric core fiber, *Electron. Lett.* **36**, pp. 1689–1691 (2000).
55. K. Thyagarajan and C. Kakkar, Fiber design for broadband, gain flattened Raman fiber amplifier, *IEEE Photon. Tech. Letts.* **15,** pp. 1701–1703 (2003).
56. P. Palai, R.K. Varshney, and K. Thyagarajan, A dispersion flattening dispersion compensating fiber design for broadband dispersion compensation, *Fib. Int. Opt.* **20,** pp. 21–27 (2001).
57. B.P. Pal and K. Pande, Optimization of a dual-core dispersion slope compensating fiber for DWDM transmission in the 1480-1610 nm band through G.652 single-mode fibers, *Opt. Comm.* **201,** pp. 335–344 (2002).
58. K. Pande and B.P. Pal, Design optimization of a dual-core dispersion compensating fiber with high figure of merit and a large effective area for dense wavelength division multiplexed transmission through standard G.655 fibers, *Appl. Opt.* **42**, pp. 3785–3791 (2003).
59. I.C. Goyal, R.K. Varshney, B.P. Pal, and A.K. Ghatak, "Design of a small residual dispersion fiber along with a matching DCF," U.S. Patent 6,650,812 B2 (2003).
60. K. Mukasa, Y. Akasaka, Y. Suzuki, and T. Kamiya, "Novel network fiber to manage dispersion at 1.55 μm with combination of 1.3 μm zero dispersion single-mode fiber," in *Proceedings of the 23rd European Conf on Opt. Commn. (ECOC)*, Edinburgh, Session MO3C, pp. 127–130 (1997).
61. K. Mukasa and T. Yagi, "Dispersion flat and low non linear optical link with new type of reverse dispersion fiber (RDF-60), in *Tech. Dig. Opt. Fib. Comm. Conference (OFC'2001),* Anaheim, CA, Paper TuH7 (2001).
62. T. Grüner-Nielsen, S.N. Knudsen, B. Edvold, P. Kristensen, T. Veng, and D. Magnussen, Dispersion compensating fibers and perspectives for future developments," in *Proceedings of the 26th European Conference Opt. Comm. (ECOC'200),* pp. 91–94 (2000).
63. Y. Miyamoto, K. Yonenaga, S. Kuwahara, M. Tomi-zawa, A. Hirano, H. Toba, K. Murata, Y. Tada, Y. Umeda, and H. Miyazawa, "1.2 Tbit/s (30x42.7 Gb/s ETDM optical channel) WDM transmission over 376 km with 125-km spacing using forward error correction and carrier suppressed RZ format," in *Tech. Dig. Opt. Fib. Comm. Conference (OFC'2000),* Baltimore, MD, Post-deadline paper, PD26 (2000).
64. L.-A. de Montmorrilon, F. Beamont, M. Gorlier, P. Nouchi, L. Fleury, P. Sillard, V. Salles, T. Sauzeau, C. Labatut, J.P. Meress, B. Dany, and O. Leclere, "Optimized Teralight™/ reverse Teralight© dispersion-managed link for 40 Gbit/s dense WDM ultra-long haul transmission systems," in *Proceedings of the 27th Europ. Conf. Opt. Comm. (ECOC'2001),* Amsterdam, pp. 464–465 (2001).
65. J. Hecht, *Understanding Fiber Optics,* 4th ed., Prentice Hall, Upper Saddle River, NJ (2002).
66. J. Ryan, Fiber considerations for metropolitan net works, *Alcatel Telecom. Review* **1**, pp. 52–56 (2002).
67. D. Culverhouse, A. Kruse, C-C. Wang, K. Ennser, and R. Vodhanel, "Corning® MetroCore™ fiber and its applications in metropolitan networks," Corning Inc. White Paper WP5078 (2000).
68. ProForum Tutorial on Polarization mode dispersion: *http://www.iec.org*
69. F. Kapron A. Dori, J. Peters, and H. Knehler, "Polarization-mode dispersion: should you be concerned?" in *Proceedings of the Nat. Fiber Optic Eng. Conf. (NFOEC'96),* pp. 756–758 (1997).
70. J.J. Refi, Optical fibers for optical networking, Bell Labs. Tech. J., **January-March,** pp. 246–261 (1999).
71. C.D. Poole and J. Nagel, "Polarization effects in light wave systems," in

Optical Fiber Telecommunications, Vol. IIIA, I.P. Kaminow and T.L. Koch (Eds.), Academic Press, San Diego, CA, pp. 114–161 (1997).

72. F.P. Kapron, "System considerations for polarization mode dispersion," in *Proceedings of the Nat. Fiber Optic Eng. Conf. (NFOEC'97)*, San Diego, CA, pp. 433–444 (1997).
73. A.C. Hart, R.G. Huff, and K.L. Walker, "Method of making a fiber having low polarization mode dispersion due to a permanent spin," U.S. Patent 5,298,047 (1994).
74. A.F. Judy, "Improved PMD stability in optical fibers and cables," in *Proceedings of the 43rd Intl. Wire and Cable Symposium*, Atlanta, GA, Nov. 14–17 (1994).

Chapter 3.2

3.2

Polymer optical fibers

Pal

3.2.1 Introduction

Glass optical fiber has received intense research and development over the past 30 years, forming the backbone of optical communication and sensing. More recently, this fiber has found applications in medical diagnosis and surgery. New types of glass optical fiber have also been developed, notably the photonic crystal fiber or holey fiber [1]. On the other hand, polymer optical fiber (POF) was developed more than 20 years ago by DuPont with poly-methal-methacrylate (PMMA) as the constitutent material. However, its attenuation was very large (>200 dB/km) compared with that of glass optical fiber. Hence, it found applications only in systems involving the use of short lengths of fiber, such as back plane interconnection and decorations. In the past 10 years, however, POF has received renewed attention mainly because of its applications in motor vehicles and home networks and also because of the development of low-loss POF. The purpose of this article is to give a brief review of the latest development of POF and its various applications.

3.2.2 Types of POFs

The most important parameter of an optical fiber for applications in communication is its attenuation. Whereas silica optical fiber has attenuation less than 0.2 dB/km at 1550 nm, polymer fiber has a much larger figure, making it suitable only for short distance application. From the attenuation point of view, there are three types of POFs that receive popular attention: PMMA fiber, deuterated PMMA fiber, and perfluorinated POF.

3.2.2.1 PMMA fiber

The most popular POF is made of PMMA material, whose chemical composition is shown in Fig. 3.2.1. The molecules in the PMMA chain oscillate with a resonant absorption wavelength between 3.3 and 3.5 μm. Its harmonics extend into the visible range of light so that the theoretical minimum attenuation of this fiber is 100vdB/km at 0.65 μm. However, the practical figure is about 200 dB/km. Thus, this fiber is used for distances of over several hundred meters. The bandwidth of the fiber depends on its refractive index profile, and multimode fiber is often used. The glass transition temperature is about 110° C.

3.2.2.2 Deuterated PMMA POF

The large attenuation in the PMMA POF occurs because the resonant wavelength of the carbon–hydrogen bond is too close to the visible wavelength of light. Hence, one way to reduce the attenuation is to somehow remove the resonant wavelength to a larger value. This has been done by replacing the hydrogen atom in the molecular chain with a heavy hydrogen atom, that is, one with two

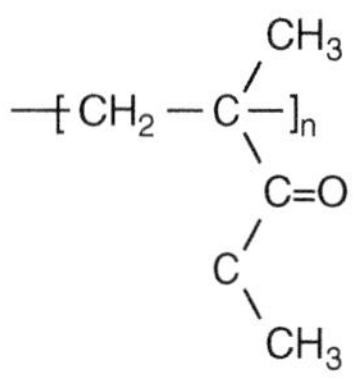

Fig. 3.2.1 Chemical composition of PMMA.

Guided Wave Optical Components and Devices; ISBN: 9780120884810

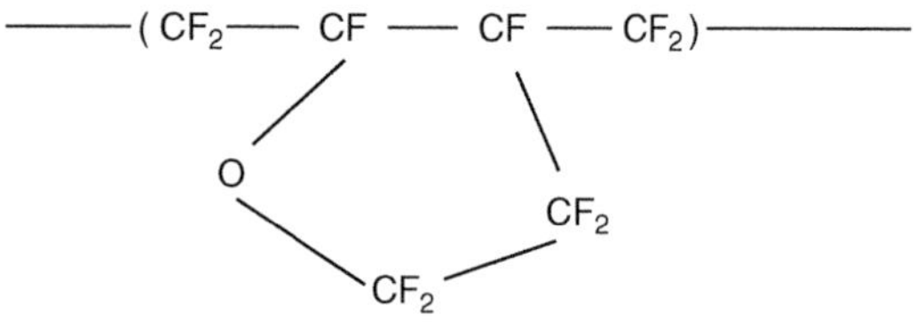

Fig. 3.2.2 Chemical composition of perfluorinated POF.

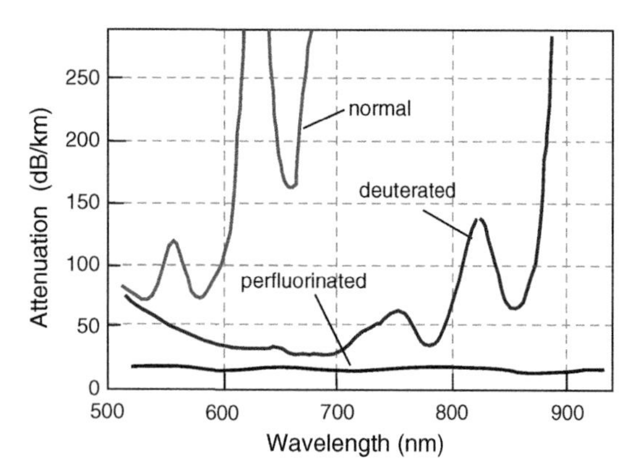

Fig. 3.2.3 Attenuation spectra of POFs.

protons in the nucleus instead of one as in the ordinary hydrogen atom. The resultant resonant wavelength then appears at 4.5 μm, and its minimum attenuation can be as low as 25 dB/km at 650 μm. However, this fiber tends to absorb water easily. In doing so, the attenuation rapidly increases. This fiber therefore is not widely used.

3.2.2.3 Perfluorinated POF

To reduce the loss further, it is necessary to remove the hydrogen atom completely from the polymer and replace it by a heavy atom. This has been achieved by means of introducing fluorine atoms into the polymer; the fiber thus fabricated is called perfluorinated POF [2]. Its chemical composition is shown in Fig. 3.2.2.

The resonant absorption wavelength of this fiber extends from 7.7 μm to 10 μm. Because this is very far from the visible range, its attenuation is correspondingly small. It has been shown [3] that its theoretical value at 1300 nm is 0.5 dB/km and at 1550 nm is 0.3 dB/km. These figures happen to coincide with those of silica fiber. The best perfluorinated POF made so far has an attenuation figure of 10 dB/km. This is a great improvement over the PMMA fiber. There are other attractive features of this fiber that are explained in the following sections.

Fig. 3.2.3 shows the attenuation spectra of the three types of POF. It is of interest to note that the perfluorinated POF has a nearly flat and low attenuation over a wide range of wavelengths.

3.2.3 Manufacture of POFs

We concentrate our discussion on the manufacture of PMMA POF. There are two methods of fabricating POF: the preform and drawing method and the extrusion method. In the extrusion method, the starting material is PMMA in powder or in pebble form that can be purchased directly from polymer companies. In the preform and drawing method, the starting material is the MMA monomer. We discuss this method first [4].

3.2.3.1 Preform and drawing method

3.2.3.1.1 Material preparation

The MMA monomer can be bought from a commercial company. Some chemicals have been added to inhibit polymerization taking place during storage and transport. Therefore, the first step in preparing the material is to remove these inhibitors. The common inhibitors are hindered phenols such as butylated hydroxy toluene. The inhibitor stops polymerization by reacting with free radicals in the monomer, as shown in Fig. 3.2.4. The inhibitor can be removed by distillation, extraction, or chromatography. Alternatively, one can simply use extra initiator to overwhelm the inhibitor.

The monomer is then degassed to remove all the air bubbles. It is also necessary to get rid of the impurities in the monomer to avoid excessive scattering loss of the

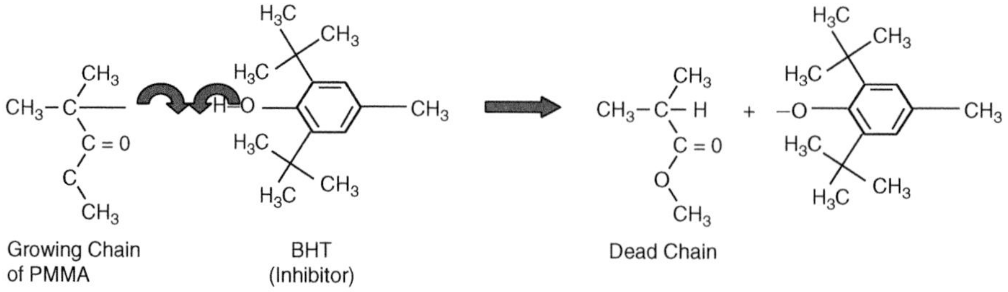

Growing Chain of PMMA — BHT (Inhibitor) — Dead Chain

Fig. 3.2.4 Function of inhibitor. BHT, butylated hydroxy toluene.

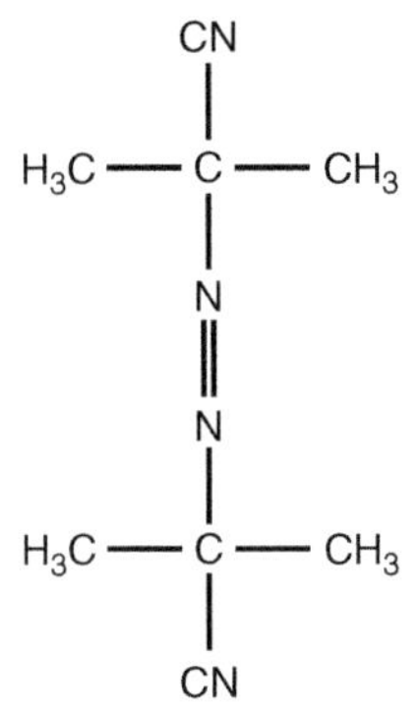

CN

H_3C — C — CH_3

N

N

H_3C — C — CH_3

CN

Fig. 3.2.5 Chemical composition of the initiator, AIBN.

resultant fiber. To start the polymerization process, we need to add to the MMA monomer with an initiator such as lauryl peroxide or AIBN (Azobisisobutyronitrile), the chemical composition of which is given in Fig. 3.2.5. In addition, we need to put in a chain transfer agent to adjust the molecular weight of the polymer. A chain transfer is a kind of termination reaction in that the growing chain radical reacts with the agent that has an even number of electrons. The growing chain is terminated, but another radical is formed. That new radical may or may not reinitiate polymerization, depending on its structure. Various compounds can act as chain transfer agents, but the most common ones are structures with an S—H bond known as mercaptans, the chemical composition of which is shown in Fig. 3.2.6, or thiols.

The growing polymer chain end radical abstracts the hydrogen atom from the mercaptan along with one of the two electrons in the S—H bond of the mercaptan. This terminates the growing chain with a hydrogen atom. One electron from the former S—H bond remains behind, creating a new radical on sulfur. This new radical can act as an initiator, reacting with additional monomer to grow a new chain. In this way, the deliberately added chain

Initiator + chain teragentransf + [—CH₂—C(CH₃)(C(=O)O—CH₃)—]$_n$ → 70° C

Monomer

Initiator + chain transfer agent — [—CH₂—C(CH₃)(C(=O)O—CH₃)—]$_n$

Polymer

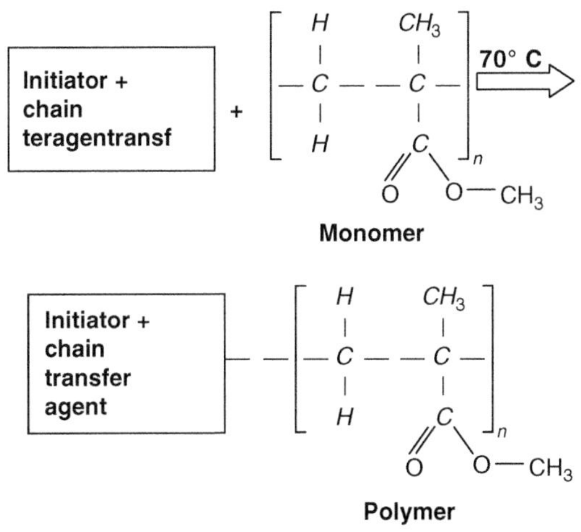

Fig. 3.2.6 Chemical composition of mercaptan.

transfer agent terminates the growing polymer chain sooner than would be the case without the agent. This reduces the molecular weight.

The MMA monomer has a refractive index of about 1.402. This is used as the cladding material of the fiber. However, for the core material, we need to raise its refractive index. This is achieved by adding an appropriate amount of dopant. The index increase is linearly proportional to the amount of dopant. There is a variety of dopants that can be used, such as benzyl methAcrylate, bromobenzen, benzyl butyl phthalate, benzyl benzoate, or diphenyl phthalate. Similarly, to reduce the index instead, one has to simply add trifluro ethyl metha-crylate.

3.2.3.1.2 Polymerization

We can summarize the polymerization process by the following chemical equation:

We start to polymerize the cladding of the preform first. The MMA monomer together with the initiator and transfer agent is poured into a glass tube that is rotated at high speed (3000 rpm) in an oven set at a temperature of 70° C. The high-speed rotation is needed to create an axial space for the core monomer to be introduced after the cladding is polymerized. The polymerization takes 1 or 2 days.

The next step after the cladding is polymerized is to start the polymerization of the core by pouring index raising dopant, initiator, and chain transfer agent into the space with prepared monomer material, that is, MMA monomer. The filled tube is again placed in an oven at 95° C and rotated at 50 rpm for 24 h. The whole preform is then polymerized and is ready for drawing into fiber. Fig. 3.2.7 shows a sample of the preform.

To fabricate a graded index preform, we rely on the interfacial gel effect [5] in which the polymerization begins from the core–cladding boundary. We also assume that the dopant molecules are larger than the MMA molecules so that the latter molecules can diffuse toward the boundary and become polymerized first. In this way, the dopant concentration gets larger toward the center.

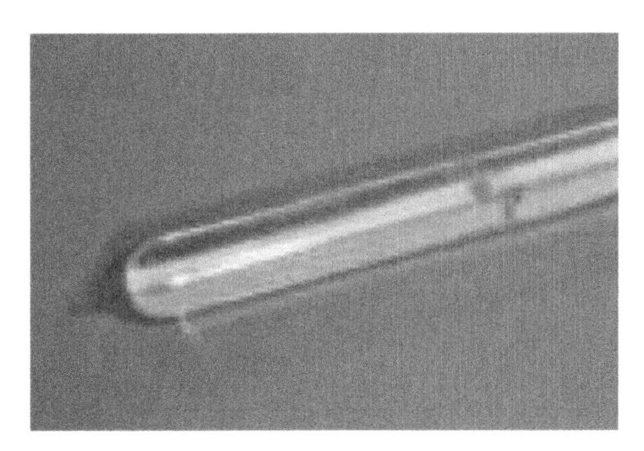

Fig. 3.2.7 POF preform.

Thus, a graded index preform is created. The exact profile is much more difficult to achieve than a silica preform by chemical vapor deposition.

3.2.3.1.3 Fiber drawing

The drawability of a POF preform is very sensitive to its glass transition temperature, which is about 110° C, and its average molecular weight, which should be less than 80,000. In the drawing process, the tip of the preform is slowly fed into the furnace under computer control at a temperature between 28 and 290° C, as shown in Fig. 3.2.8. The preform method allows us to manufacture a variety of POFs, for example, single mode and multimode fiber, twin-core fiber, dye-doped fiber, inorganic-doped fiber (e.g., erbium), and electrooptic fiber. The disadvantage of the preform method is that only a finite length of the preform is made each time, which means that a limited amount of fiber is produced. This is not commercially attractive.

3.2.3.2 Extrusion method

With this method, the starting material is not monomer but polymer in the form of powder or pebbles that can be bought directly from commercial suppliers. Fig. 3.2.9 shows the extrusion setup [6,7]. The core powder is introduced through feeder 1 while the cladding powder is fed through feeder 2. Both powders are pushed toward the output die at the exit end of the diffusion zone by a set of feed screws. There is a temperature gradient throughout the diffusion zone, reaching a melt temperature of about 280° C. This serves to help the creation of graded-index fiber. The extrusion rate varies between 93 and 245 g/hr. The length of the diffusion zone is about 6.5 cm. The die has two concentric nostrils serving to create the core and cladding of the fiber. The diameters of the nostrils determine the dimensions of the fiber core and cladding.

The advantage of the extrusion method is its ability to produce a very long length of fiber, which is commercially attractive. However, the fiber purity cannot be controlled but depends on the purity of the starting material.

3.2.4 Comparison between silica fiber and polymer fiber

3.2.4.1 Difference in diameters

The standard multimode silica fiber for communication has a core diameter of 62.5 μm and an overall fiber diameter of 125 μm. In the case of polymer fiber, the standard core diameter is 900 μm and the overall fiber diameter, 1000 μm. Thus, the cross-section of polymer fiber is much larger than that of silica fiber. Consequently, it is easier to join two polymer fibers with low loss than the silica fiber. This is significant because it leads to more economical installation of optical fiber systems.

The recent low-loss POF development [8] leads to a smaller core diameter in the order of 120 μm and a cladding diameter of 250 μm. The reason for such a reduction is the cost of the polymer, especially the perfluorinated polymer. However, such a reduction removes the advantage of the ease of joining. To keep the advantage, the fiber is jacketed with a sleeve so that the overall diameter is still about 1 mm. The multilayer structure of the fiber may introduce an undesirable effect such as the nonconcentricity of the core, leading to excess joint loss [9].

3.2.4.2 Minimum bend radius

Polymer fiber has a much smaller Young's modulus than silica, about 30 times less. It is therefore easier to negotiate bends without breakage. Hence, the minimum bend

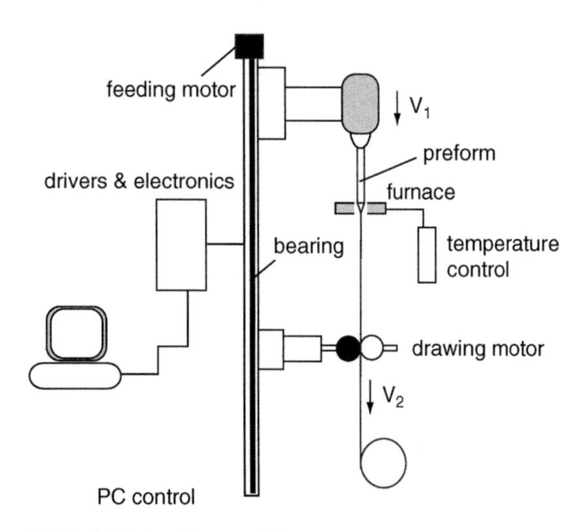

Fig. 3.2.8 POF drawing machine.

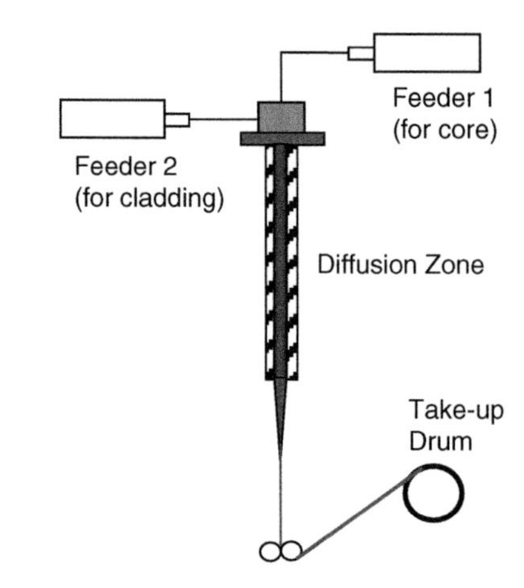

Fig. 3.2.9 Extrusion of POF.

radius of a polymer fiber is smaller than that of silica fiber for the same diameter. For example, a 125 μm overall diameter polymer fiber has a minimum bend radius of 1 cm compared with 3 cm for silica fiber. On the other hand, a 1 mm diameter polymer fiber has a minimum bend radius of 5 cm compared with 22 cm for silica fiber of the same diameter. It is noted that the bend radius increases linearly with the overall fiber diameter.

The ability to negotiate a sharp corner finds applications in very confined spaces, such as inside an automobile or a cramped compartment. This is the main reason polymer fiber is being increasingly used in automobiles [10].

3.2.4.3 Numerical aperture

The numerical aperture of an optical fiber is determined by the relative magnitudes of the refractive indices in the core and in the cladding. In the case of silica fiber, the difference between these indices cannot be too large, otherwise the residual thermal stresses in these regions will be so large the fiber will fracture. It is rare to find silica fiber with a numerical aperture larger than 0.3. However, in the case of polymer fiber, the stress build-up is less severe because of the small Young's modulus. In fact, the fiber does not fracture. It simply yields gradually if the strain is too great. Hence, it is common to find these fibers with a numerical aperture as large as 0.7 or more.

Because the optical power carrying capacity of the fiber is linearly proportional to the square of its numerical aperture, it is desirable to have a fiber with a large numerical aperture for applications in local area networks, especially in a bus configuration in which computers hang down from the fiber and drain some parts of the light.

3.2.4.4 Fiber bandwidth

Fig. 3.2.10 shows the comparison of bandwidths of the silica multimode fiber and polymer multimode fiber. Both fibers have a parabolic index profile. It can be seen that the polymer fiber has a larger bandwidth. The reason is that its material dispersion is less than that of silica fiber, as shown in Fig. 3.2.11 [11].

Fig. 3.2.11 shows that the material dispersions of all different kinds of polymer fiber approach the horizontal axis asymptotically, whereas the silica fiber crosses the axis at a wavelength near 1.3 mm and continues to increase nearly linearly, so much so that at 1.55 μm it reaches a value of 17 ps/nmkm. This shows that polymer fiber basically provides more bandwidth than silica fiber.

3.2.5 Applications of POFs

POF finds applications in situations where short length of fiber is required. Broadly, it can be divided into (1) short distance communication, (2) sensors, and (3) illumination.

3.2.5.1 Communication

PMMA fiber has an attenuation figure of about 100 dB/km. Its useful length would therefore be less than 100 m. For multimode step index fiber, its bandwidth is on the order of 10 Mb/skm. Hence, this fiber would be useful for communication systems less than 100 m with a bandwidth 100 Mb/s. It finds applications in automobile optical networks and back plane data communication. However, the low-loss perfluorinated fiber can be used for distances up to 1 km because its attenuation is now reduced to 10 dB/km. Furthermore, its bandwidth

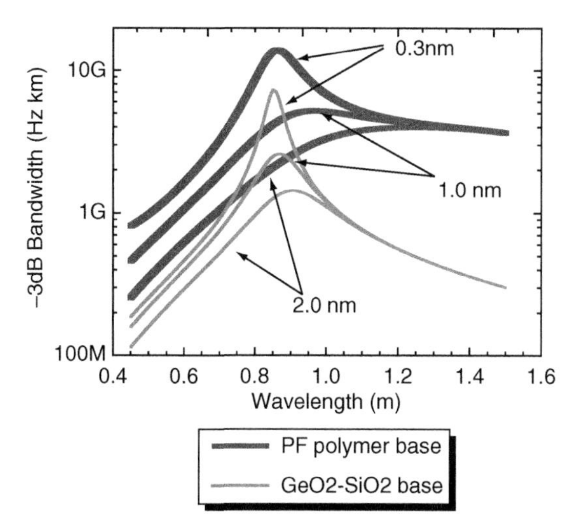

Fig. 3.2.10 Bandwidth of graded-index multimode fibers.

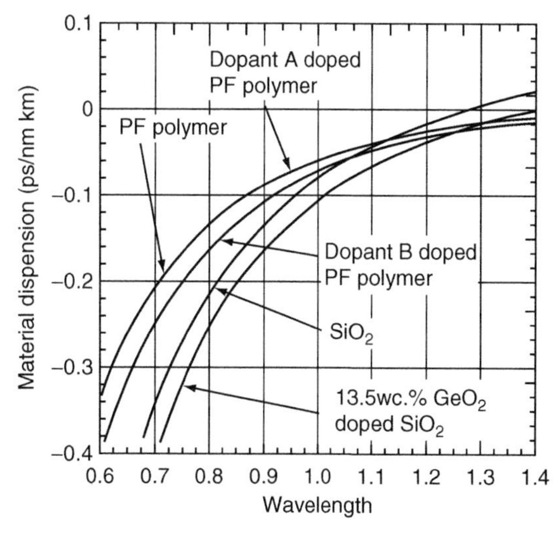

Fig. 3.2.11 Fiber material dispersion.

can reach 2.5 Gb/s. Hence, this fiber is useful for office networks and home networks.

3.2.5.1.1 Automobile optical networks

Modern motor vehicles are equipped with many communication devices, as shown in Fig. 3.2.12. If these are wired together by copper wires, there will be severe interference problems. Some of the European automobile manufacturers have come together to develop a standard communication protocol for motor cars, called MOST (media-oriented system transport) [12]. It separates the communication devices given in Fig. 3.2.12 into different networks in accordance with the bandwidth required as shown in Fig. 3.2.13.

Fig. 3.2.14 shows the network configurations for different groups of devices. Each network must satisfy the ISO/OSI seven-layer model. The total length of fiber used in the car is less than 30 m.

Fig. 3.2.15 shows the locations of different networks within the automobile.

3.2.5.1.2 Back plane interconnect

There two types of back plane interconnections: intrasystem and intersystem. In the first system, it is the connection by POF from one subsystem to another within a large system. In this case, because the system is self-contained, standards for interconnection are not needed. The fiber usually replaces the massive copper wires at the back plane as shown in Fig. 3.2.16. The fiber length is usually less than 30 m. The bit rate required at present varies from 155 to 622 Mb/s, but in the near future this will upgrade to more than 2.5 Gb/s.

In the intersystem interconnection, systems from the same vendor or different vendors are connected together. The length of fiber may be as long as 300 m. It is

Fig. 3.2.12 Different communication devices inside a modern car.

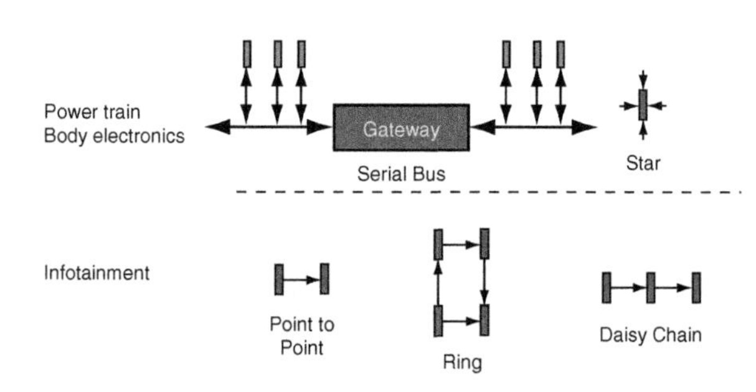

Fig. 3.2.14 Network topologies for automobiles.

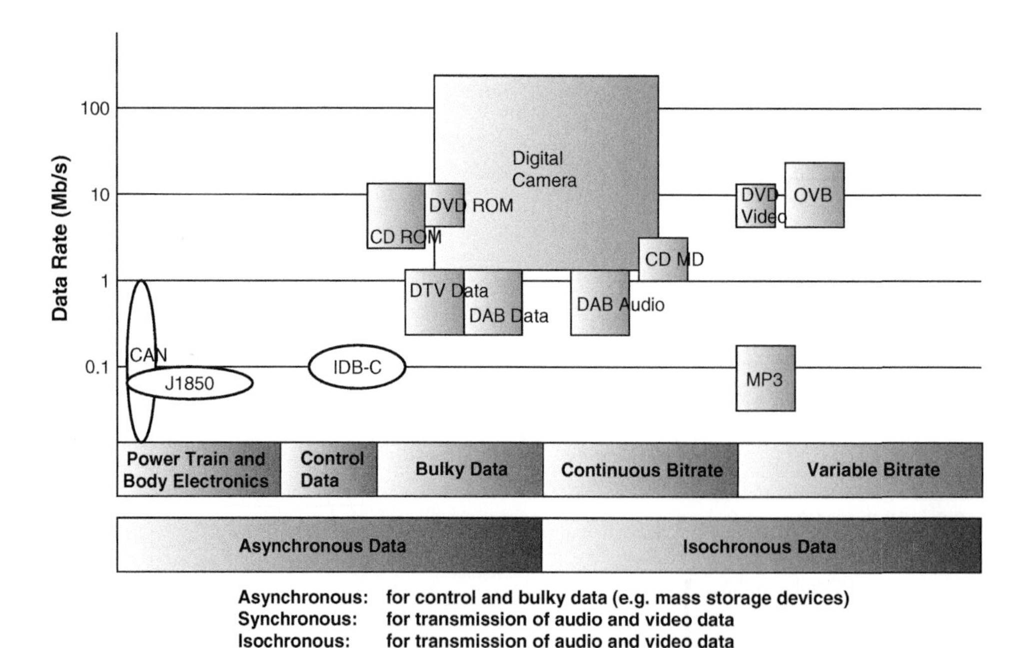

Fig. 3.2.13 Data type and data rate of automobile networks.

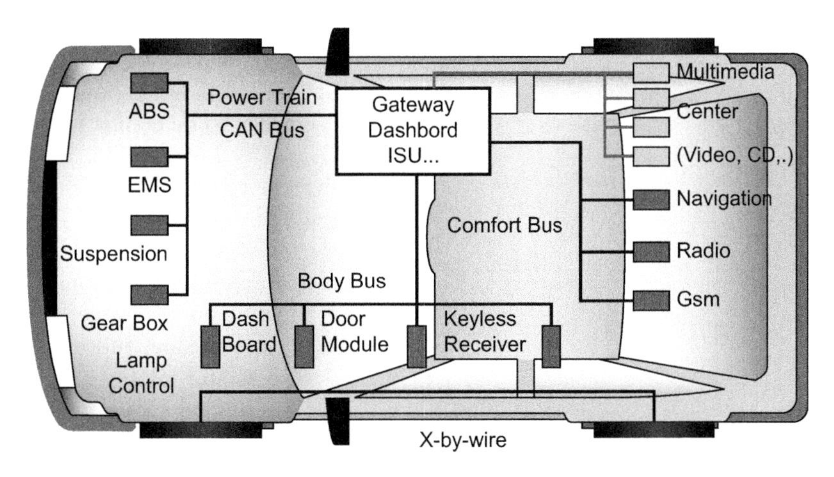

Fig. 3.2.15 POF network within the automobile.

(a) (b)

Fig. 3.2.16 (a) UTP back plane. (b) POF back plane.

important for the vendors to follow international standards, and the bit rate may reach 10 Gb/s with a bit error rate (BER) between 10^{-9} and 10^{-12}. A typical example of an intersystem connection is in the storage access network (SAN), as shown in Fig. 3.2.17.

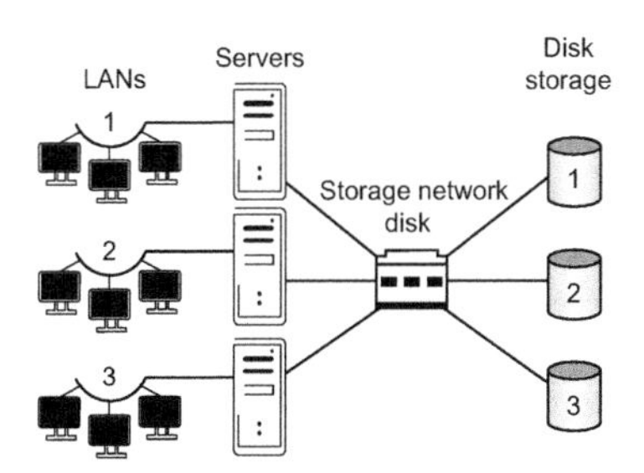

Fig. 3.2.17 Back plane interconnection in SAN.

3.2.5.1.3 Office network

The penetration of POF into office network faces challenges from UTP (untwisted pair) Cat5e and Cat6 cables and from multimode silica fiber. Cat5e can have a bandwidth of 100 Mb/s over a distance of 100 m, and Cat6 has a bandwidth of 600 Mb/s over the same distance. They are extensively used at present in most office networks. Hence, it is unlikely that POF will replace UTP for network distances less than 100 m. In fact, the POF to be used here must be graded to provide for sufficient bandwidth. The disadvantage of a UTP is that it is affected by electromagnetic interference.

Future office networks will run at a bit rate greater than 2 Gb/s. UTP will not be able to compete in this area. However, silica multimode fiber may be a strong competitor. It certainly has an advantage over POF in fiber attenuation and hence the distance used. POF still has some other advantages over silica fiber, such as its ease of connection and ability to negotiate a small bend. Finally, silica fiber is more brittle and its end may fracture

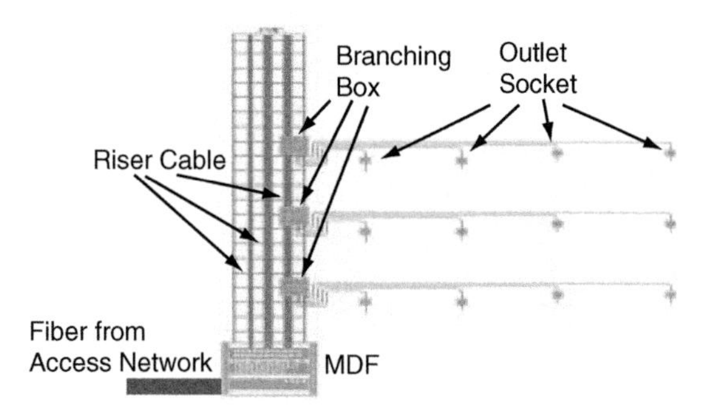

Fig. 3.2.18 Office network.

during installation, posing a danger to the installer in that the fractured glass pieces may get into the fingers.

Fig. 3.2.18 shows the wiring of UTP, silica fiber, or POF within a building leading to the office. The signal from outside the building arrives at the main distribution frame located at the basement through the access network. This signal is usually optical and, if UTP is still used inside the building, it has to be converted into an electrical signal within the main distribution frame (MDF) before it is carried upward to different floors by the riser coaxial cable. However, if the optical fiber is already installed in the office (FTTO), the MDF does not have to do any conversion but it does have to distribute the optical signal to different riser fibers for different floors. The branching box on each floor then distributes the signal to different floor fibers which are eventually terminated at some wall socket outlets.

3.2.5.1.4 Home network

The structure of the home network is similar to the office network. The MDF located at the outside wall of the house receives signals from a variety of media such as Internet, cable television, and satellite Internet and distributes them to different rooms inside the house via coaxial cables or optical fibers (silica or polymer) terminating at wall outlets as shown in Fig. 3.2.19.

The devices to be connected within the home can be summarized as shown in Fig. 3.2.20. The residential gate is equivalent to the MDF. There are two types of networks used to connect these devices depending on the quality of service required. There are some services that require guaranteed connection and transmission without loss of packets. For example, videophone, high-density television transmission, and indispensable sensors monitoring patients and infants. These services require high bit rate transmission. There are other services that do not required guaranteed service but the best of effort will do. In this category, we include voice over IP telephone, home electric appliances, sensors with redundancy, that is, more than one sensor to monitor a single event and conventional personal computers. These services require low bit rate transmission. Hence, there are two kinds of home networks: the wavelength division multiplexing (WDM)-based network which guarantees the required quality of service, and the token-based network, which offers best of effort service. Fig. 3.2.21 shows their bandwidth requirements.

For WDM networks, the coarse WDM will do because the maximum bandwidth required would not exceed a few Gb/s. It has been estimated that a total of eight wavelength channels is sufficient. However, because multimode fibers are used within the home, it is now necessary to develop CWDM over these fibers, which is a relatively new technology. On the other hand, the token-based network needs to transmit at low bit rates. The well-developed OTDM will be sufficient. The important thing to take into account is the intermodal dispersion of the fiber.

Fiber to the home is not yet popular simply because of the cost of the optical system and the need for a very

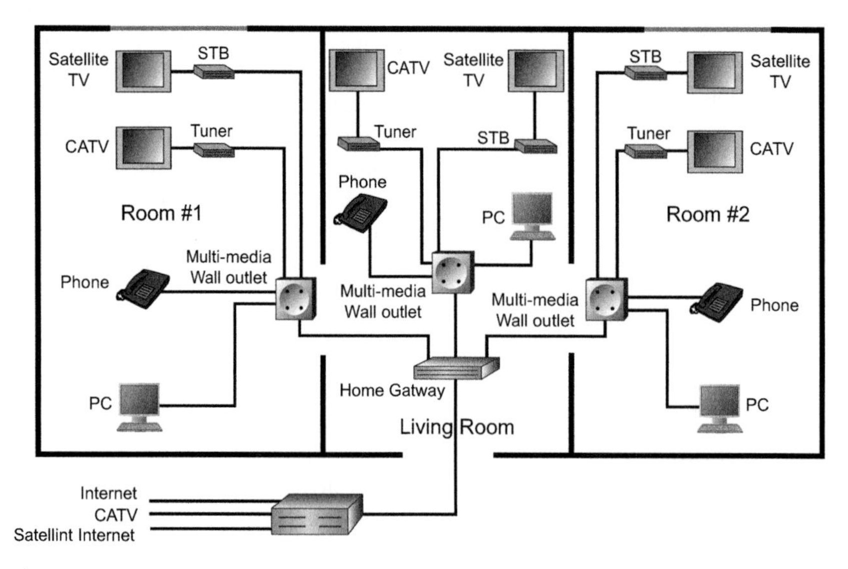

Fig. 3.2.19 Home network.

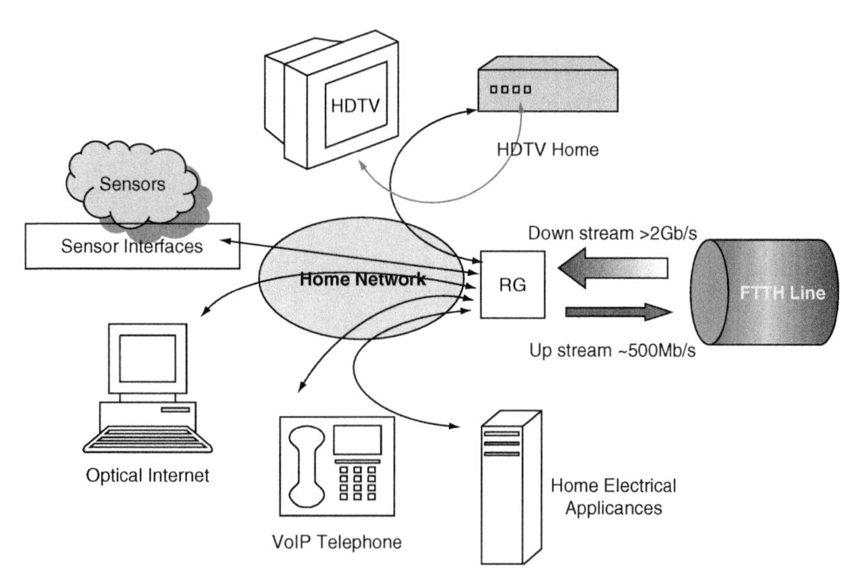

Fig. 3.2.20 Home network applications. FTTH,fiber to the home; HDTV, high-density television; RG, residential gate; VoIP, voice over IP.

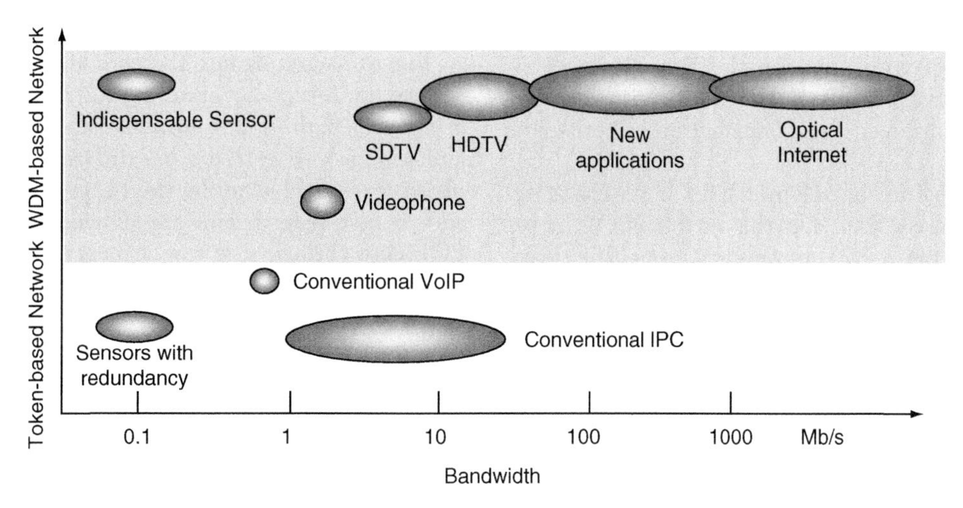

Fig. 3.2.21 Bandwidth requirements in home network. HDTV, high-density television; IPC, IP communication; SDTV, standard definition television; VoIP, voice over IP.

large bandwidth. Although optical fiber (either silica or polymer) is as cheap or even cheaper than copper cable, there is an extra cost of the transceivers at both ends of the fiber. Unless this cost is reduced to a very insignificant level, fiber to the home is unlikely to be realized. The existing copper pair joining our plain old telephone is still useful even for high bandwidth transmission if it is configured in asymmetric digital subscriber line (ADSL) or its associated format. However, it is noted that at the point of writing this chapter, a company in Germany had just produced a pair of transceivers at a market value of U.S.$5.00. This is certainly a significant step toward putting fiber into the home. In this case, polymer fiber may be a preferred candidate instead of silica fiber for the reasons mentioned above. Another competitor is wireless communication, but it is unlikely to compete strongly.

3.2.5.2 Illumination

A new application of POF is in the illumination and signage areas. Instead of making the fiber to deliver light from one end to the other for transmission of information, it is possible to design the fiber so that it leaks light uniformly from the side throughout the whole length of the fiber. This can be achieved by inserting a thin inner cladding with a refractive index in between those of core and outer cladding, as shown in Fig. 3.2.22. Fibers for illumination normally have larger diameters than communication fiber.

For the fiber illustrated in Fig. 3.2.22, the core refractive index is 1.522, for inner cladding 1.343, and for outer cladding 1.484. In the core, some insoluble particles are introduced so that light traveling in it is scattered. Those with scattering angles larger than the critical

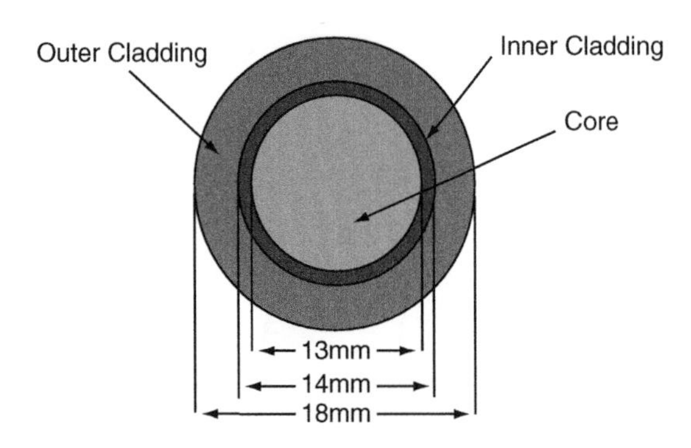

Fig. 3.2.22 Side emitting POF for illumination.

angle defined by the core and the inner cladding indices leave the core and eventually leave the fiber sideways. It is obvious that the inner cladding is used to control the rate of leakage of light. The difficulty is in ensuring a uniform leakage throughout the fiber because the intensity of the light within the core necessarily decays along its length unless the concentration of the scattering particles increases with fiber length. This obviously is difficult to achieve. However, the uniformity of light leakage may be improved if it is launched from both ends simultaneously.

The unique feature of polymer fiber is its elasticity compared with glass fiber, and this enables POF to be used for signage. Fig. 3.2.23 shows such a sign. The three letters are formed by bending a single length of POF and powered by one halogen lamp of 150 W. It is conceivable that POF will one day replace neon signs.

3.2.6 Polymer fiber gratings

Polymer is a very photosensitive material. Fig. 3.2.24 shows its change in refractive index when it is irradiated by an ultraviolet (UV) light of 488 nm at 1.84 mW [13].

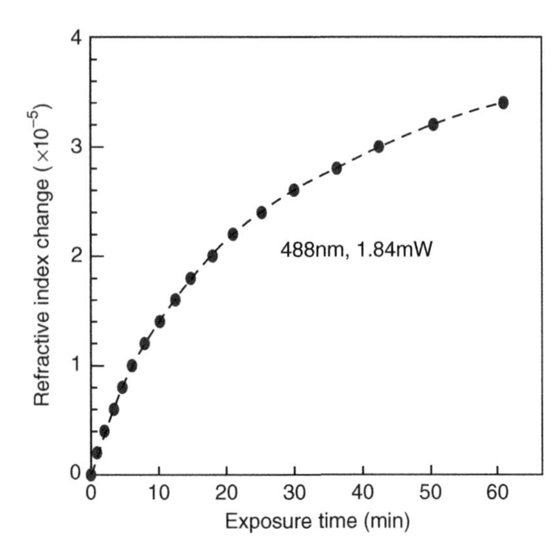

Fig. 3.2.24 Photosensitivity of PMMA fiber.

It is therefore possible to create Bragg gratings in a single-mode POF.

To write a grating into the core of a single-mode POF, we use the setup shown in Fig. 3.2.25 [14]. A UV beam at a wavelength of 355 nm irradiates the POF through a phase mask. The zero-order diffraction beam from the phase mask is blocked by the beam dumper, and the + and − first-order beams are allowed to interfere at the POF core through the ring arrangement of the prisms. The photosensitivity of the polymer creates a grating in the core. A broadband light source (amplified spontaneous emission; ASE) is connected to the POF via a coupler and a length of silica single mode fiber. The build-up of the grating during the irradiation is observed through the optical spectrum analyzer.

Fig. 3.2.26 shows that the index modulation of the core increases linearly with time until it reaches 62 min. This is called type 1 grating. After that, the modulation increases with a steep slope until 87 min. This is called

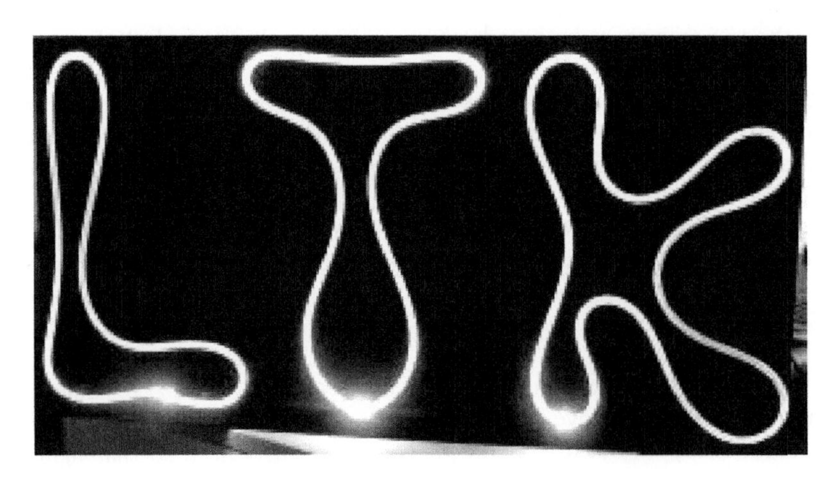

Fig. 3.2.23 POF for signage.

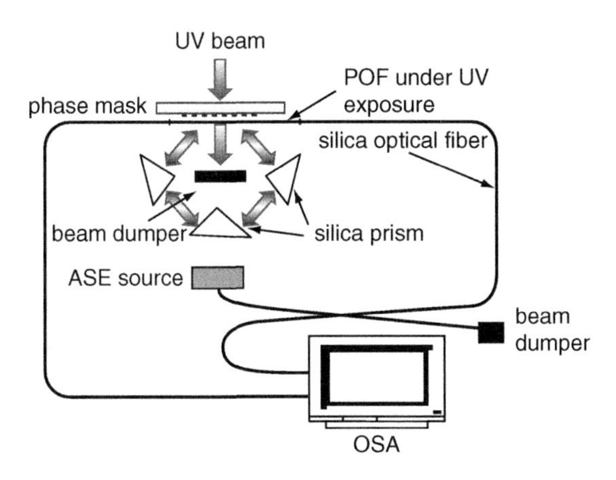

Fig. 3.2.25 Writing of Bragg gratings in single-mode POF. ASE, amplified spontaneous emission.

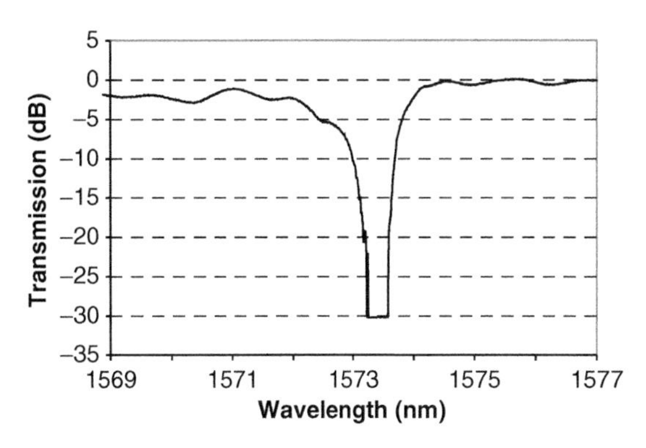

Fig. 3.2.27 Spectrum of Bragg grating in single-mode POF.

type 2 grating. The grating then remains unchanged for a longer irradiation. The energy of the irradiation beam is about 6 mJ. The spectrum of the final grating is shown in Fig. 3.2.27. The stop band is less than 0.5 nm, and the extinction ratio is 28 dB.

The unique feature of the POF grating is its large thermal and mechanical tuning ranges. Fig. 3.2.28a shows that a change of temperature over 508 C shifts the Bragg wavelength by about 8 nm, and Fig. 3.2.28b shows a mechanical strain can tune the wavelength by 10 nm [15]. More importantly, these tunings are linear, and no hysteresis is observed. These features can be utilized for constructing very sensitive sensing devices, such as current sensors and acoustic sensors.

3.2.7 Segmented cladding POF

Photonic crystal fibers (PCFs; also called holey fibers) have received intense study recently (see Chapter 3). These fibers have been made in both silica and polymer form [16]. The common feature of these fibers is that longitudinal holes are distributed either randomly or regularly over the cross-section of the fiber. One of the distinguishing characteristics of holey fiber is its large wavelength range for single-mode operation. This is based on the principle of differential leakage of the higher order modes in the fiber. As long as the effective index of the cladding is less than the mode index of the fundamental mode and larger than the mode indices of all higher order modes, all the higher order modes will eventually leak away at different rates provided the fiber is long enough, leaving the single mode in the core. Based on this principle, it is possible to design new fibers without holes that also possess large wavelength range single-mode operation. An interesting design is the segmented cladding fiber shown in Fig. 3.2.29 [17]. In this fiber, the core material fans out into the cladding in the form of spokes, thus segmenting the cladding. A polymer version of this fiber has been fabricated with four segments, as shown in Fig. 3.2.30 [18]. Its core has a diameter of 20 μm, and the overall fiber diameter is 200 μm.

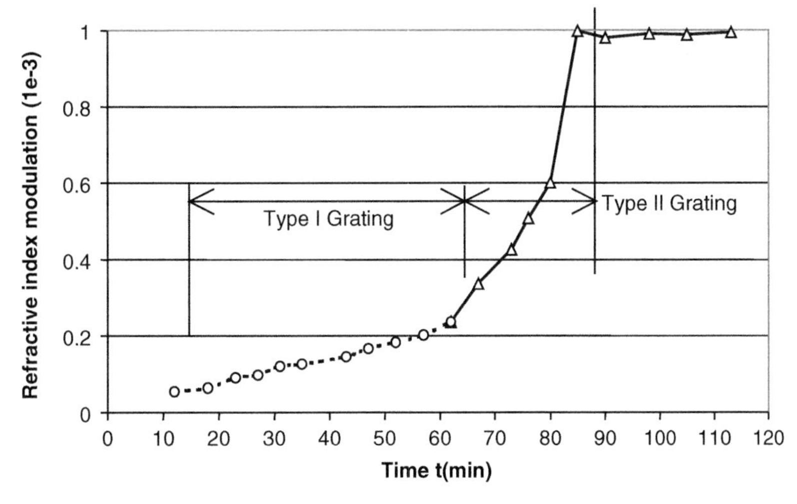

Fig. 3.2.26 Build-up of Bragg gratings in single-mode POF.

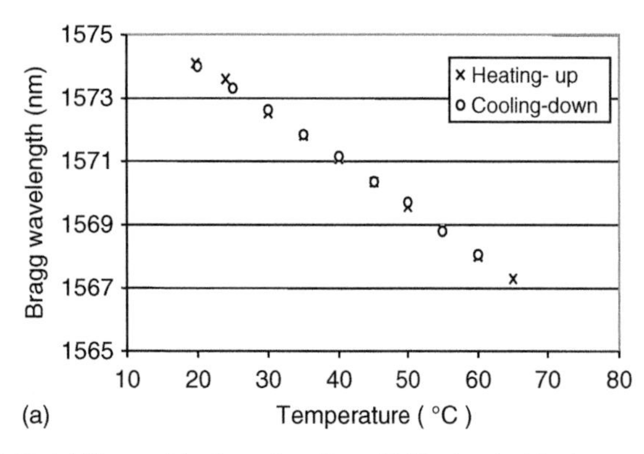

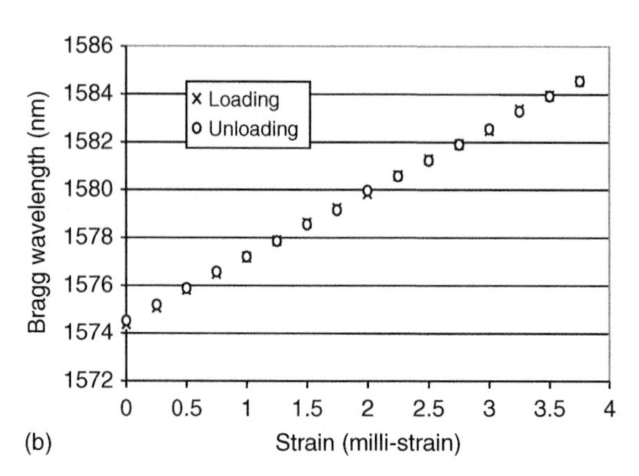

Fig. 3.2.28 (a) Thermal tuning of grating. (b) Mechanical tuning.

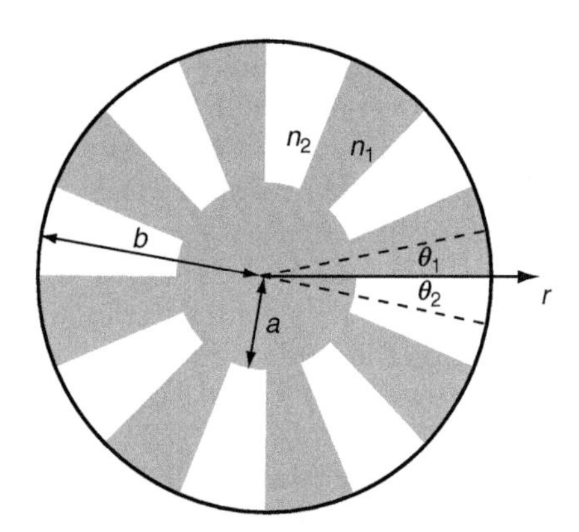

Fig. 3.2.29 Segmented cladding fiber.

With such a large core diameter, a conventional fiber would be very much multimodal even at a wavelength of 1.55 μm. However, Fig. 3.2.31 shows that this fiber is in single-mode operation at this wavelength when its length is 36.5 cm. When the signal wavelength is reduced to 0.633 μm, the fiber is few-moded as shown in Fig. 3.2.32.

Fig. 3.2.30 Four-segmented cladding fiber.

3.2.8 Dyedoped polymer fiber amplifier

Chemical dyes such as rhodamine B can be easily introduced into the MMA monomer during preform fabrication. This enables us to manufacture dye-doped polymer fibers that can function as an optical amplifier [19]. Fig. 3.2.33 shows the gain characteristic of a rhodamine B-doped polymer fiber amplifier. It can be seen that a signal gain more than 20 dB is obtained for a fiber length of 90 cm at a wavelength of about 620 nm. In this experiment, both the pump power and signal are in pulse form with a pulse width of 5 ns and a repetition rate of 10 Hz. The peak pump power is 920 W. This means that the average pump power is only 0.046 mW.

3.2.9 Conclusions

POF is gaining increasing attention from both industrialists and researchers because it has unique characteristics not available in silica fiber. In this chapter, we described how polymer fibers are manufactured and their applications in short distance communication and illumination for signage. It is expected that polymer fiber will capture the markets in these two areas. The wide tuning range of polymer fiber grating is an advantage that silica fiber grating cannot match. Hence, many tunable optical devices will be constructed from polymer fiber grating. The segmented cladding polymer fiber is a new addition to the host of holey fibers, with the advantage of the absence of holes. More study is required to ascertain unique applications of this fiber. Finally, polymer fiber amplifier opens up a new field of research, especially because dye-doped fiber can be easily fabricated. Different dyes will lead to fiber amplifiers operating at different signal wavelengths. Again, more study is needed to make these amplifiers useful in practical communication and sensing systems.

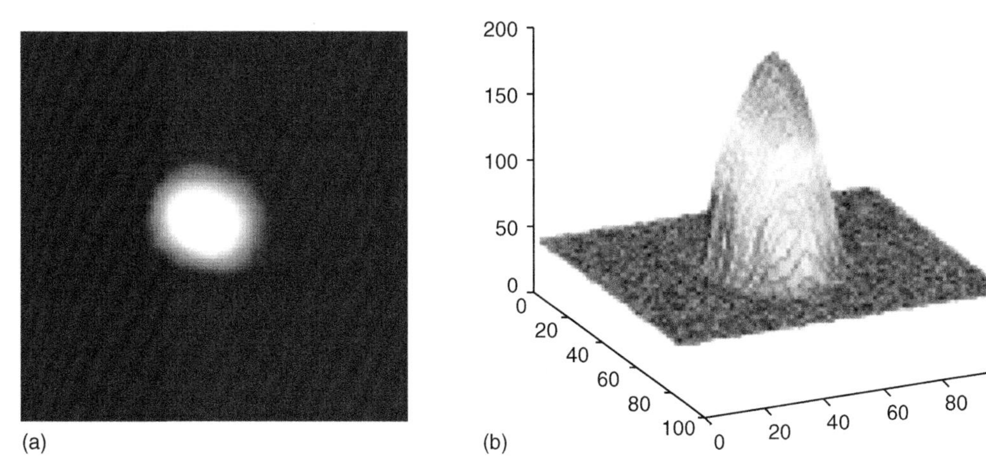

Fig. 3.2.31 (a) Far-field spot of SCF. (b) Intensity distribution.

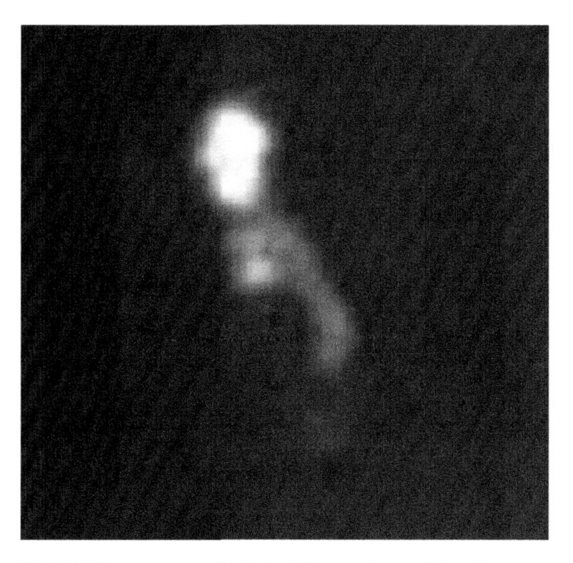

Fig. 3.2.32 Four-segmented cladding polymer fiber becomes few-moded.

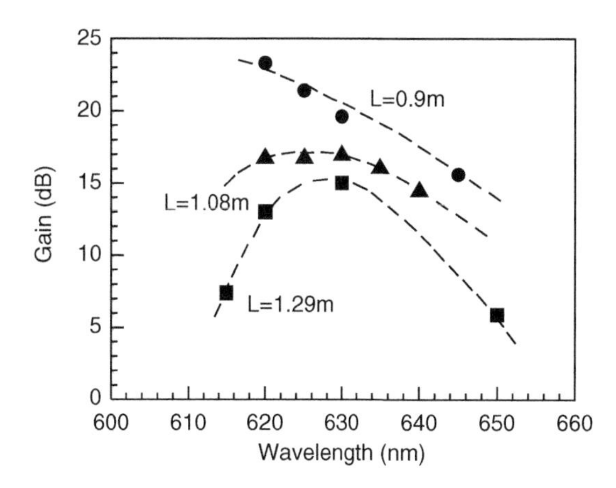

Fig. 3.2.33 Rhodamine B-doped polymer fiber amplifier.

References

1. J.C. Knight, T.A. Birks, P. St. J. Russell and D.M. Atkin, "All-silica single-mode optical fibre with photonic crystal cladding," *Optics Lett.*, **21**, No. 19, pp. 1457–1459 (1996).
2. Y. Koike, Proc. ECOC'96, paper MoB.3.1,1.41, (1996).
3. N. Tanio and Y. Koike: "What is the most transparent polymer?," Proc. POF Conf.'97, pp. 33–34, September 1997, Hawaii (1997).
4. G.D. Peng, P.L. Chu, L. Xia and R.A. Chaplin, "Fabrication and characterization of polymer optical fibres," *J. Electr. Electron. Eng. Australia*, **15**, No. 3, pp. 289–296,1995.
5. Y. Koike, T. Ishigure and E. Nihei, "High-bandwidth graded-index polymer optical fibre," *IEEE J. Lightw. Technol.*, **13**, No. 7, pp. 1475-1489, (1995).
6. B.C. Ho, J.H. Chen, W.-C. Chen, Y.-H. Chang, S.Y. YaTseng, "Gradient-index polymer fibres prepared by extrusion," *Polymer J.* **27**, No. 3, pp. 310–313, (1995).
7. I.S. Sohn, K. Yoon, T. Kang, O. Kwon and C.W. Park, "Fabrication of GI-POF by a co-extrusion method with enhanced diffusion," 11th Intern. Conf. POF, pp. 61-64, September 2002, Tokyo, Japan (2002).
8. R. Ratnagiri, M. Park, W.R. White and L.L. Blyler Jr., "Control of properties of extruded perfluorinated graded index polymer optical fibres," Proc. 12th Intern. Conf. POF, pp. 212-214, September 2003, Seattle, WA, USA (2003).
9. P.L. Chu, V. Yau and W.A. Gambling, "Effect of core non-concentricity on joint loss in polymer optical fibre," Proc. 12th Intern. Conf. POF, pp. 195–198, September 2003, Seattle, WA, USA (2003).
10. O. Ziemann, L. Giehmann, P.E. Zamzow, H. Steinberg and D. Tu, "Potential of PMMA based SI-POF for

Gbps transmission in automotive applications," Proc. Intern. Conf. POF, pp. 44-48, September 2000, Cambridge, MA, USA (2000).

11. Y. Koiko, "Progress in GI-POF: status of high speed plastic optical fibre and its future prospect," Proc. In tern. Conf. POF, pp. 1–5, September 2000, Cambridge, MA, USA (2000).
12. W. Baierl, "Evolution of automotive networks," 10th Intern. Conf. POF, pp. 161-168, September 2001, Am sterdam (2001).
13. G.D. Peng, Z. Xiong and P.L. Chu, "Photosensitivity and gratings in dye-doped polymer optical fibres," *Optical Fibre Techn.*, **5**, pp. 242–251, 1999.
14. H.Y. Liu, G.D. Peng and P.L. Chu, "Polymer fiber bragg gratings with 28dB transmission rejection," *IEEE Photon. Technol. Let.*, **14**, No. 7, pp. 935–937, 2002.
15. H.B. Liu, H.Y. Liu, G.D. Peng and P.L. Chu, "Strain and temperature sensor using a combination of polymer and silica fibre Bragg gratings," *Optics Commun* **219**, pp. 139–142, 2003.
16. G. Barton, M. van Eijkelenborg, G. Henry, N. Issa, K. F. Klein, M.Large, S. Manos, W. Padden, W. Pok and L. Poladian, "Characteristics of multimode micro- structured POF performance," Proc. 12th Intern. Conf. POF, pp. 81–84, September 2003, Seattle, WA, USA (2003).
17. V. Rastogi and K.S. Chiang, "Propagation character istics of a segmented cladding fibre," *Optics Lett.*, **26**, No. 8, pp. 491–493, 2001.
18. A. Yeung, K.S. Chiang and P.L. Chu, "Polymer segmented cladding optical fibre," 12th Intern. Conf. POF, pp. 77–80, September 2003, Seattle, WA, USA (2003).
19. G.D. Peng, P.L. Chu, Z. Xiong, T. Whitbread and R.P. Chaplin, "Broadband tunable optical amplification in rhodamine B-doped step index polymer optical fibre," *Optics Commun.*, **129**, pp. 353–357,1996.

Chapter 3.3

Microstructured optical fibers

Pal

3.3.1 Fibers with micron-scale structure

The development of core-clad silica glass optical fibers has revolutionized communications systems over the past 30 years. These "conventional" optical fibers have also made a significant impact in areas as diverse as sensing, medical imaging, laser welding, and machining and have allowed the realization of new classes of lasers and amplifiers. All these advances have been enabled by one key factor: reduction of the fiber loss. Reducing loss was a topic of intensive research and development for two decades, and dramatic improvements in the transmission of silica-based fibers in the 1.5-μm telecommunications window were achieved as a result. The widely used Corning SMF-28 fiber has a loss of less than 0.2 dB/km at 1550 nm.

In the early 1970s, when the fabrication processes for the manufacture of core-clad preforms had not yet reached maturity, Kaiser et al. proposed an alternative route to achieving low fiber losses. Kaiser et al.'s concept was to confine light within a pure (undoped) silica core by surrounding it with air [1, 2]. The core was supported by a sub-wavelength strand of silica glass and then jacketed in a silica cladding for strength. Although this new class of fibers showed promise, the fabrication methods used to produce these early single-material fibers were limited. Therefore, this new technology was quickly overtaken by improvements in the modified chemical vapor deposition process, which allowed the definition of high-quality preforms for the production of low-loss core-clad silica fibers.

In the late 1980s, work by Yablonovitch [3] on the development of three-dimensional photonic crystals identified micron-scale structuring to be a powerful means of modifying the optical characteristics of a material. The earliest photonic crystal samples were formed by drilling centimeter-scale holes to produce photonic bandgaps within which light propagation was forbidden. These samples were confirmed to have photonic bandgaps located at microwave wavelengths. In the 1990s, a number of groups worked to extend this concept to infrared and visible wavelengths by scaling down the dimensions of the photonic crystal structure to micron-scale feature sizes. The technique used most extensively for defining two-dimensional photonic crystals is electron beam lithography [4]. However, this technique is not well suited for defining structures that are truly extended in the third dimension to avoid nonuniformities in this direction modifying the properties of the photonic bandgaps. Fabricating two-dimensional photonic crystals is an engineering challenge, and although a number of approaches exist, there is a continued drive to develop cheap and flexible techniques for the large-scale production of high-quality photonic crystals.

In 1995, Birks et al. [5] proposed a novel technique for producing two-dimensionally structured silica/air photonic crystal structures by taking advantage of optical fiber manufacturing techniques. The fabrication concept was to stack macroscopic silica capillary tubes together into a hexagonal lattice to form a preform with millimeter-scale features and then to pull this preform to a fiber with micron-scale features on a drawing tower. Thus, the scale reduction and longitudinal uniformity inherent to the fiber drawing process could be utilized to

Guided Wave Optical Components and Devices; ISBN: 9780120884810

produce the first photonic crystals that could truly be considered infinite in the third dimension.

Although the index contrast between air and silica is not sufficient to form photonic bandgaps for all polarizations of light propagating within the transverse plane, by considering out-of-plane propagation, it is possible to form full photonic bandgaps in such structures [5]. Hence, it was proposed that by introducing a "defect" to the otherwise periodic transverse structure, light could be localized with the bandgap and thus guided within the defect, which acts as the fiber core. In addition, introducing an air defect raises the possibility of guiding light within an air core, something that cannot be achieved in conventional optical fibers, which guide light due to total internal reflection in a high index core material.

In 1996 the first silica/air microstructured fiber was made by this stack-and-draw technique [6]. This fiber had a hexagonal arrangement of small air holes and a central solid core, which was formed by replacing one of the capillaries within the stack with a solid rod. Although the fabrication of this fiber represented a significant breakthrough, it also raised some interesting questions. Calculations indicated that the holes within the cross-section of the fabricated fiber were too small to lead to the formation of photonic bandgaps, and thus light could not be guided within this fiber via photonic bandgap effects. Despite this, Knight and colleagues [7] demonstrated that light could be guided within the solid core of this fiber, and a number of interesting optical characteristics began to emerge within this new fiber type, most notably that they can be "endlessly single mode," guiding just a single mode at all wavelengths.

In this way, an important new class of optical fibers was discovered during the drive to produce photonic bandgap fibers (PBGFs). These "index-guiding" microstructured fibers guide light due to a modified variant of the law of total internal reflection. The arrangement of air holes acts to lower the effective refractive index in the cladding region, and so light is confined to the solid core, which has a relatively higher index. This class of fibers has attracted significant attention from both university and industrial research groups in recent years largely because they can exhibit many novel optical properties that cannot be achieved in conventional optical fibers. Note that it is not essential to use a periodic arrangement of air holes in this class of fibers [8]. The optical properties of this class of fibers are reviewed in Section 3.3.2.

A number of names have been given to these fibers, including photonic crystal fibers [6], holey fibers (HFs) [9], and microstructured optical fibers [10]. Within this chapter the term holey fiber is used to differentiate index-guiding fibers from those that guide via photonic bandgap effects, which are described as PBGFs; and the term microstructured optical fiber is used to describe all types of fibers with micron-scale transverse structural features.

In 1998, Broeng et al. [11] reported that the use of a honeycomb air hole lattice led to the formation of larger bandgaps than the triangular/hexagonal lattice both for in-plane and out-of-plane propagation. These honeycomb fibers were the first PBGFs to be fabricated [12]. The light guided by these fibers is guided within a ring-shaped mode located in the silica surrounding a central air defect in the honeycomb lattice structure. The first air-guiding PBGFs were realized in 1999 [13], 3 years after the first air/silica microstructured optical fibers were made by the stack-and-draw fabrication technique. These fibers have air holes arranged on a hexagonal lattice and a very large air-filling fraction (Fig. 3.3.1c). Research in this field is reviewed in Section 3.3.7.

In the single-material microstructured fibers described thus far, light is solely confined by the holes in the cladding. Hybrid microstructured fibers are another class of fiber that combines a doped core with a holey cladding. At one extreme, in "hole-assisted" fibers light is guided by the relatively higher index of the doped core, and the air holes located in the cladding of a conventional solid fiber act to modify properties such as dispersion [14] or bend loss. In air-clad fibers, an outer cladding with a high air-filling fraction creates a high numerical aperture (NA) inner cladding that allows the realization of cladding-pumped high power lasers [15]. Alternatively, dopants can be added to the core of an HF to create novel HF-based amplifiers and lasers (see, e.g., [16]). Work performed to date on active microstructured fibers is described in Section 3.3.5.3.

Note that the parameters Λ and d/Λ are widely used to label the feature sizes in fibers that have hexagonal hole arrangements, where Λ is the hole-to-hole spacing and d is the hole diameter. Figure 3.3.1 presents a gallery of scanning electron microscope (SEM) images of a representative selection of the microstructured optical fibers realized to date. The images presented are all of fibers made at the Optoelectronics Research Centre (ORC) at the University of Southampton, United Kingdom. Figure 3.3.1, a–d, are SEM images of silica microstructured fibers. The fabrication of silica microstructured fibers has now reached a level of maturity that allows a broad range of high-quality fiber profiles to be defined. More detailed information about fiber fabrication is given in Section 3.3.3, and work in silica fibers is described further in Section 3.3.5. There are now a number of groups (both university groups and companies) worldwide capable of producing silica fiber structures of similar quality. Figure 3.3.1, e and f, are images of two new classes of soft glass microstructured fibers. An overview of soft glass microstructured fibers is presented in Section 3.3.6.

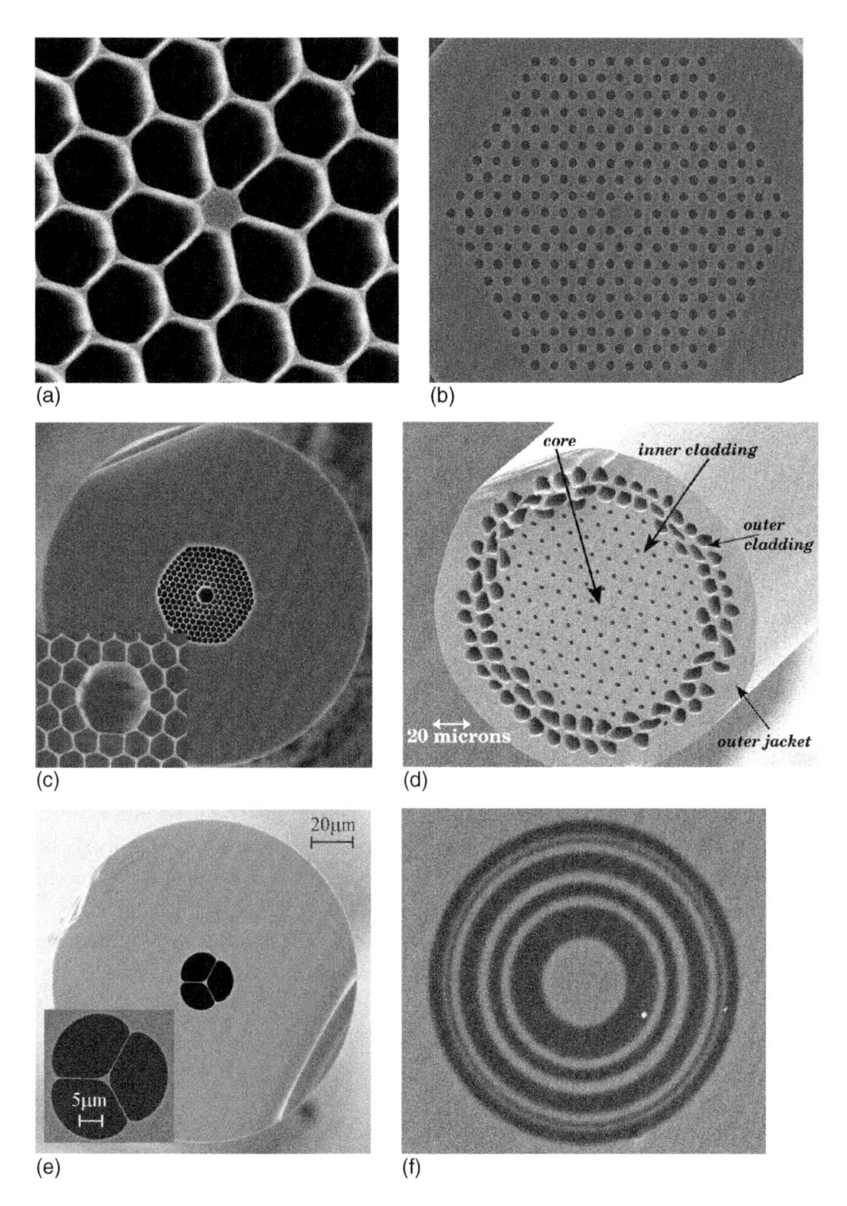

Fig. 3.3.1 A representative selection of microstructured optical fibers. The images shown are all fibers fabricated at the ORC (Southampton, UK). (a) Small-core nonlinear pure silica holey fiber. (b) Large-mode area pure silica holey fiber. (c) Air-core photonic bandgap fiber. (d) Double-clad $Yb3^{+}$-doped large-mode area fiber. (e) Extruded highly nonlinear bismuth holey fiber. (f) One-dimensional layered soft glass microstructured fiber.

3.3.2 Overview of optical properties

3.3.2.1 Introduction

This section broadly reviews the optical characteristics that have been identified thus far in both classes of microstructured optical fibers. As described in Section 3.3.1, HFs guide light due to the effective refractive index difference between the solid core and the arrangement of air holes that forms the cladding region. The effect of the holes on the fiber properties depends on the hole distribution and size(s) relative to the wavelength of light guided in the fiber. Hence, the effective refractive index of the structured cladding region (and thus the fiber's NA) can be a strong function of wavelength in these fibers. For this reason, it is possible to design fibers with spectrally unique properties not possible in conventional core-clad optical fibers. In addition, the optical properties of microstructured fibers are determined by the spatial configuration of air holes used to form the cladding, and many different arrangements can be envisaged within this flexible fiber type.

PBGFs guide light for use with a fundamentally different mechanism than conventional fibers, and so it is not surprising that the optical properties of modes guided with these fibers can be radically different from conventional fibers. It is worth noting that research in the area of PBGFs is less mature than in index-guiding fibers, and so the full range of optical properties possible in PBGFs is not yet known.

3.3.2.2 Nonlinearity tailoring

Microstructured fibers provide a powerful means of tailoring the effective nonlinearity that can be achieved in a fiber form over at least five orders of magnitude, a much wider range than can be achieved in conventional core-clad fiber designs. The effective nonlinearity (γ) of a fiber is defined as in [17] to be

$$\gamma = \frac{2\pi}{\lambda}\frac{n_2}{A_{\mathrm{eff}}}$$

where A_{eff} is the effective mode area of the fundamental guided mode (defined in [17]), λ is the wavelength of light, and n_2 is the effective nonlinear refractive index of the material. For example, the standard Corning SMF-28 fiber has $\gamma \approx 1/(\mathrm{W \cdot km})$. Observe that the effective nonlinearity of a fiber can be tailored by either modifying the mode area of the fiber or using different host materials, or both.

Fibers with high γ values are attractive for a broad range of applications in nonlinear fiber devices, and some examples are presented in Section 3.3.5.1. By modifying conventional fiber designs, values of γ as large as $20/(\mathrm{W \cdot km})$ have been achieved [18]. This is done by reducing the diameter of the fiber core and using high germanium concentrations within the core, which both increases the NA and enhances the intrinsic nonlinearity (n_2) of the material. Both modifications act to confine light more tightly within the fiber core and thus increase the nonlinearity γ by reducing the mode area A_{eff}. However, the NA that can be achieved limits the nonlinearity of conventional fiber designs.

When a large air-filling fraction is used in the fiber cladding (Fig. 3.3.1a), index-guiding HFs can offer a very high effective core-cladding refractive index contrast (and hence a high NA) relative to that which can be achieved in conventional fiber designs. Combining this with small-scale core dimensions allows this class of fibers to offer tight mode confinement. One useful way of estimating the minimum mode area (and thus maximum γ) that can be achieved in any given material system is to consider the theoretical limit of a rod of glass suspended in air. For an air-suspended silica rod, the minimum value of A_{eff} is $\sim 1.5\,\mu\mathrm{m}^2$ at 1550 nm [19]. This is illustrated in Fig. 3.3.2, which shows the effective mode area at 1550 nm for both an air-suspended rod of silica and a range of small-core silica HF designs. As Fig. 3.3.2 shows, once the core size becomes significantly smaller than the optical wavelength, the rod becomes too small to confine the light well and the mode broadens again. The fiber with the largest air-filling fraction ($d/\Lambda = 0.9$) in Fig. 3.3.2 has a minimum effective mode area of $1.7\,\mu\mathrm{m}^2$, only slightly larger than the air-suspended silica rod ($1.5\,\mu\mathrm{m}^2$).

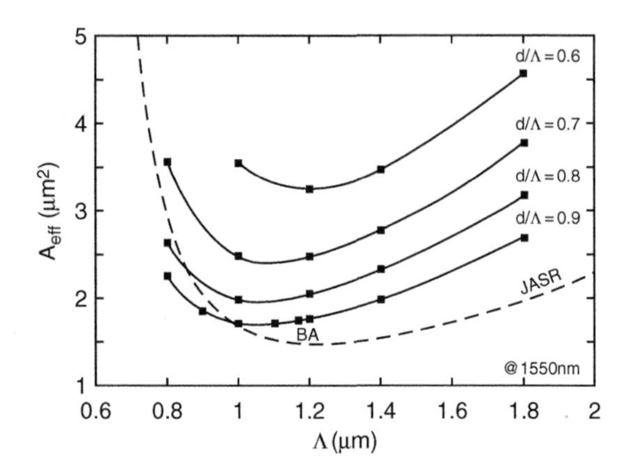

Fig. 3.3.2 Effective mode area (A_{eff}) as a function of the hole-to-hole λ spacing for a range of small-core silica holey fiber designs with air holes arranged on a hexagonal lattice. JASR corresponds to the case of an air-suspended core of diameter λ. (From ORC, Southampton, UK.)

The value of n_2 in undoped silica glass is $\sim 2.2 \times 10^{-20}$ m^2/W. Hence, the ultimate nonlinearity limit in silica air/glass fibers is $\gamma \approx 70/(\mathrm{W \cdot km})$, and fibers with effective nonlinearities close to this value have been manufactured [20]. Note that glasses with a higher linear refractive index can provide better confinement and thus smaller mode areas. Nonlinearity tailoring can also be achieved via choice of the glass host material, and significantly higher values of n_2 can be achieved using high index glass. For example, fiber nonlinearities as high as $1100/(\mathrm{W \cdot km})$ have been reported in index-guiding HFs made from bismuth oxide glass [21]. Progress in this area is reviewed in Section 3.3.6.2.

At the opposite extreme, low values of γ (and large effective mode areas) are attractive for avoiding both nonlinear effects and fiber damage for a range of high power applications. Although large-mode area fibers can be produced using conventional core-clad designs, microstructured fiber designs are attractive for a number of reasons. First, pure silica index-guiding HFs offer a number of useful features for high power delivery applications, including broadband single-mode guidance and the option of using a pure silica glass core (which offers better power handing than a doped core). Single-mode silica HFs with mode areas as large as $680\,\mu\mathrm{m}^2$ have been achieved at 1550 nm (corresponding to $\gamma \approx 0.1$) [22]. Large-mode area silica fibers are discussed in more detail in Section 3.3.5.2. PBGFs can have less than 1 % of the guided mode located in the silica within the cladding, and typical mode areas for fibers produced to date are $A_{\mathrm{eff}} \approx 20\,\mu\mathrm{m}^2$ at 1.06 μm [23]. Hence, values of γ as low as an order of 0.01 are attainable in PBGFs.

Thus, these fibers are very attractive for a broad range of high power applications. Note that for these fibers, the value of γ is a sensitive function of the air–mode overlap, and so nonlinearity measurements are a useful way of measuring this overlap. The air–mode overlap in microstructured fibers is discussed further in Section 3.3.2.5.

3.3.2.3 Dispersion

A broad range of novel dispersive properties of microstructured fibers has been predicted and observed within silica microstructured fibers. In index-guiding silica HFs, the dispersion can be particularly strongly influenced by the cladding configuration, particularly when the features within the cladding are on the scale of the wavelength of light and the core is small. In such fibers, the waveguide contribution to the total dispersion can dominate the material dispersion of silica, and so new regimes are possible, including anomalous dispersion at wavelengths down to 550 nm [24]. Such fibers have applications in new soliton-based devices (see the example in Section 3.3.5.1) and for the generation of broadband supercontinuum [25]. In both applications, the required dispersive properties are achieved using small-core designs, which offer the additional advantage of allowing nonlinear effects at low powers in short fiber lengths. Silica fibers can also be defined in which the waveguide contribution to the dispersion is finely tuned to compensate for the material dispersion over a broad wavelength range, resulting in broadband flat near-zero dispersion [26].

The impact of fiber dispersion on the nonlinear applications of these fibers is discussed in more detail in Section 3.3.5.1. Although a vast range of dispersion profiles can be achieved, it is worth noting that it is not always possible to combine all the properties that might be desired for any particular device application within the one fiber. One good example of this is the concept of development HFs for dispersion compensating devices [27]. HF designs can have very large values of normal dispersion (as large as -2000 ps/nm.km at 1550 nm), which at first appears attractive for compensating the dispersion of conventional communications fibers. However, to achieve such values, it is necessary to use fiber designs with very small effective mode areas, which results in poor integration with conventional systems, significantly higher losses, and degradation of the linear polarization of the fiber mode.

The dispersion properties of nonsilica index-guiding HFs are not yet as well mapped out as is the case for silica fibers. However, it is clear that due to the generally steeper material dispersion characteristics (as a function of wavelength) and the higher index contrast in these fibers, in general it will be more challenging to target specific values of dispersion–dispersion slope than in silica fibers.

The highly wavelength-selective nature of the guidance mechanism in PBGFs naturally gives rise to extraordinary dispersion properties in this fiber type. In other words, waveguide dispersion dominates in this class of fibers. In particular, zero dispersion can be realized within the bandgap, and in general the dispersion is anomalous across most of the bandgap [28], and the dispersion values at the edges of the bandgap can be extremely large [29, 30].

Note that the accurate prediction of dispersion properties in microstructured fibers is challenging and is in general a good test of the accuracy of any numerical method. In addition, the dispersive properties of microstructured fibers are more sensitive than most of the other fiber properties (such as mode area, number of guided modes, and birefringence) to subtle variations of or imperfections in the fiber profile.

3.3.2.4 Polarization

Microstructured optical fibers with air holes arranged on a hexagonal lattice support a degenerate pair of fundamental modes and thus are not birefringent. More generally, any fiber structure with a greater than 2-fold symmetry is not birefringent, if a perfect cross-section profile is assumed. This is proved in [31] on the basis of arguments based on group theory and symmetry classes.

However, in reality, no real fabricated fibers are perfectly symmetric. When any structural imperfections are combined with the small feature sizes and large index contrast that is possible in microstructured fiber designs, small asymmetries can translate into significant birefringence effects. For example, the fiber described in 16 has a somewhat elliptical-shaped core and has a measured birefringent beat length at 1550 nm of 0.3 mm, the record for the shortest beat length at the time it was reported. Unsurprisingly, this effect is more noticeable in highly nonlinear fibers, and typical beat lengths for such fibers are on the order of a few millimeters at 1550 nm, even for relatively symmetric structures.

Highly birefringent microstructured fibers can also be made by deliberately introducing a 2-fold symmetry to the fiber profile, and this symmetry can either be introduced in the core shape or into the cladding configuration. This can be done for either index-guiding [32] or photonic bandgap [33, 34] guidance mechanisms.

One tested way of reducing unwanted polarization effects in fibers is the process of fiber spinning, in which the fiber preform is rotated during the fiber drawing process. Recently, an order of magnitude decrease in the polarization mode dispersion of HFs was reported in spun HFs [35].

3.3.2.5 Air-light overlap

The holes in the cladding of a microstructured fiber open up new opportunities for exploiting the interaction of light with gases and liquids via evanescent field effects. For example, the concentration of pollutants in a gas could be determined by measuring the absorption that occurs as light propagates through the gas for a range of wavelengths. One attraction of the fiber geometry is that it can naturally provide extremely long optical path lengths in a compact fashion and does not require large gas volumes. Two approaches can be envisaged: either to exploit PBGFs, which guide light in air, or to use the overlap between the guided mode of an index-guiding HF and the cladding air holes.

For index-guiding HFs to be useful as evanescent field devices, a significant fraction of the modal field must be located within the holes. However, calculations indicate that there is typically only a very small overlap between the guided mode and the holes [36]. Indeed, for many HF designs, less than 1% of the guided mode's power is located in the air holes. The field distribution for the fundamental mode depends strongly on the size of the features in the HF relative to the wavelength, and so the overlap between the guided mode and air can be tailored. This leads to the requirement $\lambda < 2.2\Lambda$ for this overlap to be significant. For example, the air–mode overlap is 30% at 1.5 μm for an HF with hexagonally arranged holes with a relative hole size $d/\Lambda = 0.8$ and spacing Λ of 0.75 μm. Note that if λ is made too small, it becomes difficult to fill the holes with gas in a reasonable time. One alternative fiber type that offers the potential for easier filling is a fiber structure like the one shown in Fig. 3.3.1e, which has a small-scale core suspended on a number of thin spokelike membranes. In either case, the fiber designs required to allow significant air–mode overlap have small mode areas, which makes them more challenging to integrate with conventional systems.

PBGFs are the most promising route for evanescent field applications, because these fibers can offer nearly perfect overlap between the guided mode and the material that fills the holes. For example, Bouwmans et al. [37] reported that more than 99% of the fundamental mode is guided within the air core. This was exploited in [38], which demonstrated the measurement of acetylene and hydrogen cyanide absorptions within a PBGF.

The excellent overlap of the guided mode with air within a PBGF also makes these fibers attractive for high power delivery applications [39]. The light intensities within the silica material within the fiber cladding are low relative to the mode peak intensities, and so such fibers promise higher damage thresholds than fibers that guide light within glass. However, one important consideration with the use of these fibers for high power applications is damage of the fragile cladding structure caused by beam wander or contamination. Further work needs to be done to ascertain the relative merit of this fiber type for extending the power handling capabilities beyond those possible in more conventional fibers.

3.3.3 Fabrication approaches

3.3.3.1 Preform fabrication

The first stage in the fabrication of any microstructured optical fiber is the production of a preform, which is a macroscopic version of the structure that is to be defined in the final fiber. For most microstructured fibers, the preform fabrication stage is the most challenging step of the fiber fabrication process.

The vast majority of microstructured fibers produced to date have been made from pure silica glass, and the preforms for these fibers are usually fabricated using capillary stacking techniques. Capillary tubes are stacked in a hexagonal configuration. If the fiber to be produced is an index-guiding HF, the central capillary is typically replaced with a solid glass rod, which ultimately forms the fiber core. To produce a bandgap fiber, either 7 or 19 of the capillaries are removed to produce a large central air core. The stacking procedure is flexible: For example, active fibers can be made using rare-earth–doped core rods, and off-center or multiple core fibers can be readily made by replacing noncentral or multiple capillaries, respectively. The reproducibility that can be achieved in the transverse cross-section of the preform is determined by the length and uniformity of the capillaries that are used to form the stack.

Stacking is a flexible technique and has been used successfully by a number of groups worldwide. Note that the preforms for the fibers in Fig. 3.3.1, a–d, were all produced via capillary stacking. One significant drawback of the stacking approach is that the preform fabrication is labor intensive and the quality of the final fiber depends significantly on the craft of the fabricator in forming the preform.

Another widely applicable and flexible technique for the definition of structured preforms is drilling of bulk glass samples. Drilling has been used to produce both polymer [40] and soft glass microstructured fibers [41]. Drilling allows the definition of a broad range of hole configurations and can be applied to a very broad range of optical materials. Some drawbacks of drilling include the length of time required to produce a complex preform structure containing many holes and the reduction in preform yield with increasing complexity.

Recent attention has been focused on the fabrication of microstructured fibers in a range of nonsilica glasses (see Section 3.3.6). In general, such compound or "soft" glasses have low softening temperatures relative to silica glass, which allows new techniques to be used for the fabrication of structured preforms, such as rotational casting and extrusion. The technique used for most nonsilica glass microstructured fibers made thus far is extrusion [21, 42–49]. Casting techniques have also been used for the production of low-loss microstructured fibers in tellurite glass [50].

Extrusion techniques allow the fabrication of structured fiber preforms with millimeter-scale features directly from bulk glass billets. In this process, a glass billet is forced through a die at elevated temperatures near the softening point, whereby the die orifice determines the preform geometry. Once the optimum die geometry and process parameters have been established, the preform fabrication process can be automated. In this way good reproducibility in the preform geometry can be achieved. One advantage of the extrusion technique is that the preform for the microstructured part of a fiber can be produced in one step. In addition, extrusion allows access to a more diverse range of cladding structures, because the holes are not restricted to hexagonal arrangements.

As an example, consider the structured preform shown in Fig. 3.3.3a. In this geometry, the core (center) is attached to three long, fine, supporting struts. The outer diameter of this preform is approximately 16 mm. The preform was reduced in scale on a fiber drawing tower to a cane of approximately 1.7 mm in diameter (Fig. 3.3.3b). In the last step, the cane was inserted within an extruded jacket tube, and this assembly was drawn to the final fiber (Fig. 3.3.3c). The illustrative examples shown in Fig. 3.3.3 are made from the lead silicate glass SF57 and are described in more detail in [47]. This procedure has also been used to make nonlinear fibers from other materials, including bismuth glass [48]. A similar technique has been used to produce small-core tellurite microstructured fibers in which the core is supported by six long fine struts [44]. In [44], the preform extrusion is performed directly on a fiber drawing tower, which allows the preform to be reduced to an ~1-mm-diameter cane directly at the time of extrusion.

3.3.3.2 Fiber drawing

To produce fibers that have relatively large-scale features, such as large-mode area fibers (Fig. 3.3.1b), which typically have hole-to-hole spacing (Λ) in the range of 7–15 μm, fibers are drawn directly in a single step from the preform. To achieve micron-scale hole-to-hole spacings (Λ), it is in generally necessary to first reduce the preform to a cane of 1–2 mm in diameter, and in a second step this cane is inserted in a jacketing tube and then drawn to the final fiber.

One of the most challenging aspects of air/glass microstructured fiber fabrication is to prevent collapse of the holes and to achieve the target hole size and shape during caning and fiber drawing. The microstructured profile can be affected by the pressure inside the holes, surface tension of the glass, and temperature gradient in the preform. Hence, it is useful to have predictive tools for evaluating the best choice of fiber drawing conditions for producing any predetermined target fiber structure. Work on modeling the drawing of silica capillary tubes has led to some progress in this direction [51]. This model has produced a range of useful rules of thumb and asymptotic limits that aid the selection of appropriate drawing conditions.

A recent highlight in this area is the prediction that the air holes in microstructured fibers can survive the spinning of the preform during fiber drawing and indeed that spinning can be used as a means of controlling the geometry of the drawn fiber [52]. Indeed, it has been demonstrated in practice that the holes within the fiber can survive spinning during the fiber drawing process [35].

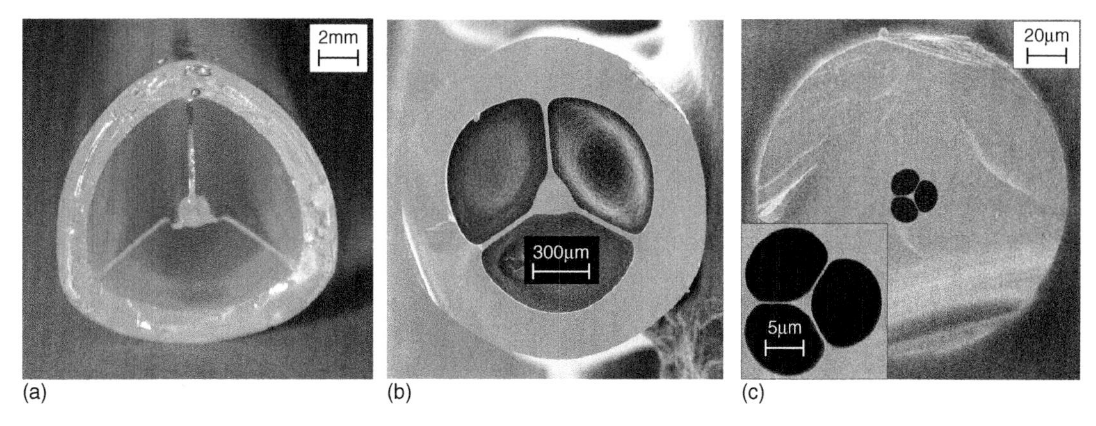

Fig. 3.3.3 (a) Photograph of a cross-section through a typical extruded SF57 glass preform. FWHM, full width at half-maximum, (b) SEM image of cane cross-section, (c) SEM image of resulting holey fiber cross-section. (Pictures from the ORC, Southampton, UK.)

Note that compared with silica glass, most non-silica glasses have significantly steeper viscosity curves, which leads to greater demands on the process control during fiber drawing. Nevertheless, a high degree of reproducibility for the HF geometry has already been demonstrated for both lead silicate glass [47] and bismuth glass [48]. The core diameter can be adjusted during fiber drawing by an appropriate choice of the external fiber diameter. Small-core dimensions are chosen to provide tight mode confinement and thus high effective non-linearity. For fiber designs of the type shown in Fig. 3.3.3, core diameters in the range of 1.7–2.3 μm correspond to struts that are typically >5 μm long and <250 nm thick. These long thin struts act to isolate the core optically from the external environment and thus ensure that confinement loss is negligible (see Section 3.3.4.3). Excellent structural reproducibility has been demonstrated using this fabrication technique.

3.3.3.3 State-of-the-art

A diverse range of high-quality transverse fiber cross-sections based on a hexagonal lattice configuration can now be fabricated in silica glass via the capillary stacking fabrication technique. Continued improvements in the fabrication procedures have reduced the losses of microstructured fibers dramatically in recent years. At the time of this writing, transmission losses of 0.28 dB/km for silica index-guiding HFs [53] and 1.7 dB/km for silica PBGFs [54] were reported at 1550 nm. It is worth noting that losses are typically larger for small-core (highly nonlinear) fibers of the type described in Section 3.3.2.2. This is principally caused by nanometer-scale surface roughness at the air–glass boundaries near the core [55]. However, this is not a significant limitation, because most applications of highly nonlinear fibers require the use of short fiber lengths.

As the numbers quoted above attest, within a decade the loss of microstructured fibers has been reduced to values close to those achieved in conventional solid transmission fibers. The prospect of microstructured silica fibers with a lower transmission loss than conventional fibers represents a tantalizing possibility with the potential to revolutionize telecommunications. Index-guiding fibers are one potential route to lower fiber losses, given that the core can be made from pure silica, which potentially offers lower loss than a doped core. Gas core bandgap fibers are another attractive potential route to lower fiber losses, given that they can have as little as 1 % of the light propogating in the silica within the cladding.

The fabrication of soft glass microstructured fibers is currently less mature. The first good quality single-mode fibers were reported in 2002 [42], and it is now possible to produce fibers in a broad range of glasses with losses on the order of 2 dB/m at 1550 nm. It is anticipated that further fabrication work should reduce the losses of these fibers to below 1 dB/m in a number of soft glass materials. Although this loss value is high relative to silica glass, it still allows the development of soft glass–based nonlinear fiber devices with better figures of merit than is possible using existing technologies (see Section 3.3.6.2).

3.3.4 Fiber design methodologies

The presence of wavelength-scale holes in microstructured fibers leads to challenges in the accurate modeling of their optical characteristics. A wide variety of fiber design techniques can be used, ranging from effective step-index fiber models to approaches that incorporate the full complexity of the transverse cross-section. Here these methods are reviewed and assessed in terms of their suitability for modeling optical properties of both index-guiding HFs and PBGFs. Some of the issues associated with designing and modeling practical fibers are highlighted. Note that when dispersion predictions are required, many of the approaches described below allow the dispersion of the material to be included ab initio through the usual Sellmeier formula.

3.3.4.1 Effective index methods

The complex nature of the cladding structure of the microstructured optical fiber does not generally allow for the direct use of analytical methods from traditional fiber theory. However, for index-guiding HFs, a scalar model based on an effective index of the cladding has proven to give a good qualitative description of their operation [7].

The fundamental idea behind this approach is to first evaluate the properties of the periodically repeated air hole lattice that forms the cladding. By solving the scalar wave equation in a hexagonal cell centered on a single air hole, the propagation constant of the lowest order mode that could propagate in the infinite cladding material is determined. In this work, the hexagonal unit cell is approximated by a circular one to simplify the analysis. This procedure allows the effective cladding index of the fundamental cladding mode, sometimes called the fundamental space-filling mode (FSM), to be determined as a function of the wavelength ($n_{FSM}(\lambda)$).

The next step of the method is then to model the fiber as a standard step-index fiber, using n_{FSM} as the cladding index. The core of this equivalent fiber is assumed to have the refractive index of pure silica with a core radius typically taken to be 0.62Λ. (This assumes that the core is created by the omission of one cladding air hole.)

Despite ignoring the spatial distribution of the refractive index profiles within HFs, the effective index method can provide some useful insight into their operation. For example, it correctly predicts the endlessly single-mode guidance regime for fibers with small d/Λ. This method has also been used as a basis for the approximate dispersion and bending analysis presented in [56]. However, this reduced model cannot accurately predict modal properties such as dispersion, birefringence, or other polarization properties that depend critically on the spatial configuration of holes within the cladding.

One difficulty that arises when using this approach is the question of how to define the properties of the equivalent step-index fiber. One method for making this choice was described above. In this work, the core radius was taken to be 0.62Λ, and the results obtained using the effective index method were made to agree well with full simulations for a particular fiber structure via the appropriate choice of this constant of proportionality. However, for different structures or wavelengths, different choices can become necessary. This restricts the usefulness of this approach, because it is typically necessary to determine the best choice of equivalent structure by referring to results from a more complete numerical model. Riishede and coworkers [57] explored the possibility of choosing the step-index fiber parameters in a more general fashion by allowing a wavelength-dependent core or cladding index. However, to date no entirely satisfactory method for ascribing parameters to the equivalent fiber has been found.

It is noteworthy that recent developments in the fabrication of structures with relatively large air holes have made it relevant to approximate the fiber by an isolated strand of glass surrounded by air (Fig. 3.3.1, a and e), and this case has been labeled the air suspended rod limit in the literature. In this limit, the step-index fiber analogy can provide useful information about the fiber properties (see, e.g., [48]).

3.3.4.2 Structural methods

To accurately model index-guiding HFs, it is typically necessary to account for the complex spatial distribution of air holes that define the cladding. A number of these techniques have also been successfully applied to PBGFs. Note that when the effective index contrast between core and cladding regions is large, the weak-guidance (scalar) approximation breaks down, leading to inaccurate results, and it is often necessary to adopt a vectorial method that includes polarization effects. This is typically necessary when the air-filling fraction of the cladding is large. Vectorial methods are also appropriate for asymmetric structures. Most of the techniques described in this section can be implemented in both scalar and vector forms.

As noted in Section 3.3.1, much of the early research in microstructured fibers was driven by the desire to fabricate a fiber operating by the photonic bandgap effect, which may be obtained in periodically structured material. It is not possible to analyze PBGFs using simple scale modeling—or effective index approaches—because the full vectorial nature of the electromagnetic waves must be accounted for to paint an accurate picture of the resulting photonic bandgaps.

In 1990, the first method for finding photonic bandgaps in photonic crystals was described [58]. The method was closely related to methods used for calculating electronic bandgaps in semiconductor crystals in that it described the magnetic field as a plane wave multiplied by a Bloch function with the two-dimensional periodicity of the photonic crystal. From Maxwell's equations, an eigenvalue equation can be formulated that is well suited for calculating the bandgaps of a periodic dielectric structure, because it describes both the field and the structure as a Bloch function. To include a core, one has to impose an artificial periodicity that is handled numerically by creating a supercell with periodically repeated core defects. This approach yields reasonable solutions if the supercell is much larger than the guided mode area [59]. Such plane-wave techniques have been used to make a broad range of useful predictions both for photonic bandgap guiding fibers and index-guiding fibers.

Beam propagation methods (BPMs) can also be used to calculate the modal properties of HFs. For example, Eggleton et al. [60] used a commercial BPM package to investigate a modified conventional germanium-doped fiber in which six large holes were added around the doped core region. A Bragg grating written in the core of the fiber was used to investigate the cladding modes of the structure, and good agreement with the BPM predictions was found. However, because BPMs calculate the modes of a fiber indirectly by propagating a light distribution along a fiber, they are relatively computationally intensive.

An alternative approach was initially suggested by Mogilevtsev et al. [61], which described the modal fields using localized functions. This technique takes advantage of mode localization and so is more efficient than the plane-wave methods; however, it cannot be accurate unless the refractive index is also represented well.

A hybrid approach that combines some of the best features of the localized function and plane-wave techniques described above was developed [9, 62–63], and some extensions to this approach are outlined briefly here. The air hole lattice was described using a plane-wave decomposition, as in the plane-wave techniques described above, and the solid core and the modal fields were described using localized functions [9, 62]. This allows for an efficient description, particularly for idealized periodic structures, because only symmetric terms need to be used in the expansions. To model HFs with asymmetric profiles or to obtain accurate predictions for higher order modes, it is necessary to extend this approach to use a complete basis set [63]. When more complex fiber profiles are considered, the advantages of describing the localized core separately from air holes is diminished, and the best combination of efficiency and accuracy is obtained by describing the entire refractive index distribution using a plane-wave expansion while using localized functions only for the modal fields. The general implementation of this hybrid approach can be used to explore the full range of both index guiding and photonic bandgap structures and modes and can predict the properties of real fibers by use of SEM photographs to define the refractive index profile (see, e.g., [64]). This allows the deviations in optical properties that are caused by the subtle changes in structure to be explored.

In this implementation, the entire transverse refractive index profile is described using a plane-wave expansion, and the Fourier coefficients are evaluated by performing overlap integrals, which only need to be calculated once for any given structure. The modal electric field is expanded into orthonormal Hermite-Gaussian functions (both even and odd functions are included). These decompositions can be used to convert the vector wave equation into a simple eigenvalue problem (as in the plane-wave method) that can be solved for the modal propagation constants and fields. To solve the system, a number of overlap integrals between the various basis functions need to be evaluated. For the choice of decompositions made here, these overlaps can be performed analytically, which is a significant advantage of this approach.

Most of the modeling done to date has considered ideal hexagonal arrangements of air holes. As discussed in Section 3.3.2.4, group theory arguments can be used to show that all symmetric structures with higher than 2-fold symmetry are not birefringent [28]. However, as the techniques described thus far perform calculations based on a Cartesian grid, they typically predict a small degree of birefringence that can be reduced (but not eliminated) by using a finer grid. However, when modeling asymmetric structures have a form birefringence that is significantly larger than this false birefringence, it is possible to make reasonably accurate predictions for fiber birefringence.

3.3.4.3 Predicting confinement loss

To design practical single-material fibers, it is important to have a means of predicting confinement loss, because all guided modes are intrinsically leaky modes. Even in index-guiding HFs with high index dopants in the core (which can have true bound modes), it can be important to understand the leakage characteristics of any cladding modes. PBGFs can also suffer significantly from confinement loss (see Section 3.3.7).

The confinement loss associated with a given mode can be extracted from the imaginary part of the modal propagation constant. Of the techniques described in Section 3.3.4.2, only the BPM can currently calculate complex propagation constants.

The finite element method is being used increasingly for analyzing microstructured fibers and is capable of calculating complex propagation constants [65, 66]. With this method, the classical Maxwell differential equations are solved for a large set of properly chosen subspaces, taking into account the continuity of the electromagnetic fields. More specifically, the modeled waveguide is split into distinct homogeneous subspaces of triangular and quadrilateral shapes. Maxwell's equations are discretized for each element, and the resulting set of elementary matrices is combined to create a global matrix system for the entire structure. This method has been shown to lead to fast and accurate numerical solutions for both classes of microstructured fibers.

Another technique that is well suited to this problem is the multipole approach [67]. This approach is suitable for studying effects caused by the finite cladding region, because it does not make use of periodic boundary conditions. Another advantage of this method is that it calculates the modal fields using decompositions that are based in each of the cladding air holes, and so it avoids the false birefringence problems associated with using a Cartesian coordinate system described above. For this reason, this method is also particularly well suited to exploring the symmetry properties of new fiber geometries. However, it cannot be used to investigate fibers with arbitrary cladding configurations and is limited to circular or at most elliptical hole shapes.

Another technique that can predict confinement loss is based on representing the refractive index distribution as a series of annular segments [68]. The algorithm uses a polar coordinate Fourier decomposition method with adjustable boundary conditions to model the outward radiating fields. The use of annular segments allows the overlap integrals between the structure and the field components to be performed directly, and so this method can be efficient. It is possible to represent arbitrary fiber profiles in this way.

3.3.4.4 Summary

A number of techniques have been adapted or developed to model microstructured optical fibers, and a range of novel guidance regimes has been identified in these fibers that promise to lead to a new generation of optical devices with tailor-made optical properties. Many of the techniques described herein complement one another and can often be used in conjunction with each other to paint a complete picture of the optical characteristics of any given microstructured fiber. The extremes that are possible in these fibers have highlighted a number of challenges in accurate modeling of their properties, and it seems likely that as the technology for fabricating these structures matures, further challenges will emerge.

Almost all the work performed to date designing and modeling microstructured optical fibers has relied on a combination of guesswork and experience to establish the fiber cross-section. By applying one or more numerical design tools, the optical characteristics of the fiber are determined, and this is often computationally intensive. In practice, it would be more useful to know what (if any) fiber cross-section(s) allows any desired optical properties to be realized. Hence, this field has reached a point where there is a great need for the application of inverse solution methodologies, and work in this area has begun to be reported. For example, genetic algorithm techniques were used to define fiber structures with optimized optical properties [69].

3.3.5 Silica HFs

3.3.5.1 Small-core fibers for nonlinear devices

3.3.5.1.1 Background

One of the most promising practical applications of index-guiding HF technology is the opportunity to develop fibers with a high optical nonlinearity per unit length [70]. Figure 3.3.1a shows a typical highly nonlinear silica HF made at the ORC (Southampton, UK). A small-core diameter combines with a large air-filling fraction in the cladding to result in a fiber that confines light tightly within the solid central core region. In this case, $\Lambda \approx 1.5$ μm and $d/\Lambda \approx 0.95$.

Even though silica is not intrinsically a highly nonlinear material, its nonlinear properties can be used if high light intensities are guided within the core, as described in Section 3.3.2.2. The breakthrough that first demonstrated the promise of nonlinear applications of microstructured fibers came with the experimental demonstration of the supercontinuum generation in microstructured silica fibers reported by Ranka et al. in 2000 [25].

The high NA that can be achieved in HFs allows the realization of nonlinearities more than 50 times higher than in standard telecommunications fiber and two times higher than the large NA conventional designs. The key principle is that modest optical powers can induce significant nonlinear effects within these fibers, and so they offer an attractive new route toward efficient, compact, fiber-based, nonlinear devices. Nonlinear effects can be used for a wide range of optical processing applications in telecommunications and beyond, and examples include optical data regeneration, wavelength conversion, optical demultiplexing, and Raman amplification. Here we review the optical properties of these small-core fiber designs and present an overview of some of the emerging device applications of this new class of fibers.

3.3.5.1.2 Design considerations

Small-core HFs can also exhibit a range of novel dispersive properties of relevance for nonlinear applications (see Section 3.3.2.3). By modifying the fiber profile, it is possible to tailor both the magnitude and the sign of the dispersion to suit a range of device applications. They can exhibit anomalous dispersion down to 550 nm, which has made soliton generation in the near-infrared (IR) and visible spectrum possible for the first time. An application of this regime was reported [16] in which the soliton self-frequency shift (SSFS) in an ytterbium-doped HF amplifier was used as the basis for a femtosecond pulse source tunable from 1.06 to 1.33 μm. Shifting the zero dispersion wavelength to regimes where there are convenient sources also allows the development of efficient supercontinuum sources [25], which are attractive for dense wavelength division multiplexing transmitters, pulse compression, and the definition of precise frequency standards. It is also possible to design nonlinear HFs with normal dispersion at 1550 nm [27]. Fibers with low values of normal dispersion are advantageous for optical thresholding devices, because normal dispersion reduces the impact of coherence degradation [71] in a nonlinear fiber device.

Highly nonlinear fibers with zero dispersion at 1.55 μm have long been pursued because these fibers are very attractive for a range of telecom applications, such as 2R regeneration [72], multiple clock recovery [73], optical parametric amplifiers [74], pulse compression [75], wavelength conversion [76], all-optical switching [77], supercontinuum-based wavelength division multiplexing telecom sources [78], and demultiplexing [79].

It is worth noting that small-core HFs pose a number of challenges for effective modeling. The high index contrast inherent in these fibers necessitates the use of a full-vectorial method. In addition, any asymmetries or imperfections in the fiber profile, when combined with this large contrast and the small structure–scale, can lead to significant form birefringence. In general, even small asymmetries can lead to noticeable birefringence for these small-core fibers. Hence, it is often necessary to use the detailed fiber profile to make accurate predictions.

When the core diameter is reduced to scales comparable with (or less than) the wavelength of light guided within the fiber, confinement loss arising from the leaky nature of the modes can contribute significantly to the overall fiber loss. Confinement loss is described in more detail in Section 3.3.4.3. Indeed, the small-core HFs fabricated to date are typically more lossy than their larger core counterparts. Here, we briefly outline some general design rules for designing low-loss high-nonlinearity HFs as described [19].

The range of effective mode areas that can be achieved using silica glass at 1550 nm is shown in Fig. 3.3.2. As Fig. 3.3.2 shows, the hole-to-hole spacing (Λ) can be chosen to minimize the value of the effective mode area, and this is true regardless of the air-filling fraction (d/Λ). However, it is not always desirable to use the structures with the smallest effective mode area, because they typically exhibit higher confinement losses [19]. In other words, in the limit of core dimensions that are much smaller than the wavelength guided by the fiber, many rings (more than six) of air holes are required to ensure low-loss operation, which increases the complexity of the fabrication process. In this small-core regime, unless many rings of holes are used, the mode can *see over* the finite cladding region. A relatively modest increase in the structure scale in this small-core regime can lead to dramatic improvements in the confinement of the mode without compromising the achievable effective nonlinearity significantly.

With careful design, it is possible to envisage practical HFs with small-core areas ($<2\ \mu m^2$) and low confinement loss (<0.2 dB/km). Note that although fiber loss limits the effective length of any nonlinear device, for highly nonlinear fibers short lengths (<10 m) are typically required, and so loss values of orders of 1 dB/km can be readily tolerated. In addition, note that reducing the core diameter to dimensions comparable with the wavelength of light generally increases the fiber loss for another reason: In relatively small-core fibers, light interacts more with the air–glass boundaries near the core, and so the effect of surface roughness can be significant [54].

3.3.5.1.3 Device demonstrations

A range of device demonstrations have now been performed using highly nonlinear silica HFs. The first such demonstration was 2R data regeneration, a function that is a crucial element in any optical network, because it allows a noisy stream of data to be regenerated optically. The first demonstration of regeneration used a silica HF with a mode area A_{eff} of just $2.8\ \mu m^2 [\gamma = 35/(W \cdot km)]$ at 1550 nm [72]. Typically, devices based on conventional fibers are ~1 km long, whereas in these early experiments just 3.3 m of HF was needed for an operating power of 15 W. Subsequent experiments used an 8.7-m-long variant of this switch for data regeneration within an optical code division multiple access system [80]. Significant improvements in system performance were obtained in this way.

The 2R regeneration scheme is reviewed briefly here as an example. A schematic of the HF-based data regenerator is shown in Fig. 3.3.4. Pulses of light propagating in a highly nonlinear fiber broaden spectrally due to self-phase modulation, and Fig. 3.3.4b shows the spectrum of 2.5-ps soliton pulses before and after propagation through the HF. Figure 3.3.4c shows the pulse power that is transmitted through a 1.0-nm narrowband filter (offset by +2.5 nm relative to the incident pulses) as a function of incident pulse peak power. The S-shaped characteristic is suitable for thresholding because at low powers the pulses do not broaden, and so transmission through the filter is negligible. This corresponds to a "0" in the data stream. For higher powers (~2 W here), substantial self-phase modulation occurs, and so transmission through the filter becomes appreciable. This corresponds to a "1." This device acts to remove noise from an incoming data stream by nullifying all noisy "0" bits and by equalizing all noisy "1" bits.

Fibers with a high effective nonlinearity also offer length/power advantages for devices based on other processes such as Brilluoin and Raman effects. The demand for increased optical bandwidth in telecommunications systems has generated enormous interest in the S- and L-bands, outside the gain band (C-band) of conventional erbium-doped fiber amplifiers (see Chapter 1). Fiber amplifiers based on Raman effects offer an attractive route to extending the range of accessible amplification bands. In addition, the fast response time (<10 fs) of the Raman effect can also be used for all-optical ultrafast signal processing applications. Despite these attractions, there is one significant drawback to Raman devices based on conventional fibers: Long lengths (~10 km) are generally required, and so Rayleigh scattering ultimately limits their performance. High nonlinearity fibers offer a method for obtaining sufficient Raman gain in a short fiber length, which eliminates this problem. For example, Yusoff et al. [81] demonstrated a 70-m fiber laser-pumped Raman amplifier. The amplifier was pumped using a pulsed fiber laser and provided gains of up to 43 dB in the L+-band for peak powers of ~7 W.

Other nonlinear device applications of HFs that have been demonstrated include a continuous-wave Raman laser [82], a wavelength division multiplexing wavelength converter [83], and pulse compression down to 20 fs [84]. Note the continuous-wave power density at the facet ($2\ W = \mu m^2$) in the continuous-wave Raman device mentioned above demonstrates that HFs can exhibit a good resilience to damage.

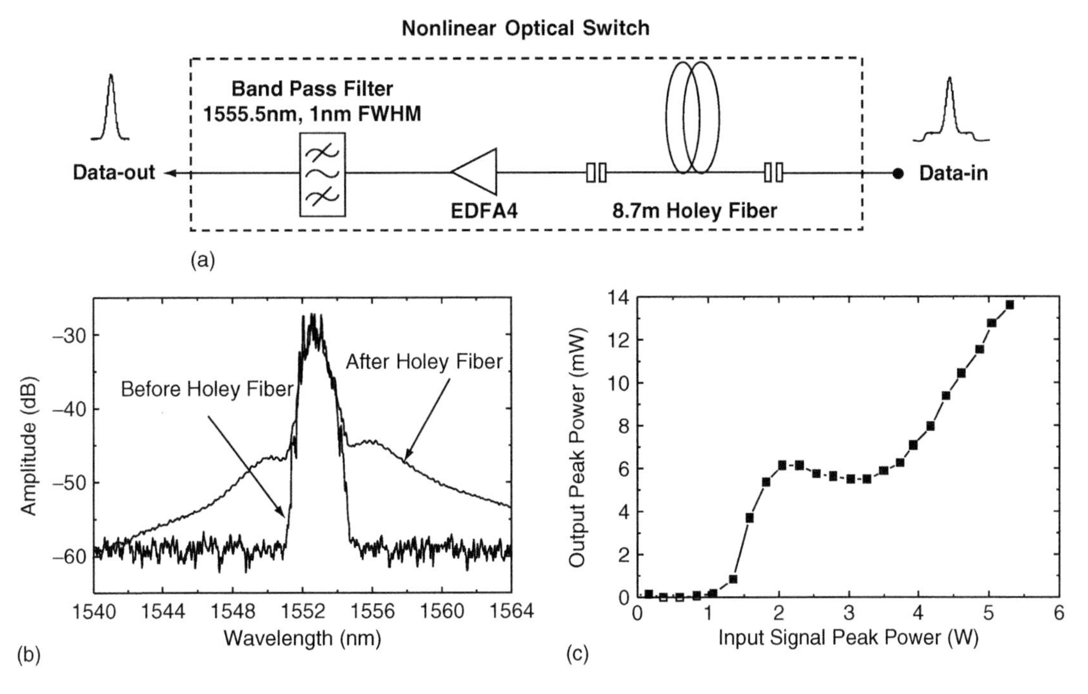

Fig. 3.3.4 (a) Schematic of thresholder. (b) Pulse spectra before and after HF. (c) Power transmitted (including offset narrowband filter). (From ORC, Southampton, UK.)

One of the applications of highly nonlinear silica HFs to be investigated most intensively is the generation of supercontinuum (see, e.g., [25, 85–88]). The continua have been used in applications including optical coherence tomography [89], spectroscopy, and frequency metrology [90]. Supercontinua covering several octaves as well as multiwatt output have been demonstrated [91]. The absence of dopants in the core of a pure silica HF has also allowed the generation and propagation of ultraviolet wavelengths in fiber [92]. Considerable effort has been made to develop a better appreciation of the complex interplay of nonlinear processes behind supercontinuum generation, and today many of the basic mechanisms (e.g., soliton fission [93, 94], self-phase modulation [95], four-wave-mixing, and stimulated Raman scattering [85]) are understood.

3.3.5.2 Large-mode area fibers for high power applications

The development of large-mode area fibers is important for a wide range of practical applications, most notably those requiring the delivery of high power optical beams. For many of these applications, spatial mode quality is a critical issue, and such fibers should preferably support just a single transverse mode. Large-moded single-mode fibers can be made using conventional fiber doping techniques such as modified chemical vapor deposition by reducing the NA of the fiber and increasing the fiber core size.

HFs are an attractive route toward such fibers [96], and single-mode HFs with effective areas as large as 1000 μm^2 have been reported [97]. Large-mode HFs (LMHFs) can be produced by designing fibers with a large hole-to-hole spacing ($\Lambda > 5$ μm) and/or small air holes ($d\Lambda <$ 0.3). A typical pure silica LMHF is shown in Fig. 3.3.1b. In addition to offering large-mode areas, LMHFs offer other unique and valuable properties; most notably they can be single-moded at all wavelengths.

The models described in Section 3.3.4 can be applied to model the optical properties of these fibers, although typically extra care is needed because of the wide range of spatial scales present. Polarization effects are typically less important in this class of fibers, and it is often sufficient to use a scalar model.

Macroscopic bend loss ultimately limits the practicality of such large-mode fibers, and so understanding bend loss is important in the design of this class of fiber. Note that single-material fibers exhibit a short wavelength bend loss edge in addition to the long wavelength loss edge found in conventional fiber designs. Generally, larger holes result in broader operational windows, whereas the hole-to-hole distance roughly determines the center position of the window (as a first approximation, the minimum bend loss occurs at a wavelength around Λ=2) [98]. Hence, standard telecommunications wavelengths fall on the short wavelength loss edge. Therefore, macrobending losses effectively limit the operational wavelength range of this class of fibers. Despite this, LMHFs with hexagonally arranged holes possess comparable bending losses to similarly sized conventional fibers at 1550 nm [99]. Note that the effective index model described in Section 3.3.4.1 is capable of predicting accurately the spectral location of the short wavelength bend loss edge [96].

Two distinct bend loss mechanisms have been identified in conventional fibers: transition loss and pure bend loss [100]. Pure bend loss occurs continually along any curved section of fiber: At some radial distance, the tails of the mode need to travel faster than the speed of light to negotiate the bend and are thus lost. The loss associated with the transition from straight to bent fiber sections is called the transition loss and is typically negligible for macrobends [99].

In any fiber, bend loss increases for decreasing values of NA and with increasing effective mode area (A_{eff}). When evaluating the relative bending losses of different fiber designs, it is thus essential to consider fibers that are equivalent in terms A_{eff} and NA at the wavelength of interest to ensure that the loss predictions for different designs can be compared. Microstructured fiber fabrication techniques permit a high level of flexibility, and bending loss can be reduced by using modified cladding hole configurations. For example, the effective mode area can be enlarged, without increasing the bending losses, by using three adjacent rods to form the fiber core instead of a single central rod [101]. This "tri-core" HF is described in more detail below.

The degree of modal distortion and the associated attenuation both increase with the severity of the bend. Baggett et al. [99] described a method for modeling the bending losses of HFs. This model uses an orthogonal function method together with a conformal transformation to obtain the distorted modal fields of the bent fiber. The bend loss is extracted by estimating the fraction of the modal field lost to radiation. This bend loss model has very few restrictions on the refractive index profile that can be considered and has been experimentally validated for LMHFs. Note that this approach does not approximate the fiber profile as a step-index fiber, as previous work does, which allows the effect of the angular orientation of the fiber relative to the bend to be explored for the first time. The results presented below show that the bend loss of HFs cannot be evaluated with effective index methods alone and that the detailed nature of the cladding configuration, which is reflected in the shape of the bent mode, is an essential consideration.

The bent mode of a standard LMHF design with $\Lambda = 12.7$ μm and $d/\Lambda = 0.45$ is shown in Fig. 3.3.5a for a bend of radius 3.4 cm. This fiber has an effective mode area of ~190 μm^2 at 1064 nm when straight. Figure 3.3.5b shows a bent mode in a tri-core LMHF chosen to have a similar A_{eff} and NA to the design shown in Fig. 3.3.5a. The tri-core design has $\Lambda = 7.4$ μm and $d/\Lambda = 0.2$. Figure 3.3.5 clearly demonstrates that the tri-core structure is better able to confine the bent mode to the core than the single-rod HF for this bend radius even though the effective indices of the two (straight) fibers are similar. The predicted critical bend radius for the tri-core fiber is ~3.0 cm, approximately 20% smaller than for traditional single-rod designs. This level of improvement is in excellent agreement with observations and corresponds to an ability to increase A_{eff} by 15% without increasing bending losses. Preliminary results for a tri-core fiber fabricated at the ORC are in excellent agreement with these predictions.

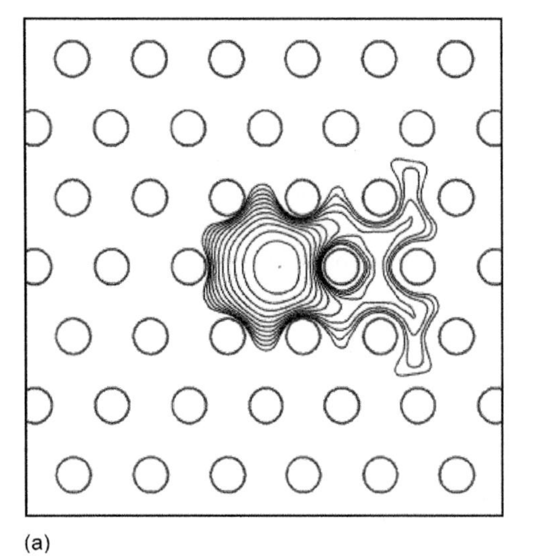
(a)

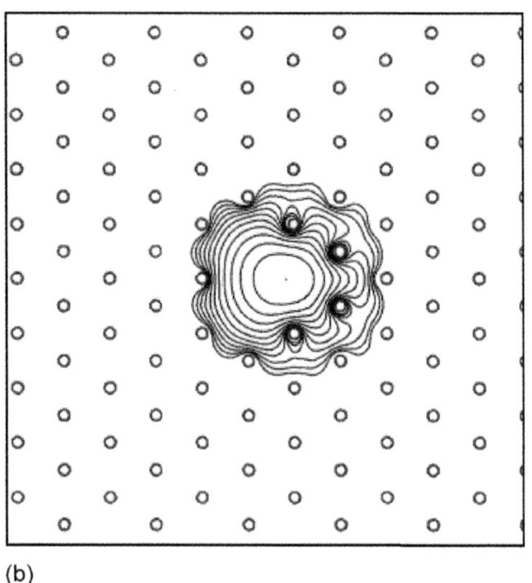
(b)

Fig. 3.3.5 Contour plots of the modal intensity and refractive index profile for (a) a standard large-mode area silica fiber and (b) a tri-core holey fiber. Contour lines are spaced by 2 dB.

3.3.5.3 Active fibers

3.3.5.3.1 Background

The capillary-stacking techniques that are generally used to make single-material silica microstructured fibers can be readily adapted to allow the production of doped fibers simply by replacing the solid silica core rod with a doped rod in the preform stack. The doped core rod is typically formed by extracting the core region from a conventional doped fiber. In the example given in [103], the starting point was an aluminosilicate ytterbium–doped rod with an NA of 0.05 produced using conventional modified chemical vapor deposition process techniques.

Many properties of the index-guiding HFs have benefits for active devices. HFs incorporating Er^{3+} [102,104] and Yb^{3+} have been fabricated, and continuous-wave [105] and mode-locked [106] Yb^{3+} HF lasers have been realized experimentally. As is the case for the passive applications of HFs, two regimes of particular practical relevance are low NA large-core active fibers and high NA small-core active fibers. Large-mode areas offer the potential for high power generation without nonlinear effects or fiber damage and in addition allow broadband single-mode laser output. At the other extreme, nonlinear fibers allow the realization of very high gain efficiency (>8.5 dB/mW was reported in [104]).

Below one active device of each type is reviewed: a tunable fs-soliton source based on highly nonlinear fiber and a cladding pumped laser based on double-clad large-mode area fiber.

3.3.5.3.2 Tunable fs-soliton source

Fibers with anomalous dispersion extending down through the visible regions of the spectrum allow the prospect of extending soliton laser techniques that have been developed for 1550 nm to Yb^{3+}-doped and Nd^{3+}-doped fiber lasers. Wavelength tuneable femtosecond (fs) pulse sources have applications in areas such as ultrafast spectroscopy, materials processing, and nonlinear optics. Traditional fs pulse sources based on bulk crystals offer a limited wavelength range, particularly above 1.1 μm. The SSFS in fibers [107] has opened up the possibility of wavelength tuneable fs-soliton fiber sources. To use the SSFS effect, the frequency shifting fiber must exhibit anomalous dispersion at the seed wavelength and across the required tuning range. Although conventional fibers can have anomalous dispersion beyond 1.3 μm, as discussed in Section 3.3.2.3, small-core, large, air-filling fraction HFs extend this range to 550 nm. Such fibers also have a high effective non-linearity, which allows soliton formation with just pJ pulse energies in meter long fiber lengths [108].

Here we outline a continuously tuneable soliton source operating from 1.06 to 1.33 μm based on HF technology. A fuller exposition of this work can be found in [16]. This source is seeded by a diode-pumped 1.06-μm Yb^{3p}-doped silica fiber laser and relies on SSFS effects in a Yb 3^{+}-doped holey fiber amplifier. The fiber used in these experiments was single mode for all wavelengths considered here and has an effective mode area of just $A_{eff} \approx 2.5\ \mu m^2$ at 1550 nm (an SEM image of this fiber is shown in the inset to Fig. 3.3.6).

The mode-locked seed laser produces pulses at 1.06 μm with a positive linear chirp, which are launched into Yb 3^{+}-doped HF amplifiers, together with a pump beam from a diode laser to control the gain, as shown by Fig. 3.3.6. Because of the amplification and nonlinear evolution of the pulses as they pass through the amplifier, Raman solitons form and are continuously wavelength shifted via the SSFS. Because the nonlinear evolution of the pulses depends on the pulse peak power, the wavelength of the Raman solitons at the amplifier output can be tuned by varying the gain in the amplifier. Using ~5 m of amplifier fiber, monocolor soliton output pulses have been wavelength tuned from 1.06 to 1.33 μm, as shown at the bottom of Fig. 3.3.6. At higher pump powers, the change in gain distribution causes the Raman solitons to form earlier within the amplifier, thereby leaving them a greater length of fiber within which to walk-off to longer wavelengths through the SSFS. The final central wavelength of the pulses varies in an almost linear fashion with the level of incident pump power. The maximum wavelength shift of the Raman soliton increases with the length of fiber and is ultimately only limited by the absorption of silica near 2.3 μm.

3.3.5.3.3 Cladding pumped fiber laser

Capillary stacking techniques can be extended to allow the production of all-glass double-clad fibers, and an example is shown in Fig. 3.3.1d. This approach is attractive for high power active fiber devices, because it allows the use of cladding pumping, and for such applications large-mode area fiber designs are of particular interest (see Section 3.3.5.2 for a review of passive LMHFs). However, the use of rare-earth dopants in LMHFs is challenging. The presence of dopants (and associated co-dopants such as germanium, aluminium, and boron that are required to incorporate the rare-earth ions at reasonable concentrations and to maintain laser efficiency) modifies the refractive index of the host glass. This affects the NA of the fiber and can lead to the loss of some of the most attractive LMHF features, such as broadband single-mode guidance, unless care is taken in the fiber designs.

Here we review laser development based on double-clad HFs [103]. The ultimate advantages of these fibers relative to polymer-coated dual-clad fibers are that they allow for all-glass structures, with inner cladding NAs in

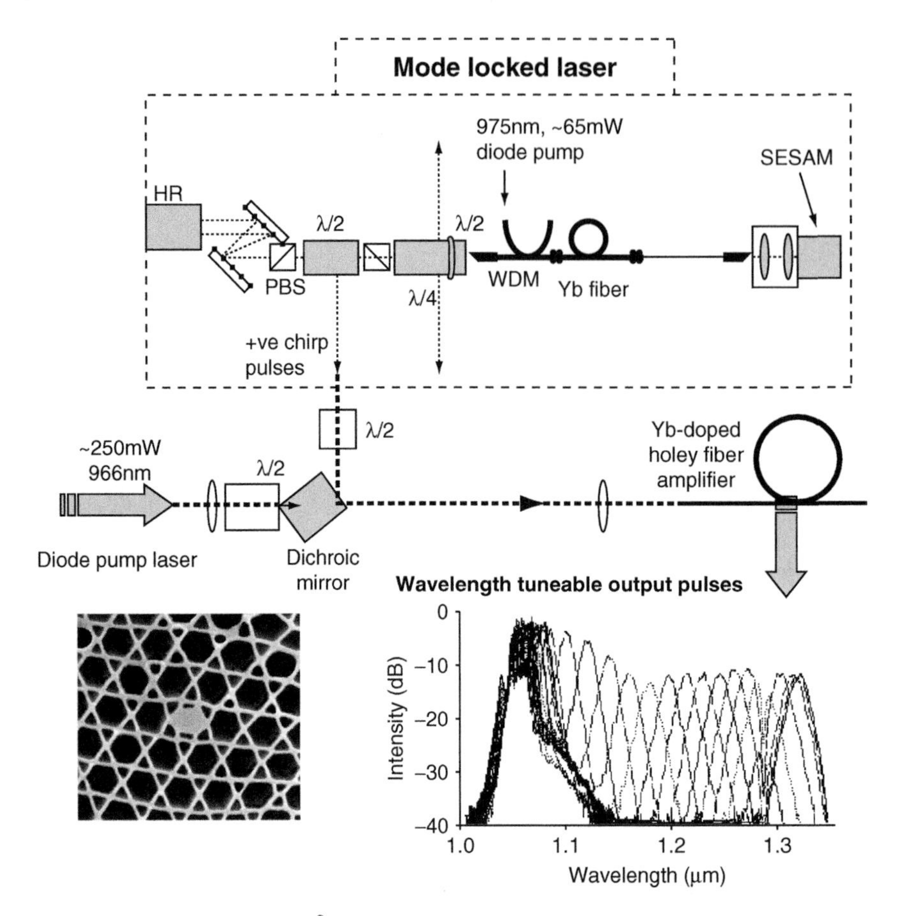

Fig. 3.3.6 Experimental setup of the mode-locked Yb^{3+} fiber seed laser (diode-pumped), the launch arrangement for seeding the pulses and pump laser to the Yb^{3+}-doped holey fiber amplifier, and the wavelength tunable output pulses. The Yb^{3+}-doped holey fiber used here is shown in the inset. PBS, polarizing beam splitter; SESAM, semiconductor saturable absorber mirror; WDM, wavelength division multiplexing.

excess of 0.5, and good pump/mode mixing. In addition, they offer the combination of single-mode guidance and large-core dimensions. In terms of device, these features translate to the possibility of higher coupled diode powers, shorter device lengths, and extended tuning ranges. Consequently, HF technology represents a most interesting proposition for application in the future power scaling of both continuous-wave and pulsed fiber laser systems.

The fiber in Fig. 3.3.1d has a Yb^{3+}-doped core surrounded by an inner cladding consisting of five rings of small ($d \approx 2.7$ μm) holes separated by $\Lambda \approx 9.7$ μm. The inner cladding NA was measured to be 0.3 in a ~10-cm fiber length. Two rings of larger holes define the outer cladding. Note that the core is offset from the center of the fiber, which breaks the cladding symmetry and hence enhances the pump absorption.

Slope efficiencies as high as 82% were recorded in a 4-m fiber length, comparable with the best conventional ytterbium fiber lasers. As expected, the output beam was robustly single mode. Because of the low NA of the inner core/cladding structure, a significant fraction of the pump was launched into the inner cladding, and so this laser acts as a hybrid core/cladding pumped fiber laser. A cladding pumped laser was realized using a low brightness 915-nm fiber-coupled laser diode. In this first demonstration, average powers in excess of 1 W were achieved at a 7.5-m fiber length, with a measured slope efficiency of 70% [103]. Both Q-switched and mode-locked operations were demonstrated in this cladding-pumped HF laser. In the mode-locking experiments, fundamental mode locking was obtained over a wavelength tuning range in excess of 60 nm. The pulse duration was estimated to ~100 ps. An output power of more than 500 mW was achieved for a pump power of 1.33 W.

Further progress in this class of fibers was reported [109], which presents an HF laser generating up to 80 W of output power with a slope efficiency of 78%. Single transverse mode operation is achieved using a fiber with a mode-field area of 350 μm^2. No thermo-optical limitations are observed at the extracted power level of 35 W/m, which implies that such fibers allow scaling to even higher powers. Limpert et al. [110] explored the issue of thermal management of air-clad microstructured fibers in comparison with conventional double-clad fibers. This work revealed that the temperature in the fiber core is determined primarily by heat transport through the outer surface of the fiber. If the dimensions of the air-cladding region are properly designed, the temperature profile can be comparable with a conventional double-clad fiber. Hence, the air-clad region is not necessarily a limitation to the power scaling capabilities of microstructured fiber lasers, and therefore air-clad microstructure fibers are likely to be scalable to power levels of several kilowatts.

3.3.6 Soft glass fibers

3.3.6.1 Background

The combination of the microstructured fiber concept with nonsilica glasses is an emerging field that promises the development of a host of new fibers and operational regimes not achievable with existing fiber technology. For example, recently, compound or "soft" glasses with high intrinsic optical nonlinearity (such as lead silicate, tellurite, and bismuth oxide) have been used to produce HFs with extreme values of effective fiber nonlinearity [21,42–49]. Work on high nonlinearity soft glass HFs is reviewed within this section. A number of nonsilica glasses transmit at wavelengths substantially beyond silica into the mid-IR (e.g., chalcogenide glasses), and microstructured fibers made from such materials promise a new means of providing power delivery at these wavelengths. Other classes of microstructured fiber have also begun to emerge, such as microstructured fibers with solid cladding designs, and these fibers are also described in this section.

The fabrication of soft glass microstructured fibers was reviewed briefly in Section 3.3.3. Note that in contrast to conventional fibers, microstructured fibers can be made from a single material, which eliminates the problems induced by the requirement that the core/cladding glasses are thermally and chemically matched. This opens up the prospect of using micro-structured fiber technology as a tool for realizing optical fibers from an extremely broad range of optical materials.

3.3.6.2 Extreme nonlinearity

As described in Section 3.3.2.2, the combination of a highly nonlinear glass composition and small-core/high NA HF geometry allows a dramatic increase of the fiber nonlinearity relative to conventional fibers. Soft glasses typically exhibit a higher linear refractive index than silica and thus can provide better confinement and smaller mode areas, in addition to higher material nonlinearity. For example, an air-suspended rod of bismuth oxide glass (which has a refractive index of 2.02 [111]) has a minimum $A_{eff} \approx 0.6\ \mu m^2$ at 1550 nm. When this is combined with the nonlinear refractive index of bismuth oxide glass, which is $3.2 \times 10^{-19}\ m^2/W$ [111], the maximum γ that could be achieved in an HF made from this material is ~2200/(W·km), a factor of 2 higher than the record nonlinearity achieved thus far for a chalcogenide fiber with conventional solid cladding [112].

Lead silicate glasses are one promising family of glasses for use in highly nonlinear index-guiding HFs. Although their intrinsic material nonlinearity (and indeed linear refractive index) is lower than chalcogenide and heavy metal oxide glasses [113], they offer higher thermal and crystallization stability and less steep viscosity–temperature curves while exhibiting low softening temperatures [114]. Among commercially available lead silicate glasses, the Schott glass SF57 exhibits the highest nonlinearity. The softening temperature of SF57 is 520°C [115], the nonlinear refractive index of this glass was measured to be $4.1 \times 10^{-19} rn^2/W$ at 1060 nm [116], and the linear refractive index of SF57 is ~1.8 at 1550 nm [115]. The zero dispersion wavelength for this glass is 1970 nm, and the material dispersion is strongly normal at 1550 nm.

Some key results that have been achieved in microstructured lead silicate fibers include effective nonlinearity coefficients (γ) as high as 640/(W·km) with a loss ~2.6 dB/m at 1550 nm, anomalous dispersion at 1550 nm, and Raman soliton generation [46]. Observations of the SSFS and pulse compression have also been made in SF57 HF [46]. Super-continuum generation has been observed in SF6 glass HF [43].

Bismuth oxide–based glasses are also attractive materials for nonlinear devices. It shows high nonlinearity but without containing toxic elements such as Pb, As, Se, and Te [117]. Moreover, the bismuth-based glass exhibits good mechanical, chemical, and thermal stability, which allows easy fiber fabrication process. A nonlinear fiber [118] and a short Er-doped fiber amplifier with broadband emission [119] have been developed from this glass. In addition, bismuth oxide-based fibers can be fusion spliced to a silica fiber [120], which offers easy integration to silica-based networks. Because of the high bismuth content, the glass exhibits a high linear and nonlinear refractive index of $n = 2.02$ and $n_2 = 3.2 \times 10^{-19} m^2/W$ at 1550 nm, respectively [118], and has a softening temperature of 550°C.

Bismuth oxide–based glass HFs with effective fiber nonlinearities as high as $\gamma = 1100/(W.km)$ have been realized [21]. In addition, the splicing of small-core bismuth HFs to conventional fibers has been reported [49]. To reduce the overall mode-mismatch loss, two intermediate buffer stages were used. The splices were mechanically strong with respect to strain in the axial direction. Although the total splicing losses achieved to date are still quite high (5.8 dB), they can largely be accounted for by mode mismatches at the buffer fiber interfaces. The introduction of an additional buffer stage should help to reduce the mode mismatch. Splicing of bismuth glass HF to silica fiber has resulted in two benefits in performance. One is the reduction of coupling losses by 0.9 dB relative to butt coupling. The other is the achievement of single-mode guidance in the bismuth HF at 1550 nm, although the fiber can support more than one mode in the case of free-space coupling.

Tellurite glasses, like lead silicate and bismuth glasses, offer high refractive index and high optical nonlinearity ($n_2 = 2.5 \times 10^{-19} m^2/W$) [44]. In addition, tellurite glass has good infrared transmittance and has a low phonon energy relative to other oxide glasses [121]. Furthermore, tellurite glasses are more stable than fluoride glasses, have higher rare-earth solubilities than chalcogenide glasses [121], and have an order of magnitude larger Raman gain peak than fused silica [122]. Tellurite glasses have a low softening temperature around 350° C [121].

Tellurite HFs with $\gamma = 48/(W \cdot km)$ and a loss of 5 dB/m at 1550 nm were reported [44]. First- and second-order Stokes stimulated Raman scattering were observed in 1 m of this tellurite HF. More recently, low-loss index-guiding tellurite fibers with a low of just 0.18 dB/m at 1550 nm and $\gamma \approx 675/(W \cdot km)$ have been reported [50]. This is the lowest low reported to date in a nonsilica micro-structured fiber. In addition, this fiber also has a zero-dispersion wavelength that has been shifted to near 1550 nm.

Both the effective fiber nonlinearity (γ) and effective fiber length (L_{eff}) determine the performance of a nonlinear device. The effective length of a fiber depends on the fiber's propagation loss via $L_{eff} = [1-\exp(-\alpha L)]/\alpha$. Small-core high-NA HFs based on bismuth oxide and lead silicate glasses have clearly higher fiber nonlinearity but also higher propagation loss compared with silica HFs or conventional nonlinear fiber. However, for short devices using ≤ 1 m fiber length, fiber losses of ≤ 2 dB/m can generally be tolerated. When the propagation loss is less than 2 dB/m, the effective fiber length is more than 80% of the real fiber length of ≤ 1 m, whereas the nonlinearity of nonsilica fibers can be up to 10 times higher than that of silica HFs. In other words, in compact devices using ≤ 1 m fiber length with ≤ 2 dB/m loss, the increase in fiber nonlinearity obtained by using nonlinear glass compositions clearly outweighs the decrease of the effective fiber length due to higher propagation losses. Thus, provided that relatively low-loss fibers can be produced, highly nonlinear compound glass HFs provide a route to better nonlinear performance than existing silica fibers (in terms of lower power consumption and/or shorter fiber length).

Most soft glasses have a high normal material dispersion at 1550 nm, which tends to dominate the overall dispersion of fibers with a conventional solid cladding structure, and for most glass compositions, near-zero dispersion at a wavelength of about 1550 nm cannot readily be achieved using conventional technologies. However, for many nonlinear device applications, anomalous or near-zero dispersion is required. Fortunately, the cladding geometry of a microstructured fiber can result in a large enough waveguide dispersion to allow the highly normal material dispersion to be overcome. (Indeed, as mentioned above, for example, anomalously dispersive lead silicate HFs at 1550 nm have been demonstrated [47].) The fact that fiber dispersion is anomalous also enables us to exploit soliton effects [46].

3.3.6.2 New transmission fibers

A range of nonsilica glasses can exhibit properties such as transparency in the mid-IR region and high solubility of rare-earth ions that are not available in silica glass. Conventional fiber approaches have thus far made limited progress in developing low-loss single-mode soft glass fibers due to difficulties in finding compatible core/cladding materials. For this reason, single material nonsilica index-guiding HFs are of particular promise for applications in mid-IR region and active devices.

One challenge associated with the use of soft glass–based fibers for high power applications is the onset of intensity-dependent nonlinear effects and optical damage. The use of single-mode LMHF designs [96] is one means of minimizing nonlinear effects. Because the nonlinear refractive index (n_2) of nonsilica glass is typically in the range of (1–50) x $10^{-19} m^2/W$, higher than that of silica glass by 1–2 orders of magnitude, large-mode area designs ($\gg 100\ \mu m^2$) are required to reduce the effective nonlinearity (g) even to the $60/(W \cdot km)$ level. In contrast to the nonlinear fiber designs described in Section 3.3.6.2, an LMHF cladding needs to contain a high feature count to provide low NA guidance (i.e., by containing a large number of relatively small air holes).

An HF with a mode area of 40 μm^2 at 800 nm fabricated from the Schott lead silicate glass SF6 using the conventional capillary-stacking technique was reported [123]. This fiber had a four-ring micro-structured cladding with a hole-to-hole spacing Λ of 4.3 μm. Robust single-mode guidance at 800 nm was observed. Although, as described above, this value of mode area is still not large enough to avoid intensity-dependent nonlinear effects, this work demonstrates that the capillary-stacking technique could be exploited to fabricate soft glass HFs with a complex holey cladding. It should be feasible to fabricate single-mode fibers with larger-mode areas in high-index glasses using this technique in the future.

3.3.6.4 Solid microstructured fibers

Research to date on silica microstructured fibers has shown that the combination of wavelength-scale features and a large refractive index contrast is a powerful means of obtaining fibers with a broad range of useful optical properties. However, there are some practical drawbacks associated with the use of air/ glass fibers. When compared with solid fibers, air/ glass microstructured fibers are challenging to splice, polish, and taper, and when the cross-section is largely composed of air, they can be fragile. In addition, it can be challenging to fabricate kilometer-scale HFs with identical and controllable cladding configurations. This is because the transverse profile of a drawn microstructured fiber is sensitive to the effect of pressure inside the holes, surface tension at the air-glass boundaries, and temperature gradients present during the fiber drawing process. To prevent the collapse of the holey microstructure during drawing, the holes within the preform are often sealed, and consequently the air pressure inside the structure changes during fiber drawing. Hence, the precise details of the final fiber microstructure are typically time dependent as well as dimension dependent. Note that the optical characteristics of microstructured fibers can be sensitive to the cladding configuration, and even minor changes in the microstructure can cause noticeable deviations in properties such as dispersion.

The development of microstructured fibers that have a solid microstructured cladding promises to eliminate these drawbacks. These fibers combine the practicality of a solid cladding with the design flexibility provided by the transverse microstructure. Such fibers have the potential to advance the development of both index-guiding and photonic band-gap-guiding microstructured fibers. Two different geometric implementations of this solid microstructured fiber concept have recently been realized.

The first approach is to replace the air holes in the transverse structure of an air/glass microstructured preform with low index glass regions to produce an index-guiding solid variant of an HF [41]. Another approach is to form a fiber in which the cladding is defined by a series of thin-nested concentric layers of two or more different glasses [124]. Such fibers are often referred to as Bragg fibers (see Chapter 4). In both approaches, the basic requirement is the identification of materials that are thermally and chemically matched so that they can be drawn into optical fiber and provide a sufficient refractive index contrast to allow light to be confined either by index-guiding or photonic bandgap effects without requiring unfeasibly large numbers of cladding features. Note that although all-solid microstructured fibers may have some practical advantages, as suggested above, this approach clearly restricts the range of materials that can be used relative to single-material fiber designs. Here an example of each of the four possible classes of solid fiber is briefly reviewed.

Feng et al. [41] described the development of all-solid index-guiding HF based on two thermally matched silicate glasses with a large index contrast. A borosilicate glass with a refractive index of $n = 1.76$ at 1550 nm was selected as the background material for this fiber. A glass with $n = 1.53$ at 1550 nm was used as the material to fill the holes. These glasses were selected because of their mechanical, rheological, thermodynamic, and chemical compatibility. Calculations indicate that it is possible to design fibers from these two materials with negligible confinement loss provided that the d/Λ ratio is sufficiently large, as is the case for the fibers described below [41].

Rods of the low index glass were inserted into the tubes made from the higher index glass, both of which were drilled from bulk glass samples. These rod–tube structures were caned on a fiber drawing tower and stacked around a core rod of the high index glass within a high index jacket tube. This preform was then drawn using a two-stage procedure to produce two fibers: one with an outer diameter of 440 μm and the other with an outer diameter of 220 μm. Figure 3.3.7 compares the cladding configurations of the 1-mm cane, 440-μm fiber, and 220-μm fiber. Note that all the low index (black) regions retain their circularity, regardless of the draw-down ratio. In addition, the value of the d/Λ ratio is ~0.81 regardless of the fiber outer diameter. These fibers thus provide a practical way of avoiding structural deformations during fiber drawing.

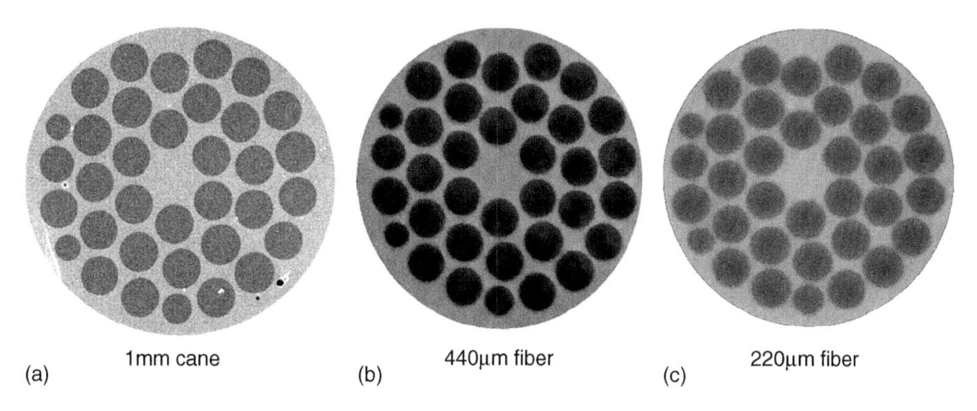

Fig. 3.3.7 SEM imags of microstructured cladding in (a) 1-mm cane, (b) 440-μm outer diameter (OD) fiber with A = 4 μm, and (c) 220-μm OD fibor with $\Lambda = 2$ μm.

Even though the index contrast between silica/air leads to a similar minimum mode area as this combination of materials, the significantly larger material nonlinearity (n_2) of these soft glass materials results in a dramatic improvement in the non-linearity. Hence, although the maximum nonlinearity that can be achieved in a silica/air HF is ~60/(W·km), more than 500/(W·km) is possible in using this combination of high index glasses. The fiber in Fig. 3.3.7c has a measured nonlinearity of $\gamma \approx 230/(W\cdot km)$.

Using the commercial Schott glasses LLF1 and SF6, Luan et al. [125] reported the fabrication of an all-solid PBGF. The periodic cladding in this fiber was formed by stacking LLF1/SF6 tubes in a hexagonal pattern, as was done in the index-guiding fiber described above. In this case, light was guided via photonic bandgap effects within the relatively lower index core region made from LLF1 glass. One advantage of this approach relative to air/silica PBGFs is that it appears to avoid the problem of surface modes with the bandgaps, which can restrict the practically useful wavelength range in these fibers.

Index-guiding fibers with a one-dimensionally multilayered structured cladding and a solid core have also been realized [126], and an example is shown in Fig. 3.3.1f (dark regions represent the low index regions). In this case, the structured preform was formed by extruding an alternating stack of thin films into a preform with a macroscopic ring structure. Single-mode guidance and high effective fiber nonlinearity [$\gamma \approx 260/(W\cdot km)$] have been observed in this fiber.

The first reported example of a fiber with a high index-contrast solid multilayered cladding guided light in an air core via photonic bandgap effects [124]. To achieve high index contrast in the layered portion of the fiber, a chalcogenide glass with a refractive index of 2.8 (arsenic triselenide) was combined with a high glass-transition temperature thermoplastic polymer with a refractive index of 1.55, poly(ether sulfone). The glass layers were thermally evaporated onto a polymer film and the coated film rolled to form a hollow multilayered tubular preform, which was drawn into a fiber with submicron layer thicknesses. Note that the polymer material is not transparent at the operating wavelength, and fiber losses below 1 dB/m were made possible by the short penetration depths of electromagnetic waves in the high refractive index contrast photonic crystal structure. Using a fiber with a fundamental photonic bandgap centered near 10.6 μm and a hollow core with a diameter of 700 μm, CO_2 laser emission was successfully transmitted. No damage to the fibers was reported when a laser power density of ~300 W=cm^2 was coupled into the hollow fiber core.

3.3.7 PBGFs

The index-guiding HFs discussed in Sections 3.3.5 and 3.3.6 can to a first approximation be regarded as a variant of the traditional step-index fiber with a larger and wavelength dependent index contrast between core and cladding. In contrast, with PBGFs, light is localized at a defect placed in a photonic bandgap material. The periodically structured material acts acts to suppress the transverse propagation of light at the frequency of the guided mode. Here progress in the development of PBGFs is reviewed for the two fiber types that have been realized experimentally: *honeycomb fibers,* whose properties in many respects resemble those of index-guiding microstructured fibers, and *air-guiding fibers,* which can only be realized through the photonic bandgap effect.

For the photonic bandgap-guiding mechanism to operate, the bandgap must extend over the whole plane perpendicular to the direction of propagation. In silica/air structures such a gap can only be achieved for a finite (non-zero) longitudinal propagation constant, the minimum value required being dependent on the fiber structure. The simple hexagonal arrangement of cladding holes commonly used for making index-guiding microstructured fibers only provides a complete bandgap for air-filling fractions of 30% or higher. In contrast, arranging the holes on a honeycomb lattice makes it possible to open up complete gaps at much lower air-filling fractions (<1%) [127]. For this reason, the first PBGF experimentally fabricated was based on the honeycomb lattice. The perfect honeycomb lattice can be thought of as an array of silica rods, separated by rings of air holes. The core region of the waveguide is created by introducing an extra air hole in the central silica rod, thereby lowering the effective index of this region.

It is a general feature of PBGFs that the central defect is typically created by lowering the effective index of the core region (whereas for index-guiding fibers the effective core index is raised to trap the guided mode). The magnitude and position of the bandgaps are controlled by the radius of the cladding holes. Depending on the size of the central core hole, the fundamental defect mode may be pushed into either the first or a higher order bandgap. In principle, any of the bandgaps may support a number of guided modes. By proper tuning of the core and cladding hole configurations, one can create fibers that are either single or multimode. Note that this class of fibers allows the possibility of guiding higher order modes without guiding the fundamental mode. For example, the honeycomb PBGF reported in [128] guided third-order mode only. These provided the first proof-of-principle of the photonic bandgap waveguiding mechanism for honeycomb fiber designs.

Somewhat surprisingly, modeling results imply that the honeycomb PBGFs have a similar potential for dispersion engineering as do index-guiding fibers [129]. For example, chromatic group-velocity dispersion curves may be flattened over large wavelength intervals, zero dispersion may be obtained at short wavelengths, and fibers can have large anomalous group-velocity dispersion coefficients (several hundred ps/nmkm). As is the case for index-guiding fibers, by breaking the 6-fold symmetry of the structure, it is possible to produce polarization-maintaining PBGFs [130]. This can be done either by introducing an asymmetric core defect or an asymmetric cladding configuration. One unique possibility birefringent PBGFs offer is a means of realizing fibers that support a single polarization (i.e., one of the polarization states can be pushed out of the photonic bandgap).

One of the greatest attractions of PBGF technology is that it can allow light to be guided in an air core. This occurs when modes with an effective index of ~1 fall within the bandgap. In other words, for light to be guided in air, the effective index of the cladding mode at the lower boundary of the gap falls above the light line. As mentioned above, the hexagonal lattice structure has this property for sufficiently large air-filling fractions. As an example, Fig. 3.3.8 shows the bandgap diagram for a structure with a filling fraction of 70% [131]. It can be seen that several bandgaps cross the air line. To confine an air-guided mode within these bandgaps, a core defect consisting of a rather large air hole must be introduced. In the inset of Fig. 3.3.8, a structure is shown in which the core defect has been obtained by replacing seven holes by a single air hole. This structure was found to support both a fundamental and a second-order guided mode, though not at the same frequency. The traces of the two modes are shown in the upper inset of Fig. 3.3.8. Inside the bandgap the modes are bound, whereas outside they appear as leaky resonances within the bands of cladding modes. For both modes, guidance only takes place within a rather narrow frequency interval. Furthermore, the guidance properties of the structure depend strongly on the radius of the core defect.

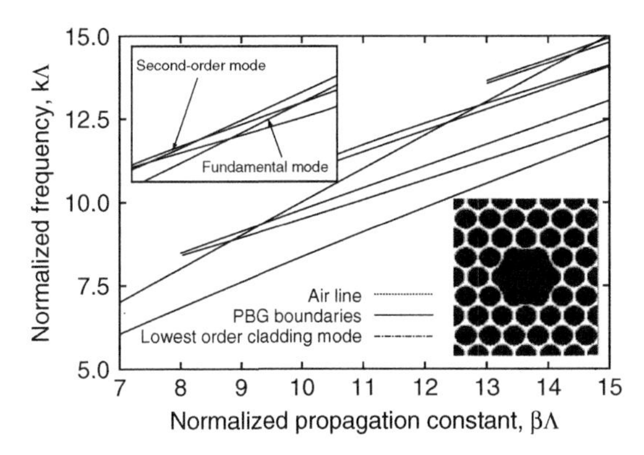

Fig. 3.3.8 Photonic bandgap (PBG) boundaries and defect mode traces (upper inset) for the air-guiding PBGF discussed in the text. The defect modes are confined when they fall inside the PBG (and leaky elsewhere). The fiber microstructure is shown in the lower inset. (From [131].)

The fabrication of air-guiding PBGFs was first reported in [13]. In these first fibers, structures with $\Lambda \approx 5\ \mu m$ and an air-filling fraction of 30–50% resulted in several air-guiding transmission bands over a fiber length of a few centimeters, in good agreement with predictions. Since this first report, considerable progress has been made to reduce the loss of PBGFs, and losses of 1.7 dB/km can now be achieved at 1550 nm [54]. As noted in Section 3.3.3.1, to produce a bandgap fiber preform, the fiber core can be formed in a number of ways, and the two cases explored in the most detail thus far are 7-cell and 19-cell air cores. The typical air-core PBGF, seen in Fig. 3.3.1c, has a 7-cell core.

In addition to using microstructured optical fibers as light pipes, they can be used to manipulate light propagating across the fiber [132]. Fundamental and higher order bandgaps have been observed experimentally via the spectral features that result from the periodic nature of the fiber microstructure in the transverse direction.

The dispersion properties of PBGFs were reviewed briefly in Section 3.3.2.3. It is important to note that the dispersion of PBGFs is strongly wavelength dependent. Typically, negative group-velocity dispersion values (normal dispersion) are found at the short wavelength edge of the bandgap, passing through zero dispersion near the center of the band-gap with positive values (anomalous dispersion) at longer wavelengths. The dispersion slope increases rapidly near the upper band edge and decreases rapidly near the lower band edge [133, 134], and values of more than ± 2000 ps/nmkm are typical. In effect, the dispersion of the fiber can be tuned by tuning the operating wavelength with respect to the location of the bandgap (or vice versa). The dispersion slope can be significantly modified by changing the geometry of the core region [133].

The loss, dispersion, and leakage properties of the fundamental mode of a PBGF can all be affected by the presence of surface modes within the bandgap [135]. This occurs because the air-guided modes can couple to surface modes within the bandgap. Surface modes can be supported within the glass region surrounding the air core, and the quantity and configuration of glass surrounding the core determines the properties of these states. Some fiber designs do not have surface states within the bandgap, and this is desirable because surface modes can reduce the effective bandgap width and increase the fiber loss. On a more positive note, surface modes can lead to narrowband regions of strong anomalous dispersion (via avoided crossings) that may find use in novel device applications [136,137].

Now that high-quality, low-loss, air-guiding fibers can be made, many new applications can be envisaged. These include low-loss/high-power transmission at wavelengths where the absorption in silica is high, in gas-sensing applications, and for spectral filtering exploiting the narrowband nature of the guided-mode transmission bands. Applications for laser-assisted atom transport or even as particle accelerators have also been proposed [138].

One recently demonstrated example of an application of air-guiding PBGF is all-fiber chirped pulse amplification [137]. Pulses from a wavelength- and duration-tunable femtosecond/picosecond source at 10 GHz were dispersed in 100 m of dispersion compensating fiber before being amplified in an erbium-doped fiber amplifier and subsequently recompressed in 10 m of the anomalously dispersive PBGF. Pulses as short as 1.1 ps were obtained. The advantage of air-core fibers for this application is that they present negligible nonlinearity and thus can potentially be used to obtain ultra-high pulse peak powers. The novel dispersion properties of air-core PBGFs have also been used for dispersion compensation within a fiber laser [139].

Hollow-core PBGFs are also ideally suited to deliver high power laser beams. Recent work in this area has concluded that 7-unit cell cores are currently most suitable for transmission of femtosecond and sub-picosecond pulses, whereas larger cores (e.g., 19-cell cores) are better for delivering nanosecond pulses and continuous-wave beams [140].

3.3.8 Conclusion and the future

Microstructured optical fibers have now developed to the point where they are not only of interest from a research perspective, but are also becoming available commercially. As this chapter demonstrated, good quality index-guiding HFs and PBGFs based on a cladding structure with a triangular lattice of air holes embedded in pure silica glass can now routinely be fabricated over a wide parameter range. More complicated structures, such as double-clad fibers and active fibers, have also been demonstrated.

The unique properties and design flexibility of these fibers opens up a wide range of possible applications as functional components in fiber communication networks (including devices for amplification or dispersion compensation), in novel broadband sources, or for high-power transmission, to name a few. One area in which significant progress has been made recently is the field of highly nonlinear index-guiding HFs. The generation of broad super-continuum spectra and all-optical data regeneration are examples of the significant advances that have resulted from this silica-based HF technology. Moreover, nonsilica index-guiding HFs with extremely high nonlinearity can now be routinely fabricated [21], and such fibers promise to offer nonlinear fiber devices at unprecedentedly low operating powers (1–10 mW) and short device lengths (0.1–1 m).

By filling the holes of these fibers with a range of materials, it should be possible to significantly extend their functionality. One example is the filling of index-guiding HFs with a high index liquid to form tunable PBGFs [141].

The synergy of novel nonsilica glass materials and microstructured fiber technology promises a broad range of new and potentially useful optical fibers. The large index contrast possible in nonsilica micro-structured fibers is promising for the development of fibers based on photonic bandgap effects. However, it has been demonstrated that increasing the refractive index contrast beyond that available in air/silica does not necessarily broaden the photonic bandgaps that are available [142]. One particularly promising application of this new fiber type will be the development of new air-guiding fibers for broadband high power IR transmission.

It was demonstrated that solid microstructured fibers can be produced using a relatively small number of cladding features. It is expected that the use of a solid fiber structure may lead to a number of practical advantages relative to air/glass fibers. For example, edge polishing, angle polishing, and splicing should all be more straightforward in solid microstructured fibers. By combining a solid cladding with an air core, it is possible to demonstrate that low fiber attenuation can be achieved through structural design rather than high-transparency material selection [124]. As the fabrication techniques used to produce structured preforms and to draw high quality fibers continue to improve, it is anticipated that more novel optical properties and promising applications will continue to emerge.

Acknowledgments

I thank a number of colleagues at the ORC, University of Southampton who have made many important contributions to the research described within this chapter. In particular, warm thanks to Joanne Baggett, Heike Ebendorff-Heidepriem, Xian Feng, Vittoria Finazzi, Kentaro Furusawa, John Hayes, Ju Han Lee, Periklis Petropoulous, Jonathan Price, and David Richardson. I also acknowledge the support of a Royal Society University Research Fellowship.

References

1. P. Kasier, E.A.J. Marcatili and S.E. Miller, "A new optical fiber," *Bell Sys. Tech. J.* **52**, 265–269 (1973).
2. P. Kaiser and H.W Astle, "Low-loss single-material fibers made from pure fused silica," *Bell Sys. Tech. J.* **54**, 1021–1039 (1974).
3. E. Yablonovitch, "Inhibited spontaneous emission in solid-state physics and electronics," *Physical Rev. Lett.* **58**, 2059–2062 (1987).
4. M.A. McCord and R. Michael J., "Electron beam lith ography," in P.R. Coudry (Ed.), *Handbook of Micro-lithography, Micromachining and Microfabrication,* **1**, SPIE Press, Bellingham, WA (1997).
5. T.A. Birks, P.J. Roberts, P.St.J. Russell, D.M. Atkin and T.J. Shepherd, "Full 2-D photonic bandgaps in silica/air structures," *Electron. Lett.* **31**,1941–1943 (1995).
6. J.C. Knight, T.A. Birks, P.St.J. Russell and D.M. Atkins, "All-silica single-mode optical fiber with photonic crystal cladding," *Optics Lett.* **21**, 1547–1549 (1996).
7. T.A. Birks, J.C. Knight, P.St.J. Russell, "Endlessly single-mode photonic crystal fiber," *Opt. Lett.* **22**, 961–963 (1997).
8. T.M Monro, P.J. Bennett, N.G.R. Broderick and D.J. Richardson, "Holey fibers with random cladding distributions," *Opt. Lett.* **25**, 206–208 (2000).
9. T.M. Monro, D.J. Richardson, N.G.R. Broderick and P.J. Bennett, "Holey optical fibers: an efficient modal model," *J. Lightwave Technol.* **17**, 1093–1102 (1999).
10. B.J. Eggleton, C. Kerbage, P. Westbrook, R.S. Winde- ler and A. Hale, "Microstructured optical fiber de vices," *Opt. Express* **9**, 698–713 (2001).
11. J. Broeng, S.E. Barkou, A Bjarklev, J.C. Knight, T.A. Birks and P.St.J. Russell, "Highly increased photonic bandgaps in silica/air structures," *Opt. Commun.* **156**, 240–244 (1998).
12. J.C Knight, J. Broeng, T.A. Birks and P.St.J. Russell, "Photonic bandgap guidance in optical fibers," *Science* **282**, 1476–1478 (1998).
13. R.F. Cregan, B.J. Mangan, J.C. Knight, T.A. Birks, P.St.J. Russell, P.J. Roberts and D.C. Allan, "Single- mode photonic bandgap guidance of light in air," *Science* **285**, 1537–1549 (1999).
14. T. Hasegawa, E. Sasaoka, M. Onishi, M. Nishimura, Y. Tsuji and M. Koshiba, "Hole-assisted lightguide fiber for large anomalous dispersion and low optical loss," *Opt. Express* **9**, pp. 681–686 (2001).
15. J.K. Sahu, C.C. Renaud, K. Furusawa, R. Selvas, J.A. Alvarez-Chavez, D.J. Richardson and J. Nilsson, "Jacketed air-clad cladding pumped ytterbium-doped fibre laser with wide tuning range," *Electron. Lett.* **37**, pp. 1116–1117 (2001).
16. J.H.V. Price, K. Furusawa, T.M. Monro, L. Lefort and D.J. Richardson, "Tunable, femtosecond pulse source operating in the range 1.06-1.33mm based on an Yb3þ - doped holey fiber amplifier," *J. Opt. Soc. Amer.* B **19**, pp. 1286–1294 (2002).
17. G. P. Agrawal, *Nonlinear Fiber Optics*, Academic Press, Boston (2001).
18. T. Okuno, M. Onishi, T. Kashiwada, S. Ishikawa and M. Nishimura, "Silica-based functional fibers with enhanced nonlinearity and their applications," *IEEE J. Sel. Top. Quant.* **5**, 1385–1391 (1999).
19. V. Finazzi, T.M. Monro and D.J. Richardson, "Small core silica holey fibers: nonlinearity and confinement loss trade-offs," *J. Opt. Soc. Amer. B* **20**, 1427–1436 (2003).
20. W. Belardi, J.H. Lee, K. Furusawa, Z. Yusoff, P. Petropoulos, M. Ibsen, T. M. Monro and D.J. Richardson, "A 10Gbit/s tuneable wavelength converter based on four-wave mixing in highly non linear holey fibre," in Proc. ECOC, Copenhagen, Denmark, September 2002. Postdeadline paper PD1.2.
21. H. Ebendorff-Heidepriem, P. Petropoulos, S. Asimakis, V. Finazzi, R. C. Moore, K. Frampton, F. Koizumi, D. J. Richardson and T. M. Monro, "Bismuth glass holey fibers with high nonlinearity," *Opt. Express* **12**, 5082–5087 (2004).
22. J.C. Baggett, T.M. Monro, K. Furusawa and D.J. Richardson, "Comparative study of large mode holey and conventional fibers," *Opt. Lett.* **26**, 1045–1047 (2001).
23. M. N. Petrorich, V. Finazzi, T. M. Monro, D. J. Richardson, E. J. O'Driscoll, M. A. Watson, T. Del-mark, "Photonic bandgap fibers for infrared applica tions," 2nd EMRS DTC Conference, Edinburgh, UK (2005).
24. J.C. Knight, J. Arriaga, T.A. Birks, A. Ortigosa-Blanch, J.W. Wadsworth and P.St.J. Russell, "Anomalous dispersion in photonic crystal fiber," *IEEE Photon. Technol. Lett.*, **12**, 807–809 (2000).
25. J.K. Ranka, R. S. Windeler and A.J. Stentz, "Visible continuum generation in air silica microstructure optical fibers with anomalous dispersion at 800 nm," *Opt. Lett.* **25**, 25–27 (2000).
26. K. Saitoh, M. Koshiba, T. Hasegawa and E. Sasaoka, "Chromatic dispersion control in photonic crystal fibers: application to ultra-flattened dispersion," *Opt. Express* **11**, 843–852 (2003).
27. T.A. Birks, D. Mogilevtsev, J.C. Knight and P.St.J. Russell, "Dispersion compensation using single material fibers," *IEEE Photon. Technol. Lett.* **11**, 674–676 (1999).
28. D.G. Ouzounov, F.R. Ahmed, D. Muller, N. Venkataraman, M.T. Gallagher, M.G. Thomas, J. Silcox, K.W. Koch and A.L. Gaeta, "Generation of megawatt optical solitons in hollow core photonic band-gap fibers," *Science* **301**, 1702–1704 (2003).
29. K. Saitoh and M. Koshiba, "Leakage loss and group velocity dispersion in air-core photonic bandgap fibers," *Opt. Express***11**, 3100–3109 (2003).
30. H. Lim and F. W. Wise, "Control of dispersion in a femtosecond ytterbium laser by use of hollow-core photonic bandgap fiber," *Opt. Express* **12**, 2231–2235 (2004).
31. M.J. Steel, T.P. White, C.M. de Sterke, R.C. McPhedran and L. C. Botten, "Symmetry and degeneracy in microstructured optical fibers," *Opt. Lett.* **26**, 488–490 (2001).
32. T. P. Hansen, J. Broeng, E. B. Libori, E. Knudsen, A. Bjarklev, J. R. Jensen and H. Simonsen, "Highly birefringent index-guiding photonic crystal fibers," *IEEE Photon. Technol. Lett.* **13**, 588–590 (2001).
33. A. Bjarklev, J. Broeng, S. E. Barkou, E. Knudsen, T. Søndergaard, T. W. Berg and M. G. Dyndgaard, "Polarization properties of honeycomb-structured photonic bandgap fibres," *J. Opt.* A **2**, 584–588 (2000).
34. X. Chen, M. Li, N. Venkataraman, M. T. Gallagher, W. A. Wood, A. M. Crowley, J. P. Carberry, L. A. Zenteno and K. W. Koch, "Highly birefringent hollow-core photonic bandgap fiber," *Opt. Express* **12**, 3888–3893 (2004).

35. M. Fuochi, J.R. Hayes, K. Furusawa, W. Belardi, J.C. Baggett, T.M. Monro and D.J. Richardson, "Polariza tion mode dispersion reduction in spun large mode area silica holey fibres," *Opt. Express* **12**,1972–1977 (2004).
36. T.M. Monro, D. J. Richardson and P. J. Bennett, "Developing holey fibres for evanescent field devices," *Electron. Lett.* **35**, 1188–1189 (1999).
37. G. Bouwmans, F. Luan, J. C. Knight, P. St. J. Russell, L. Farr, B. J. Mangan and H. Sabert, "Properties of a hollow-core photonic bandgap fiber at 850nm wave length," *Opt. Express* **11**, 1613–1620 (2003).
38. T. Ritari, J. Tuominen, H. Ludvigsen, J. C. Petersen, T. Sørensen, T. P. Hansen and H. R. Simonsen, "Gas sensing using air-guiding photonic bandgap fibers," *Opt. Express* **12**, 4080–4087 (2004).
39. G. Humbert, J. C. Knight, G. Bouwmans, P. S. J. Russell, D. P. Williams, P. J. Roberts and B. J. Mangan, "Hollow core photonic crystal fibers for beam delivery," *Opt. Express* **12**, 1477–1484 (2004).
40. M. van Eijkelenborg, M. Large, A. Argyros, J. Zagari, S. Manos, N.A. Issa, I.M. Bassett, S.C. Fleming, R.C. McPhedran, C.M. deSterke and N.A. P. Nicorovici, "Microstructured polymer optical fibre," *Opt. Express* **9**, 319 (2001).
41. X. Feng, T.M. Monro, P. Petropoulos, V. Finazzi and D.W. Hewak, "Solid microstructured optical fiber," *Opt. Express* **11**, pp. 2225–2230 (2003).
42. K.M. Kiang, K. Frampton, T.M. Monro, R. Moore, J. Trucknott, D.W. Hewak and D.J. Richardson, "Extruded singlemode non-silica glass holey optical fibres," *Electron. Lett.* **38**, 546–547 (2002).
43. V.V.R.K. Kumar, A.K. George, W.H. Reeves, J.C. Knight and P.S.J. Russell, "Extruded soft glass photonic crystal fiber for ultrabroadband superconti- nuum generation," *Opt. Express***10**, 1520–1525 (2002).
44. V.V.R.K. Kumar, A.K. George, J.C. Knight and P.S.J. Russell, "Tellurite photonic crystal fiber," *Opt. Express* **11**, pp. 2641–2645 (2003).
45. T.M. Monro, K.M. Kiang, J.H. Lee, K. Frampton, Z. Yusoff, R. Moore, J. Trucknott, D.W. Hewak, H.N. Rutt and D.J. Richardson, "High nonlinearity extruded single-mode holey optical fibers," Proc. Optical Fiber Communications Conference (OFC 2002), Anaheim, California, postdeadline paper FA1-1 (2002).
46. P. Petropoulos, T.M. Monro, H. Ebendorff-Heide- priem, K. Frampton, R.C. Moore, H.N. Rutt, D. J. Richardson, "Soliton-self-frequency-shift effects and pulse compression in anomalously dispersive high nonlinearity lead silicate holey fiber," Proc. Optical Fiber Communications Conference (OFC), Atlanta, Georgia, postdeadline paper PD03 (2003).
47. P. Petropoulos, H. Ebendorff-Heidepriem, V. Finazzi, R.C. Moore, K. Frampton, D.J. Richardson and T.M. Monro, "Highly nonlinear and anomalously dispersive lead silicate glass holey fibers," *Opt. Express* **11**, 3568–3573 (2003).
48. H. Ebendorff-Heidepriem, P. Petropoulos, V. Finazzi, K. Frampton, R.C. Moore, D.J. Richardson and T.M. Monro, "Highly nonlinear bismuth-oxide-based glass holey fiber," Proc. Optical Fiber Communications Conference (OFC), Los Angeles, California, paper ThA4 (2004).
49. P. Petropoulos, H. Ebendorff-Heidepriem, T. Kogure, K. Furusawa, V. Finazzi, T.M. Monro and D.J. Richardson, "A spliced and connectorized highly non linear and anomalously dispersive bismuth-oxide glass holey fiber," Proc. Conference on Lasers and Electro- Optics (CLEO), San Francisco, California, paper CTuD (2004).
50. A. Mori, K. Shikano, K. Enbutsu, K. Oikawa, K. Naganuma, M. Kato and S. Aozasa, "1.5 μm band zero-dispersion shifted tellurite photonic crystal fibre with a nonlinear coefficient *g* of 675 W"[1]km"[1]," Proc. European Conference on Optical Communications (ECOC) (2004).
51. A. D. Fitt, K. Furusawa, T. M. Monro, C. P. Please and D. J. Richardson, "The mathematical modelling of capillary drawing for holey fibre manufacture," *J. Eng. Math.* **43**, 201–227 (2002).
52. C. J. Voyce, A. D. Fitt and T. M. Monro, "Mathemat ical model of the spinning of microstructured fibres," *Opt. Express* **12**, 5810–5820 (2004).
53. K. Tajima, J. Zhou, K. Kurokawa and K. Nakajima, "Low water peak photonic crystal fibres," Proc. European Conference on Optical Communication (ECOC), Rimini, Italy, postdeadline paper Th4.1.6 (2003).
54. B. Mangan, L. Farr, A. Langford, P.J. Roberts, D.P. Williams, F. Couny, M. Lawman, M. Mason, S. Coupland, R. Flea, H. Sabert, T.A. Birks, J.C. Knight and P.St.J. Russell, "Low loss (1.7 dB/km) hollow core photonic bandgap fiber," Proc. Optical Fiber Communications (OFC), Anaheim, paper PDP24 (2004).
55. L. Farr, J.C. Knight, B.J. Mangan and P.J. Roberts, "Low loss photonic crystal fibre," in Proc ECOC, paper PD1.3, Copenhagen, Denmark (2002).
56. A. Bjarklev, J. Broeng, S.E. Barkou and K. Dridi., "Dispersion properties of photonic crystal fibres," 24th European Conference on Optical Communication (ECOC), Madrid 1,135–136 (1998).
57. J. Riishede, S.B. Libori, A. Bjarklev, J. Broeng and E. Knudsen, "Photonic crystal fibers and effective index approaches," Proceedings of the 27th European Conference on Optical Communications (ECOC), Th.A.1.5 (2001).
58. K.M. Ho, C.T. Chan and C.M. Soukoulis, "Existence of a photonic gap in periodic dielectric structures," *Phys. Rev. Lett.* **65**, 3152–3155 (1990).
59. R.D. Meade, A.M. Rappe. K.D. Brommer, J.D. Joan- nopoulos and O.L. Alerhand, "Accurate theoretical analysis of photonic band-gap materials," *Phys. Rev. B* **48**, 8434–8437 (1993).
60. B.J. Eggleton, P.S. Westbrook, R.S. Windeler, S. Spalter and T.A Strasser, "Grating resonances in air/silica microstructured optical fibers," *Opt. Lett.* **24**, 1460–1462 (1999).
61. D. Mogilevtsev, T.A. Birks and P.St.J. Russell, "Group-velocity dispersion in photonic crystal fibers," *Opt. Lett.* **23**, 1662–1664 (1998).
62. T.M. Monro, D.J. Richardson, N.G.R. Broderick and P.J. Bennett, "Modelling large air fraction holey op tical fibers," *J. Lightwave Technol.* **18**, 50–56 (2000).
63. T.M. Monro, N.G.R. Broderick and D.J. Richardson, "Exploring the optical properties of holey fibres," NATO Summer School on Nanoscale Linear and Nonlinear Optics, Erice, Sicily, July (2000).
64. P.J. Bennett, T.M. Monro and D.J. Richardson, "Towards practical holey fibre technology: fabrication splicing modeling and characterization," *Opt. Lett.* **24**, 1203–1205 (1999).
65. F. Brechet, J. Marcou, D. Pagnoux and P. Roy, "Complete analysis of the characteristics of propagation into

photonic crystal fibers, by the finite element method," *Opt. Fiber Technol.* **6**, 181–191 (2000).

66. H.P. Uranus and H. Hoekstra, "Modelling of micro- structured waveguides using a finite-element-based vectorial mode solver with transparent boundary conditions," *Opt. Express* **12**, 2795–2809 (2004).
67. T.P. White, R.C. McPhedran, C.M. deSterke, L.C. Botten and M.J. Steel, "Confinement losses in micro-structured optical fibers," *Opt. Lett.* **26**, 1660–1662 (2001).
68. L. Poladian, N.A. Issa and T.M. Monro, "Fourier decomposition algorithm for leaky modes of fibres with arbitrary geometry," *Opt. Express* **10**, 449–454 (2002).
69. E. Kerrinckx, L. Bigot, M. Douay and Y. Quiquempois, "Photonic crystal fiber design by means of a genetic algorithm," *Opt. Express* **12**, 1990–1995 (2004).
70. N.G.R. Broderick, T.M. Monro, P.J. Bennett and D.J. Richardson, "Nonlinearity in holey optical fibers: measurement and future opportunities," *Opt. Lett.* **24**, 1395–1397 (1999).
71. N. Nakazawa, H. Kubota and K. Tamura, "Random evolution and coherence degradation of a high-order optical soliton train in the presence of noise," *Opt. Lett.* **24**, 318–320 (1999).
72. P. Petropoulos, T.M. Monro, W. Belardi, K. Furusawa, J.H. Lee and D. J. Richardson, "2R-regenerative all-optical switch based on a highly nonlinear holey fiber," *Opt. Lett.* **26**, 1233–1235 (2001).
73. F. Futami, S. Watanabe and T. Chikama, "Simultan eous recovery of 20 × 20 GHz WDM optical clock using supercontinuum in a nonlinear fiber," in Proc. ECOC, (2000).
74. J. Hansryd and P.A. Andrekson, "Broad-band continuous-wave-pumped fiber optical parametric amplifier with 49-dB gain and wavelength-conversion efficiency," *Photon. Technol. Lett.* **13**, 194–196 (2001).
75. F. Druon, N. Sanner, G. Lucas-Leclin, P. Georges, R. Gaumé, B. Viana, K.P. Hansen and A. Petersson, "Self-compression of 1-um femtosecond pulses in a photonic crystal fiber," in Proc. CLEO (2002).
76. J.H. Lee, Z. Yusoff, W. Belardi, M. Ibsen, T.M. Monro, B. Thomsen and D.J. Richardson, "A holey fiber based WDM wavelength converter incorporating an apo- dized fiber Bragg grating filter," in Proc. CLEO (2002).
77. J.E. Sharping, M. Fiorentino, P. Kumar and R.S. Windeler, "All optical switching based on cross-phase modulation in microstructure fiber," *Photon. Technol. Lett.***14**, 77–79 (2002).
78. H. Takara, T. Ohara, K. Mori, K. Sato, E. Yamada, K. Jinguji, Y. Inoue, T. Shibata, T. Morioka and K.-I. Sato, "Over 1000 channel optical frequency chain generation from a single supercontinuum source with 12.5 GHz channel spacing for DWDM and frequency standards," in Proc. ECOC (2000).
79. K.P. Hansen, J.R. Jensen, C. Jacobsen, H.R. Simonsen, J. Broeng, P.M.W. Skovgaard, A. Petersson and A. Bjarklev, "Highly nonlinear photonic crystal fiber with zero-dispersion at 1. 55 mm," OFC '02 postdeadline paper (2002).
80. J.H. Lee, P.C. Teh, Z. Yusoff, M. Ibsen, W. Belardi, T.M. Monro and D.J. Richardson, "A holey fiber- based nonlinear thresholding device for optical CDMA receiver performance enhancement," *IEEE Photon. Technol. Lett.* **14**, 876–878 (2002).
81. Z. Yusoff, J.H. Lee, W. Belardi, T.M. Monro, P.C. the and D.J. Richardson, "Raman effects in a highly nonlinear holey fiber: amplification and modulation," *Opt. Lett.* **27**, 424–426 (2002).
82. J. Nilsson, R. Selvas, W. Belardi, J.H. Lee, Z. Yusoff, T.M. Monro, D.J. Richardson, K.D. Park, P.H. Kim and N. Park, "Continuous-wave pumped holey fiber Raman laser," in Proc. OFC, OSA Technical Digest, Anaheim, California, (2002).
83. J.H. Lee, Z. Yusoff, W. Belardi, M. Ibsen, T.M. Monro, B. Thomsen and D.J. Richardson, "A holey fiber based WDM wavelength converter incorporating an apodized fiber Bragg grating filter," CLEO/QELS 2002, Long Beach, California, (2002).
84. F. Druon and P. Georges, "Pulse-compression down to 20 fs using a photonic crystal fiber seeded by a diode-pumped Yb:SYS laser at 1070 nm," *Opt. Express* **12**, 3383–3396 (2004).
85. S. Coen, A.H.L. Chau, R. Leonhardt and J.D. Harvey, "Supercontinuum generation via stimulated Raman scattering and parametric four-wave mixing in photo nic crystal fibers," *J. Opt. Soc. Amer. B* (2002).
86. K.P. Hansen, J. J. Larsen, J.R. Jensen, S. Keiding, J. Broeng, H.R. Simonsen and A. Bjarklev, "Super continuum generation at 800 nm in highly nonlinear photonic crystal fibers with normal dispersion," in Proc. LEOS (2001).
87. K. M. Hilligsøe, T. V. Andersen, H. N. Paulsen, C. K. Nielsen, K. Mølmer, S. Keiding, R. Kristiansen, K. P. Hansen and J. J. Larsen, "Supercontinuum generation in a photonic crystal fiber with two zero dispersion wavelengths," *Opt. Express* **12**, 1045–1054 (2004).
88. S.G. Leon-Saval, T.A. Birks, W.J. Wadsworth, P.St.J. Russell and M.W. Mason, "Supercontinuum generation in submicron fibre waveguides," *Opt. Express* **12**, 2864–2869 (2004).
89. I. Hartl, X.D. Li, C. Chudoba, R.K. Ghanta, T.H. Ko and J.G. Fujimoto, "Ultrahigh-resolution optical co herence tomography using continuum generation in an air-silica microstructure optical fiber," *Opt. Lett.* **26**, 608–610 (2001).
90. R.E. Drullinger, S.A. Diddams, K.R. Vogel, C.W. Oates, E. A. Curtis, W.D. Lee, W.M. Itano, L. Hollberg and J.C. Bergquist, "All-optical atomic clocks," Inter national Frequency Control Symposium and PDA Ex hibition, 69–75 (2001).
91. P.A. Champert, S.V. Popov and J.R. Taylor, "Gener ation of multiwatt, broadband continua in holey fibers," *Opt. Lett.* **27**, 122–124 (2002).
92. J.H.V. Price, T.M. Monro, K. Furusawa, W. Belardi, J.C. Baggett, S. Coyle, C. Netti, J.J. Baumberg, R. Paschotta and D.J. Richardson, "UV generation in a pure silica holey fiber," *Appl. Phys.* B **77**, 291–298 (2003).
93. A.V. Husakou and J. Herrmann, "Supercontinuum generation of higher-order solitons by fission in photonic crystal fibers," *Phys. Rev. Lett.* **87**, (2001).
94. J. Herrmann, U. Griebner, N. Zhavoronkov, A. Husakou, D. Nickel, J.C. Knight, W.J. Wadsworth, P.St.J. Russell, and G. Korn, "Experimental evidence for supercontinuum generation by fission of higher-order solitons in photonic fibers," *Phys. Rev. Lett.* **88**, 173–201 (2002).
95. K.P. Hansen, J.R. Jensen, D. Birkedal, J.M. Hvam and A. Bjarklev, "Pumping wavelength dependence of super continuum generation in photonic crystal fibers," in Proc. Conference on Optical Fiber Communication (2002).

96. J.C. Knight, T.A. Birks, R.F. Cregan, P.St.J. Russell and J.-P. De Sandro, "Large mode area photonic crys tal fibre," *Electron. Lett.* **34**, 1347–1348 (1998).
97. J. Limpert, A. Liem, M. Reich, T. Schreiber, S. Nolte, H. Zellmer, A. Tunnermann, J. Broeng, A. Petersson and C. Jakobsen, "Low-nonlinearity single-transverse-mode ytterbium-doped photonic crystal fiber amplifier," *Opt. Express* **12**, 1313–1319 (2004).
98. T. Sørensen, J. Broeng, A. Bjarklev, E. Knudsen, S.E. Barkou, H.R. Simonsen and J. Riis Jensen, "Macrobending loss properties of photonic crystal fibres with different air filling fractions," in Proceed ings of ECOC'2001, Amsterdam, The Netherlands (2001).
99. J.C. Baggett, T.M. Monro, K. Furusawa and D.J. Richardson, "Understanding bending losses in holey optical fibers," *Opt. Commun.* **227**, 317–335 (2003).
100. A. Gambling, H. Matsumura, C.M. Ragdale, R.A. Sammut, "Measurement of radiation loss in curved single-mode fibres," *Microwaves Opt. Acoust.* **2**, 134–140 (1978).
101. N.A. Mortensen, M.D. Nielsen, J.R. Folkenberg, A. Petersson and H.R. Simonsen, "Improved large-mode-area endlessly single-mode photonic crystal fibers," *Opt. Lett.* **28**, 393–395 (2003).
102. R.F. Cregan, J.C. Knight, P.St.J. Russell and P.J. Roberts, "Distribution of spontaneous emission from an Er^{3+}-doped photonic crystal fiber," *IEEE J. Lightwave. Technol.* **17**, 2138–2141 (1999).
103. K. Furusawa, A. Malinowski, J.H.V. Price, T.M. Monro, J.K. Sahu, J. Nilsson and D.J. Richardson, "A cladding pumped Ytterbium-doped fiber laser with holey inner and outer cladding," *Opt. Express* **9**, 714–720 (2001).
104. K. Furusawa, T. Kogure, T.M. Monro and D.J. Richardson, "High gain efficiency amplifier based on an erbium doped aluminosilicate holey fiber," *Opt. Express* **12**, 3452–3458 (2004).
105. W.J. Wadsworth, J.C. Knight, W.H. Reeves, P.St.J. Russell and J. Arriaga, "Yb^{3+}-doped photonic crystal fiber laser," *Electron. Lett.* **36**, 1452–1454 (2000).
106. K. Furusawa, T.M. Monro, P. Petropoulos and D.J. Richardson, "A mode-locked laser based on ytter bium doped holey fiber," *Electron. Lett.* **37**, 560–561 (2001).
107. F.M. Mitschke and L.F. Mollenauer, "Discovery of the soliton self-frequency shift," *Opt. Lett.* **11**, 659–661 (1986).
108. J.H.V Price, W. Belardi, L. Lefort, T.M. Monro and D.J. Richardson, "Nonlinear pulse compression, dis persion compensation, and soliton propagation in holey fiber at 1 micron," in Proc Nonlinear Guided Waves and Their Applications (NLGW), paper WB1- 2 (2001).
109. J. Limpert, T. Schreiber, S. Nolte, H. Zellmer, T. Tunnermann, R. Iliew, F. Lederer, J. Broeng, G. Vienne, A. Petersson and C. Jakobsen, "High-power air-clad large-mode-area photonic crystal fiber laser," *Opt. Express* **11**, 818–823 (2003).
110. J. Limpert, T. Schreiber, A. Liem, S. Nolte,H. Zellmer, T. Peschel, V. Guyenot and A. Tünnermann, "Thermo-optical properties of air-clad photonic crystal fiber lasers in high power operation," *Opt. Express* **11**, 2982–2990 (2003).
111. K. Kikuchi, K. Taira and N. Sugimoto, "Highly nonlinear bismuth oxide-based glass fibers for all- optical signal processing," *Electron. Lett.* **38**, 166–167 (2002).
112. R.E. Slusher, J.S. Sanghera, L.B. Shaw and I.D. Aggarwal, "Nonlinear optical properties of As-Se fiber," in Proceedings of OSA Topical meeting on Nonlinear Guided Waves and their Applications, Stresa, Italy, 1–4 Sept. (2003).
113. E.M. Vogel, M.J. Weber and D.M. Krol, "Nonlinear optical phenomena in glass," *Phys. Chem. Glasses* **32**, 231–254 (1991).
114. S. Fujino, H. Ijiri, F. Shimizu and K. Morinaga, "Measurement of viscosity of multi-component glasses in the wide range for fiber drawing," *J. Jpn. Inst. Met.* **62**, 106–110 (1998).
115. Schott Glass Catalogue, 2003. Available from: www.us.schott.com/optics-devices/english/download
116. S.R. Fribergand and P.W. Smith, "Nonlinear optical- glasses for ultrafast optical switches," *IEEE J. Quant. Electron.* **23**, 2089–2094 (1987).
117. N. Sugimoto, H. Kanbara, S. Fujiwara, K. Tanaka, Y. Shimizugawa and K. Hirao, "Third-order optical nonlinearities and their ultrafast response in Bi_2O3-B_2O3-SiO_2 glasses," *J. Opt. Soc. Am. B* **16**, 1904–1908 (1999).
118. K. Kikuchi, K. Taira and N. Sugimoto, "Highly non linear bismuth oxide-based glass fibers for all-optical signal processing," *Electron. Lett.* **38**, 166–167 (2002).
119. N. Sugimoto, Y. Kuroiwa, K. Ochiai, S. Ohara, Y. Furusawa, S. Ito, S. Tanabe and T. Hanada, "Novel short-length EDF for CþL band amplification," in Proceedings of Optical Amplifiers and their Applica tions, Quebec City, Canada (2000).
120. Y. Kuroiwa, N. Sugimoto, K. Ochiai, S. Ohara, Y. Furusawa, S. Ito, S. Tanabe and T. Hanada, "Fusion spliceable and high efficient Bi2O3-based EDF for short length and broadband application pumped at 1480 nm," in Proc. Optical Fiber Communications (OFC), Anaheim, California, paper TuI5 (2001).
121. J.S. Wang, E.M. Vogel and E. Snitzer, "Tellurite glass: a new candidate for fiber devices," *Opt. Mat.* **3**, 187–203 (1994).
122. R. Stegeman, L. Jankovic, H. Kim, C. Rivero, G. Tegeman, K. Richardson, P. Delfyett, Y. Guo, A. Schulte and T. Cardinal, "Tellurite glasses with peak absolute Raman gain coefficients up to 30 times that of fused silica," *Opt. Lett.* **28**, 1126–1128 (2003).
123. X. Feng, A.K. Mairaj, D.W. Hewak and T.M. Monro, "Towards high-index glass based monomode holey fiber with large mode area," *Electron. Lett.* **40**, 167–169 (2004).
124. B. Temelkuran, S.D. Hart, G. Benoit, J.D. Joanno- poulos and Y. Fink, "Wavelength-scalable hollow optical fibres with large photonic bandgaps for CO_2 laser transmission," *Nature* **420**, 650–653 (2002).
125. F. Luan, A.K. George, T.D. Hedley, G.J. Pearce, D.M. Bird, J.C. Knight and P.St.J. Russell, "All-solid photonic bandgap fiber," *Opt. Lett.* **29**, 2369–2371 (2004).
126. X. Feng, T.M. Monro, P. Petropoulos, V. Finazzi and D.J. Richardson, "Single-mode high-index-core one-dimensional microstructured fiber with high nonlinearity," Proc. Optical Fiber Communications (OFC), 2005.
127. S.E. Barkou, J. Broeng and A. Bjarklev, "Silica-air photonic crystal fiber design that permits waveguiding by a true photonic bandgap effect," *Opt. Lett.* **24**, 46–48 (1999).
128. J.C. Knight, J. Broeng, T.A. Birks and P.St.J. Russell, "Photonic bandgap guidance in optical fibers," *Science* **282**, 1476–1478 (1998).

129. S.E. Barkou, J. Broeng and A. Bjarklev, "Dispersion properties of photonic bandgap guiding fibers," in Proc. Optical Fiber Communications (OFC), paper FG5, pp. 117–119 (1999).
130. A. Bjarklev, J. Broeng, S.E. Barkou, E. Knudsen, T. Søndergaard, T.W. Berg and M.G. Dyndgaard, "Polarization properties of honeycomb-structured photonic bandgap fibres," *J. Opt. A* **2**, 584–588 (2000).
131. J. Broeng, S.E. Barkou, T. Søndergaard and A. Bjarklev, "Analysis of air-guiding photonic bandgap fibers," *Opt. Lett.* **25**, 96–98 (2000).
132. H.C. Nguyen, P. Domachuk, B.J. Eggleton, M.J. Steel, M. Straub, M. Gu and M. Sumetsky, "A new slant on photonic crystal fibers," *Opt. Express* **12**, 1528–1539 (2004).
133. K. Saitoh and M. Koshiba, "Leakage loss and group velocity dispersion in air-core photonic bandgap fibers," *Opt. Express* **11**, 3100–3109 (2003).
134. J.A. West, N. Venkataraman, C.M. Smith and M.T. Gallagher, "Photonic crystal fibers," in Proc. European Conference on Optical Communications (ECOC), Th.A.2.2 (2001).
135. K. Saitoh, N.A. Mortensen and M. Koshiba, "Air- core photonic band-gap fibers: the impact of surface modes," *Opt. Express* **12**, 394–400 (2004).
136. D.G. Ouzounov, F.R. Ahmad, D. Müller, N. Venkataraman, M.T. Gallagher, M.G. Thomas, J. Silcox, K. W. Koch and A.L. Gaeta, "Generation of megawatt optical solitons in hollow-core photonic band-gap fibers," *Science* **301**, 1702–1704 (2003).
137. C.J.S. de Matos, J.R. Taylor, T.P. Hansen, K.P. Hansen and J. Broeng, "All-fiber chirped pulse amplification using highly-dispersive air-core photonic bandgap fiber," *Opt. Express* **11**, 2832–2837 (2003).
138. N. Venkataraman, M.T. Gallagher, C. M. Smith, D. Müller, J.A. West, K.W. Koch and J.C. Fajardo, "Low loss (13 dB/km) air core photonic band-gap fibre," in Proc. European Conference on Optical Communications (ECOC) Copenhagen, Denmark, postdeadline paper PD1.1 (2002).
139. H. Lim and F.W. Wise, "Control of dispersion in a femtosecond ytterbium laser by use of hollow-core photonic bandgap fiber," *Opt. Express* **12**, 2231–2235 (2004).
140. G. Humbert, J.C. Knight, G. Bouwmans, P.S.J. Russell, D.P. Williams, P.J. Roberts and B.J. Mangan, "Hollow core photonic crystal fibers for beam delivery," *Opt. Express* **12**, 1477–1484 (2004).
141. N.M. Litchinitser, S.C. Dunn, P.E. Steinvurzel, B.J. Eggleton, T.P. White, R.C. McPhedran and C.M. de Sterke, "Application of an ARROW model for designing tunable photonic devices," *Opt. Express* **12**, 1540–1550 (2004).
142. L.B. Shaw, J.S. Sanghera, I.D. Aggarwal and F.H. Hung, "As-S and As-Se based photonic bandgap fiber for IR laser transmission," *Opt. Express* **11**, 3455–3460 (2003).

Chapter 3.4

3.4

Photonic bandgap–guided Bragg fibers

Pal

3.4.1 Introduction

Consequent to the mind-boggling progress in highspeed optical telecommunications witnessed in recent times, it appeared that it would only be a matter of time before the huge theoretical bandwidth of 53 THz offered by low-loss transmission windows (extending from 1280 nm [235 THz] to 1650 nm [182 THz]) in OH^--free high-silica optical fibers would be tapped for telecommunication through dense wavelength division multiplexing techniques! In spite of this possibility, there has been a considerable resurgence of interest among researchers to develop specialty fibers, that is, fibers in which transmission loss of the material would not be a limiting factor and in which nonlinearity or dispersion properties could be conveniently tailored to achieve transmission characteristics that are otherwise almost impossible to realize in conventional high-silica fibers. Research targeted at such fiber designs gave rise to a new class of fibers, known as *microstructured* optical fibers (see Chapter 3). One category of such microstructured fibers is known as *photonic bandgap* fibers (PBGFs). In a conventional optical fiber, light is guided by total internal reflection because of the refractive index contrast that exists between a finite-sized cylindrical core and the cladding of lower refractive index that surrounds it. On the other hand, in a PBGF, light of *certain frequencies* cannot propagate along directions perpendicular to the fiber axis but instead are free to propagate along its length confined to the fiber core. This phenomenon that forbids the propagation of photons (of certain frequencies) transverse to the axis of microstructured fibers, led to the christening of these specialty fibers as photonic bandgap-guided optical fibers, in analogy with the electronic bandgaps encountered by electrons in semiconductors. In contrast to the electronic bandgap, which is the consequence of periodic arrangement of atoms/molecules in a semiconductor crystal lattice, photonic bandgap arises due to a *periodic distribution of refractive index* in certain dielectric structures, generically referred to as *photonic crystals* (reminiscent of semiconductor crystals in solid-state physics) [1]. If the frequency of incident light happens to fall within the photonic bandgap, which is characteristic of the photonic crystal, then light propagation is forbidden in it. Depending on the number of dimensions in which periodicity in refractive index exists, photonic crystals are classified as one-, two-, or three-dimensional photonic crystals (Fig. 3.4.1) [2]. Opals are naturally occurring photonic crystals consisting of a three-dimensional lattice of dielectric spheres, and *opalescence* is a direct consequence of their photonic crystal nature. The iridescent wings of some butterflies and beetles are also essentially the result of naturally occurring photonic crystal effects that evolved on their surface with time. In 1987, Yablonovitch first proposed the possibility of controlling properties of light through the photonic bandgap effect in man-made photonic crystals.

PBGFs essentially consist of a core surrounded by a periodic cladding having a photonic crystal-like structure (i.e., a periodic refractive index distribution). As explained above, the inherent periodicity of the cladding results in a photonic bandgap, which forbids propagation

Guided Wave Optical Components and Devices; ISBN: 9780120884810

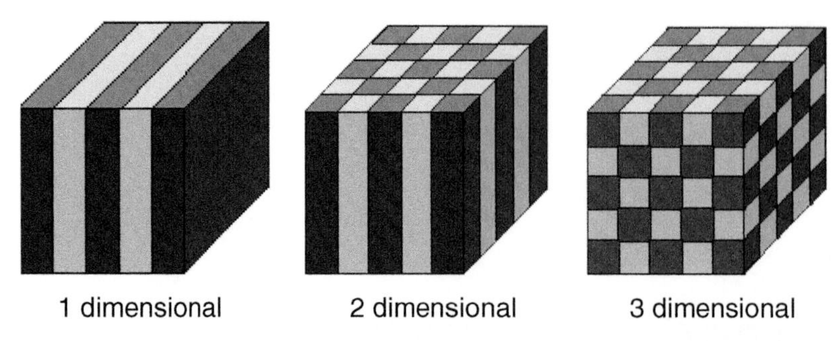

Fig. 3.4.1 Schematic representations of photonic crystals in one, two, and three dimensions. (After [2].)

of certain frequencies of light into the cladding. Consequently, if light of these frequencies is launched into the core of the micro-structured fiber, it could be confined within this region and be guided along the fiber length. PBGFs can be broadly classified as either *Bragg fibers* [4] or *photonic crystal fibers* [5, 6]. The former consists of a one-dimensional cladding that is periodic along the radial direction, whereas the latter consists of a two-dimensional periodic cladding extending along the directions radial as well as azimuthal. In contrast to conventional fibers, a PBGF can guide light through the low-index core (air) and hence offers several advantages as compared with light transmission through conventional fibers. In air-core PBGFs, because light is guided in air, two major benefits gained are much lower transmission loss and reduced sensitivity to nonlinear optical impairments such as four-wave mixing, cross-phase modulation, and so on. Following the same reasoning, the power-handling capability of these fibers would be much better and material dispersion effects would be much reduced. In addition, because of a multitude of physical parameters that can be altered independently, their propagation characteristics can be tuned with ease to control and maneuver light guidance for a variety of applications, ranging from telecommunications to sensors [7]. In this chapter our focus will remain only on Bragg fibers.

3.4.2 Bragg fibers

Bragg fibers consist of a low-index core surrounded by periodic multilayer cladding (concentric layers) of alternate high and low refractive index materials (each of which has a refractive index higher than that of the core). As mentioned above, the inherent periodicity in the cladding forms a photonic bandgap does not allow light of certain frequencies from leaking through the cladding and thereby confines it within the core region. Fig. 3.4.2 shows the cross-sectional view of a conventional optical fiber (a) and a Bragg fiber (b) (air core). The Bragg fiber was proposed as early as 1978 [4]. However, because of the lack of advanced fabrication techniques and doubts related to their applicability [8], there was hardly any further research in this field for almost a decade. Interest revived again in the late 1990s in the context of a flurry of research on photonic crystals, and several articles reported successful fabrication of broadband, low-loss, hollow-core fiber waveguides suitable for various ranges of wavelengths [9,10].

3.4.2.1 Bandgap in one-dimensional periodic medium

The multilayer cladding of the Bragg fiber is periodic along the radial direction and results in a one-dimensional photonic bandgap. It is functionally similar to a multilayer planar stack that consists of thin films of alternate refractive indices, n_1 and n_2, having thickness l_1 and l_2, respectively, such that $\Lambda = l_1 + l_2$ (Fig. 3.4.3). Accordingly, wave guidance in a Bragg fiber can be conveniently understood in terms of the physics that underlies the formation of the bandgap and the concept of decaying Bloch waves in a multilayer planar stack. Following the

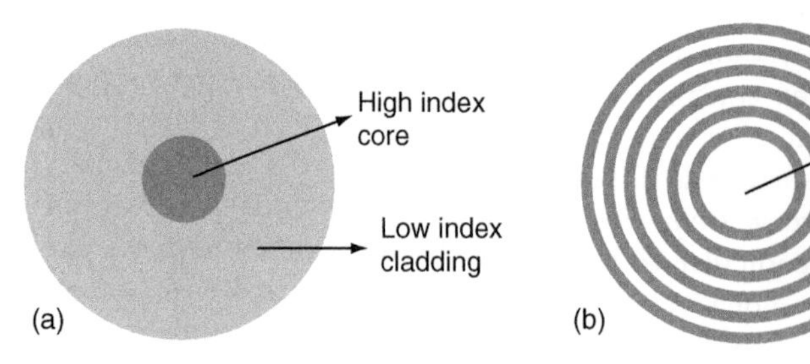

Fig. 3.4.2 (a) Cross-sectional view of a conventional fiber. (b) Cross-sectional view of a Bragg fiber.

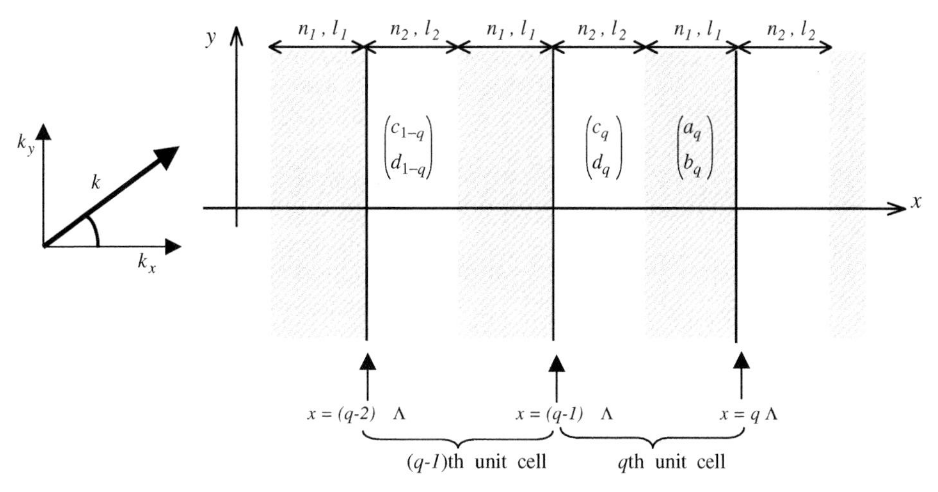

Fig. 3.4.3 Planar Bragg structure (Adapted from (11); © John Woley & Sons, 1991).

analysis presented in [11], the electric field of the wave propagating in the x–y plane in the qth unit cell of a planar stack (periodic in the x-direction, Fig. 3.4.3) may be written as the superposition of a forward and a backward propagating plane wave as follows:

$$\begin{aligned} E_z(x,y) &= \left[a_q e^{-ik_{1x}(x-q\Lambda)} + b_q e^{-ik_{1x}(x-q\Lambda)}\right] e^{-ik_y y} \\ E_z(x,y) &= \left[c_q e^{-ik_{2x}^{(x-q\Lambda)}} + b_q e^{-ik_{2x}^{(x-q\Lambda)}}\right] e^{-ik_y^y}, \end{aligned} \tag{3.4.1}$$

where $(a_q,\ b_q)$ and $(c_q,\ d_q)$ are the coefficients of the forward and backward propagating fields in the layers having refractive indices n_1 and n_2, respectively; k_y is the y-component of the propagation vector k, and $k_{jx} = \left[\left(\frac{n_j\omega}{c}\right)^2 - k_y^2\right]^{1/2}, j = 1, 2$. In general, the Bloch theorem can be used to obtain the normal modes of such a periodic medium. According to this theorem, the normal modes of a one-dimensional periodic structure (having a periodicity ^ along the x-direction) can be written as the product of a periodic Bloch wave function and a plane wave function:

$$\begin{aligned} E_z(x,y) &= \left[E_K(x)e^{-iKx}\right]e^{-ik_y y} \\ H_z(x,y) &= \left[H_K(x)e^{-iKx}\right]e^{-ik_y y}, \end{aligned} \tag{3.4.2}$$

where K is the periodic Bloch wave vector; and $E_K(x)$ and $H_K(x)$ satisfy the following periodicity conditions:

$$\begin{aligned} E_K(x) &= E_K(x+\Lambda) \\ II_K(x) &= II_K(x+\Lambda), \end{aligned} \tag{3.4.3}$$

Combining Eqs. (3.4.1) and (3.4.2) along with (3.4.3), the coefficients of electric field amplitudes in two consecutive layers can be related through

$$\begin{pmatrix} a_{q-1} \\ b_{q-1} \end{pmatrix} = e^{iK\Lambda}\begin{pmatrix} a_q \\ b_q \end{pmatrix}. \tag{3.4.4}$$

Alternatively, the complex amplitudes of the plane waves in a layer characterized by a refractive index n_1 in a certain unit cell can be related to that of the equivalent layer in the next unit cell by the unit cell *translation matrix* (see Eq. (3.4.3, 3.4.4, 3.4.5)). The continuity conditions for the transverse electric and magnetic field components E_z and H_y (for TE modes) and H_z and E_y (for TM modes) at the interfaces and $x = (q-1)\Lambda$ and $x = (q-1)\Lambda + l_2$ yield the translation matrix as

$$\begin{pmatrix} a_{q-1} \\ b_{q-1} \end{pmatrix} = \begin{pmatrix} A_{TE/TM} & B_{TE/TM} \\ C_{TE/TM} & D_{TE/TM} \end{pmatrix}\begin{pmatrix} a_q \\ b_q \end{pmatrix}, \tag{3.4.5}$$

where

$$\begin{aligned} A_{TE} &= e^{ik_{1x}l_1}\left[\cos(k_{2x}l_2) + \frac{i}{2}\left(\frac{k_{2x}}{k_{1x}} + \frac{k_{1x}}{k_{2x}}\right)\sin(k_{2x}l_2)\right] \\ B_{TE} &= e^{-ik_x l_1}\left[\frac{i}{2}\left(\frac{k_{2x}}{k_{1x}} - \frac{k_{1x}}{k_{2x}}\right)\sin(k_{2x}l_2)\right] \\ C_{TE} &= e^{ik_{1x}l_1}\left[-\frac{i}{2}\left(\frac{k_{2x}}{k_{1x}} - \frac{k_{1x}}{k_{2x}}\right)\sin(k_{2x}l_2)\right] \\ D_{TE} &= e^{-ik_{1x}l_1}\left[\cos(k_{2x}l_2) - \frac{i}{2}\left(\frac{k_{2x}}{k_{1x}} + \frac{k_{1x}}{k_{2x}}\right)\sin(k_{2x}l_2)\right]. \end{aligned} \tag{3.4.6}$$

$$A_{TM} = e^{ik_{1x}l_1}\left[\cos(k_{2x}l_2) + \frac{i}{2}\left(\frac{n_2^2 k_{1x}}{n_1^2 k_{2x}} + \frac{n_1^2 k_{2x}}{n_2^2 k_{1x}}\right)\sin(k_{2x}l_2)\right]$$
$$B_{TM} = e^{-ik_{1x}l_1}\left[\frac{i}{2}\left(\frac{n_2^2 k_{1x}}{n_1^2 k_{2x}} - \frac{n_1^2 k_{2x}}{n_2^2 k_{1x}}\right)\sin(k_{2x}l_2)\right]$$
$$C_{TM} = e^{ik_{1x}l_1}\left[-\frac{i}{2}\left(\frac{n_2^2 k_{1x}}{n_1^2 k_{2x}} - \frac{n_1^2 k_{2x}}{n_2^2 k_{1x}}\right)\sin(k_{2x}l_2)\right]$$
$$D_{TM} = e^{-ik_{1x}l_1}\left[\cos(k_{2x}l_2) - \frac{i}{2}\left(\frac{n_2^2 k_{1x}}{n_1^2 k_{2x}} + \frac{n_1^2 k_{2x}}{n_2^2 k_{1x}}\right)\sin(k_{2x}l_2)\right] \quad (3.4.7)$$

The eigenvalue equation that governs the photonic bandgap, is obtained by substituting Eq. (3.4.5) into Eq. (3.4.4) as

$$e^{iK\Lambda} = \frac{1}{2}(A+D) \pm \left[\left\{\frac{1}{2}(A+D)\right\}^2 - 1\right]^{1/2}. \quad (3.4.8)$$

Subscripts TM and TE have been dropped for simplicity. Note that if $\left|\frac{1}{2}(A+D)\right| < 1$, it would imply that the Bloch wave vector K is real, which would correspond to a propagating Bloch wave. However, if $\left|\frac{1}{2}(A+D)\right| > 1$, K would be complex and this condition would correspond to an evanescent Bloch wave, whose amplitude decays exponentially along the direction of periodicity. Thus, light frequencies that satisfy the condition $\left|\frac{1}{2}(A+D)\right| > 1$ constitute the "forbidden" bands of the periodic medium.

Fig. 3.4.4 shows the typical band structure of such a one-dimensional periodic medium. The white (blank) regions in the figure correspond to decaying Bloch waves, and light frequencies that fall within this region are forbidden to propagate in the medium. However, the one-dimensional bandgap is not *omnidirectional* (*omni* means *all* in Latin) because, in general, such a structure is unable to reflect light at arbitrary incident angles. An omnidirectional bandgap can confine light within a certain frequency range irrespective of the incident angle and state of polarization of light [12]. It occurs if the forbidden frequencies in the bandgaps overlap along all possible directions in space. Intuitively, it might seem that this would require a system that is periodic along the three orthogonal directions (three-dimensional photonic crystal). However, omnidirectional reflection can be achieved even from a one-dimensional photonic crystal (one-dimensional planar stack, in this case) if the wave incident from the ambient medium does not couple to the propagating states of the photonic crystal [12,13].

To comprehend this feature, we first explain the concept of *light line*. Fig. 3.4.5a shows the dispersion relation for light waves in free space. The solid line corresponds to $\omega = ck$, where ω is the frequency of light, k is the magnitude of the propagation wave vector, and c is the speed of light in a vacuum. We assume the x–y plane as the plane of propagation of the light wave. In general, the propagation vector k would have both x- and y-components, such that to $\omega = c(k_y^2+k_x^2)^{1/2}$ and $\omega \geq ck_y$. Again, referring to the planar stack in Fig. 3.4.3, $k_y = k \sin\theta$, where θ is the angle of incidence on the planar stack such that $-90° \leq \theta \leq 90°$. Hence, the range of all possible values of k_y would be $-k \leq k_y \leq k$. Fig. 3.4.5b shows the variation of k_y with ω, wherein any horizontal line (e.g., the dashed line) would correspond to all possible values of the incident wave vector for a given frequency. The y-axis corresponds to $k_y = 0$ and the *light line* corresponds to $\omega = ck_y = ck$ ($k_x = 0$, $0 = \pm 90°$). The entire region above the light line satisfies $\omega > ck_y$, and hence all allowed states of the ambient medium lie in this region. To obtain the condition of omnidirectionality, we assume

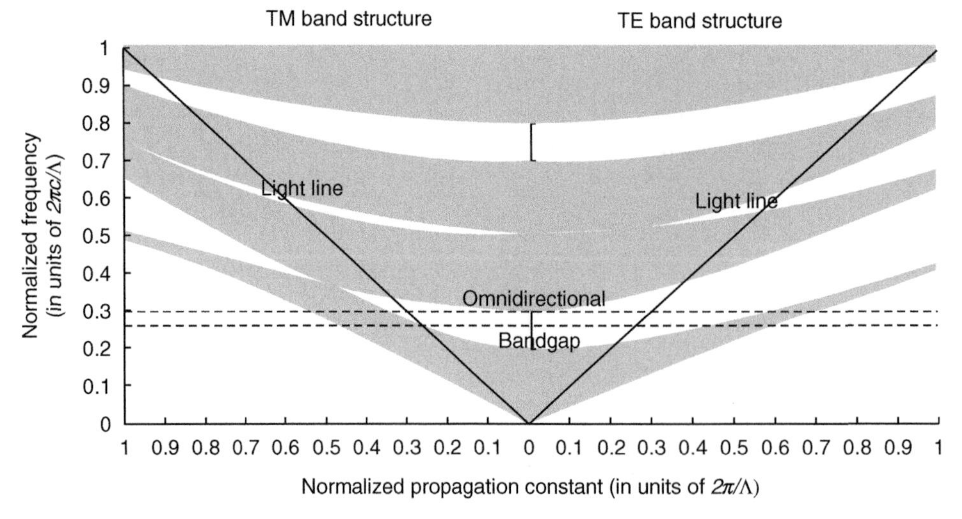

Fig. 3.4.4 Photonic bandgap structure for TE and TM polarization states for a one-dimensional planar stack.

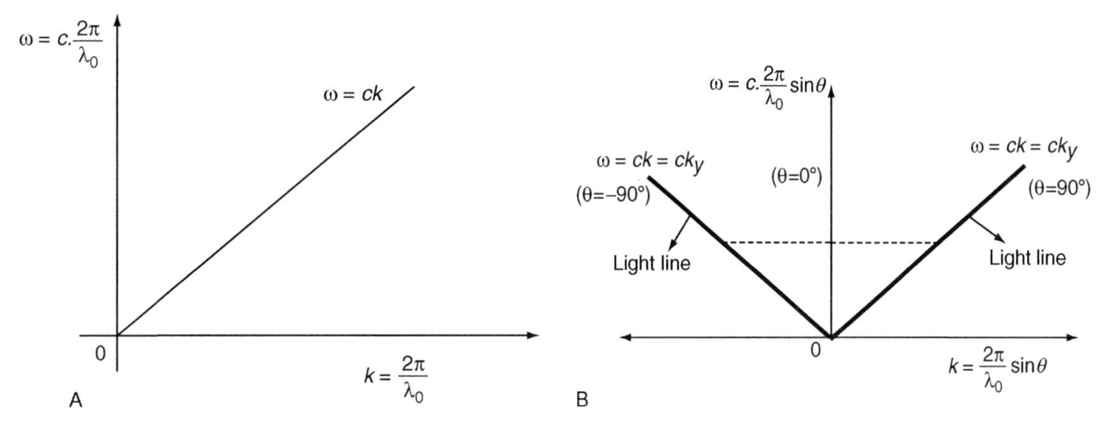

Fig. 3.4.5 (a) Variation of ω vs. k (dispersion relation of light wave) in free space. (b) Variation of ω vs. k_y (dispersion relation of light wave) in free space; θ is the angle at which the wave vector is incident on the planar stack in Fig. 4.3.

Fig. 3.4.5b to be superimposed on the band structure of the planar stack (Fig. 3.4.4). If there now exists a range of "forbidden" frequencies within the region $\omega > ck_y$, the incident wave from the ambient medium at those frequencies would encounter the bandgap for all possible angles of incidence. Consequently, the incident wave would be unable to couple to the propagating states of the periodic structure, and hence omnidirectional reflection could be achieved [12]. In Fig. 3.4.4, the frequencies lying within the two horizontal dashed lines constitute the omnidirectional bandgap. For any further details on band structure and reflection characteristics of periodic planar stacks, we refer readers to [11].

3.4.2.2 Light propagation in Bragg fibers

Propagation of electromagnetic waves in Bragg fibers is governed by the well-known Maxwell's equations in charge-free space as

$$\begin{aligned} \nabla \cdot D &= 0 \\ \nabla \times E &= -\frac{\partial B}{\partial t} \\ \nabla \cdot B &= 0 \\ \nabla \times H &= -\frac{\partial D}{\partial t}. \end{aligned} \quad (3.4.9)$$

The electric and magnetic fields associated with the guided modes propagating in the z-direction (in cylindrical coordinates) can be assumed to have the following form:

$$\Psi(r,\theta,z,t) = \psi(r,\theta)e^{-i(\Omega t-\beta z)}, \quad (3.4.10)$$

where $\Psi(r, \theta)$ could be any of the field components: E_z, E_r, E_θ, H_z,H_r, and H_θ. The field components E_z and H_z satisfy the following wave equation:

$$(\nabla_t^2 + (\omega^2\mu\epsilon - \beta^2))\begin{pmatrix} E_z \\ H_z \end{pmatrix} = 0, \quad (3.4.11)$$

where $\nabla_t^2 = \nabla^2 - \frac{\partial^2}{\partial z^2}$ is the transverse Laplacian operator. Bragg fibers possess cylindrical symmetry, and the general solutions of the field components E_z and H_z, that satisfy the wave equation, can be written as follows (see, e.g., [3]):

$$\begin{aligned} E_z &= [A_jJ_l(k_jr) + B_jY_l(k_jr)]\cos(l\theta + \Phi_j)j = 1,2 \\ H_z &= [C_jJ_l(k_jr) + D_jY_l(k_jr)]\cos(l\theta + \chi_j)j = 1,2, \end{aligned} \quad (3.4.12)$$

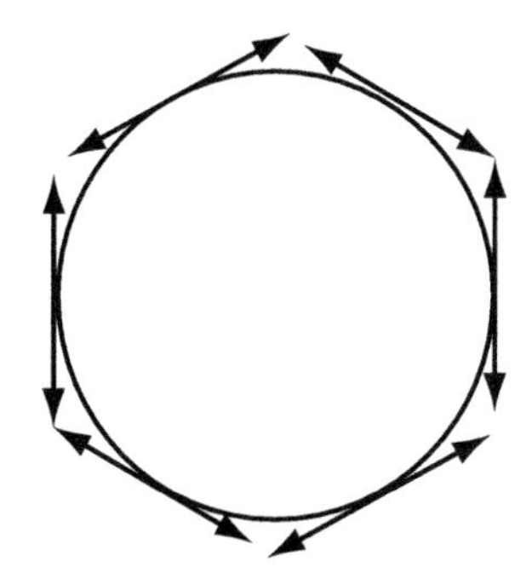

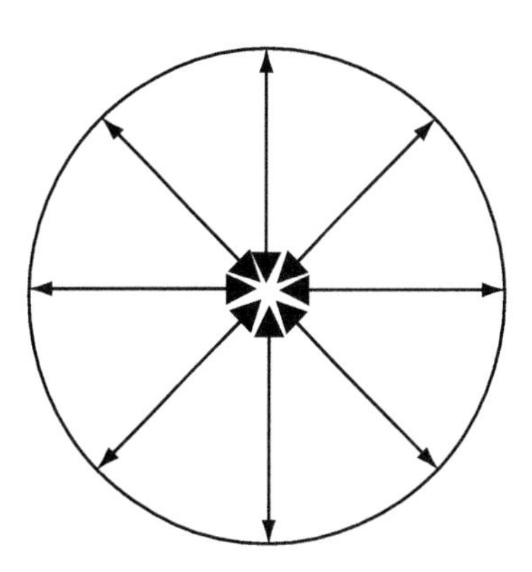

Fig. 3.4.6 Schematic of electric field components of TE and TM modes in a Bragg fiber.

where $k_j = \left[\left(\frac{n_j\omega}{c}\right)^2 - \beta^2\right]^{1/2}$ and $j = 1, 2$ corresponds to the cladding layers of refractive index n_1 and n_2 and thickness l_1 and l_2. The other transverse field components of the guided modes (E_θ, E_r, H_θ, H_r) can be written in terms of E_z and H_z as [15]

$$\begin{aligned} E_r &= \frac{i\beta}{\omega^2\mu\varepsilon-\beta^2}\left(\frac{\partial}{\partial r}E_z + \frac{\omega\mu}{\beta}\frac{\partial}{r\partial\theta}H_z\right) \\ E_\theta &= \frac{i\beta}{\omega^2\mu\varepsilon-\beta^2}\left(\frac{\partial}{r\partial\theta}E_z - \frac{\omega\mu}{\beta}\frac{\partial}{\partial r}H_z\right) \\ H_r &= \frac{i\beta}{\omega^2\mu\varepsilon-\beta^2}\left(\frac{\partial}{\partial r}H_z - \frac{\omega\epsilon}{\beta}\frac{\partial}{r\partial\theta}E_z\right) \\ H_\theta &= \frac{i\beta}{\omega^2\mu\varepsilon-\beta^2}\left(\frac{\partial}{r\partial\theta}H_z + \frac{\omega\epsilon}{\beta}\frac{\partial}{\partial r}E_z\right). \end{aligned} \quad (3.4.13)$$

3.4.2.2.1 Minimization procedure

In the case of conventional optical fibers, the effective index of a mode can be easily obtained by using the boundary condition that the transverse field decays exponentially to zero, outside the cladding-air interface. However, this is not applicable to Bragg fibers because they support oscillatory solutions even in the cladding layers. Moreover, the aperiodic nature of Bessel functions does not allow the Bloch theorem to be applied to the fields in the multilayer cladding. Hence, to obtain the effective index of a guided mode in such fibers, the outward flowing power flux has to be minimized by appropriately choosing the thickness of the cladding layers [4]. The radial component of the net Poynting vector is

$$S_r = \left(\frac{1}{2}\right)\mathrm{Re}[E_\theta H_z^*]. \quad (3.4.14)$$

S_r is zero for any Bragg fiber structure with an infinite or finite number of cladding layers if an infinitely extended source is used to excite the modes. However, in a real-world situation, the net outward flow of power is finite and constitutes the propagation loss. Optimum confinement is achieved when the cladding bilayers are of "half-wave" thickness, implying that each layer satisfies the following *quarter-wave thickness* condition:

$$\mathrm{k}_1 l_1 = \mathrm{k}_2 l_2 = \pi/2. \quad (3.4.15)$$

Because the zeroes of Bessel functions are not exactly periodic, theoretically the cladding layers of a Bragg fiber that satisfy the quarter-wave thickness condition cannot be of equal thickness!

3.4.2.2.2 Asymptotic matrix theory

The minimization procedure, explained above, is relatively easy for pure TE and TM modes but becomes tedious and time-consuming for hybrid modes. However, Xu et al. proposed an *asymptotic matrix theory* that yielded an analytical equation for obtaining propagation constants of the guided modes in a Bragg fiber, and it required much less computation time [15]. It was based on the fact that in the asymptotic limit, Bessel functions essentially resemble the plane waves with an additional amplitude factor of $\sqrt{r}$ in its denominator. In this method, the field in the core of the Bragg fiber and up to a certain finite number of cladding layers (say N_1) are assumed to be exact solutions of Maxwell's equations (in the form of Bessel functions). In the subsequent cladding layers beyond the (N_1)th layer, the field is assumed to follow the asymptotic form of Bessel functions as follows [15]:

$$\left.\begin{aligned} E_z &= A_{j'}J_l(\mathrm{k}_{j'}r) + B_{j'}Y_l(\mathrm{k}_{j'}r) \\ H_z &= C_{j'}J_l(\mathrm{k}_{j'}r) + D_{j'}Y_l(\mathrm{k}_{j'}r) \end{aligned}\right\} j' = 1, 2, \ldots, N_1,$$

where $\mathrm{k}_{j'} = \left[\left(\frac{n_{j'}\omega}{c}\right)^2 - \beta^2\right]^{1/2}$ and n_j' is the refractive index in the (j')th layer.

$$\left.\begin{aligned} E_z &= \frac{a_j^q e^{i\mathrm{k}_q(r-\rho_j^q)} + b_j^q e^{-i\mathrm{k}_q(r-\rho_j)}}{\sqrt{\mathrm{k}_q r}} \\ H_z &= \frac{c_j^q e^{i\mathrm{k}_q(r-\rho_j)} + d_j^q e^{-i\mathrm{k}_q(r-\rho_j)}}{\sqrt{\mathrm{k}_q r}} \end{aligned}\right\} \rho_j^q \le r < \rho_j^q + l_q$$
$$j = 1, 2, \ldots N - N_1, q = 1, 2 \quad (3.4.16)$$

where $\mathrm{k}_q = \left[\left(\frac{n_q\omega}{c}\right)^2 - \beta^2\right]^{1/2}$ and where $q = 1, 2$ corresponds to the cladding layers having refractive index n_1 and n_2, and N is the total number of cladding layers. The eigenvalue equation for the guided modes is obtained by applying the Bloch theorem to the field in the outer cladding layers and using the continuity conditions for the electric and magnetic fields (i.e., E_z, E_θ, H_z, H_θ) at the layer interfaces. The accuracy of the solutions could be improved by increasing the number of inner cladding layers for which the solutions were assumed to be (rigorously correct) Bessel functions. The asymptotic approximation is valid only if $\mathrm{k}_j r \gg 1$ [16]. The condition that $\mathrm{k}_j r \gg 1$ may not be true if the fiber has a very small core radius or if the refractive index of the cladding layers is not sufficiently large relative to the refractive index of air. Other numerical techniques, like the finite difference time domain method [17] and perturbation method [18, 19], are also widely used to obtain the propagation constants of guided modes in Bragg fibers.

3.4.2.3 Modal characteristics

In general, bandgap structures for the TE and TM polarizations do not overlap. However, a large refractive index contrast in the cladding layers can result in an omnidirectional bandgap [20]. TE and TM modes are the fundamental modes of a Bragg fiber, for which the electric fields are purely parallel (TE) and normal (TM) to the layer interfaces (Fig. 3.4.6). Unlike the LP modes of a conventional fiber, the TE and TM modes of an air-core Bragg fiber are nondegenerate. Hence, Bragg fibers, which support a single TE or TM mode, are inherently free of polarization mode dispersion. Though omnidirectionality is not essential for a mode to be guided along the Bragg fiber, the TE_{01} modal properties of a dielectric waveguide with a large index contrast in the cladding closely resemble the modal structure of hollow metallic waveguides if the TE_{01} mode lies within the omnidirectional bandgap [21]. Similar to metallic waveguides, TE_{01} mode is the lowest-loss mode out of the several (including TM and hybrid) modes supported in Bragg fibers, all of which exhibit much larger differential loss relative to the TE_{01} mode. In view of this feature, single-mode propagation of TE_{01} mode could effectively be realized even in relatively large-core Bragg fibers [19]. Accordingly, the TE_{01} mode is relatively more tightly confined inside the core as compared with other modes (Fig. 3.4.7). In addition to the bandgap-guided modes (TE, TM, and hybrid), Bragg fibers also support other index-guided modes (which decay exponentially in the air regions but extend throughout the dielectric cladding) and surface modes (which decay exponentially in both air and dielectric layers).

3.4.2.3.1 Propagation loss

Propagation loss of Bragg fibers primarily arises from material absorption and scattering losses and waveguide-related radiation loss. The material-related losses are dictated by the choice of materials used for the core and the cladding layers, uniformity of the layer interfaces and penetration of the field into the cladding. However, because a large fraction of the TE_{01} modal energy essentially propagates through the hollow core in an air-core Bragg fiber, material absorption losses of this mode are not very significant [22]. The radiation loss of the propagating modes depends on the refractive index contrast and number of cladding layers. The guiding region (air core) of a Bragg fiber has a refractive index that is lower than that of all the cladding layers and it supports only leaky modes. Ideally, a Bragg fiber with infinitely extended cladding layers would have zero radiation loss. However, in practice, a fabricated Bragg fiber has only a finite number of cladding layers, and consequently, the modes supported by it would invariably suffer a finite radiation loss. Hence, the modes supported by a Bragg fiber are termed "quasi-modes." However, due to the low loss of the TE_{01} mode relative to higher order modes, these would leak out within a characteristic length (L_{SM}) of the fiber, beyond which the fiber becomes effectively single moded. Furthermore, because the guided mode (i.e., TE_{01}) itself is also leaky by nature, there exists a characteristic fiber length L_{max} beyond which the power of the guided mode reduces below useful limits [23]. The quarter-wave stack condition (Eq. (3.4.15)) is an optimization condition that optimizes the radiation loss of TE modes in large-core Bragg fibers, and the radiation loss is proportional to the square of the field amplitude in the cladding. Modal loss also depends on the refractive index contrast in the cladding layers and core radius of the fiber. For TE_{01} modes, it is proportional

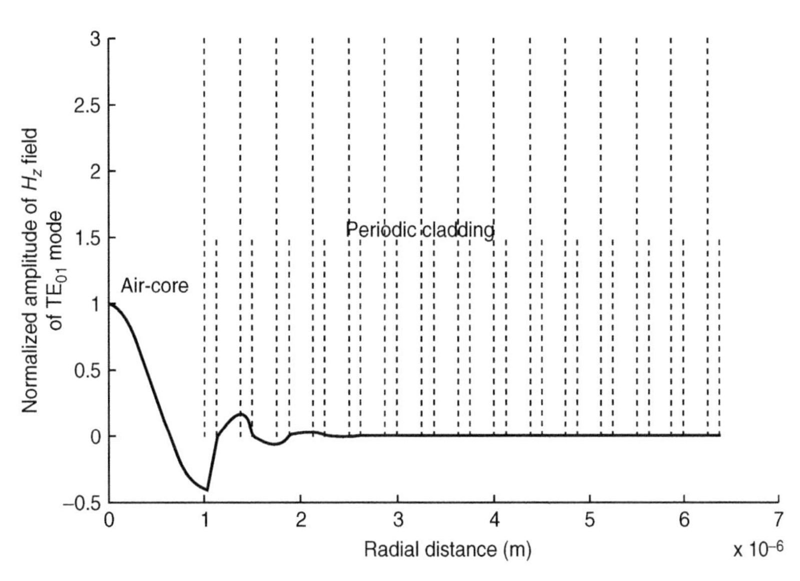

Fig. 3.4.7 Typical variation of Hz field of the TE_{01} mode along the radial direction. Note that the field rapidly decays to zero within a few cladding layers. Vertical dashed lines correspond to the cladding layer interfaces; the length of the lines is proportional to the refractive index in the particular layer.

to $1/r_{co}^3$, whereas for TM_{01} modes, the loss is proportional to $1/r_{co}$, where r_{co} is the core radius of the fiber [22]. Besides the refractive index contrast, modal radiation loss of the Bragg fiber also depends on the number of cladding layers, N, and scales as Δ^N, where Δ is a constant whose value depends on the mode being considered [22].

3.4.2.3.2 Dispersion

Because of the multitude of index profile parameters involved, Bragg fibers offer fiber designers wide latitude in the choice of parameters to tailor their dispersion characteristics for diverse functionality. Judicious choice of the core radius, cladding layer thickness, and refractive index contrast in the cladding layers can yield dispersion characteristics that are virtually unique to Bragg fibers and almost unachievable in conventional single-mode fibers. For example, single-mode light propagation at 1.06 μm with zero dispersion at a wavelength ≈ 1 μm through a silica core Bragg fiber was experimentally demonstrated in 2000 [9]. Subsequently, various designs have been proposed to achieve multiple zero dispersion [24] and high negative dispersion for dispersion compensation purposes [24, 25] and single polarization single-mode propagation [26].

3.4.3 Dispersion compensating Bragg fiber

Dispersion compensators are an integral part of any long-haul fiber optic telecommunications link. Conventional high-silica single-mode fibers exhibit positive temporal dispersion beyond a wavelength of 1310 nm. Consequently, a signal carrying the LP_{01} mode acquires positive dispersion as it propagates through a conventional single-mode transmission fiber, which unless canceled through insertion of a component to combat it would limit signal transmission capacity of the fiber. This is achieved by inserting a dispersion compensating module, which in most cases is a dispersion compensating fiber (DCF; see Chapter 1). A DCF is normally designed to exhibit a large negative dispersion coefficient (D) across the signal wavelength band such that a relatively short length (l_C) of the DCF could cancel the dispersion that the signal acquires while propagating through much longer lengths of the transmission fiber (l_{Tx}). Equation (3.4.17) determines the length of the DCF required for compensating the dispersion accumulated in a transmission fiber.

$$D_C l_C + D_{Tx} l_{Tx} = 0, \tag{3.4.17}$$

where D_C and D_{Tx} are the dispersion coefficients of the DCF and the transmission fiber, respectively. Besides high negative dispersion, an efficient DCF must also exhibit low loss in the operating wavelength range. The overall efficiency of a DCF is measured in terms of an integral parameter known as figure of merit (FOM), which is defined as the ratio of dispersion to loss of the DCF:

$$\text{FOM} = D/\alpha, \tag{3.4.18}$$

where the loss is measured in dB/km, so that FOM is expressed in units of ps/nm·dB. The larger the value of FOM, the more efficient is the DCF. One can achieve large negative dispersion in conventional silica fibers through suitable tailoring of its refractive index profile. The largest dispersion coefficient of −1800 ps/nm km demonstrated to date was based on a dual-core DCF design [27, 28]. However, the material loss of silica limits the FOM that can be achieved, and typically FOM ranges from ~300 to 400 ps/nm·dB in commercially available high FOM DCFs. Moreover, typical refractive index profiles required to generate a large waveguide (negative) dispersion to substantially offset the positive material dispersion (within the gain spectrum of an erbium-doped fiber amplifier) are necessarily characterized with a relatively small mode effective area (A_{eff}) ≈15–20 μm^2, which makes these DCFs sensitive to detrimental nonlinear effects unless the launched signal power into the DCF is restricted.

Bragg fibers have evolved as an attractive alternative to conventional DCFs, and various designs have been proposed for realizing high negative dispersion fibers with a relatively large A_{eff}. One of the earliest designs reported an estimated D up to approximately −25,000 ps/nm·km through the hybrid HE_{11} mode of a Bragg fiber [25]. However, the hybrid nature and higher loss of HE_{11} mode (as compared with TE_{01} mode) and small core radius were major issues that limited the practical use of this otherwise attractive design. Subsequently, it was reported that high negative D could be achieved for the TE_{01} mode as well by intentionally incorporating a defect layer within the periodic cladding layers [24]. The fiber is so designed that the fundamental (TE_{01}) mode of the defect-free Bragg fiber and the defect mode supported by the defect layer are in weak resonance, analogous to phase matching of the supermodes in a fiber directional coupler at a particular wavelength. Near the resonance wavelength, the fiber exhibits very large negative dispersion. The thickness and localization of the defect layer were crucial in determining precise dispersion characteristics of such Bragg fibers. These parameters can be altered suitably to match the dispersion and dispersion-slope for achieving broadband dispersion compensation characteristics. A fundamental design rule for such Bragg fiber–based DCFs with an intentional defect layer(s) is that its modal field must penetrate sufficiently into the

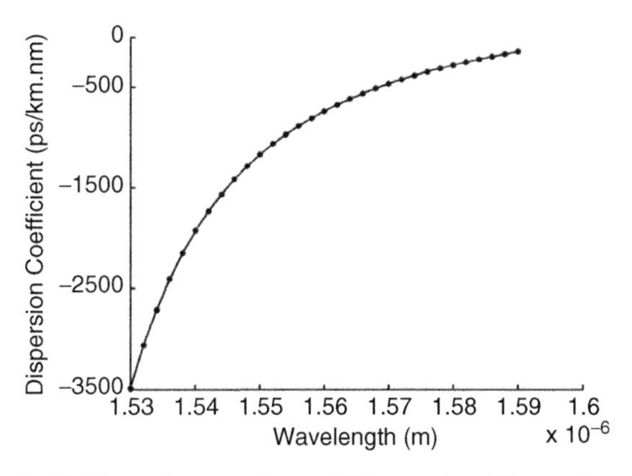

Fig. 3.4.8 Dispersion spectrum of TE_{01} mode of Bragg fiber based DCF.

cladding for interactions with the defect layer so as to generate large negative dispersion. This procedure, however, inadvertently increases the material-related loss of the structure besides making the modal field sensitive to nonuniformities in cladding thickness. An alternate design was recently proposed to realize a DCF with a high FOM [29] in which the quarter-wave stack condition (Eq. (3.4.15)) was modified to

$$\frac{2\pi}{\lambda_0}n_1l_1 = \frac{2\pi}{\lambda_0}n_2l_2 = n\pi/2, \tag{3.4.19}$$

where n is an odd integer. The multiple quarter-wave stack condition (Eq. (3.4.19) with $n = 3$) was found to confine the TE_{01} mode tightly within the core with a low radiation loss while at the same time exhibiting high negative dispersion. The dispersion spectrum of one such designed DCF is shown in Fig. 3.4.8. It can be seen from Fig. 3.4.8 that the so designed DCF's high negative D extends over a wavelength range of about 60 nm. The design was targeted to minimize the radiation loss

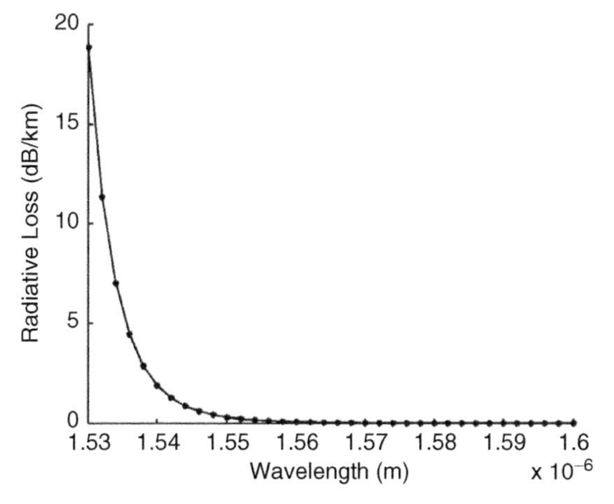

Fig. 3.4.9 Radiation loss spectrum of TE_{01} mode of Bragg fiber based DCF.

(Fig. 3.4.9), thereby yielding a very high estimate for the FOM (~4000 ps/nm·dB at 1550 nm) for the designed fiber. Though the proposed design supports more than one mode in the operating wavelength band, the differential losses suffered by the higher order modes with respect to the TE_{01} mode were found to be an order of magnitude higher, thus effectively making the fiber single moded.

3.4.4 Bragg fibers for metro networks

With the growth in Internet usage and increasing demand in data services and network efficiency, metro optical networks are increasingly evolving as *intelligent* and *data-centric* networks. Metropolitan networks bridge the gap between local access networks and long distance telecommunications networks. As a rule of thumb, metro networks are required to be suitably designed to address features like low installation cost, high degree of scalability, dynamism that is capable of accommodating unpredicted traffic growth besides flexibility to add/drop individual signals at any central office in the network, and interoperability to support protocols such as cell phones, synchronous optical networks/synchronous digital hierarchy, and legacy equipment. The low cost requirement implies that metro-specific fiber designs are aimed at minimizing use of components such as amplifiers, dispersion compensators, gain flattening filters, and so on. In any case, dispersion as well as loss spectra are the dictating factors in estimating maximum signal reach in a network.

Reported designs for metro fibers are based on their operation either as negative dispersion fibers [30, 31] or positive dispersion fibers across the C-band [32], so as to achieve a span length of ~100 km, without the need for a dispersion compensating device. However, these networks still need an amplifier after approximately every 80 km. Because Bragg fibers offer the potentiality of realizing ultra-low-loss transmission, deployment of Bragg fibers could minimize investment cost on amplifier head and provide a useful platform for dispersion tailoring, both of which should be very attractive for a metro network. Dispersion spectrum of one such Bragg fiber designed in-house having an air-core radius of 10.0 μm with 15 cladding bilayers and with a refractive index contrast of 2.0 is shown in Fig. 3.4.10. Loss suffered by higher order modes of the fiber is much larger as compared with the TE_{01} mode, thus making the fiber effectively single moded. It can be seen from Fig. 3.4.10 that the average dispersion of the fiber across the C-band is ~9.9 ps/nm·km, with a dispersion slope of 0.23 ps/nm^2·km. This should enable a dispersion-limited fiber

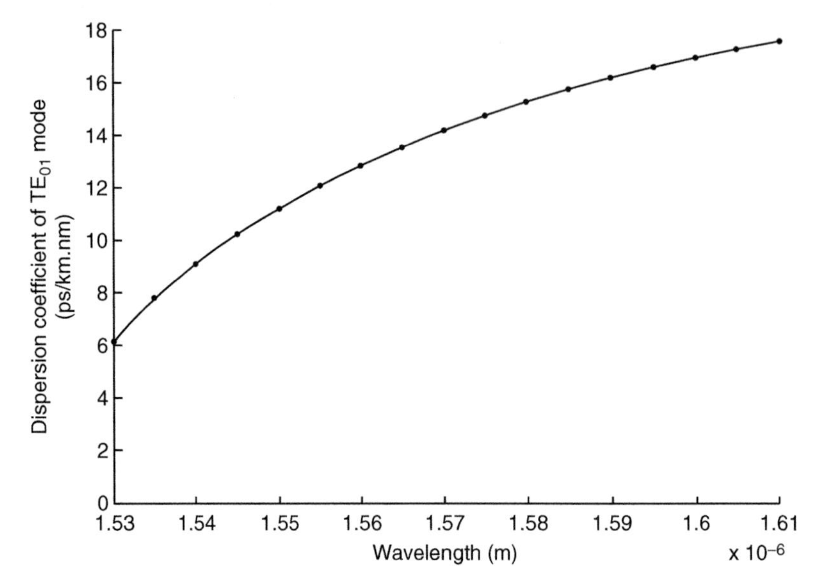

Fig. 3.4.10 Dispersion spectrum of TE_{01} mode of metro-centric Bragg fiber.

length of ~100 km in the C-band at 10 Gb/s. The average dispersion across the L-band is ~16.0 ps/nm·km, with a dispersion slope of 0.07 ps = nm^2 · km at 1.59 μm. Because the TE_{01} mode is very well confined within the air core, the material absorption losses in the cladding could be ignored as compared with the radiation loss. The radiation loss of the TE_{01} mode across the C- and L-bands is tabulated in Table 3.4.1. To achieve sufficient coupling into the TE_{01} mode from a laser pigtailed source, the fiber may be tapered to attain significant modal overlap between the TE_{01} mode and the approximate Gaussian field of the single-mode fiber pigtail.

3.4.5 Fabrication

In spite of their many appealing features, Bragg fibers did not gain much popularity until the late 1990s because of the unavailability of suitable fabrication techniques. Recently, development in fabrication technologies has enabled the realization of broadband, low-loss, omnidirectional Bragg fibers. In the year 2000, the classical modified chemical vapor deposition technique was used to manufacture the first "depressed core index photonic bandgap" fiber, which allowed single-mode propagation of light at 1060 nm [9]. The core of the depressed core index PBGF was made of slightly F-doped silica and the cladding layers were made of Ge-doped silica such that the refractive index contrast of consecutive layers was 0.009. In 1999, a combination of techniques that included dip coating on a capillary, serving as the core, and thermal evaporation and deposition was used to fabricate a broadband, low-loss, hollow waveguide in the 10 μm range [10]. The transmission through the waveguide around a 90° bend was also measured to demonstrate the low-loss features of such structures even in the presence of bends in the spectral band corresponding to the photonic bandgap. In 2003, a large effective area Bragg fiber with a silica core was fabricated using the modified chemical vapor deposition process [33], which supported a single HE11 mode with a mode effective area of ~526 μm^2 (core radius ≈17.5 μm) and exhibited losses of 0.4 dB/m at 1550 nm with only three cladding bilayers. Air-core Bragg fibers with four pairs of Si/Si_3N_4 cladding layers have also been fabricated, yielding propagation loss below 1 dB/cm [22]. The experimental loss figures are a consequence of additional loss in the fiber due to the formation of gas inlet ports in the fiber during the chemical vapor depositing fabrication process. Very recently, an air-silica Bragg fiber supporting the TE_{01} mode was demonstrated for the first time [34] (see Fig. 3.4.11). The air-core Bragg fiber consisted of silica rings separated by 2.3-μm-thick air rings and 45-nm-thick support

Table 3.4.1 Variation of radiation loss of the metro-centric Bragg fiber with wavelength.

Wavelength (nm)	1530	1540	1550	1560	1570	1580	1590	1600	1610
Radiation loss (× 10^{-2} dB/km)	10.5	4.03	1.69	0.76	0.37	0.19	0.10	0.006	0.003

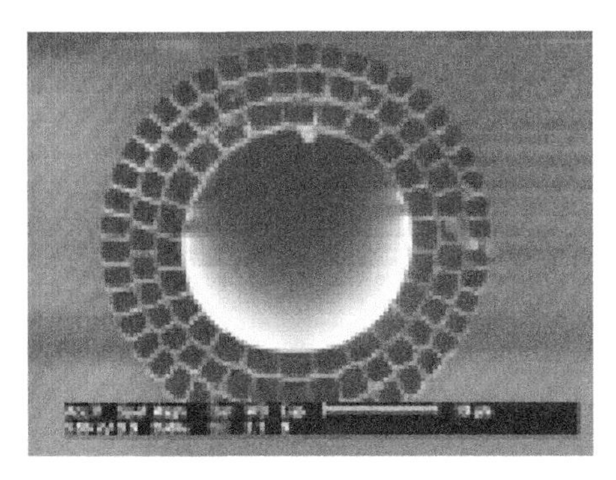

Fig. 3.4.11 Air-silica Bragg fiber (reproduced from [34] © Optical Society of America).

bridges and exhibited a propagation loss of ~1.5 dB/m at the transmission peak.

Though the transmission loss of Bragg fibers fabricated to date is orders of magnitude higher than conventional silica fibers (~0.2 dB/km), further improvement in fabrication techniques to achieve more uniform layer thickness and larger number of cladding layers should enable propagation loss of Bragg fibers to be brought down to well below the 0.2-dB/km figure. Additionally, the low loss of these fibers would not be restricted to the 1550-nm window. These distinct advantages in wavelength tun-ability for loss and dispersion should open up avenues for a wide range of potentially newer applications and devices.

3.4.6 Conclusion

In this chapter we attempted to describe relatively recent history, underlying physics, modeling of propagation characteristics, design issues, and fabrication of one-dimensional photonic bandgap-guided Bragg fibers. Bragg fibers have indeed emerged as a specialty fiber with the potential for several attractive applications.

References

1. J.D. Joannopoulos, R.D. Meade, and J.N. Winn, *Photonic Crystals: Molding the Flow of Light*, Princeton University Press, Princeton, NJ (1995).
2. http://ab-initio.mit.edu/photons/tutorial/photonic-intro.pdf
3. E. Yablonovitch, "Inhibited spontaneous emission in solid-state physics and electronics," *Phys. Rev. Lett.*, **58**, 2059 (1987).
4. P. Yeh, A. Yariv, and E. Marom, "Theory of Bragg fiber," *J. Opt. Soc. Amer.*, **68**,1196 (1978).
5. J.C. Knight, "Photonic crystal fibers," *Nature*, **424**, 847 (2003).
6. T.A. Birks, J.C. Knight, B.J. Mangan, and P.St.J. Russel, "Photonic crystal fibers: an endless variety," *IEICE Trans Electron.*, **E84-C**, 585 (2001).
7. B.J. Eggleton, C. Kerbage, P.S. Westbrook, R.S. Windeler, and A. Hale, "Microstructured optical fiber devices," *Opt. Exp.*, **9**, 698 (2001).
8 N.J. Doran and K.J. Blow, "Cylindrical Bragg fibers: a design and feasibility study for optical communications," *J. Lightwave Technol.*, **LT-1**, 588 (1983).
9. F. Brechet, P. Roy, J. Marcou, and D. Pagnoux, "Single-mode propagation into depressed-core-index photonic bandgap fiber designed for zero-dispersion propagation at short wavelengths," *Electron. Lett.*, **36**, 514 (2000).
10. Y. Fink, D.J. Ripin, S. Fan, C. Chen, J.D. Joannopoulos, and E.L. Thomas, "Guiding optical light in air using an all-dielectric structure," *J. Lightwave Technol.*, **17**, 2039 (1999).
11. P. Yeh, *Optical Waves in layered media*, John Wiley & Sons. Inc., Singapore (1991).
12. J.N. Winn, Y. Fink, S. Fan, and J.D. Joannopoulos, "Omnidirectional reflection from a one-dimensional photonic crystal," *Opt. Lett.*, **23**, 1573 (1998).
13. Y. Fink, J.N. Winn, S. Fan, C. Chen, J. Michel, J.D. Joannopoulos, and E.L. Thomas, "A dielectric omnidirectional reflector," *Science*, **282**, 1679 (1998).
14. J.A. Stratton, *Electromagnetic Theory*, McGraw-Hill, New York (1941).
15. Y. Xu, G.X. Ouyang, R.K. Lee, and A. Yariv, "Asymptotic matrix theory of Bragg fibers," *J. Lightwave Technol.*, **20**, 428 (2002).
16. S. Guo, S. Albin, and R.S. Rogowski, "Comparative analysis of Bragg fibers," *Opt. Exp.*, **12**,198 (2004).
17. A. Taflove and S.C. Hagness, *Computational Electro dynamics: The Finite-Difference-Time-Domain Method*, Artech House, Boston (2000).
18. D.Q. Chowdhary and D.A. Nolan, "Perturbation model for computing optical fiber birefringence from a two-dimensional refractive-index profile," *Opt. Lett.*, **20**,1973 (1995).
19. S.G. Johnson, M. Ibanescu, M. Skorobogatiy, O. Weisberg, T.D. Engeness, M. Soljačić, S.A. Jacobs, J.D. Joannopoulos, and Y. Fink, "Low-loss asymptotically single-mode propagation in large-core OmniGuide fibers," *Opt. Exp.*, **9**, 748 (2001).
20. S.D. Hart, G.R. Maskaly, B. Temelkuran, P.H. Prideaux, J.D. Joannopoulos, and Y. Fink, "External reflection from omnidirectional dielectric mirror fibers," *Science*, **296,** 510 (2002).
21. M. Ibanescu, S.G. Johnson, M. Soljačić, J.D. Joannopoulos, Y. Fink, O. Wiesberg, T.D. Engeness, S.A. Jacobs, and M. Skorobogatiy, "Analysis of mode structure in hollow dielectric waveguide fibers," *Phys. Rev. E*, **67**, e046608 (2003).
22. Y. Xu, A. Yariv, J.G. Fleming, and S.Y. Lin, "Asymptotic analysis of silicon based Bragg fibers," *Opt. Exp.*, **11**, 1039 (2003).
23. A. Argyros, "Guided modes and loss in Bragg fibres," *Opt. Exp.*, **10**, 1411 (2002).
24. T.D. Engeness, M. Ibanescu, S.G. Johnson, O. Weisberg, M. Skorobogatiy, S. Jacobs, and Y. Fink, "Dispersion tailoring and compensation by modal interactions in omniguide fibers," *Opt. Exp.*, **11**, 1175 (2003).
25. G. Ouyang, Y. Xu, and A. Yariv, "Theoretical study on dispersion

compensation in air core Bragg fibers," *Opt. Exp.*, **10**, 899 (2002).

26. G.R. Hadley, J. G. Fleming, and S.Y. Lin, "Bragg fiber design for linear polarization," *Opt. Lett.*, **29,** 809 (2004).
27. K. Thyagarajan, R.K. Varshney, P. Palai, A.K. Ghatak, and I.C. Goyal, "A novel design of a dispersion compensating fiber," *IEEE Photon. Technol. Lett.*, **8,** 1510 (1996).
28. J. L. Auguste, R. Jindal, J. M. Blondy, J. Marcou, B. Dussardier, G. Monnom, D.B. Ostrowsky, B.P. Pal, and K. Thyagarajan, "−1800 (ps/nm)/km chromatic dispersion at 1.55 μm in dual concentric core fiber," *Electron. Lett.*, **36**, 1689 (2000).
29. S. Dasgupta, B.P. Pal, and M.R. Shenoy, "Design of a low loss bragg fiber with high negative dispersion for the TE01 mode," *2004 Frontiers in Optics*, Pres. no. FWH49, Rochester, New York, USA (2004).
30. I. Tomkos, B. Hallock, I. Roudas, R. Hesse, A. Boskovic, J. Nakano, and R. Vodhanel, "10-Gb/s transmission of 1.55-μm directly modulated signal over 100 km of negative dispersion fiber," *IEEE Photon. Technol. Lett.*, **13**, 735 (2001).
31. T. Okuno, H. Hatayama, K. Soma, T. Sasaki, M. Onishi, and M. Shigematsu, "Negative dispersion-flattened fiber suitable for 10 Gbit/s directly modulated signal transmission in whole telecommunication band," *Electron. Lett.*, **40**, 723 (2004).
32. D. Culverhouse, A. Kruse, C. Wang, K. Ennser, and R. Vodhanel, "Corning® MetroCor™ fiber and its application in metropolitan networks," 2000 Corning Incorporated, White paper at http://www.corning.com/docs/opticalfiber/wp5078_7-00.pdf
33. S. Férvier, P. Viale, F. Gérôme, P. Leproux, P. Roy, J. M. Blondy, B. Dussardier, and G. Monnom, "Very large effective area singlemode photonic bandgap fibers," *Electron. Lett.*, **39**, 1240 (2003).
34. G. Vienne, Y. Xu, C. Jakobsen, H.J. Deyerl, T.P. Hansen, B.H. Larsen, J.B. Jensen, T. Sorensen, M. Terrel, Y. Huang, R. Lee, N.A. Mortensen, J. Broeng, H. Simonsen, A. Bjarklev, and A. Yariv, "First demonstration of air-silica Bragg fiber," OFC 2004 Post deadline paper, **PDP25,** Los Angeles, California, USA (2004).

Section Four

Optical transmitters and receivers

Chapter 4.1

4.1

Optical transmitters

DeCusatis

In this chapter we will discuss optical transmitters for fiber optic systems. This includes basic fiber optic communications light sources, such as LEDs, different types and wavelengths of lasers, and optical modulators. Our focus in this chapter is primarily on communication and sensor systems. Certain types of optical transmitters have become optimized for fiber optic communications, and other sources used for illumination and imaging have very different specifications and operating principles. While we discussed the many types of optical fiber in a single chapter, because a number of different applications use the same type of fiber, we cannot do the same with light sources; a fiber optic link is fundamentally different from a physician's fiberscope. This chapter describes the type of transmitters used in fiber optic links; sources for medical, industrial and lighting applications have different characteristics and will be treated in a separate chapter.

Following a review of basic terminology, we will discuss LEDs, lasers, and optical modulators. LEDs are frequently used for multimode fiber optic links. The two types of LEDs, surface emitting LEDs and edge emitting LEDs (ELEDs), have high **divergence** of the output light (meaning that the light beams emerging from these devices can spread out over 120° or more); slow rise times (>1 ns) that limit communication to speeds of less than a few hundred MHz; and output powers on the order of 0.1–3 mW. However, LEDs are less temperature sensitive than lasers and are very reliable.

For higher data rates, laser sources at short wavelengths (780–850 nm) are typically used with multimode fiber, and longer wavelengths (1.3–1.5 microns) with single-mode fiber. Typical lasers used in communications systems are **edge emitting** lasers (light emerges from the side of the device), including double heterostructure (DH), quantum well (QW), strained layer (SL), distributed feedback (DFB), and distributed Bragg reflector (DBR). Vertical cavity **surface emitting** lasers (VCSELs) have been introduced for short wavelength applications, and are under development for longer wavelengths; fibers may couple more easily to these surface emitting sources. Compared with LEDs, laser sources have much higher powers (3–100 mW or more), narrower spectral widths (<10 nm), smaller beam divergences (5–10 degrees), and faster modulation rates (hundreds of MHz to several GHz or more). They are also more sensitive to temperature fluctuations and other effects such as back reflections of light into the laser cavity.

While direct modulation is common for LEDs, it causes various problems with lasers, including turn-on delay, relaxation oscillation, mode hopping, and frequency chirping. External modulators used with continuous wave (CW) diode lasers are one possible solution (e.g., **lithium niobate modulators** based on the electro-optic effect, used in a Y-branch interferometric configuration). **Electroabsorption modulators,** made from III-V semiconductors, are also used because they allow for more compact designs and monolithic integration. **Electro-optic** and **electrorefractive** modulators are another option, which use phase changes, rather than absorption changes, to modulate light.

Fiber Optic Essentials; ISBN: 9780122084317

4.1.1 Basic transmitter specification terminology

There are several common features which we will find in a typical optical transmitter specification, regardless of the type of communication system involved. Every transmitter specification should include a physical description with dimensions and construction details (e.g., plastic or metal housing) and optical connector style. Specifications should also include a list of environmental characteristics (temperature ranges for operation and storage, humidity), mechanical specifications, and power requirements. Device characteristic curves may also be included, such as optical power output vs. forward current, or spectral output vs. wavelength.

A table of **absolute maximum ratings** will list properties such as the highest temperature for the case and device, as well as the maximum temperature to be used during soldering, the maximum current in forward or reverse bias, and the maximum operation optical power and current. Care should be taken with more recent vintage devices, which may use environmentally friendly lead-free solders and thus require higher soldering temperatures. Temperature is a very significant factor in determining the operation and lifetime (reliability) of a semiconductor optical source. This is one reason why many optical transceivers for communication systems are being developed as modular, pluggable components which do not require soldering to a printed circuit card.

The **spectral width, $\Delta\lambda$,** is a measure of the range of optical frequencies or the spectral distribution emitted by the source, given in nanometers. This may be calculated from the spectrum as the full width at half maximum (FWHM), which happens to exactly define the **spectral bandwidth.** Other measurements may include the root mean square (RMS) spectral width, which is a statistical value. LEDs will have a very broad spectrum compared with lasers; short wavelength lasers will have a broader spectrum than long wavelength lasers (this is another way of saying that the short wavelength lasers are less **coherent).**

The **center wavelength** is the midpoint of the spectral distribution (the wavelength that divides the distribution into two parts of equal energy). It is given in nanometers or micrometers (microns). The **peak wavelength, λp,** is the spectral line with the highest output optical power. The center wavelength and the peak wavelength are usually both listed on most specifications, although occasionally red LEDs which were designed to be coupled with plastic fiber leave out the central wavelength. The terminology **spectral bandwidth** and **spectral width** are used interchangeably in these specifications.

The **temperature coefficient of wavelength** measures how the temperature affects the output power at a specific wavelength. It is expressed in units of nanometers per degree Celsius. The **temperature coefficient of the optical power** is a broadband measurement, which is expressed in a percentage per degree Celsius.

The **coupling efficiency** between a source and a fiber is the ratio of the input power (the available power from the source), to the power transferred to the fiber or other component, converted to percent. A related value is the **fiber coupled power, Poc,** which is the coupling efficiency multiplied by the radiant power; this is given in units of either mW or dBm. Different values for the fiber coupled power are normally listed for different sized cores, especially the standard multimode fiber core diameters of 50 micron, 62.5 micron and 100 micron or a standard single-mode fiber of 9 micron.

The rise/fall time is a measurement of how fast the source can be directly modulated, or how fast the light intensity can be ramped up and down, usually in nanoseconds. A typical measurement is the 10/90 rise or fall time, which is the time during which the source output rises from 10 to 90 percent of its peak value, or falls from 90 to 10 percent. Sometimes the 80/20 rise and fall times are specified instead. The **data rate** for serial transmission is simply the number of bits per second which can be transmitted. This should not be confused with the modulation frequency, measured in cycles per second, which is half the bit rate. For example, consider a square wave which is being transmitted by an optical source; if the frequency is 100 MHz, then a complete square wave is transmitted every 0.01 microseconds. However, this is enough time to transmit 2 data bits, so the bit rate would be 200 Mbit/second. The time required to transmit one bit is called the **unit interval,** as expressed in bits/second; it is often specified in communication standards.

4.1.1.1 Laser safety

Generally speaking, fiber optic systems are no more dangerous to work with than any other type of electronics system if the user is aware of the potential hazards and takes nominal precautions to avoid them. Whenever designing or working with optical sources, eye safety is a significant concern. All types of optical sources, including lasers, LEDs, and even white light sources can cause vision problems if used incorrectly (e.g., if they are viewed directly through a magnifying lens or eye loop). In particular, fiber optic systems typically use infrared light which is invisible to the unaided eye; under the wrong conditions, an infrared source of less than 1 mW power can cause serious eye injury. Optical safety is not limited to viewing the source directly, but may also include viewing light emerging from a open connector, especially if a lens is used as noted earlier. In the United States, the

Table 4.1.1 ANSI Laser safety classifications.

Class	Description
Class 1	• inherently safe • no viewing hazard during normal use or maintenance • no controls or label requirements (typically 0.4 microwatts or less output power)
Class 1M	• inherently safe if not viewed through collecting optics • designed to allow for arrays of optical sources which may be viewed at the same time
Class 2	• normal human eye blink response or aversion response is sufficient to protect the user • low power visible lasers ($<$1 mW continuous operation) • in pulsed operation, warning labels are required if power levels exceed the class 1 acceptable exposure limits for the exposure duration, but do not exceed class 1 limits for a 0.25 second exposure
Class 2A	• warning labels are required • low power visible lasers that do not exceed class 1 acceptable exposure limits for 1000 seconds or less and systems which are not designed for intentional viewing of the beam
Class 3A	• normal human eye blink response or aversion response is sufficient to protect the user, unless laser is viewed through collecting optics • requires warning labels, enclosure/interlocks on laser system, and warning signs at room entrance where the laser is housed • typically 1–5 mW power
Class 3B	• direct viewing is a hazard, and specular reflections may pose a hazard • same warning label requirements as class 3A plus power actuated warning light when laser is in operation • typically 5–500 mW continuous output power, $<$ 10 joules per square centimeter pulsed operation for $<$ 0.25 seconds
Class 4	• direct viewing is a hazard, and specular or diffuse reflections may pose a hazard • skin protection and fire protection are concerns • same warning label requirements as class 3B plus a locked door, door actuated power kill switch or door actuated optical filter, shutter, or equivalent • typically $>$ 500 mW continuous output power, $>$ 10 joules per square centimeter pulsed operation

Department of Health and Human Services (DHHS) of the Occupational Safety and Health Administration (OSHA) and the U.S. Food and Drug Administration (FDA) defines a set of eye safety standards and classifications for different types of lasers (FDA/CDRH 21 CFR subchapter J). International standards (outside North America) are defined by the International Electrotechnical Commission (IEC/CEI 825) and may differ from the standards used in North America. The basic standard for the safe use of lasers was first issued in 1998 by the American National Standards Institute; since then, several updates and additional standards have been released (ANSI standard Z136.X, where X refers to one of six existing standards for lasers in different environments). Copies of these standards are available from various sources, including the Laser Institute of America (LIA), which also offers laser safety training. There may be additional state or local regulations governing safe use of lasers as well (e.g., New York State Code Rule 50 requires all laser manufacturers to track their primary components using a state-issued serial number). Anyone working with fiber optics on a regular basis should be familiar with the different types of laser safety classifications and how they apply to a specific product. Laser safety training is required by law in order to work with anything other than class 1 products (see Table 4.1.1 for a list of laser safety classifications). All lasers other than class 1 (inherently safe) require a safety label; their classification depends on factors including the optical wavelength, power level, pulse duration, and safety interlocks built into their packaging. Fiber optic practitioners should be familiar with the safety requirements for systems they will be using.

In addition, practitioners should be aware of all standard safety precautions for handling fiber optic components. For example, cleaving and splicing of glass fibers can produce very small, thin shards of glass which can lodge in the skin. Tools for fusion splicing produce sufficient heat to melt glass fibers, and either the equipment or a freshly heated fiber splice can cause burns. Chemical safety precautions should be used when handling index matching gels, acetone, or optical cleaning chemicals. Alcohol is sometimes used for cleaning fibers and connectors, but can be flammable if used improperly. Compressed air canisters used for dusting optical components should not be punctured or shipped by air freight to avoid explosive decompression. Normal safety precautions for working with electronics equipment and high voltage or current levels may also apply. In general, only a certified, trained professional should be allowed to work unsupervised with optical fiber, optoelectronics, or similar equipment.

4.1.2 Light emitting diodes

Light emitting diodes used in data communications are solid state semiconductor devices; to understand their function, we must first describe a bit of semiconductor

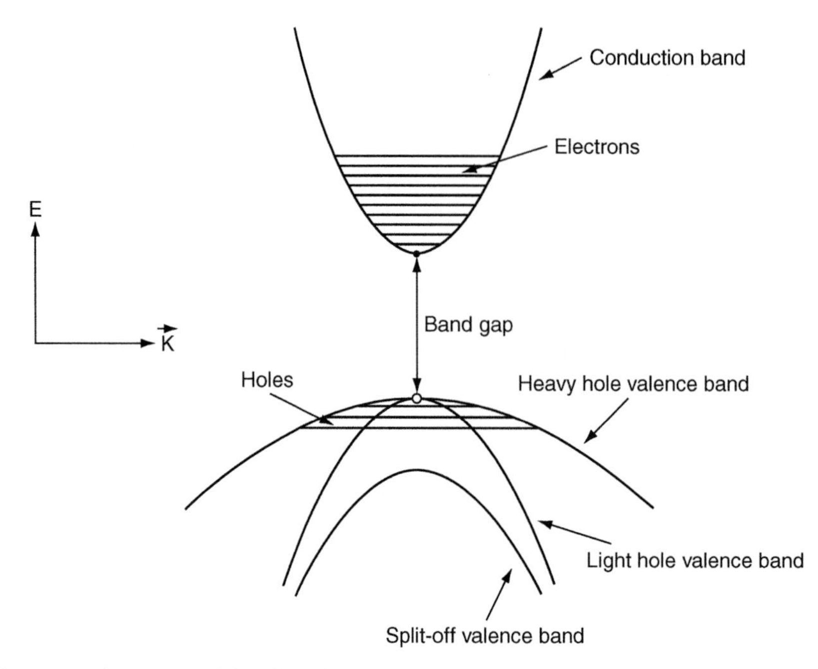

Fig. 4.1.1 Energy band structure (energy vs. *k*) for direct band gap semiconductor.

physics. For the interested reader, other introductory references to solid state physics, semiconductors, and condensed matter are available [1,2]. In a solid state device, we can speak of charge carriers as either electrons or holes (the "spaces" left behind in the material when electrons are absent). The carriers are limited to occupying certain energy states; the electron potential can be described in terms of **conduction bands** and **valence bands**, rather than individual potential wells (Fig. 4.1.1). The highest energy level containing electrons is called the **Fermi Level**. If a material is a **conductor**, the conduction and valence bands overlap and charge carriers (electrons or holes) flow freely; the material carries an electrical current. An **insulator** is a material for which there is a large enough gap between the conduction and valence bands to prohibit the flow of carriers; the Fermi level lies in the middle of the forbidden region between bands, called the **band gap**. A **semiconductor** is a material for which the band gap is small enough that carriers can be excited into the conduction band with some stimulus; the Fermi level lies at the edge of the valence band (if the majority of carriers are holes) or at the edge of the conduction band (if the majority carriers are electrons). The first case is called a **p-type** semiconductor (since the carriers are positively charged holes), the second is called **n-type** (since the carriers are negatively charged electrons). When a p-type and an n-type layer are sandwiched together, they form a PN junction (Fig. 4.1.2). PN junctions are useful for generating light because when they are forward biased, electrons are injected from the

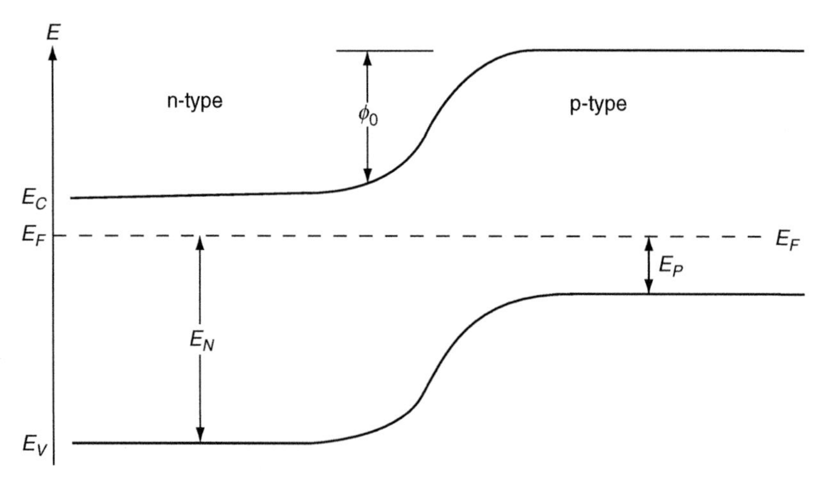

Fig. 4.1.2 Conduction and valence bands of a semiconductor.

N region and holes are injected from the P region into the active region. Free electrons and free holes can recombine across the band gap and emit photons (called **spontaneous emission**) with energy near the band gap, resulting in an LED. Like any other diode, an LED has a characteristic current-voltage response and will pass current only in one direction. Its optical output power depends on the current density, which in turn depends on the applied voltage; below a certain threshold current, the optical power is negligible.

In fiber communications systems, LEDs are used for low-cost, high reliability sources, typically operating with 62.5/125 micron graded-index multimode fiber at data rates up to a few hundred Mb/s or perhaps as high as 1 Gbit/s in some cases. For short distances, GalliumArsenide (GaAs) based LEDs operating near 850 nm are used because of their low cost and low temperature dependence. For distances up to ~10 km, more complex LEDs of InGAAsP grown on InP and emitting light at 1.3 microns wavelengths are often used [2]. The switching speed of an LED is proportional to the electron-hole recombination rate, R, given by

$$R = \frac{J}{de}, \tag{4.1.1}$$

where J is the current density in A/m^2, d is the thickness of the recombination region, and e is the charge of an electron. The optical output power of an LED is proportional to the drive current, I, according to

$$P_{out} = I\left(\frac{Qhc}{e\lambda}\right), \tag{4.1.2}$$

where Q is the quantum efficiency, h is Planck's constant, and λ is the wavelength of light.

Spontaneous emission results in light with a broad spectrum which is distributed in all directions inside the active layer. The wavelength range emitted by the LED widens as its absolute junction temperature, T, increases, according to

$$\Delta\lambda = 3.3\left(\frac{kT}{h}\right)\left(\frac{\lambda^2}{c}\right), \tag{4.1.3}$$

where k is Boltzmann's constant and c is the speed of light. As the LED temperature rises, its spectrum shifts to longer wavelengths and decreases in amplitude.

There are two types of LEDs used in communications, surface emitting LEDs and ELEDs. Thus, there are two ways to couple as much light as possible from the LED into the fiber. Butt coupling a multimode fiber to the LED surface (as shown in Fig. 4.1.3) is the simplest method, since a surface emitting geometry allows light to come out the top (or bottom) of the semiconductor structure. The alternative is to cleave the LED, and collect the light that is emitted from the edge, thus ELEDs are formed (Fig. 4.1.4). While surface emitting LEDs are typically less expensive, ELEDs have the advantage that they can be used with single-mode fiber. Both can be modulated up to around 622 Mb/s.

For a surface emitting LED, the cone of spontaneous emission that can be accepted by a typical fiber (numerical aperture NA = 0.25) is only ~0.2 percent. This coupling is made even smaller by reflection losses (such as Fresnel reflection). To improve the coupling efficiency, lenses and mirrors are used in several different geometries (e.g., see Fig. 4.1.5). Typical operating specifications for a surface emitting LED at 1.3 microns pigtailed to a 62.5/125 micron core graded-index fiber might be 16 µW at 100 mA input current, for ~0.02 percent overall efficiency with a modulation capability of 622 Mb/s. They are usually packaged in lensed TO-18 or TO-46 cans, although SMA and ST connectors are also used.

Edge emitting LEDs have greater coupling efficiencies because their source area is smaller. Their geometry is

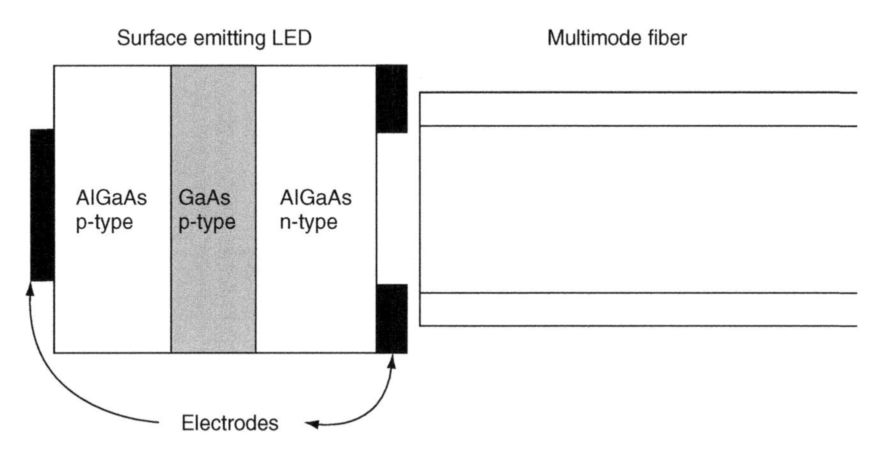

Figure 4.1.3 Butt-coupling between a surface emitting LED and a multimode fiber.

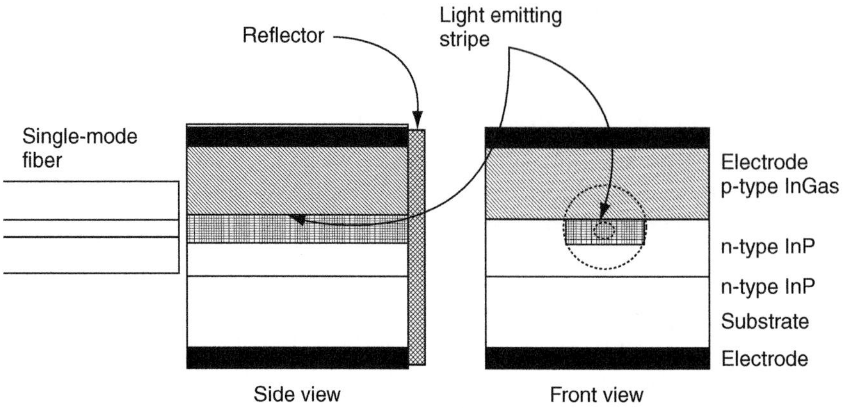

Reflector
Light emitting stripe
Single-mode fiber
Electrode
p-type InGas
n-type InP
n-type InP
Substrate
Electrode
Side view
Front view

Fig. 4.1.4 An edge emitting LED.

similar to a conventional diode laser, because light travels back and forth in the plane of the active region and is emitted from one anti-reflection coated cleaved end. These LEDs differ from lasers in that they do not have a feedback cavity. Because of their smaller source size, they can be coupled into single mode fiber with modest efficiency ($\sim$0.04%). ELEDs are typically used as low-coherence sources for fiber sensor applications, rather than in communications, because of their broad emission spectrum.

Device characteristic curves for LEDs include the forward current vs. applied voltage, the light emission or luminous intensity vs. forward current, the spectral distribution and the luminous intensity vs. temperature and angle. An example of these is shown in Fig. 4.1.6. Since LEDs are chosen for low cost applications, they are modulated by varying the drive current, rather than using an external modulator.

4.1.3 Lasers

There are many different types of lasers, only some of which are useful for communication systems. The term "**laser**" is an acronym for Light Amplification by Stimulated Emission of Radiation. In order to form a basic laser system, we require a media which can produce light (photons) through **stimulated emission.** This means that electrons in the media are excited to a higher than normal energy state, then decay into a lower energy state, releasing photons in the process. These photons will have the same wavelength, or frequency, as well as the same phase; thus, they will add up to create a **coherent** optical beam which emerges from the laser. Although we speak of laser light as being **monochromatic**, or having a single wavelength, in reality a laser produces a narrow spread of different wavelengths, or an emission spectra. The material which creates photons in this way, called the **gain media**, is most commonly a gas (in the case of medical lasers) or a semiconductor material (in the case of some medical devices and almost all communication systems). Electrons are excited by supplying energy from some external source, a process called **pumping** the laser. Most lasers are pumped with an electrical voltage, though another optical beam can sometimes be used to provide the necessary energy. Since the electrons are normally found in their lower energy state, pumping them into a higher energy state creates what is known as a **population inversion**. The final element in creating a laser is an optical feedback mechanism, such as a pair of

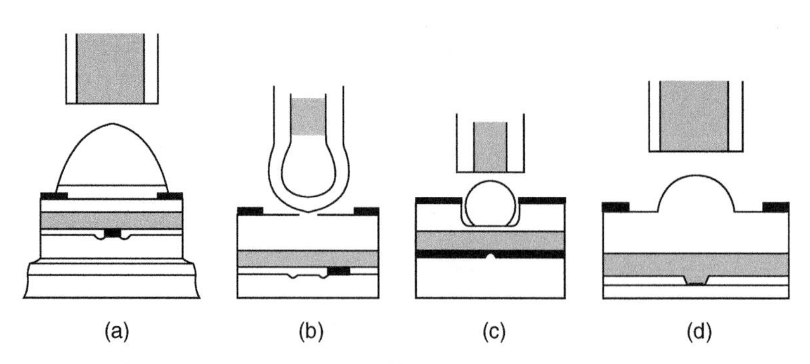

Fig. 4.1.5 Typical geometries for coupling from LEDs into fibers: (a) hemispherical lens attached with encapsulating plastic; (b) lensed fiber tip; (c) microlens aligned through use of an etched well; and (d) spherical semiconductor surface formed on the substrate side of the LED.

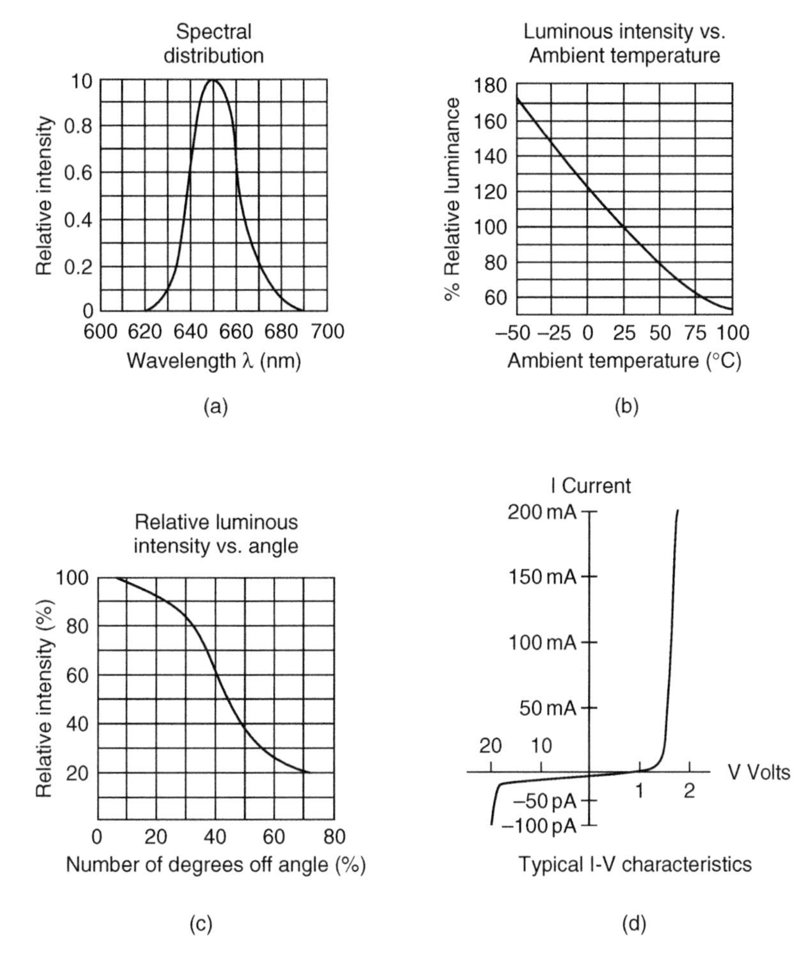

Fig. 4.1.6 Characteristics of an LED: spectral distribution, luminous intensity vs. temperature and angle, and I vs. V characteristics.

mirrors, to reflect light many times through the gain media and produce a stronger output beam. Laser light is coherent, meaning that all the photons are at the same wavelength (frequency) and all are in phase with each other. Lasers have very low divergence (a laser spot will spread out very little over long distances).

In most fiber optic communication systems, we are interested in the type of lasers based on solid state devices, similar to LEDs in that they use PN heterojunctions to emit photons. In LEDs, PN junctions are used for generating light because when they are forward biased, electrons are injected from the N region and holes are injected from the P region into the active region. Free electrons and free holes can recombine across the band gap and emit photons (called spontaneous emission) with energy near the band gap. Lasers are different from LEDs because the light is produced by **stimulated emission** of radiation which results in light of a very narrow, precise wavelength. As noted above, the semiconductor laser diode requires three essential operation elements:

1. Optical feedback from mirrors on either side of the laser cavity (originally provided by a cleaved facet or multilayer Fresnel reflector, more recently provided by wide band-gap layers of cladding).
2. A gain medium (the PN heterojunction, with spontaneous emission providing optical gain).
3. A pumping source (applied current).

The optical cavity used for some types of lasers is based on a **Fabry-Perot (FP)** device, commonly described in basic physics, in which the length of a mirrored cavity is an integral multiple of the laser wavelength. Compared with LED emission, laser sources exhibit higher output power (3–100 mW or more), spread over a narrower band of wavelengths (<10 nm), are more directional (beam divergence 5–10 degrees), have a greater modulation bandwidth (hundreds of MHz to several GHz or more), and are coherent (emit light at a single frequency and phase) [3].

The minimum requirement for lasing action to occur is that the light intensity after one complete trip through the cavity must at least be equal to its starting intensity, so that there is enough gain in the optical signal to overcome losses in the cavity. This **threshold condition** is given by the relationship

$$R_1R_2\exp(2(g-\alpha)L) = 1, \tag{4.1.4}$$

where the reflectivity of the cavity mirrors is R_1 and R_2, the lasing cavity is of length L, and the material has a gain of g and a loss of α.

The **threshold current** is the minimum current to produce laser light. This is a measured operating characteristic and for double heterojunction, lasers should be under ~ 500 A/cm^2 at 1.3 μm and ~ 1000 A/cm^2 at 1.55 μm.

The laser emits photons whose energy, E, and wavelength, λ (or frequency, f) are related by

$$E = hf = \frac{hc}{\lambda}, \tag{4.1.5}$$

where h is Planck's constant. For semiconductor laser diodes, photons are produced when electrons that were pumped into a higher energy state drop back into a lower energy state. The difference in energy states is called the **band gap**, Eg, and the wavelength of light produced is given by

$$\lambda(\text{microns}) = \frac{Eg(\text{electron volts})}{hc} = \frac{Eg}{1.24}. \tag{4.1.6}$$

When we discuss photodetectors in a later chapter, we will see that the reverse process is also useful; incident light which illuminates a semiconductor device can create an electrical current by exciting electrons across the band gap. The band gap in many semiconductor compounds can be changed by doping the material to produce different wavelengths of light.

Given that a laser diode with an active area of width w and length L will have a current $I = JwL$, and the current density is related to the carrier density through $J = eNd/\tau$, where τ is the lifetime of the electron/hole pairs, N is the carrier density, and d is the layer thickness, the **threshold current** required to achieve laser operation can be calculated from

$$I_{th} = I_{tr} + ewN(1/2\ \text{In}(1/R_1R_2) + \alpha L)d\frac{(1+2/V^2)}{\tau a_L}, \tag{4.1.7}$$

where I_{tr} is the transparency current, a_L is the proportionality constant between the gain and the carrier density near transparency and the waveguide V parameter is calculated from the requirement that a waveguide must be thinner than the cutoff value for higher order modes for proper optical confinement,

$$V = dk\sqrt{(n_g^2 - n_c^2)}, \tag{4.1.8}$$

where n_g is the refractive index of the waveguide area, n_c is the refractive index of the cladding, d is the thickness of the waveguide layer and $k = 2\pi/\lambda$.

Measured curves of light output vs. current are also shown on most specifications. When operating above threshold, the output power can be calculated from

$$P = (hv/e)(\alpha/(\alpha+\alpha_i))(I - I_{th} - I_L)\eta, \tag{4.1.9}$$

where α_i is any internal losses for the laser mode, and η is the quantum efficiency. Calculating the quantum efficiency is beyond the scope of this book, but can be found in reference [4].

Multiple PN junctions can be used in a single laser device, and there are many different types of lasers which can be made in this way. Structurally, a semiconductor DH laser is similar to ELEDs, except that photons and carriers are confined by growing the active region as a thin layer and surrounding it with layers of wide band-gap material. When the active layer is as thin as a few tens of nanometers, the free electron and hole energy levels become quantized, and the active layer becomes a **quantum well**. Because of their high gain, one or more quantum wells are often used as the active layer. While the double heterostructure and quantum well lasers use cleaved facets to provide optical feedback, distributed feedback and distributed Bragg reflector lasers replace one or more facet reflectors with a waveguide grating located outside of the active region to fine-tune the operating wavelength.

Vertical cavity surface emitting lasers have significantly different architecture, which has advantages for low cost data transmission. The light is emitted directly from the surface, not from the edge, which means fiber can be directly butt-coupled. Existing commercial VCSELs emit light exclusively at short wavelengths (near 850 nm). Long wavelength VCSELs (1.3 and 1.55 μm) have been demonstrated in laboratories but are not yet commercially available – there are a number of technical issues to be addressed, including reliability, poor high temperature characteristics and low reflectivity because of their InP/InGaAsP Bragg mirrors. There are solutions to these problems under development, such as providing different types of dielectric mirrors, wafer fusion, metamorphic Bragg reflectors or wafer fusing GaAs/AsGaAs Bragg mirrors to the InP lasers.

4.1.3.1 Double heterostructure laser diodes

The standard telecommunications source is an edge-emitting laser, grown with an active layer that has a band gap near either 1.5 or 1.3 microns. The layers are made of InGaAsP crystals, grown so they are lattice matched to InP for long wavelengths, or $Al_yGa_{1-y}As/Al_xGa_{1-x}As$ ($x > y > 0$) for near infrared wavelength operation. The typical geometry is shown in Fig. 4.1.7.

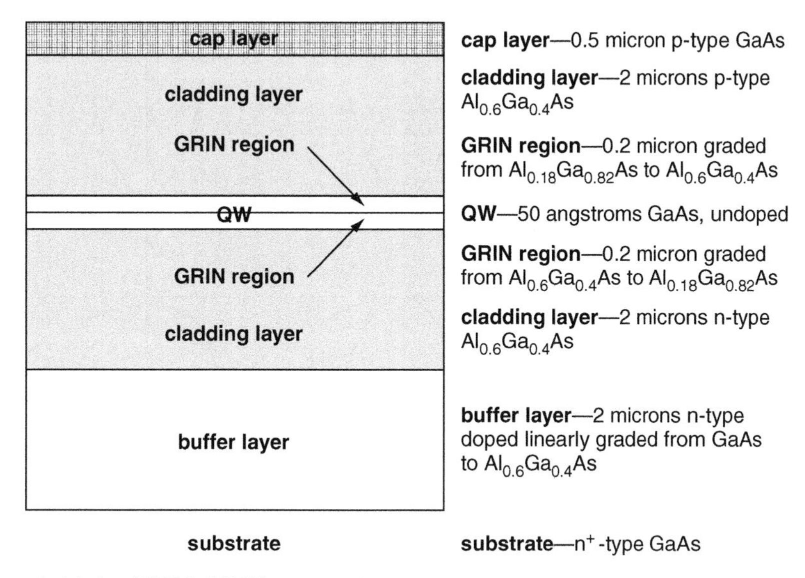

Fig. 4.1.7 A GaAs/AlAs graded-index (GRIN) SQW laser structure.

The current is confined to a stripe region which is the width of the optically active area, for efficiency. While the simplest laser diode structures do not specifically confine light, except as a result of their stripe geometry and carrier injection direction (known as **gain-guided**), these are not usually used for telecommunications applications. Higher quality lasers confine the light using buried heterostructures (BH) or ridge waveguides (RWG). The DH semiconductor laser represents the single largest constituent of today's semiconductor laser production because of its application in compact disk (CD) players and CD data storage.

4.1.3.2 Quantum well (QW) and strained layer (SL) quantum well lasers

When the active layer of a semiconductor laser is so thin that the confined carriers have quantized energies, it becomes a QW. The main advantage to this design is higher gain. In practice, either multiple QWs must be used, or else we require a guided wave structure that focuses the light on a single quantum well (SQW) (Fig. 4.1.8). Strained layer (SL) quantum wells have advanced the performance of InP lasers. In a bulk semiconductor, there are two valence bands (called heavy hole and light hole) which are at the same level as the potential well minimum. QWs separate these levels, which allow population inversion to become more efficient. Strain moves these levels even further apart, making them perform even more efficiently (Fig. 4.1.9). Typical 1.3 micron and 1.5 micron InP lasers use 5 to 15 internally strained QWs.

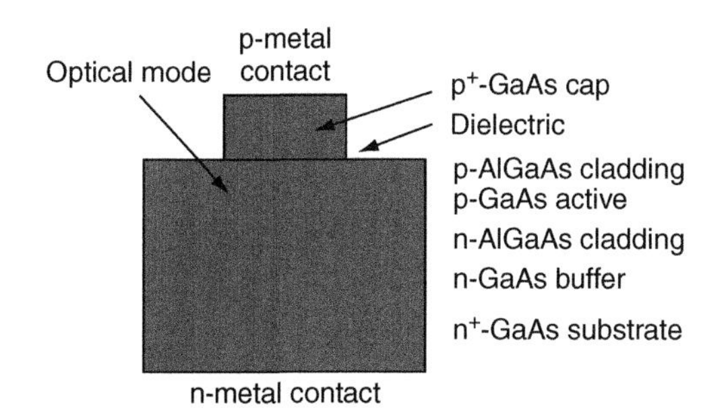

Fig. 4.1.8 Schematic diagram of an ridge waveguide GaAs/AlGaAs double-heterojunction (DH) laser.

4.1.3.3 Distributed Bragg reflector (DBR) and distributed feedback (DFB) lasers

The DBR replaces the facet reflectors with a diffraction grating outside of the active region (see Fig. 4.1.10). The Bragg reflector will have wavelength dependent reflectivity with resonance properties. The resulting devices behave like FP lasers, except that the narrow resonance gives them single-mode operation at all excitation levels.

Distributed feedback lasers are similar to DBR lasers, except that the Bragg reflector is placed directly on the active region or its cladding (Fig. 4.1.11). There is no on-resonance solution to the threshold condition in this geometry. This is resolved by either detuning to off-resonance solutions, or adding a quarter wavelength shifted grating.

Once the DFB laser is optimized, it will be single mode at essentially all power levels and under all modulation conditions. This is a significant advantage; over

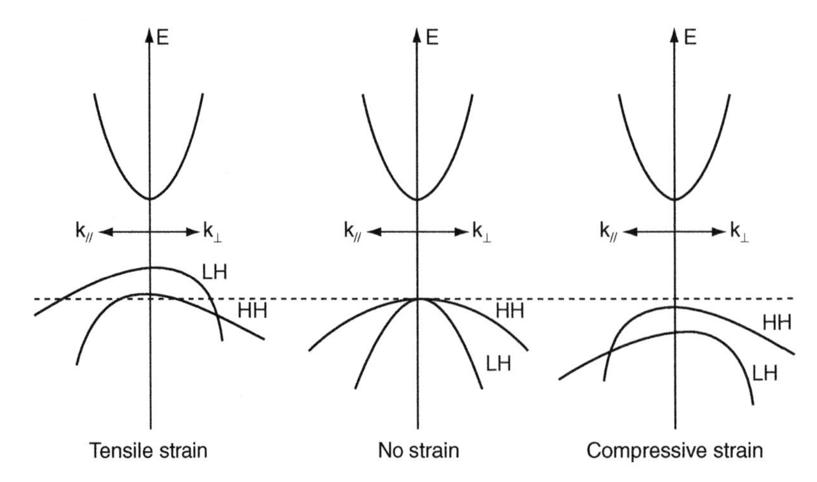

Fig. 4.1.9 Schematic energy band diagram in *k* space showing the removal of degeneracy between heavy hole (HH) valence band edge and light hole valence band edge for both compressive and tensile strained InGaAs gain medium.

the past few years, many telecommunication applications have adopted strained QW-distributed feedback lasers at 1.3 micron or 1.55 micron.

4.1.3.4 Vertical cavity surface emitting lasers

Vertical cavity surface-emitting lasers incorporate many of the advances described in the previous sections, such as QWs and DBR, into a surface emitting geometry that has many theoretical and practical advantages. While this design is difficult to achieve for long wavelengths (1.3 and 1.55 micron), it is very popular in 850-nm versions which have found wide applications in consumer electronics. These sources are low cost and compatible with inexpensive silicon detectors.

A diagram of a typical VCSEL is shown in Fig. 4.1.12. The VCSEL has mirrors on the top and bottom of the active layer, forming a vertical cavity. DBRs (multiplayer quarter wavelength dielectric stacks) form high reflectance mirrors. A typical active region might consist of three QWs each with 10 nm thickness, separated by about 10 nm. It is a technical challenge to inject carriers from the top electrode down through the Bragg reflector, because the GaAs layers provide potential wells that trap carriers. The operating voltage is two to three times that of edge emitting lasers. The light emitted from VCSELs will have similar performance to edge emitting lasers, with a more graceful turn-on due to more spontaneous emission. More information about the operating characteristics can be found in reference [4].

4.1.4 Modulators

In an optical communication system, an electrical signal is converted to an optical signal by **modulating** the source of light. Turning the source on and off is **direct modulation**, but often a **modulator** is used to switch a constant source of light. As noted previously, some optical sources can be directly modulated, while others require external modulation. External modulation is primarily used in telecommunication systems for more efficient operation

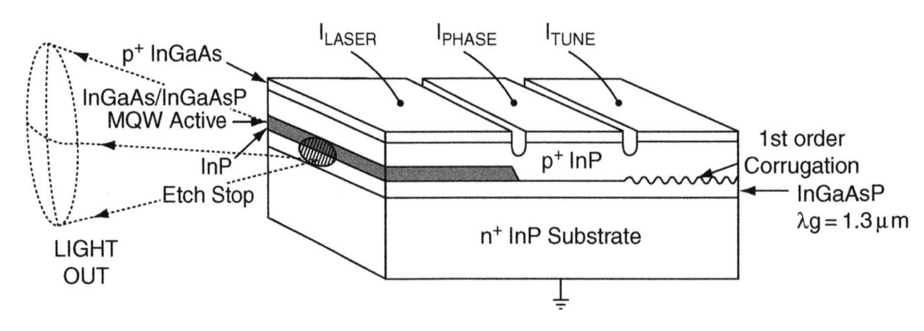

Fig. 4.1.10 Schematic for DBR laser configuration in a geometry that includes a phase portion for phase tuning and a tunable DBR grating. Fixed-wavelength DBR lasers do not require this tuning region. Designed for 1.55 μm output, light is waveguided in the transparent layer below the MQW that has a bandgap at a wavelength of 1.3 μm. The guided wave reflects from the rear grating, sees gain in the MQW active region, and is partially emitted and partially reflected from the cleaved front facet. Fully planar integration is possible if the front cleave is replaced by another DBR grating [5].

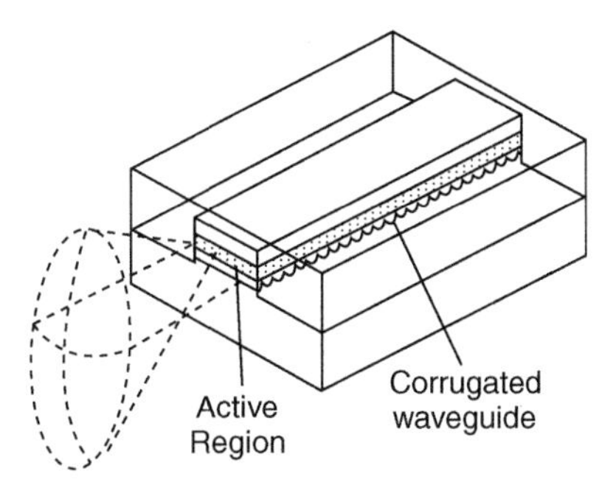

Fig. 4.1.11 Geometry for a DFB laser, showing a buried grating waveguide that forms the separate confinement heterostructure laser, which was grown on top of a grating etched substrate. The cross-hatched region contains the MQW active layer. A stripe mesa is etched and regrown to provide a buried heterostructure laser. Reflection from the cleaved facets must be suppressed by means of an antireflection coating.

at higher data rates. In a CW laser, the optical output is constant over time, and the laser light is produced in an uninterrupted fashion. The alternative is pulsed or modulated lasers. While LEDs may directly modulate their signal by turning on and off their electrical power, designs for much higher data rates may use CW lasers in conjunction with an external optical modulator. This is done to reduce effects such as turn-on delay, relaxation oscillation, mode hopping and/or frequency chirping.

Turn-on delay occurs in semiconductor lasers because it takes time for the carrier density to reach its threshold value and for the photon density to build up to a critical value. There is thus a small delay between application of a voltage to the laser and the resulting optical output (see Fig. 4.1.13).

Relaxation oscillations are overshoots of the desired signal level, which occur as the photon dynamics and carrier dynamics are coming into equilibrium. This creates oscillations in the light output for short periods of time following a rise or fall in optical power (again, see Fig. 4.1.13).

Mode hopping is the sudden shift of the laser diode output beam from one longitudinal mode to another

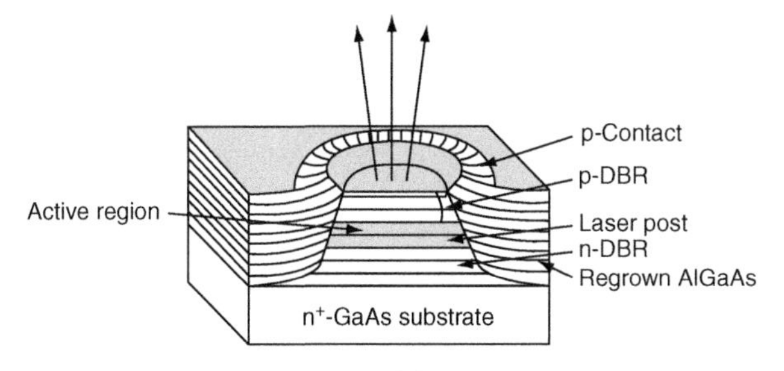

Fig. 4.1.12 One example of an VCSEL geometry. This is a passive antiguide region (PAR) VCSEL. Light is reflected up and down through the active region by the two DBR mirrors. After the laser post is etched, regrowth in the region outside the mesa provides a high refractive index AlGaAs*nipi* region to stop current flow and to provide excess loss to higher order modes.

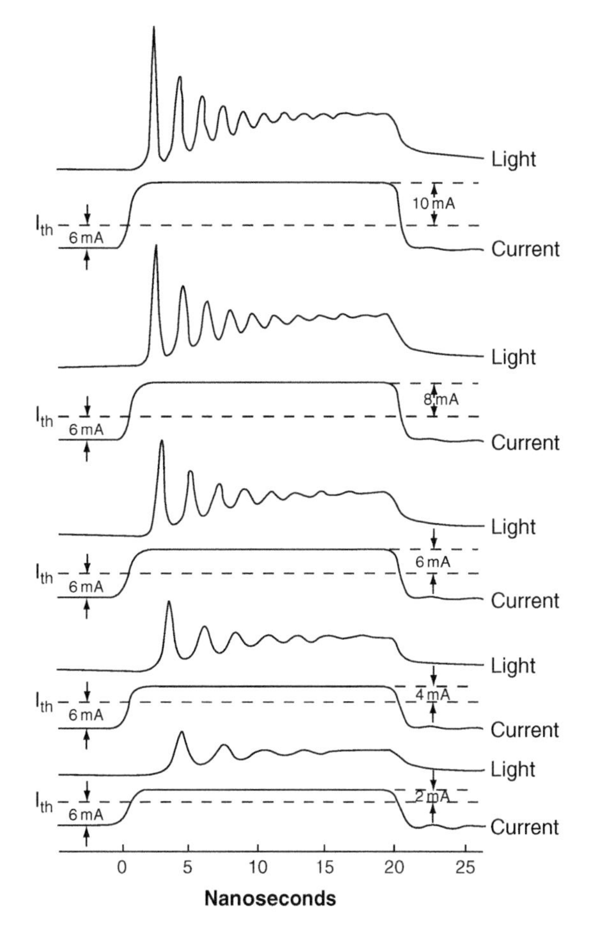

Fig. 4.1.13 Experimental example of turn-on delay and relaxation oscillations in a laser diode when the operating current is suddenly switched from 6 mA below the threshold current of 177 mA to varying levels above threshold (from 2 to 10 mA). The GaAs laser diode was 50 μm long, with a SiO_2 defined stripe 20 μm wide. Light output and current pulse are shown for each case.

during operation. This may cause excessive modal noise in an optical link.

Chirping occurs when the carrier density in the active region is rapidly changed, causing a shift of the index of refraction with time. This modifies the frequency of the output light (the term "chirp" refers to a linear change in frequency over time) and affects dispersion and other link noise sources. Chirping is most often observed in pulsed sources.

The **extinction ratio** of a modulator is the ratio of the optical power level generated when the light source is "on", compared to when the light source is "off". It is typically expressed as a fraction at a specific wavelength, or in dB. Put another way, for **intensity modulators** we can define the **modulation depth (MD)** in terms of the intensity when no data is transmitted, I_0, to the intensity when data is transmitted, I;

$$MD = \frac{(I_0 - I)}{I_0} \qquad (4.1.10)$$

Then extinction ratio is the maximum value of the ratio MD(max) to MD when the intensity of the transmitted beam is at its minimum. For **phase modulators** it is defined in a similar way, provided light intensity is related to phase. For **frequency modulators** we speak instead of the **maximum frequency deviation**, D(max), given by

$$D(\max) = \frac{|f_m - f_0|}{f_0}, \qquad (4.1.11)$$

where f_m is the maximum frequency shift of the carrier frequency f_0. Some specification sheets define the **degree of isolation**, in dB, which is given by 10 times the log of either the extinction ratio or the maximum frequency deviation.

4.1.4.1 Lithium niobate modulators

Certain types of materials, such as lithium noibate crystals, can experience a change in their refractive index when an electric field is applied. This is known as the **electro-optic effect**. In the case of lithium niobate, choosing the crystal orientation to maximize the change in the index of refraction, this change can be calculated from

$$\Delta n_z = -n_z^3 r_{33} E_z \Gamma/2 = -n_z r_{33} V\Gamma/G^2, \qquad (4.1.12)$$

where n_z is the index of refraction along the crystal's z-axis, r_{33} is a coefficient describing the magnitude of the electro-optic effect along the new axis of the crystal (in m/V), E_z is the electric field applied in the z direction, V is the applied voltage, G is the gap between the electrodes, and Γ is a filling factor (also known as an optical-electrical field overlap parameter). The filling factor corrects for the fact that the applied field may not be uniform as it overlaps the waveguide, effectively diminishing the applied field.

The electro-optic effect produces a phase shift in the modulated light, which is given by

$$\Delta\phi = \Delta n_2 kL, \qquad (4.1.13)$$

where k is $\frac{2\pi}{\lambda}$ and the electrode length is L.

Considering the index of refraction for bulk lithium niobate at a wavelength of 1.55 microns, when G = 10 microns and $V = 5$ volts, a phase shift of π is expected in a length $L \sim 1$ cm. Since Δn_z is directly proportional to the voltage, the electro-optic phase shift depends on the product of the voltage and the length, which leads engineers to use this as a figure of merit. This example would describe a 5 V-cm figure of merit.

Lithium niobate's electro-optic phase shift allows it be used to modulate intensity using interference in Y-branch **interferometric modulators**, as shown in Fig. 4.1.14. The light entering the waveguide is evenly split at the Y junction, and then goes through oppositely biased lithium niobate crystals. The fraction of light transmitted is

$$\eta = [\cos(\Delta\phi/2)]^2. \qquad (4.1.14)$$

If a phase difference of π is induced, then destructive interference occurs when the light recombines, and there is no light output from the device. Otherwise there is no destructive interference, and light emerges from the device. In this way, the interferometric modulator can switch light on or off.

Modulators are used to avoid or minimize turn-on delay, relaxation oscillation, mode hopping and/or chirping. Commercially available lithium niobate modulators operate at frequencies up to 8 GHz at 1.55 microns. Direct coupling from a laser or polarization maintaining fiber is necessary, because their phase shifting design is inherently polarization dependent.

4.1.4.2 Electroabsorption modulators

Electroabsorption refers to the electric field dependence of the absorption near the band edge of a semiconductor. This dependence is particularly strong in QWs, where it is called the quantum-confined Stark effect. There is a peak in the QW absorption spectrum known as the exciton resonance. This can form the basis for an external optical modulator. It is desirable to build modulators from III-V semiconductors, so they can be integrated directly on the same chip as the laser, although they may also be external, coupled by means of a microlens or a fiber pigtail.

The III-V semiconductors are a good material to use in QWs, which can be designed to take advantage of electroabsorption.

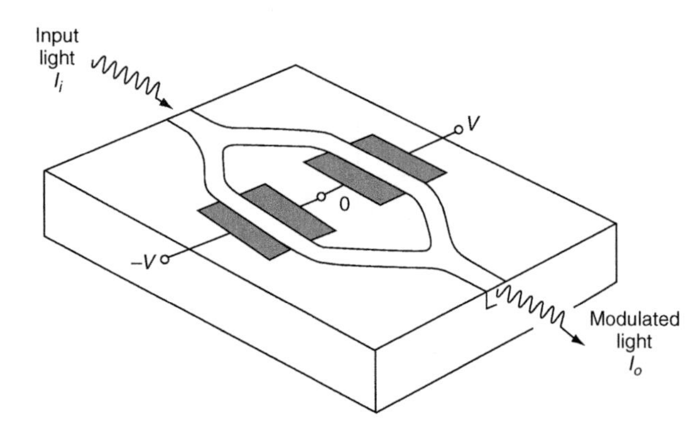

Figure 4.1.14 *Y*-branch interferometric modulator in the "push-pull" configuration. Center electrodes are grounded. Light is modulated by applying positive or negative voltage to the outer electrodes.

When light passes through the QW material, the transmission is a function of applied voltage. The usual geometry involves several QWs in a waveguide to prevent excessive light loss from diffraction. Performance is usually characterized by two quantities, contrast ratio (CR) and insertion loss. **Contrast ratio** is given by

$$CR = \frac{T_{high}}{T_{low}} = \exp(\delta\alpha L), \tag{4.1.15}$$

where T is transmission, and $\delta\alpha$ is the change in $\alpha(E)$, the QW absorption multiplied by the filling factor of the QWs in the waveguide, as it changes from the field E for a low signal to a high signal. The **insertion loss** is given by

$$A = 1 - T_{high} = 1 - \exp(-\alpha_{low}L) \sim \alpha_{low}L. \tag{4.1.16}$$

Modulator lengths are designed so the $L \sim 1/\alpha_{low}$, to keep the insertion loss low, despite the associated limitations on the contrast ratio. This results in an CR which varies as $CR = \exp(\delta\alpha/\alpha_{low})$. CR may reach as high as 10/1 or more with <2 V applied.

4.1.4.3 Electro optic and electrorefractive semiconductor modulators

The electro-optic effect is not limited to lithium niobate. Some III-V semiconductors, which are easier to package because they can be integrated on the same substrate as a laser, also experience a degree of anisotropy when an electric field is applied. The advantage of this kind of system over electroabsorption is that they offer reduced chirp (frequency broadening due to time-varying refractive index).

Semiconductor modulators can be designed based on another basic property called **electrorefraction.** The change in the electric field dependence of the absorption near the band edge, electroabsorption, is closely connected to a change in the index of refraction, which can be calculated using the Kramers-Kronig relations [4]. As with electroabsorption modulators, a field is applied across a semiconductor PIN junction. The difference is that the light is refracted, rather than absorbed. This can result in a reduction in length and drive voltages required for phase modulation in waveguides.

References

[1] G. Burns, *Solid State Physics*, Chapter 10, pp. 243–349, Orlando, Fl.: Academic Press (1985).

[2] C. Kittel, *Introduction to Solid State Physics*, Chapter 8, pp. 81–214, Wiley (1986).

[3] W. Jiang, and Lebby S. Michael "Optical Sources: Light Emitting Diodes and Laser Technology", in *Handbook of Fiber Optic Data Communication*, C. DeCusatis (ed.), Chapter 1, pp. 41–81 (2002).

[4] E. Garmire, "Sources, Modulators and Detectors for Fiber-Optic Communications Systems", *Handbook of Optics*, M. Bass, *et al.* (ed.), pp. 4. 1–4.78, McGraw-Hill (2001).

[5] T.L. Koch, and U. Koran, *IEEE J. Quantum Electron.*, 27, p. 641 (1991).

Chapter 4.2

4.2

Optical detectors and receivers

DeCusatis

In this chapter we will review optical detectors and receivers for optical communication systems. An optical receiver includes both the optical detectors, or **photodetectors** (the solid state device that convert an optical signal to an electrical signal), and various electronic circuits such as amplifiers or clock recovery circuits. We will discuss semiconductor devices such as PIN photodiodes, avalanche photodiodes (APD), photo-diode arrays, Schottky-Barrier photodiodes, metal-semiconductor-metal (MSM) detectors, resonant-cavity photodetectors, and interferometric devices. APDs are used in high sensitivity applications, MSM photoconductive detectors are sometimes used because of their ease of fabrication, and Schottky-barrier photodiodes are used for high speed applications. Resonant-cavity photodetectors are not commonly used today, but have potential for future development. Other types of detectors are also becoming more common in applications such as wavelength division multiplexing (WDM) and parallel optical interconnects (POI). After a brief discussion of optical receiver circuits, we will conclude with a discussion of noise sources and their impact on the detection electronics.

Despite the fact that they both detect and measure light, optical fiber **sensors** should not be confused with fiber optic **receivers.** Unlike receivers in communication systems, fiber sensors do not directly demodulate light; rather they often use diffraction gratings to measure interference effects in systems employing optical fibers, and thereby measure environmental factors such as temperature, vibration, and strain. Often these devices require optical spectral analysis or other techniques, rather than the more simple optical power or flux measurements provided by PIN diodes. **Interferometric sensors** are discussed briefly with other detectors.

It is worth noting that fiber optic systems typically package transmitters (discussed in Chapter 2) and receivers in a single housing known as transceivers (TRX). These are used for most two-fiber bi-directional communication links. Depending on the application, more or less electronics may be packaged within a transceiver. As before, the focus in this chapter will be mostly on communications systems; there are many types of optical detectors used in photometry, imaging systems, and other applications which are beyond the scope of this book.

Detectors are an integral part of all fiber optic communication systems. They convert a received optical signal into an electrical signal, demodulating the optical signal that was modulated at the transmitter. This initial signal may be subsequently amplified or further processed. Optical detectors must satisfy stringent requirements such as high sensitivity at the operating wavelengths, fast response speed, and minimal noise. In addition, the detector should be compact in size, use low biasing voltages and be reliable under all operating conditions. There are many different types of optical detectors available, which are sensitive over the entire optical spectra, the infrared and the ultraviolet, including such devices as photomultiplier tubes, charge coupled devices (CCDs), thermal and pho-toconductive detectors [1]. However, fiber optic communication systems

Fiber Optic Essentials; ISBN: 9780122084317

most commonly use solid state semiconductor devices, such as the PIN photodiode.

A general semiconductor detector has basically three processes: first, carriers (electron-hole pairs) are generated by the incident light; second, carriers are transported through the detection device and the gain may be multiplied by whatever current-gain mechanism may be present; and third, the current interacts with the external circuit to provide the output signal. In this chapter, we will begin with an overview of detector terminology. This is followed by a detailed description of the types of detectors used extensively in fiber optics. The PIN photodiode is the most commonly used fiber optic detector, and almost exclusively used in transceivers.

4.2.1 Basic detector specification terminology

Every detector specification should include a physical description, including dimensions and construction details (such as metal or plastic housing). We have tried to be inclusive in the following list of terms, which means that not all of these quantities will apply to every detector specification. Since specifications are not standardized, it is impossible to include all possible terms used; however, most detectors are described by certain standard figures of merit which will be discussed in this section. It is important to consider the manufacturer's context for all values; a detector designed for a specific application may not be appropriate for a different application even though the specification seems appropriate.

There are several figures of merit used to characterize the performance of different detectors. **Responsivity**, or Response, is the sensitivity of the detector to input flux. It is given by

$$R(\lambda) = \frac{I}{\phi(\lambda)}. \tag{4.2.1}$$

Where I is the detector output current (in amperes) and ϕ is the incident optical power on the detector (in watts). Thus, the units of responsivity are amperes per watt (A/W). Even when the detector is not illuminated, some current will flow; this **dark current** may be subtracted from the detector output signal when determining detector performance. Dark current is the thermally generated current in a photodiode under a completely dark environment; it depends on the material, doping and structure of the photodiode. It is the lowest level of thermal noise. Dark current in photodiodes limits the sensitivity (minimum detectable power). The reduction of dark current is important for the improvement of the minimum detectable power. It is usually simply measured and then subtracted from the flux, like background noise, in most specifications. However, the dark current is temperature dependent, so care must be taken to evaluate it over the expected operating conditions. If the anticipated signal is only a fraction of the dark current, then RMS noise in the dark current may mask the signal. Responsivity is defined at a specific wavelength; the term **spectral responsivity** is used to describe the variation at different wavelengths. Responsivity vs. wavelength is often included in a specification as a graph, as well as placed in a performance chart at a specified wavelength.

Quantum efficiency (QE) is the ratio of the number of electron-hole pairs collected at the detector electrodes to the number of photons in the incident light. It depends on the material from which the detector is made, and is primarily determined by reflectivity, absorption coefficient, and carrier diffusion length considerations. As the absorption coefficient is dependent on the incident light wavelength, the quantum efficiency has a **spectral response** as well. QE is the fundamental efficiency of the diode for converting photons into electron-hole pairs. For example, the QE of a PIN diode can be calculated by

$$\mathrm{QE} = (1 - R)T(1 - e^{-\alpha W}), \tag{4.2.2}$$

where R is the surface reflectivity, T is the transmission of any lossy electrode layers, W is the thickness of the absorbing layer, and α is the absorption coefficient.

Quantum efficiency affects detector performance through the responsivity (R) which can be calculated

$$R(\lambda) = \frac{\mathrm{QE}\lambda q}{\mathrm{hc}}, \tag{4.2.3}$$

where q is the charge of an electron (1.6×10^{-19} coulomb), λ is the wavelength of the incident photon, h is Planck's constant (6.626×10^{-34} W), and c is the velocity of light (3×10^{8} m/s). If wavelength is in microns and R is responsivity flux, then the units of quantum efficiency are A/W. Responsivity is the ratio of the diode's output current to input optical power and is given in A/W. A PIN photodiode typically has a responsivity of 0.6–0.8 A/W. A responsivity of 0.8 A/W means that incident light having 50 microwatts of power results in 40 microamps of current, in other words

$$\mathrm{I} = 50\mu\mathrm{W} \times 0.8\mathrm{A/W} = 40\mu\mathrm{A}, \tag{4.2.4}$$

where I is the photodiode current. For an APD, a typical responsivity is 80 A/W. The same 50 microwatts of optical power now produces 4 mA of current;

$$\mathrm{I} = 50\mu\mathrm{W} \times 80\mathrm{A/W} = 4\mathrm{mA}. \tag{4.2.5}$$

The minimum power detectable by the photodiode determines the lowest level of incident optical power that the photodiode can detect. It is related to the **dark current** in the diode, or the current which flows when no light is present; the dark current will set the lower limit. Other noise sources will be discussed later and are also included in detector specifications, including those associated with the diode and those associated with the receiver. The **noise floor** of a photodiode, which tell us the minimum detectable power, is the ratio of noise current to responsivity.

$$\text{Noise floor} = \frac{\text{noise}}{\text{responsivity}}. \tag{4.2.6}$$

For initial evaluation of a photodiode, we can use the dark current to estimate the noise floor. Consider a photodiode with R = 0.8 A/W and a dark current of 2 nA. The minimum detectable power is

$$\text{Noise floor} = \frac{2 \text{ nA}}{0.8 \text{ nA/nW}} = 2.5 \text{ nW}. \tag{4.2.7}$$

More precise estimates must include other noise sources, such as thermal and shot noise. As discussed, the noise depends on current, load resistance, temperature, and bandwidth.

Response time is the time required for the photodiode to respond to an incoming optical signal and produce an external current. Similarly to a source, response time is usually specified as a **rise time** and a **fall time**, measured between the 10 and 90% points of amplitude (other specifications may measure rise and fall times at the 20-80% point), or when the signal rises or falls to $1/e$ of its initial value.

As we have seen in prior chapters, **bandwidth**, or the difference between the highest and the lowest frequency that can be transmitted (see Section 1.2 and Section 2.1, etc.) is an essential feature of any fiber-optic communications system. The bandwidth of a photodiode can be limited by its **timing response** (its rise time and fall time) or its RC time constant, whichever results in the slower speed or bandwidth. The bandwidth of a circuit limited by the RC time constant is

$$B = \frac{1}{2\pi RC}, \tag{4.2.8}$$

where R is the load resistance and C is the diode capacitance. Fig. 4.2.1 shows the equivalent circuit model of a photodiode. It consists of a current source in parallel with a resistance and a capacitance. It appears as a low-pass filter, a resistor-capacitor network that passes low frequencies and attenuates high frequencies. The **cutoff frequency** is the highest frequency (or wavelength) for which the photodetector is sensitive; in practice, it is often defined as the frequency for which the signal is attenuated 3 dB (this is also called the **3 dB bandwidth**). Photodiodes for high speed operation must have a very low capacitance. The capacitance in a photodiode is mainly the junction capacitance formed at the PN junction, as well as any capacitance contributed by the packaging.

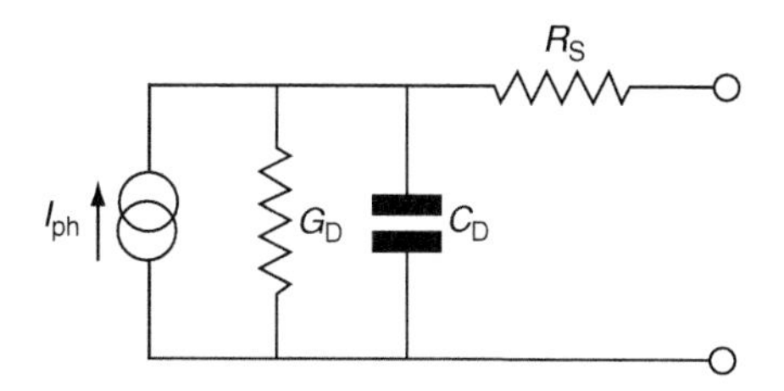

Fig. 4.2.1 Small-signal equivalent circuit for a reversed biased photodiode.

Bias voltage refers to an external voltage applied to the detector, and will be more fully described in the following section. Photodiodes require bias voltages ranging from as low as 0 V for some PIN photodiodes to several hundred volts for APDs. Bias voltage significantly affects operation, since dark current, responsivity, and response time all increase with bias voltage. APDs are usually biased near their avalanche breakdown point to ensure fast response.

Active Area and **Effective Sensing Area** are fairly straightforward; they measure the size of the detecting surface of the detection element. The **uniformity** of response refers to the percentage change of the sensitivity across the active area. Operating temperature is the temperature range over which a detector is accurate and will not be damaged by being powered. However, there may be changes in sensitivity and dark current which must be taken into account for specific applications. Storage temperature will have a considerably larger range; basically, it describes the temperature range under which the detector will not melt, freeze or otherwise be damaged or lose its operating characteristics.

Noise Equivalent Power (NEP) is the amount of flux that would create a signal of the same strength as the RMS detector noise. In other words, it is a measure of the minimum detectable signal; for this reason, it is the most commonly used version of the more generic figure of merit, Noise Equivalent Detector Input. More formally, it may be defined as the optical power (of a given wavelength or spectral content) required to produce a detector current equal to RMS noise in a unit bandwidth of 1 Hz.

$$\text{NEP}(\lambda) = \frac{i_n(\lambda)}{R(\lambda)}, \tag{4.2.9}$$

where i_n is the RMS noise current and R is the responsivity, defined above. It can be shown [2] that to a good approximation,

$$\text{NEP} = \frac{2hc}{\text{QE}\lambda}, \tag{4.2.10}$$

where this expression gives the NEP of an ideal diode when QE = 1. If the dark current is large, this expression may be approximated by

$$\text{NEP} = \frac{hc\sqrt{2qI}}{\text{QE}q\lambda}, \tag{4.2.11}$$

where I is the detector current. Sometimes it is easier to work with **detectivity**, which is the reciprocal of NEP. The higher the detectivity, the smaller the signal a detector can measure; this is a convenient way to characterize more sensitive detectors. Detectivity and NEP vary with the inverse of the square of active area of the detector, as well as with temperature, wavelength, modulation frequency, signal voltage, and bandwidth. For a photodiode detecting monochromatic light and dominated by dark current, detectivity is given by

$$D = \frac{\text{QE}q\lambda}{hc\sqrt{2qI}}. \tag{4.2.12}$$

The quantity-specific detectivity accounts for the fact that dark current is often proportional to detector area, A; it is defined by

$$D^* = D\sqrt{A}. \tag{4.2.13}$$

Normalized detectivity is detectivity multiplied by the square root of the product of active area and bandwidth; this product is usually constant, and allows comparison of different detector types independent of size and bandwidth limits. This is because most detector noise is white noise (Gaussian power spectra), and white noise power is proportional to the bandwidth of the detector electronics; thus the noise signal is proportional to the square root of bandwidth. Also, note that electrical noise power is usually proportional to detector area and the voltage which provides a measure of that noise is proportional to the square root of power. Normalized detectivity is given by:

$$D_\text{n} = D\sqrt{AB} = \frac{\sqrt{AB}}{\text{NEP}}, \tag{4.2.14}$$

where B is the bandwidth. The units are $\sqrt{\text{cm Hz}}\,\text{W}^{-1}$. Normalized detectivity is a function of wavelength and spectral responsivity; it is often quoted as normalized spectral responsivity.

Bandwidth, B, is the range of frequencies over which a particular instrument is designed to function within specified limits (see Eq. 4.2.8). Bandwidth is often adjusted to limit noise; in some specifications it is chosen as 1 Hz, so NEP is quoted in watts/Hz. Wide bandwidth detectors required in optical data communication often operate at a low resistance and require a minimal signal current much larger than the dark current; use of NEP, D, D^*, and D_n inappropriate for characterizing these applications.

Linear range is the range of incident radiant flux over which the signal output is a linear function of the input. The lower limit of linearity is NEP, and the upper limit is saturation. **Saturation** occurs when the detector begins to form less signal output for the same increase of input flux; increasing the input signal no longer results in an increase in the detector output. When a detector begins to saturate, it has reached the end of its linear range. **Dynamic range** can be used to describe non-linear detectors, like the human eye. Although communication systems do not typically use filters on the detector elements, neutral density filters can be used to increase the dynamic range of a detector system by creating islands of linearity, whose actual flux is determined by dividing output signals of the detector by the transmission of the filter. Without filtering, the dynamic range would be limited to the linear range of the detector, which would be less because the detector would saturate without the filter to limit the incident flux. The units of linear range are incident radiant flux or power (watts or irradiance).

Measuring the response of a detector to flux is known as **calibration**. Some detectors can be self-calibrated, others require manufacturer calibration. Calibration certificates are supplied by most manufacturers for fiber optic test instrumentation; they are dated, and have certain time limits. The gain, also known as the amplification, is the ratio of electron hole pairs generated per incident photon. Sometimes detector electronics allows the user to adjust the gain. Wiring and pin output diagrams tell the user how to operate the equipment, by schematically showing how to connect the input and output leads.

4.2.2 PN photodiode

We will limit our discussion in this chapter to photodiode detectors used in data communications. These are solid state devices; to understand their function, we must first describe a bit of semiconductor physics, as done in Chapter 2. For the interested reader, other introductory references to solid state physics, semiconductors, and condensed matter are available [2]. In a solid state device, the electron potential can be described in terms of conduction bands and valence bands, rather than

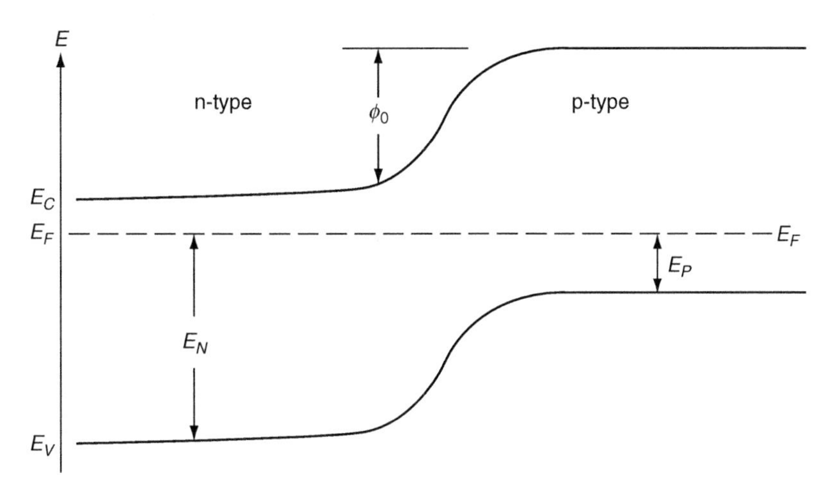

Fig. 4.2.2 Conduction and valence bands of a semiconductor.

individual potential wells (Fig. 4.2.2). The highest energy level containing electrons is called the Fermi level. If a material is a conductor, the conduction and valence bands overlap and charge carriers (electrons or holes) flow freely; the material carries an electrical current. An insulator is a material for which there is a large enough gap between the conduction and valence bands to prohibit the flow of carriers; the Fermi level lies in the middle of the forbidden region between bands, called the band gap. A semiconductor is a material for which the band gap is small enough that carriers can be excited into the conduction band with some stimulus; the Fermi level lies at the edge of the valence band (if the majority of carriers are holes) or at the edge of the conduction band (if the majority carriers are electrons). The first case is called a p-type semiconductor, the second is called n-type. These materials are useful for optical detection because incident light can excite electrons across the band gap and generate a photocurrent.

The simplest photodiode is the **PN photodiode** shown in Fig. 4.2.3. Although this type of detector is not widely used in fiber optics, it serves the purpose of illustrating the basic ideas of semiconductor photodetection. Other devices—the PIN and APDs - are designed to overcome the limitations of the PN diode. When the PN photodiode is reverse biased (negative battery or power supply potential connected to p-type material), very little current flows. The applied electric field creates a depletion region on either side of the PN junction. Carriers—free electrons and holes—leave the junction area. In other words, electrons migrate toward the negative terminal of the device and holes toward the positive terminal. Because the depletion region has no carriers, its resistance is very high, and most of the voltage drop occurs across the junction. As a result, electrical fields are high in this region and negligible elsewhere. An incident photon absorbed by the diode gives a bound electron sufficient energy to move from the valence band to the conduction band, creating a free electron and a hole. If this creation of carriers occurs in the depletion region, the carriers quickly separate and drift rapidly toward their respective regions. This movement sets an electron flowing as current in the external circuit. When the carriers reach the edge of the depletion region, where electrical fields are small, their movement, however, relies on diffusion mechanism. When electron-hole creation occurs outside of the depletion region, the carriers move slowly toward their respective regions. Many carriers recombine before reaching it. Their contribution to the total current is negligible. Those carriers remaining and reaching the depleted area are swiftly swept across the junction by the large electrical fields in the region to produce an external electrical current. This current, however, is delayed with respect to the absorption of the photon that created the

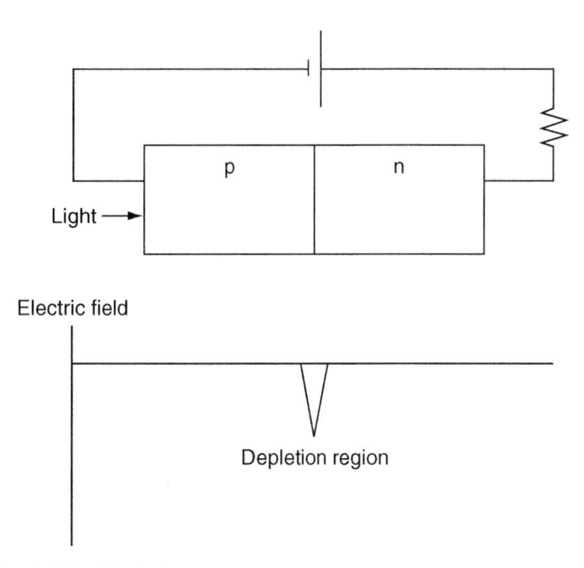

Fig. 4.2.3 PN diode.

carriers because of the initial slow movement of carriers toward their respective regions. Current, then, will continue to flow after the light is removed. This slow response, due to slow diffusion of carriers, is called slow tail response.

Two characteristics of the PN diode make it unsuitable for most fiber optic applications. First, because the depletion area is a relatively small portion of the diode's total volume, many of the absorbed photons do not contribute to the external current. The created free electrons and holes recombine before they contribute significantly to the external current. Second, the slow tail response from slow diffusion makes the diode too slow for high speed applications. This slow response limits operations to the kHz range.

4.2.3 PIN photodiode

The structure of the PIN diode is designed to overcome the deficiencies of its PN counterpart. The **PIN diode** is a photoconductive device formed from a sandwich of three layers of crystal, each layer with different band structures caused by adding impurities (doping) to the base material, usually indium gallium arsenide, silicon, or germanium. The layers are doped in this arrangement: p-type (or positive) on top, intrinsic, meaning undoped, in a thin middle layer, and n-type (or negative) on the bottom. For a silicon crystal, a typical p-type impurity would be boron, and indium would be a p-type impurity for germanium [2–6]. Actually, the intrinsic layer may also be lightly doped, although not enough to make it either p-type or n-type. The change in potential at the interface has the effect of influencing the direction of current flow, creating a diode; obviously, the name PIN diode comes from the sandwich of p-type, intrinsic, and n-type layers.

The structure of a typical PIN photodiode is shown in Fig. 4.2.4. The p-type and n-type silicon form a potential at the intrinsic region; this potential gradient depletes the junction region of charge carriers, both electrons and holes, and results in the conduction band bending. The intrinsic region has no free carriers, and thus exhibits high resistance. The junction drives holes into the p-type material and electrons into the n-type material. The difference in potential of the two materials determines the energy an electron must have to flow through the junction. When photons fall on the active area of the device they generate carriers near the junction, resulting in a voltage difference between the p-type and the n-type regions. If the diode is connected to external circuitry, a current will flow which is proportional to the illumination. The PIN diode structure addresses the main problem with PN diodes, namely providing a large depletion region for the absorption of photons. There is a tradeoff involved in the design of PIN diodes. Since most of the photons are absorbed in the intrinsic region, a thick intrinsic layer is desirable to improve photon-carrier conversion efficiency (to increase the probability of a photon being absorbed in the intrinsic region). On the other hand, a thin intrinsic region is desirable for high speed devices, since it reduces the transit time of photogenerated carriers. These two conditions must be balanced in the design of PIN diodes.

Photodiodes can be operated either with or without a bias voltage. Unbiased operation is called the photovoltaic mode; certain types of noise, including $1/f$ noise, are lower and the NEP is better at low frequencies. Signal-to-noise ratio (SNR) is superior to the biased mode of operation for frequencies below about 100 kHz [6]. Biasing (connecting a voltage potential to the two sides of the junction) will sweep carriers out of the junction region faster and change the energy requirement for carrier generation to a limited extent. Biased operation (photoconductive mode) can be either forward or reverse bias. Reverse bias of the junction (positive potential connected to the n-side and negative connected to the p-side) reduces junction capacitance and improves response time; for this reason it is the preferred operation mode for pulsed detectors. A PIN diode used for photodetection may also be forward biased (the positive potential connected to the p-side and the negative to the n-side of the junction) to make the potential scaled for current to flow less, or in other words to increase the sensitivity of the detector (Fig. 4.2.5).

An advantage of the PIN structure is that the operating wavelength and voltage, diode capacitance, and frequency response may all be predetermined during the manufacturing process. For a diode whose intrinsic layer thickness is w with an applied bias voltage of V, the

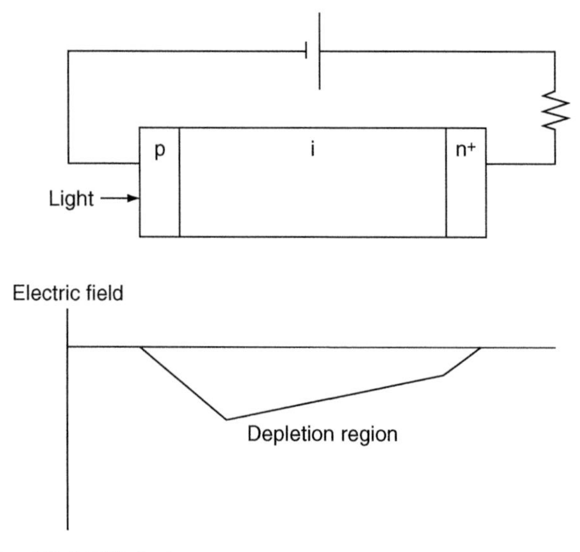

Fig. 4.2.4 PIN diode.

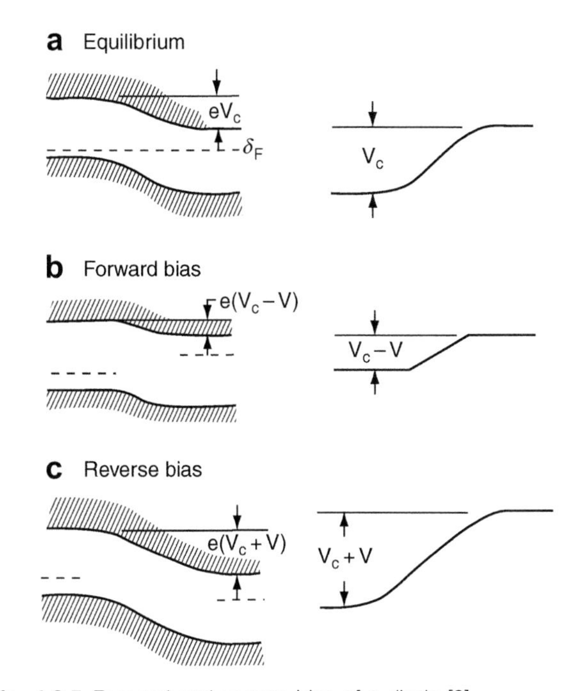

Fig. 4.2.5 Forward and reverse bias of a diode [2].

self-capacitance of the diode, C, approaches that of a parallel plate capacitor,

$$C = \frac{\varepsilon_0 \varepsilon_1 A_0}{w}, \tag{4.2.15}$$

where A_o is the junction area, ε_0 the free space permittivity (8.849×10^{-12} farad/m), and ε_l the relative permittivity. Taking typical values of $\varepsilon_l = 12$, $w = 50$ microns, and $A_o = 10^{-7}$ m^2, C = 0.2 pF. QEs of 0.8 or higher can be achieved at wavelengths of 0.8–0.9 microns, with dark currents less than 1 nA at room temperature. The sensitivity of a PIN diode can vary widely by quality of manufacture. A typical PIN diode size ranges from 5 mm × 5 mm to 25 mm × 25 mm. Ideally, the detection surface will be uniformly sensitive (at the National Institute for Standards and Technology (NIST) there is a detector profiler which, by using extremely well focused light sources, can determine the sensitivity of a detector's surface [3]). For most applications, it is required that the detector is uniformly illuminated or overfilled. The spectral responsivity of an uncorrected silicon photodiode is shown in Fig. 4.2.6. The typical QE curve is also shown for comparison. An ideal silicon detector would have zero responsivity and QE for photons whose energies are less than the band gap, or wavelengths much longer than about 1.1 micron. Just below the long wave limit, this ideal diode would have 100% QE and responsivity close to 1 A/W; responsivity vs. wavelength would be expected to follow the intrinsic spectral response of the material. In practice, this does not happen;

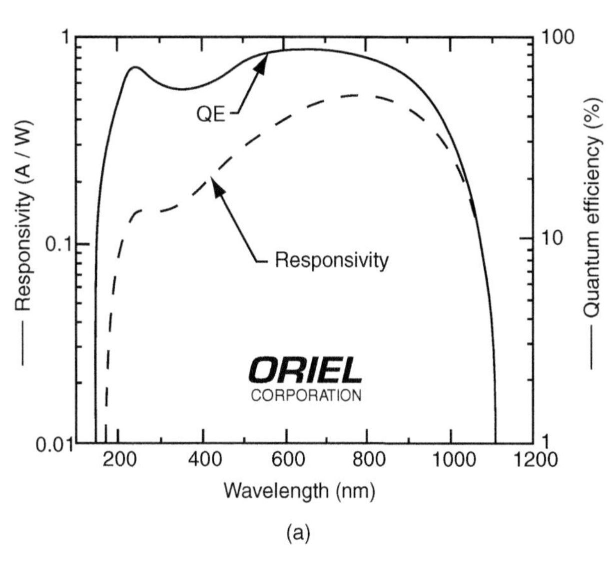

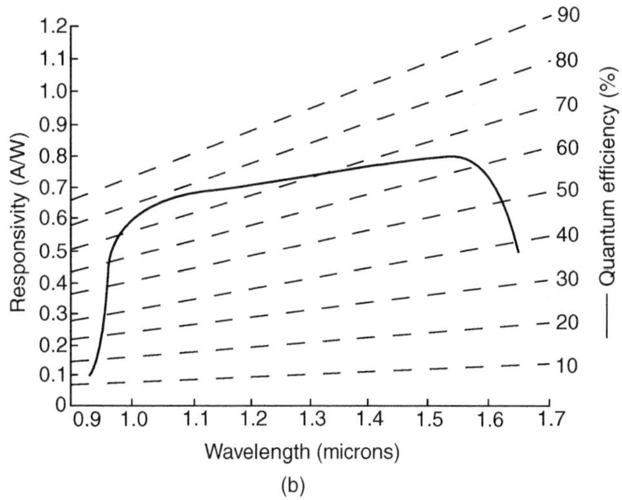

Fig. 4.2.6 PIN diode spectral response and quantum efficiency: (a) silicon and (b) InGaAs.

these detectors are less sensitive in the blue region, which can sometimes be enhanced by clever doping, but not more than an order of magnitude. This lack of sensitivity is because there are fewer short wavelength photons per watt, so responsivity in terms of power drops off, and because more energetic blue photons may not be absorbed in the junction region. For color-sensitive applications such as photometry, filters are used so detectors will respond photometrically, or to the standardized CIE color coordinates; however, the lack of overall sensitivity in the blue region can potentially create noise problems when measuring a low intensity blue signal. In the deep ultraviolet, photons are often absorbed before they reach the sensitive region by detector windows or surface coatings on the semiconductor. The departure from 100% QE in real devices is typically due to Fresnel reflections from the detector surface. The long wavelength cutoff is more gradual than expected for an ideal device because the absorption coefficient decreases at long wavelengths, so more photons pass through the photosensitive layers and do not contribute to the QE. As a result, QE tends to roll off gradually near the band gap limit. This response is typical for silicon devices, which make excellent detectors in the wavelength range 0.8-0.9 microns. A common material for fiber optic applications is InGaAs, which is most sensitive in the near infrared (0.8–1.7 microns). Other PIN diode materials include HgCdZnTe for wavelengths of 2–12 microns. In the 1940s, a popular photoconductive material, lead sulfide (PbS), for infrared solid state detectors was introduced. This material is still commonly used in the region from 1 to 4 microns [6].

PIN diode detectors are not very sensitive to temperature ($-25\,°C - +80\,°C$) or shock and vibration, making them an ideal choice for a data communications transceiver. It is very important to keep the surface of any detector clean. This becomes an issue with PIN diode detectors because they are sufficiently rugged than they can be brought into applications which expose them to contamination. Both transceivers and optical connectors should be cleaned regularly during use to avoid dust and dirt buildup.

A sample specification for a photodiode is given in Table 4.2.1. Bias voltage (V) is the voltage applied to a silicon photodiode to change the potential photoelectrons must scale to become part of the signal. Bias voltage is basically a set operating characteristic in a pre-packaged detector, in this case −24 V. Shunt Resistance is the resistance of a silicon photodiode when not biased. Junction Capacitance is the capacitance of a silicon photodiode when not biased. Breakdown Voltage is the voltage applied as a bias which is large enough to create signal on its own. When this happens, the contribution of photoelectrons is minimal, so the detector cannot function. However, once the incorrect bias is removed the detector should return to normal.

To increase response speed and QE, a variation on the PIN diode, known as the Schottky barrier diode, can be used. This approach will be treated in a later Section (4.2.4.3). For wavelengths longer than about 0.8 microns, a heterojunction diode may be used for this same reason. Heterojunction diodes retain the PIN sandwich structure, but the surface layer is doped to have a wider band gap and thus reduced absorption. In this case, absorption is strongest in the narrower bandgap region at the heterojunction, where the electric field is maximum; hence, good QEs can be obtained. The most common material systems for heterojunction diodes are InGaAsP on an InP substrate, or GaAlAsSb on a GaSb substrate.

A typical circuit for biased operation of a photodiode is shown in Fig. 4.2.7, where C_A and R_A represent the input impedance of a post-amplifier. The photodiode impulse response will differ from an ideal square wave for several reasons, including transit time resulting from the drift of carriers across the depletion layer, delay caused by the diffusion of carriers generated outside the depletion layer, and the RC time constant of the diode and its load. If we return to the photodiode equivalent circuit of Fig. 4.2.1 and insert it into this typical bias circuit, we can make the approximation that R_s is much smaller than R_A to arrive at the equivalent small signal circuit model of Fig. 4.2.8. In this model the resistance R is approximately equal to R_A, and the capacitance C is the sum of the diode capacitance, amplifier capacitance, and some distributed stray capacitance. In response to an optical pulse falling on the detector, the load voltage V_{in} will rise and fall exponentially with a time constant RC. In response to a photocurrent I_p which varies sinusoidally at the angular frequency

$$\omega = 2\pi f, \tag{4.2.16}$$

the response of the load voltage will be given by

$$\frac{V_{in}(f)}{I_P(f)} = \frac{R}{1 + j2\pi fCR}. \tag{4.2.17}$$

To obtain good high frequency response, C must be kept as low as possible; as discussed earlier, the photodiode contribution can normally be kept well below 1 pF. There is ongoing debate concerning the best approach to improve high frequency response; we must either reduce R or provide high frequency equalization. As a rule of thumb, no equalization is needed if

$$R < 1/2\pi C\Delta f, \tag{4.2.18}$$

where Δf is the frequency bandwidth of interest. We can also avoid the need for equalization by using a transimpedance feedback amplifier, which is often employed in commercial optical datacom receivers. These apparently simple receiver circuits can exhibit very complex behavior, and it is not always intuitive how to design the optimal detector circuit for a given application. A detailed analysis of receiver response, including the relative noise contributions and tradeoffs between different types of photodiodes, is beyond the scope of this chapter; the interested reader is referred to several good references on this subject [4–12].

4.2.4 Other detectors

The PIN photodiode is definitely the most commonly used type of detector in the fiber optic communication industry. But there are a number of detectors that offer

Table 4.2.1 Sample specifications for a PIN diode.

Parameter	Symbol	Test conditions	Min.	Typical	Max.	Units
Receiver section						
Data rate (NRZ)	B	–	10	–	156	Mb/s
Sensitivity (avg)	P_{DH}	62.5 μm fiber	−32.5	–	−14.0	dBm
		0.275 NA, BER $\leq 10^{-10}$				
Optical wavelength	λ_{th}	–	1270	–	1380	nm
Duty cycle	–	–	25	50	75	%
Output risetime	t_{TLH}	20–80%	0.5	–	2.5	ns
		50 Ω to V_{CC} − 2 V				
Output falltime	t_{THL}	80-20%	0.5	–	2.5	ns
		50 Ω to V_{CC} − 2 V				
Output voltage	V_{OL}	–	V_{CC} − 1.025	–	V_{CC} − 0.88	V
	V_{OH}		V_{CC} − 1.81	–	V_{CC} − 1.62	V
Signal detect	V_A	$P_{IN} > P_A$	V_{CC} − 1.025	–	V_{CC} − 0.88	V
	V_D	$P_{IN} < P_D$	V_{CC} − 1.81	–	V_{CC} − 1.62	V
P_{IN} power levels:						
Deassert	P_B	–	−39.0 or P_B	–	–32.5	dBm
Assert	P_A	–	−38.0	–	–30.0	dBm
Hysteresis		–	1.5	2.0	–	dB
Signal detect delay time:						
Deassert	–	–	–	–	50	μs
Assert	–	–	–	–	50	μs
Power supply voltage	V_{CC}–V_{EE}	–	4.75	5.0	5.25	V
Power supply current	I_{CC} or I_{EE}	–	–	–	150	mA
Operating temperature	T_A	–	0	–	70	°C
Absolute maximum ratings: transceiver						
Storage temperature	–	–	−40	–	100	°C
Lead soldering limits	–	–	–	–	240/10	°C/s
Supply voltage	V_{CC}–V_{EE}	–	−0.2	–	7.00	V

advantages in more complex situations, or for specialized types of fiber signaling; we will discuss them in the following sections. For example, POIs require detector arrays, which can be formed from either PIN diodes or MSM photoreceivers. In WDM, the optical signal traveling over one fiber optic cable is wavelength-separated, coded, and recombined. To properly receive and interpret this combined signal there is a need for more sensitive detectors, such as APDs, as well as position-sensitive detectors, such as photodiode arrays and MSM photoreceiver arrays, and also wavelength-sensitive detectors, such resonant-cavity enhanced photodetectors. The rising importance of WDM hints at the importance of growing data rates. Schottky-barrier photodiodes have always been considered for situations which required faster response than PIN diodes (100 Ghz modulation has been reported), but less sensitivity. MSM detectors, which is a Schottky diode based planar detectors, are

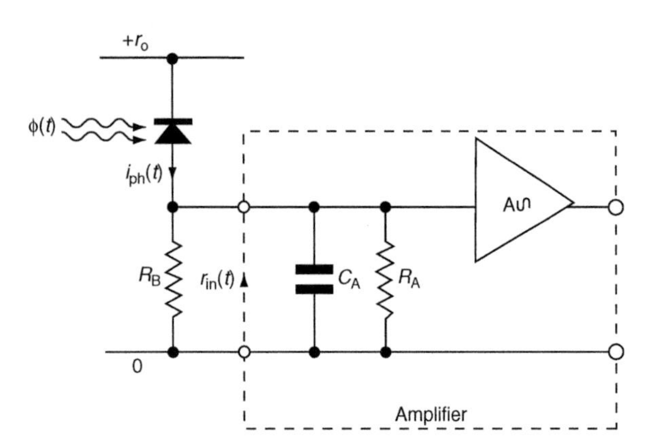

Fig. 4.2.7 PIN diode and amplifier equivalent circuit.

cheap and easy to fabricate, making them desirable for some low cost applications where there are a number of parallel channels and dense integration. Resonant-cavity enhancement (RECAP) can increase the signal from Schottky-barrier and MSM detectors, as well as making high speed, thin i layer, PIN diode detectors feasible.

4.2.4.1 Avalanche photodiode

PIN diodes can be used to detect light because when photon flux irradiates the junction, the light creates electron hole pairs with their energy determined by the wavelength of the light. Current will flow if the energy is sufficient to scale the potential created by the PIN junction. This is known as the photovoltaic effect [2, 4–6]. It should be mentioned that the material from which the top layers of a PIN diode is constructed must be transparent (and clean!) to allow the passage of light to the junction. If the bias voltage is increased significantly, the photogenerated carriers have enough energy to start an avalanche process, knocking more electrons free from the lattice which contribute to amplification of the signal. This is known as an **avalanche photodiode (APD)**; it provides higher responsivity, especially in the near infrared, but also produces higher noise due to the electron avalanche process. For a PIN photodiode, each absorbed photon ideally creates one electron-hole pair, which, in turn, sets one electron flowing in the external circuit. In this sense, we can loosely compare it to an LED. There is basically a one-to-one relationship between photons and carriers and current. Extending this comparison allows us to say that an APD resembles a laser, where the relationship is not one-to-one. In a laser, a few primary carriers result in many emitted photons. In an APD, a few incident photons result in many carriers and appreciable external current.

The structure of the APD, shown in Fig. 4.2.9, creates a very strong electrical field in a portion of the depletion region. Primary carriers— the free electrons and holes created by absorbed photons—within this field are accelerated by the field, thereby gaining several electron volts of kinetic energy. A collision of these fast carriers with neutral atoms causes the accelerated carrier to use some of its energy to raise a bound electron from the valence band to the conduction band. A free electron-hole pair is thus created. Carriers created in this way,

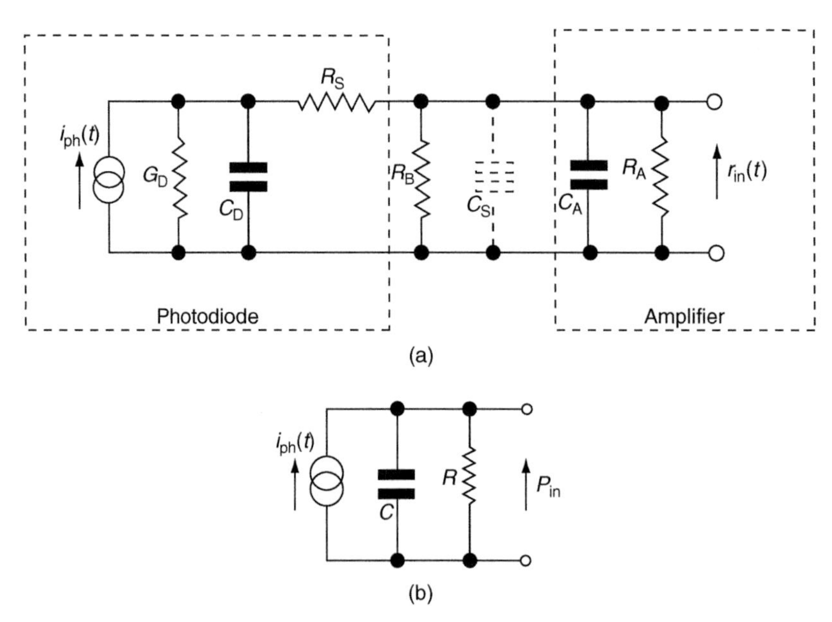

Fig. 4.2.8 Small-signal equivalent circuit for a normally biased photodiode and amplifier: (a) complete circuit and (b) reduced circuit obtained by neglecting R_s and lumping together the parallel components.

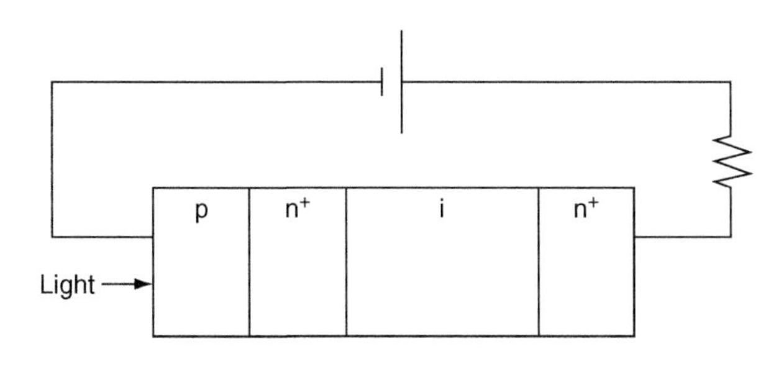

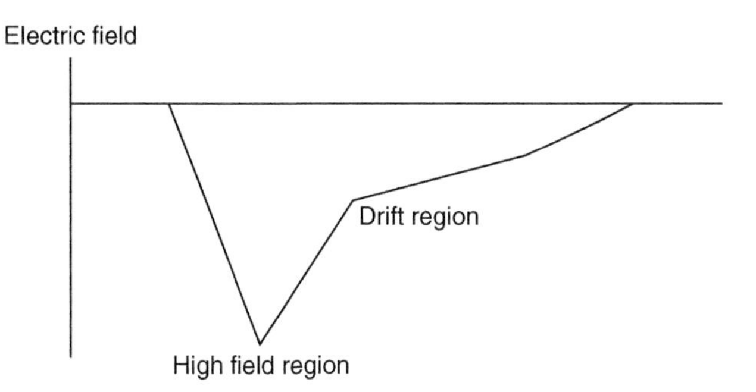

Fig. 4.2.9 Avalanche photodiode.

through collision with a primary carrier, are called secondary carriers. This process of creating secondary carriers is known as collision ionization. A primary carrier can create several new secondary carriers, and secondary carriers themselves can accelerate and create new carriers. The whole process is called **photomultiplication**, which is a form of gain process. The number of electrons set flowing in the external circuit by each absorbed photon depends on the APD's multiplication factor. Typical multiplication ranges in tens and hundreds. A multiplication factor of 50 means that, on the average, 50 external electrons flow for each photon. The phrase "on the average" is important. The multiplication factor is an average, a statistical mean. Each primary carrier created by a photon may create more or less secondary carriers and therefore external current. For an APD with a multiplication factor of 50, for example, any given primary carrier may actually create 44 secondary carriers or 53 secondary carriers. This variation is one source of noise that limits the sensitivity of a receiver using an APD.

The multiplication factor M_{DC} varies with the bias voltage as described in Equation (3.19)

$$M_{DC} = \frac{1}{(1 - V/V_B)^n}, \tag{4.2.19}$$

where V_B is the breakdown voltage, and n varies between 3 and 6, depending on the semiconductor.

The photocurrent is multiplied by the multiplication factor

$$I = \frac{MQEq\phi(\lambda)\lambda}{hc}, \tag{4.2.20}$$

as is the responsivity

$$R(\lambda) = \frac{MQE\lambda q}{hc}. \tag{4.2.21}$$

The shot noise in an APD is that of a PIN diode multiplied by M times an excess noise factor, denoted by the square root of F, where

$$F(M) = \beta M + (1 - \beta)\left(2 - \frac{1}{M}\right). \tag{4.2.22}$$

In this case, β is the ratio of the ionization coefficient of the holes divided by the ionization coefficient of the electrons. In III-V semiconductors $F = M$. It is also important to remember that the dark current of APDs is also multiplied, according to the same equations as the shot noise.

Because the accelerating forces must be strong enough to impart energies to the carriers, high bias voltages (several hundred volts in many cases) are required to create the high-field region. At lower voltages, the APD operates like a PIN diode and exhibits no internal gain. The avalanche breakdown voltage of an APD is the voltage at which collision ionization begins. An APD biased above the breakdown point will produce current in the absence of optical power. The voltage itself is sufficient to create carriers and cause collision ionization. The APD is often biased just below the breakdown point, so any optical power will create a fast response and strong output. The tradeoffs are that dark current (the current resulting from generation of electron-hole pairs even in the absence of absorbed photons) increases with bias voltage, and a high-voltage power supply is needed. Additionally, as one might expect, the avalanche breakdown process is temperature sensitive, and most APDs will require temperature compensation in datacom applications. For these reasons, APDs are not as commonly used in datacom applications as PIN diodes, despite the potentially greater sensitivity of the APD. However, they are currently being used in applications where sensitivity is very important, such as WDM.

4.2.4.2 Photodiode array

Photodiode arrays are beginning to find applications as detectors in parallel optics for supercomputers and enhanced backplane or clustering interconnects for telecommunication products. They are also used for spectrometers (along with CCD arrays), which makes them useful to test fiber optic systems. Photodiode arrays are also being considered for WDM applications.

In the previous sections (Sections 4.2.2 and 4.2.3), we have discussed the physics guiding photodiode operation. In a photodiode array, the individual diode elements

Fig. 4.2.10 An optical demultiplexing receiver array, where data channels are focused onto an array of high-speed InGaAs photodiodes. The signals are then amplified using a trans-impedance amplifier (TIA) array.

respond to incident flux by producing photocurrents, which charge individual storage capacitors. InGaAs photodiode arrays are the materials used in spectroscopic applications. Cross-talk, or signal leakage between neighboring pixels, is normally a concern in array systems. In InGaAs photodiode arrays it is limited to nearest-neighbor interactions. However, the erbium doped fiber amplifiers (EDFA) do exhibit cross-talk, so there is some discussion of using different amplifiers, such as trans-impedance amplifiers (TIA arrays). It is expected that this type of detector will be able to operate at speeds up to 10Gbit/s (Fig. 4.2.10) [13].

While some WDMs filter out the wavelength dependent signal, another common way to separate the signal involves using a grating to place different wavelengths at different physical positions, as is done in spectrometers. In ultra-dense WDM, as many as 100 optical channels can be used. Photodiode arrays are an obvious detector choice for these applications. In addition to detecting the signal, they offer performance feedback to the tunable lasers. At this time, photodiode arrays have not been commercially used in WDM, but their use would be similar to that in spectrometer design.

4.2.4.3 Schottky-barrier photodiodes

A variation on the PIN diode structure is shown in Fig. 4.2.11. This is known as a **Schottky-barrier diode**; the top layer of semiconductor material has been eliminated in favor of a reverse biased, MSM contact. The metal layer must be thin enough to be transparent to incident light, about 10 nm; alternate structures using interdigital metal transducers are also possible. The advantage of this approach is improved QE, because there is no recombination of carriers in the surface layer before

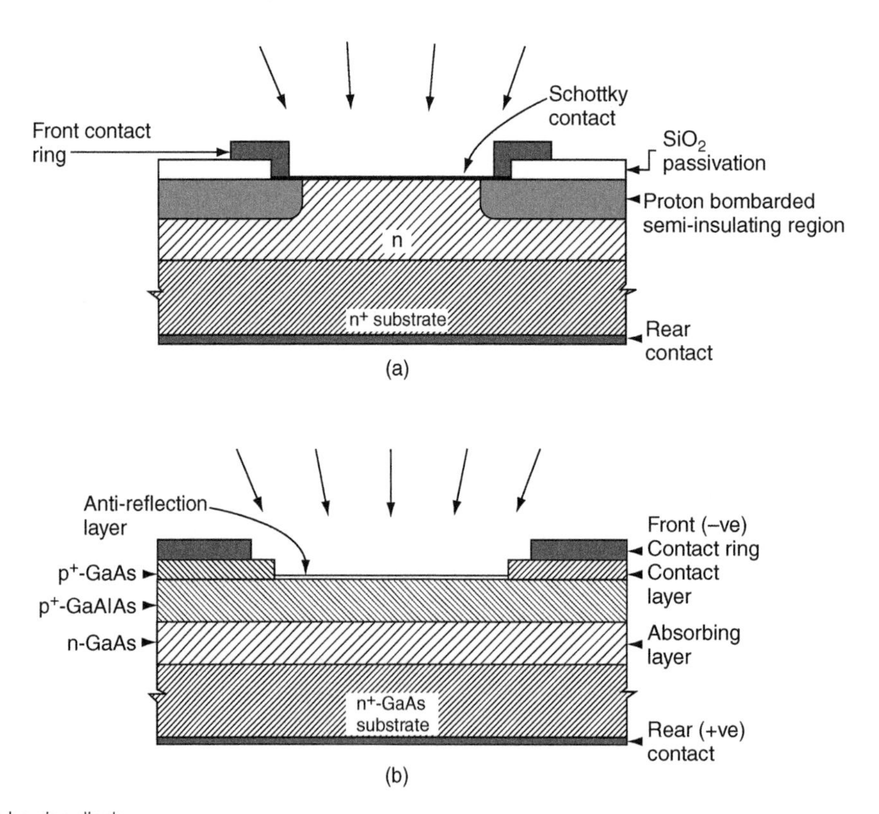

Fig. 4.2.11 Schottky-barrier diode.

they can diffuse to either the ohmic contacts or the depletion region.

In order to make a PIN detector respond faster, we need to make the inductor layer thinner. This decreases the signal. Some architectures have been designed to try to get around this problem, by using resonant cavities (see Section 4.2.4.5), or by placing a number of PIN junctions into an optical waveguide, known as traveling wave photodetectors. But Schottky-Barrier photodiodes have been another traditional alternative.

Schottky-Barrier photodiodes are not PIN based, unlike the prior detectors discussed. Instead, a thin metal layer replaces half of the PN junction. However, it does result in the same voltage characteristics— when an incident photon hits the metal layer, carriers are created in the depletion region, and their movement again sets current flowing. The similarity in the band structure can be seen in comparing the band structure of a forward biased PIN diode in Fig. 4.2.5 to the band structure of a forward biased n-type Schottky-Barrier diode in Fig. 4.2.12. The voltage and field relations derived for PIN junctions can be applied to Schottky-Barrier diodes by treating the metal layer as if it were an extremely heavily doped semiconductor.

The metal layer is a good conductor, and so electrons leave the junction immediately. This results in faster operation—up to 100 Ghz modulation has been reported [14]. It also lowers the chance of recombination, increasing efficiency, and increases the types of semiconductors which can be used, since you are less limited by lattice-matching fabrication constraints. However, their sensitivity is lower than PIN diodes with the usual sized intrinsic layer. Resonant enhancement (see Section 4.2.4.5.) has been experimented with to increase sensitivity.

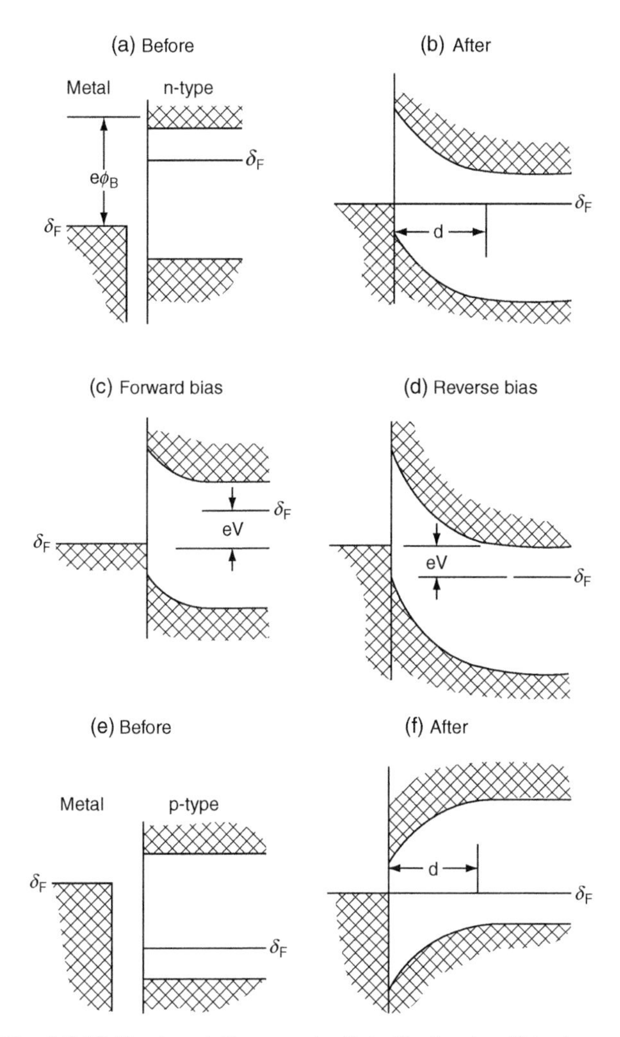

Fig. 4.2.12 The band diagram of a Schottky-barrier. This shows the metal-semiconductor ohmic contact with an n-type semiconductor (a) before contact (b) after contact (c) forward bias (d) reverse bias (e) and (f) are the same as (a) and (b), but the semiconductor is p-type [2].

4.2.4.4 Metal-semiconductor-metal (MSM) detectors

A **MSM detector** is made by forming two Schottky diodes on top of a semi-conductor layer. The top metal layer has two contact pads and a series of "interdigitated fingers" (Fig. 4.2.13), which form the active area of the device. One diode gets forward biased, and the other diode gets reverse biased. When illuminated, these detectors create a time-varying electrical signal. MSM photoreceiver arrays have been used as POIs. They are also being considered for use in smart pixels.

Like Schottky diodes, speed is an advantage of MSM detectors. The response speed of MSM detectors can be increased by reducing the finger spacing, and thus the carrier transit times. However, for very small finger spacing, the thickness of the absorption layer becomes comparable to the finger spacing and limits the high speed performance.

The other advantages of MSM detectors are their ease in fabrication and integration into electrical systems. That is why they are being placed in systems like arrays and smart pixels.

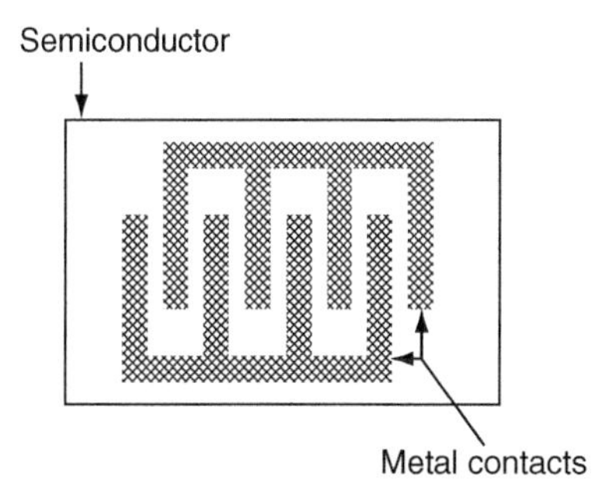

Fig. 4.2.13 A diagram of an MSM detector.

MSM detectors have the same disadvantages of other Schottky based systems—weak signal and noise. They have some advantages in systems with a number of parallel channels and dense integration of detectors, which could eventually be applied to WDM systems.

As with Schottky-Barrier photodiodes, experiments have been done to increase their signal using resonant enhancement (see Section 4.2.4.5).

For a MSM detector deposited on a photoconductive semiconductor with a distance L between the electrodes, the photocurrent and time-varying resistance can be calculated as follows. Let us assume the detector is irradiated by a input optical flux P, at a photon energy $h\nu$.

$$Iph = \frac{CPQEq\phi(\lambda)\lambda}{hc}, \tag{4.2.23}$$

where QE is the quantum efficiency, and G is the photoconductive gain, which is the ratio of the carrier lifetime to the carrier transit time. It can be seen that increasing the carrier lifetime decreases the speed, but increases the sensitivity, which is what you would expect.

The time varying resistance $R(t)$, is dependent on the photo induced carrier density $N(t)$.

$$N(t) = \frac{QE\phi(\lambda)\lambda}{hc}, \tag{4.2.24}$$

$$R(t) = L/(eN(t)\mu wd). \tag{4.2.25}$$

Where μ is the sum of the electron and whole mobilites, w is the length along the electrodes excited by the light, and d is the effective absorption depth into the semiconductor. The specifications for a typical MSM detector are shown in Table 4.2.2.

Table 4.2.2 Sample specifications for an MSM detector.

Detector type	MSM (metal-semiconductor-metal)
Active material	InGaAs
Bandwidth	(−3 dB electrical) 20 GHz resp. 35 GHz
Rise time	(10–90%) <11ps
Pulse width (FWHM)	<18ps
Wavelength range	400 nm-1.6 μm
Responsivity	0.21 A/W at 670 nm, 0.24 A/W at 810 nm, 0.19 A/W at 1.5 μm
Bias voltage	2-10 V
Bias input connector	SMC male
RF signal ouput connector	K-type female
Optical input connector	FC/PC on 9 μm single-mode fiber
Maximum optical peak input power	200 mW at 20 ps, 1.3 μm
Maximum optical average input power	2 mW at 1.3 μm
Options	
Bias input connector	MSM battery case with cable
RF signal output connector	K-type male
Optical input connector on request	SMA, SC, ST, etc.
Optical input	50 μm multimode fiber, free space with collimation optics

4.2.4.5 Resonant-cavity enhanced photodetectors (recap)

The principle of **RECAP** involves placing a fast photodetector into an FP cavity to enhance the signal magnitude (Fig. 4.2.14). In practice, the FP cavity's end mirrors are made of quarter-wave stacks of GaAs/AlAs, InGaAs/InAlAs or any other semiconductor or insulator materials which have the desired refractive index contrast. A typical example of a fast photodetector is a PIN photodetector with a very thin i-layer. By making the inductor layer thinner, you increase the speed of the detector, but decrease the signal. Schottky-Barrier photodiodes (see Section 4.2.4.3), and MSM detectors (see Section 4.2.4.4) can also have their signal magnitude increased using RECAP techniques.

By placing the detector in an FP cavity, you pass the light through it several times, which results in an increase of signal. However, due to the resonance properties of the FP cavity, this detector is highly wavelength selective. It is also highly sensitive to the position of the detector in the FP cavity, since it is a resonance effect.

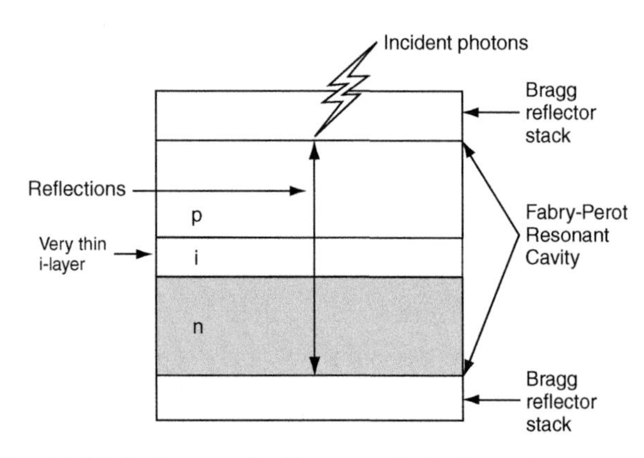

Fig. 4.2.14 A diagram of a Resonant Cavity Photodiode.

The maximum QE, QE_{max}, of a RECAP detector can be calculated by

$$QE_{max} = \left\{\left(1 + R_2 e^{-ad}\right)(1 - R_1)/\left(1 - R_1 R_2 e^{-ad}\right)^2\right\}^* (1 - e^{-ad}), \quad (4.2.26)$$

where R_1 and R_2 are the reflectivity of the top and bottom mirrors, α is the absorption of the active area of the detector, and d is thickness of the active area of the detector. If the reflection of the bottom mirror $R_2 = 0$, this reduces to the QE of a photodiode, only with an ideal total transmission through the top electrode (see Equation 4.2.2).

$$QE = (1 - R)T\left(1 - e^{-\alpha W}\right). \quad (4.2.2)$$

The resonant cavity enhancement, RE, can allow RCE detectors to have quantum efficiencies of nearly unity at their peak wavelength.

$$RE = \left\{\left(1 + R_2 e^{-\alpha d}\right)/\left(1 - \sqrt{(R_1 R_2) e^{-\alpha d}}\right)^2\right\}. \quad (4.2.27)$$

The theoretical wavelength dependence of the quantum efficiency is shown in Fig. 4.2.15.

Since WDM applications are by their very nature wavelength dependent, the wavelength selectivity of RECAP may turn out to be a useful design feature. RECAP is not yet being used in the field, but reported experimental quantum efficiencies have reached 82% for PIN diodes, and 50% improvements of photocurrent for Schottky-Barrier photodiodes [15].

4.2.4.6 Interferometric sensors

Optical fiber sensors are different from communications circuits in that they measure environmental factors rather than transmitting signals. In Section 1.3, we discussed the most commonly used optical sensor, which is used for sensing strain and temperature in remote locations such as oil wells. FP interferometric sensors are another sensor technology. We have chosen to treat them in this chapter because of their similarity to detectors, although they could also be classified as a type of specialty fiber.

In **FP interferometric sensors** a single-mode laser diode is attached to a coupler, where one side illuminates an FP cavity and the other side a reference, index matching gel. The FP cavity is formed between an input single-mode fiber and a reflecting target element that may be a fiber (Fig. 4.2.16). The interference between the reflected signals from the FP cavity and the reference allows us to measure strain and displacement in environments where temperature is not anticipated to change.

Fabry-Perot interferometric sensors can be either extrinsic or intrinsic. The difference between the two lies in the design of the FP cavity. Extrinsic sensors have the cavity around the fiber, and intrinsic sensors form the cavity within the optical fiber (Fig. 4.2.17). Interferometric sensors have a high sensitivity and bandwidth, but are limited by nonlinearity in their output signals.

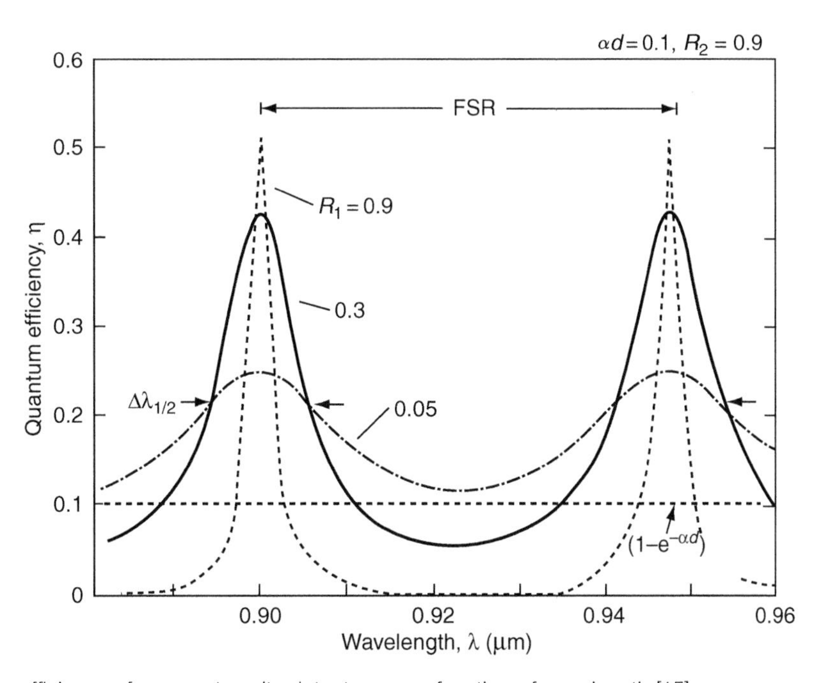

Fig. 4.2.15 The quantum efficiency of resonant cavity detectors as a function of wavelength [15].

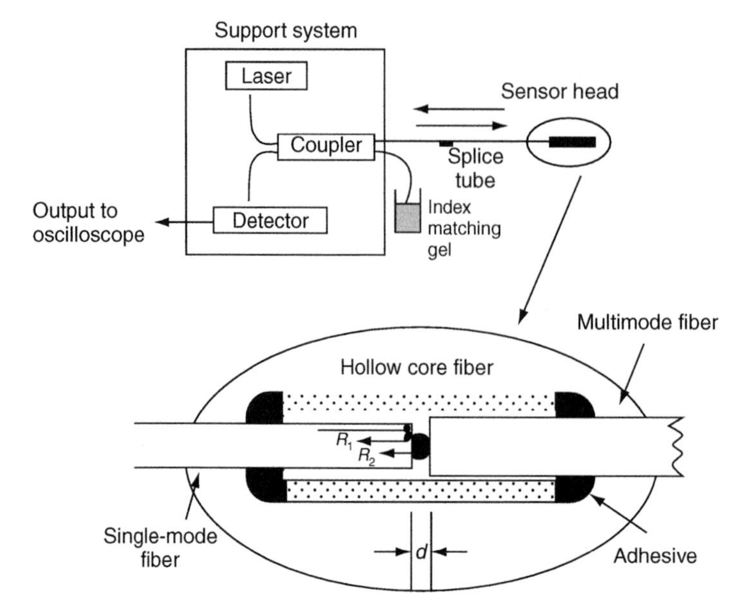

Fig. 4.2.16 Extrinsic Fabry-Perot interferometric (EFPI) sensor and system.

4.2.5 Noise

Any optical detection or communication system is subject to various types of noise. There can be noise in the signal, noise created by the detector, and noise in the electronics. A complete discussion of noise sources has already filled several good reference books; since this is a chapter on detectors, we will briefly discuss the noise created by detectors. For a more complete discussion, the reader is referred to treatments by Dereniak and Crowe [6], who have categorized the major noise sources. The purpose of the detector is to create an electrical current in response to incident photons. It must accept highly attenuated optical energy and produce an electrical current. This current is usually feeble because of the low levels of optical power involved, often only in the order of nanowatts. Subsequent stages of the receiver amplify and possibly reshape the signal from the detector. Noise is an serious problem that limits the detector's performance. Broadly speaking, noise is any electrical or optical energy apart from the signal itself.

Although noise can and does occur in every part of a communication system, it is of greatest concern in the receiver input. The reason is that receiver works with very weak signals that have been attenuated during transmission. Although very small compared to the signal levels in most circuits, the noise level is significant in relation to the weak detected signals. The same noise level in a transmitter is usually insignificant because signal levels are very strong in comparison. Indeed, the very limit of the diode's sensitivity is the noise. An optical signal that is too weak cannot be distinguished from the noise. To detect such a signal, we must either reduce the noise level or increase the power level of the signal. In the following sections, we will describe in detail several different noise sources; in practice, it is often assumed that the noise in a detection system has a constant frequency spectrum over the measurement range of interest; this is the so-called "white noise" or "Gaussian noise," and is often a combination of the effects we will describe here.

There are two kinds of amplifiers discussed in fiber optics. The first is **electronic amplifiers** that amplify the

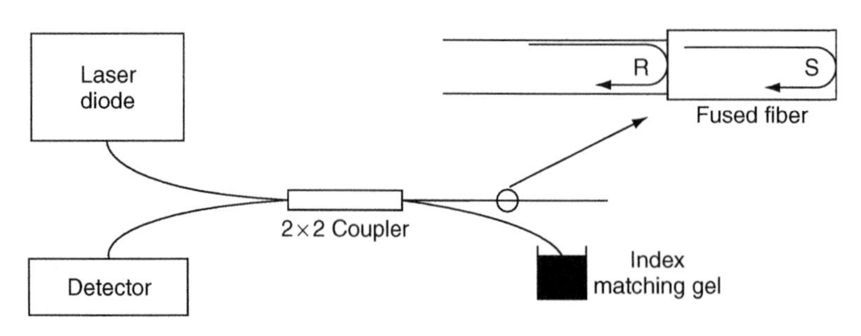

Fig. 4.2.17 The intrinsic Fabry-Perot interferometric (IFPI) sensor.

detector signal, which we will treat here. The second is **optical amplifiers,** which like repeaters and optical repeaters are placed standalone in the link to amplify the transmitter signal and extend networks to very long distances. Optical amplifiers are discussed in a later chapter.

Earlier in this chapter, we considered several circuit diagrams of detectors, including Fig. 4.2.8, which has a PIN diode and amplifier equivalent circuit. Since we do not have all-optical communications systems, it is necessary to consider the electronic aspects of the signal-recovery process at the receiver.

Fiber optic systems typically send digital information, by which we mean a stream of binary data is transmitted by modulating the optical source such that the energy emitted during each bit period is at one of two levels. The high corresponds to 1 and the low corresponds to 0, even though the low is probably set to a non-zero level (as compared to "no signal"). A pre-amplifier will typically convert the current signal from the photodiode into a voltage signal, which may be filtered to clean it up a bit. A pulse regenerator with a decision circuit will convert this into a digital signal that can be used by electronics (Fig. 4.2.18).

There are three major sources of regeneration error. The first is simply noise. The receiver noise level increases with the bandwidth.

The second source of regeneration error is timing errors. The decision circuit samples the signal, and any timing variations which affect the moment of arrival of the signal, or the moment of the sampling within the decision circuit may cause the waveform to be sampled at a point other than at its maximum amplitude. The proper point in the pulse cycle to sample the signal is determined by a **clock recovery circuitry** using phase locked loops. The narrower the pulse, the more serious such variations are.

The third source of regeneration error is intersymbol interference, which occurs if the power received during one bit period affects the signal amplitude during any other bit period. The lower the amplifier bandwidth, the more likely this is, because the pulse response is more spread out, and extends into adjacent bit periods. Electronic filters can be designed to minimize this problem.

The amplification stages of the receiver amplify both the signal and the noise. There are various ways to minimize the noise by narrowing the receiver bandwidth or performing signal averaging, including boxcar integration, multichannel scaling, pulse height analysis, and phase-sensitive detection. For example, boxcar averaging takes its name from gating the signal detection time into a repetitive train of N pulse intervals, or "boxes," during which the signal is present. Since noise which would have been accumulated during times when the gating is off is eliminated, this process improves the SNR by a factor of the square root of N for white noise, Johnson noise, or shot noise. (This is because the integrated signal contribution increases as N, while the noise contribution increases only as the square root of N.) [6] Narrowing the bandwidth of a fiber optic receiver can also have beneficial effects in controlling RIN noise (see Section 4.3.3) and modal noise (see Section 4.3.5). The use of differential signaling is also common in datacom receiver circuits, although they typically do not use lock-in amps but rather solid state electronics such as operational amplifiers.

4.2.5.1 Shot noise

Shot noise occurs in all types of radiation detectors. It is due to the quantum nature of photoelectrons. Since individual photoelectrons are created by absorbed photons at random intervals, the resulting signal has some variation with time. The variation of detector current with time appears as noise; this can be due to either the desired signal photons or by background flux (in the latter case, the detector is said to operate in a **Background Limited in Performance**, or BLIP, mode). To study the shot noise in a photodiode, we will consider the photodetection process. An optical signal and background radiation are absorbed by the photodiode, whereby electron-hole pairs are generated. These electrons and holes are then separated by the electric field and drift toward the opposite sides of the p-n junction. In the process, a displacement current is induced in the external load resister. The photocurrent generated by the optical signal is I_p. The current generated by the background radiation is I_b. The current generated by the thermal generation of electron-hole pairs under completely dark environment is I_d. Because of the randomness of the

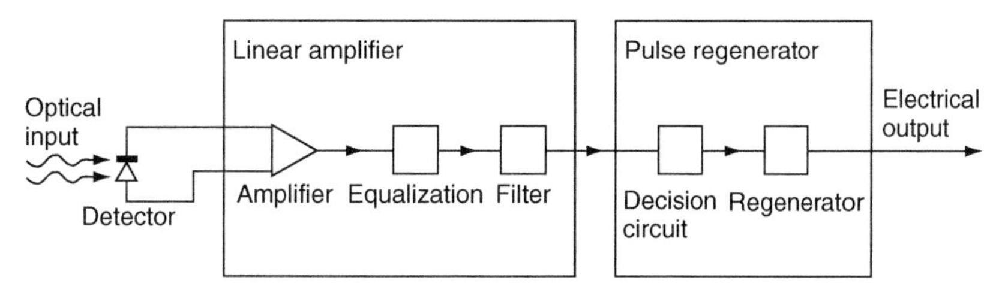

Fig. 4.2.18 Block diagram of a digital optical communication receiver.

generation of all these currents, they contribute shot noise given by a mean square current variation of

$$I_s^2 = 2q(I_p + I_b + I_d)B, \quad (4.2.28)$$

where q is the charge of an electron (1.6×10^{-19} coulomb) and B is the bandwidth. The equation shows that shot noise increases with current and with bandwidth. Shot noise is at its minimum when only dark current exists, and it increases with the current resulting from optical input.

4.2.5.2 Thermal noise

Also known as **Johnson** or **Nyquist noise**, thermal noise is caused by randomness in carriers generation and recombination due to thermal excitation in a conductor; it results in fluctuations in the detector's internal resistance, or in any resistance in series with the detector. These resistances consist of R_j the junction resistance, R_s the series resistance, R_l the load resistance, and R_i the input resistance of the following amplifier. All the resistances contribute additional thermal noise to the system.

The series resistance R_s is usually much smaller than the other resistance and can be neglected. The thermal noise is given by

$$I_t^2 = 4kTB\left(\frac{1}{R_j} + \frac{1}{R_l} + \frac{1}{R_i}\right), \quad (4.2.29)$$

where k is Boltzmann's constant (1.38×10^{-23} J/K), T is absolute temperature (kelvin scale), and B is the bandwidth.

4.2.5.3 Other noise sources

Generation-recombination noise and ***1/f* noise** are particular to photo-conductors. Absorbed photons can produce both positive and negative charge carriers, some of which may recombine before being collected. Generation recombination noise is due to the randomness in the creation and cancellation of individual charge carriers. It can be shown [4] that the magnitude of this noise is given by

$$I_{gr} = 2I\sqrt{(\tau B/N(1 + (2\pi f\tau)^2)} = 2qG\sqrt{\varepsilon EAB}, \quad (4.2.30)$$

where I is the average current due to all sources of carriers (not just photocarriers), τ is the carrier lifetime, N is the total number of free carriers, f is the frequency at which the noise is measured, G is the photoconductive gain (number of electrons generated per photogenerated electron), E is the photon irradiance, and A is the detector active area.

Flicker or the so-called "$1/f$ noise" is particular to biased conductors. Its cause is not well understood, but it is thought to be connected to the imperfect conductive contact at detector electrodes. It can be measured to follow a curve of $1/f^\beta$, where β is a constant which varies between 0.8 and 1.2; the rapid falloff with 1/f gives rise to the name. Lack of good ohmic contact increases this noise, but it is not known if any particular type of electrical contact will eliminate this noise. The empirical expression for the noise current is

$$I_f = \alpha\left(\frac{iB}{f\beta}\right)^{1/2}, \quad (4.2.31)$$

where α is a proportionality constant, i is the current through the detector, and the exponents are empirically estimated to be $\alpha = 2$ and $\beta = 1$. Note this is only an empirical expression for a poorly understood phenomena; the noise current does not become infinite as f approaches zero (DC operation). There may be other noise sources in the detector circuitry, as well; these can also be modeled as equivalent currents. The noise sources described here are uncorrelated and thus must be summed as RMS values rather than a linear summation (put another way, they add in quadrature or use vector addition), so that the total noise is given by

$$I_{tot}^2 = I_f^2 + I_{gr}^2 + I_s^2 + I_t^2. \quad (4.2.32)$$

Usually, one of the components in the above expression will be the dominant noise source in a given application. When designing a data link, one must keep in mind likely sources of noise, their expected contribution, and how to best reduce them. Choose a detector so that the signal will be significantly larger than the detector's expected noise. In a laboratory environment, cooling some detectors will minimize the dark current; this is not practical in most applications. There are other rules of thumb which can be applied to specific detectors, as well. For example, although Johnson noise cannot be eliminated, it can be minimized. In photodiodes, the shot noise is approximately 3X greater than Johnson noise if the DC voltage generated through a transimpedance amplifier is more than about 500 mV. This results in a higher degree of linearity in the measurements while minimizing thermal noise. The same rule of thumb applies to PbS and PbSe-based detectors, though care should be taken not to exceed the maximum bias voltage of these devices or catastrophic breakdown will occur. For the measured values of NEP, detectivity, and specific detectivity to be meaningful, the detector should be operating in a high impedance mode so that the principle source of noise is the shot noise associated with the dark current and the signal current. Although it is possible to use electronic circuits to filter out some types of noise, it is better to have the signal much stronger than the noise by either having a strong signal level or a low noise level. Several types of noise are associated with the photodiode itself and with the receiver; for example, we have already mentioned multiplication noise in an APD, which arises because multiplication varies around a statistical mean.

4.2.5.4 Signal-to-noise ratio

Signal-to-noise ratio is a way to describe the quality of signals in a communication system. SNR is simply the ratio of the average signal power, S, to the average noise power, N, from all noise sources.

$$\mathrm{SNR} = \frac{S}{N}. \qquad (4.2.33)$$

SNR can also be written in decibels as

$$\mathrm{SNR} = 10\log_{10}\left(\frac{S}{N}\right). \qquad (4.2.34)$$

If the signal power is 20 mW and the noise power is 20 nW, the SNR ratio is 1000, or 30 dB. A large SNR means that the signal is much larger than the noise. The signal power depends on the power of the incoming optical power. The specification for SNR is dependent on the application requirements.

For digital systems, **bit error rate (BER)** usually replaces SNR as a performance indicator of system quality. BER is the ratio of incorrectly transmitted bits to correctly transmitted bits. A ratio of 10^{-10} means that one wrong bit is received for every 10 billion bits transmitted. Similarly with SNR, the specification for BER is also dependent on the application requirements. SNR and BER are related. In a ideal system, a better SNR should also have a better BER. However, BER also depends on data-encoding formats and receiver designs. There are techniques to detect and correct bit errors. We can not easily calculate the BER from the SNR, because the relationship depends on several factors, including circuit design and bit error correction techniques. For a more complete overview of BER and other sources of error in a fiber optic data link, see Chapter 7.

References

[1] C.J. Sher DeCusatis, "Fundamentals of Detectors", in *Handbook of Applied Photometry*", C. DeCusatis (ed.), Chapter 3, pp. 101–132, AIP Press: N.Y. (1997).

[2] G. Burns, *Solid State Physics*, Chapter 10, pp. 323–324, Academic Press: Orlando, Fl. (1985).

[3] NBS Special Publication SP250–17, "The NBS Photodetector Spectral Response Calibration Transfer Program".

[4] T.P. Lee, and T. Li, "Photodetectors", in *Optical Fiber Communications*, S.E. Miller and A.G. Chynoweth (eds) Chapter 18, Academic Press: N.Y. (1979).

[5] J. Gowar, *Optical Communication Systems*, Prentice Hall, Englewood Cliffs, N.J.

[6] E. Dereniak, and D. Crowe, *Optical Radiation Detectors*, John Wiley & Sons: N.Y. (1984).

[7] J. Graeme, "Divide and conquer noise in photodiode amplifiers", *Electronic Design*, pp. 10–26, 27 June (1994).

[8] J. Graeme, "Filtering cuts noise in photodiode amplifiers", *Electronic Design*, pp. 9–22, 7 November (1994).

[9] J. Graeme, "FET op amps convert photodiode outputs to usable signals", EDN, p. 205, 29 October (1987).

[10] J. Graeme, "Phase compensation optimizes photodiode bandwidth", EDN, p. 177, 7 May (1982).

[11] D.A. Bell, *Noise and the Solid State*, John Wiley & Sons: N.Y. (1985).

[12] R. Burt, and R. Stitt, "Circuit lowers photodiode amplifier noise", EDN, p. 203, 1 September (1988).

[13] E. Garmire, "Sources, Modulators and Detectors for Fiber Optic Communication Systems", *Handbook of Optics IV*, McGraw-Hill (2001).

[14] J. Marshall Cohen, "Photodiode arrays help meet demand for WDM", *Optoelectronics World*, Supplement to Laser Focus World, August (2000).

[15] M. Selim Ünlü and Samuel Strite "Resonant Cavity Enhanced (RCE) Photonic Devices", at http://photon.bu.edu/selim/papers/apr-95/node1.html. (1995).

Chapter 4.3

4.3

Fiber optic link design

DeCusatis

In this chapter, we will examine the technical requirements for designing fiber optic communication systems. We begin by defining some figures of merit to characterize the system performance. Then, concentrating on digital optical communication systems, we will describe how to design an optical link loss budget and how to account for various types of noise sources in the link.

4.3.1 Figures of merit

There are several possible figures of merit which may be used to characterize the performance of an optical communication system. Different figures of merit may be more suitable for different applications, such as analog or digital transmission. Even if we ignore the practical considerations of laser eye safety standards, an optical transmitter is capable of launching a limited amount of optical power into a fiber; similarly, there is a limit as to how weak a signal can be detected by the receiver in the presence of noise. Thus, a fundamental consideration in optical communication systems design is the optical link **power budget**, or the difference between the transmitted and the received optical power levels. Some power will be lost due to connections, splices, and attenuation in the fiber; there may also be optical power penalties due to dispersion or other effects which we will describe later. The optical power levels define the **SNR** at the receiver, which is often used to characterize the performance of analog communication systems (such as some types of cable television). For digital transmission, the most common figure of merit is the **BER**, which is the ratio of received bit errors to the total number of transmitted bits. SNR is related to the BER by the equation

$$BER = \frac{1}{\sqrt{2\pi}} \int_{Q}^{\infty} e^{-\frac{Q^2}{2}} dQ \cong; \frac{1}{Q\sqrt{2\pi}} e^{-\frac{Q^2}{2}} \qquad (4.3.1)$$

where Q represents the SNR for simplicity of notation [1–4]. A plot of BER vs. received optical power yields a straight line on a semilog scale, as illustrated in Fig. 4.3.1. Some effects, such as fiber attenuation, can be overcome by increasing the received optical power, subject to constraints on maximum optical power (laser eye safety) and the limits of receiver sensitivity. Other types of noise sources are independent of signal strength. When such noise is present, no amount of increase in transmitted signal strength will affect the BER; a noise floor is produced, as shown by curve B in Fig. 4.3.1. This type of noise can be a serious limitation on link performance. If we plot BER vs. receiver sensitivity for increasing optical power, we obtain a curve similar to Fig. 4.3.2 which shows that for very high power levels, the receiver will go into saturation. The characteristic "bathtub" shaped curve illustrates a window of operation with both upper and lower limits on the received power.

In the design of some analog optical communication systems, as well as some digital television systems (e.g., those based on 64 bit Quadrature Amplitude Modulation), another possible figure of merit is the **modulation error ratio (MER)**. To understand this metric, consider the standard definition given by the Digital Video Broadcasting (DVB) Measurements Group [5]. First, the video receiver captures a time record of *N* received signal coordinate pairs, representing the position of information on a two-dimensional screen. The ideal

Fiber Optic Essentials; ISBN: 9780122084317

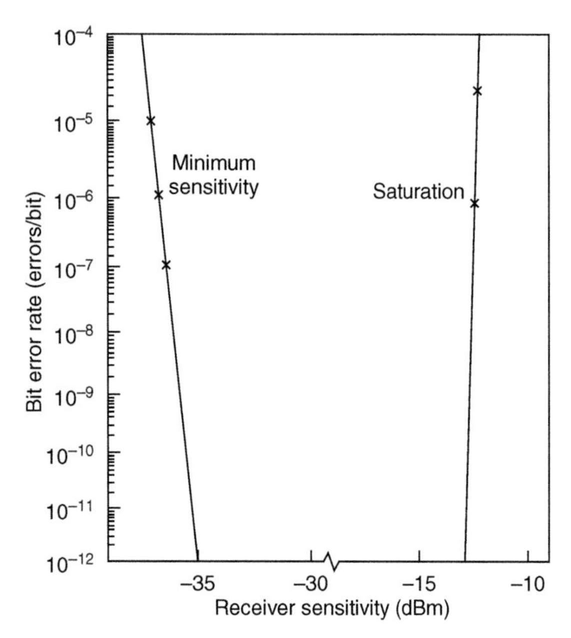

Fig. 4.3.1 Bit error rate as a function of received optical power illustrating range of operation from minimum sensitivity to saturation [3].

position coordinates are given by the vector (X_j, Y_j). For each received symbol, a decision is made as to which symbol was transmitted, and an error vector $(\Delta X_j, \Delta Y_j)$ is defined as the distance from the ideal position to the actual position of the received symbol. The MER is then defined as the sum of the squares of the magnitudes of the ideal symbol vector divided by the sum of the squares of the magnitudes of the symbol error vectors:

$$MER = 10\log \frac{\sum_{j=1}^{N}(X_j^2 + Y_j^2)}{\sum_{j=1}^{N}(\Delta X_j^2 + \Delta Y_j^2)} \text{ dB.} \qquad (4.3.2)$$

When the signal vectors are corrupted by noise, they can be treated as random variables. The denominator becomes an estimate of the average power of the error vector and contains all signal degradation due to noise, reflections, etc. If the only significant source of signal degradation is additive white Gaussian noise, then MER and SNR are equivalent. For communication systems which contain other noise sources, MER offers some advantages; in particular, for some digital transmission systems there may be a very sharp change in BER as a function of SNR (the so-called "cliff effect") which means that BER alone cannot be used as an early predictor of system failures. MER, on the other hand, can be used to measure signal-to-interference ratios accurately for such systems. Because MER is a statistical measurement, its accuracy is directly related to the number of vectors, N, used in the computation; good accuracy can

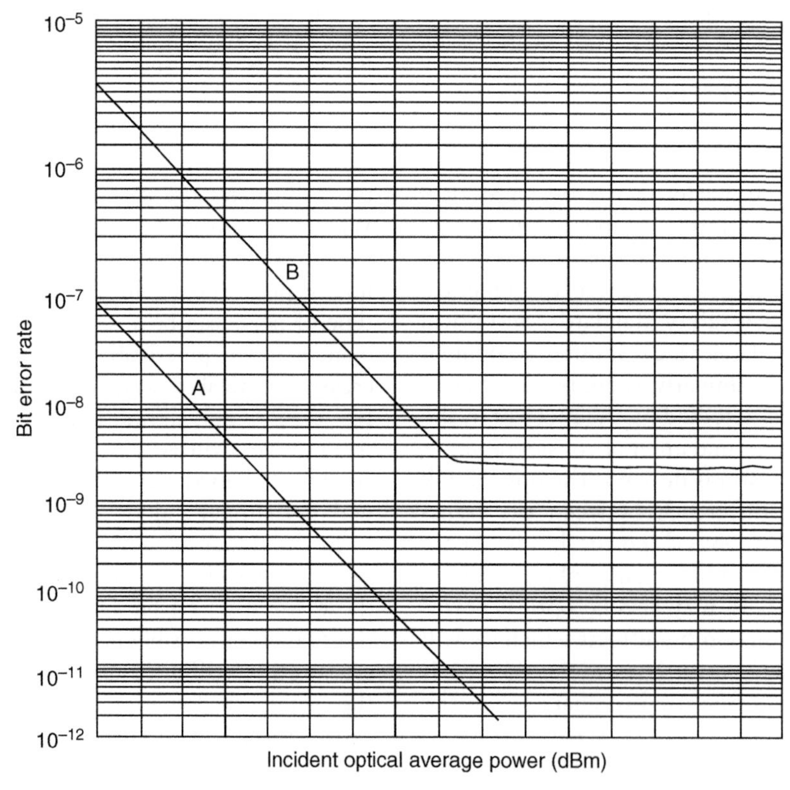

Fig. 4.3.2 Bit error rate as a function of received optical power. Curve A shows typical performance, whereas curve B shows a BER floor [3].

be obtained with $N = 10{,}000$, which would require about 2 ms to accumulate at the industry standard digital video rate of 5.057 Msymbols/s.

Analog fiber optic links are in general nonlinear. That is, if the input electrical information is a harmonic signal of frequency f_0, the output electrical signal will contain the fundamental frequency f_0 as well as high-order harmonics of frequencies nf_0 $(n > 2)$. These high-order harmonics comprise the harmonic distortions of analog fiber optic links [6]. The nonlinear behavior is caused by nonlinearities in the transmitter, the fiber, and the receiver.

The same sources of nonlinearities in the fiber optic links lead to **Inter-Modulation Distortions (IMD)**, which can be best illustrated in a two-tone transmission scenario. If the input electrical information is a superposition of two harmonic signals of frequencies f_1 and f_2, the output electrical signal will contain second-order intermodulation at frequencies $f_1 + f_2$ and $f_1 - f_2$ as well as third-order intermodulation at frequencies $2f_1 - f_2$ and $2f_2 - f_1$.

Most analog fiber optic links require bandwidth of less than one octave $(f_{max} < 2f_{min})$. As a result harmonic distortions as well as second-order IMD products are not important as they can be filtered out electronically. However, third-order IMD products are in the same frequency range (between f_{min} and f_{max}) as the signal itself and therefore appear in the output signal as the spurious response. Thus the linearity of analog fiber optic links is determined by the level of third-order IMD products. In the case of analog links where third-order IMD is eliminated through linearization circuitry, the lowest odd-order IMD determines the linearity of the link.

To quantify IMD distortions, a two-tone experiment (or simulation) is usually conducted where the input radio frequency (RF) powers of the two tones are equal. There are several important features to bear in mind when interpreting the linear and nonlinear power transfer functions (the output RF power of each of two input tones and the second or third-order IMD product as a function of the input RF power of each input harmonic signal). When plotted on a log-log scale, the fundamental power transfer function should be a line with a slope of unity. The second- (third-) order power transfer function should be a line with a slope of two (three). The intersections of the power transfer functions are called second- and third-order intercept points, respectively. Because of the fixed slopes of the power transfer functions, the intercept points can be calculated from measurements obtained at a single input power level. At a certain input level, the output power of each of the two fundamental tones, the second-order IMD product, and third-order IMD products are P_1, P_2, and P_3, respectively. When the power levels are in units of dB or dBm, the second-order and third-order intercept points are

$$IP_2 = 2P_1 - P_2 \tag{4.3.3}$$

and

$$IP_3 = \frac{3P_1 - P_3}{2}. \tag{4.3.4}$$

The dynamic range is a measure of the ability of an analog fiber optic link to faithfully transmit signals at various power levels. At the low input power end, the analog link can fail due to insufficient power level so that the output power is below the noise level. At the high input power end, the analog link can fail due to the fact that the IMD products become the dominant source of signal degradation. In terms of the output power, the dynamic range (of the output power) is defined as the ratio of the fundamental output to the noise power. However, it should be noted that the third-order IMD products increase three times faster than the fundamental signal. After the third-order IMD products exceeds the noise floor, the ratio of the fundamental output to the noise power is meaningless as the dominant degradation of the output signal comes from IMD products. So a more meaningful definition of the dynamic range is the so-called **spurious-free dynamic range (SFDR)** [6, 7], which is the ratio of the fundamental output to the noise power at the point where the IMD products is at the noise level. The SFDR is then practically the maximum dynamic range. Since the noise floor depends on the bandwidth of interest, the unit for SFDR should be $dBHz^{2/3}$. The dynamic range decreases as the bandwidth of the system is increased. The SFDR is also often defined with reference to the input power, which corresponds to SFDR with reference to the output power if there is no gain compression.

4.3.2 Link budget analysis

In order to design a proper optical data link, the contribution of different types of noise sources should be assessed when developing a link budget. There are two basic approaches to link budget modeling. One method is to design the link to operate at the desired BER when all the individual link components assume their worst case performance. This conservative approach is desirable when very high performance is required, or when it is difficult or inconvenient to replace failing components near the end of their useful lifetimes. The resulting design has a high safety margin; in some cases, it may be overdesigned for the required level of performance. Since it is very unlikely that all the elements of the link will assume their worst case performance at the same time, an alternative is to model the link budget statistically. For this method, distributions of transmitter power output, receiver sensitivity, and other parameters are either measured or estimated. They are then combined statistically using an approach such as the Monte Carlo method, in which many possible link combinations are simulated to generate an overall distribution of the available link optical power. A typical approach is the **3-sigma design,** in which the combined variations of all link components are not allowed to extend more than 3 standard deviations from the average performance target in either direction. The statistical approach results in greater design flexibility, and generally increased distance compared with a worst-case model at the same BER.

4.3.2.1 Installation loss

It is convenient to break down the link budget into two areas: installation loss and available power. **Installation loss** refers to optical losses associated with the fiber cable, such as connector loss and splice loss. **Available optical power** is the difference between the transmitter output and the receiver input powers, minus additional losses due to noise sources. With this approach, the installation loss budget may be treated statistically and the available power budget as worst case. First, we consider the installation loss budget, which can be broken down into three areas, namely transmission loss, fiber attenuation as a function of wavelength, and connector or splice losses.

4.3.2.2 Transmission loss

Transmission loss is perhaps the most important property of an optical fiber; it affects the link budget and maximum unrepeated distance. Since the maximum optical power launched into an optical fiber is determined by international laser eye safety standards [8], the number and separation between optical repeaters and regenerators is largely determined by this loss. The mechanisms responsible for this loss include material absorption as well as both linear and nonlinear scattering of light from impurities in the fiber [1–5]. Typical loss for single-mode optical fiber is about 2–3 dB/km near 800 nm wavelength, 0.5 dB/km near 1300 nm, and 0.25dB/km near 1550 nm. Multimode fiber loss is slightly higher, and bending loss will only increase the link attenuation further.

4.3.2.3 Attenuation vs. wavelength

Since fiber loss varies with wavelength, changes in the source wavelength or use of sources with a spectrum of wavelengths will produce additional loss. Transmission loss is minimized near the 1550 nm wavelength band, which unfortunately does not correspond with the dispersion minimum at around 1310 nm. An accurate model for fiber loss as a function of wavelength is quite complex, and must include effects such as linear scattering, macrobending, material absorption due to ultraviolet and infrared band edges, hydroxide (OH) absorption, and absorption from common impurities such as phosphorous. Typical loss due to wavelength dependent attenuation for laser sources on single-mode fiber is specified by the manufacturer and can be held below 0.1 dB/km.

4.3.2.4 Connector and splice loss

There are also installation losses associated with fiber optic connectors and splices; both of these are inherently statistical in nature. There are many different kinds of standardized optical connectors, and different models which have been published for estimating connection loss due to fiber misalignment [9, 10]; most of these treat loss due to misalignment of fiber cores, offset of fibers on either side of the connector, and angular misalignment of fibers. There is no general model available to treat all types of connectors, but typical connector loss values average about 0.5 dB worst case for multimode, slightly higher for single mode.

Optical splices are required for longer links, since fiber is usually available in spools of 1–5 km, or to repair broken fibers. There are two basic types: mechanical splices (which involve placing the two fiber ends in a receptacle that holds them close together, usually with epoxy) and the more commonly used fusion splices (in which the fibers are aligned, then heated sufficiently to fuse the two ends together). Typical splice loss values can be kept well below 0.2 dB.

4.3.3 Optical power penalties

Next, we will consider the assembly loss budget, which is the difference between the transmitter output and the receiver input powers, allowing for optical power penalties due to noise sources in the link. We will follow the standard convention of assuming a digital optical communication link which is best characterized by its BER. Contributing factors to link performance include the following:

- Dispersion (modal and chromatic) or intersymbol interference
- Mode partition noise
- Multipath interference
- Relative intensity noise
- Timing jitter
- Radiation induced darkening
- Modal noise.

Higher order, nonlinear effects including Stimulated Raman and Brillouin scattering and frequency chirping are important for long haul links or optical amplifiers, but are quite complex and beyond the scope of this book.

4.3.3.1 Dispersion

The most important fiber characteristic after transmission loss is dispersion, or intersymbol interference. This refers to the broadening of optical pulses as they propagate along the fiber. As pulses broaden, they tend to interfere with adjacent pulses; this limits the maximum achievable data rate. In multimode fibers, there are two dominant kinds of dispersion, modal and chromatic. **Modal dispersion** refers to the fact that different modes will travel at different velocities and cause pulse broadening. The fiber's **modal bandwidth** in units of MHz-km is specified according to the expression

$$BW_{modal} = BW_1/L^{\gamma}, \tag{4.3.5}$$

where BW_{modal} is the modal bandwidth for a length L of fiber, BW_1 is the manufacturer-specified modal bandwidth of a 1 km section of fiber, and γ is a constant known as the modal bandwidth concatenation length scaling factor. The term γ usually assumes a value between 0.5 and 1, depending on details of the fiber manufacturing and design as well as the operating wavelength; it is conservative to take $\gamma = 1.0$.

The other major contribution is chromatic dispersion, BW_{chrom}, which occurs because different wavelengths of light propagate at different velocities in the fiber. For multimode fiber, this is given by

$$BW_{chrom} = \frac{L^{\gamma_c}}{\left(\sqrt{\lambda_w}(a_0 + a_1|\lambda_c - \lambda_{eff}|\right)}, \tag{4.3.6}$$

where L is the fiber length in km; λ_c is the center wavelength of the source in nm; λ_w is the source FWHM spectral width in nm; γ_c is the chromatic bandwidth length scaling coefficient, a constant; λ_{eff} is the effective wavelength, which combines the effects of the fiber zero dispersion wavelength and spectral loss signature; and the constants a_1 and a_0 are determined by a regression fit of measured data. The chromatic bandwidth for 62.5/125 micron fiber is empirically given by [11]

$$BW_{chrom} = \frac{10^4 L^{-0.69}}{\sqrt{\lambda_w}(1.1 + 0.0189|\lambda_c - 1370|)}. \tag{4.3.7}$$

For this expression, the center wavelength was 1335 nm and λ_{eff} was chosen midway between λ_c and the water absorption peak at 1390 nm; although λ_{eff} was estimated in this case, the expression still provides a good fit to the data. For 50/125 micron fiber, the expression becomes

$$BW_{chrom} = \frac{10^4 L^{-0.65}}{\sqrt{\lambda_w}(1.01 + 0.0177|\lambda_c - 1330|)}. \tag{4.3.8}$$

For this case, λ_c was 1313 nm and the chromatic bandwidth peaked at $\lambda_{eff} = 1330$ nm. Recall that this is only one possible model for fiber bandwidth [1]. The total bandwidth capacity of multimode fiber BW_t is obtained by combining the modal and chromatic dispersion contributions, according to

$$\frac{1}{BW_t^2} = \frac{1}{BW_{chrom}^2} + \frac{1}{BW_{modal}^2}. \tag{4.3.9}$$

Once the total bandwidth is known, the dispersion penalty can be calculated for a given data rate. One expression for the dispersion penalty in dB is

$$P_d = 1.22\left[\frac{Bit\ Rate\ (Mb/s)}{BW_t(MHz)}\right]^2. \tag{4.3.10}$$

For typical telecommunication grade fiber, the dispersion penalty for a 20km link is about 0.5 dB.

Dispersion is usually minimized at wavelengths near 1310 nm; special types of fiber have been developed which manipulate the index profile across the core to achieve minimal dispersion near 1550 nm, which is also the wavelength region of minimal transmission loss. Unfortunately, this dispersion-shifted fiber suffers from some practical drawbacks, including susceptibility to certain kinds of nonlinear noise and increased interference between adjacent channels in a wavelength multiplexing environment. There is a new type of fiber which minimizes dispersion while reducing the unwanted crosstalk effects, called **dispersion optimized fiber**. By using a very sophisticated fiber profile, it is possible to minimize dispersion over the entire wavelength range from 1300 to 1550 nm, at the expense of very high loss (around 2dB/km); this is known as **dispersion flattened fiber**. Yet another approach is called **dispersion compensating fiber**; this fiber is designed with negative dispersion characteristics, so that when used in series with conventional fiber it will "undisperse" the signal. Dispersion compensating fiber has a much narrower core than standard single-mode fiber, which makes it susceptible to nonlinear effects; it is also birefringent and suffers from polarization mode dispersion, in which different states of polarized light propagate with very different group velocities.

By definition, single-mode fiber does not suffer modal dispersion. **Chromatic dispersion** is an important effect, though, even given the relatively narrow spectral width of most laser diodes. The dispersion of single-mode fiber is given by

$$D = \frac{d\tau_g}{d\lambda} = \frac{S_o}{4}\left(\lambda_c - \frac{\lambda_o^4}{\lambda_c^3}\right), \quad (4.3.11)$$

where D is the dispersion in ps/(km-nm) and λ_c is the laser center wavelength. The fiber is characterized by its zero dispersion wavelength, λ_o, and zero dispersion slope, S_o. Usually, both center wavelength and zero dispersion wavelength are specified over a range of values; it is necessary to consider both upper and lower bounds in order to determine the worst case dispersion penalty. Once the dispersion is determined, the intersymbol interference penalty as a function of link length, L, can be determined to a good approximation from [12]:

$$P_d = 5\log(1 + 2\pi(BD\Delta\lambda)^2L^2), \quad (4.3.12)$$

where B is the bit rate and $\Delta\lambda$ is the RMS spectral width of the source. By maintaining a close match between the operating and the zero dispersion wavelengths, this penalty can be kept to a tolerable 0.5–1.0 dB in most cases.

4.3.3.2 Mode partition noise

This penalty is related to the properties of a FP type laser diode cavity; although the total optical power output from the laser may remain constant, the optical power distribution among the laser's longitudinal modes will fluctuate. We must be careful to distinguish this behavior of the instantaneous laser spectrum, which varies with time, from the time-averaged spectrum which is normally observed experimentally. The light propagates through a fiber with wavelength dependent dispersion or attenuation, which deforms the pulse shape. Each mode is delayed by a different amount due to group velocity dispersion in the fiber; this leads to additional signal degradation at the receiver, known as mode partition noise; it is capable of generating BER floors, such that additional optical power into the receiver will not improve the link BER. The power penalty due to mode partition noise can be estimated from [13]:

$$P_{mp} = 5\log(1 - Q^2\sigma_{mp}^2), \quad (4.3.13)$$

where

$$\sigma_{mp}^2 \frac{1}{2} k^2(\pi B)^4 \left[A_1^4\Delta\lambda^4 + 42A_1^2A_2^2\Delta\lambda^6 + 48A_2^4\Delta\lambda^8\right], \quad (4.3.14)$$

$$A_1 = DL, \quad (4.3.15)$$

and

$$A_2 = \frac{A_1}{2(\lambda_c - \lambda_o)}. \quad (4.3.16)$$

The mode partition coefficient, k, is a number between 0 and 1 which describes how much of the optical power is randomly shared between modes; it summarizes the statistical nature of mode partition noise. While there are significantly more complex models available and work in this area is ongoing, a practical rule of thumb is to keep the mode partition noise penalty less than 1.0 dB maximum, provided that this penalty is far away from any noise floors.

4.3.3.3 Relative intensity noise

Stray light reflected back into a FP type laser diode gives rise to intensity fluctuations in the laser output. This is a complicated phenomena, strongly dependent on the type of laser; it is called either reflection-induced intensity noise or RIN. This effect is important since it can also generate BER floors. The power penalty due to RIN is the subject of ongoing research; since the reflected light is measured at a specified signal level, RIN is data dependent although it is independent of link length. Since many laser diodes are packaged in windowed containers, it is difficult to correlate the RIN measurements on an unpackaged laser with those of a commercial product. There have been several detailed attempts to characterize RIN [14, 15]; typically, the RIN noise is assumed Gaussian in amplitude and uniform in frequency over the receiver bandwidth of interest. The RIN value is specified for a given laser by measuring changes in the optical power when a controlled amount of light is fed back into the laser; it is signal dependent, and is also influenced by temperature, bias voltage, laser structure, and other factors which typically influence laser output power [15]. If we assume that the effect of RIN is to produce an equivalent noise current at the receiver, then the additional receiver noise σ_r may be modeled as

$$\sigma_r = \gamma^2 S^{2g} B, \quad (4.3.17)$$

where S is the signal level during a bit period, B is the bit rate, and g is a noise exponent which defines the amount of signal dependent noise. If $g = 0$, noise power is independent of the signal, while for $g = 1$ noise power is proportional to the square of the signal strength. The coefficient γ is given by

$$\gamma^2 = S_i^{2(1-g)} 10^{(RIN_i/10)}, \quad (4.3.18)$$

where RIN_i is the measured RIN value at the average signal level S_i, including worst case backreflection conditions and operating temperatures. One approximation for the RIN power penalty is given by

$$P_{rin} = -5\log\left[1 - Q^2(BW)(1+M_r)^{2g}\right.$$
$$\left.(10^{RIN/10})\left(\frac{1}{M_r}\right)^2\right] \quad (4.3.19)$$

where the RIN value is specified in dB/Hz, BW is the receiver bandwidth, M_r is the receiver modulation index, and the exponent g is a constant varying between 0 and 1 which relates the magnitude of RIN noise to the optical power level. The maximum RIN noise penalty in a link can usually be kept to below 0.5 dB.

4.3.3.4 Jitter

Although it is not strictly an optical phenomena, another important area in link design deals with the effects of timing jitter on the optical signal. In a typical optical link, a clock is extracted from the incoming data signal which is used to retime and reshape the received digital pulse; the received pulse is then compared with a threshold to determine if a digital "1" or "0" was transmitted. So far, we have discussed BER testing with the implicit assumption that the measurement was made in the center of the received data bit; to achieve this, a clock transition at the center of the bit is required. When the clock is generated from a receiver timing recovery circuit, it will have some variation in time and the exact location of the clock edge will be uncertain. Even if the clock is positioned at the center of the bit, its position may drift over time. There will be a region of the bit interval, or eye, in the time domain where the BER is acceptable; this region is defined as the eye width [1–3]. Eye width measurements are an important parameter for evaluation of fiber optic links; they are intimately related to the BER, as well as the acceptable clock drift, pulse width distortion, and optical power. At low optical power levels, the receiver SNR is reduced; increased noise causes amplitude variations in the received signal. These amplitude variations are translated into time domain variations in the receiver decision circuitry, which narrows the eyewidth. At the other extreme, an optical receiver may become saturated at high optical power, reducing the eyewidth and making the system more sensitive to timing jitter. This behavior results in the typical "bathtub" curve shown in Fig. 4.3.2; for this measurement, the clock is delayed from one end of the bit cell to the other, with the BER calculated at each position. Near the ends of the cell, a large number of errors occur; towards the center of the cell, the BER decreases to its true value. The eye opening may be defined as the portion of the eye for which the BER remains constant; pulse width distortion occurs near the edges of the eye, which denotes the limits of the valid clock timing. Uncertainty in the data pulse arrival times causes errors to occur by closing the eye window and causing the eye pattern to be sampled away from the center. This is one of the fundamental problems of optical and digital signal processing, and a large body of work has been done in this area [16,17]. In general, multiple jitter sources will be present in a link; these will tend to be uncorrelated.

Industry standards bodies have adopted a definition of jitter [17] as short-term variations of the significant instants (rising or falling edges) of a digital signal from their ideal position in time. Longer-term variations are described as wander; in terms of frequency, the distinction between jitter and wander is somewhat unclear. Each component of the optical link (data source, serializer, transmitter, encoder, fiber, receiver, retiming/clock recovery/deserialization, decision circuit) will contribute some fraction of the total system jitter. If we consider the link to be a "black box" (but not necessarily a linear system) then we can measure the level of output jitter in the absence of input jitter; this is known as the "intrinsic jitter" of the link. The relative importance of jitter from different sources may be evaluated by measuring the spectral density of the jitter. Another approach is the maximum tolerable input jitter (MTIJ) for the link. Finally, since jitter is essentially a stochastic process, we may attempt to characterize the jitter transfer function (JTF) of the link, or estimate the probability density function of the jitter. The problem of jitter accumulation in a chain of repeaters becomes increasingly complex; however, we can state some general rules of thumb. For well designed practical networks, the basic results of jitter modeling remain valid for N nominally identical repeaters transmitting random data; systematic jitter accumulates in proportion to $N^{1/2}$ and random jitter accumulates in proportion to $N^{1/4}$. For most applications, the maximum timing jitter should be kept below about 30% of the maximum receiver eye opening.

4.3.3.5 Modal noise

Because high capacity optical links tend to use highly coherent laser transmitters, random coupling between fiber modes causes fluctuations in the optical power coupled through splices and connectors; this phenomena is known as modal noise [18]. As one might expect, modal noise is worst when using laser sources in conjunction with multimode fiber; recent industry standards have allowed the use of short-wave lasers (750–850 nm) on 50 micron fiber which may experience this problem. Modal noise is usually considered to be nonexistent in single-mode systems. However, modal noise in single-mode fibers can arise when higher order modes are generated at imperfect connections or splices. If the lossy mode is not completely attenuated before it reaches the next connection, interference with the dominant mode may occur. For N sections of fiber, each of length L in a single-mode link, the worst case sigma for modal noise can be given by

$$\sigma_m = \sqrt{2}N\eta(1-\eta)e^{-\alpha L}, \quad (4.3.20)$$

where α is the attenuation coefficient of the LP_{11} mode, and η is the splice transmission efficiency, given by

$$\eta = 10^{-(\eta_o/10)}, \quad (4.3.21)$$

where η_o is the mean splice loss (typically, splice transmission efficiency will exceed 90%). The corresponding optical power penalty due to modal noise is given by

$$P = -5\log(1 - Q^2\sigma_m^2), \quad (4.3.22)$$

where Q corresponds to the desired BER. This power penalty should be kept to less than 0.5 dB.

4.3.3.6 Radiation induced loss

Another important environmental factor as mentioned earlier is exposure of the fiber to ionizing radiation damage. There is a large body of literature concerning the effects of ionizing radiation on fiber links [19, 20]. There are many factors which can affect the radiation susceptibility of optical fiber, including the type of fiber, type of radiation (gamma radiation is usually assumed to be representative), total dose, dose rate (important only for higher exposure levels), prior irradiation history of the fiber, temperature, wavelength, and data rate. Almost all commercial fiber is intentionally doped to control the refractive index of the core and cladding, as well as dispersion properties. Because of the many factors involved, there does not exist a comprehensive theory to model radiation damage in optical fibers, although the basic physics of the interaction has been described. There are two dominant mechanisms: radiation induced darkening and scintillation. First, high energy radiation can interact with dopants, impurities, or defects in the glass structure to produce color centers which absorb strongly at the operating wavelength. Carriers can also be freed by radiolytic or photochemical processes; some of these become trapped at defect sites, which modifies the band structure of the fiber and causes strong absorption at infrared wavelengths. This radiation induced darkening increases the fiber attenuation; in some cases, it is partially reversible when the radiation is removed, although high levels or prolonged exposure will permanently damage the fiber. The second effect, scintillation, is caused if the radiation interacts with impurities to produce stray light. This light is generally broadband, but will tend to degrade the BER at the receiver; scintillation is a weaker effect than radiation-induced darkening. These effects will degrade the BER of a link; they can be prevented by shielding the fiber, or partially overcome by a third mechanism, called photobleaching. The presence of intense light at the proper wavelength can partially reverse the effects of darkening in a fiber. It is also possible to treat silica core fibers by briefly exposing them to controlled levels of radiation at controlled temperatures; this increases the fiber loss, but makes the fiber less susceptible to future irradiation. These so-called "radiation hardened fibers" are often used in environments where radiation is anticipated to play an important role. The loss due to normal background radiation exposure over a typical link lifetime can be held below about 0.5 dB.

References

[1] S.E. Miller, and A.G. Chynoweth, (eds) *Optical Fiber Telecommunications*, Academic Press, Inc.: New York (1979).

[2] J. Gowar, *Optical Communication Systems*, Prentice Hall, Englewood Cliffs: N.J. (1984).

[3] C. DeCusatis, E. Maass, D. Clement, and R. Lasky, (eds) *Handbook of Fiber Optic Data Communication*, Academic Press, NY; see also *Optical Engineering*, special issue on optical data communication (December 1998).

[4] R. Lasky, U. Osterberg, and D. Stigliani, (eds) *Optoelectronics for Data Communication*, Academic Press, NY (1995).

[5] Digital video broadcasting (DVB), Measurement Guidelines for DVB systems, European Telecommunications Standards Institute ETSI Technical Report ETR 290, May 1997; Digital Multi-Programme Systems for Television Sound and Data Services for Cable Distribution, International Telecommunications Union ITU-T Recommendation J. 83, 1995; Digital Broadcasting System for Television, Sound and Data Services; Framing Structure, Channel Coding and Modulation for Cable Systems, European Telecommunications Standards Institute ETSI 300 429, 1994.

[6] W.E. Stephens, and T.R. Joseph, "System Characteristics of Direct Modulated and Externally Modulated RF Fiber-Optic Links", *IEEE J. Lightwave Technol.*, vol. LT-5 (3), pp. 380–387 (1987).

[7] C.H. Cox III, and E.I. Ackerman, "Some limits on the performance of an analog optical link", Proceedings of the SPIE-The International Society for Optical Engineering, vol. 3463, pp. 2-7 (1999).

[8] United States laser safety standards are regulated by the Dept. of Health

and Human Services (DHHS), Occupational Safety and Health Administration (OSHA), Food and Drug Administration (FDA), Code of Radiological Health (CDRH), 21 Code of Federal Regulations (CFR) subchapter J; the relevant standards are ANSI Z136.1, "Standard for the safe use of lasers" (1993 revision) and ANSI Z136.2, "Standard for the safe use of optical fiber communication systems utilizing laser diodes and LED sources" (1996-97 revision); elsewhere in the world, the relevant standard is International Electrotechnical Commission (IEC/CEI) 825 (1993 revision).

[9] D. Gloge, "Propagation effects in optical fibers", *IEEE Trans. Microw. Theor. Tech.*, vol. MTT-23, pp. 106–120 (1975).

[10] P.M. Shanker, "Effect of modal noise on single-mode fiber optic network", *Opt. Comm.* 64, pp. 347–350 (1988).

[11] J.J. Refi, "LED bandwidth of multimode fiber as a function of source bandwidth and LED spectral characteristics", *IEEE J. Lightwave Tech.*, vol. LT-14, pp. 265–272 (1986).

[12] G.P. Agrawal, *et al.* "Dispersion penalty for 1.3 micron lightwave systems with multimode semiconductor lasers", *IEEE J. Lightwave Tech.*, vol. 6, pp. 620–625 (1988).

[13] K. Ogawa, "Analysis of mode partition noise in laser transmission systems", *IEEE J. Quantum Elec.*, vol. QE-18, pp. 849–855 (1982).

[14] J. Radcliffe, "Fiber optic link performance in the presence of internal noise sources", IBM Technical Report, Glendale Labs, Endicott, NY (1989).

[15] L.L. Xiao, C.B. Su, and R.B. Lauer, "Increase in laser RIN due to asymmetric nonlinear gain, fiber dispersion, and modulation", *IEEE Photon. Tech. Lett.*, vol. 4, pp. 774–777 (1992).

[16] P. Trischitta, and P. Sannuti, "The accumulation of pattern dependent jitter for a chain of fiber optic regenerators", *IEEE Trans. Comm.*, vol. 36, pp. 761–765 (1988).

[17] CCITT Recommendations G.824, G. 823, O.171, and G.703 on timing jitter in digital systems (1984).

[18] D. Marcuse, and H.M. Presby, "Mode coupling in an optical fiber with core distortion", *Bell Sys. Tech. J.*, vol. 1, p. 3 (1975).

[19] E.J. Frieble, *et al.* "Effect of low dose rate irradiation on doped silica core optical fibers", *App. Opt.*, vol. 23, pp. 4202–4208 (1984).

[20] J.B. Haber, *et al.* "Assessment of radiation induced loss for AT&T fiber optic transmission systems in the terrestrial environment", *IEEE J. Lightwave Tech.*, vol. 6, pp. 150–154 (1988).

Section Five

Fiber optic data communication

Chapter 5.1

5.1

History of fiber optical communication

DeCusatis

In this review of the history of communication via fiber optics, we examine this relatively recent advancement within the context of communication through history. We also offer projections of where this continuing advancement in communication technology may lead us over the next half century.

5.1.1 Earliest civilization to the printing press

5.1.1.1 Communication through the ages

All species communicate within their group. The evolution of the human species, however, appears to have been much more rapid and dramatic than the evolution of other species. This human advancement has coincided with an increasingly rapid advancement in communication capability. Is this merely a coincidence, or is there a causal relationship?

The earliest human communication, we assume, was vocal; a capability shared by numerous other species. Archaeological information, however, indicates that, tens of thousands of years ago, humans also began to communicate via stored information in addition to the vocal mode. Cave paintings and cliffside carvings have survived over time, to now, conveying information that at the time was useful. Findings also indicate signal fires existed in those early times, to transmit (via light) information, presumably the sighting of the approach of other humans, the appearance of game animals, or other intelligence. Smoke signal communication emerged along with nautical signal flags, followed by light-beam and flag semaphores.

As civilization advanced (and humans apparently became much more numerous), communication became increasingly complex. Symbols to represent items of interest were conceived and adopted. Techniques were developed to carve these symbols in stone, or to paint them onto media such as walls or sheets made from papyrus reeds — the early communication storage media. The papyrus-enscribed messages were especially significant, in that they were transportable — the early telecommunication ("communication at a distance").

While the development of symbols and media was a major advancement, there were still some major handicaps. Carved messages, in particular, had very low portability. A more general problem was that forming the symbols into the media was a high-level skill that required years of training. Kings and common people could not write (and, in general, could not read). Beyond the limited number of scribes available, and the relatively high cost per message inscribed, was the time required to complete a message; hours to days for a simple scroll; lifetimes for stone carvings. Also, each copy, if wanted, required as much effort and time as the original. These general techniques, however, did not change dramatically over a span of thousands of years. A degree of "shorthand" symbols were developed for commercial messages, and the language became richer through development of more and increasingly refined symbols. Still, it remained a slow form of communication, limited to royalty, wealthy merchants, military leaders, and scholars.

As the need for copies of messages, such as distribution of proclamations, increased, entrepreneurs developed the technique of transferring a symbolic message from the original by applying ink and transferring the

Fiber Optic Data Communication; ISBN: 9780122078927

message to another surface. Printing! Naturally, as this technique evolved, message originators also evolved to sending out more copies. There also naturally evolved a tendency to create longer, more complex messages. So, although making multiple copies became feasible, crafting the original print master remained the role of a master craftsman and, as messages became longer, more time was required.

Within this period, some messages became long enough to be "books." Creating the print master for a book occupied a crew of engravers for many years. Although communication certainly was advancing, it remained expensive and slow to initiate in transportable, storable form.

5.1.2 The next 500 years: Printing press to year 2000

5.1.2.1 Printing press changes the rules

The invention of the movable-type printing press by J. Gutenberg, circa 1450, was a major breakthrough. By this time, the language of communication had evolved from pictorial symbols to words formed from a set of characters or other symbols. These were laboriously engraved into printing plates, requiring days to years per plate. With the availability of movable type pieces that could be arranged to construct a clamped-together plate, the time to create a plate was reduced by orders of magnitude; from days to minutes. Of equal importance, the plates could now be constructed by a technician having relatively modest training, instead of by a skilled artisan with years of training and apprenticeship. With the Gutenberg press, the cost of books could be greatly reduced, becoming financially available to a much larger segment of the populace. Over the ensuing 400 years, instruction books became widely available to all students, current news publication flourished, and entertainment books emerged.

5.1.2.2 The cascade of invention

With the evolution of the printing press, the worldwide exchange of information between scholars, inventors, and other innovators accelerated. Especially over the most recent two centuries, significant inventions cascaded, often standing on the shoulders of earlier inventions. Some of the key inventions related to the advancement of communication are noted in Fig. 5.1.1. Signal transmission through space by electromagnetics (Marconi), electrical conductance principles (Maxwell), mechanized digital computing (Babbage), the telephone (Bell) were landmark inventions that set the platforms for the just-completed Magnetic Century. Vacuum tube amplifiers and rectifiers emerged, making radio transmission and reception feasible (and, ultimately, ubiquitous and affordable). Electronic computing evolved, mid-century, from an interesting intellectual concept to become a tool, albeit very expensive, for controlling massive electrical power grids and for tackling otherwise overwhelmingly challenging scientific calculations. (It was visualized that several of these machines, perhaps dozens, might ultimately be useful worldwide; Thomas J. Watson, International Business Machines Chairman, postulated a potential worldwide market for perhaps five of their computing machines.)

Over the 1633–1882 span, mechanical computation machines were of continuing interest, with concepts developed by Pascal, Leibnitz, and Schickhard, culminating in the first serious effort to build a mechanical calculator machine (by Charles Babbage, in 1882). The first working electromechanical calculator was built by IBM engineers in 1930 (the IBM Automatic Sequence Controlled Calculator, Mark I), under the direction of Professor Aiken of Harvard University [1]. The first electronic calculator, ENIAC, was built by Eckert and Mauchly, of the University of Pennsylvania, in 1946.

The Strowger switch, invented within Bell Laboratories, illustrates a significant point that keeps recurring in the evolution of communication (and in other fields): When a problem evolves and advances to the point that it threatens the continuing evolution of an important field, inventive minds find a feasible solution. The early wire-line telephone systems required switching, to connect a specific originating telephone to the desired other telephone instrument. This was done by an operator who received verbal instructions from the originator, then plugged a connection cord between the two appropriate receptacles on the switchboard. As the number of subscribers and the number of calls per subscriber steadily increased, it became apparent that within a relatively few years it would no longer be feasible to recruit enough operators to do the switching. Thus, the Strowger switch, doing the same task based on telephone-number-based electrical signals, was developed. This switch occupied a lot less physical space, and did the task faster, at less cost, and with higher 24-hour-per-day dependability and accuracy. The Strowger switch, introduced in the late 1800s, bridged the transition into the Electromagnetic Century.

The advancement of telecommunication technology and facilities was especially dramatic through the first half of the 20th century. Telephone communication advanced from two-wire lines to hundreds of parallel voice grade lines, as illustrated in Fig. 5.1.2, colliding with another roadblock. The number of open, uninsulated

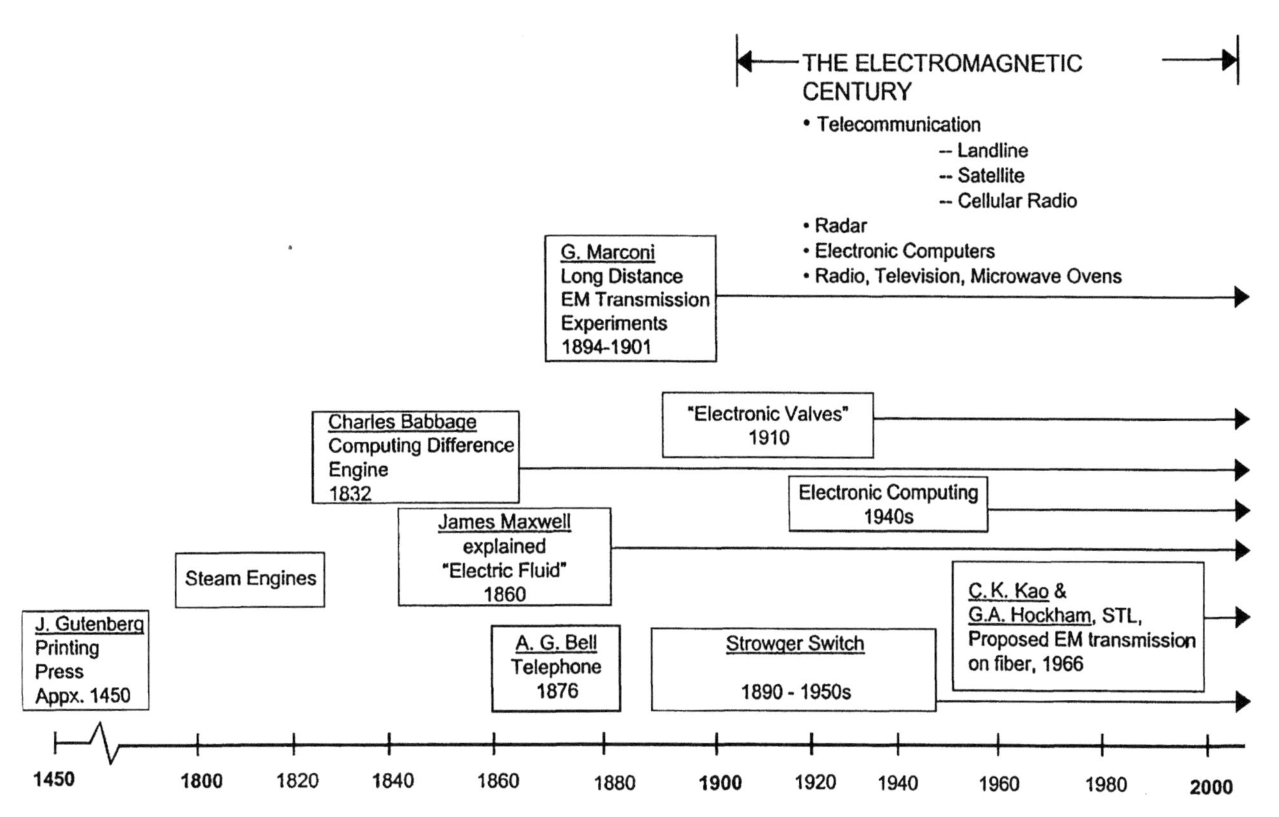

Fig. 5.1.1 The electromagnetic century.

lines routed along city streets and into major office buildings approached the physical space limits. This drove network developers to evolve to "twisted pair" insulated copper wires that greatly reduced the space required for transmission lines. (This was followed by the development of coaxial cables, which could transmit hundreds of voice signals multiplexed onto a single cable.)

This evolved to large cables, "flexible as a sewer pipe," enclosing hundreds of twisted pairs plus several coaxial cables. Most of this cable, installed from about 1930 to date, is still in operation, mainly in metropolitan access networks in North America, Europe, and Japan, and still used for long-haul trunk lines in less developed countries.

World War II interrupted the deployment of civilian communication networks, especially through 1940–1945. Paradoxically, however, this global conflict accelerated the technology of microwave technology, deployed initially primarily in radar systems. Rocket vehicle technology also advanced dramatically during this

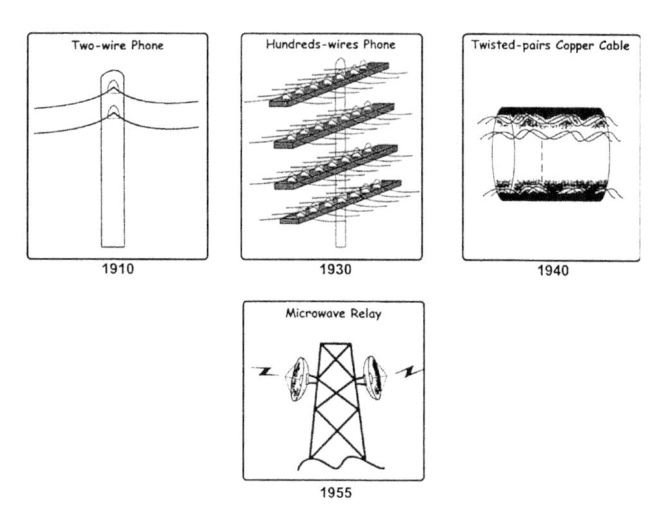

Fig. 5.1.2 Communication transport 19th-century evolution.

relatively brief interval. With the return to peacetime priorities, point-to-point microwave communication relay products evolved from the radar components base. Led by AT&T, General Electric, and RCA, microwave picked up a large and rapidly increasing share of the long-haul transcontinental telecommunication transport.

Terrestrial microwave relay communication expanded rapidly, but was limited by radio-frequency spectrum space availability, and also by the requirement for line-of-sight transmission. The microwave free space transmission beams also spread, as a function of the antenna dimensions (in wavelengths) and of the distance traversed, limiting networks to links of a few tens of miles. Fortunately, as terrestrial microwave relay became increasingly limited in expansion, especially in the most-developed regions, microwave technology was combined with the continuing rocket launch vehicle development to bring satellite microwave communication to commercial reality.

Satellite microwave communication was a fierce competitor to fiber optic communication, especially through the 1975–1990 era, for long-haul high-volume communication. For transoceanic communication, in particular, before the optical amplifier became commercially available, fiber optic cable was not feasible, and satellite microwave took a major share of this market away from underseas copper telecommunication cable. Satellite microwave also was a strong competitor for terrestrial long haul, overcoming the line-of-sight limitations of terrestrial point-to-point microwave relay. At 2000, satellite microwave remains a major element of long-haul communication, business voice and data transport as well as residential television.

5.1.3 Fiber optic communication advancement, 1950–2000

Communication technology and facilities advanced rapidly through the first half of the 20th century, evolving from dual open space wires to hundreds of open space wires, then to hundreds of twisted pair wires in cables, then augmented by terrestrial microwave relay, followed by satellite microwave. The expansion of microwave transmission, however, is limited by radio frequency (RF) spectrum availability (although advancing modulation technologies such as CDMA have greatly extended these limits). Copper wire transmission has a severe distance-times-bandwidth limitation. Fiber optic waveguide has become the next-generation transmission media, initially for long-distance, high-data-rate transport. As technologies and production volumes have advanced, with a dramatic fall in cost per gigabit-kilometer of transport, fiber optic networks have now also become the most economical solution for short/medium distance, modest-data-rate transport in new installations, such as residential and business access, displacing copper.

The evolution of fiber-optic-based communication was built upon many different initial concepts that were then advanced through the years by succeeding scientists.The most significant of these were:

- Transmission of light through a confining media, with very low loss per length;
- A light source, at wavelengths corresponding to low media loss; modulatable at high data rates and having practical lifetime;
- An amplifier of the light signal in the media.

These three legs of the fiber optic communication stool evolved somewhat in parallel, with impetus in one field coming from advancements in another. These primary advancements were augmented in later years, as the industry evolved to networks of much greater complexity, by advancements in digital computing and in rapidly accessable memory storage.

Serious signal transmission by lightwave was preceded by microwave signal transmission. Thus, it is not surprising that, through the past half century, many of the developers of lightwave communication technology, products, and application have moved over from the microwave field.

5.1.3.1 Long road to low-loss fiber

The path to the current low-loss optical glass fiber has had many entrance points. Probably the most significant was the perception by Charles K. Kao, a native Chinese engineer working for Standard Telecommunications Laboratories (STL), UK, that eliminating impurities in glass could yield glass having very low light transmission loss, less than 20 dB per kilometer, building on the concept of total internal reflectance of light in a glass fiber core with a glass cladding of lower index of refraction. Working with microwave engineer George Hockham (1964–65), supporting data was collected, results were published in 1966, and application to long-distance communication over single-mode fiber was proposed. This effort was in the STL optical communications laboratory, headed by Antoni E. Karbowiak, who resigned in 1964 and was succeeded by Charles Kao. (Dr. Kao later retired from STL to return to China, joining the staff of a major university, where he has been a major influence in the current strength of China in the fiber optic communications field.)

The research and initiative by Charles Kao leveraged from earlier work by Elias Snitzer at American Optical Corporation, and Wilbur Hicks (who started and developed the optical fiber components program at AO in 1953). Will Hicks later started his own nearby company, Mosaic Fabrications, which evolved into Galileo. He subsequently started Incom, which was acquired by Polaroid. Snitzer and Hicks demonstrated, in 1961, waveguiding characteristics in glass fiber, total internal reflectance, comparable to theory developed earlier in microwave dielectric waveguides. Elias Snitzer also is credited with being first to propose the principal of the optical fiber amplifier, in 1961 at AO, with first publication in 1964. The concept had to remain on the shelf, however, until an adequate light pump became available.

Will Hicks was an early proponent of Raman amplifiers, and of wavelength-division multiplexers (WDM) based on circular resonant cavities as the wavelength-selective element.

The low-loss glass fiber concept, demonstration, and promotion by Charles Kao was followed by the development of commercially feasible optical fiber production in 1970 by Donald Keck, Robert Maurer, and Peter Schultz at Corning Glass Works. High-purity glass core was achieved by depositing solids from gasses inside a heated quartz tube to form a rod from which fiber was drawn. This was soon followed by low-loss fiber producibility demonstrated at AT&T Bell Laboratories, by John MacChesney and staff.

The development of optical communication fiber also drew from various glass fiber and rod experiments through the first half of the 20th century. Clarence W. Hansell in the United States, and John L. Baird in the UK, developed and patented the use of transparent rods or hollow pipes for image transmission. This was followed by experiments and results reported by Heinrich Lamm (Germany) of image transmission by bundles of glass fiber. This was followed, in 1954, by fiber bundle imaging research reported by Abraham van Heel, Technical University of Delft, Holland, and by Harold H. Hopkins and Narinder Kapany at Imperial College, UK. This in turn was followed, in the late 1950s, by glass-clad fiber bundle imaging reporting by Lawrence Curtiss, University of Michigan.

5.1.3.2 Light power to the core

For practical communication transport over single-mode fiber, the light signal must be coupled into the fiber core (which is only a few microns diameter), must be modulatable at high data rates (initially, a few megabits per second; will reach 40 gigabits per second, commercially, in 2001), and must have a long lifetime (to ensure high reliability of the network). Early experiments used flash lamps, and the earliest optical communication links (generally in industrial or other specialized applications, rather than telecommunication) used light-emitting diodes (LEDs) with relatively large core multimode fiber. There was broad acceptance of the concept that the semiconductor laser diode was the most promising long-term candidate, but its availability lagged behind the optical fiber. Theodore Maiman, of the Hughes Research Laboratory of Hughes Aircraft (U.S.), built the first laser, based on synthetic ruby, in 1960. Reliable solid state lasers still had many years of development ahead.

Laser concepts go back to quantum mechanics theory, outlined in 1900 by Max Planck and advanced in 1905 by Albert Einstein, introducing the photon concept of light propagation and the ability of electrons to absorb and emit photons. This was followed by Einstein's discovery of stimulated emission, evolving from Niels Bohr's 1913 publication of atomic model theory.

Microwave research during and after World War II set the stage for the next phase of laser development. Charles Townes, in 1951, as head of the Columbia University Radiation Laboratory, pursuing microwave physics research, working with James Gordon and Herbert Zeiger, built a molecular-based microwave oscillator to operate in the submillimeter range (evolving from mechanical-based centimeter-wavelength oscillators). Townes named this MASER (Microwave Amplification by Stimulated Emission of Radiation). (Succeeding laboratories, pursuing government-funded research and development contracts, were heard to refer to this as Money Acquisition Scheme for Extended Research.) Further experimentation by Townes determined the concept could be extended into the lightwave region; for which, "Light" was substituted for "Microwave"; thus, LASER. Results were published in 1958, leading to the 1960 ruby laser announcement by Theodore Maiman.

Early research and development of gas and crystal lasers used flashlamp pumps. Meanwhile, semiconductor device development was proceeding, sparked by the transistor invention in 1948, at AT&T Bell Laboratories, by William Schockley, Walter Brattain, and John Bardeen. Heinrich Welker, of Siemens, Germany, in 1952 suggested that semiconductors based on III-V compounds (from columns III and V of the periodic table) could be useful semiconductors. Gallium arsenide, in particular, appeared promising as a base for a communication semiconductor laser. Work on GaAs lasers progressed on several fronts, leading to operational GaAs lasers demonstrated in 1962 by General Electric, IBM, and Lincoln Laboratory of Massachusetts Institute of Technology.

These early GaAs lasers, however, had very short life; seconds, evolving to hours, due to creation of excessive heat in operation. Cooling attempts were inadequate to solve the problem. The solution was to confine the laser action to a thin active layer, as proposed in 1963 by Herbert Kroemer, University of Colorado. This led to a multilayered crystal modified GaAs structure, doped with aluminum, suggested in 1967 by Morton Panish and Izuo Hayashi at AT&T Bell Laboratories. This led, after years of research and development by several leading laboratories, to operational trials of semiconductor communication lasers by AT&T, Atlanta (1976), and the first commercial laser-driven fiber communication deployment, in Chicago (1977).

A major contribution to the lifetime of communication semiconductor laser diodes was the development of molecular-beam epitaxy (MBE) crystal growth by J. R. Arthur and A. Y. Cho at AT&T Bell Laboratories. This achieved much greater precision of layer thickness, permitting higher operational efficiency, thus less heat and longer life (one million hours).

Early deployment of telecommunication fiber links progressed slowly. The 1977–1978 deployment totaled only about 600 miles. The first significant thrust was the AT&T Northeast Corridor, 611 miles Boston to Washington, DC (later extended to New York City); plans submitted in 1980, deployment completed and service "turned up" in 1984. The Northeast Corridor was planned for 45 Mbps; upgraded to 90 Mbps during deployment.

5.1.3.3 Light amplification for long reach

Early fiber communication systems, powered by laser diode transmitters, achieved relatively short link distances between regenerators (which detected the electronic signals from the light beam modulation, reconstituted the pulses, and re-transmitted). Higher power transmitters were not an early alternative, due to linearity and life problems. The regenerators were (and still are) expensive and a source of system failure. Thus, there was a need for periodic amplification of the light power level, along the trunk line. The concept of the optical fiber amplifier had been proposed earlier by Elias Snitzer at American Optical, and in 1985 the concept of using erbium-doped optical fiber as the pumped amplification medium was discovered by S. B. Poole at the University of Southampton, UK. The short length of erbium-doped fiber functions as an externally pumped fiber laser. This concept was developed into a demonstration laser by David Payne and P. J. Mears at the University of Southampton and by Emmanuel Desurvire at AT&T Bell Laboratories. This was demonstrated by Bell Laboratories in 1991. These efforts led to the further development, rigorous life testing, and ultimate deployment of ultra-reliable optical amplifiers in the AT&T/KDD joint transpacific submarine cable.

The preceding summary highlights only a few key efforts that contributed to the dynamic advancement of fiber optic communication. The *City of Light*, by Jeff Hecht, details many more of the breakthroughs and the recognized researchers. Behind this recognition were thousands of unrecognized researchers, worldwide, who moved this technology to the marketplace. (For a detailed, linear chronology of the development and commercial realization of communication-grade optical fiber, the reader is referred to the excellent, thoroughly researched book, *City of Light*, by Jeff Hecht [2]). The following discussion will highlight some of the key elements of this progression, and place them in context with other, parallel developments. The historical (and continuing) development of optical communication fiber is, indeed, impressive. Its commercial feasibility, however, has also benefited greatly from the serendipitous development of lasers (particularly, laser diodes), which supported optical amplifiers, as well as the development of magnetic data storage, semiconductor integrated circuits, and other technical advancements. In the final analysis, also, all of these technical advancements have been greatly accelerated by the fact that they could be used to enable attractive returns on invested capital.

The manufacture of glass began thousands of years ago; initially, a precious commodity for decorative purposes, evolving very slowly (until the middle of the last millennia) to commercial use. Techniques for producing glass fibers emerged hundreds of years ago, and they were applied to practical light transmission for various illumination applications by the late 1800s. The concept and principals of using a bundle of glass fibers for image transmission was outlined by Clarence W. Hansell in 1926, laying the basis for a thriving imaging product industry. American Optical (AO), in Massachusetts, was an early leader in this field, with production accelerating from the early 1950s. AO was an early base for Will Hicks, a scientist/entrepreneur who has subsequently boosted fiber optic communication development in several contexts. He was a founder of the first fiber optics company, Mosaic Fabrications, in 1958, but continued to cooperate with AO, particularly in developing single-mode fiber, which subsequently was advanced by Elias Snitzer of AO.

As already discussed, the post-World War II era brought the realization, by AT&T Bell Laboratories, Standard Telecommunication Laboratories (STL), and other major communication firms, that twisted-pair copper cable was approaching its limit as an economically viable long-haul transmission media. Driven by economics, numerous alternatives were explored; cylindrical millimeter microwave waveguide, satellite microwave,... and optical fiber. Charles K. Kao and George Hockham, in the STL optical communication program, were early pioneers, particularly through the late 1960s, in advancement of the technical and economic arguments for optical fiber communication.

Moving into the 1970s, the commercial feasibility arguments advanced by Kao/Hockham at STL (later acquired by Northern Telecom), and others, convinced Corning to support the development of commercially producible low-loss optical fiber. A team of Robert Maurer, Donald Keck, Peter Schultz, and Frank Zimar achieved rapid successive breakthroughs in low-loss fiber development, especially through the early 1970s.

Meanwhile, there remained the practical realization that low-loss optical transmission of signals was only an intellectual exercise, unless a light source capable of high-speed modulation could be developed. (The photo-detector capability was also evolving, with less drama.) The primary candidate was the laser (first demonstrated by Theodore Maiman, of Hughes Research Laboratories, in 1960). More specifically, semiconductor diode lasers; but, early laser diodes had almost zero lifetime. Numerous parallel laser diode development programs proceeded, with Robert N. Hall's group at General Electric first to demonstrate operation, in 1962 (but with short life, and only by operating in liquid nitrogen temperature). STL demonstrated 1 gigabit per second (Gbps) laser diode modulation in 1972. Bell Labs demonstrated 1000 hours laser diode lifetime in 1973, and Laser Diode Labs (a spinoff from RCA Sarnoff Labs) demonstrated room-temperature operation of a commercial CW laser diode in 1975. In 1976, Bell Labs demonstrated 100,000-hour life of selected laser diodes, at room temperature. Also in 1976, Bell Labs demonstrated 45 megabits per second (Mbps) modulation of laser diodes, coupled with graded-index optical fiber.

Thus, driven by major economic imperatives, the development of optical fiber and laser diodes advanced dramatically through the 1960–1975 span. This laid the base for the dawn of commercial deployment of optical fiber communication networks, starting with the AT&T Northeast Corridor project (Boston-New York-Washington, DC; initially planned as 45 Mbps transmission; deployed as 90 Mbps). (The first independent consultant forecasts of fiber optic communication deployment were published in 1976, led by *Fiber Optic and Laser Communication Forecast*, Jeff D. Montgomery and Helmut F. Wolf, Gnostic Concepts, Inc.)

The last quarter century, 1975–2000, has seen the explosive development of technology and commercial realization of fiber optic communication. The global consumption of fiber optic cable and other components, for example, advanced by about 5 orders of magnitude, from $2.5 million in 1975 to $15.8 billion in 2000. Component development has proceeded through hundreds of laboratories, handing off to hundreds of factories, large and small. Fiber loss continued to drop, laser modulation speeds increased; an old concept, wavelength division multiplexing, found economic justification and catapulted into the marketplace.

Much of this advancement, however, would not have occurred without support from the sidelines: the optical fiber amplifier. The optical fiber amplifier concept was first outlined by Elias Snitzer, in 1961, but for many years it went nowhere, for two reasons: (1) little commercial need was seen; (2) pump laser diodes, an essential component of the amplifier, were not available. The travails of the transmitter laser diode were previously discussed. The pump diode experienced similar difficulties. It needed to operate at much higher peak powers than the transmit diodes of that time; thus, lifetime was an even more severe problem. Also, it needed to operate at a significantly different wavelength than the transmit diodes, so it could benefit less from the earlier diode development.

Early developmental pump diodes had lifetimes of milliseconds; gradually expanded to minutes, then hours, then thousands of hours. The evolution of the optical fiber amplifier, in the context of other related components, is illustrated in Fig. 5.1.3.

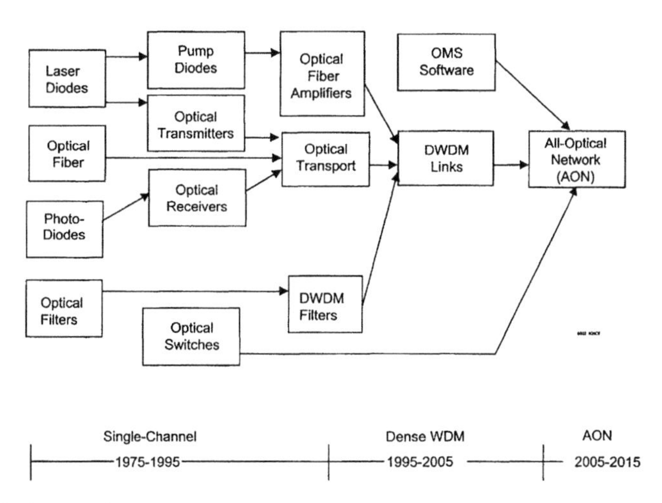

Fig. 5.1.3 Evolving to the all-optical network.

As with many other breakthroughs, the optical fiber amplifier became a commercial product because of its apparent economic payoff. Commercial realization was retarded by the pump lifetime problem previously mentioned, plus the very high cost of final development and life test/demonstration, plus the expected high cost of production, after demonstration of technical feasibility. This barrier was broken by a partnership of AT&T Submarine Cable Systems and KDD (Japan), in development of a transpacific submarine fiber cable. Calculations indicated that, if the amplifiers met specifications and had sufficient lifetime, they could substantially boost the cable performance/cost ratio. The team funded the design, production, and life test of about 200 amplifiers, at an estimated cost of $40–50 million (an impressive amount at the time). The amplifiers were produced, demonstrated long life, and were deployed.

This demonstration led other network developers, both submarine and terrestrial long haul, to consider optical amplifiers. Although they were initially quite expensive, they could eliminate a substantial share of the also-expensive optical/electrical/optical regenerator nodes in the network, so deployment accelerated. With increased production, costs dropped, opening additional markets.

A point of this is, without the optical fiber amplifier, dense wavelength division multiplexing (DWDM) over interexchange (long-haul) networks would not be feasible. So, although DWDM was not a dominant element in the initial amplifier development, it benefited and opened up a major new market.

With DWDM, plus evolutionary developments (such as 40 Gbps transmission per wavelength), terabit-per-fiber transmission becomes feasible, as shown in Fig. 5.1.4. With cables now commercially available with over 1000 fibers, this provides a base for petabit-per-cable systems.

5.1.4 Communication storage and retrieval

It is important to recognize that modern communication depends greatly on the storage of messages and other information, as well as on the technology of transmitting this intelligence from one location to another. With the early storage of communication by carving symbols into stone, transport was essentially impossible, the recipient had to travel to the message. Storage by inscription onto papyrus sheets was transportable, but required a lot of time to create and, generally, a lot more time (weeks to months) to deliver the scroll to the recipient. The printing press drastically shortened the printing time, and the parallel evolution of physical transportation shortened delivery time. Still, however, delivery of this stored message typically required hours to weeks.

The telephone was the dramatic answer to shortening delivery time; essentially instantaneous. However, it simultaneously lost the storage capability. It was necessary for the recipient to handwrite the perceived message, typically in abbreviated and possibly erroneous format, or else have no storage at all. So, the telephone was a very useful advance for the transmission of personal viewpoints and general information, but problematic for conveying precise business data.

As electronic computers emerged, integral data storage was essential for their operation. Earlier mechanical computers used exotic mechanisms for rudimentary storage, but this was infeasible for major machines. This requirement drove the early development of the individual-switch magnetic wound core memory and the development of magnetic tape memory.

The early computers (with much less capability than year 2000 electronic pocket calculators) depended on wound core and tape memory, vacuum tube switches and

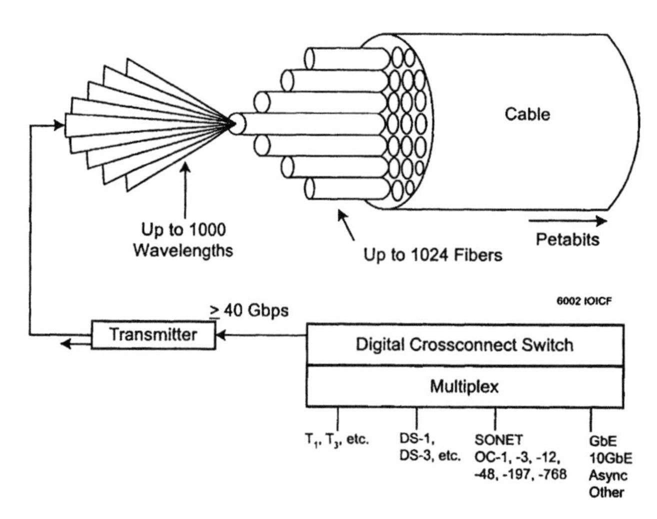

Fig. 5.1.4 Future fiber data transport.

punched card instructions. This was adequate for the perceived demand of a few additional machines per year. Thus, there was relatively little pressure to pursue other than evolutionary advancement of storage (or switching, or input instruction) technologies.

Quite unrelated, AT&T encountered a serious and steadily increasing problem with vacuum tubes. These tubes were key elements in the microwave relay transmit/receive equipment that, leveraging from military-developed microwave technology during World War II, were used for transporting much higher volume of communication than had been feasible over the earlier wire line. Thousands of tubes were in operation in a system, and the average lifetime of the tubes was a few thousand hours, so, statistically, a tube failure in the system could be expected at an average hourly interval but, at the statistical edges, any minute. The solution to this problem was to deploy crews of technicians that continually replaced tubes in the system with new tubes, statistically far in advance of their expected failure time. It could be extrapolated that this, like the earlier switchboard operator roadblock, would become an impossible handicap within a few years. Substantial research was applied to improving tube lifetime, but with limited results.

This dilemma drove AT&T Bell Labs to approach the signal amplification problem from a new base: semiconductor effects. This, within a relatively short period, emerged as the transistor. The transistor, serendipitously, turned out to be much more than the solution to the amplifier tube problem. Early on, it superseded the vacuum tubes in next-generation computers. The long operating lifetime of the transistor, its much smaller size, and its much lower cost in high-volume production, led to developing data storage based on the transistor.

As transmission data rates inside computers and other digital machines move up to gigabits per channel, interconnect links are evolving from copper to optical. Guided wave internal optical interconnect links were widely used in digital crossconnect switches, servers, and other machines in 2000, and terabit free space links are being developed, under US DARPA sponsorship, for military/aerospace ultracompact systems.

Historically, most computer internal interconnect has been from digital signal processors (DSPs) to memory, over copper; the various data streams combined into a single TDM stream by a serializer IC chip, and separated at the other end by a deserializer chip. As DSPs have progressed from 4-bit to 64-bit chips, and data rates per pin have advanced from a few megabits to gigabit level, the cost of serializer/deserializer sets has increased exponentially, and the reach of the copper link falls inversely proportional to data rates. With the cost per gigabit of optical links falling rapidly, optical links will dominate future short reach communication, as well as long haul.

The early transistor, though a major advancement over the vacuum tube for most applications, still in its early years was an individually packaged and socketed device, with multiple transistors connected by discrete wiring, evolving to conductive patterns printed on "printed wiring" boards. A significant supporting technical advancement was the perception, by a team (led by Jack St. Clair Kilby) at Texas Instruments, that it might be useful to process the interconnection between transistors on the parent silicon wafer itself, rather than separating the individual transistor chips out of the wafer, connecting wire leads to the transistor, then connecting these leads to printed wiring board conductors to reach another transistor, etc. This interesting, demonstrated concept attracted little interest or enthusiasm for (in retrospect) a very long time, but it finally burst onto the commercial market as the "integrated circuit." It found an early home in computers (which by this time had progressed beyond the market concept of a few units per year). Since then, the number of interconnected transistors has been doubling about every 18 months. This wasn't very impressive in the early years (few equipment designers could envisage a need for 64 transistors in a single package). The interconnected transistors per chip, however, in 2000 had advanced well beyond the million-device level, and continuing.

As fiber optic communication links advance to production of tens of millions per year, including internal interconnects numbering hundreds per equipment, pressure is increasing for both reduction of physical space per transceiver and reduction of cost per gigabit. In 2000, several major fiber component producers shifted into automated packaging and test of transceivers, optical amplifiers, photonic switches, and other components, achieving volume reduction of 75 to 99 percent. Parallel links also have evolved to production (by Infineon, in Germany); 12 transmitters per module. These trends demand major advancements in package design to meet heat transfer, optical, and electrical isolation requirements in micron-tolerance low-cost packages, as well as advanced assembly/test equipment. These trends are evolving to hybrid optoelectronic integrated circuit (HOEIC) packaging, which in turn will evolve to monolithic optoelectronic ICs over the 2000–2010 decade, supporting hundreds of optical channels per module.

Closing the loop back to communication: about the same time that the transistor phased into widespread application and production, and the integrated circuit (IC) began its market multiplication, AT&T approached another "switchboard operator" roadblock with the Strowger switch. Telephone company central offices had large rooms filled with these electromechanical marvels, clanking away. Projections of switching demand indicated a new solution was needed, and soon. The integrated circuit became the key to telephone signal switching; quite similar technology to computer applications.

Parallel to the application of the IC to telephone signal switching, computers (and other digital machines) found that digital data storage could be accomplished in ICs. The earliest application was as a replacement of the wound core memory, but this soon evolved for use also for archival storage. Also in parallel, however, magnetic disk memory superseded magnetic tape memory, with lower cost per memory bit, much less physical space, and faster store/retrieve. The IC-based memory has become a key element in year 2000 communication equipment, as well as some computer sections. The magnetic disk, however, through aggressive increases in storage efficiency (bits per square inch), access speed, and size options, has maintained a strong commercial position in computers.

5.1.4.1 Task networking augmented by fiber

Along with the continuing advancement of digital signal processor speed, and the trend to harnessing many DSPs to build a mainframe computer, a parallel trend of computer networking has emerged. Networking has taken two forms:

1. Synchronized interconnection of a number of separately located mainframes, for simultaneous processing of different elements of a single problem;
2. Digital communication between computers; data transfer, analogous to telephone voice networks.

Several interconnected high-end workstation computers can provide computing power matching supercomputers. This has evolved from concept to practice, 1965–2000, in both government laboratories and commercial organizations. Optical fiber is the only transmission media that is practical for this interconnection, due to the increasingly high data rates and long connection distances.

The advancement of task-sharing computer networking owes much to the funding of research and development in this field by the U.S. Advanced Research Project Agency (ARPA; now DARPA), starting in the late 1960s. This led to the 1970 inauguration of the ARPAnet, precursor to the Internet, interconnecting four U.S. west coast universities. In 1972, the first International Conference on Computer Communication (ICCC) was held in Washington, DC, to discuss progress of these early efforts. The chairman of this conference was Vinton Cerf, who, along with Robert Kahn, would release the standard Internet protocol TCP/IP just four years later. It was also Cerf who proposed linking ARPAnet with the National Science Foundation's CSNet via a TCP/IP gateway in 1980, which some consider to be the birth of the modern Internet.

The National Physics Laboratory in the United Kingdom and the Societe Internationale de Telecommunications Aeronotique in France, in the 1960s, also explored similar concepts.

5.1.5 Future of fiber optic communications, 2000–2050

Fiber optic networks in 2000 were transmitting several hundred gigabits per fiber; terabit per fiber capability had been demonstrated. The throughput of fiber cables deployed in 2000 could be increased by typically two orders of magnitude by a combination of increased DWDM (more wavelengths per fiber) within currently developed spectral bands plus higher data rate modulation. Beyond this, the available spectral bandwidth will probably be expanded by at least 10X over the 2000–2010 span. Beyond 2010, new fibers can extend the low-loss spectrum by hundreds of nanometers. And, of course, it will always be feasible to deploy additional cables.

Unlike RF/microwave communication, which is limited by available spectrum, and copper cable transmission, which is limited by low data rate capability, high cost per transmitted bit, and the large physical space consumption, future fiber optic communication expansion is relatively unlimited. The key question is: who needs it? Is there a long-term commercial demand for rapidly increasing bandwidth per subscriber, at a rapidly falling price per transmitted bit? To those with a long-term background in integrated circuits and in computers, this is a familiar and long-since-answered question.

To phrase the question differently: will there be new services, enabled by greater bandwidth, at declining cost per gigabit-kilometer, for which subscribers will see economic justification for purchase? Can higher data rate global communication, at little additional cost, enable a business to increase revenues; decrease costs; better negotiate business cycles through greater agility? Will there be bandwidth-dependent residential services offering greater subscriber satisfaction, at little additional cost? Will business and residential interests tend to merge? The history of commercial and residential communication services over the past 50 years, and especially the past 15, supports an affirmative response.

E-business is now emerging, with fits and starts, but the underlying logic seems clear; there are numerous ways it can reduce costs and risks. E-business will require increasingly voluminous instant global transfer of numeric and graphic data. Global subscriber-to-subscriber internets will proliferate, with rapidly falling communication costs per bit.

Personal videoconferencing, on the "back burner" for personal communication as an augmentation of the telephone over the past 40 years, is now entering through the back door; on the personal computer terminal, which is rapidly evolving to an internet data/voice/video communication terminal. This will evolve to higher resolution, color video, using orders of magnitude more bandwidth compared to voice grade lines. It will be affordable, because costs will drop rapidly with increasing volume use, and it will make communication more effective, thus of greater value.

Residential entertainment video is still in an early stage of development. With analog broadcast TV, the TV set designer has been limited to whatever functions she could accomplish within about 4.5 GHz of bandwidth. Very impressive advancements have been made within this restriction; evolution to color TV, and increased definition and/or larger screens. Another restriction has been the vacuum tube display screen; limited in size, although advancements have been achieved. With virtually unlimited bandwidth at very low cost, and with continuing evolution of high-resolution flat panel displays, life-sized three-dimensional "virtual window" displays will become technically and economically feasible. Coupled with entertainment content improvements made feasible by computer graphics/animation, 3D effects, and other innovations, a major future digital TV market appears likely; requiring megabits now; tens of megabits (per receiver) by 2010; gigabits within 50 years.

5.1.5.1 Dynamic, credible video creation

Major advancements were made, over the 1990–2000 decade, in the perception of quality of created (versus recorded/reality) video. Much of this was driven by the economic returns from video games; their increased perception of reality added to popularity and profits. In broadcast entertainment and advertising, also, major advances were made in three-dimensional effects, animation, and nuance. This impressive advancement, however, was restricted by the high cost of massive computational power, size, and resolution limitations of video display screens, and the television receiver bandwidth restriction imposed by the requirement to transmit all program information within an imposed narrow segment of the radio frequency (RF) spectrum. Associated with this was the inevitable time required for the learning process, each incremental step building upon the accumulated knowledge. These restrictions are yielding to the continuing advancements of technology on many fronts.

5.1.5.2 Animation technology advancing

Within the 2000–2050 span, animators will create libraries of "actors" of virtually unlimited scope, with totally lifelike visual and voice replication of a Marilyn Monroe, John Wayne, or any other then-known or fantasy actor. These animatons will be programmable for seamless performance. They will not be temperamental or temporary, and will not demand huge salaries. They will be instantly, continually available to program creators. Copyright laws will expand to protect the rights of both "real" and created actors.

5.1.5.3 Surge of video screen development

The capability of "flat panel" video screens advanced impressively, 1990–2000, while the cost/performance ratio drastically dropped, driven by the economic returns available from improved portable computers, cellular phones, and other instruments. (This followed the path of semiconductors, which were accelerated into wide usage by Sony's high-volume sales of semiconductor-based portable personal radios.)

5.1.5.4 Fiber bandwidth supports receiver revolution

The use of a substantial segment of the RF spectrum for entertainment video transmission will be displaced, within the 2000–2050 period, in the more industrially advanced nations, by signal transmission over fiber. (Regions that cannot economically be served by fiber will receive TV broadcasts via satellite microwave and by local, low-power terrestrial microwave transmitters.) This will free television receiver designers from the current restriction of accomplishing all signal conversion and application within a bandwidth of about 4.5 MHz. By 2010, over 100 million homes globally will receive television entertainment over fiber at data rates up to 622 Mbps. By 2050, this will stretch to over one billion homes at up to 10 Gbps. This will enable the design and profitable supply of receivers presenting apparent life-size, three-dimensional television via instant selection from remote storage servers offering tens of thousands of programs.

While television viewers continue to show strong acceptance of fictional/fantasy presentations, there also is rapidly increasing demand for reality-based programs. The evolving technology of video program creation using computer-enabled script and character creation will make feasible lifelike evening programs based on same-day morning news.

5.1.5.5 Major advancement of non-entertainment video

The 2000–2050 advancements in video content creation, including animation plus 3D screen presentation, enabled by broadband signal fiber transmission to homes and businesses, will support strong growth in video instruction for both formal education and instructional "how to" markets. The rapid (and accelerating) advancement of technology on all fronts means a person cannot complete their formal education and then remain proficient in their chosen field for their entire career through only workplace exposure. Continuing education will be required; fiber-enabled broadband video increasingly will become the most economical medium to deliver this education. Instruction manuals will be displaced by video clips.

5.1.5.6 Dynamic future of communication fiber optics industry

The economics-driven growth of global, regional, and local fiber communication networks over the past 25 years has supported, and been enabled by, dynamic expansion of fiber network deployment. The global consumption of fiber optic components in communication networks exploded from only $2.5 million in 1975 to $15.8 billion in 2000. Continued growth to $739 billion in 2025 is forecast [3]. The value of fiber optic communication equipment and trunk lines incorporating these components in 2000 was more than twice the component value. While year-to-year growth rates varied significantly over the past quarter century, the overall trend was a pattern of gradually slowing investment growth rate, while communication transport capacity expanded at a relatively steady exponential rate.

5.1.5.7 Standards, integration, mass production keys to fiber ubiquity

Transistors, in their early market, cost more than vacuum tubes. If that price relationship had continued, the advantages of semiconductors nevertheless would have supported a growing market. However, over the past 50 years the average price of a transistor has dropped by about six orders of magnitude. This enabled transistors to take market share away from vacuum tubes but, much more importantly, supported the design and economic production of a dazzling range of products that would not have been technically or economically feasible with vacuum tube technology. This includes nearly all of the computer market, portable electronics, most vehicular and space electronics, and much more. Keys to the unprecedented price drop achieved by transistors have included increasing integration (2001 laboratory demonstration: 400 million transistors on a chip) and related miniaturization, plus mass production enabled by substantial standardization and automation of processes, packaging, and testing. The production of fiber optic cable has become relatively mature by 2000, comparable to the copper communication cable of 1975. Other fiber optic components are much less mature; many new components entered the market 2000–2001, and many are yet to be conceived and developed.

5.1.5.8 Components are far from mature

Fiber optic active and integratable products in 2000 were at a maturity stage comparable to semiconductor integrated circuits circa 1970 (a few transistors on a chip). Integration of active photonic components is now evolving aggressively in hybrid format, including multiple parallel channels, with production of over one million units projected for 2001. This will expand to over ten million units in 2010. Monolithically integrated optoelectronic circuits are now mainly developmental, but will exceed one million production units by 2010. Wavelength-division multiplexing (WDM) and parallel channel integration will combine to achieve ten-terabit interconnection fiber links, terminated in single small transmit/receive modules, well before 2050.

Fiber optic component production has evolved, 1975–2000, from small quantities assembled by engineers plus semi-skilled labor, to quantities of hundreds per day assembled by highly skilled, tooling-assisted labor (mostly in low labor cost regions). The industry, in 2000, was at a very early stage of high-volume automated assembly, packaging, and testing along the path pioneered circa 1975 by the semiconductor industry. With the evolutionary progress through 2000, the price of optoelectronic transmitters and receivers (in current dollars), measured in megabits per dollar, dropped by about four orders of magnitude. An even larger decrease will occur over the 2000–2050 span. To achieve this, automation of alignment and attachment, especially of optical beam transmission components, to sub-micron tolerances is required. Standardized packages in several formats, metallic and nonmetallic, will be necessary, along with precision pick-and-place machines. Heat transfer plus both optical and electrical crosstalk challenges must be met.

The historic growth of the economically significant fiber optic component categories 1975–2000, and forecasted growth 2000–2050, are presented in Table 5.1.1. The total global consumption of fiber optic components in 2050 is forecast to reach about $28 trillion. While fiber optic cable has dominated component consumption

Table 5.1.1 Global fiber optic component consumption growth rates, by function.

	1975		2000		2025		2050		Average annual growth rate %		
Component type	**$Million**	**%**	**$Billion**	**%**	**$Billion**	**%**	**$Billion**	**%**	**1975–2000**	**2000–2025**	**2025–2050**
Fiber Optic Cable	1.78	**71**	25.20	70	**430**	**40**	3750	13	**47**	**12**	**9**
Active Components	0.40	**16**	7.40	20	**490**	**45**	20090	72	**48**	**21**	**16**
Passive Components	0.16	**6**	2.45	7	**124**	**11**	3290	12	**47**	**19**	**14**
Other Components	0.15	**6**	1.20	3	**39**	**4**	665	2	**43**	**15**	**12**
Total Consumption	2.49	100	36.25	100	**1083**	100	27795	100	**47**	**15**	**14**

value to date, active components will steadily move to dominance over the next half century as many passive functions are integrated into monolithic OEICs and as consumption quantities are dominated by much shorter links, requiring little fiber, compared to year 2000 mix.

5.1.5.9 Advancement will be evolutionary

The advancement of fiber optic components over the next 25 and 50 years will be evolutionary, in a pattern resembling the evolution of the semiconductor device industry over the past half century. The underlying trends will be toward

- Higher performance;
- Higher integration (more functions per component);
- Miniaturization; function/volume ratio increased by more than four orders of magnitude;
- Lower cost per function, by more than four orders of magnitude.

5.1.5.10 Optical packet switches by 2010

Amazing progress along this trendline occurred over the 1998–2001 span, with two orders of magnitude size reduction along with one order of cost reduction in commercially available transmitters and receivers of fixed performance. Components with further dramatic advancements were demonstrated, in 2001, in numerous industry laboratories and will become commercially available 2002–2003. These include dozens of transmitter/receiver pairs in a single small package; dozens of optical amplifiers on a single small chip; photonic transparent nonblocking matrix switches with 10,000 × 10,000 port capability. The low-loss fiber transport spectrum, supported by available and emerging fiber and other components, will expand by more than an order of magnitude over the 2000–2025 span. Optical packet switching capable of 40 Gbps per channel, terabits per fiber throughput, supported by holographic memory and sub-nanosecond semiconductor optical switches, will be deployed in long-haul network packet switches by 2010.

The anticipated trend of fiber optic communication over the 2025–2050 span becomes more hazy. It is highly probable that the basic function will continue to grow; rapidly in terms of capacity (throughput; gigabits x kilometers), and at a declining but still-positive constant dollar (i.e., after deducting inflation effects) investment rate. Throughput of a fiber cable will continue to increase, while the cost per gigabit-kilometer continues to drop. Expanded services that consume much more bandwidth will continue to emerge. Miniaturization, upward integration, and quantity increases will continue. Standard monolithic optoelectronic integrated circuits (MOEICs) will become the mainstream standard component category, but a major market in application-specific MOEICs also will emerge. Technical breakthroughs that have already been conceived, and perhaps demonstrated, as well as inventions through the 2000–2025 span, will have a major impact, not now identifiable, on 2025–2050 components.

Who, in 1950, could project that transistors would largely supersede vacuum tubes, with many orders of magnitude per-function cost reduction, thus opening the way for major new, useful applications requiring multi-billion-dollar annual equipment production? Who, even in 1975, could forecast that millions of transistors could be processed on a single tiny chip selling for a few dollars each, by 2000? (Indeed, there were scholarly treatises demonstrating this would be impossible.) Fiber optic communication will not progress in a competitive vacuum. Although copper cable will decline in relative significance (as have vacuum tubes over the past 50 years), it will still be deployed in some situations. More competitive will be radio-frequency wireless and unguided optical spacebeam communication, strongly supported by semiconductor-based encoding that greatly expands the available channels and usable bandwidth within a served area. By 2050, satellite-supported wireless will make video and high data transfer rates accessible to miniature portable units having immense memory, anywhere in the world, at a cost that is economically attractive to businesses and to professional people.

5.1.5.11 Communication evolution will continue

Communication with, between, and among professional individuals in 2050 will extend commonplace technology beyond the 2000 perceived outer limits. At an economically justifiable cost, the professional, wherever he or she travels in the world, can continuously send and receive voice, data, graphic, and video communication. This will include transfer of massive data and text, which can be stored in the personal terminal. Verbal language translation will be automatic. An immense library of public and private information will be instantly accessible. Voice-to-hardcopy translation in any commonly used language can immediately be printed.

All known prospective contacts can be contacted/connected in seconds, and desired but unidentified contacts can quickly be identified and connected. This will be accomplished by the global communication network's combination of fiber optic, satellite, and cellular transport.

If these 2000–2050 forecasts seem aggressive, they should be judged in the context of advancement over the past 500 years, not the past 50. (This is the rule, mentioned earlier, that over half of the progress of any period has occurred during the last 10 percent of the period.) They should be judged alongside the probability that wealthy individuals (of which there will be millions) in 2050 can (and some will) have several hundred children, with or without genetic relationship. The probability that nuclear fusion power (the 20-year future breakthrough forecast since 1960) can provide virtually unlimited low-cost clean electrical energy that can displace fossil fuels in transportation, heating, and industrial processes; and can achieve low-cost water purification and transport, supporting major greening of the planet and alleviation of hunger. The probability of space travel to solar planets. The probability of dramatic advancement in the treatment of common diseases. These and many other possibilities, viewed from 2000, appear more likely than the world of 2000 appeared to the thinkers of 1500.

Communication is not a fad. Modern society considers it a necessity, and will see economic justification for better, more responsive, ubiquitous communication. Fiber optics technologies and related markets will continue growth over the next 50 years, and beyond.

References

1. C. DeCusatis, *Handbook of Fiber Optic Data Communication*. San Diego: Academic Press (1998).
2. J. Hecht, *City of Light, The Story of Fiber Optics*. Oxford, UK: Oxford University Press (1999).
3. J. D. Montgomery, *Fifty Years of Fiber Optics*. Nashua, N.H.: Lightwave/PennWell Corp (1999).

Chapter 5.2

5.2

Small form factor fiber optic connectors

DeCusatis

In this chapter, we will describe the similarities and differences between four of the industry's Small Form Factor connectors. Although there are certainly more than four connectors that have been developed to meet the criteria of a smaller form optical interface, the MT-RJ, SC-DC, VF-45, and LC connectors are the focus of this chapter based on their popularity and general acceptance across the industry. The connectors' physical characteristics, coupling techniques, and optical performances will be discussed in detail in order to provide a comprehensive comparison of these predominant connector types.

5.2.1 Introduction

The natural progression of technology will inevitably drive new product designs to be faster, smaller, and less expensive. This is also the case in the world of fiber optic interconnects, and a new generation of connectors, collectively known as Small Form Factor connectors, have now been developed with these goals in mind. Although the MT-RJ, SC-DC, VF-45, and LC connectors have all significantly reduced the size and cost of optical interfaces, in comparison to the standard SC duplex connector, each design approach is unique and therefore there are significant physical and functional differences that can make the connectors application dependent.

Various types of next-generation SFF optical interfaces have been proposed to the Electronics Industry Association/Telecommunications Industry Association (EIA/TIA), for inclusion in developing standards such as the Commercial Cabling Standard TIA-568-B. While no single connector has been selected for inclusion in this standard, it requires that connectors be defined by a reference document called a Fiber Optic Connector Inter-matability Standard (FOCIS), which defines the connector geometry so that the same connector build by different manufacturers will be mechanically compatible. The EIA/TIA also requires connectors to meet some minimal performance levels, independent of connector design; test methodologies are defined by the EIA/TIA Fiber Optic Test Procedures (FOTPs). Other industry specifications are also relevant to these connectors; for example, Bellcore spec. GR-326-CORE defines fiber protrusion from a ferrule to ensure physical contact and prevent back reflections in the connector. The relevant FOCIS documents defined for connectors that have currently been proposed to the TIA are given in Table 5.2.1.

5.2.2 MT-RJ connector

The MT-RJ connector utilizes the same rectangular plastic ferrule technology as the MTP array-style connector first developed by NTT, with a single ferrule body housing two fibers at a 750-μm pitch (Fig. 5.2.1). These ferrules are available in both single-mode and multimode tolerances, with the lower-cost multimode version typically comprised of a glass-filled thermoplastic and the critically tolerance single-mode version comprised of

Fiber Optic Data Communication; ISBN: 9780122078927

Table 5.2.1 EIA/TIA proposed SFF connector standards.

Connector type	FOCIS document number	FOCIS author*
MT-RJ	12	Amp
SC-DC[a]	11	Siecor
LC	10	Lucent
SG (VF-45[b])	7	3 M
Fiber Jack	6	Panduit

Note. The authoring company listed does not necessarily support or manufacture only one connector type, nor does this table include all of the supporters or manufacturers of each connector type.
[a] SC-DC is a trademark of Siecor.
[b] VF-45 is a trademark of 3 M.

a glass-filled thermoset material. Unlike the thermoplastic multimode ferrules, which can be manufactured using the standard injection mold process, the thermoset single-mode ferrules must be transfer molded, which is generally a slower but more accurate process.

By design the alignment of two MT-RJ ferrules is achieved by mating a pair of metal guide pins with a corresponding pair of holes in the receptacle (Fig. 5.2.2). This feature makes the MT-RJ the only Small Form Factor connector with a distinct male and female connector. As a general rule, wall outlets, transceivers, and internal patch panel connectors will retain the guide pins (thus making their gender male) and the interconnecting jumpers will have no pins (female). In the event that two jumper assemblies require mating mid-span a special cable assembly with one male end and one female end must be used. However, some unique designs do allow for the insertion and extraction of guide pins in the field, affording the user the ability to change the connector's gender as required.

Latching of the MT-RJ connector is modeled after the copper RJ-45 connector, whereby a single latch arm

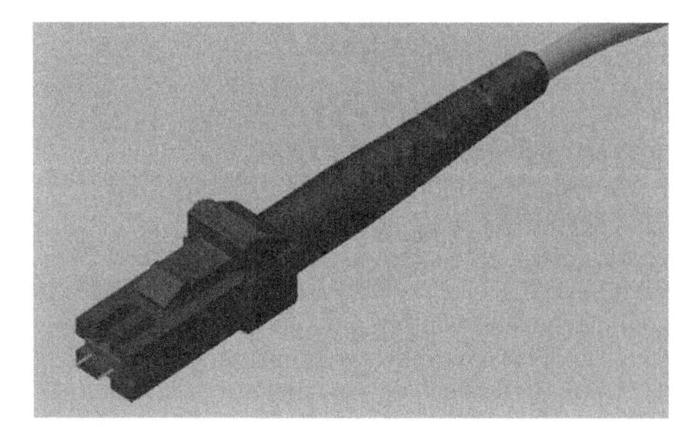

Fig. 5.2.1 Standard MT-RJ male connector.

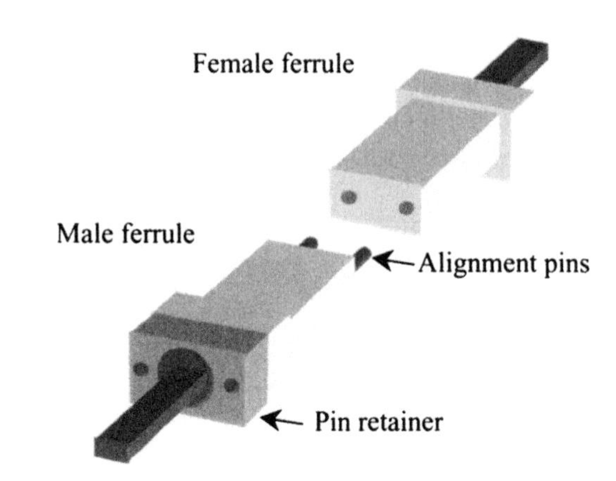

Fig. 5.2.2 MT-RJ ferrule alignment method.

positioned at the top of the connector housing is positively latched into the coupler or transceiver window. Although this latch design is similar in all the MT-RJ connector designs, individual latch pull strengths may vary depending on the connector material, arm deflection, and the relief angles built into the mating receptacles. For this reason it is recommended to evaluate connector pull strengths as a complete interface, depending on the specific manufacturer's connector, coupler, or transceiver design, as the coupling performances may vary.

MT-RJ connectors are typically assembled on 2.8 mm round jacketed cable housing two optical fibers in one of three internal configurations. The first construction style consists of the two optical fibers encapsulated within a ribbon at a 750 μm pitch (Fig. 5.2.3). This approach is unique to the MT-RJ connector and designed specifically to match the fiber spacing to the pitch of the ferrule for ease of fiber insertion. Although this construction style may be ideal for a MT-RJ termination, it can cause some difficulty when manufacturing a hybrid assembly, and availability may also be an issue based on its uniqueness. A second design which is more universal, utilizes a single 900 μm buffer to house two 250 μm fibers (Fig. 5.2.4). This construction is more conducive to hybrid cable manufacturing but the fibers will naturally maintain a 250 μm pitch, thus making fiber insertion rather difficult. The third design is considered a standard construction and is used across the industry (Fig. 5.2.5). In

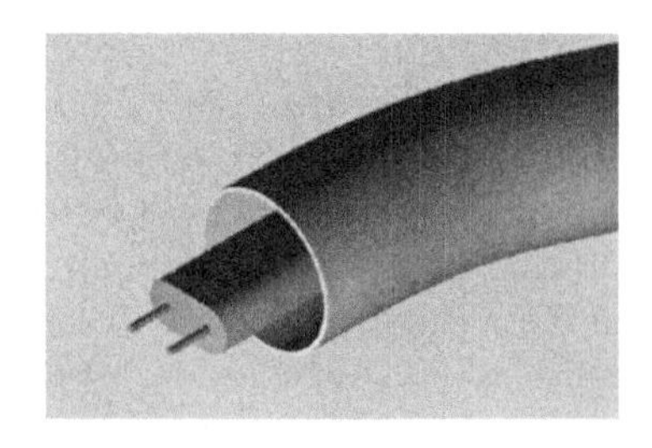

Fig. 5.2.3 Two fiber ribbon construction.

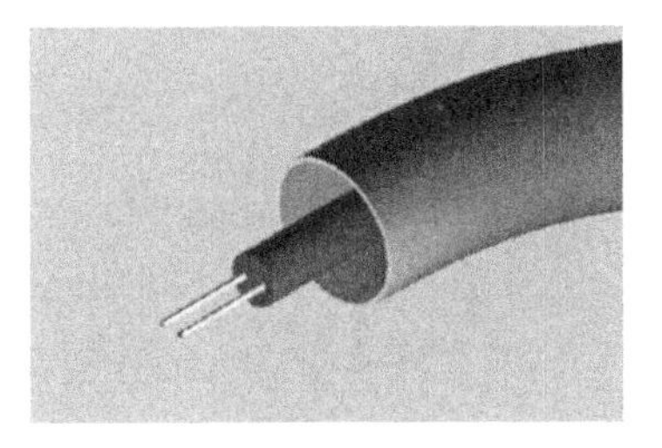

Fig. 5.2.4 Dual 250 micron construction.

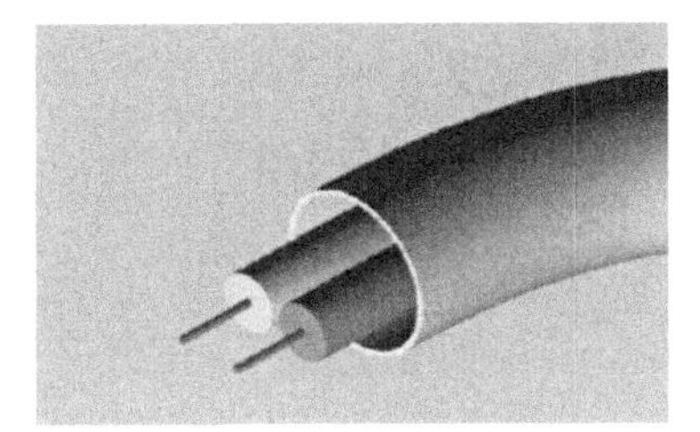

Fig. 5.2.5 Dual 900 micron construction.

this configuration each individual fiber is buffered with a PVC coating. The coating thickness is typically 900 μm, but as in the previous case this does cause a mismatch of the fiber to ferrule pitch. To compensate for this, some connector designs incorporate a fiber transition boot, which gradually reduces the fiber pitch to 750 μm, while others simply use a non-standard buffer coating of 750 μm.

In general the assembly and polish of the MT-RJ factory-style connector is considerably more difficult than the other small form factor connectors. Typical MT-RJ designs have a minimum of eight individual components that must be assembled after the ferrule has been polished, allowing for a number of handling concerns (Fig. 5.2.6). As with the case of most MT-style ferrules, the perpendicularity or flatness of the ferrule endface with reference to the ferrule's inner shoulder is critical and this cannot be accomplished if the connector is pre-assembled. Another unique requirement of the MT-RJ polish involves fiber protrusion. Although the ferrule endface is considered to be flat, depending on the polishing equipment, fixtures, and even contamination some angularity may occur. Therefore it is recommended that the fibers themselves protrude 1.2 to 3.0 μms from the ferrule surface in order to guarantee fiber-to-fiber contact.

For the reasons previously mentioned, the factory-style MT-RJ connector is not a good candidate for field assembly and polishing, so a number of similar yet unique Field Installable Connectors have been developed. All of these field solutions utilize a pre-polished ferrule assembly, which is mated with two cleaved fibers and aligned by v-grooves, with the entire interface filled with index matching gel to compensate for any possible air gaps. One of the major differences between the various solutions is in the mechanism used to open and close the spring clip that maintains constant pressure on the two sandwiched halves of the channeled interface. In one design a cam, which is integrated into the connector body, is used to separate the halves while in other designs a separate hand tool is required. The other general difference between the field solutions is the application design — a number of the connectors are designed to be just that — a field connector with a distinct gender that can be terminated onto distribution style cable — while others are designed to be male receptacles only that must be wall or cabinet mounted. Because of the inherent difference between the MT-RJ fields solutions one solution may be better suited to a given application than another.

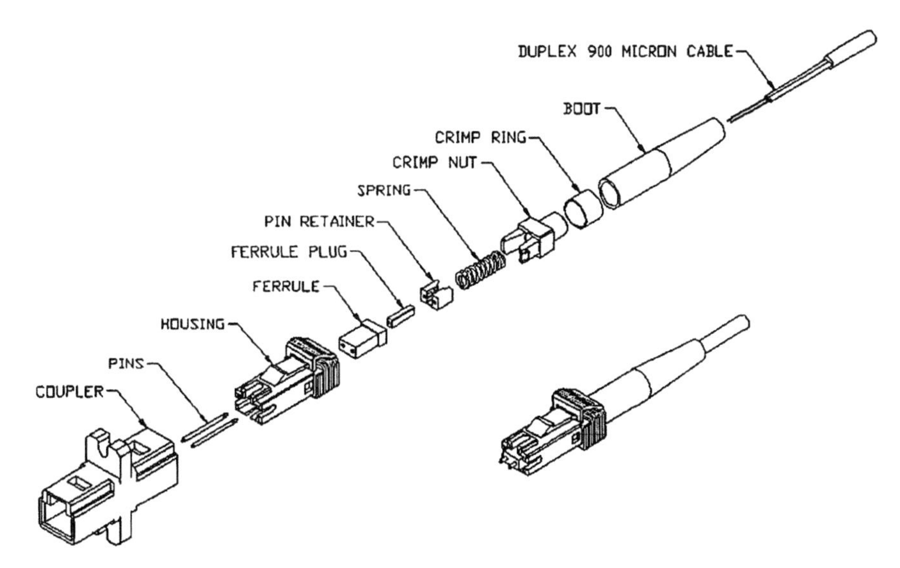

Fig. 5.2.6 Exploded view of the MT-RJ connector components and coupler.

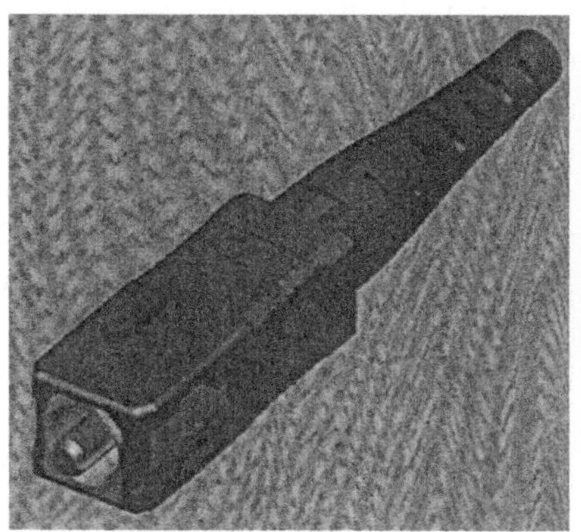
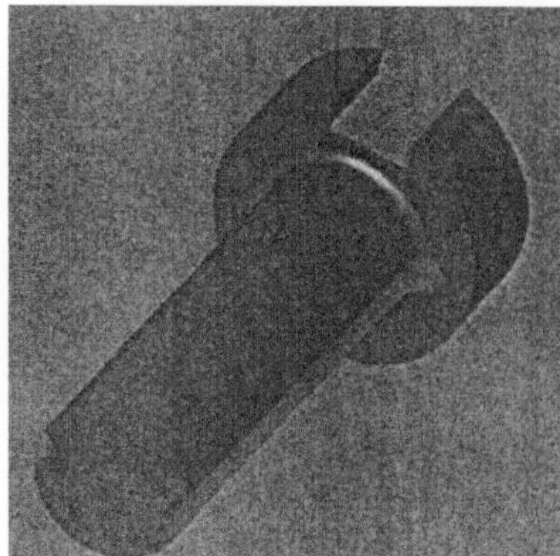

Fig. 5.2.7 Siecor SC-DC/SC-QC connector and ferrule assembly.

5.2.3 SC-DC Connector

The SC-DC (dual contact) or SC-QC (Quattro contact), developed by Siecor, has a connector body design resembling a SC simplex connector with a round thermoset molded ferrule that is capable of handling either two or four optical fibers (Fig. 5.2.7). The connector is designed to support both single-mode and multimode cabling applications. However, the SC-DC is one of the few connectors that is used exclusively for cable interconnecting and therefore has no transceiver support.

As in the case of the MT-RJ connector the SC-DC fibers are on a 750 μm pitch while the SC-QC fibers are on a 250 μm pitch. The same ferrule and housings are used in both connectors so the only feature that distinguishes between the two is simply the number of fibers used. The four ferrule holes of approximately 126 μm diameter are placed on 250 μm centers along the ferrule center line to form a SC-QC connector. By using only the two outermost ferrule holes and leaving the inner two empty the SC-DC is created.

Although the ferrule composition is very similar to that used across the MT technologies, the geometry and alignment methods are very different. The SC-DC ferrule has a standard round shape with a 2.5 mm diameter, but unlike its ceramic counterparts there are two semicircular grooves with a 350 μm radius positioned along each side of the ferrule at a 180° interval. This feature provides the ferrule-to-ferrule alignment when mated with the corresponding ribs of the coupler. The design of this feature within the couplers is different for a single-mode and multimode connection. The multimode coupler uses a one-piece all-composite alignment insert with molded ribs, while the single-mode coupler incorporates two precision alignment pins captured inside the sleeve (Fig. 5.2.8). By design the ribs of the coupler and grooves in the ferrule are on the same 2.6 mm pitch as the pins of a male MT-RJ connector to allow for possible hybrid mating of the two connector types.

The latching mechanism of the SC-DC connector is the same push-pull style used on the industry standard SC simplex connector and the outer housing dimensions are identical. Because the two connectors physically appear the same but are not functionally interchangeable, the housing alignment key of the SC-DC is offset to help distinguish between the two connector types and prevent any confusion in the field. By following the same basic footprint as the SC connector this new SFF optical connector already has a familiar look and feel with proven plug reliability.

SC-DC connectors and cable assemblies are solely manufactured and sold by Corning, so the variability of

Fig. 5.2.8 Siecor SC-DC/SC-QC multimode and single-mode coupling devices.

design and cable material is not an issue as is the case with other SFF connectors. The connectors are typically assembled onto 2.8 mm round jacketed fiber incorporating a single-ribbon fiber populated with either two or four fibers, which assists in the alignment of the fiber to the ferrule holes. The internal ferrule geometry can accommodate the standard single-fiber cables and this may be an option for the fabrication of hybrid cable assemblies.

As previously stated the size and shape of the SC-DC ferrule is similar to the standard ceramic and therefore can be polished in much the same manner and still produce endface geometries akin to the MT-RJ. A flat endface polish is the desired result, much like with the MT-style ferrules, however the perpendicularity is referenced to the sides of the ferrule rather than to an internal feature, so the SC-DC connector can be preassembled and polished as a complete connector. The ability to pre-assemble a connector significantly reduces complexity of manufacturing and typically results in better yields.

The SC-DC connector is available in a field installable version, the SC-DC UniCam, which utilizes prepolished fiber stubs much like other SFF solutions. Alignment of the fiber stubs to the in-field cleaved fiber is achieved through a gel-filled mechanical splice element. The splice element is opened and closed with a mechanical cam and is retained within a connector housing. Termination of the standard SC-DC connector can also be accomplished in the field by using conventional equipment and methods much like the field termination of the SC-style connector.

5.2.4 VF-45 connector

The VF-45 connector, developed by 3 M, is perhaps the most innovative SFF connector design, in that it eliminates the need for precision ferrules and sleeves altogether. The overall look and feel of the "Plug-to-socket" design closely resembles the standard telephony RJ-45 "Connector-to-jack" system whereby the cable assembly mates directly to a terminated socket, reducing the need for couplers (Fig. 5.2.9). Although this concept has been around for years in the copper industry, the creation of a bare fiber optical interface, using alignment grooves and no index matching gels, requires some revolutionary techniques.

The VF-45 connector incorporates two 125 μm optical fibers, suspended in free space on a 4.5 mm pitch protected by a RJ-45 style housing with a retractable front door designed to protect the fibers. The connector design supports both single-mode and multimode tolerances by relying on the inherent precision of the optical fibers within the two injected molded v-grooves of either the transceiver or a VF-45 socket. The design of the interconnect allows the natural spring forces of the optical fibers to align the fibers within the v-grooves as well as ensuring physical fiber-to-fiber contact.

Because of the uniqueness of this interconnect the geometry of the end-face polish of both the plug and receptacle fibers has been modified to provide optimum performance. The VF-45 optical connection relies on the spring force created by the bowing of the optical fibers to provide a physical contact force of approximately 0.1 N and this force coupled with an 8-degree angle polished endfaces produces the optimum connection and return loss results. The tips of the plug fibers are also beveled at 35 degrees, allowing a 90 μm contact area and providing a relief for the fiber to slide into the v-grooves with no damage to the core region (Fig. 5.2.10). This chamfer is not required on the receptacle fibers since they remain stationary while the plug fibers may have to endure multiple insertions (Fig. 5.2.11). As previously mentioned, the contact force created at the optical interface directly influences the optical performance of the plug-socket connection. This downward compressive force is generated when the two fibers of the plug engage with the resident fibers of the socket and cause a slight "bow" in the plug fibers (Fig. 5.2.12). Because of the constant stress on these fibers, long-term reliability on standard

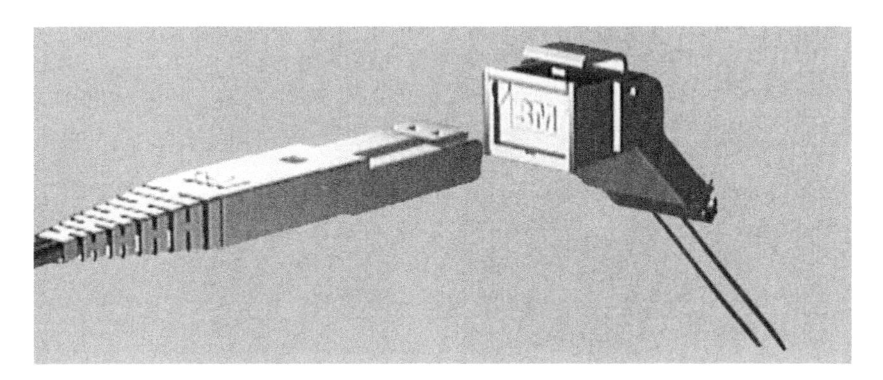

Fig. 5.2.9 3M VF-45 "Connector-to jack" system.

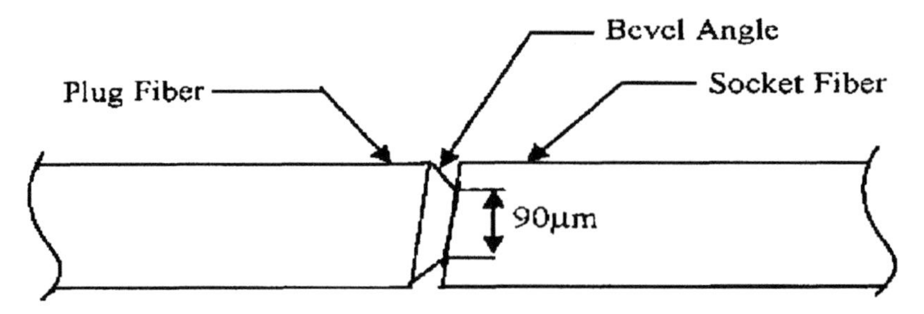

Fig. 5.2.10 VF-45 plug fiber to socket fiber interface.

optical fibers became a concern and therefore a specialized high-strength optical fiber was developed for this application, called GGP (glass-glass-polymer) fiber. GGP fiber consists of 100μm glass fiber with a polymeric coating applied to bring the outer diameter to 125 μm. By reducing the outer diameter of the glass the tensile stress on the fiber is minimized and the additional coating also provides protection against abrasions from the v-grooves and reduces the chance of damage to the glass during the mechanical stripping process used in the termination process.

The factory termination of the VF-45 jumper plugs is considerably different than the conventional ferrule-based connectors. The process of threading a 125 μm fiber into a precision ferrule hole filled with epoxy is now eliminated and replaced with a mechanical fiber holder that grips the fibers in place. The fibers are then cleaved and polished to the endface geometry previously described and the cable strain relief slid into place. The fibers and holder are then placed into a protective shroud and the front door cover installed. The relative simplicity of this manufacturing process makes the VF-45 connector one of the best candidates for a fully automated production line.

The socket of the VF-45 was specifically designed for termination in the field with minimal effort and training. After preparing the fibers for termination by removing outer buffer material, they are inserted into a mechanical fiber holder that retains the fiber by gripping them inside a deformable aluminum crimp. The fibers are then cleaved and hand-polished to an 8-degree angle with a slight radius generated by the durometer of the polishing pad. The fiber holder with the polished fibers is guided into the socket v-grooves and the housing plate snapped into place (Fig. 5.2.13). Although this method of field termination does vary from the other pre-polished SFF connectors, the total termination time and the complexity of the process is very similar.

5.2.5 LC connector

The LC connector developed by Lucent Technologies is a more evolutionary approach to achieving the goals of a SFF connector. The LC connector utilizes the traditional components of a SC duplex connector having independent ceramic ferrules and housings with the overall size scaled down by one half (Fig. 5.2.14). The LC family of connectors includes a stand-alone simplex design; a "behind the wall" (BTW) connector; and the duplex connector available in both single-mode and multimode tolerances, all designed using the RJ-style latch.

The outward appearance and physical size of the LC connector varies slightly depending on the application

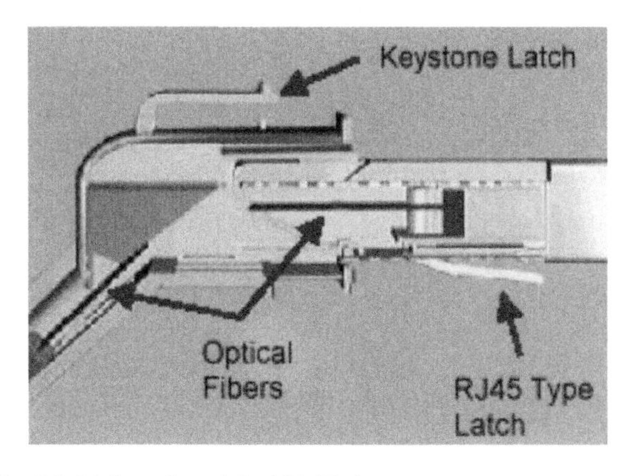

Fig. 5.2.11 Insertion of the VF-45 plug.

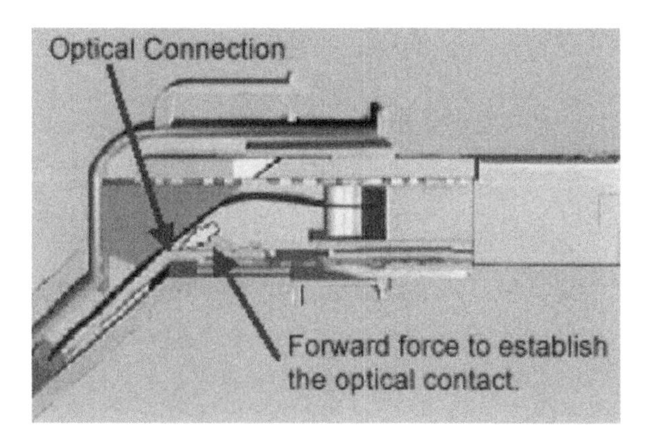

Fig. 5.2.12 VF-45 optical connection.

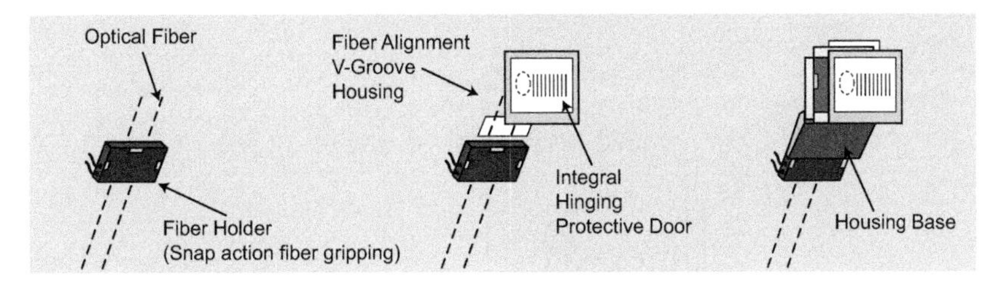

Fig. 5.2.13 Field termination of the 3M VF-45 socket.

and vendor preference. Although all the connectors in the LC family have similar latch styles modeled after the copper RJ latch, the simplex version of the connector has a slightly longer body than either the duplex or BTW version and the latch has an additional latch actuator arm that is designed to assist in plugging as well to prevent snagging in the field. The BTW connector is the smallest of the LC family and designed as a field or board mountable connector using 900 μm buffered fiber and in some cases has a slightly extended latch for extraction purposes. The duplex version of this connector has a modified body to accept the duplexing clip that joins the two connector bodies together and actuates the two latches as one (Fig 5.2.15). Finally, even the duplex clip itself has variations depending on the vendor. In some cases the duplex clip is a solid one-piece design and must be placed on the cable prior to connectorization, while other designs have slots built into each side to allow the clip to be installed after connectorization. In conclusion, all LC connectors are not created equal and depending on style and manufacturer's preference there may be attributes that make one connector more suitable for a specific application than another.

The LC duplex connector incorporates two round ceramic ferrules with outer diameters of 1.25 mm and a duplex pitch of 6.25 mm. Alignment of these ferrules is achieved through the traditional couplers and bores using precision ceramic split or solid sleeves. In an attempt to improve the optical performance to better than 0.10 db at these interfaces most of the ferrule and backbone assemblies are designed to allow the cable manufacturer to tune them. Tuning of the LC connector simply consists of rotating the ferrule to one of four available positions dictated by the backbone design. The concept is basically to align the concentricity offset of each ferrule to a single quadrant at 12:00; in effect, if all the cores are slightly offset in the same direction, the probability of a core-to-core alignment is increased and optimum performance can be achieved. Although this concept has its merits, it is yet another costly step in the manufacturing process and in the case where a tuned connector is mated with an untuned connector, the performance increase may not be realized.

Typically the LC duplex connectors are terminated onto a new reduced-size zipcord referred to as mini-zip; however, as the product matures and the applications expand it may be found on a number of different cordages. The mini-zip cord is one of the smallest in the industry with an outer diameter of 1.6 mm compared with the standard zipcord for SC style product of 3.0 mm. Although this cable has passed industry standard testing there are some issues raised by the cable manufacturers concerning the ability of the 900 μm fibers to move freely inside 1.6 mm jacket and others involving the overall crimped pull strengths. For these reasons some end users and cable manufactures are opting for a larger 2.0 mm, 2.4 mm or even the standard 3.0 mm zipcord. In applications where the fiber is either protected within a wall

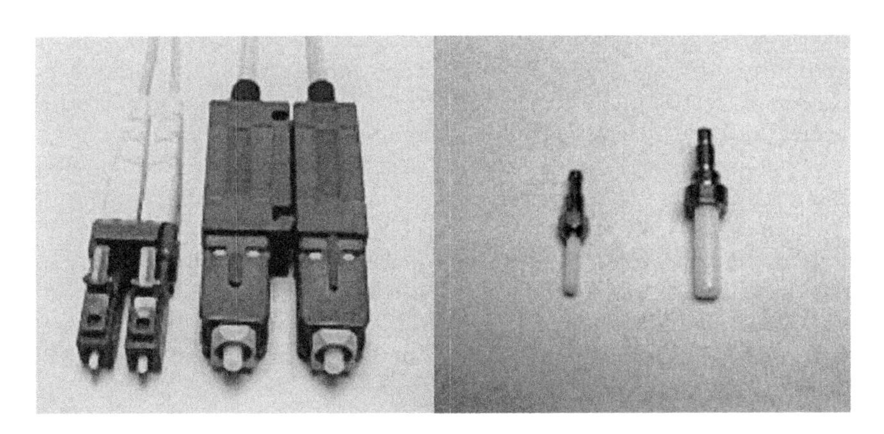

Fig. 5.2.14 LC duplex connector and ferrule compared with a SC duplex connector and ferrule.

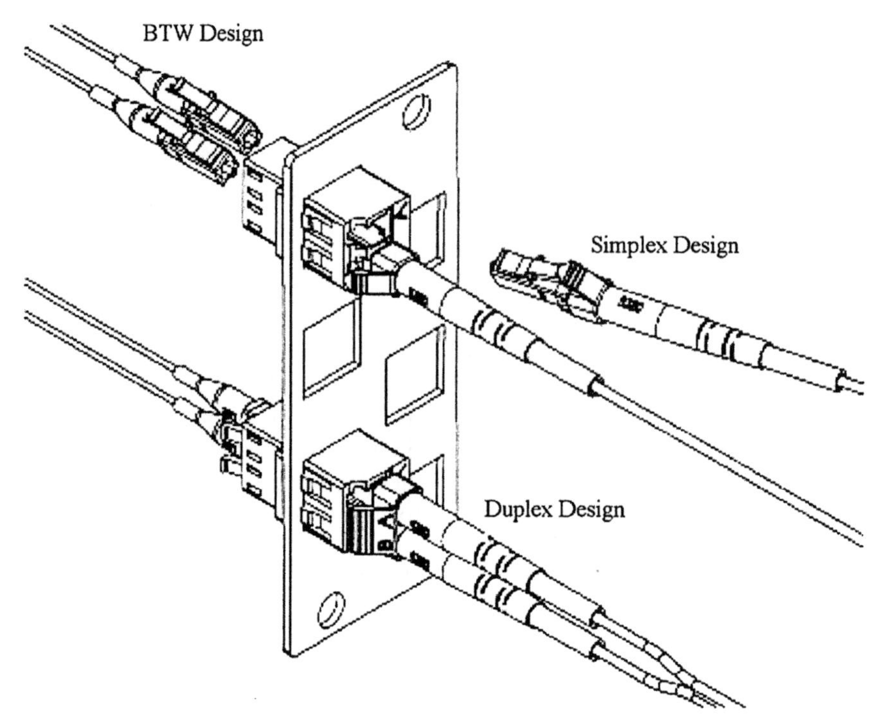

Fig. 5.2.15 The family of LC-style connectors.

outlet or cabinet the BTW connector is used and terminated directly onto the 900 μm buffers with no jacket protection.

The factory termination of the LC cable assemblies is very similar to other ceramic-based ferrules using the standard pot and polish processes with a few minor differences. The one-piece design of the connector minimizes production handling and helps to increase process yields when compared with other SFF and standard connector types. Because of the smaller diameter ferrule, the polishing times for an LC ferrule may be slightly lower than the standard 2.5 mm connectors, but the real production advantage is realized in the increased number of connectors that can be polished at one time in a mass polisher. For the reasons mentioned above and the fact that the process is familiar to most manufacturers, the LC connector may be considered one of the easiest SFF connectors to factory terminate.

Field termination of the LC connector has typically been accomplished through the standard pot and polish techniques using the BTW connector. However, a pre-polished, crimp and cleave connector is also available. The LCQuick Light field-mountable BTW-style connector made by Lucent Technologies is a one-piece design with a factory polished ferrule and an internal cleaved fiber stub. Unlike other pre-polished SFF connectors previously discussed, the LCQuick light secures the inserted field cleaved fiber to a factory polished stub by crimping or collapsing the metallic entry tube onto the buffered portion. This is accomplished by using a special crimp tool designed not to damage the fibers. However, this means the installer has but one chance for a good connection. The LCQuick light is designed specifically for use in protected environments such as cabinets and wall outlets and has no provision for outer jacket or Kevlar protection.

5.2.6 Other types of SFF connectors

Following a renewed interest in fiber optic cabling and transceiver footprint reduction, there have been many types of small form factor optical interfaces proposed. Some of these are no longer widely used; for example, the mini-MT connector was originally proposed as a 2-fiber version of the MTP connector, but was largely supplanted by the MT-RJ design. In this chapter, we have concentrated on optical interfaces for next-generation transceivers and cable plant infrastructures, and have limited our treatment so as not to include the MTP/MPO connectors, the SMC parallel optical connector, or other emerging multi-fiber interfaces for Infiniband. However, there are some other SFF interfaces currently in use besides the 4 major types discussed earlier; we will briefly describe them here.

The Fiber-Jack (FJ) is the non-trademarked name for the Opti-Jack connector developed by Panduit Corporation. As shown in Fig. 5.2.16, this connector incorporates two industry standard SC duplex ceramic ferrules, each 2.5 mm diameter. However, the spacing

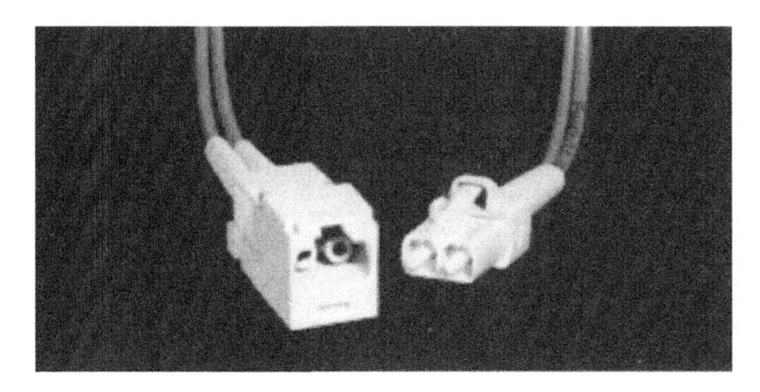

Fig. 5.2.16 Fiber-Jack.

between ferrules has been reduced from 1.27 mm (0.5 inches) as in a standard SC Duplex connector to only 0.63 mm (0.25 inches). The ferrules are independently spring loaded, and are aligned in a receptacle by standard split-sleeve mechanical techniques. While this simplifies the connector design, it also means that the connector must incur the full cost of 2 ceramic ferrules; how this will eventually compare with other SFF ferrule costs has yet to be determined. The connector latch is modeled after the industry standard RJ-45 wall jack, and has found many initial applications in building wiring to wall outlets. The connector is available in both single-mode and multimode versions, which preserve the TIA industry standard color coding on the plug body and the termination cap on the jack. In addition, the FJ connector is available with color coding to identify different networks, applications, areas of the building, or portions of the cable infrastructure, to facilitate network adminstration. Following the category 5 wiring conventions, the FJ housing and plugs may be color coded in black, blue, gray, orange, red, beige, or white. The FJ was also among the first SFF connectors to be specified by a TIA FOCIS document for use with plastic optical fiber; because it employs standard ceramic ferrules, the FJ can accommodate plastic fiber with an outer diameter up to 1 mm. The FJ supports standard duplex jumper cables, couplers, and adapters; while FJ transceivers are not widely available, some development work in this area has been reported.

Another type of SFF connector, developed by NTT to serve both as a standard fiber optic patch cable and as an optical backplane interface, is the multi-termination unibody or MU connector. This connector is also available from various sources under slightly different names; for example, the version manufactured by Sanwa Corporation is called the SMU. As shown in Fig. 5.2.17, the basic MU connector is a simplex design measuring 6.6 mm wide and 4.4 mm high, with a center-to-center spacing of 4.5 mm in duplex or multi-fiber applications. It has been standardized by IEC 61754-6, "Interface standards type MU connector family," in 1997; other standard bodies including JIS and IEEE 1355 Heterogeneous Interconnect (a future bus architecture) have endorsed the MU as well. A backplane version of the MU is available, which measures 13 mm wide and 42 mm high. Its small size is achieved by using a ceramic ferrule 1.25 mm in diameter, roughly half the size of a standard SC connector ferrule. Consequently, a different type of physical contact polishing was developed to accommodate the smaller ferrules. Smaller diameter zirconia split-sleeves have also been developed to support duplex couplers, adapters, and similar jumper cable applications. A self-retention mechanism is employed, similar in design to a miniature push-pull SC latch; indeed, the MU is sometimes referred to as a "mini-SC" connector because of the similarities in look and feel. This is consistent with the intended applications, as it permits blind mating of the connector in a printed circuit board backplane more readily than an RJ-45 style latch. The MU has found some applications as a front panel patch cable on optical switches or multiplexers that have a large amount of fiber connections; these interfaces are expected to be incorporated into optical backplanes on next generation equipment. The plug and jack structure of MU connectors can be easily cascaded to produce multi-fiber parallel

Fig. 5.2.17 MU.

Table 5.2.2 Comparison of SFF connector features.

	LC	MT-RJ	SC-DC™	VF-45[ix]
Fiber spacing	6.25 mm	0.75 mm	0.75 mm	4.5 mm
# of ferrules	2	1	1	0
Ferrule material	Ceramic	Plastic	Plastic	None
Alignment	Bore and ferrule	Pin and ferrule	Rail and ferrule	V-groove
Ferrule size	ϕ 1.25 mm	2.5 mm × 4.4 mm	ϕ2.5 mm	None
Trx opening: (width × height × length)	11.1 mm × 5.7 mm × 14.6 mm	7.2 mm × 5.7 mm × 14 mm	11 mm × 7.5 mm × 12.7 mm	12.1 mm × 8 mm × 21 mm
Fiber cable	duplex	duplex or ribbon	duplex or ribbon	GGP polymer coated
Field term: Plug	pot and polish	pre-polished stub	pre-polished stub	not available
Field term: Socket	plug + coupler	plug + coupler and socket	plug + coupler	cleave and polish socket
Latch	RJ-top2 latch coupled	RJ-top latch	SC push pull	RJ-top latch

interface designs, it has been suggested that future designs using the MU could accommodate hundreds or even thousands of fibers when used in this configuration. Transceivers for the MU interface are not yet widely available, although some development work is under way in this area.

5.2.7 Transceivers

As noted earlier, optical transceivers are widely available from multiple vendors for the **MT-RJ** and LC interfaces, and to a lesser degree for the VF-45 and FJ interfaces. While there is some ongoing development of transceivers for the other SFF interfaces, they are not widely available at this time. The industry trend has been to adopt the **MT-RJ** interface for low data rate (below 1 Gbit/s), multimode applications, and the LC for high data rate (above 1 Gbit/s) applications, both single-mode and multi-mode (mainly using SX transceivers). Transceiver development has been facilitated by ad hoc industry standards or multi-source agreements (MSAs) which govern transceiver package dimensions, electrical interfaces and host board layouts, card bezel design, mechanical specifications (including insertion, extraction, and retention forces), and transceiver labeling. The original MSA for SFF transceivers was supported by 15 companies including Agilent, IBM, Lucent, Siemens/Infineon, Amp/Tyco, and others. It defined a pin through hole device with 2 rows of 5 pins each, using the signal definitions listed in Table 5.2.2.

Recently, a second MSA has been approved by the member companies, which defines a pluggable transceiver that mates with a surface mountable card receptacle. These small form factor pluggable (SFP) transceivers make it possible to change the optical interface at the last step of card manufacturing, or even in the field, to accommodate different connector interfaces or a mix of SX and LX transceivers. This should make it easier to adjust optical interface characteristics on future system designs, in much the same way that the GBIC transceiver did for the SC duplex interface (in fact, the SFP is sometimes known as a "mini-GBIC"). The SFP has twenty signal connections and provides three additional functions in addition to the original 10 SFF signal pins; these new functions include module definition pins which specify a serial ID indicating the type of transceiver function (such as LX vs SX transmitters), a data rate select function (such as 1 Gbit/s vs 2 Gbit/s), and a transmitter fault signal. The signal definitions are provided in Table 5.2.2.

A metal receptacle, sometimes called a cage or a garage, is surface mounted to the printed circuit board to accept the pluggable transceivers. In addition to providing easy replacement and reconfiguration of the transceiver interface, this offers several other advantages. The transceiver is often the only pin through hole component on a modern card design; the SFP cage allows the elimination of extra manufacturing processing steps and potentially reduces cost. By removing the optical components from the soldering process, the SFP should provide improved reliability of the optics, and permit the use of higher soldering temperatures (this may be important as future lead-free solders with higher process temperatures are required by environmental regulations).

5.2.8 SFF comparison

Conventional duplex fiber optic connectors, such as the SC duplex defined by the ANSI Fibre Channel Standard [1], achieve the required alignment tolerances by threading each optical fiber through a precision ceramic ferrule. The ferrules have an outer diameter of 2.5 mm, and the resulting fiber-to-fiber spacing (or pitch) of a duplex connector is approximately 12.5 mm. Because the outer diameter of an optical fiber is only 125 μm, it should be possible to design a significantly smaller optical connector. Smaller connectors with fewer precision parts could dramatically reduce manufacturing costs and have the potential to open up new applications such as fiber to the desktop. Smaller connectors and transceivers would also permit more ports to be added to enterprise servers, fiber optic switches, and communications equipment without increasing the size and cost of these devices[2]. Recently, a new class of small form factor (SFF) fiber optic connectors has been introduced with the goal of reducing the size of a fiber optic connector to one-half that of an SC Duplex connector while maintaining or reducing the cost[4], namely the LC[5], MT-RJ[6], SC-DC[7], andVF-45[8].

Table 5.2.2 gives a comparison of the different features of the four major SFF connectors. A brief description of each connector and its alignment method is given, followed by a discussion of the distinguishing characteristics and their impact on the connector and transceiver.

There are several different design approaches to reducing the dimensions of a fiber optic connector. One approach is to use a single ferrule with multiple fibers; this is the concept behind the SC-DC and MT-RJ connectors. The SC-DC (dual connect) and SC-QC (quad connect) use a standard SC connector body and latching mechanism with an offset key, but a new round plastic ferrule design that incorporates either 2 fibers (750 μm pitch) or 4 fibers (250 μm pitch) in a linear array. Alignment is provided by semicircular grooves in the sides of the SC-DC ferrule, which mate with corresponding ribs in the receptacle. This connector has been used by IBM Global Services as part of the Fiber Quick Connect system; it is currently limited to applications in patch panels and the cable infrastructure. The most radical, and innovative, approach for a smaller connector is to eliminate ferrules altogether; this is the case for the VF-45 connector. In this connector, a pair of optical fibers is aligned using injection-molded thermoplastic v-grooves; the fibers are cantilevered in free space on 4.5mm pitch, and protected by the connector outer body. When plugged into a receptacle, the fibers bend slightly in order to achieve physical contact; better performance is achieved when using optical fibers that have a special strength coating in addition to the outer jacket. The MT-RJ connector uses the same rectangular plastic ferrule concept as the multifiber MTP connector, with 2 fibers on 750 μm pitch and a latching mechanism based on the RJ-45 connector. Alignment in this case is provided by a pair of metal guide pins in the connector, which mate with a corresponding pair of holes in the receptacle; this feature makes the MT-RJ the only small form factor connector with distinct male and female connector ends. A more evolutionary approach to designing SFF connectors involves simply shrinking the standard SC duplex connector, maintaining a single fiber in each of the ceramic ferrules and using conventional alignment techniques applied to the ferrules. The LC connector uses this approach and shrinks the ferrules to 1.25 mm in diameter with a fiber pitch of 6.25 mm (duplex). LC is the only small form factor connector that can be either simplex or duplex.

In a comparative analysis of the different connectors the first feature with striking differences is the fiber pitch. The connectors can be broken up into two classes: small (0.75 mm) and large (>4.5 mm) fiber pitch. The small pitch presents challenges to the transceiver design for cross talk and space transforms for use with opto-electronic devices packaged in standard f 5.4 mm TO cans. Suppliers are also breaking away from the hermetic TO can to non-hermetic silicon optical bench (SiOB) technology to address these packaging concerns. Small pitch connectors such as the MT-RJ also have both fibers in a single ferrule while large pitch connectors such as the LC have the fibers in separate ferrules (or without ferrules in separate v-grooves as in the VF-45). The number of ferrules is significant because the ferrules are precision-made parts (mm tolerances) and are traditionally the most costly part in the connector. Ferrule material as well as the quantity of material is an issue. The plastic ferrule connectors should have a cost advantage over the ceramic ones and this advantage should increase as the manufacturing volume increases. However, the ceramic ferrule technology is more mature and the prices are currently falling due to market pressures. New lower cost glass ceramic ferrules are currently being produced. The plastic ferrules are newer and are currently controlled by a limited number of suppliers, resulting in current prices being as much as twice the cost of two of the standard ceramic ferrules. The plastic ferrules are both made with the same glass-filled, thermoset material, which must be transfer molded. Transfer molding is generally slower than injection molding, but more accurate. Some ferrule suppliers are beginning development of low-cost injection-molded ferrules.

The alignment schemes vary for the connector types and this reflects on the complexity of the transceiver package design. For the LC connector each ferrule requires a precision bore (5 mm/<1 mm tolerance for multimode/single-mode fibers) on the transceiver, which may add cost to the transceiver. The MT-RJ single ferrule connector requires two precision pins (0.25 mm tolerance) placed at a precision (3 mm tolerance) separation, which may eliminate any cost advantage to the transceiver from having one fewer bore than the LC. The alignment process in the assembly of the transceiver optical coupler(s) can also be a source of differentiation for the small- and large-pitched connectors. Connectors with a single ferrule may have a potential cost advantage in that a single optical alignment may be possible. With a small-pitched single ferrule connector such as the MT-RJ, it may be possible to reduce the number of parts in the transceiver by placing both transmitter and receiver in a single package and aligning them simultaneously with a single X,Y,Z,q alignment. Large pitch connectors such as the LC do not have the fibers placed accurately with respect to each other and therefore require two separate packages for transmitter and receiver and two separate alignments (X,Y,Z). It should be recognized that the alignments may in some cases be reduced to passive alignment, which may reduce the cost benefit of having one ferrule.

The transceiver opening is a major concern because of the small form factor transceiver dimensional constraints of 14 mm width, 9.8 mm height, and 31 mm length. All SFF transceivers must comply with an industry consortium multiple source agreement (MSA), which governs the maximum outside dimensions of the transceiver body and the size of a transceiver opening in a card bezel. The MT-RJ connector requires the least volume in the transceiver; use of the space around the connector is questionable so a key dimension is the length into the transceiver. MT-RJ and LC are very close in length of their transceivers (VF-45 can be longer). The single ferrule connector has an added space burden. Part of the transceiver volume may be necessary to space apart the devices from the small fiber pitch to avoid electrical and optical cross talk in silicon optical bench packages and to accommodate devices packaged in conventional TO cans.

Field termination can be an important issue for some applications. Ceramic ferrule connectors such as the LC have standard pot and polish field termination kits so the time and cost to terminate the connectors should not change from standard SC duplex. The single-ferrule MT-RJ field termination is accomplished using a pre-polished fiber stub connector with index matching gel. This approach should offer installation time at least equivalent to the SC duplex, and potentially shorter.

High mechanical accuracy of the ferrule alignment surfaces relative to the fiber position is a necessary condition for a connector that uses a ferrule. Generally, suppliers of the MT-RJ have begun to offer slight amounts of fiber protrusion in order to ensure physical contact between two ferrules in a duplex coupler or between the ferrule and transceiver. Some designs of the MT-RJ duplex coupler have encountered a stubbing problem when the connector pins did not properly align with holes in the receptacle. Early coupler designs attempted to address this problem by including alignment tabs or cross-hairs inside the coupler, designed to provide coarse alignment of the ferrules; subsequent design refinements of the MT-RJ connector have shown that this feature may not be necessary to produce adequate performance. The correct amount of protrusion and the acceptable effects of environment and plugging on these protrusions remains an ongoing development item for the industry.

Although the SFF connectors are tested for resistance to both axial and off-axis pull forces, the computer equipment that uses these connectors must still provide some form of strain relief to prevent cable pull forces from being transmitted back to the optical transceivers. This is particularly true for connectors such as the LC and MT-RJ, which only provide latching mechanisms on one side of the connector body; if the cable assembly is pulled in a preferential direction opposite to the latch, excessive losses may result (SFF connectors are generally more susceptible to pull forces because of their reduced size). Some current designs for fiber cable strain relief are based on serpentine (S-shaped) grooves in a plastic housing, which are capable of absorbing very high forces from fiber cables. However, these current designs are not well suited to more than a dozen fibers at a time; the greatly increased density provided by SFF packaging may require a new form of optical cable strain relief. It is desirable for the new strain relief to also be more user-friendly; the current design requires fiber to be manually placed around grooves and tabs, and is rather cumbersome. As a result, it is not used in real-world applications as widely as might be desired. Furthermore, the older strain relief was designed for a larger cable outer jacket and would not accommodate SFF cables as well. In new computer environments that employ many SFF transceivers on a single-channel card, it is particularly desirable to avoid any failure mechanisms due to mechanical damage on the connectors, as this requires replacement of the entire card and disruption of a large number of channels. Although there are industry standard FOTPs that address many of the connector functional requirements, some important application tests do not have a corresponding FOTP procedure (for example, random connection loss, insertion/withdrawal force testing, temperature cycle, and off axis pull testing). In addition, comparison of different connector types is not always possible, since some of the FOTPs are not applicable to connectors with male and female ends (such as MT-RJ) or without a duplex coupler (such as VF-45). These additional tests simulate common stresses that are found when connectors are used in the field under raised floors, in racks of equipement, and even on the wall outlets. The limits of these tests are set by the application specifications; performance that is adequate for fiber-to-the-desktop, for example, may not be adequate for a mainframe or supercomputer environment.

The intent of many of the SFF connectors was to replace the SC Duplex and thus to have the same or better quality and performance as the SC Duplex but half the size. Thus it is reasonable to use the SC Duplex as a benchmark against which to measure the new SFF connectors. Recent independent testing has shown that the SFF interfaces are still evolving, and performance is likely to continue improving over the coming years; however, at present some versions of the LC and MT-RJ have demonstrated equivalent performance to SC duplex connectors. As the assembly procedure has matured and become more standardized, the variability across suppliers has decreased, at least for the larger or first-tier suppliers. The improved performance of LC connectors, for example, may be partially attributed to a novel polishing technique for the smaller ceramic ferrules, the use of low eccentricity standard bulk fiber, and a new manufacturing process of "tuning" the ferrules by rotating them at small angular increments to obtain the orientation with minimal eccentricity prior to assembling the ferrule into the cable connector. In addition, fine tuning of the MT-RJ design has taken place, including changes in the connector body outer dimensions which facilitate better engagement with the RJ-45 latch. A small chamfer has been added around the ferrule alignment holes; this avoids chipping of the hole edge as the guide pins are inserted, and makes for a more re-peatable connection with reduced plug force. Since the MT-RJ employs a novel rectangular ferrule, polishing techniques also have needed to be refined. A conventional ferrule polishing machine intended for round end-faces and symmetrical, cylindrical ferrules may not provide uniform polish on the MT-RJ; in fact, it can polish the endface with an angular bias along one axis. It was originally thought that this would not affect connector performance, since the fiber pitch is so small that only a relatively small area of the ferrule needed to have a high degree of polish; however, it has subsequently been shown that polishing is critical for the ferrule area near the alignment pins as well. Improper handling can result in a ferrule endface polished with multiple compound angles, which interferes with the guide pin alignment and resulting optical performance. More recent polishing jigs developed specifically for the MT-RJ have reduced or eliminated this problem.

Axial pull requirements are also a very interesting area. Some SFF connectors such as the SC-DC and VF-45 connectors are designed to disengage from their receptacles under an applied pull force above 45 N; this is to prevent damage to the connector, and induce an obvious optical failure in the link (it is problematic to diagnose a link failure if the connector remains engaged but exhibits high optical losses under stress). Another design approach used by LC and MT-RJ is to retain the connector in the receptacle under much higher forces without excessive optical loss. The magnitude of the pull force is a requirement that will vary depending on the application; however, it is essential for all the applications that the cable unplug or mechanically fail before the loss increases. End users should be cautioned that despite improvements in standardized manufacturing processes, there is still a broad range of connector performance currently available from second- and third-tier suppliers. Differences in assembly procedures result in different performance, thus representing low cost/performance options arising from connector assembly procedures. The connectors continue to undergo minor design revisions to improve performance and tighten up the assembly procedures across multiple suppliers. With properly chosen suppliers, however, some SFF connectors have matured to the point where they can be considered adequate replacements for SC duplex.

References

1. ANSI Fibre Channel - physical and signaling interface (FC-PH) X3.230 rev. 4.3 (1994).
2. J. Trewhella, C. DeCusatis, and J. Fox. "Performance comparison of small form factor fiber optic connectors." *IEEE Transactions on Advanced Packaging* **23**(2): 188–196 (2000).
3. C. DeCusatis, "Small form factor fiber optics for enterprise computing applications." IBM Journ. Research and Development (To be publ).
4. C. Schwantes, "Small-form factors herald the next generation of optical components." *Lightwave* pp. 65–68 (1998).
5. M. A., Shahid, et al. "Small and Efficient Connector System." *Proc. 49th Electronic Components and Technology Conference*, San Diego, Calif., June 1999 and TIA FOICS 10.
6. Y. Tamaki, et al. "Compact and Durable MT-RJ Connector." *Proc. 49th Elec tronic Components and Technology Conference*, San Diego, Calif., June 1999 and TIA FOICS 12.
7. K. Wagner, "SC-DC/SC-QC connector." *Optical Engineering* **37**(12):3129 and TIA FOICS 11.
8. R. Selli, et al. "A novel v-groove based interconnect technology." *Optical Engineering* **37**(12):3134 and TIA FOICS 7.

Chapter 5.3

5.3

Specialty fiber optic cables

DeCusatis

5.3.1 Introduction

In this chapter, we will describe different types of fiber optic cable that have been developed for a wide range of applications, including enhanced distance, optical amplification and attenuation, dispersion and polarization management, and other areas. Future optical fiber designs, still several years away from commercial use, will also be discussed. While it isn't possible to provide a comprehensive list of every type of specialty fiber currently available, we will describe the most common types and their uses. To begin, we need to review the manufacturing process for the more common graded index single-mode and multimode fiber types, which have been discussed in previous chapters.

5.3.2 Fabrication of conventional fiber cables

The fabrication of conventional low loss silica fiber optic cables [1] involves precision control of the glass composition to both control impurities and to ensure accurate refractive index profiles. High purity materials and well-controlled tolerances are important to the manufacturing process. The desired refractive index profile is first fabricated in a large glass preform, typically several centimeters in diameter and about a meter long, which maintains the relative dimensions and doping profiles for the core and cladding. The glass is selectively doped to provide the desired index gradients. This preform, or boule, is later heated in an electric resistance furnace until it reaches its melting point over the entire cross-section; it can take up to an hour to establish uniform heating of the preform. Thin glass fibers are then drawn upward from the preform in a drawing tower, as illustrated in Fig. 5.3.1a; the fiber cools and solidifies very quickly, within a few centimeters of the furnace. The pulling force controls the rate of fiber production, and hence the fiber diameter, which is monitored by a laser interferometer. Bare glass fiber is then drawn through a vat of polymer and receives a protective coating extruded over it to a diameter of about 250 microns; this must be done as soon as possible after drawing to avoid water contamination in the fiber. Finally, the fibers are spooled evenly onto a mandrel about 20 cm in diameter, to avoid microbending. If the preform is uniformly heated (and therefore has a uniform viscosity) then the cross-section and index profile of the drawn fiber will be exactly the same as in the preform. In this manner, fibers with very complex refractive index profiles can be produced. In order to facilitate easy manufacturing of the glass fibers, it is easier to use fairly large preforms and doping methods with a fast deposition rate. The primary technology used in fiber preform manufacturing is chemical vapor deposition (CVD), in which submicron silica particles are produced through one or both of the following chemical reactions, carried out at temperatures of around 1800–2000 C:

$$\begin{aligned} SiCl_4 + O_2 &\rightarrow SiO_2 + 2Cl_2 \\ SiCl_4 + 2H_2O_2 &\rightarrow SiO_2 + HCL \end{aligned} \qquad (5.3.1)$$

Fiber Optic Data Communication; ISBN: 9780122078927

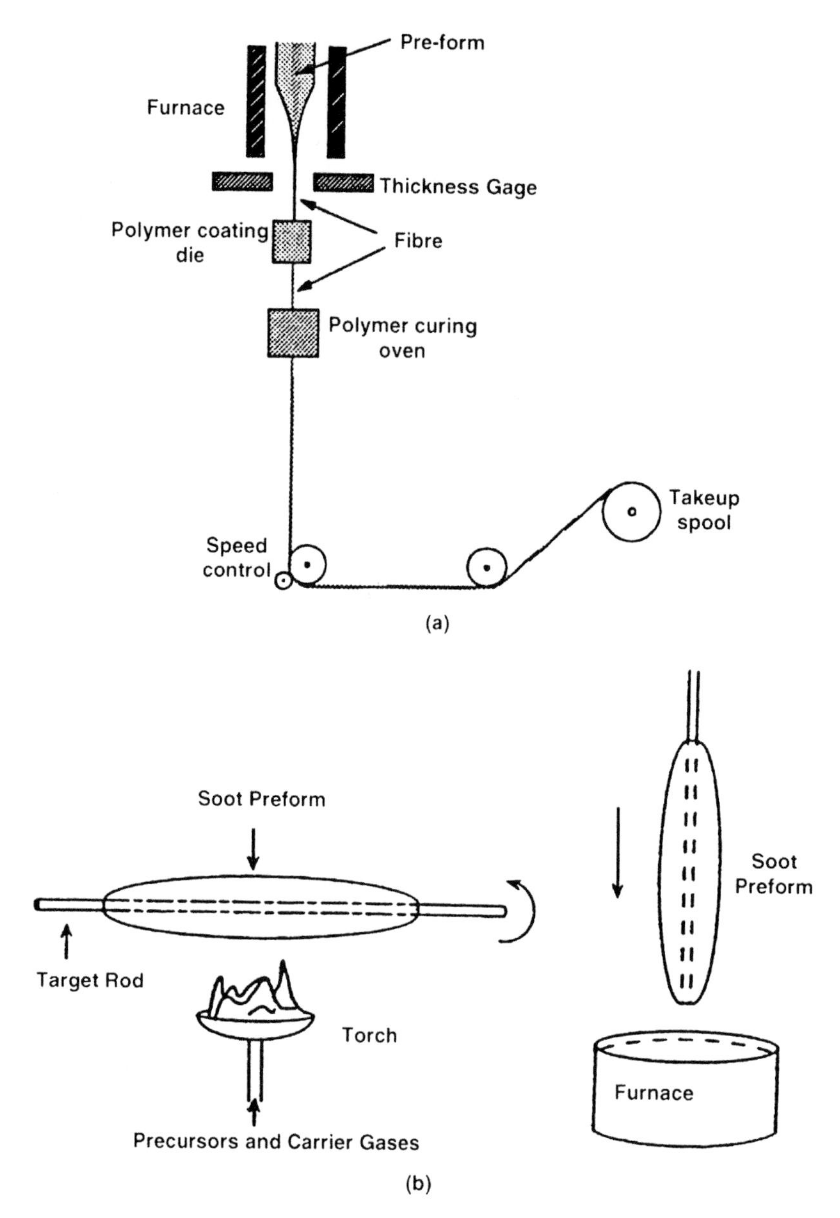

Fig. 5.3.1 (a) Typical optical fiber drawing tower. (b) Preform fabrication with a torch and carrier gasses.

This deposition produces a high purity silica soot which is then sintered to form optical quality glass. There are two basic manufacturing techniques commonly used. In the so-called "inside process," a rotating silica substrate tube is subjected to an internal flow of reactive gasses. There are two variations on this approach, modified chemical vapor deposition (MCVD) and plasma-assisted chemical vapor deposition (PCVD). In both cases, layers of material are successively deposited, controlling the composition at each step, in order to reach the desired refractive index. MCVD accomplishes this deposition by application of a heat source, such as a torch, over a small area on the outside of the silica tube. This heat is necessary for sintering the deposited SiO_2 and for the oxidation reactions shown in the equation. Submicron particles are deposited at the leading edge of the heat source; as the heat moves over these particles, they are sintered into a layered, glassy deposit. This requires fairly precise control over the temperature gradients in the tube, but has the advantage of accomplishing the sintering and deposition in one step. MCVD accounts for a large portion of the fiber produced today, especially in Europe and America. By contrast, the PCVD process provides the necessary energy for the chemical reactions by direct RF excitation of a microwave-generated plasma. Because the microwave field can be moved very quickly along the tube (since it heats the plasma directly, not the silica tube itself) it is possible to traverse the tube thousands of times and deposit very thin layers at each pass, which makes for very precise control of the preform index profile. A separate step is then required for sintering of the glass. In both cases, the preforms require a final heating to around 2150 °C in a furnace to collapse the preform into a state from which glass is ready to be drawn. All inside vapor deposition (IVD) processes require a tube to be used as a preform; minor flaws in the tube can induce corresponding dips and peaks in the fiber index profile. This can be a problem in wavelength multiplexing systems, which may require only a few parts per million tolerance on the optical components in a link. This also affects the fiber's suitability for transmission of high data rates; the use of legacy fiber for Gigabit Ethernet links requires careful evaluation of these distortions in the core of IVD fibers.

In the so-called "outside process," a rotating, thin cylindrical mandrel is used as the substrate for subsequent CVD; the mandrel is then removed before the preform is sintered. An external torch fed by carrier gasses is used to supply the chemical components for the reaction, as well as to provide the necessary heat for the reaction to occur. Two outside processes have been widely used — the outside vapor deposition (OVD) and the vapor axial deposition (VAD) methods (today, Corning exclusively uses the OVD process for standard commercial fiber). Much of the control in these techniques lies in the construction of the torch. For example, OVD is basically a flame hydrolysis process in which the torch consists of discrete holes in a pattern of concentric rings which each provide a different constitutent element for the chemical reactions; a stream of oxygen is used between successive rings to act as a shield between the different chemicals. The torch is moved back and forth along the rotating preform and the dopants in the flame are dynamically controlled to generate the desired index profile, as illustrated in Fig. 5.3.1b. OVD is not widely used any more, although it was among the first processes developed, because of its high cost (preforms are limited in size, as it is a batch process) and technical problems such as difficulty in removing all the water (OH groups) from the formed glass, a tendency for the fibers to have a large depression in refractive index near the core.

The VAD process is similar in concept, using a set of concentric annular apertures in the torch; in this case, the preform is pulled slowly across the stationary torch. It was found that by mixing $GeCL_4$ as a dopant into the $SiCl_4$-O_2 feed, the proportion of germania (GeO_2) deposited with the silica varies with the temperature of the flame; if a wide flame is used, the temperature gradient produces a graded portion of germania deposit. This process is still used, particularly in Japan, for commercial fiber production.

Silica glass has absorption bands in both the ultraviolet (UV) and mid infrared (IR) wavelength ranges, which provides a fundamental limit to the attenuation that can be achieved. This occurs despite the fact that the Raleigh scattering contribution decreases inversly as the fourth power of the wavelength, and the UV Urbach absorption edge decreases even faster with increasing wavelength. The infrared absorption increases at long wavelengths, becoming dominant at wavelengths greater than about 1.6 microns, which results in the minimal loss for wavelengths around 1.55 microns. Note that for many years, optical fiber attenuation was limited by a strong hydroxide (OH) absorption band near wavelengths of 1.4 microns; this has been steadily reduced over time with improved fabrication methods, until the loss minimum near 1.55 microns was brought close to the Raleigh limit. The best dopants for altering the refractive index of silica glass are those that provide a weak index change without inducing a large shift in the ultraviolet (UV) absorption edge. These include GeO_2 and P_2O_5 (which increase the refractive index in the fiber core), B_2O_3 and SiF_4 (which decrease the refractive index in the cladding). The chemical reactions for the general process often use high vapor pressure liquids such as $GeCl_4$, $POCl_3$, or SIF_4 along with oxygen as a carrier gas; these reactions are well documented [2–4]. It is more difficult to introduce more exotic dopants, such as the rare earth elements used in optical fiber amplifiers, and there is not a single widely used technique at this time. There are also other preform fabrication and fiber drawing techniques that are not generally used for telecommunication grade optical fiber, but which can be important for other material systems and applications. These include bulk casting of the preform and the non-CVD method of "rod and tube" casting (in which the core and cladding are cast separately and combined in a final melting step). There are also preform-free drawing techniques such as the "double crucible" method, in which the core and cladding are formed separately in a pair of platinum crucibles and combined in the drawing process itself. This method was important in the past, but is not widely used today because it does not provide the same precision control as the drawing tower process.

There has been a great deal of research into fiber materials with better transparency in the infrared. Although none of these materials has yet proven to be a serious competitor with doped silica, they may prove useful for other applications that do not have the strict requirements of telecommunication systems, such as CO_2 laser transmission, medical applications, or remote sensing and imaging [5]. For example, bulk infrared optics and fibers can be made from sulfide, selenide, and telluride glasses; collectively known as chalcogenide fibers, they exhibit transmission loss on the order of 1 dB/meter in the wavelength range 5–7 microns [6]. Another example is the heavy-metal fluroide fibers [7], which hold some interest for communication systems because their theoretical limit for Rayleigh scattering is much lower than for silica (this is due to a high-energy ultraviolet absorption edge, and better infrared transparency). However, excess absorption has proven difficult to reduce, and current state-of-the-art fluroide fibers continue to exhibit losses near 1 dB/km, well above the Rayleigh limit. Fabrication of short lengths of fluroide fiber has been somewhat more successful in transmitting longer wavelengths, with reported losses as low as 0.025 dB/km at wavelengths of 2.55 microns [8]. The residual loss mechanisms in longer fibers remain the subject of ongoing research, and are thought to be due to extrinsic impurities or defects such as platinum particles from the fabrication crucibles, bubbles in the core preform and at the core-cladding boundry, and fluoride microcrystals.

Many types of optical fiber are available for different environments, including undersea cables, outdoors (with integrated strength members to facilitate hanging cables from telephone poles), and indoors. The properties of optical fiber cables are governed by various industry standards including G.652 (with a maximum bandwidth of 50 GHz) and G.655 (with a maximum bandwidth of several THz). Fiber jackets are typically rated as either riser, plenum, low smoke, low/zero halogen, or a dual-rated combination of the above; the IEC 1034 specification provides the dual-rated jacket specifications, while the zero halogen jackets are typically free of chlorine, fluroine, and bromides. New cable jacket types are also emerging from the national fire and electrical codes, such as the "limited combustibility" designation for some types of plenum cables. Some cable types are rated for use under a computer room raised floor, others for installation via air blowers in plenum ducts. Many types of multifiber connectors, structured cabling systems, and cable pullers, conduits, and patch panels are available. Special designs of conventional optical connectors are available, such as the so-called "elite MT" connectors, a customized version of the 12 fiber MT ferrule that is sorted to guarantee less than 0.35 dB maximum loss per fiber. New types of multifiber connectors are also being developed using two-dimensional ferrules to accommodate 24 or more fibers in a single plugging operation. In addition, many companies manufacture custom optical fibers to a user-specified refractive index profile, doping, core or cladding geometry, numerical aperture, cutoff wavelength, or other characteristics. Short sections of optical fiber may be packaged as loopbacks or wrap plugs for transceiver testing, while long haul spools may require special shielding for mechanical or structural reasons. Specialty fibers with coatings to increase their mechanical performance under bending are used today in small form factor VF-45 style connectors; there are many types of fiber optic connectors available, including so-called "no polish" field installable connectors intended for quick and low-cost installation by untrained personnel, physical and non-physical contact connectors, flat ferrules and angle polished ferrules for low back reflectance, connectors with built-in variable attenuators or mode scramblers, metallized fibers that are soldered into place, and many, many other variants. While it isn't practical to give a comprehensive list of every specialty fiber type in this chapter, we will provide an overview of some major fiber types that are commonly encountered, as well as a synopsis of emerging fiber types that may become important in the future.

5.3.2.1 Optical cable, couplers, and splitters

There are many different types of fiber optic cable; as shown in Fig. 5.3.2, it is possible to package multiple fibers into a single unibody cable or into a ribbon or zipcord structure. Bundles of fibers whose ends are bound together, ground, and polished can form flexible light pipes. Of course, it is possible to bundle the fibers in such a way that there is no fixed relationship between the location of an input fiber and an output fiber; the principle purpose of such structures is to conduct light from one location to another, for illumination as an example; these are sometimes referred to as incoherent bundles [9], although they have little to do with optical coherence theory. A more interesting case is when the fibers are carefully arranged so that they occupy the same relative positions at both ends of the bundle; such bundles are said to be coherent. A coherent bundle of single-mode fiber is capable of conducting a high-quality image even when the bundle is made highly flexible; such fiber arrays have many applications in remote vision systems, and are used in fiber optic endoscopes for medical applications. Not all fiber arrays are made flexible; fused, rigid bundles or mosaics can be used to replace low-resolution sheet glass in cathode ray tubes. Mosaics consisting of hundreds to millions of individual fibers with their claddings fused together have mechanical properties very much like homogeneous glass [10]. Another common application of mosaics is as a field flattener.

If the image formed by a lens system falls on a curved surface, it is often desirable to reshape it into a plane, for example to match a photographic film plate. A mosaic can be ground and polished on one end surface to correspond with the contours of the image, and on the other surface to match the configuration of the detector. Similarly, a sheet of fused tapered fibers can be used to either magnify an image or miniaturize an image, depending on whether the light enters the smaller or larger end of the fibers.

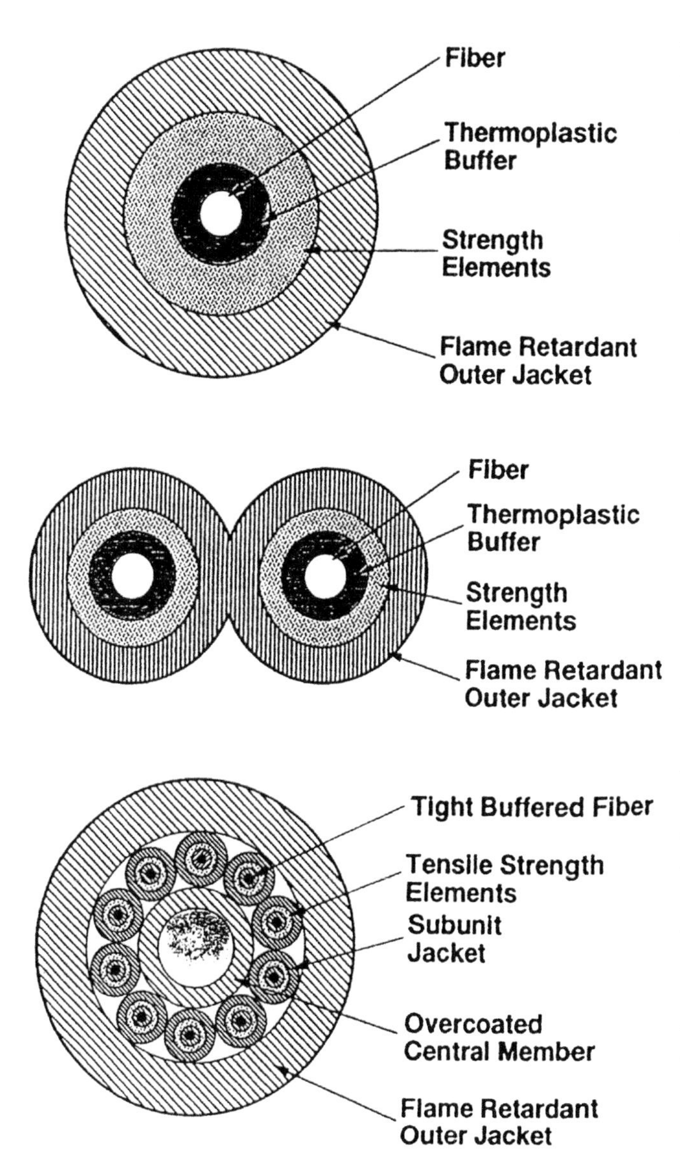

Fig. 5.3.2 Different types of fiber optic cable cross-sections.

Many simple devices such as fiber optic splitters, couplers, and combiners have been manufactured; the most common techniques include fiber tapering and fusion splicing [11–14], etching [15], and polishing [16–18]. Other fabrication techniques can also be used, including micro-optics and integrated optical components; however, optical fiber devices are particularly useful because they can be inserted into existing networks as just another piece of cable. One of the most common devices is a tapered fiber optic power splitter, often implemented in single-mode fiber [19]. In this process, two glass fibers with their protective jackets removed are brought close together and parallel to each other, then fused and stretched using a torch or similar heat source. Light that is initially launched into only one fiber will be partially coupled into the adjacent fiber as it propagates through the tapered region. Light propagating in the single-mode fiber is not confined to the core but extends into the surrounding cladding. In the case of a fiber taper it has been shown [20] that light propagating through the input fiber core is initially transferred to the cladding interface as it enters the tapered region, then to the core-cladding mode of the adjacent fiber. The light transfers back to the core modes as it exits the tapered region. This is known as a cladding mode coupling device. Light that is transferred to a higher order mode of the core-cladding structure is readily stripped away by the higher refractive index of the fiber coating, resulting in excess attenuation. The simplest case of light coupling from the cladding of one fiber into another through a fused taper can be described to a good approximation by the scalar wave equation and first-order perturbation theory [21]; if light is propagating along the z axis, then the exchange of optical power, P, is given by

$$P = \sin^2(kz), \tag{5.3.2}$$

where z is propagation distance and k is a complex function of the optical wavelength, refractive indices of the core and cladding, material properties, and the overlap distance between the two fibers [22]. Although this is only an approximation and neglects higher order terms, it does reflect the sinusoidal dependence of coupled power on wavelength and the dependence of power transfer on cladding diameter and other effects. Tapered couplers can be used to separate wavelengths using this dependence; by proper choice of the device length and taper ratio, two wavelengths can be made to emerge from two different output ports. Some applications include filters for wavelength division multiplexing (WDM) systems, or multiplexing signal and pump beams in an erbium-doped fiber amplifier. In some cases, such as optical power splitters, it is more desirable to remove the dependence of coupled power on wavelength; acromatic couplers can be fabricated by using two fibers with different propagation constants. These are known as dissimilar fibers; in most cases, fibers are made dissimilar by changing their cladding diameters or cladding indices. In this case, the preceding equation for coupled power must be modified and the power vs. distance is not simply sinusoidal, but becomes much more complex [22].

Other approaches are also possible, such as tapering the device such that the modes expand well beyond the cladding boundaries [23], or encapsulating the fibers in a third material with a different refractive index [24–25]. Often, it is desirable to taper multiple fibers together so that an input signal is split between many output fibers. Typically, a single input is split into 2^M outputs (i.e., 2, 4, 8, or 16), where the configuration of fibers in the tapered region affects the output power distribution; care must be taken to achieve a uniform optical power distribution among the output fibers [26]. The optical power coupled from one fiber into another can also be changed by bending the tapered device at its midpoint; this frustrates coupled power transfer. For example, displacing one end of a 1 cm long taper by only 1 mm can change the coupled power by over 30 dB [11]. Applications for this effect include variable optical attenuators and optical switches.

5.3.3 Fiber transport services

Given the many different types of fiber optic data links in a modern enterprise data center, the design of an optical cable infrastructure that will accommodate both current and future needs has become increasingly complicated. For example, IBM Site and Connectivity Services has developed structured cabling systems to support multigigabit cable plants. In this section, we briefly describe several recent innovations in fiber optic cable and connector technology for the IBM structured cabling solution, known as Fiber Transport Services (FTS) or Fiber Quick Connect (FQC).

A central concept of FTS is the use of multifiber trunks, rather than collections of 2-fiber jumper cables, to interconnect the various elements of a large data center [27–30]. FTS provides up to 144 fibers in a common trunk, which greatly simplifies cable management and reduces installation time. Cable congestion has become a significant problem in large data centers, with up to 256 ESCON channels on a large director or host processor. With the introduction of smaller, air-cooled CMOS-based processors and the extended distance provided by optical fiber attachments, it is increasingly common for data processing equipment to be rearranged and moved to different locations, sometimes on a daily basis. It can be time consuming to reroute 256 individual jumper cables without making any connection errors or accidentally damaging the cables. To relieve this problem, this year FTS and S/390 have introduced the Fiber Quick Connect system for multifiber trunks. The trunks are terminated with a special 12-fiber optical connector known as a Multifiber Termination Push-on (MTP) connector, as shown in Fig. 5.3.3. Each MTP contains 12 fibers or 6 duplex channels in a connector smaller than most duplex connections in use today (barely 0.5 inches wide). In this way, a 72-fiber trunk cable can be terminated with 6 MTP connectors; relocating a 256-channel ESCON director now requires only re-plugging 43 connections. Trunk cables terminated with multiple MTP connectors are available in 4 versions, either 12-fiber/6 channels, 36-fiber/18 channels, 72-fiber/36 channels, or 144-fiber/72 channels. Optical alignment is facilitated by a pair of metal guide pins in the ferrule of a male MTP connector, which mate with corresponding holes in the female MTP connector. Under the covers of a director or enterprise host processor, the MTP connectors attach to a coupler bracket (similar to a miniature patch panel); from there, a cable harness fans out each MTP into 6 duplex connectors which mate with the fiber optic transceivers as shown in Fig. 5.3.4. Since the qualification of the cable harness, under the covers patch panel, and trunk cable strain relief for FTS are all done in collaboration with the mainframe server development organization, the FTS solution functions as an integral part of the applications.

At the other end of the FTS trunk, individual fiber channels are fanned out at a patch panel or main distribution facility (MDF), where duplex fiber connectors are used to re-configure individual channels to different destinations. These fanouts are available for different fiber optic connector types, although ESCON and Subscriber Connection (SC) duplex are most common for multimode and SC duplex for single-mode. Fanning out the duplex fiber connections at an MDF also offers the advantage of being able to arrange the MDF connections in consecutive order of the channel identifiers on the host machine, greatly simplifying link reconfigurations. As the

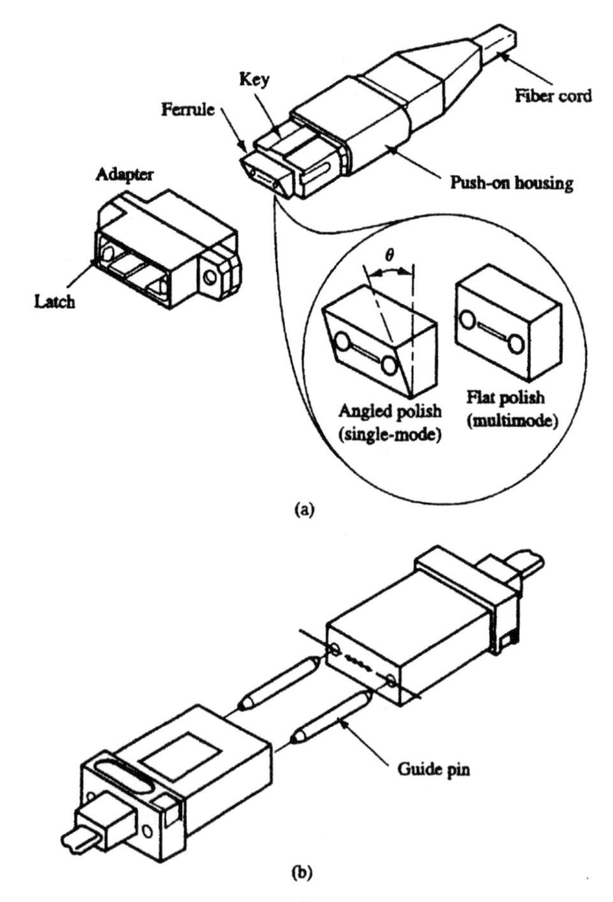

Fig. 5.3.3 (a) MTP connector with flat or angle polished ferrule; (b) detail of alignment pins.

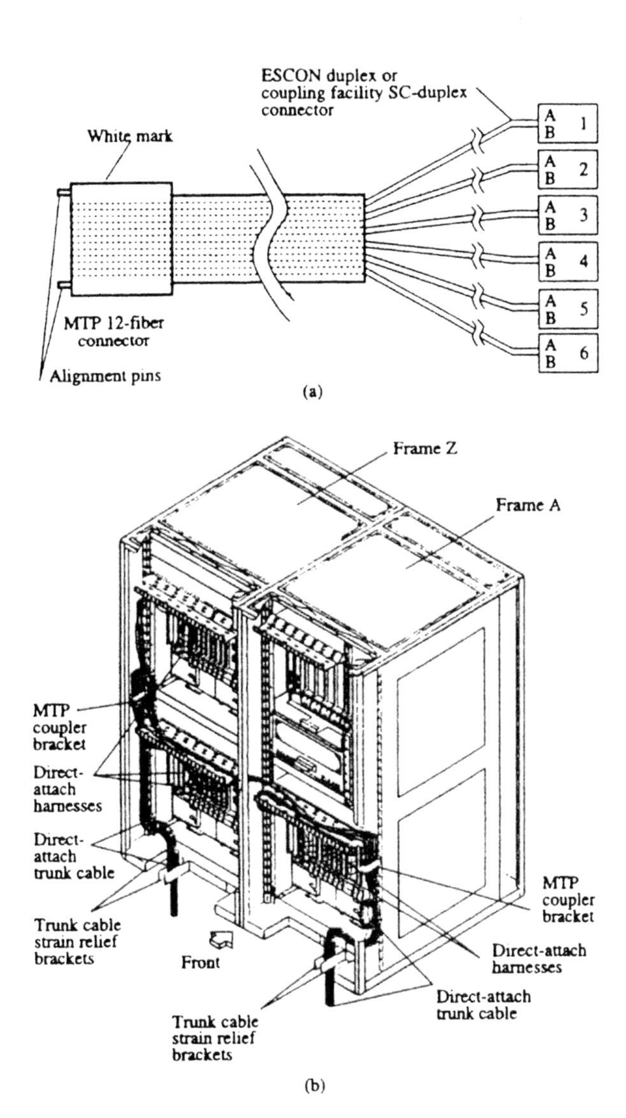

Fig. 5.3.4 Cable harness using MTP connector fanout.

size of the servers has been reduced and the number of channels has increased, the size of the MDF soon became a limiting factor in many installations. In order to keep the MDF from occupying more floor space than the processors, a more dense optical connector technology was required for the Fiber Quick Connect system. To meet this need, IBM Global Services has adopted a new small form factor fiber optic connector as the preferred interconnect for multimode patch panels, the SC-DC. Structured cabling solutions similar to this are available from other companies as well; they may include overhead or underfloor cable trays and raceways, as well as cabinets or rack-mounted enclosures (standard units are compatible with either 19- or 23-inch-wide equipment racks, with heights between 1 and 7 U[1] tall).

5.3.3.1 Attenuated cables for WDM and cable TV

Some applications require optical attenuators to control or limit the optical power on a fiber link. One example is wavelength division multiplexing (WDM) equipment, which uses fixed attenuators to allow a common transceiver to intemperate with many different physical layers. Another is the fiber links in the cable television industry.

Fixed attenuators can be expensive, and must be incorporated into the design of the cable system; this means that duplex cables cannot be used if attenuation is required in only one path of a duplex fiber link. Separating duplex connectors defeats the keying that prevents the connector from being improperly inserted into a receptacle or transceiver. The attenuators also provide an extra connection point in the link, which must be cleaned and may be susceptible to mechanical vibration that will tend to dislodge connectors in data communication products. Instead of using fixed, pluggable attenuators, it is possible to manufacture in-line optical attenuators as part of the fiber optic cable assembly. Several approaches can be used. For multimode attenuators, a short piece of single-mode fiber can be spliced into the cable; by controlling the alignment between the single-mode fiber stub and the multimode cable on either side, as well as the length of the stub, various levels of attenuation can be achieved. The stub may be actively aligned during cable maufacturing, then protected by an external sheath or package to protect it from mechanical shock, vibration, and cable flexing. Similar effects can be achieved by deliberate misalignment of a multimode or single-mode fiber fusion splice. In most cases, arbitrary attenuation values from 0.5 dB to over 20 dB can be realized with a tolerance of less than 0.5 dB. The resulting attenuated cables are quite robust, and in many cases they serve a dual purpose, since the application would have required jumper cables anyway to adapt to different styles of optical connectors or connect with subtended equipment.

For other applications in which optical power must be controlled, specialty fibers are available with high attenuation that is flat over a certain spectral region. With so much design effort directed toward reducing the fiber attenuation to improve link budgets and distances, it can be easy to forget that optical fiber can just as easily be designed for high attenuation. Using the same precision-controlled manufacturing techniques that consistently yield low loss fiber, it is possible to dope the fiber in such a way that very consistent, high attenuation is provided over a wide range of operating wavelengths. This can be

[1] The Electronics Industry Association (EIA) defines a standard height of 1 U as equivalent to 1.75 inches for a data center equipment rack.

an advantage in designing attenuators for WDM systems. As with the offset spliced cable attenuators, these fibers tend to be more reliable and robust than conventional airgap attenuators. They have the additional advantage that a controlled amount of attenuation can be selected by simply cutting and splicing a desired length of the fiber. Short sections of high attenuation fiber are also being integrated in some types of fiber optic components, and are finding applications in optical test and measurement systems. Typical fibers are available with attenuations ranging from 0.5 dB/meter to 30 dB/meter in 0.5 dB increments, or from 0.25 dB/cm to 25 dB/cm in 0.25 dB increments.

5.3.3.2 Encapsulated fiber and flex circuits

Recently, there has been increasing interest in using optical interconnections within the backplanes of computer systems, or for connections between equipment frames or racks. Various types of embedded surface waveguides have been proposed, which could be patterned using standard photolithographic techniques to produce an optical wiring plane on a standard multilayer printed circuit board or within a multichip module. However, many of these proposals require special materials or handling procedures, and it is difficult to make waveguide materials capable of withstanding high processing temperatures involved in circuit board rework without becoming opaque. Consequently, there have been several proposals and commercial product offerings dealing with embedding optical fibers into printed circuit boards. One approach is to embed conventional fibers into a flexible polymer laminate; in this way, optical circuit boards can be designed using existing computer drafting techniques, with the fibers being treated as if they were wires on a separate circuit board plane. The fibers can be terminated with standard optical connectors, or special blind-mateable multifiber backplane connectors, which can co-exist with copper interconnects. This has the advantage of reducing the size and complexity of optical interconnect designs, making them less susceptible to mechanical or thermal degradation, as well as allowing implementation of time domain filters and optical delay lines. One example is the OptiFlex[2] material developed by Lucent; acrylate or polyimide-coated fibers can be embedded in a 0.5 mm-thick polymer up to 75 square cm in area [31]; the resulting material sheet can withstand operating temperatures from −20 to +85 C, and accommodates all standard types of single-mode and multimode fibers.

5.3.3.3 Next-generation multimode fiber

Conventional datacom links use single-mode fiber for long-distance, highspeed links and multimode fiber for shorter links. The recent deployment of short wave (780–850 nm) laser transceivers has made it possible to use multimode fiber at data rates of 1.25 Gbit/s at distances of a few hundred meters. Because the cost of short wave transceivers is presently lower than for long wave transceivers, there remains some question as to the preferred fiber to install for some applications, and the best mixture of 62.5 micron and 50 micron multimode fiber. The IEEE has recently recommended using 62.5micron multimode fiber in building backbones for distances up to 100 meters, and 50 micron fiber for distances between 100 and 300 meters. Care must be taken not to mix different types of multimode fiber in the same cable plant, as the resulting mismatch in core size and numerical aperture creates high losses. This can make it difficult to administer a mixed cable plant, as there is no industry standard connector keying to prevent misplugging different types of multimode fiber into the wrong location. In general, 50 micron fiber has been widely deployed in Europe and Japan, while North America has primarily used 62.5 micron multimode fiber.

Because there is such a large amount of legacy multimode fiber installed, new industry standards have attempted to accommodate multimode fiber even at higher data rates. Much of the interest in 50 micron fiber has been a result of its higher bandwidth and longer distances that can be achieved for shortwave laser links. While the idea of backward compatibility works reasonably well up to 1 Gbit/s (distances of a few hundred meters can be achieved), it begins to break down at higher data rates when the achievable distance is reduced even further. Designing a future-proof cable infrastructure under these conditions becomes increasingly difficult; at some point, new fiber needs to replace the legacy multimode fiber. An alternative to single-mode is the emerging "next-generation" multimode fiber, a 50 micron fiber optimized for 850 nm transmission, which can achieve distances up to at least 300 meters at 10 Gbit/s data rates. This would allow less expensive VCSEL lasers to support 10 Gigabit Ethernet, Fibre Channel, and telecom data rates, rather than more expensive long-wave lasers over single-mode fiber. The cost tradeoff at a system level is less clear at this point, but this concept has been jointly presented to various standards bodies by Corning and Lucent. An example is the Systimax LazrSPEED[3] fiber recently introduced by Lucent; backward compatible with existing multimode

[2] OptiFlex is a trademark of Lucent Corporation.
[3] Systimax and LazrSPEED are trademarks of Lucent Corporation.

systems and requiring no special installation tools or skills, this fiber uses a green jacket to distinguish it from existing multimode (orange), single-mode (yellow), and dispersion-managed (purple) fiber cables. Attenuation is about 3.5 dB/km at 850 nm and 1.5 dB/km at 1300 nm; bandwidth is 2200 MHz-km at 850 nm (500 MHz-km overfilled) and 500 MHz-km at 1300 nm (no change when overfilled). Another example is the Corning InfiniCore[4] fiber; the CL 1000 line consists of 62.5 micron fiber made with an outside vapor deposition process that achieves 500 meter distances at 850 nm and 1 km at 1300 nm. Similarly, the CL 2000 line of 50 micron fiber supports 600 meter distance at 850 nm, and 2 km at 1300 nm. Additional details on modal dispersion in multimode fiber can be found in the Telecommunication Industry Association (TIA) task group on modal dependencies of bandwidth (TIA FO-2.2).

5.3.3.4 Optical mode conditioners

Because of the bandwidth limitations of multimode optical fiber, future multi-gigabit fiber optic interconnects will be based on single-mode fiber cables. For this reason, most new fiber installations include at least some single-mode fiber in the cable infrastructure. However, many applications continue to use multimode fiber extensively; a recent survey of building premise cable installers reported that most LAN infrastructures currently installed are composed of about 90% multimode fiber [32]. As the fiber cable plant is upgraded to support higher data rates on single-mode fiber, we must also provide a migration path that continues to reuse the installed multimode cable plant for as long as possible. The need to migrate from multimode to single-mode fiber affects many important datacom applications [33]:

- I/O applications currently using multimode fiber for ESCON will need to migrate the cable plant to single-mode fiber in order to take full advantage of the higher bandwidth of FICON links. Future FICON enhancements that extend this protocol to multi-gigabit data rates will also require single-mode fiber.
- Networking applications such as ATM have traditionally used different adapter cards to support multimode and single-mode fiber. Gigabit Ethernet standard (IEEE 802.3z) is the first industry standard to propose the use of both fiber types with the same adapter card.
- Parallel Sysplex links were originally offered as either 50 Mbyte/s data rates over multimode fiber or 100 Mbyte/s data rates over single-mode fiber. With the announcement of more recent servers, support for multimode fiber has been withdrawn as a standard feature and is now available only on special request. There is a need to support 100 Mbyte/s adapter cards over installed multimode fiber to facilitiate migration of those customers who have been using the 50 Mbyte/s option.

In order to address these concerns, special fiber optic adapter cables have been developed, known as mode conditioning patch cables (MCP). This cable contains both single-mode and multimode fibers, and should be inserted on both ends of a link to interface between a single-mode adapter card and a multimode cable plant. This allows the maximum achievable distance for multimode fiber (550 meters) and enables some applications to continue using the installed multimode cable plant. The MCPs for parallel sysplex links, Gigabit Ethernet, Fibre Channel, and many other applications are available today.

Next, let us describe the technical issues associated with this approach. The bandwidth of an optical fiber is typically measured using an over-filled launch condition, which results in equal optical power being launched into all fiber modes [32]. This is also known as a mode scrambled launch, and is approximately equivalent to the conditions achieved when using a Lambertian source such as an LED. By contrast, laser sources being more highly collimated tend to produce an under-filled launch condition; this can result in either larger or smaller effective bandwidth relative to an overfilled launch, and is sensitive to small changes in the fiber's refractive index profile. As discovered in recent gigabit link tests [34], bandwidth measured using over-filled launch conditions is not always a good indication of link performance for laser applications over multimode fiber. As illustrated in Fig. 5.3.5, when a fast rise time laser pulse is applied to multimode fiber, significant pulse broadening occurs due to the difference in propagation times of different modes within the fiber. This pulse broadening is known as differential mode delay (DMD); it is observed as an additional contribution to timing jitter (measured in ps/m) and can be large enough to render a gigabit link inoperable. DMD values are unique to the modal weighting of a source, the modal delay and mode group separation properties of the fiber, mode-specific attenuation in the fiber, and the launch conditions of the test. DMD is made worse by the excitation of relatively few modes with similar power levels in widely spaced mode groups and a high percentage of modal power concentrated in lower order modes. The impact of DMD increases with link length. There is, unfortunately, not a simple relationship between the industry specified over-fill launch measured

[4] InfiniCore CL 1000 and CL 2000 are trademarks of Corning Corporation.

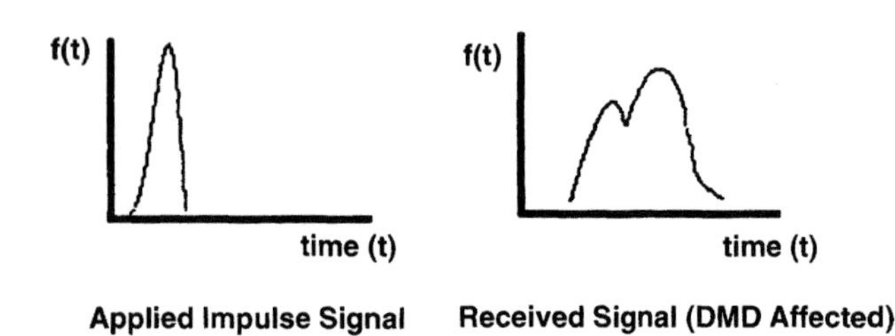

Fig. 5.3.5 Effect of DMD on signals passing through multimode fiber.

bandwidths of the fiber and the effective bandwidth due to DMD.

The radial over-fill launch method was developed as a way to establish consistent and repeatable modal bandwidth measurement of a given fiber coupled with a given source [34]. A radial over-fill launch is obtained when a laser spot is projected onto the core of the multimode fiber, symmetric about the core center with the optic axis of the source and fiber aligned; the laser spot must be larger than the fiber core, and the laser divergence angle must be less than the fiber's numerical aperture. When these conditions are satisfied, the worst case modal bandwidth of the link is taken to be the worse of the over-fill and radial over-fill launch condition measurements (although for most applications, the radial over-fill launch will be the worst case). There is a good correlation between the radial over-fill launch bandwidth and the DMD limited bandwidth of a fiber; thus, high-speed laser links implemented over multimode fiber will likely experience bandwidth values closer to the radial over-fill launch method rather than the more commonly specified over-fill launch method.

To allow for laser transmitters to operate at gigabit rates over multimode fiber without being unduly limited by DMD, a special type of fiber optic jumper cable was developed to "condition" the laser launch and obtain an effective bandwidth closer to that measured by the over-fill launch method. The intent is to excite a large number of modes in the fiber, weighted in the mode groups that are highly excited by over-fill launch conditions, and to avoid exciting widely separated mode groups with similar power levels. This is accomplished by launching the laser light into a conventional single-mode fiber, then coupling into a multimode fiber that is off-center relative to the single-mode core, as shown in Fig. 5.3.6. There are two ways in which the offset launch can be introduced. One version requires manufacturing a splice between the single-mode and multimode fiber with a controlled amount of lateral offset between the fiber cores. A tolerance analysis of this approach revealed that some installations could experience unacceptable variability in the splice elements, resulting in poor alignment and ineffective mode conditioning. For this and other reasons, the preferred embodiment uses standard ceramic ferrule technology with an offset in the ferrule alignment. Different offsets are required for 50.0 and 62.5 micron multimode fiber cores. Evaluations conducted by the Gigabit Ethernet Task Force, Modal Bandwidth Investigation Group, have verified that single-mode to 62.5 micron multimode MCPs with lateral offsets in the 17–23 micron range can achieve an effective modal bandwidth equivalent to the over-fill launch method across 99% of the installed multimode fiber infrastructure. Similar work has shown that single-mode to 50 micron multimode offset launch cables with lateral offsets in the 10–16 micron range will achieve similar results.

The MCP is illustrated in Fig. 5.3.7; its form factor is similar to a standard 2 meter jumper cable, except that it contains both single-mode and multimode fibers and includes a small package for the offset ferrules near one end. During the manufacturing process, the offset ferrules are actively aligned and then sealed with a potting

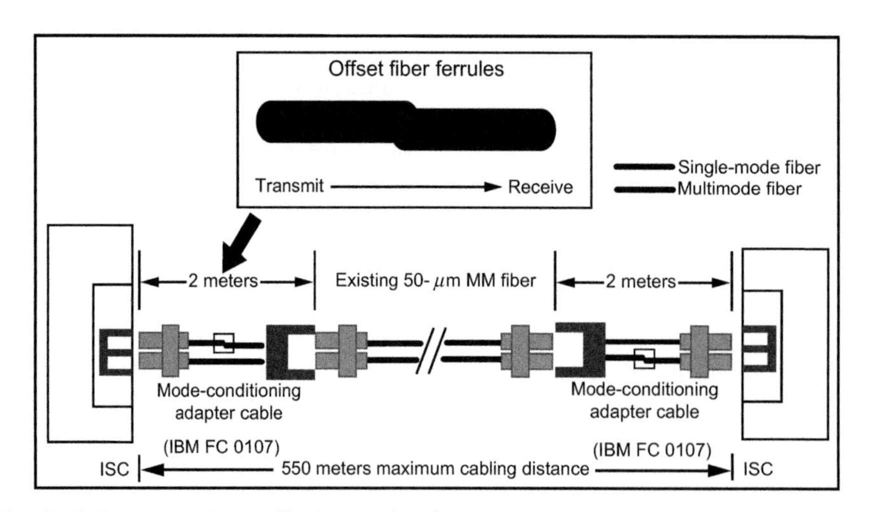

Fig. 5.3.6 Off-center ferrule design for mode conditioning patch cables.

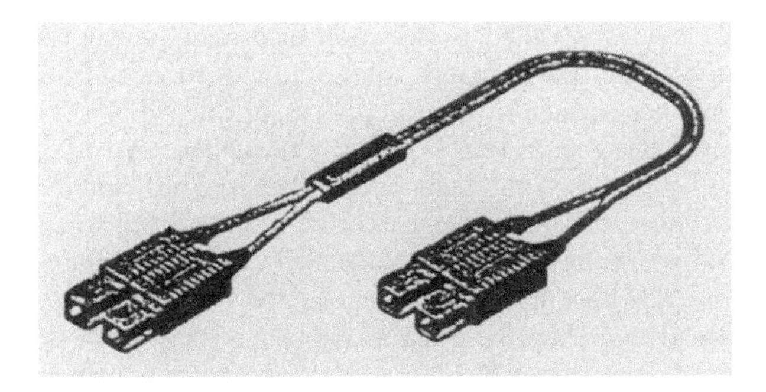

Fig. 5.3.7 Mode conditioning patch cable (MCP). The small box just behind one connector contains the offset ferrules; the transmit fiber from the optical transceiver is singlemode, all other fibers are multimode.

compound to provide thermal and mechanical stability. The active alignment apparatus is shown in Fig. 5.3.8; a wide field charge-coupled device (CCD) camera is used to measure the two-dimensional spatial distribution of optical power at the output of the MCE. Typical results of this measurement are shown in Fig. 5.3.9(a) and (b); the first plot illustrates the optical power distribution for a long wavelength laser source coupled directly into single-mode fiber, and the second plot shows the same laser launched into an MCP and then into a 3-meter multimode jumper cable. It can be seen from these figures that the MCP-conditioned launch provides a uniform distribution of optical power among all the modes of the multimode fiber, and that the MCP-conditioned launch is virtually indistinguishable from the laser launch into single-mode fiber. Once the ferrules have been aligned and sealed into their optimal position, a simple assembly loss-type measurement can be used to evaluate the MCP performance, rather than measuring the complete two-dimensional coupled power profile. As shown in Fig. 5.3.10, there is a good correlation between the connection loss of the offset ferrules and the coupled power ratio. Note that alternate designs for MCPs have also been proposed, in which the offset launch condition

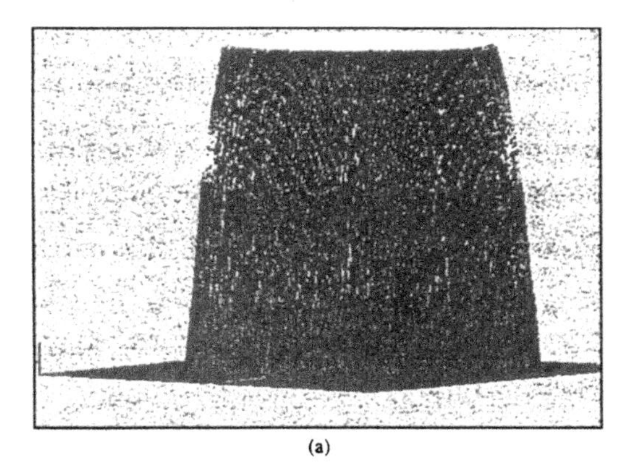

(a)

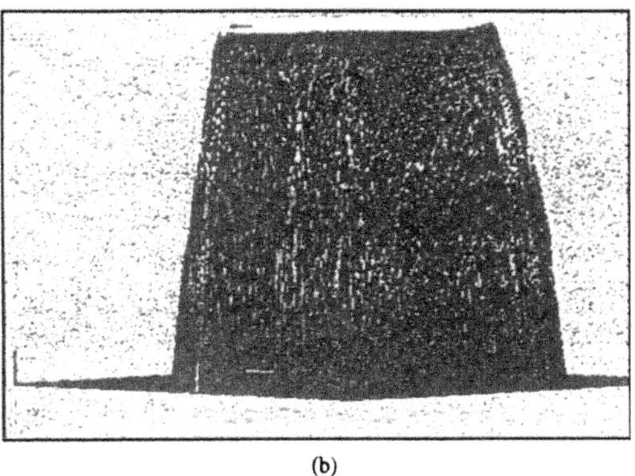

(b)

Fig. 5.3.9 MCP optical power profiles (a) single-mode fiber (b) multimode fiber.

is replaced by special optics that convert the laser spot into a donut-shaped launch. This acts to minimize power in the fundamental mode of the fiber, to achieve the same effect as an offset launch. MCPs have been tested under a variety of stressful conditions [32, 33].

5.3.4 Polarization controlling fibers

In the design of fiber optic systems it is important to know how many modes can propagate in the fiber, the

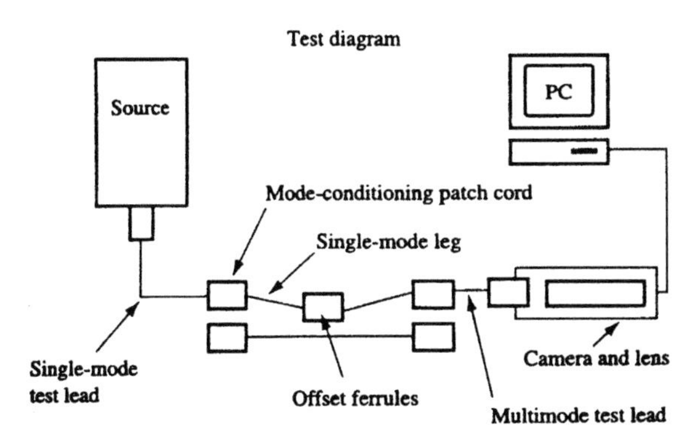

Fig. 5.3.8 Schematic of alignment apparatus for MCP cable manufacturing.

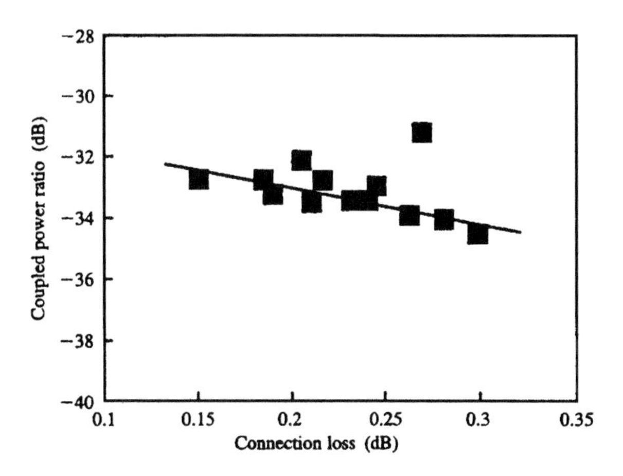

Fig. 5.3.10 Coupled optical power vs MCP connection loss.

phase constants of the different modes, and their spatial profiles. To do this we need to solve the wave equation for a particular fiber geometry. The solution depends on the specific refractive index profile of the fiber. For the case of step index fiber profiles, a complete set of analytical solutions exists [35]; these can be grouped into three different types of modes depending on the direction of the electric field vector relative to the direction of propagation. They are called transverse electric (TE), transverse magnetic (TM), and hybrid modes. The hybrid modes can be further separated into two classes depending on whether the electric field, E, or magnetic field, H, is larger in the transverse direction; these are called EH and HE modes, respectively. In practice, the refractive index difference between the core and cladding of an optical fiber is so small (about 0.002 to 0.009) that most of these modes are degenerate and it is sufficient to use a single notation for all modes, called the linearly polarized or LP notation. An LP mode is denoted by 2 subscripts, which refer to the radial and azimuthal zeros of the particular mode; for example, the fundamental mode is the LP_{01} mode. This is the only mode that will propagate in a single-mode fiber.

The cylindrical symmetry of an optical fiber leads to a natural decoupling of the radial and tangential components of the electric field vector; hence, standard single-mode fiber does not maintain the polarization state of the light when it is launched. However, these two polarizations are nearly degenerate, and a fiber with circular symmetry is most often described in terms of orthogonal linear polarizations. This near-degeneracy of the two polarization modes is easily broken by any imperfections in the cylindrical symmetric geometry of the fiber, including mechanical stresses on the fibers. These effects can either be introduced intentionally during the fiber manufacturing process, or they may arise inadvertently after the optical fiber has been installed. The effect is known as birefringence; it results in two orthogonally polarized modes with slightly different propagation constants (note that the two modes need not be linearly polarized, and in general they will be elliptical polarizations). Because each mode experiences a slightly different refractive index, the modes will drift in phase relative to each other; at any point in time, the light in the fiber exists in a state of polarization that is a superposition of the two orthogonally polarized modes. Birefringence of a fiber may be specified as the difference in refractive index between the two modes of propagation. The net polarization evolves as the light propagates through various states of ellipticity and orientation; after some distance, the two modes will differ in phase by a multiple of 2π, resulting in a state of polarization identical to that at the fiber input. This characteristic length is known as the beat length, and is a measure of the intrinsic material birefringence in the fiber; the time delay between the two modes is called polarization dispersion, and it can impact the performance of communication links in a manner similar to intermodal dispersion [9]. For example, if the time delay is less than the coherence time of the light source, the light in the fiber remains coherent and fully polarized. For sources of wide spectral width, however, this condition is reversed and light emerges from the fiber in a partially polarized and unpolarized state (the orthogonal polarizations have little or no statistical correlation). Links producing an unpolarized output can experience a 3 dB power penalty when passing through a polarizing optical element at the output of the fiber.

A stable polarization state can be ensured by deliberately introducing birefringence into an optical fiber; this is known as polarization preserving fiber or polarization maintaining fiber (PMF). Fibers with an asymmetric core profile will be strongly birefringent, having a different refractive index and group velocity for the two orthogonal polarizations (this is sometimes known as loss discrimination between modes). Such fibers are useful in some types of systems that require control of the transmitted light polarization. There are many possible core configurations as shown in Fig. 5.3.11; for example, elliptical cores provide a simple form of PMF by using very high levels of dopant in the core. These so-called "high birefringence" fibers also experience high attenuation because of the elevated dopant levels in the core. A double-core geometry (not shown in the figure) will also introduce a large birefringence. Another approach is to create mechanical stress within the fiber, such as in the bow tie configuration, by inserting stress-inducing members near the fiber core. Note that in all of these examples, polarization is only preserved if the initial signal is polarized along one of the preferred directions in the PMF; otherwise, the polarization of the signal will continue to drift as light propagates along the fiber.

Another example is the Polarization Maintaining and Absorption Reducing (PANDA) fiber; the areas highlighted in Fig. 5.3.12 show parts of the fiber core doped to create an area with a different coefficient of expansion than that of the cladding. In manufacturing, as this fiber cools, stresses are set up due to this difference, which in turn modifies the refractive index without requiring high

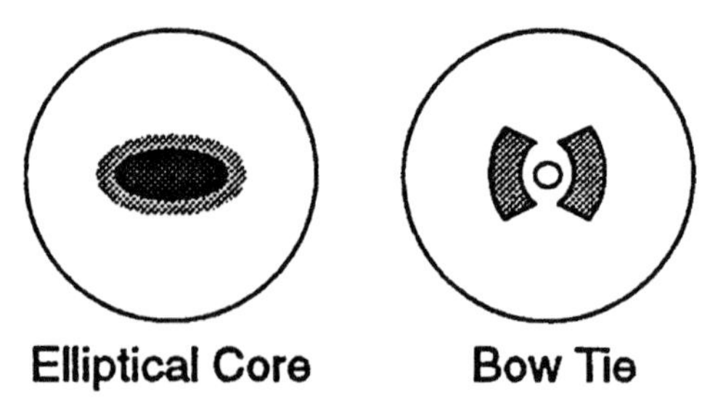

Fig. 5.3.11 Cross sections of polarization maintaining fibers.

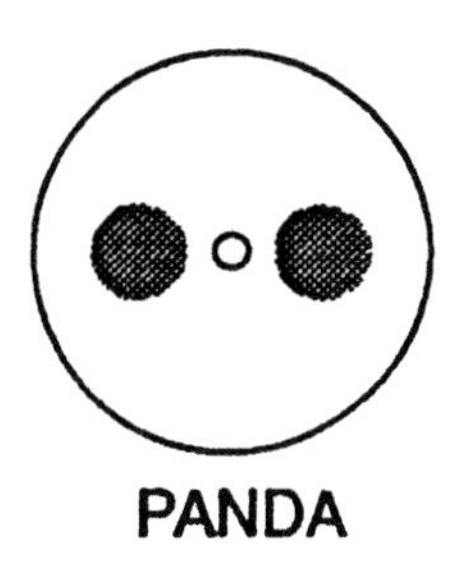

Fig. 5.3.12 Cross-section of PANDA fiber, showing mechanical members that apply strain to the fiber.

levels of dopants in the core. These stress-applying members are present along the entire length of the fiber; such fiber is commercially available.

There are other ways to make PMF, although they are not widely used in commercial products [36]. For example, so-called "low birefringence" fibers can be made by very carefully controlling the fiber profile, since there is no reason for power to couple between orthogonally polarized modes if there are no irregularities in a perfectly circular fiber. Another way is to twist the fiber during manufacturing, or deliberately make the core off-center, so that the two polarization modes become circular in opposite directions and power coupling cannot take place. Yet another variation is called "spun fiber"; a PANDA fiber pre-form is spun while the fiber is being drawn, producing a full revolution about every 5 mm. Spun fiber has no polarization dependence at all, but is very difficult to make successfully at lengths much beyond about 200 meters and is very expensive; consequently it is not commonly used for communication systems.

Polarization effects can manifest themselves in a number of ways. For example, standard erbium-doped fiber amplifiers (EDFAs) exhibit two forms of birefringence, which are usually considered trivial but may build up in systems with many optical amplifiers. First, polarization dependent loss (PDL) refers to the fact that most EDFAs exhibit higher gain for one polarization state than for the orthogonal state. Because the arriving signal is composed of a superposition of states, the gain changes slowly (over a timescale of minutes); however, the amplified spontaneous emission noise is unpolarized and experiences a fixed gain. Hence, there is variation in the signal-to-noise ratio over time. Note that this effect accumulates as the root mean square of all the amps in a chain, rather than as a straight summation. Second, polarization dependent gain (PDG) is a saturation effect in the EDFA itself; the amplifier exhibits a higher gain in the polarization state orthogonal to that of the signal. One way to combat these effects is by polarization scrambling of the input signals. Another is to design a polarization-insensitive EDFA; this can be done by using a circulator to direct light into a Faraday polarization rotating mirror such that the light makes two trips through the EDFA. The PDL and PDG effects introduced on one polarization state in the first pass through the EDFA are also induced in the second polarization state during the second pass through the EDFA, such that the emerging light has uniform gain across both polarization states.

Recently, new types of polarization maintaining fibers have been announced for high capacity fiber systems. The Lucent TruePhase family of polarization-maintaining fibers is offered in application-specific wavelengths. A new nonzero-dispersion fiber from Pirelli, the Advanced Free-Light fiber features a reduction of 50% in its polarization mode dispersion ratio over previously available fibers. As this is a rapidly evolving field, we can expect many new fiber types to be introduced in the coming years with improved properties.

5.3.5 Dispersion controlling fibers

Multimode optical fibers are subject to modal dispersion, while both multimode and single-mode fibers experience a combination of material (or chromatic) dispersion and waveguide dispersion. It was also noted that chromatic and waveguide dispersion have opposite signs so they may cancel each other out; this is why conventional silica fiber has a dispersion minima around 1300-nm wavelength. We may group together the collective effects of all these factors under the term group velocity dispersion (GVD). Standard single-mode fiber can exhibit either normal or anomalous dispersion. Under normal dispersion, long wavelengths have a higher group velocity than short wavelengths; if a wide spectrum of light is launched into this fiber, the red wavelengths will emerge first, followed by the blue wavelengths (this is also known as a positive frequency chirp or up-chirp). For anomalous dispersion, the situation is reversed; short wavelengths travel faster than long wavelengths, and blue light will emerge from the fiber before red light (this is called a negative frequency chirp or down-chirp).[5] Because modulation of a signal necessarily increases its bandwidth, and all practical light sources have some finite spectral width, dispersion effects occur in all communication systems.

As shown earlier, a typical silica optical fiber has an attenuation minimum around wavelengths of 1.55 microns, and a zero dispersion point at 1.3 microns. This

[5] Note that while this terminology is consistent with most other reference books, in some engineering texts the meaning is reversed; the definition given here for normal dispertion is called anomalous, and the definition given here for anomalous is called normal.

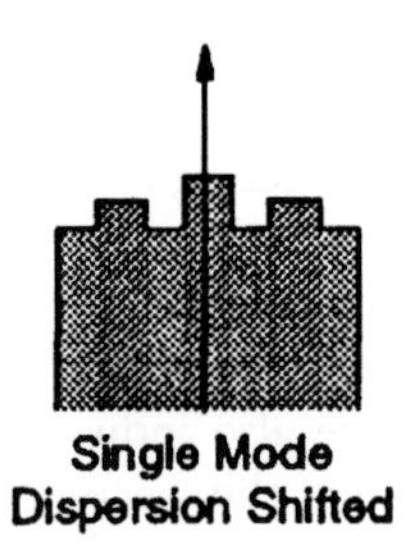

Fig. 5.3.13 Index profile of dispersion-shifted fiber.

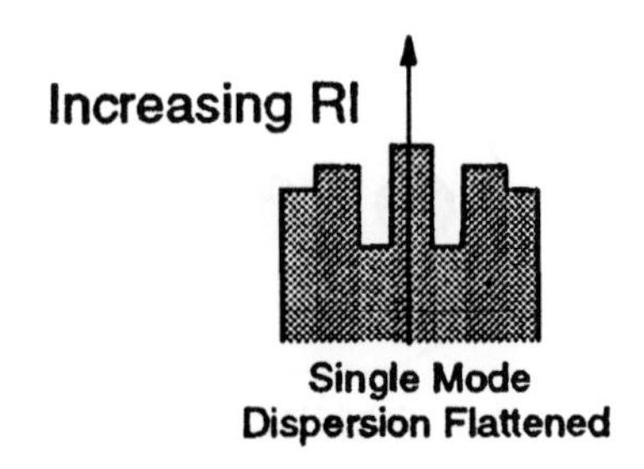

Fig. 5.3.14 Index profile of dispersion-flattened fiber.

represented a fundamental tradeoff in fiber link design, depending on whether the optical links were intended to be loss-limited or dispersion-limited; shorter distance data communication systems such as ESCON typically chose to operate at 1.3 microns, while links designed for long-haul communications were designed around 1.55 microns. Both the loss minimum and material dispersion are inherent physical properties of the silica fiber materials. However, waveguide dispersion can be affected by the refractive index profile design [36]. The profile shown in Fig. 5.3.13 has been successfully used to shift the zero dispersion point 1.55 microns; this is known as dispersion shifted fiber (DSF). Conventional fiber that has not been treated in this manner is called non-zero dispersion shifted fiber (NZDSF). Currently available, DSF has a number of practical problems. It is more prone to some forms of signal nonlinearity, especially because its slightly smaller mode field diameter concentrates electromagnetic field more strongly in the core. WDM systems can also experience strong interchannel interference, or so-called near-end crosstalk (NEXT). For example, nonlinear effects such as four-wave mixing (FWM), also known as four-photon mixing, can be a serious design issue for WDM systems. FWM is strongly influenced by the wavelength channel spacing and by the fiber dispersion. In order for FWM to occur, each channel must stay in phase with its adjacent channel for a considerable distance. Thus, if fiber dispersion is high (as with standard NDSF in the 1550 nm band, typically around 17 ps-nm-km) the effects of FWM are minimal for channel spacings greater than about 25 Ghz (if channel spacing is reduced below about 15 GHz, the effect of FWM can be severe even on standard fiber). If DSF is used (dispersion less than 1 ps-nm-km), then FWM effects are maximized; the effect causes degradation at channel spacings of less than 80 GHz, and at a channel spacing of 25 GHz around 80% of the optical energy in the original two signals will be transferred into either sum or difference frequencies. So, one might ask why all fiber isn't dispersion-shifted to take advantage of the minimal dispersion properties; part of the answer is

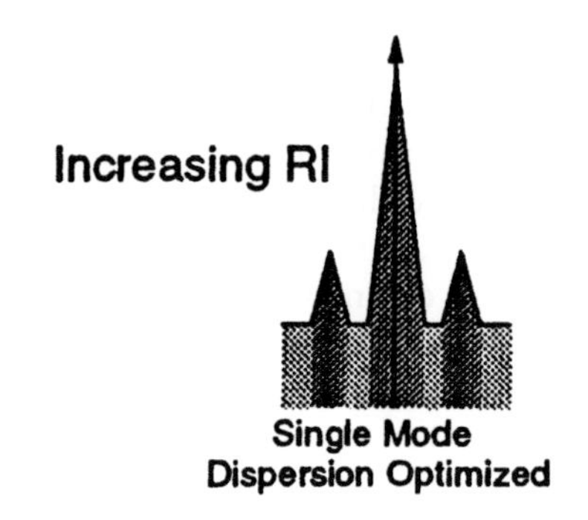

Fig. 5.3.15 Index profile of dispersion-optimized fiber.

that DSF significantly increases problems like FWM. Other nonlinear effects, such as stimulated Raman scattering are also worse when operating over DSF.

In order to address these problems, there are new types of fiber available, which essentially guarantee a certain level of dispersion (around 4 ps-nm-km), although these are currently very expensive. Other variants known as dispersion optimized fiber are also available, with a refractive index profile as shown in Fig. 5.3.14. This guarantees around 4 ps-nm-km dispersion in the 1530–1570 nm wavelength range; it is available under various brand names, including Tru-Wave[6] fiber from AT&T and so-called SMF-LS fibers from Corning.

In addition, more sophisticated dispersion compensation can be achieved by adding several core and cladding layers to the fiber design. The index profile shown in Fig. 5.3.15 is quite complex, but it is possible to realize dispersion less than 3 ps-nm-km over the entire wavelength range 1300 nm–1700 nm using this approach. This is known as dispersion-flattened fiber; it was intended to allow users to easily migrate from 1300 nm systems to 1550 nm systems without changing the installed fiber. This fiber also has potential applications to broadband WDM applications, for which fiber dispersion must be kept as uniform as possible over a range of operating wavelengths. The principle drawback is its high loss, around 2 dB/km, which prevents general use in the wide area network.

[6] Tru-Wave is a trademark of AT&T.

It is also possible to construct a fiber index profile for which the total dispersion is over 100 ps-nm-km in the opposite direction to the material dispersion; this can be used to reverse the effects of conventional fiber dispersion. So-called dispersion compensating fiber (DCF) is commercially available with an attenuation of around 0.5 dB/km. DCF has a much narrower core than standard single-mode fiber, which accentuates nonlinear effects; it is also typically birefringent and suffers from polarization mode dispersion.

Different fiber designs intended for use in WDM environments have recently been introduced. For example, Corning has recently introduced its large effective area fiber, known as LEAF, for use in the 1550 nm window (both C-band and L-band wavelengths) at data rates up to 10 Gbit/s and beyond. The LEAF fiber is a single-mode, non-zero dispersion-shifted fiber with a larger effective area in which optical power can be transmitted; typically this fiber can accommodate 2 dB more than conventional fibers without introducing nonlinear effects that can arise because of high power levels in the core, especially at the high power levels associated with multi-wavelength WDM systems. LEAF fiber has enhanced bandwidth, and effectively quadruples the information-carrying capacity of the fiber. This has made LEAF fiber particularly well-suited to long distance carriers and datacom service providers. A variation on this technology is the so-called MetroCor[7] fiber from Corning, a single-mode NZDSF compatible with industry standard G.655, which is designed to handle both C-band and L-band transmission in metropolitan area networks (1280 to 1625 nm). Similarly, the Allwave[8] fiber from Lucent provides a 50% larger spectrum than conventional fiber, lowering the attenuation between the 1300 nm and 1550 nm windows while maintaining low dispersion at 1300 nm. To protect against bending loss on single-mode WDM systems, special fiber is available such as the Blue Tiger[9] cables from Lucent. Identified by a blue cable jacket, this fiber is specially designed to support very tight bend radius applications (only 0.3 dB loss on a 10 mm bend, as compared with 1 to 1.5 dB for a conventional fiber under these conditions). Because a sharp fiber bend can induce enough loss to force WDM equipment to protection switch, this type of fiber may be important as WDM finds increasing applications in the metropolitan-area network.

Dispersion-flattened fibers have recently been investigated in soliton propagation systems for long distance communication without optical repeaters. We mention this for completeness, since datacom systems typically do not use soliton transmission; this is presently reserved for long haul telecommunication links. The nonzero dispersion slope in the fiber causes different wavelengths to experience different average dispersions; this can be a significant limitation in classical soliton long distance WDM transmission. By combining different types of commercially available fiber with different signs of dispersion and dispersion slopes, it is possible to design paths with essentially zero dispersion slope. For example, this was demonstrated in a recent experiment in which almost flat average dispersion ($D = 0.3$ ps/nm-km) was achieved by combining standard, dispersion compensated, and Tru-Wave fibers; this enabled soliton transmission of twenty-seven WDM channels, each carrying 10 Gbit/s, over more than 9000 km. Soliton experiments may lead to other types of specialty fibers; for example, adiabatic soliton compression can be obtained by using a fiber whose dispersion slowly decreases with distance (so-called dispersion tapered fiber).

5.3.6 Photosensitive fibers

Many types of glass are sensitive to ultraviolet light, which can induce a permanent change in their refractive index. These are known as photosensitive or photorefractive materials; in this case, the refractive index profile of the fiber can be changed by light that either propagates along the fiber length or illuminates an unjacketed fiber from the side. Many different types of materials can be used for this effect [37]; standard glass fiber can be made photosensitive by doping it with hydrogen, for example, while other fiber types do not require the hydrogenation process. If the fiber is illuminated through a transmission mask, or by an interference pattern created by two light beams, then the photorefractive effect can be used to write a diffraction grating into the fiber index variations. Recently, a class of devices known as in-line fiber Bragg gratings (FBGs) have been developed which hold great promise for many applications such as optical filters, wavelength add/drop multiplexers, and dispersion compensators. Writing a fiber Bragg grating requires a larger full width at half maximum response due to saturation of the index change. Due to coupling of radiation modes, undesirable side lobes of the grating can also be created at shorter wavelengths; however, this can be controlled with more recently developed types of fiber and grating writing techniques. For example, a grating stronger than 30 dB can be written in a few minutes using specialty fibers, with sidelobes kept below 0.1 dB.

[7] MetroCor is a trademark of Corning Corporation.
[8] Allwave is a trademark of Lucent.
[9] Blue Tiger is a trademark of Lucent.

Photosensitive fibers can also be used to write chirped fiber Bragg gratings for dispersion compensation. Recently, it has been demonstrated that the delay curve of grating-based dispersion compensators can be designed for a nearly perfectly linear response. This has caused increased interest in the use of FBGs for dispersion compensation in WDM systems. The FBG does have a certain amount of intrinsic polarization mode dispersion, but this can be overcome by coupling them with a suitable PMD compensator. This type of dispersion compensation is one of the many alternatives to dispersion shifted, flattened, and compensated fibers discussed earlier; with the increased interest in high bit rates (10 to 40 Gbit/s and beyond) over long haul distances (hundreds of km), the market for dispersion compensation is growing rapidly, and there will likely be many types of specialized optical fiber designed for different applications.

5.3.7 Plastic optical fiber

In an effort to reduce cost, many of the metal and ceramic components in optical fiber connectors are being replaced by plastic or polymer materials. Eliminating metal from fiber optic connectors has the advantage of improving electromagnetic radiation susceptibility, since the other components of a typical fiber cable are non-conductive. For example, radiation is generated by electronic equipment, which can escape from the receptacle of a fiber optic transceiver; metallic elements in the fiber connector act like antennas to re-radiate these emissions, possibly causing interference with either the original circuit board or with nearby unshielded electrical cables. In this manner, a product with many transceivers such as a fiber optic switch may pass electromagnetic noise emission testing with no fiber cables attached, but may fail if fiber cables with metal elements are plugged into the transceiver ports.

While ferrules in most optical connectors are traditionally made of ceramic, there are only a few companies in the world that manufacture these ferrules to the precision tolerances required. This can lead to capacity problems when there is a high demand for optical fiber; it also means that mul-timode ferrules may become scarce, as production lines migrate to higher volume (and higher margin) single-mode parts. With improvements in the single-mode ferrule manufacturing process, yields are high and there is very little product that can be recycled into multimode ferrules with their less restrictive tolerances. As a result, many companies are investigating plastic or polymer ferrules to address high-volume requirements and to reduce the cost of optical connector manufacturing. New polishing techniques need to be developed for plastic ferrules; there are also concerns with reliability and damage to the ferrule after hundreds of mating cycles. Plastic ferrules are just emerging as a viable alternative to ceramics, and are expected to play an increasingly important role in future cable systems.

This effort begs the question of whether the fiber itself could be replaced by plastic, at least for applications that can use transmission wavelengths not highly attenuated by the new material. Although most optical fibers are made of doped silica glass, plastic optical fibers are also available; these are commonly used in applications that do not require long transmission distances, such as medical instrumentation, automobile and aircraft control systems, and consumer electronics. In fact, short pieces of plastic fiber with large cores (about 900 microns) have already been used in optical loopbacks and wrap plugs, due to their low cost and ease of alignment with both multimode and single-mode transceivers. The short lengths required for a loopback mean that attenuation is not an issue, and in some cases the high attenuation of short plastic fibers is an advantage because it prevents saturation of the transceiver. However, there is some interest in using plastic fiber for very low cost data communication links, especially for the small office/home office (SOHO) environment. The combination of simplified alignment with optical sources and detectors (due to the large core diameter of plastic fibers) as well as the low cost of visible optical sources makes plastic optical fiber cost competitive for some applications. Plastic fibers are also very easy to connectorize; with minimal training and very simple tools, an amateur can connectorize bare plastic fiber in a few minutes. By contrast, glass fiber requires highly trained technicians and expensive equipment; this is a major difference, and one of the reasons for interest in plastic fiber for the do-it-yourself installations of homes and small offices; a significant inhibitor to wider use of optical links in this environment is the high cost of installation compared with the more reasonable cost of hardware and raw materials. Plastic fiber links are also used with visible light sources around 570 to 650 nm wavelength; this makes alignment of the fibers easier to perform, and high-power visible sources are readily available at low cost. A major drawback to plastic fiber is the difficulty in creating fusion splices with acceptable attenuation; typical splice losses are about 5 dB.

Both step index and graded index plastic fiber are available, although only step index is considered a commercial product at this time. While there are many potential plastics that could be considered, the most commonly used is Poly(Methyl MethylAcrylate) or PMMA. Doping the PMMA fibers can dramatically change their transmission properties; for example, Fig. 5.3.16 shows typical plastic fiber attenuation vs. wavelength for both standard PMMA fiber and a more recently introduced fiber in which deuterium replaces

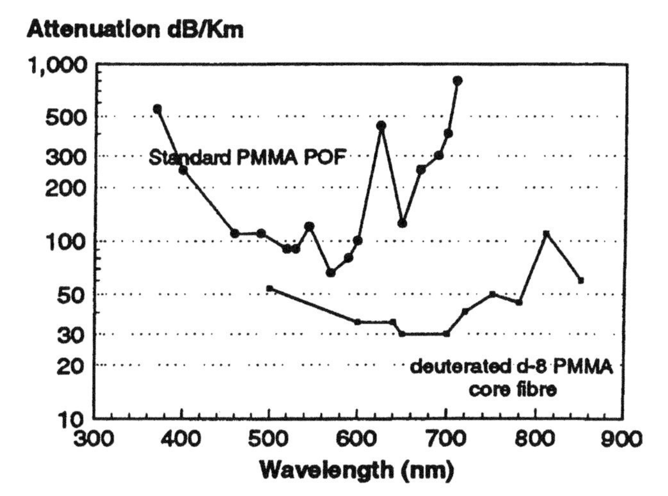

Fig. 5.3.16 Attenuation vs Wavelength for different types of plastic optical fiber.

hydrogen in some parts of the polymer molecules. Attenuation is very high compared with glass fiber at all wavelengths; transmission windows using visible light near 570 nm and 650 nm are feasible for short distances (up to 100 meters). At this distance, step index fiber has a bandwidth of about 125 MHz, while for graded index fibers bandwidths of better than 500 MHz have been reported [9]. Table 5.3.1 gives some typical properties of plastic optical fiber; note the large core diameter, which is 100 times bigger than single-mode fiber. Although plastic fiber has been adopted outside the United State, most notably in Japan, there are not yet established standards for its use; some proposals suggest that plastic fiber could be implemented for links up to 50 meters and 50 to 100 Mbit/s data rates. Various types of plastic fiber and connectors have been proposed for standardization by ATM and other groups. One example is the so-called Lucina[10] graded index plastic optical fiber developed by Asahi Glass Company; it is made of a transparent fluorpolymer, CYTOP, which is proprietary to Asahi Glass at this time. Available in both single-mode and multimode versions, including up to 200 micron core diameters, this fiber can potentially support over 1 Gbit/s up to 500 meters; attenuation is about 50 dB/km at 850 nm wavelength, bandwidth is around 200–300 MHz-km.

Other combinations of fiber materials are the subject of ongoing research, in particular Hard Polymer Clad (glass) Fiber or HPCF. This uses a relatively thick glass fiber core with a step-index hard plastic cladding; it is claimed to offer significantly less attenuation than conventional plastic fiber at around the same cost. It is also thinner than POF and consequently suffers from less modal dispersion. Although there are no general standards yet for HPCF fiber, the specification in Table 5.3.2 has been approved by the ATM Forum for use at 155 Mbit/s up to 100 meters. Since the fiber is using wavelengths around 650 nm, attenuation is fairly high; this wavelength is used to keep HPCF compatible with plastic fiber links.

Table 5.3.1 Typical specifications of plastic optical fiber.

Core Diameter	980 microns
Cladding Diameter	1000 microns (1 mm)
Jacket Diameter	2.2 mm
Attenuation (at 850 nm)	<18 dB per 100 meters (180 dB/km)
Numerical Aperture	0.30
Bandwidth (at 100 meters)	Step index: 125 MHz Graded Index: 500 MHz

Table 5.3.2 Properties of typical HPCF fiber.

Core Diameter	200 microns
Cladding Diameter	225 microns
Buffer Diameter	500 microns
Oncore diameter	2.2 mm
Attenuation (at 650 nm)	0.8 dB/100 meters
Numerical Aperture	0.3 max.
Bend Radius	2 cm, loss of 0.05 dB
Bandwidth	10MHz-km

5.3.8 Optical amplifiers

In order to increase the maximum transmission distance of a loss limited optical link, various types of optical amplifiers have been proposed. This topic can easily consume an entire series of books, and many excellent references are available [38,39]; we will note here only a few important points related to the design of such systems. Some types of optical amplifiers are based on semiconductor devices, which act much like optically pumped lasers in the fiber link. Similarly, optical amps made by selectively doping fibers can be thought of as an optically pumped light source, and such fibers have been proposed for just such applications. Some types of rare earth elements have transition bands that correspond to the near infrared

[10] Lucina is a trademark of Ashai Glass Corporation.

optical communication spectrum. When inserted into a glass matrix, these materials can absorb light from a pumping wavelength, storing energy that can subsequently be used to amplify incident light. The most common type used today is erbium-doped fiber amplifiers (EDFAs), which operate in the 1550 nm band, and are transparent to protocol, bit rate, and bit format.

In order to amplify signals at other wavelengths, including the 1300 nm band commonly used for data communication, other rare earth dopants may be used. Because glass is amorphous rather than crystalline, it offers a large gain spectrum when doped with a rare earth element; this can be shared equally between many different wavelength channels, making doped fiber a good candidate for WDM systems. Praseodymium (Pr) doped amplifiers are one solution; this is based on implanting Pr^{3+} ions in the fiber material. Silica glass cannot be used for this type of amplifier; instead, Fluorozir-conate or ZBLAN glasses are used with a very narrow (2 micron diameter) core to concentrate the pump light. This creates significant losses when attempting to couple into standard optical fiber links. These amps can be practically pumped at wavelengths of either 1017 nm (using a semiconductor InGaAs laser) or 1047 nm (using a Nd:YLF crystal laser). Recently, there has been a good deal of research into new material systems using Pr-doped Chalcogenide fibers with much higher gains (around 24 dB). Another possible dopant for the 1300 nm band is Neodymium (Nd), which can amplify over wavelength ranges of 1310 to 1360 in ZBLAN glass and 1360 to 1400 nm in silica. Efficient pump wavelengths for Nd-doped amplifiers are either 795 nm or 810 nm. Recent research has also been done in amplifiers using plastic fiber [40], especially as part of compact integrated optical systems, or as part of a hybrid glass-plastic long-haul communication system (for longer distance applications using plastic fiber, it may be easier to use silica links rather than to develop a plastic fiber amplifier). The main interest in this area lies in the fact that the gain medium can be an organic compound, which is introduced into the plastic at a relatively low temperature; this is much easier than the process of doping standard glass or ZBLAN with rare earth elements. Organic dyes cannot be used as a gain medium in glass fiber, since they tend to break down around silica's melting point of 2000 °C. One example is a plastic fiber amplifier using Rhodamine B doped PMMA, with a reported gain window between 610 and 640 nm, a pump efficiency of 33%, and a gain of 24 dB [40].

One way to get higher pump power in a rare-earth-doped fiber is by using broad area laser diodes or arrays. Although these devices emit very high pump power, they cannot be efficiently coupled into the single-mode fiber used for rare-earth-doped amplifiers. An alternative is to combine a multi-mode pump and a single-mode signal into one fiber, the so-called double-clad optical amplifier. The single-mode core is placed inside the multimode pump region; if the pump light overlaps with the core, it will be absorbed over a certain section of the fiber. To optimize this overlap, the circular symmetry of the fiber must be broken using either an off-center core or a pump region that is triangular, rectangular, polygonal, or D-shaped. Absorption into the doped core must be made very strong in order to reach the bleaching power (gain region) with relatively low pump power densities. One method of achieving this is co-doping — a first rare-earth dopant is used as an absorber that in turn transfers its energy to a second dopant, which then provides gain in the desired wavelength band. A common example is the use of ytterbium in combination with erbium in a double-clad fiber amplifier or fiber booster.

5.3.9 Futures

In this chapter, we have discussed many types of optical fibers and cable assemblies in use today. We conclude with a short description of the new types of optical fibers currently under development; while these fibers are not yet commercially available, they are promising for long-term applications in optical data communications.

5.3.9.1 Photonic crystal fibers

Another type of optical fiber that holds promise for future applications is the so-called photonic crystal fiber, also known as microstructured fiber, photonic bandgap fiber, or holey fiber. This technology was first demonstrated by Phillip Russel at the University of Bath, England, in 1997. Rather than guiding light through silica using the principle of total internal reflection, these fibers consist of a microstructure cross-section along their length (long capillaries filled with air). A typical configuration consists of air holes surrounding a silica core as shown in Fig. 5.3.17. The fiber is made by assembling a small (7 mm diameter) bundle of silica capillary tubes surrounding a solid silica rod in a hexagonal packing arrangement. This structure is first drawn into a 1 mm diameter strand; then it is inserted into an 8 m diameter borosilicate jacket and drawn to a diameter of 260 microns. The fiber is then covered with a protective polymer coating. Light is guided in the hollow fiber core using the same quantum confinement principles of a photonic bandgap structure in semiconductor materials. There are two basic types of photonic crystal fibers, namely high-index fibers (using a solid silica core and micro-pores in the cladding) and low-index fibers (using a hollow core or a core with micro-pores). The micro-holes in the fiber are typically about 1.9 microns diameter

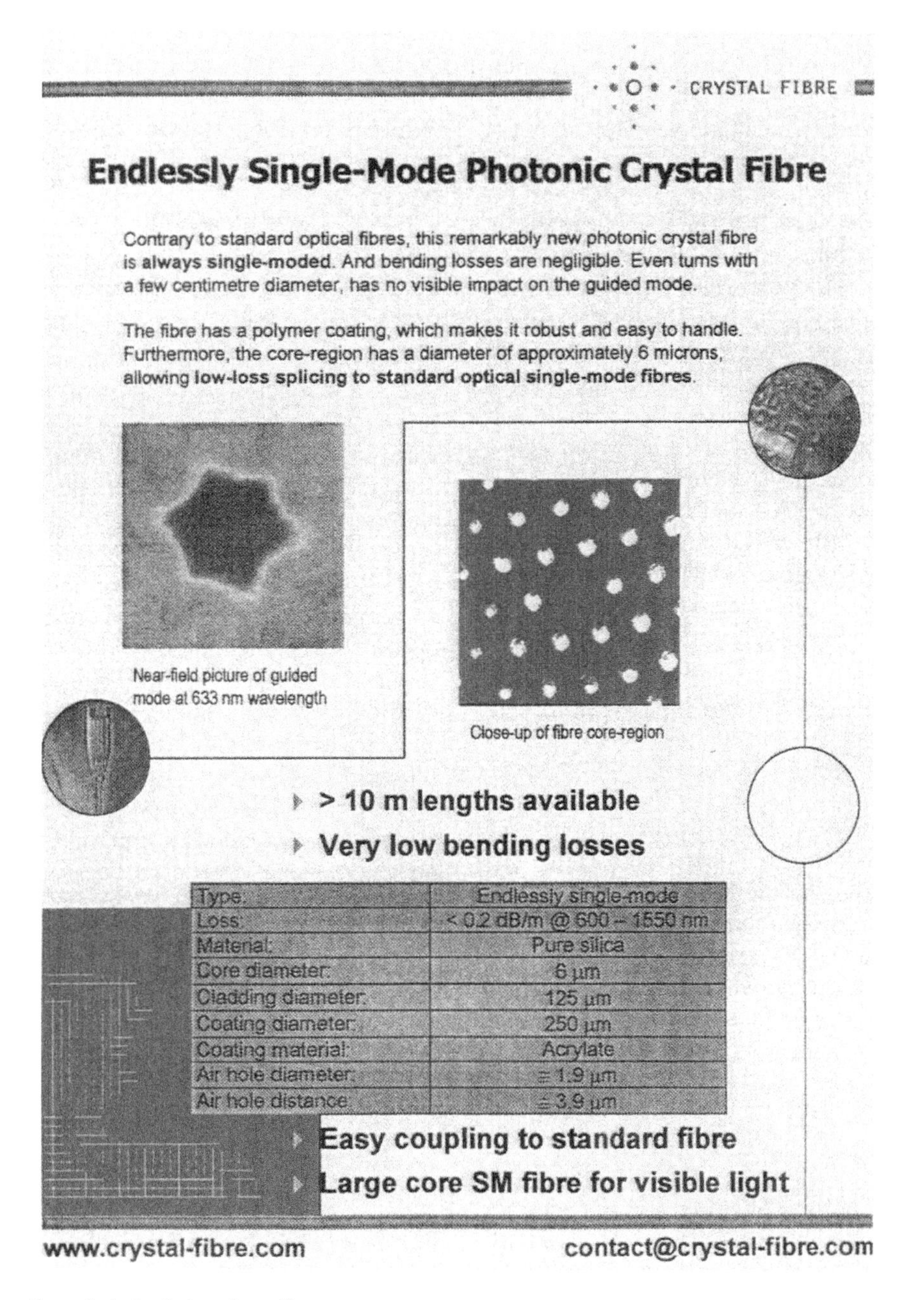

Fig. 5.3.17 Cross-section of photonic bandgap fiber.

on a 3.9 micron spacing. This fiber has several unique properties; for example, in theory there is no minimum propagation distance before the light becomes single-moded (this has led to the design of "endlessly single-mode" fibers with 6 micron core and 125 micron acrylate cladding). Bending losses should be negligible, even for a few centimeters bend radius.

Although bandgap effects are present in these fibers, it has also been found that light is guided by a secondary mechanism, namely the difference in average refractive index between the core and cladding. The large index difference between glass and air allows for much stronger confinement of the modes than with conventional fibers. Because the light is effectively guided through air, very low transmission losses are predicted; current fibers exhibit around 0.2 dB/meter at wavelengths between 600 and 1550 nm. Group velocity dispersion of about 50 picoseconds/nm/km was measured between wavelengths of 1480–1590 nm using a 6.5 meter long sample of fiber. These fibers also offer very small effective mode field area, and a selectable zero dispersion point; together, these features mean that the fiber can experience nonlinear effects over very short lengths (less than 1 meter) as opposed to the minimum interaction lengths required in silica fibers for effects such as four-wave mixing. This has potential applications to wavelength conversion, dispersion compensation, soliton generation, and white-light fiber lasers. These fibers also facilitate the construction of double-core or multi-core fibers, with selective optical coupling between modes; these are most often used for optical remote sensors, but may find applications in multichannel communication systems as well. The fibers can also have doped cores, and even use ytterbium doping to build photonic crystal fiber lasers. They have applications as polarization-preserving fibers or mode conditioners/converters. Furthermore, different wavelengths in the fiber experience a different number of holes and hence a different refractive index; having an effective index strongly dependent on wavelength opens up possible applications in wavelength multiplexing (the fiber could be used as part of a laser system to generate a wide band of wavelengths). The area in which modes propagate is very large, and can be controlled over three orders of magnitude; if the effective mode field diameter is made very large, higher optical power can be transmitted without nonlinear effects. Recently, it was shown that the fiber can be designed with anomalous dispersion at short wavelengths (780 nm), opening up possible applications in dispersion management; soliton transmission at short wavelengths may even be possible. However, photonic crystal fiber is still in the research phase and considerable work remains to be done before it is widely available as a commercial product; researchers at the Technical University of Denmark and NKT Group have collaborated to form Crystal Fibre company, while researchers in the United Kingdom have started Blaze Photonics, to further development of this technology.

5.3.9.2 Free space optical links

For some applications, optical fiber may not be necessary at all; in cases where it is difficult or expensive to install fibers, free space optical communication may provide an alternative. For example, free space optics have been proposed for communication between buildings in areas where fiber cannot be readily installed underground or between tall office buildings in metropolitan areas [41], between ships at dock and their mooring station, or to bridge the last mile between private homes and a service provider network [42]. Most of these systems consist of some form of telescope optics to collimate a tight beam for long-distance transmission, or to receive the dispersed light at a remote location. As one might expect, these approaches must deal with pointing accuracy and do not work as well under adverse weather conditions, since the infrared light signal is strongly absorbed by moisture in the atmosphere. On a smaller scale, free space optical interconnects have also been proposed for intra-machine communications, using micro-optics and lenses to collimate and focus light beams within a computer system or between adjacent equipment racks. While many such systems have been proposed, commercial products for use within a computer system are not yet available.

5.3.9.3 Omnidirectional fibers

In 1998, researchers at MIT developed a new type of mirror that reflects light from all angles and polarizations, just like metallic mirrors, but can also be as low loss as dielectric mirrors. Prior to this breakthrough, it was widely thought that a mirror could only be designed to offer high reflectivity for light with certain polarizations and angles of incidence. There are many potential applications for this so-called "perfect mirror" technology, one of which is fabricating the new mirrors into a tube with a hollow air-filled core to create an omnidirectional optical waveguide. Because this approach guides light through air rather than through silica glass, it should theoretically result in lower attenuation per km, as well as increased optical power per channel, polarization insensitivity, and improved dispersion characteristics. Also, unlike conventional waveguides, which are subject to loss at tight bend radius, the omnidirectional waveguide can make very small bends without incurring optical loss. This could enable new applications for integrated optics by allowing the miniaturization of many optical components. It has also been suggested that the monodirectional reflective material could be fabricated into a conventional coaxial cable structure with a metal or dielectric core, capable of performing as either an electrical or optical cable.

5.3.9.4 Superluminal waveguides

Finally, researchers at NEC corporation have recently published [43] the first evidence for optical signals propagating faster than the speed of light in a vacuum. It's a well-known principle of physics and Fourier optics that a pulsed signal (finite duration in time) can be created by adding up an infinite number of waves with infinite duration at different frequencies. The shorter the desired pulse, the larger the bandwidth of frequencies that must be used. All pulses of light or microwaves are thus formed by a packet of waves, each of which has a different frequency, amplitude, and phase. There is a distinction between the speeds of the individual frequency components, called the phase velocity, and the speed at which the wave packet propagates, called the group velocity. Conventional physics allows the phase velocity to exceed the speed of light, c, although the group velocity (the speed at which information is actually conveyed) must be less than *c*. It's possible to have a medium in which the phase and group velocities are not only different, but in opposite directions (one positive, the other negative). While the idea of some wave components moving backward while the wave packet only moves forward is non-intuitive, these so-called backward propagating modes are not new, and have been observed under many circumstances. In a medium with negative group velocity or so-called anomalous dispersion, it is possible in theory to propagate optical pulses faster than *c*. There are some materials that exhibit negative group velocities, typically near an atomic absorption frequency; unfortunately, this region also corresponds to very high optical attenuation and nonlinearities, which have made many prior experiments inconclusive. However, recent experiments have shown evidence of microwave pulses propagating about 7% faster than *c* over distances of about 1 meter — about ten times the wavelength of the radiation itself. Subsequently, researchers at NEC have demonstrated [43] superluminal propagation of optical signals, again over distances much larger than the wavelength of light. Working with a medium that exhibits gain, or optical amplification, near the region of negative group velocity, a narrow optical pulse (3.7 microseconds) was launched into a chamber 6 cm long that was filled with excited atomic cesium vapor. If the chamber had been filled with vacuum, the optical pulse should have taken about 0.02 nanoseconds to pass through; instead, the leading edge of the optical pulse appeared at the chamber's exit 62 nanoseconds before the bulk of the pulse entered the chamber. In other words, the wave packet had traveled nearly 20 meters away from the chamber exit before the incoming pulse entered. The optical pulse was shifted forward in time by about 1.7% of its width, giving a group velocity inside the chamber of about —*c*/330. The optical pulse emerging from the chamber was identical in shape to the one that entered. The mechanism for this is not completely understood, although it does not appear to violate the causality principle since the leading edge of the input pulse contains all the information necessary to reconstruct the peak of the output pulse, so the entire pulse does not need to enter the chamber before it exists at the opposite side. The output pulse may be just a clever interference effect, shaping newly created photons from the gain region into an exact copy of the incident light pulse. The original incoming pulse is subsequently canceled out by a backward propagating mode that travels from the chamber exit to its entrance with a phase velocity about 330 times faster than *c*. While we have little hope of developing a superluminal optical data link any time soon, research like this into the basic physics of optical propagation may lead to very practical benefits in future communication systems.

References

1. D. Nolan, "Tapered fiber couplers, mux and demux." Chapter 8 in *Handbook of Optics* vol. III and IV. Optical Society of America (2000).
2. K. Hill, "Fiber bragg gratings." Chapter 9 in *Handbook of Optics* vol. III and IV, Optical Society of America (2000).
3. MJ. Adams, *An Introduction to Optical Waveguides*. Chichester, England: John Wiley and Sons (1981).
4. A.W. Carlisle, "Small size high performance lightguide connectors for LANs." *Proc. Opt. Fiber Comm.* Paper TUQ 18: 74–75 (1985).
5. S.E. Miller, and A.G. Chynoweth. *Optical Fiber Telecommunications*, New York: Academic Press (1979).
6. J. Nishii, et.al., "Recent advances and trends in chalcogenide glass fiber technology: a review." *Journ. Non-crystalline Solids* 140: 199–208 (1992).
7. J.S. Sanghera, B.B. Harbison, and I.D. Agrawal. "Challenges in obtaining low loss fluoride glass fibers," *Journ. Non-crystalline Solids* 140: 146–149 (1992).
8. S. Takahashi, "Prospects for ultra-low loss using fluoride glass optical fiber," *Journ. Non-Crystalline Solids* 140: 172–178 (1992).
9. C. DeCusatis, "Design and engineering of fiber optic systems." In *The Optical Engineer's Desk Reference*, E. Wolfe, ed. Optical Society of America (2001).
10. E. Hect, and A. Zajac. *Optics*. New York: Addison Wesley (1979).
11. D. Nolan, "Tapered fiber couplers, mux and demux." Chapter 8 in *Hand book of Optics* vol. IV. OSA Press (2000).
12. T. Ozeki, and B.S. Kawaski. "New star coupler compatible with singe multimode fiber links." *Elec. Lett.* 12: 151–152 (1976).
13. B.S. Kawaski, and K.O. Hill. "Low loss access coupler for multimode optical fiber distribution networks." *App. Optcs* 16: 1794–1795 (1977).
14. G.E. Rawson, and M.D. Bailey. "Bitaper star couplers with up to 100 fiber channels." *Elec. Lett.* 15: 432–433 (1975).
15. S.K. Sheem, and T.G. Giallorenzi. "Singlemode fiber optical power divided; encapsulated teching technique." *Opt. Lett.* 4: 31 (1979).
16. Y. Tsujimoto, et al. "Fabrication of low loss 3 dB couplers with multi- mode optical fibers." *Elec. Lett.* 14: 157–158 (1978).
17. R.A. Bergh, G. Kotler, and HJ. Shaw. "Singlemode fiber optic direc tional coupler." *Elec. Lett.* 16: 260–261 (1980).
18. O. Parriaux, S. Gidon, and A. Kuznetsov. "Distributed coupler on polished singlemode fiber." App. *Opt.* 20: 2420–2423 (1981).
19. B.S. Kawaski, K.O. Hill, and R.G. Lamont. "Biconical-tapered single-mode fiber coupler." *Opt. Lett.* 6: 327 (1981).
20. R.G. Lamont, D.C. Johnson, and K.O. Hill. *App. Opt.* 24: 327–332 (1984).
21. A. Snyder, and J.D. Love. *Optical Waveguide Theory.* New York: Chapman and Hall (1983).
22. T. Brown, "Optical fibers and fiber optic communications." Chapter 1 in *Handbook of Optics* vol. III and IV. Optical Society of America (2000).

23. D.L. Weidman, "Achromat overclad coupler." US Patent no. 5,268,979 (Dec 1993).
24. C.M. Truesdale, and D.A. Nolan. "Core-clad mode coupling in a new three-index structure." European Conference On Optical Communications, Barcelona, Spain (1986).
25. D.B. Keck, A.J. Morrow, D. A. Nolan, and D. A. Thompson. *Journ. Lightwave Tech.* 7: 1623–1633 (1989).
26. W.J. Miller, D.A. Nolan, and G.E. Williams. "Method of making a 1 X N coupler." U.S. Patent no. 5,017,206 (Apr 1991).
27. "Fiber Transport Services Physical and Configuration Planning." (IBM document number GA22-7234), IBM Corp., Mechanicsburg, Pa (1998).
28. "Planning for Fiber Optic Channel Links." (IBM document number GA23-0367), IBM Corp., Mechanicsburg, Pa (1993).
29. "Maintenance Information for Fiber Optic Channel Links." (IBM document number SY27-2597), IBM Corp., Mechanicsburg, Pa (1993).
30. C. DeCusatis, "Fiber Optic Data Communication: Overview and Future Directions," *Optical Engineering*, special issue on Optical Data Communication (Dec 1998).
31. See Lucent product information at www.lucent.com.
32. T. Giles, J. Fox, and A. MacGregor. "Bandwidth reduction in gigabit ethernet transmission over multimode fiber and recovery through laser trans mitter mode conditioning." *Opt. Eng.* 37: 3156–3161 (1998).
33. C. DeCusatis, D. Stigliani, W. Mostowy, M. Lewis, D. Petersen, and N. Dhondy. "Fiber optic interconnects for the IBM Generation 5 parallel enterprise server." *IBM Journal of Research and Development*. **43**(5/6): 807–828 (Sept–Nov 1999).
34. C.R. Giles, "Lightwave applications of fibre Bragg gratings." *IEEE Journ. Lightwave Tech.* **15**(8): 1391–1404 (1997).
35. T. Okoshi, *Optical Fibers*. New York: Academic Press (1982).
36. H. Dutton, *Optical Communications*. New York: Academic Press (1999).
37. J. Buck, "Nonlinear effects in optical fibers." Chapter 3 in *Handbook of Optics* vol. III and IV, Optical Society of America (2000).
38. J-M P. Delavaux, "Multi-stage erbium doped fiber amplifier designs." *IEEE Journ. Lightwave Tech.* **13**(5): 703–720 (1995).
39. N.A. Olsen, "Lightwave systems with optical amplifiers." *IEEE Journ. Lightwave Tech.* **13**(7): 1071–1082 (1989).
40. G.D. Peng, PL. Chu, Z. Xiong, T. Whitbread, and R.P. Chaplin. "Broad band tunable optical amplification in Rhodamine B-doped step-index polymer fibre." *Opt. Comm.* 129: 353–357 (1996).
41. I. Kim, R. Steiger, J. Koontz, C. Moursund, M. Barclay, P. Adhikari, J. Schuster, E. Korevaar, R. Ruigrok, and C. DeCusatis. "Wireless optical transmission of fast ethernet, FDDI, ATM, and ESCON protocol data using the TerraLink laser communication system." *Opt. Eng.* 37: 3143–3156 (1988).
42. P. Humbert, and W. Weller. "Mesh algorithms enable the free space laser revolution." *Lightwave* **17**(13): 170–178 (Dec 2000).
43. L.J. Wang, A. Kuzmich, and A. Dogariu. "Gain-assisted su- perluminal light propagation." *Nature* 406: 277–279 (July 20, 2000).

Chapter 5.4

5.4

Optical wavelength division multiplexing

DeCusatis

5.4.1 Introduction and background

Although fiber optic communication systems have seen widespread commercial use for some time, in recent years there has been increasing use of fiber in computer and data communication networks, as compared with their applications in voice and telephone communications. There are several unique requirements that distinguish the sub-field of optical data communication. Datacom systems must maintain very low bit error rates, typically between 10^{-12} and 10^{-15}, since the consequences of a single bit error can be very serious in a computer system; by contrast, background static in voice communications, such as cellular phones, can often be tolerated by the listener. Optical data links also face a tradeoff between optical power and unrepeated distance. Computer applications such as distributed computing or real-time remote backup of data for disaster recovery require fiber optic links with relatively long unrepeated distances (10–50 km). At the same time, because computer equipment is often located in areas with unrestricted access, the optical links must comply with international laser safety regulations, which limit the transmitter output power [1]. Telecom links may stretch hundreds of kilometers or more using low jitter signal regenerators; by contrast, most datacom systems cannot use mid-span repeaters in long links. This makes link budget analysis a critical element for the datacorn system designer; it also implies that there are a much larger number of datacom transceivers per km of fiber than in telecom applications, making the datacom market very sensitive to the cost of optoelectronics. This drives a further tradeoff between low-cost and high-reliability components. Datacom systems require rugged components because data centers require continuous availability of the computer applications, but they are a hostile environment for optical devices; connectors are not cleaned regularly, cable reconfigurations are frequent, cable strain relief and bend radius are often not managed properly, and transceivers must withstand large numbers of reconnections, high pull forces on fiber cables, and other environmental stress.

Despite these challenges, the use of optical networking has brought about some convergence between telecommunication and data communication, at least at the physical layer. The need for ever-increasing bandwidth, data rates, and distance have made fiber optic links an integral part of computer system architectures over the past 10 years, particularly in high-performance applications, which have traditionally driven the need for higher bandwidth and consequently been the first computer applications to employ new technologies. A typical example is the increasing use of fiber optic data links on the IBM System/390 Enterprise Servers (mainframe computers). This continues to be an active area of development; mainframes and large servers, long recognized for their high security, reliability, and continuous availability, remain the data repository for major Fortune 1000 companies and contain an estimated 70% of the world's data. We can categorize the applications driving high-end optical data links into three areas:

(1) input/output (I/O) devices (such as tape or magnetic disk storage);

(2) clustered parallel computer processors (such as the IBM Parallel Sysplex architecture);

(3) inter-networking in the local, metropolitan, or wide area network (LAN/WAN/MAN).

Fiber Optic Data Communication; ISBN: 9780122078927

The bandwidth requirements for networking environments using asynchronous transfer mode (ATM) over synchronous optical network (SONET) or similar protocols are well known [1]. However, Internet traffic has recently been growing at a rate of about 150% per year, and the amount of data traffic carried on the public telephone network now exceeds the amount of voice traffic by many estimates. This traffic has somewhat different properties than voice traffic; for example, the average connection distance for Internet traffic is about 3000 km, or about 5 times larger than for voice traffic. This implies that future optical transport networks (OTNs) can reduce costs by using recently developed technologies such as long-range optical transmission, amplification, and switching. Internet traffic also tends to come in bursts, with a longer average connection duration than voice traffic, which makes capacity planning more challenging. However, telecommunications remains an important area for fiber optic networking. Virtually all telephone systems, including wireless and satellite communications, rely on a backbone of fiber optic networks. Most voice traffic is carried over protocols such as ATM/SONET; switching equipment designed for these protocols provides features such as fast protection switching so that calls are uninterrupted in the event of a fiber break or equipment failure. They also help ensure high quality of service (QOS) by providing error-free transmission of voice signals, sometimes by retiming them for increased fidelity. However, these features have until recently been available only on very expensive telecom switching equipment designed for a phone company's central office, and have applied to ATM/SONET protocols only. Users were forced to provide a separate overlay network for data communications, often at high cost, which did not offer the same features for voice and data. This required the construction of protocol-specific overlay networks, as shown in Fig. 5.4.1, which are expensive and do not scale well. Recent DWDM equipment provides QOS functions such as protection switching across all protocols; this allows users to save on leased optical fiber cost by combining voice and data over a single network, and for the first time makes it cost effective to build so-called virtual private networks (VPNs) to handle all the communication requirements of a single organization. Furthermore, DWDM can be used by telecom and datacom service providers as the basis for a service offering. Some telecom companies currently lease networking services in the same way that they lease telephone lines for voice traffic; this is a first step toward making bandwidth a commodity and enabling so-called all-optical networks (AON). The ability to manage an entire network from one location at the service provider central office is critical, as well as the ability to manage remotely (for example, one customer has products installed in Phoenix, Arizona, which are monitored through their call center in Minneapolis, Minnesota). DWDM devices also allow telecom carriers to migrate from their legacy voice-only, SONET-based equipment to new,

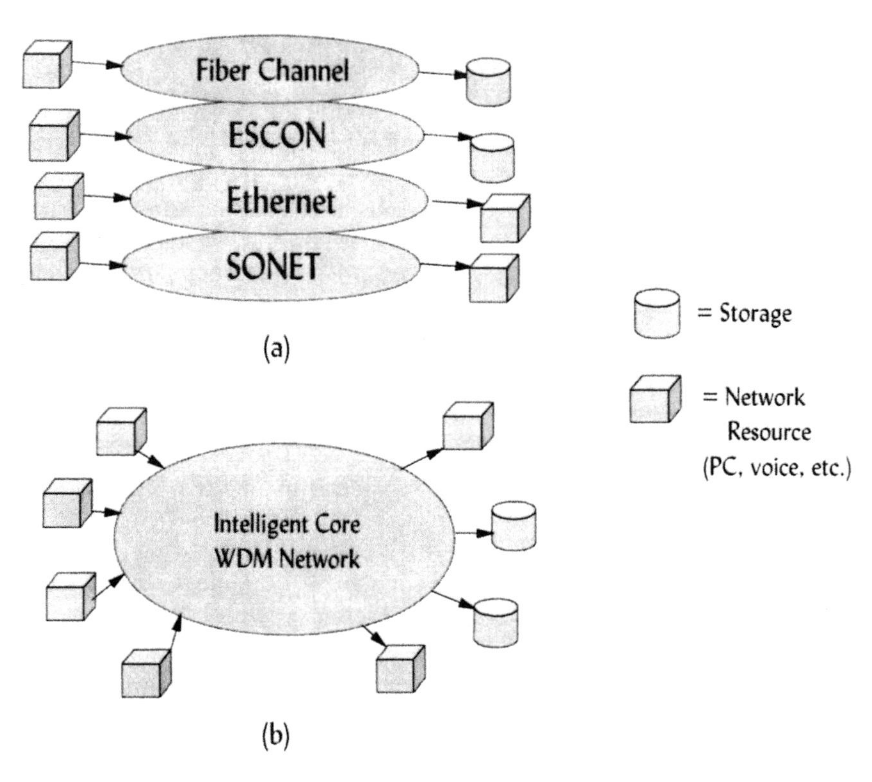

Fig. 5.4.1 (a) Existing service specific overlay network (b) future service transparent optical network with DWDM core.

more cost efficient network types, without sacrificing performance.

Transport networks in the MAN continue to evolve at a rapid pace. Datacom applications such as virtual private networks, Internet Protocol (IP) over WDM, electronic commerce, and the exponential growth of the Internet and Web continue to drive the need for increased bandwidth in the MAN (indeed, growth of the "optical internet" is among the largest driving factors in the deployment of WDM solutions). In this environment, the amount of installed fiber is being consumed quickly by new service offerings, and many networks face congested routes or fiber exhaust problems which limit their scalability and constrain new service offerings. Installation of new fiber, upgrading to higher line rates, or adding new terminal equipment as a workaround to fiber exhaust can be prohibitively expensive and may disrupt existing services. As fiber becomes scarce, there is increasing pressure to decommission legacy services in order to free fiber for higher data rate systems; a migration path is needed to maintain use of existing critical services and protect the installed service base, while maximizing use of the existing fiber plant. Network topologies within the MAN also continue to evolve; there is strong interest in sharing data among many widely distributed locations, and converging the data network with voice, video, and other services. Because the underlying protocols for ESCON and Parallel Sysplex were designed for point-to-point systems, and switches to enable fabrics of FICON and Fibre Channel devices are not yet commercially available, new MAN topologies are being driven by the capability of next-generation DWDM systems to support more than point-to-point configurations. While first-generation DWDM systems were strictly point-to-point, next-generation systems include self-healing hubbed ring and meshed ring configurations.

Large servers require dedicated storage area networks (SANs) to interconnect with various types of magnetic disk, tape, and optical storage devices, printers, and other equipment; the dramatic increase in use of fiber optics for these applications has led to the term "Optical Data Center" (ODC). A SAN is a network of computers and storage devices (magnetic disk and tape, optical disk, printers, etc.) interconnected by a switching device to provide continuous access to all the data in the event that any single communication path fails. Examples of SANs include networks of ESCON, FICON, Fibre Channel, or Gigabit Ethernet devices, which can require tens to hundreds of fiber optic links. For example, large organizations such as airlines, banks, credit card companies, the federal government, and others are currently using huge SANs with many petabytes of storage; in the near future, exabyte SANs will likely emerge. A closely related concept is network attached storage (NAS), which refers to any type of storage device that is not locally connected to a server but attached only via a network interface. The cost and manageability of these systems is greatly improved by using wavelength multiplexing technology to reduce the number of fibers required; because it is protocol independent, it also provides the ability to work with different communication protocols and equipment from many different companies. Large SANs also benefit from the ability to construct protocol independent DWDM rings with QOS features such as "1 + 1" fast protection switching on individual channels for all protocols, not just those used for telecom voice communication. New applications such as data mining and requirements to archive data in real time at a remote location for disaster recovery with minimal service interruptions continue to drive bandwidth demands in this area. This has led to the emergence of client-server-based networks employing either circuit or packet switching, and the emergence of network-centric computing models in which a high bandwidth, open protocol network is the most critical resource in a computer system, surpassing even the processor speed in its importance to overall performance.

The recent trend toward clustered, parallel processors to enhance performance has also driven the requirement for high bandwidth fiber optic coupling links between computers. For example, large water-cooled mainframe computers using bipolar silicon processors are being replaced by smaller, air-cooled servers using complimentary metal oxide semiconductor (CMOS) processors [1, 2]. These new processors can far surpass the performance of older systems because of their ability to couple together many central processing units (CPUs) in parallel. There are many examples of parallel coupled processor architectures one important example is a Parallel Sysplex, an ad hoc standard developed by IBM, which allows mainframe computers to be clustered together using optical fiber links, and to work in parallel as if they were a single system. In this manner, the advantages of parallel processing can be used to increase the performance of the computer system without increasing the speed of the microprocessors, and to increase system reliability to better than 99.999%. A Parallel Sysplex requires a minimum of 30 to 60 duplex fiber optic channels for a small installation, and larger installations can require hundreds of channels. For disaster recovery purposes, the elements of a Parallel Sysplex are often split among physical locations up to 40 km or more apart; this is known as a Geographically Dispersed Parallel Sysplex (GDPS). The high cost of leased optical fiber over these distances ($300 per mile per month for 1 channel) makes it cost prohibitive for many users to implement Parallel Sysplex without using wavelength multiplexing. However, a Parallel Sysplex has unique requirements for the fiber optic channels and the wavelength multiplexer. In particular, a Sysplex requires some links (known as InterSystem

Channels) to support the ANSI Open Fiber Control (OFC) protocols; this protocol specifies point-to-point channels only, and does not describe how to include repeaters or multiplexers in the link without violating the proper channel operation, or how to extend an OFC channel beyond distances of 20 km as required for many Sysplexes. Furthermore, a Sysplex requires that the computer's clock be distributed to remote locations up to 40 km apart, which causes a variety of latency and timing concerns when a multiplexer is included in the link. For these and other technical reasons, very few WDM solutions are able to support GDPS. The combined effects of these application areas has made high-end computer systems a near term application for multi-terabit communication networks incorporating wavelength division multiplexing. In the remainder of this chapter, we will first describe the dominant approaches to data storage and coupled Parallel Sysplex processors using optical fiber connectivity and datacom protocols. We will then discuss how dense wavelength division multiplexing (DWDM) is being used in large data communication systems, including specific examples of current and next-generation DWDM devices and systems. Technical requirements, network management and security, fault tolerant systems, new network topologies, and the role of time division multiplexing in the network will be presented. Finally, we describe future directions for this technology.

5.4.2 Wavelength multiplexing

Multiplexing wavelengths is a way to take advantage of the high bandwidth of fiber optic cables without requiring extremely high modulation rates at the transceiver. With an available bandwidth of about 25 THz, a single optical fiber could carry all the telephone traffic in the United States on the busiest day of the year (recently, Mother's Day has been slightly exceeded by Valentine's Day). This technology represents an estimated US $1.6 B market with over 50% annual growth; wavelength multiplexing systems may be classified according to their wavelength spacing and number of channels as follows [3]:

- Coarse WDM systems typically employed only 2–3 wavelengths widely spaced, for example 1300 nm and 1550 nm. Applications of this technology in data communications are limited, although recently coarse WDM systems with 4 to 8 channels have been used in small networks.
- Wide Spectrum WDM (WWDM) systems can support up to 16 channels, using wavelengths that are spaced relatively far apart; there is no standardized wavelength spacing currently defined for such systems, although spacing of 1 to 30 nm has been employed. These systems are meant to serve as a low-cost alternative to dense wavelength division multiplexing (DWDM) for applications that do not require large numbers of channels on a single fiber path, and are being considered as an option for the emerging 10 Gbit/s Ethernet standard [4]. These are also known as Sparse WDM (SWDM) systems.
- Dense WDM systems (DWDM) employed wavelengths spaced much closer together, typically following multiples of the International Telecommunications Union (ITU) industry standard grid [5] with wavelengths near 1550 nm and a minimum wavelength spacing of 0.8 nm (100 GHz). This may be further subdivided as follows:
- First-generation DWDM systems typically employed up to 8 full duplex channels multiplexed into a single duplex channel.
- Second-generation DWDM systems employ up to 16 channels.
- Third-generation DWDM systems employ up to 32 channels; this is the largest system currently in commercial production for data communication applications.
- Fourth-generation or Ultra-dense WDM are expected to employ 40 channels or more and may deviate from the current ITU grid wavelengths; channel spacing as small as 0.4 nm (50 GHz) have been proposed [6]. These systems are not yet commercially available, and are the subject of much ongoing research.

Normally, one communication channel requires two optical fibers, one to transmit and the other to receive data; a multiplexer provides the means to run many independent data or voice channels over a single pair of fibers. This device takes advantage of the fact that different wavelengths of light will not interfere with each other when they are carried over the same optical fiber; this principle is known as wavelength division multiplexing (WDM). The concept is similar to frequency multiplexing used by FM radio, except that the carrier "frequencies" are in the optical portion of the spectrum (around 1550 nanometers wavelength, or 2×10^{14} Hertz). Thus, by placing each data channel on a different wavelength (frequency) of light, it is possible to send many channels of data over the same fiber. More data channels can be carried if the wavelengths are spaced closer together; this is known as dense wavelength division multiplexing (DWDM). Following standards set by the International Telecommunications Union (ITU) [5], the wavelength spacing for DWDM products is a minimum of 0.8 nm, or about 100 GHz; in practice, many products use a slightly broader spacing such as 1.6 nm, or about 200 GHz, to simplify the design and lower overall product cost. A list of ITU grid standard wavelengths for DWDM is shown in Table 5.4.1. The concept of combining multiple data channels

Table 5.4.1 ITU grid standard wavelengths for dense WDM systems. C-band (or blue band) extends from about 1528.77nm (196.1 THz) to 1556.31nm (191.4 THz), and L-band (or red band) extends from 1565.48nm (191.5 THz) to 1605.73nm (186.7 THz). The minimum channel spacing is 0.8nm (100 GHz) Anchored to a 193.1 THz reference (after ITU standard G.MCS Annex A of COM15-R 67-E).

Wavelength (nm)	Frequency (THz)	Wavelength (nm)	Frequency (THz)
1528.77	194.6	1554.13	192.9
1542.14	194.4	1554.94	192.8
1543.73	194.2	1555.75	192.7
1545.32	194.0	1556.55	192.6
1546.92	193.8	1557.36	192.5
1532.68	193.6	1558.17	192.4
1533.47	193.4	1558.98	192.3
1551.72	193.2	1559.79	192.2
1553.33	193.0	1560.61	192.1
1554.94	192.8	1561.42	192.0
1536.61	195.1	1562.23	191.9
1537.40	195.0	1563.05	191.8
1538.19	194.9	1563.86	191.7
1538.98	194.8	1564.67	191.6
1539.77	194.7	1565.48	191.5
1540.56	194.6	1566.31	191.4
1541.35	194.5	1567.13	191.3
1542.14	194.4	1567.94	191.2
1542.94	194.3	1568.77	191.1
1543.73	194.2	1569.59	191.0
1544.53	194.1	1570.42	190.9
1545.32	194.0	1571.24	190.8
1546.12	193.9	1572.06	190.7
1546.92	193.8	1572.05	190.6
1547.72	193.7	1573.71	190.5
1548.51	193.6	1574.54	190.4
1549.32	193.5	1575.37	190.3
1550.12	193.4	1576.19	190.2
1550.92	193.3	1577.03	190.1
1551.72	193.2	1577.85	190.0
1552.52	193.1	1578.69	189.9
1553.33	193.0	1579.51	189.8

Continued

Table 5.4.1 ITU grid standard wavelengths for dense WDM systems. C-band (or blue band) extends from about 1528.77nm (196.1 THz) to 1556.31nm (191.4 THz), and L-band (or red band) extends from 1565.48nm (191.5 THz) to 1605.73nm (186.7 THz). The minimum channel spacing is 0.8nm (100 GHz) Anchored to a 193.1 THz reference (after ITU standard G.MCS Annex A of COM15-R 67-E). — *cont'd*

Wavelength (nm)	Frequency (THz)	Wavelength (nm)	Frequency (THz)
1580.35	189.7	1593.80	188.1
1581.18	189.6	1594.64	188.0
1582.02	189.5	1595.49	187.9
1582.85	189.4	1596.34	187.8
1583.69	189.3	1597.19	187.7
1584.52	189.2	1598.04	187.6
1585.36	189.1	1598.89	187.5
1586.19	189.0	1599.74	187.4
1587.04	188.9	1600.60	187.3
1587.88	188.8	1601.45	187.2
1588.73	187.7	1602.31	187.1
1598.89	187.5	1603.16	187.0
1590.41	187.3	1604.02	186.9
1591.25	187.1	1604.88	186.8
1592.10	186.9	1605.73	186.7
1592.94	186.7		

over a common fiber (physical media) is illustrated in Fig. 5.4.2. Note that the process is in principle protocol independent; it provides a selection of fiber optic interfaces to attach any type of voice or data communication channel. Input data channels are converted from optical to electrical signals, routed to an appropriate output port, converted into optical DWDM signals, and then combined into a single channel. The wavelengths may be combined in many ways; for example, a diffraction grating or prism may be used. Both of these components act as dispersive optical elements for the wavelengths of interest; they can separate or re-combine different wavelengths of light. The prism or grating can be packaged with fiber optic pigtails and integrated optical lenses to focus the light from multiple optical fibers into a single optical fiber; demultiplexing reverses

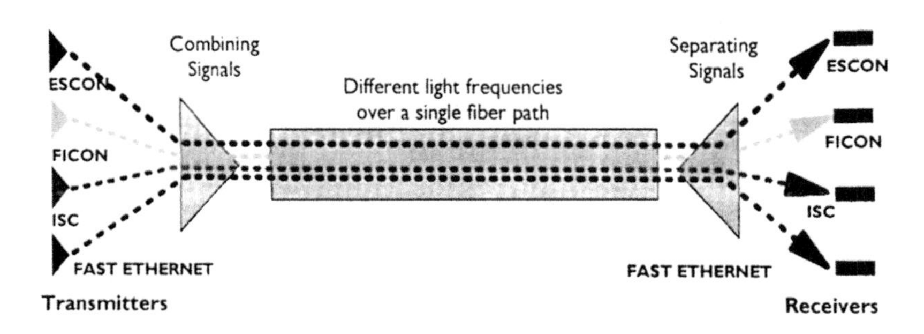

Fig. 5.4.2 Multiple data channels and protocols transmitted over a single optical fiber by WDM.

the process. The grating or prism can be quite small, and may be suitable for integration within a coarse WDM transceiver package. Such components are typically fabricated from a glass material with low coefficient of thermal expansion, since the diffraction properties change with temperature. For coarse WDM applications, this is not an issue because of the relatively wide spacing between wavelengths. For DWDM systems, the optical components must often be temperature compensated using heat sinks or thermoelectric coolers to maintain good wavelength separation over a wide range of ambient operating temperatures. Because of the precision tolerances required to fabricate these parts and the complex systems to keep them protected and temperature stable, this solution can become expensive, especially in a network with many add/drop locations (at least one grating or prism would be required for each add/drop location).

Another form of multiplexing element is the array waveguide grating (AWG); this is basically just a group of optical fibers or waveguides of slightly different lengths, which controls the optical delay of different wavelengths and thereby acts like a diffraction grating. AWGs can be fabricated as integrated optical devices either in glass or silicon substrates, which are thermally stable and offer proven reliability. These devices also provide low insertion loss, low polarization sensitivity, narrow, accurate wavelength channel spacing, and do not require hermetic packaging; one example of a commercial product is the Lucent Lightby 40 channel AWG mux/demux. Recently, the technology for fabricating in-fiber Bragg gratings has shown promise as an alternative approach; this will be discussed later in the chapter. Another promising new technology announced recently by Mitel Corporation's Semiconductor division implements an Echelle grating by etching deep, vertical grooves into a silica substrate; this process is not only compatible with standard silicon processing technology, which may offer the ability to integrate it with other semiconductor WDM components, but also offers a footprint for a 40-channel grating up to 5 times smaller than most commercially available devices.

However, in many commercially available DWDM devices a more common approach is to use thin film interference filters on a glass or other transparent substrate, as illustrated in Fig. 5.4.3. These multilayer filters can selectively pass or block a narrow range of wavelengths; by pigtailing optical fibers to the filters, it is possible to either combine many wavelengths into a single channel or split apart individual wavelengths from a common fiber. Note that many filters may be required to accommodate a system with a large number of wavelengths, and each filter has some insertion and absorption loss associated with it; this can affect the link budget in a large WDM network. For example, a typical 4-channel thin film add/drop filter can have as much as 3–4 dB loss for wavelengths that are not added or dropped; a cascade of many such filters can reduce the effective link budget and distance of a network by 10–20 km or more. The combined wavelengths are carried over a single pair of fibers; another multiplexer at the far end of the fiber link reverses the process and provides the original data streams.

There are many important characteristics to consider when designing a DWDM system. One of the most obvious design points is the largest total number of channels (largest total amount of data) supported over the multiplexed fiber optic network. Typically, one wavelength is required to support a data stream; duplex data streams may require two different wavelengths in each direction or may use the same wavelength for bi-directional transmission. As we will discuss shortly, additional channel capacity may be added to the network using a combination of WDM and other features,

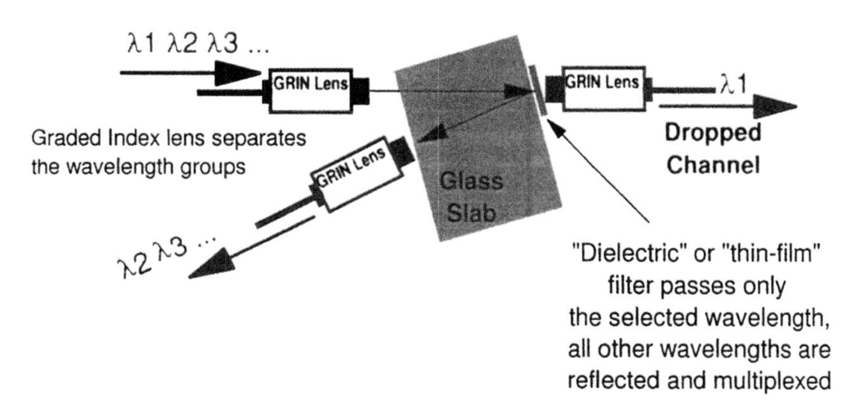

Fig. 5.4.3 Dielectric thin film optical filters for wavelength multiplexing.

including TDM and wavelength reuse. Wavelength reuse refers to the product's ability to reuse the same wavelength channel for communication between multiple locations; this increases the number of channels in the network. For example, consider a ring with 3 different locations A, B, and C. Without wavelength reuse, the network would require one wavelength to communicate between sites A and C, and another wavelength to communicate between sites B and C, or two wavelength channels total. With wavelength reuse, the first and second sites may communicate over one wavelength, then the second and third sites may reuse the same wavelength to communicate rather than requiring a new wavelength. Thus, the second wavelength is now available to carry other traffic in the network. Wavelength reuse is desirable because it allows the product to increase the number of physical locations or data channels supported on a ring without increasing the number of wavelengths required; a tradeoff is that systems with wavelength reuse cannot offer protection switching on the reused channels.

Time Division Multiplexing (TDM) is another way in which some WDM products increase the number of channels on the network. Multiple data streams share a common fiber path by dividing it into time slots, which are then interleaved onto the fiber as illustrated in Fig. 5.4.4. TDM acts as a front-end for WDM by combining several low data rate channels into a higher data rate channel; because the higher data rate channel only requires a single wavelength of the WDM, this method provides for increased numbers of low-speed channels. As an example, if the maximum data rate on a WDM channel is 1 Gbit/s then it should be possible to TDM up to 4 channels over this wavelength, each with a bit rate of 200 Mbit/s, and still have some margin for channel overhead and other features. The TDM function may be offered as part of a separate product, such as a data switch, that interoperates with the wavelength multiplexer; preferably, it would be integrated into the WDM design. Some products only support TDM for selected telecommunications protocols such as SONET; in fact, the telecom protocols are designed to function in a TDM-only network, and can easily be concatenated at successively faster data rates. However, since WDM technology can be made protocol independent, it is desirable for the TDM to also be bit-rate and protocol-independent, or at least be able to accommodate other than SONET-based protocols. This is sometimes referred to as being "frequency agile." Note that while a pure TDM network requires that the maximum bit rate continue to increase in order to support more traffic, a WDM network does not require the individual channel bit rates to increase. Depending on the type of network being used, a hybrid TDM and WDM solution may offer the best overall cost performance; however, TDM alone does not scale as well as WDM. A comparison of the two multiplexing approaches is given in Fig. 5.4.5.

Another way to measure the capacity of the multiplexer is by its maximum bandwidth, which refers to the product of the maximum number of channels and the maximum data rate per channel. For example, a product that supports up to 16 channels, each with a maximum data rate of 1.25 Gigabit per second (Gbps), has a maximum bandwidth of 20 Gbps. Note that the best way to measure bandwidth is in terms of the protocol-independent channels supported on the device; some multiplexers may offer very large bandwidth, but only when carrying well-behaved protocols such as SONET or SDH, not when fully configured with a mixture of datacom and telecom protocol adapters. Another way to measure the multiplexer's performance is in terms of the maximum number of protocol-independent, full-duplex wavelength channels that can be reduced to a single channel using wavelength multiplexing only. Some products offer either greater or fewer numbers of channels when used with options such as a fiber optic switch.

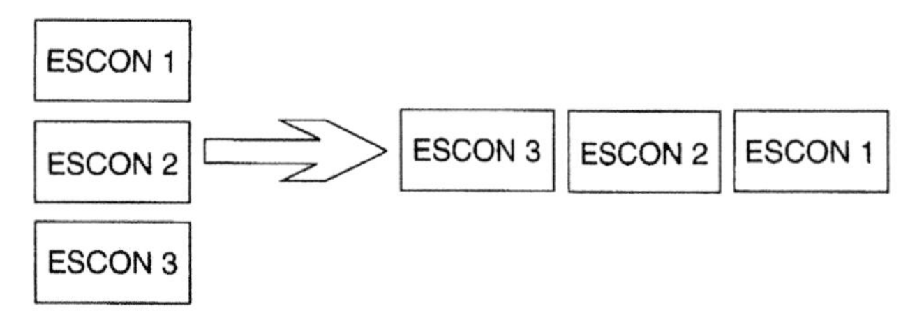

Fig. 5.4.4 Example of time division multiplexing 3 ESCON data frames.

5.4.2.1 WDM design considerations

Another important consideration in the design of WDM equipment is the number of multiplexing stages (or cards) required. It is desirable to have the smallest

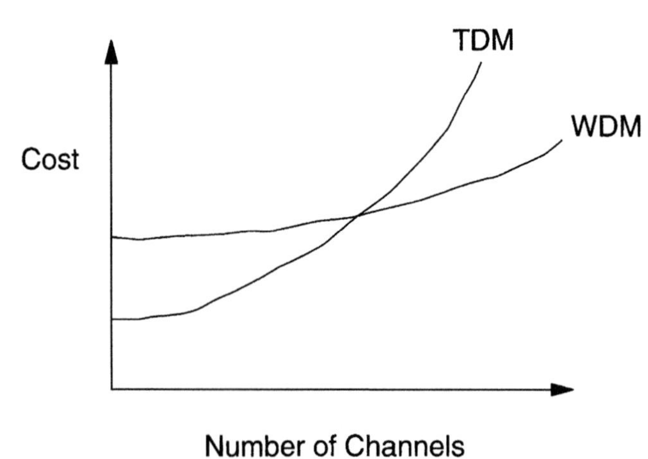

Fig. 5.4.5 Comparison of TDM and WDM scaling potential.

number of cards supporting a full range of datacom and telecom protocols. Generally speaking, a WDM device contains 2 optical interfaces, one for attachment of input or client signals (which may be protocol specific) and one for attachment of the WDM signals. Each client interface may require a unique adapter card; for example, some protocols require a physical layer that is based on an LED transmitter operating over 62.5 micron multimode fiber, others use short-wavelength lasers with 50 micron multimode fiber, and still others require long wavelength lasers with single-mode fiber. Likewise, each channel on the WDM interface uses a different wavelength laser transmitter tuned to an ITU grid wavelength, and therefore requires a unique adapter card. Some designs place these 2 interfaces on a single card, which means that more cards are required to support the system; as an example, a product with 16 wavelength channels may require 16 cards to support ESCON, 16 more to support ATM, and in general to support N channels with M protocols would require N * M cards. Typically N = 16 to 32 channels and M = 10 to 15 protocols, so this translates into greater total cost for a large system, greater cost in tracking more part numbers and carrying more spare cards in inventory, and possibly lower reliability (since the card with both features can be quite complex).

This common card is sometimes known as a transponder, especially if it offers only optical input and output interfaces. An advantage of the transponder design is that all connections to the product are made with optical fiber; the backplane does not carry high-speed signals, and upgrades to the system can be made more easily by swapping adapter cards. However, a cross-connected high-bandwidth backplane is still a desirable feature to avoid backplane bus congestion at higher data rates and larger channel counts; furthermore, it provides the possibility of extending the backplane into rack-to-rack type interconnections using parallel optical interconnects. Another possible design point places the client interface and WDM interface on separate cards, and uses a common card on the client side to support many protocol types, so only 1 card for each protocol is required. Continuing our example, a product with 16 channels may support 10 different protocols; using the first design point discussed above this requires 160 cards, while the second design point requires only 16 cards, a significant simplification and cost savings. The tradeoff with this second design point is that more total cards may be required to populate the system, since 2 cards are required for every transponder, or in other words a larger footprint for the same number of wavelength channels. In practice, a common client interface card may be configurable to support many different protocols by using optical adapters and attenuators at the interface, and making features such as retiming programmable for different data rates. This means that significantly less than 1 card per protocol is required; a maximum of 2–5 cards should be able to support the full range of networking protocols listed in Table 5.4.2. Note that WDM devices that do not offer native attachment of all protocols may require separate optical patch panels, strain relief, or other protection for the optical fibers; this may require additional installation space and cost.

Features such as adapters or patch panels that must be field installed also implies that a particular configuration cannot be fully tested before it reaches the end user location; field-installed components also tend to have lower reliability than factory built and installed components. For example, if a product supports ESCON protocols, a user should be able to plug an

ESCON duplex connector directly into the product as delivered, without requiring adapters for the optical connectors. Note that some ad hoc industry standards such as the Parallel Sysplex architecture for coupling mainframe computers have very specific performance requirements, and may not be supported on all DWDM platforms; recently published technical data from

Table 5.4.2 Protocols supported by WDM, including native physical layer specifications and attachment distances; MM = multimode fiber, SM = single-mode fiber, TX = transmitter output, RX = receiver input, LX = long wavelength transmitter, SX = short wavelength transmitter.

Protocol type	Physical layer specification	Native attach distance
ESCON/SBCON MM and	TX: −15 to −20.5	3 km
Sysplex Timer MM	RX: −14 to −29	
ESCON/SBCON SM	TX: −3 to −8 RX: −3 to −28	20 km
FICON SM	TX: −3 to −8.5 RX: −3 to −22	10 km
ATM 155MM	TX: −14 to −19 RX: −14 to −30	2 km
ATM 155 SM	TX: −8 to −15 RX: −8 to −32.5	10 km
FDDI MM	TX: −14 to −19 RX: −14 to −31.8	2 km
Gigabit Ethernet LX SM	TX: −14 to −20 RX: −17 to −31	5 km
Gig. EN SX MM	TX: −4 to −10	550 meters
(850 nm)	RX: −17 to −31	
HiPerLinks for Parallel	TX: −3 to −11	10vkm
Sysplex and GDPS	RX: −3 to −20	

a refereed journal is a good reference to determine which protocols have been tested in a given application; there may also be other network design considerations, such as configuring a total network solution in which the properties of the subtended equipment are as well understood as those of the DWDM solution.

There are several emerging technologies that may help address these design points in the near future. One example is wavelength tunable or agile lasers, whose output wavelength can be adjusted to cover several possible wavelengths on the ITU grid. This would mean that fewer long wavelength adapter cards or transponders would be required; even if the wavelength agile laser could be tuned over only 2 ITU grid wavelengths, this would still cut the total number of cards in half. Another potentially useful technology is pluggable optical transceivers, such as the gigabit interface converter (GBIC) package or emerging pluggable small form factor transceivers. This could mean that an adapter card would be upgradable in the field, or could be more easily repaired by simply changing the optical interface; this reduces the requirement to keep large numbers of cards as field spares in case of failure. Also, a pluggable interface may be able to support the full physical layer of some protocols, including the maximum distance, without the need for optical attenuators, patch panels, or other connections; the tradeoff for this native attachment is that changing the card protocol would require swapping the optical transceiver on the adapter card.

Using currently available technology, it is much easier to implement ITU grid lasers and optical amplifiers at C-band wavelengths than at L-band wavelengths. Hence, some implementations use tighter wavelength spacing in order to fit 32 or more channels in C-band; the scalability of this approach remains open to question. Other approaches available today make use of L-band wavelength and larger inter-wavelength spacing, and can be more easily scaled to larger wavelength counts in the future. Performance of the multiplexer's optical transfer function (OTF), or characterization of the allowable optical power ripple as a function of wavelength, is an important parameter in both C-band and L-band systems.

5.4.2.2 Network topologies

Conventional SONET networks are designed for the WAN and are based on reconfigurable ring topologies, while most datacom networks function as switched networks in the LAN and point-to-point in the WAN or MAN. There has been a great deal of work done on optimizing nationwide WANs for performance and scalability, and interfacing them with suitable LAN and MAN topologies; WDM plays a key role at all three network levels, and various traffic engineering approaches will be discussed later in this chapter. Despite the protocol independent nature of WDM technology, many WDM products targeted at telecom carriers or local exchange carriers (LECs) were designed to carry only SONET or SDH compatible traffic. Prior to the introduction of WDM, it was not possible to run other protocols over a ring unless they were compatible with SONET frames; WDM has made it possible to construct new types of protocol-independent network topologies. For the first time, datacom protocols such as ESCON may be configured into WDM rings, including hubbed rings (a central node communicating with multiple remote nodes), dual hubbed rings (the same as a single hub ring except that the hub is mirrored into another backup location), meshed rings (any-to-any or peer-to-peer communication between nodes on a ring), and linear optical add/drop multiplexing (OADM) or so-called "opened rings" (point-to-point systems with add/drop of channels at intermediate points along the link in addition to the endpoints). These topologies are illustrated in Fig. 5.4.6. Note that since the DWDM network is protocol independent, care must be taken to construct networks that are functionally compatible with the attached equipment; as an example, it is possible to build a DWDM ring with attached Fibre Channel equipment that does not comply with recommended configurations such as Fibre Channel Arbitrated Loop. Other network implementations are also possible; for example, some metro WDM equipment offers a 2-tier ring consisting of a dual fiber ring and a separate, dedicated fiber link between each node on the ring to facilitate network management and configuration flexibility (also known as a "dual homing" architecture).

5.4.2.3 Distance and repeaters

It is desirable to support the longest distance possible without repeaters between nodes in a WDM network. Note that the total supported distance for a WDM system may depend on the number of channels in use; adding more channels requires additional wavelength multiplexing stages, and the optical fibers can reduce the available link budget. The available distance is also a function of the network topology; WDM filters may need to be configured differently, depending on whether they form an optical seam (configuration that does not allow a set of wavelengths to propagate into the next stage of the network) or optical bypass (configuration that permits wavelengths to pass through into the rest of the network). Thus, the total distance and available link loss budget in a point-to-point network may be different from the distance in a ring network. The available distance is typically independent of data rate up to around 2 Gbit/s; at higher data rates, dispersion may limit the achievable distances. This should be kept in mind when installing a new WDM system that is planned to be upgraded to significantly higher data rates in the future. The maximum available distance and link loss may also be reduced if optional optical switches are included in the network for protection purposes. In some cases, it is possible to concatenate or cascade WDM networks together to achieve longer total distances; for example, by daisy chaining two point-to-point networks the effective distance can be doubled (if there is a suitable location in the middle of the link). Better performance in concatenated applications is usually achieved with channels that retime the data; data retiming is a desirable property, because it improves the signal fidelity and reduces noise and jitter. There are three levels of functionality, namely Retiming (removes timing jitter to improve clock recovery at the receiver), Reshaping (removes pulse shape distortion such as that caused by dispersion), and Regeneration (ensures the outgoing signal has sufficient power to reach its next destination). Devices that support only the first two are known as "2R" repeaters, while devices that support all three are called "3R" repeaters. Generally speaking, longer distances and better data fidelity are possible using 3R repeaters; however, this class of repeaters must be configured in either software, hardware, or both to recognize at least the data rate on the link. Care must be taken to keep the advantages of a protocol-independent design when configuring a 3R repeater.

Some WDM devices may also support longer distances using optical amplifiers in either pre-amp, post-amp, or mid-span configurations. Recently, record-breaking terabit per second point-to-point transmission in systems over more than 2000 km has been reported [7] based on dispersion-managed nonlinear transmission techniques, without the need for signal regeneration devices, using only linear optical amplifiers. However, if we extend our discussion to consider terrestrial photonic networks with dynamic routing capability, more sophisticated regeneration schemes could be necessary in order to compensate for signal quality discrepancies between high data rate WDM channels, which could be routed at different times over variable distances. Although classical

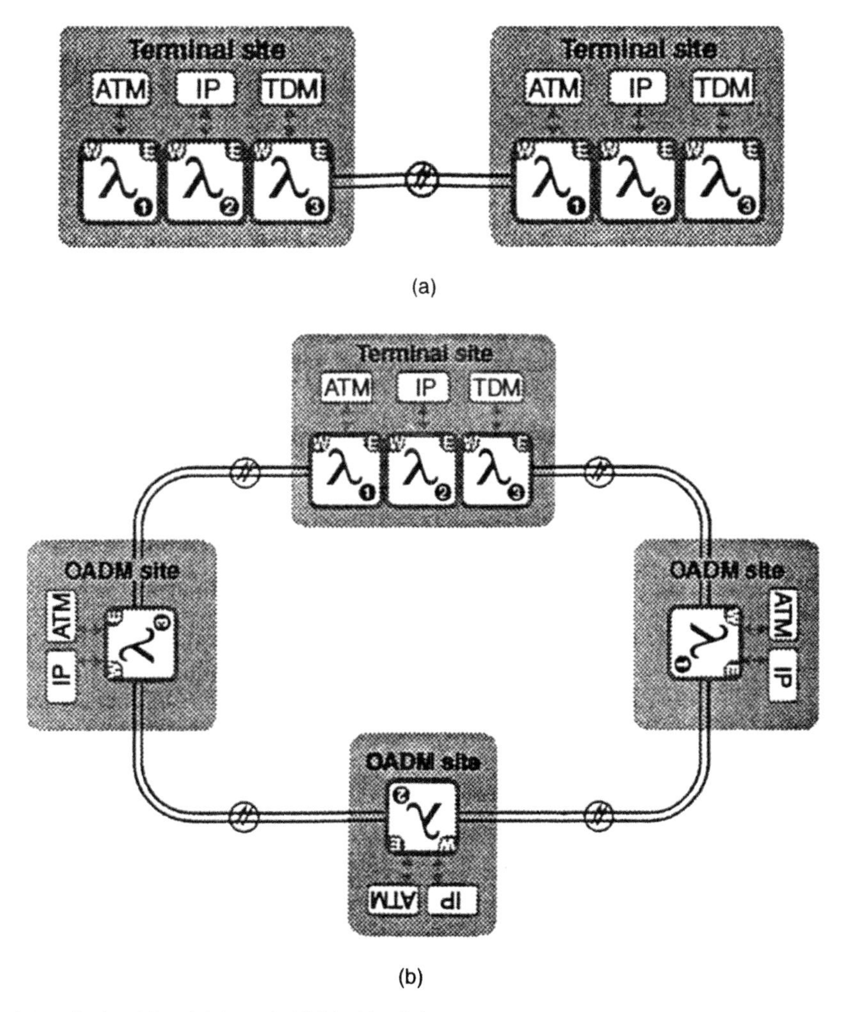

Fig. 5.4.6 WDM network topologies (a) point-to-point (b) hubbed ring.

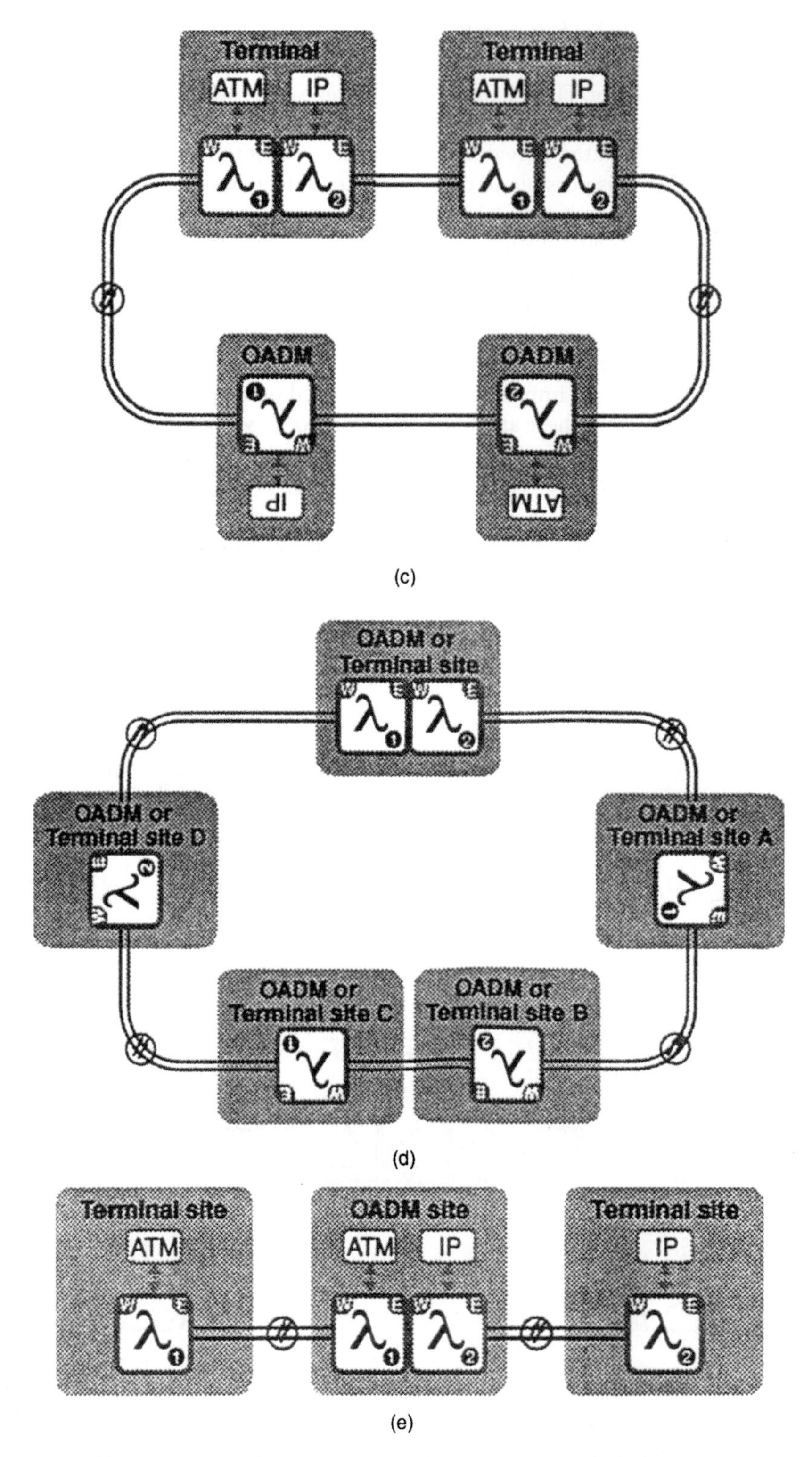

Fig. 5.4.6 (c) dual hubbed ring (d) meshed ring. (e) linear optical add/drop multiplexer (OADM).

optoelectronic regenerators as used in the modern telecommunication industry constitute an attractive solution to this problem for single-channel, single-data-rate transmission, it is not yet clear whether they could also serve as a cost-effective solution for a rapidly growing broadband WDM network. All-optical regenerators based on components that can process low as well as high bit rate signals could be an interesting alternative.

Signal degeneration in fiber systems arises from various sources, including amplified spontaneous emission (ASE) due to optical amplifiers, pulse spreading due to group velocity dispersion (GVD) (which can be corrected in principle through passive dispersion compensation schemes), polarization mode dispersion (PMD), and other nonlinear effects such as Kerr effect signal distortion and jitter for data rates above 10 Gbit/s. If timing jitter is negligible, simple amplification and reshaping processes are usually enough to maintain signal quality over long distances by preventing the accumulation of noise and distortion. Many repeaters convert the WDM optical signal to an electrical signal, then back into an optical signal for re-transmission; however, various schemes for all-optical 2R and 3R repeaters have been suggested. A 2R regenerator consists mainly of a linear amplifier (which may be an optical amplifier) followed by a data-driven nonlinear optical gate (NLOG), which modulates a low-noise continuous light source. If the gate transmission vs. signal intensity characteristics yield a thresholding or limiting behavior, then the signal extinction ratio can be improved and ASE noise can be partially reduced; in addition, accumulated signal chirp can be compensated. In some cases timing jitter is also a concern, for example, due to cross-phase modulation in WDM systems or pulse edge distortions due to the finite response times of nonlinear analog signal processing devices (wavelength converters). In these cases, 3R regeneration may be necessary. The basic structure of an optical 3R regenerator consists of an amplifier (which may be optical), a clock recovery function to provide a jitter-free short pulse clock signal, and a data-driven nonlinear optical gate that modulates this clock signal. The core function of the optical regenerators is thus a nonlinear gate featuring signal extinction ratio enhancement and noise reduction. Such gates can be either optical in nature (as in the case of proposed all-optical repeaters) or electronic (as in the case of hybrid optical-to-electrical conversion-based devices). Semiconductor-based gates are much more compact than fiber-based devices. It is also possible to further subdivide this class of semiconductor devices into passive devices, such as saturable absorbers, and active devices, such as semiconductor optical amplifiers that require an electrical power supply. For optical 3R regeneration, a synchronous jitter-free clock stream must be recovered from the incident signal [8]. Many solutions to this problem have been reported, and it would be beyond the scope of this chapter to describe them all. Although 2R regenerators are attractive because of their relative simplicity, it is not clear whether they will be adequate above 10 Gbit/s data rates since this would require components with very short transition time responses.

5.4.2.4 Latency

DWDM devices also function as channel extenders, allowing many data-corn protocols to reach previously impossible distances (50–100 km or more). Combined with optical amplifier technology, this has led some industry analysts to proclaim "the death of distance," meaning connection distances should no longer pose a serious limitation in optical network design. However, in many real-world applications, it is not sufficient to simply extend a physical connection; performance of the attached datacom equipment must also be considered. Latency, or propagation delay due to extended distances, remains a formidable problem for optical data communication. The effects of latency are often protocol specific or device specific. For example, using DWDM technology it is possible to extend an ESCON channel to well over 50 km. However, many ESCON control units and DASD are synchronous, and exhibit timing problems at distances beyond about 43 km. Some types of asynchronous DASD overcome this limitation; however, performance of the ESCON protocol also degrades with distance. Due to factors such as the buffer size on an ESCON channel interface card and the relatively large number of acknowledgments or handshakes required to complete a data block transfer (up to 6 or more), ESCON begins to exhibit performance drop at around 9 km on a typical channel, which grows progressively worse at longer distances. For example, at 23 km a typical ESCON channel has degraded from a maximum throughput of 17.5 MByte/s to about 10 MByte/s; if the application is a SAN trying to back up a petabyte database, the time required to complete a full backup operation increases significantly. This problem can be addressed to some degree by using more channels (possibly driving the need for a multiplexer to avoid fiber exhaust) or larger data block sizes, and is also somewhat application dependent. Other protocols, such as FICON, can be designed to perform much better at extended distances.

A further consideration is the performance of the attached computer equipment. For example, consider the effect of a 10 km duplex channel with 100 microseconds round-trip latency. A fast PC running at 500 MHz clock rate would expend 50,000 clock cycles waiting for the attached device to respond, while a mainframe executing 1000 MIPS could expend over 100,000 instruction cycles in the same amount of time. The effect this may have on the end user depends on factors such as the application software; some of the effects this latency can have on the performance of a Parallel Sysplex under similar conditions. As another example, there is tremendous activity in burst mode routing and control traffic for next-generation Internet (NGI) applications.

In many cases, it is desirable for data to be transported from one point in the network to one or more other points in the least possible time. For some applications the time sensitivity is so important that minimum delay is the overriding factor for all protocol and equipment design decisions. A number of schemes have been proposed to meet this requirement, including both signaling based and equipment intensive solutions employing network interface units (NIUs) and optical crossbar switches (OXBS). In order to appreciate the impetus for designing burst mode switching networks, it is useful to consider the delays that are encountered in wide area data networks such as the Internet. For example, it takes 20.5 ms for light to travel from San Francisco to New York City in a straight line through an optical fiber, without considering any intervening equipment or a realistic route. Many transmission protocols require that data packets traverse the network only after a circuit has been established; the setup phase in TCP/IP, for example, involves 3 network traversals (sending a S YN packet, responding with an ACK, and completing the procedure with another ACK) before any data packets can be sent. While this helps ensure reliable data transport, it also guarantees a minimum network delay of more than 80 ms before data can be received. For some applications this may be an unnecessary overhead to impose.

One way to mitigate the impact of network setup time on packet latency is to pipeline the signaling messages. Two methods for accomplishing these time savings have been proposed [9, 10] which allow the first packet delay to be reduced from 2 round trips to 1 ½ round trips; these schemes are currently undergoing field trials. Another paradigm for providing a burst mode switching capability using conventional equipment is to send a signaling packet prior to the actual data packet transmission, and transmitting the data while the setup is in progress. Some WDM systems have proposed using this approach [11], using a dedicated wavelength for signal control information, a burst storage unit in order to buffer packets when necessary, and an implementation-friendly link scheduling algorithm to ensure efficient utilization of the WDM channels. Another alternative is commonly known as "spoofing" the channel; the attached channel extension equipment will be configured to send acknowledgments prior to actually delivering the data. While this reduces latency and improves performance, it also makes the assumption that data can be delivered very reliably on the optical link; if there is a transmission problem after the attached channels have received their acknowledgment of successful delivery, the error recovery problem becomes quite difficult. Although this approach has been implemented in commercial devices, it has not been widely accepted because of the potential data integrity exposures inherent in the design.

In many optical networks, cross-connect switches are employed, which act like electronically reconfigurable patch panels. The cross-connect changes its switching state in response to external control information, such as from an outband signal in the data network. The alternative to signaled data transmission is data switching using header information. This is somewhat different from a typical ESCON or Gigabit Ethernet switch, which performs optical to electrical conversion, reads the data header, sets the switch accordingly, and then reconverts the data to the optical domain for transmission (sometimes retiming the signal to remove jitter in the process). An example of outband switching networks has been demonstrated [12]; in the approach, a data packet is sent simultaneously with a header that contains routing information. The headers are carried along with the data, but out of band, using subcarrier multiplexing or different wavelengths. At switch nodes inside the network, the optical signal is sampled just prior to entering a short fiber delay line. While most of the signal is being delayed, a fast packet processing engine determines the correct state of the switching fabric, based on the incoming header and a local forwarding table. The switch fabric is commanded to enter a net state just before the packet exits the delay line and enters the switch. This method provides the lowest possible latency for packet transmission and removes the task of pre-calculating the arrival time of burst mode data packets.

5.4.2.5 Protection and restoration

Backup fiber protection or restoration refers to the multiplexer's ability to support a secondary fiber path for redundancy in case of a fiber break or equipment failure. The most desirable is so-called "1 + 1" SONET-type protection switching, the standard used by the telecom industry, in which the data is transmitted along both the primary and backup paths simultaneously, and the data switches from the primary to the backup path within 50 ms. (This is the SONET industry standard for voice communications; while it has been commonly adopted by protocol-independent WDM devices, the effects of switching time on the attached equipment depends on the application, as in the previous discussion of latency effects.) There are different meanings for protection, depending on whether the fiber itself or the fiber and electronics are redundant. In general, a fully protected system includes dual redundant cards and electrical paths for the data within the multiplexer, so that traffic is switched from the primary path to the secondary path not only if the fiber breaks, but also if a piece of equipment in the multiplexer fails. A less sophisticated but more cost-effective option is the so-called fiber trunk switch, which simply switches from the primary path to the backup path if the primary path breaks. Not all trunk switches monitor the backup path as 1 + 1 switches do, so they can run the risk of switching traffic to a path that is not intact (some trunk switches provide a so-called "heartbeat" function, sending a light pulse down the backup path every second or so to establish that the backup link remains available). Trunk switches are also slower, typically taking from 100 ms to as much as 2 seconds to perform this switchover; this can be disruptive to data traffic. There are different types of switches; a unidirectional switch will only switch the broken fiber to its backup path (for example, if the transmit fiber breaks, then the receive fiber will not switch at the same time). By contrast, a bidirectional switch will move both the transmit and receive links if only one of the fibers is cut. Some types of datacom protocols can only function properly in a bidirectional switches environment because of timing dependencies on the attached equipment; the link may be required to maintain a constant delay for both the transmit and received signal in a synchronous computer system, for example. Some switches will toggle between the primary and secondary paths searching for a complete link, while others are non-revertive and will not return to the primary path once they have switched over. It is desirable to have the ability to switch a network on demand from the network management console, or to lock the data onto a single path and prevent switching (for example, during link maintenance).

Another desirable switching feature is the ability to protect individual channels on a per channel basis as the application requires; this is preferable to the "all or none" approach of trunk switches, which require that either all channels be protected, or no channels be protected. Hybrid schemes using both trunk switching and 1 + 1 protection are generally not used, as they require some means of establishing apriority of which protection mechanism will switch first and they defeat the purpose of the lower cost trunk switch. Other features such as dual redundant power supplies and cooling units with concurrent maintenance (so-called "hot swappable" components, which can be replaced without powering down the device) should also be part of a high-reliability installation.

Another desirable property is self-healing, which means that in the event of a fiber break or equipment failure the surviving network will continue to operate uninterrupted. This may be accomplished by re-routing traffic around the failed link elements; some form of protection switching or bypass switching can restore a network in this manner. In a larger network consisting of multiple cross-connects or add/drop multiplexers, there are two approaches to producing optical self-healing networks [13]. One is to configure a physical-mesh topology network with optical cross connects (OXCs), and the other is to configure a physical ring topology network with optical add/drop multiplexers (OADMs, or WDM ADMs). Ring topology based optical self-healing networks are generally preferred as a first step because of the lower cost of OADMs compared with OXCs; also the protection speed in a ring topology is much faster, and the OADM is more transparent to data rate and format. Eventually, in future photonic networks, multiple optical self-healing rings may be constructed with emerging large-scale OXCs. New architectures have also been proposed, such as bi-directional wavelength path-switched ring (BWPSR), which uses bidirectional wavelength-based protection and a wavelength-based protection trigger self-healing ring network [14].

5.4.2.6 Network management

Some DWDM devices offer minimal network management capabilities, limited to a bank of colored lamps on the front panel; others offer sophisticated IP management and are configured similar to a router or switch. IP devices normally require attached PCs for setup (defining IP addresses, etc.) and maintenance; some can have a "dumb terminal" attachment to an IP device which simplifies the setup, but offers no backup or redundancy if the IP site fails. Some devices offer minimal information about the network; others offer an in-band or out-band service channel that carries management traffic for the entire network. When using IP management, it is important to consider the number of network gateway devices (typically routers or switches) that may be attached to the DWDM product and used to send management data and alarms to a remote location. More gateways are desirable for greater flexibility in managing the product. Also, router protocols such as Open Shortest Path First (OSPF) and Border Gateway Protocol (BGP) are desirable because they are more flexible than "static" routers; OSPF can be used to dynamically route information to multiple destinations.

Many types of network management software are available — in data-com applications, these are often based on standard SNMP protocols supported by many applications such as HP Openview, CA Unicenter, and Tivoli Netview. IP network management can be somewhat complex, involving considerations such as the number of IP addresses, number of gateway elements, resolving address conflicts in an IP network, and others. Telecom environments will often use the TL-1 standard command codes, either menu-driven or from a command line interface, and may require other management features to support a legacy network environment; this may include CLEI codes compatible with the TERKS system used by the telecommunications industry and administered by Telcordia Corp., formerly Bell Labs [15]. User-friendly network management is important for large DWDM networks, and facilitates network troubleshooting and installation. Many systems require one or more personal computers to run network management applications; network management software may be provided with the DWDM, or may be required from another source. It is desirable to have the PC software pre-loaded to reduce time to installation (TTI), pre-commissioned and pre-provisioned with the user's configuration data, and tested prior to shipment to ensure it will arrive in working order. Finally, note that some protocol-specific WDM implementations also collect network traffic statistics, such as reporting the number of frames, SCSI read/write operations, Mbytes payload per frame, loss of sync conditions, and code violations (data errors). Optical management of WDM may also include various forms of monitoring the physical layer, including average optical power per channel and power spectral density, in order to proactively detect near end-of-life components or optimize performance in amplified WDM networks.

5.4.2.7 Nonlinear effects and optical amplifiers for WDM

The interaction of light with the optical fiber material is typically very small, particularly at low optical power levels. However, as the level of optical power in the fiber is increased nonlinear effects can become significant; this is especially important for long-distance fiber links, which provide the opportunity for smaller effects to build up over distance. Because WDM involves transmitting many optical signals over a common fiber, nonlinear effects in the link can become significant. In an optical communication system, nonlinear effects can induce transmission errors that place fundamental limits on system performance, in much the same fashion as attenuation or dispersion effects. At the same time, some important components such as amplifiers for extended distance WDM systems rely on nonlinear effects for their operation.

One of the most common nonlinear interactions is known as "four wave mixing" (FWM), which occurs when two or more optical signals propagate in the same direction along a common single-mode fiber. As illustrated in Fig. 5.4.7, optical signals in the fiber can mix to produce new signals at wavelengths that are spaced at the same intervals as the original signals. The effect can also occur between three or more signals, making the overall effect quite complex. FWM increases exponentially with signal power, and becomes greater as the channel spacing is reduced; in particular, it is a concern with dense wavelength division multiplexing systems. If WDM channels are evenly spaced, then the spurious FWM signals will appear in adjacent wavelength channels and act as noise. One method of dealing with this problem is to space the channels unevenly to reduce the effect of added noise on adjacent channels; however, FWM still removes some optical power from the desired signal levels. Because FWM is caused by signals that remain in phase with each other over a significant propagation distance, the effect is stronger for lasers with a long coherence length. Also, FWM is strongly influenced by chromatic dispersion — because dispersion ensures that different signals do not stay in phase with each other for very long, it acts to reduce the effect of FWM.

Another nonlinear optical phenomena is known as frequency chirping, or inducing a linear frequency sweep in an optical pulse. Until now, we have assumed that the refractive index of the fiber core is a constant, independent of the optical power. Actually, sufficiently high optical power levels can affect the material properties of the glass and induce small changes in the refractive index. The frequency chirp is generated by self-phase modulation, which arises from the interaction of the propagating light and the intensity dependent portion of the fiber's refractive index [17]. Because these effects are caused by the propagating signal itself, they are known as carrier-induced phase modulation (CIP). The fiber's refractive index can be expressed as follows:

$$n_1 = n_{10} + n_{12}I(t), \tag{5.4.1}$$

where n_{10} is the refractive index under low optical power conditions (for this case $n_1 = n_{10}$), $I(t)$ is the intensity

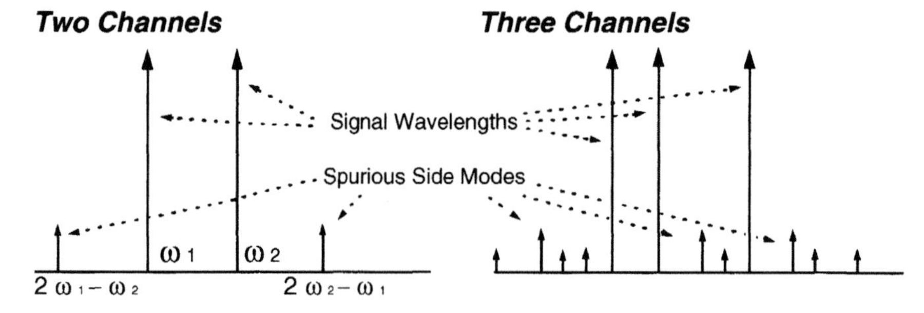

Fig. 5.4.7 Example of four wave mixing (FWM) in WDM networks with 2 channels and with 3 channels.

profile of the propagating light, and n_{12} is a positive material constant [17]. The propagation constant is then given by

$$k_1 = \omega_0 n_{10}/c + \Gamma I(t), \tag{5.4.2}$$

where the constant Γ represents a collection of terms [17]. The phase of the optical pulse now becomes

$$\Psi = \omega_0 t - \omega_0 n_{10} z/c - \Gamma I(t), \tag{5.4.3}$$

where z is the propagation distance. The instantaneous frequency of the light is thus proportional to the negative time derivative of the intensity profile,

$$\omega i = d\Psi/dt = \omega_0 - \Gamma dI(t)/dt, \tag{5.4.4}$$

and the properties of the resulting chirp depend on the time-varying light intensity; for many pulse shapes including Gaussian, this results in a nonuni-form frequency chirp. Another property of interest is the group velocity dispersion (GVD); this is calculated by expanding the propagation constant, $k(\omega)$, about its center frequency and retaining the second-order derivative [18]. Optical fibers with positive GVD cause the frequency components to spread out as the light propagates along the fiber; by contrast, in a fiber with negative GVD the frequency components move closer together as the light propagates. The specific material properties of the fiber and the wavelength of light determine whether the GVD is positive or negative. The combined effects of self-phase modulation and GVD are called dispersive self-phase modulation; this effect has applications in optical pulse compression systems [18]. When there are multiple signals at different wavelengths in the same fiber, nonlinear phase modulation in one signal can induce phase modulation of the other signals. This is known as cross-phase modulation; in contrast with other nonlinear effects, it does not involve power transfer between signals. Cross-phase modulation can introduce asymmetric spectral broadening and distortion of the pulse shapes.

At higher optical power levels, nonlinear scattering may limit the behavior of a fiber optic link. The dominant effects are stimulated Raman and Brillouin scattering. When incident optical power exceeds a threshold value, significant amounts of light may be scattered from small imperfections in the fiber core or by mechanical (acoustic) vibrations in the transmission media. These vibrations can be caused by the high-intensity electromagnetic fields of light concentrated in the core of a single-mode fiber. Because the scattering process also involves the generation of photons, the scattered light can be frequency shifted [16]. Put another way, we can think of the high-intensity light as generating a regular pattern of very slight differences in the fiber refractive index; this creates a moving diffraction grating in the fiber core, and the scattered light from this grating is Doppler shifted in frequency by about 11 GHz. This effect is known as stimulated Brillouin scattering (SBS); under these conditions, the output light intensity becomes nonlinear as well. Stimulated Brillouin scattering will not occur below a critical optical power threshold. Brillouin scattering has been observed in single-mode fibers at wavelengths greater than cutoff with optical power as low as 5 mW; it can be a serious problem in long-distance communication systems when the span between amplifiers is low and the bit rate is less than about 2 Gbit/s, in WDM systems up to about 10 Gbit/s when the spectral width of the signal is very narrow, or in remote pumping of some types of optical amplifiers [16]. In general, SBS is worse for narrow laser linewidths (and is generally not a problem for channel bandwidth greater than 100 MHz), wavelengths used in WDM (SBS is worse near 1550 nm than near 1300 nm), and higher signal power per unit area in the fiber core. In cases where SBS could be a problem, the source linewidth can be intentionally broadened by using an external modulator or additional RF modulation on the laser injection current. However, this is a tradeoff against long-distance transmission, since broadening the linewidth also increases the effects of chromatic dispersion.

When the scattered light experiences frequency shifts outside the acoustic phonon range, due instead to modulation by impurities or molecular vibrations in the fiber core, the effect is known as stimulated Raman scattering (SRS). The mechanism is similar to SBS, and scattered light can occur in both the forward and backward directions along the fiber; the effect will not occur below a threshold optical power level. As a rule of thumb, the optical power threshold for Raman scattering is about three times larger than for Brillouin scattering. Another good rule of thumb is that SRS can be kept to acceptable levels if the product of total power and total optical bandwidth is less than 500 GHz-W. This is quite a lot. For example, consider a 10-channel DWDM system with standard wavelength spacing of 1.6 nm (200 GHz). The bandwidth becomes $200 \times 10 = 2000$ GHz, so the total power in all 10 channels would be limited to 250 mW in this case (in most DWDM systems, each channel will be well below 10 mW for other reasons such as laser safety considerations). In single-mode fiber, typical thresholds for Brillouin scattering are about 10 mW and for Raman scattering about 35 mW; these effects rarely occur in multimode fiber, where the thresholds are about 150 mW and 450 mW, respectively. In general, the effect of SRS becomes greater as the signals are moved further apart in wavelength (within some limits); this introduces a tradeoff with FWM, which is reduced as the signal spacing increases.

Optical amplifiers can also be constructed using the principle of SRS; if a pump signal with relatively high power (half a watt or more) and a frequency 13.2 THz higher than the signal frequency is coupled into a sufficiently long length of fiber (about 1 km), then amplification of the signal will occur. Unfortunately, more efficient amplifiers require that the signal and pump wavelengths be spaced by almost exactly the Raman shift of 13.2 THz, otherwise the amplification effect is greatly reduced. It is not possible to build high-power lasers at arbitrary signal wavelengths; one possible solution is to build a pump laser at a convenient wavelength, then wavelength shift the signal by the desired amount [16]. However, another good alternative to SRS amplifiers is the widely used erbium-doped fiber amplifiers (EDFA). These allow the amplification of optical signals along their direction of travel in a fiber, without the need to convert back and forth from the electrical domain. While there are other types of optical amplifiers based on other rare earth elements such as praseodymium (Pd) or neodymium (Nd), and even some optical amplifiers based on semiconductor devices, the erbium-doped amplifiers are the most widely used because of their maturity and good performance at wavelengths of interest near 1550 nm.

An EDFA operates on the same principle as an optically pumped laser; it consists of a relatively short (about 10 meters) section of fiber doped with a controlled amount of erbium ions. When this fiber is pumped at high power (10 to 300 mW) with light at the proper wavelength (either 980 nm or 1480 nm) the erbium ions absorb the light and are excited to a higher energy state. Another incident photon around 1550 nm wavelength will cause stimulated emission of light at the same wavelength, phase, and direction of travel as the incident signal. EDFAs are often characterized by their gain coefficient, defined as the small signal gain divided by the pump power. As the input power is increased, the total gain of the EDFA will slowly decrease; at some point, the EDFA enters gain saturation, and further increases to the input power cease to result in any increase in output power. Because the EDFA does not distort the signal, unlike electronic amplifiers, it is often used in gain saturation. The gain curve of a typical EDFA as a function of wavelength is shown in Fig. 5.4.8; note that the gain at 1560 nm is about twice as large as the gain at 1540 nm. This can be a problem when operating WDM systems; some channels will be strongly amplified and dominate over other channels that are lost in the noise. Furthermore, a significant complication with EDFAs is that their gain profile changes with input signal power levels; so, for example, in a WDM system the amplifier response may become nonuniform (different channels have different effective gain) when channels are added or dropped from the fiber. This requires some form of equalization to achieve a flat gain across all channels. There has been

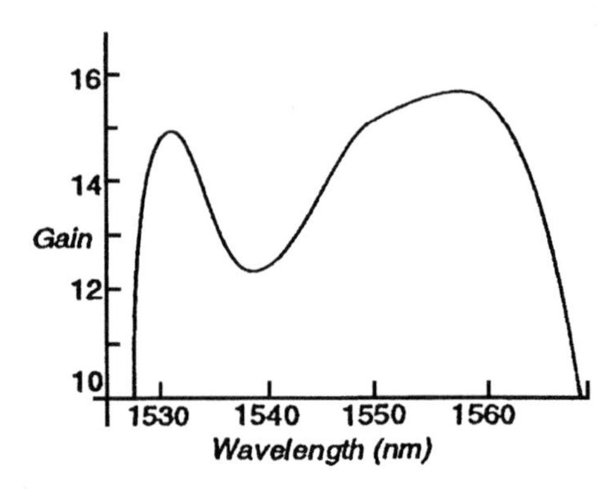

Fig. 5.4.8 Gain vs. wavelength for a typical erbium-doped optical amplifier.

a great deal of research in this area; some proposals include adding an extra WDM channel locally at the EDFA to absorb excess power (gain clamping), and manipulating either the fiber doping or core structure. Another concern with EDFAs is that some of the excited erbium undergoes spontaneous emission, which can create light propagating in the same direction as the desired signal. This random light is amplified and acts as background noise on the fiber link; the effect is known as amplified spontaneous emission (ASE). Since ASE can be at the same wavelength as the desired signal, it may be difficult to filter out; furthermore, ASE accumulates in systems with multiple amplifier stages and is proportional to the amplifier gain.

Other nonlinear effects can be used to produce useful devices by changing the properties of the fiber itself; one of the most common examples is fiber Bragg gratings. When an optical fiber is exposed to ultraviolet light, the fiber's refractive index is changed; if the fiber is then heated or annealed for a few hours, the index changes can become permanent. This phenomena is called photosensitivity [19, 20]. The magnitude of the index change depends on many factors, including the irradiation wavelength, intensity, and total dose, the composition and doping of the fiber core, and any materials processing done either prior or subsequent to irradiation. In germanium-doped single-mode fibers, index differences between 10^{-3} and 10^{-5} have been obtained. Using this effect, periodic diffraction gratings can be written in the core of an optical fiber. This was first achieved by interference between light propagating along the fiber and its own reflection from the fiber endface [21]; this is known as the internal writing technique and the resulting gratings are known as Hill gratings. Another approach is the transverse holographic technique, in which the fiber is irradiated from the side by two beams which intersect at an angle within the fiber core. Gratings can also be written in the fiber core by irradiating the fiber through a phase mask with a periodic structure. These techniques can be used to write fiber Bragg gratings in the fiber core;

such gratings reflect light in a narrow bandwidth centered around the Bragg wavelength, λ_B, which is given by

$$\lambda_B = 2N_{eff}\Lambda, \quad (5.4.5)$$

where Λ is the spatial period, or pitch, of the periodic index variations and N_{eff} is the effective refractive index for light propagating in the fiber core. There are many applications for fiber Bragg gratings in optical communications and optical sensors, such as tapped optical delay lines, filters, multiplexers, optical strain gauges, and others (an extensive review is provided in ref. [22,23]). Fiber Bragg gratings function in reflection, while many applications require a transmission effect; this conversion is accomplished using an optical circulator, a Michaelson or Mach-Zender interferometer, or a Sagnac loop [24]. Fiber Bragg gratings can be used to multiplex and demultiplex wavelengths in a WDM system, or to fabricate add/drop filters within the optical fibers that offer very low insertion loss; they can also be used in various dispersion compensation schemes. These devices represent a promising new technology for future commercial WDM applications.

5.4.3 Commercial WDM systems

Many commercial DWDM products in use today have been developed for the telecommunication and data communications market; there are also a number of testbeds and service trials underway, and new products or technologies are being proposed at a rapid pace. Some examples include the MultiWave WDM terminal from Ciena Corp., which accommodates up to 16 °C-48 channels and will soon be upgradable to more than 32 channels, and the WaveMux from Pirelli, which handles up to 10 °C-192 channels or 32 °C-48 channels. There is also a unique implementation of WDM using free-space optics, the Lucent OptiAir (a free space 4-channel wavelength multiplexed communication link, for applications where physical fiber connections are not practical, such as communication with ships at dock in a seaport). Several other corporations are also investigating applications of WDM, including the ESPRIT program in Europe, the Optical Network Technology Consortium, the All Optical Network Consortium, the MONET consortium led by AT&T, and others [25].

In particular, the MONET project is a multi-vendor government-sponsored network consortium and testbed for future WDM systems [26]. While its capabilities and applications continue to evolve, we can discuss a recent snapshot of the network as an example of how WDM is being evaluated for future networking applications. The MONET network in Washington, D.C. consists of a pair of two fiber rings denoted as the east and west ring. The east ring is provided and maintained by Lucent Technologies. This ring consists of two wavelength selective cross-connects located at the Laboratory for Telecommunication Sciences at the National Security Agency and the Naval Research Laboratory, one wavelength add/drop multiplexer at the National Aeronautics and Space Administration, and two wavelength amplifiers in the Bell Atlantic central office. The west ring is provided by Tellium and managed by Telcordia Technologies. This ring consists of three wavelength ADMs at the Defense Advanced Research Projects Agency/Information Systems Institute, Defense Information Systems Agency, and Defense Intelligence Agency, as well as four wavelength amplifiers in the Bell Atlantic central offices. Bell Atlantic provides the in-ground fiber infrastructure for the network. The whole network is controlled and managed by Telcordia Technologies' CORBA-based network control and management (NC&M) system, which runs on the ATM network at OC-3c rates over a supervisory wavelength channel at 1510 nm. The Lucent switching fabric is based on lithium niobate devices and is thus protocol transparent. The Tellium ADMs are based on wavelength transponders and transceivers using O/E/O conversion and an electrical switching fabric. Thus, the west ring is opaque and has wavelength interchange capability.

The west ring has 4 modes of transmission; OC-3, OC-12, OC-48, and optical signal (OSIG) mode, which is used to transmit non-SONET rate signals up to 2.5 Gbit/s without timing recovery during the O/E/O conversion. Many experiments have been conducted on this network; for example, recently it was used to test the long-distance transmission capabilities of Gigabit Ethernet in OSIG mode [26]. This was accomplished using two Pentium II workstations as test hosts running short wave Gigabit Ethernet links to a switch, which was in turn configured into two virtual LANs (VLANs) to prevent the hosts from bridging packets directly between them. Packets were thus forced over the Gigabit Ethernet LX ports into the WDM network. This work has demonstrated for the first time transport of Gigabit Ethernet packets directly over WDM for 1062 km; this suggests that this protocol, and probably also the Fibre Channel physical layer, can be used for WAN backbone transmission, bypassing the intermediate ATM/SONET protocol layers. Of course, performance issues remain to be resolved; using the jumbo frames defined in the Gigabit Ethernet standard, a maximum of about 9 Kbytes data block size is permitted, which is not large compared with Fibre Channel or ESCON block sizes. Thus, the role of Gigabit Ethernet in the SAN remains to be determined.

MONET is an example of an active WDM WAN employing a multiven-dor environment. Some of the other WDM networking efforts currently underway deal with passive optical networks (PON) [27]; other research programs are underway to investigate new architectures for client-independent OTNs including IP over DWDM; one example is the EURESCOM project P918 [28]. Similarly, commercial WDM products offer a wide range of options. Coarse WDM products may consist of individual transceivers or multichip modules with built-in optical multiplexing functions. Most DWDM products are based on a fundamental building block, such as a card cage, which can be daisy-chained or scaled to accommodate more wavelength channels. Some building blocks are mounted in a standard 19- or 23-inch equipment rack, which is assembled in the field; others are pre-packaged as standalone boxes, which vary from large systems to smaller packages (the size of a large PC); the physical size of a WDM system depends on many factors, including the choice of design point, transponders vs. separate client/network cards, use of small form factor optical transceivers, etc. In some cases, a smaller "satellite box" with only a few wavelengths or even a coarse WDM solution is deployed close to the end user premises, which feeds into a larger DWDM node some distance away on the WAN. Many different types of network management are offered, which may consist of a proprietary graphic user interface with protocol-specific performance monitoring to an industry standard SNMP approach or a TL-1 (per EIA standard 232) and CMISE interface which allows telco standard operations, administration, maintenance, and provisioning functions (OAM&P). Use of sub-rate TDM or CDMA, supported network topologies with different bandwidths depending on the configuration, number of channels over a single fiber or pair of fibers, and many other design choices contribute to the differences between commercially available WDM products. To further complicate matters, many WDM products are resold or rebranded by different companies offering a range of technical support and maintenance options.

Given the rapid growth of this technology, it is not possible to comprehensively list all of the commercial products being introduced, or all of the research efforts currently underway to support future products. A sample of selected WDM companies and WAN backbone carriers is given in Table 5.4.3. Product marketing literature and claims of supported features should always be examined with care; potential users should evaluate the most recent technical information from prospective companies to find the best solution for a given application. Instead of attempting to give a detailed technical description of all the major commercial WDM product offerings, we will select one representative product to illustrate first-generation and third-generation devices (fourth-generation ultra-dense WDM devices are not yet commercially available). For each of these typical devices, we will provide a detailed description of its components, packaging, and functionality; similar building blocks are used by other commercial products.

5.4.3.1 Firstgeneration WDM: The IBM 9729 optical wavelength division multiplexer

One example of a first-generation DWDM system is the IBM 9729 Optical Wavelength Division Multiplexer [1, 29, 30]. The first WDM product developed specifically for the datacom industry, it became available as a special request product in 1993 and was released as a commercial product in 1996. The device is shown in Fig. 5.4.9; it allows the transmission of up to 10 full duplex links (20 independent data streams) over a single fiber. Using different adapter cards, a mixture of ESCON, Fibre Channel, FDDI, and ATM links can be plugged into the device, which is protocol independent. The data is remodulated using distributed feedback laser diodes with wavelengths spaced 1 nm apart near the 1550 nm region, in C-band only. The optical signals are then combined using a diffraction grating with embedded fiber pigtails, and coupled into a single-mode fiber; another unit at the far end of the link demultiplexes the signals.

The maximum unrepeated distance for the 9729 is 50 km for data in the 200 Mbit/s range (such as ESCON or OC-3) with a 15 dB link budget, and 40 km for data in the 1 Gbit/s range (such as HiPerLinks) with a 12 dB link budget. Thus, the 9729 functions as a channel extender for most protocols, including ESCON, ETR, CLO, and HiPerLinks. Note that despite the low data rate of ETR links, they are limited to maximum distances of 40 km because of timing considerations. Proprietary signaling is used between a pair of 9729s to support OFC propagation beyond the 20 km limits imposed by the Fibre Channel Standard. For protection against a broken optical link between the units, an optional dual-fiber switch card is available, which consists of an optical switch and a second fiber link. The units automatically detect if the primary fiber link is broken, and switch operation to the secondary fiber within 2 seconds; this is not intended to provide continuous operation of the attached systems, only to restore the broken link capacity more quickly. The 9729s are managed through a serial data port, and can report their status or receive simple commands, such as switching to the second fiber link, from a personal computer or workstation running a software management package.

A typical GDPS installation can be implemented by routing all point-to-point links between site 1 and site 2 over the 9729; testing and performance of this system have been described previously [31]. Today, many large systems use point-to-point wavelength division multiplexing to reduce the total number of inter-site fiber optic links. However, there is considerable interest in extending this architecture into ring topologies, with additional features for data protection and management. In the following section, we describe an example of

Table 5.4.3 WDM device and equipment providers.
(a) partial, non-comprehensive list of commercial WDM device and equipment providers; for latest information on resale agreements and brand names, contact the companies listed below. Many new companies enter the WDM industry each year; this is not intended to be a comprehensive list, and does not imply any endorsement of the product or companies listed below; this information is provided for reference purposes only. All brand names listed below are registered trademarks of their respective parent companies.
Adva Fiber Service Platform (also offered under various OEM and resale agreements over the past several years, including those with Canoga-Perkins (Lambda-Access or WA 8/16), Controlware, Hitachi, Centron, CNT or Computer Network Technologies (UltraNet Wave Multiplexer), Inrange, formerly General Signal Networks (OptiMux 9000), and Cisco (Metro 1500)). The initial product offering was known as the OCM-8 (Optical Channel Multiplexer).
Alcatel Networks WDM platform (backbone and OXC equipment)
Astral Point ON 5000 optical access transmission system and Optical Services Architecture
Avanex PowerMux DWDM devices
BrightLink Networks (OXC equipment)
Centerpoint Broadband Technologies (telecommunications equipment)
Ciena LightWorks product line; MultiWave 1600, MultiWave Sentry 1600/4000, MultiWave CoreStream, MultiWave CoreDirector, MultiWave EdgeDirector
Cisco Metro 1500 product line, WaveMux 6400, TeraMux, Wavelength Router (ONS 15900 series)
Corvis CorWave product family; Optical Network Gateway, Optical Amplifier, Optical Routing Switch, CorManager system
Ericsson ERION (Ericsson Optical Networking) product line, including ERION Linear and FlexRing
Finisar OptiCity metro DWDM product line
Fujitsu (backbone and OXC equipment)
IBM 2029 Fiber Saver (replaces the IBM 9729 optical wavelength division multiplexer; the 2029 is part of a joint development agreement with Nortel Networks)
Juniper Networks (backbone equipment)
Kestrel Solutions TalonMX optical frequency division multiplexer
Lucent WaveStar product line; OLS 40 G, OLS 80 G, OLS 400 G, TDM 2.5 G, TDM 10 G, ADM 16/1, DACS 4/4/1, DVS, Bandwidth Manager, LambdaRouter
LuxN WavStation product line and Multiplex Channel Module (MCM)
NEC (backbone and OXC equipment)
Nortel Networks OPTera Metro 5200 MultiService Platform, OPTera LH product line, and OPTera Connect
Osicom GigaMux (a subsidiary, Sorrento Networks, offers the EPC sub-rate TDM solution)
Pirelli TeraMux product line
Sycamore Networks SN 8000 Metro Core system, Intelligent Optical Network Node SN 8000, optical switch SN 16000
Tellabs Titan 6100 series WDM platform
Tellium Aurora 32 and 512 product lines (backbone and OXC equipment)
(b) major North American optical fiber network backbone carriers
AT&T
Nortel
MCI WorldCom

Continued

Table 5.4.3 WDM device and equipment providers. —*cont'd*
Sprint
Global Crossing
Williams
Qwest
IXC
Level 3
GTE
Enton
(c) major European optical fiber network backbone carriers
Interoute
GTS Group
Viatel
Teleglobe Communications
Energis
COLT
Carrier 1

Fig. 5.4.9 A pair of IBM 9729 optical wavelength division multiplexers.

a next-generation DWDM system recently developed for GDPS and other datacom applications.

5.4.3.2 Thirdgeneration WDM: The IBM 2029 fiber saver (Nortel optera metro 5200 multiservice platform)

In February 2000, IBM announced a third-generation DWDM solution, the 2029 Fiber Saver, as a follow-on to the first generation 9729 technology [32, 33]. This product is the result of a joint development relationship with Nortel Networks, and the 2029 is based on the same building blocks used in the Nortel Optera Metro 5200 MultiService Platform. The two products share a common set of hardware and software, although the 2029 supports only a pre-tested level of Optera hardware and code. While the Optera is provided as a standard telecom service provider package, the 2029 is repackaged by IBM with additional features, including turnkey installation with a pre-tested and configured PC, a class 1 laser eye safe cabinet for enterprise applications, integrated patch panels for native attachment of all datacom interfaces, and standard dual AC power supplies. For the sake of brevity, we will describe the design features of both products in this section, referring to the 2029 for our examples, and note those areas in which the two designs may be different.

The 2029 is a fundamentally different network architecture from the 9729 and offers many additional features that make it a more modular, scaleable approach to DWDM. Each 2029 model contains up to 2 shelves mounted on a 19-inch rack inside a standard-size datacom cabinet, as illustrated in Fig. 5.4.10. The cabinet also contains standard dual redundant power supplies, an optical patch panel, and (if required) an Ethernet hub for managing inter-shelf communications. We will discuss each of these functions in detail. We will discuss the link budgets and distances in more detail shortly; for now, we note that they allow a maximum distance of 50 km in a point-to-point configuration. Although the individual ITU grid lasers used to achieve this distance meet international Class 1 laser safety requirements, the multiplexed link carrying more than 16 wavelengths is Class 1 in North America only; elsewhere it is a Class 3A device per IEC 825 standards. (For a discussion of laser safety standards.) This is a standard requirement for all WDM systems supporting more than 16 wavelengths at this distance; the only way to avoid this and achieve worldwide class 1 operation of the entire link would be to either reduce the supported distance, reduce the number of wavelength channels, or implement some form of open fiber control interlock on all channels. (This use of OFC would not be industry standardized, and may introduce additional design tradeoffs related to loss of light propagation across the WDM network, especially for ring topologies.) Thus, an Optera system installed outside North America must be located in a locked room or similar environment with access restricted to individuals with appropriate laser safety training. However, the 2029 provides restricted access to this interface by means of a lockable cabinet, screw-down covers over the

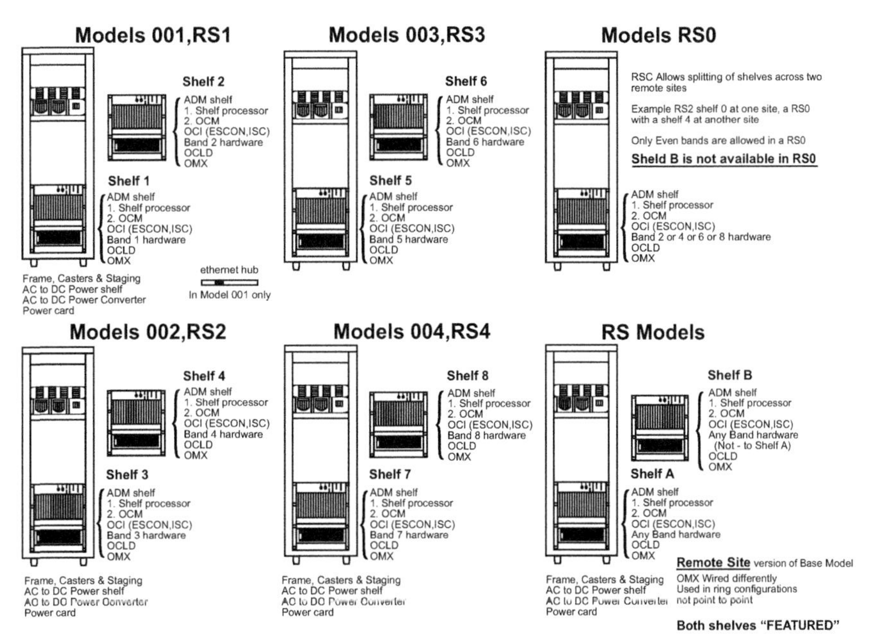

Fig. 5.4.10 The IBM 2029 Fiber Saver.

multiplexed fiber connections, appropriate safety labeling on the product, and supporting safety materials in the documentation; thus, the 2029 provides a self-contained laser safe environment and there are no restrictions on its installation outside North America. All of the interfaces for subtended equipment meet Class 1 laser safety requirements; a single client interface card supports many different protocols, so optical attenuators and adapters or hybrid fiber cables are required to attach datacom channels to the WDM device. The Optera provides plug-in optical attenuators for this purpose, which are configured to a separate rack mountable patch panel. The 2029 uses attenuated fiber optic cables with the appropriate connector types available on a patch panel integrated into the enterprise cabinet. There is a 1 to 1 mapping of interfaces on the patch panel to adapter cards in the shelves.

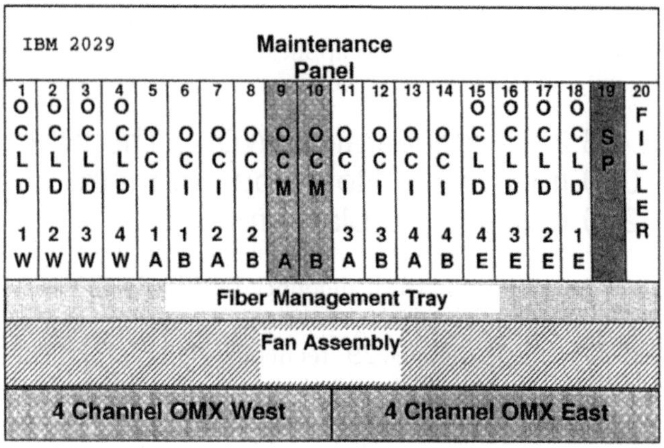

IBM 2029 — Maintenance Panel

1	2	3	4	5	6	7	8	9	10	11	12	13	14	15	16	17	18	19	20
OCLD	OCLD	OCLD	OCLD	OCI	OCI	OCI	OCI	OCM	OCM	OCI	OCI	OCI	OCI	OCLD	OCLD	OCLD	OCLD	SP	FILLER
1W	2W	3W	4W	1A	1B	2A	2B	A	B	3A	3B	4A	4B	4E	3E	2E	1E		

Fiber Management Tray

Fan Assembly

4 Channel OMX West	4 Channel OMX East

Fig. 5.4.12 Card types in a 2029 shelf.

A detailed view of the shelf is given in Fig. 5.4.11; it consists of a card cage that holds different types of adapters, a maintenance panel with power supply breakers and connections for monitoring and telemetry, some fiber slack management, a dual redundant cooling unit, and a tray containing the optical multiplexing (OMX) modules. Each OMX module is a passive device that can multiplex up to 4 optical wavelengths into one fiber path; the OMX modules are wavelength specific and are identified by their band (1 to 8). As shown in Fig. 5.4.12, the shelf contains up to 20 active circuit cards of 4 different types:

- 1 Shelf Processor (SP) card; this is a programmable processor that does not handle data, but provides management information and IP addressing for the shelf. The SP card monitors all circuit cards in the shelf, using feedback from each card to provide performance monitoring, software and configuration management, and alarm reporting. It has only an electrical connection to the shelf backplane, and always occupies slot 19.
- 8 Optical Channel Interface (OCI) cards; these connect to the equipment to be multiplexed or demultiplexed through optical fiber attachments to the patch panel, and perform optical to electrical conversion of the data prior to multiplexing and electrical to optical conversion after demultiplexing. There are different types of OCI cards available; a low-speed card that supports single-mode protocols only up to a maximum data rate of 622 Mbit/s, a high-speed card that supports both multimode and single-mode protocols up to 1.25 Gbit/s, another high speed card that extends this range to 2.5 Gbit/s, a 4:1 TDM card that puts up to 4 signals at data rates of 270 Mbit/s or less over a single 1.25 Gbit/s wavelength channel, and a protocol-specific card for Parallel Sysplex coupling links. All of these operate at 1300 nm wavelength; a separate

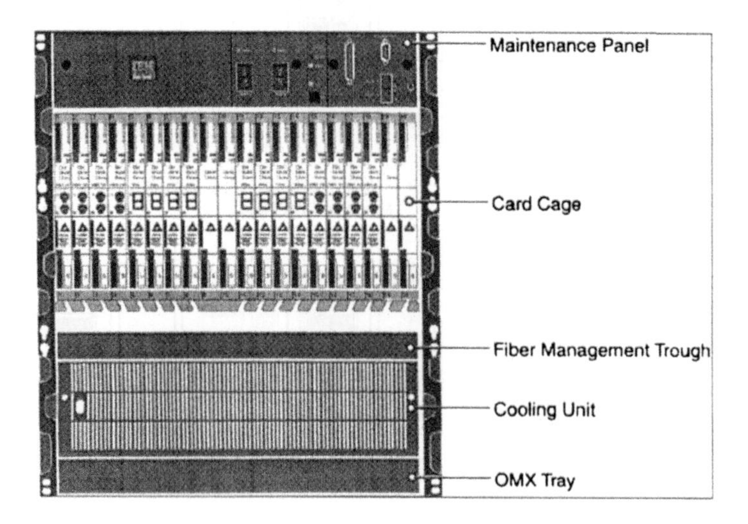

Fig. 5.4.11 Detail of a 2029 shelf.

OCI card is available for short-wavelength (850 nm) operation. These cards are located in slots 5 to 8 and 11 to 14; they typically use SC duplex connectors, except for the 4TDM card, which uses MT-RJ interfaces.

- 8 Optical Channel Laser and Detector (OCLD) cards; these perform electrical to optical and optical to electrical conversion of the data onto the ITU grid long-wavelength lasers. There is a different OCLD for each wavelength; they are identified by wavelength band (1 to 8) and by channel (1 to 4). They must correspond to the wavelength of the OMX filters in the shelf. Specific channels have fixed locations in the shelf, occupying slots 1 to 4 and 18 to 15; they use a pair of FC optical connectors to attach to the OMX modules.
- 2 Optical Channel Manager (OCM) cards; these are dual redundant, and provide switching functions from the fully cross-connected backplane for backup protection switching of data. All of the data to be multiplexed flows through one or both OCM cards, which allow any OCI card to map to any OCLD in the same shelf. The OCM stores configuration and provisioning data, as well as copies of the IP address information for the shelf. The OCMs have fixed positions in slots 9 and 10, and have only an electrical connection to the shelf backplane.

The shelf can be divided roughly in half, with channels corresponding to either the "east" or "west" side (while this designation has been adopted by many telecom service providers, it does not carry any inherent meaning; the two paths can just as readily be referred to as North/South, Red/Blue, etc.). The two halves of the shelf act as dual redundant data paths in protected or high availability mode. In unprotected or base mode, the data flow is as shown in Fig. 5.4.13; there are up to 8 duplex channels (OCI cards) per shelf, each with a corresponding OCLD. The output of each shelf is a single fiber link on the east and west sides, carrying a multiplex of up to 4 wavelengths. If there is a card failure, that channel is lost; if there is a fiber cut on either the east or west side, 4 channels are lost. By contrast, the protected or high availability configuration is shown in Fig. 5.4.14; in this case, only 4 channels are used, but the data is split after passing through the OCI card and travels over dual-redundant OCLD cards and fiber paths. There is no single point of failure in this configuration; an equipment failure or fiber break results in the data being switched to the redundant path within 50 ms. The system is a complete "3R" repeater (repeats, retimes, and regenerates the signal). A single shelf thus supports up to 8 high availability or 4 base channels, without using TDM; with TDM, the capacity is increased by a factor of 4 (note that the 4TDM channels are treated as a single wavelength for protection purposes). Up to 4 models (8 shelves) can be daisy-chained together with optical fibers, which allows a maximum of 32 full duplex links to be multiplexed over a single pair of optical fibers (the 32 wavelengths are compliant with the ITU grid, half in C-band and half in L-band, and for convenience are grouped into 8 bands of 4 wavelengths each). In this way, a 2029 network contains up to 32 high availability or 64 base channels, without TDM; using the 4TDM OCI card increases this capacity by a factor of 4. Note that the 2029 is compatible with external TDM devices as well; for example, as shown in Fig. 5.4.15, a 9032-5 ESCON Director with the FICON Bridge feature can be used, so that up to 8 ESCON channels occupy only a single WDM wavelength; in this manner, the total capacity of a 2029 network can be increased to 256 ESCON channels, or the full capacity of an S/390 mainframe computer.

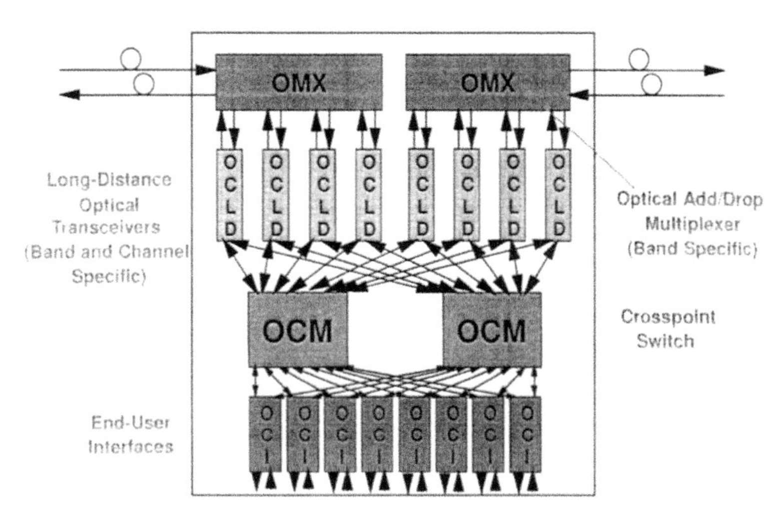

Fig. 5.4.13 2029 shelf traffic flow in base or unprotected mode.

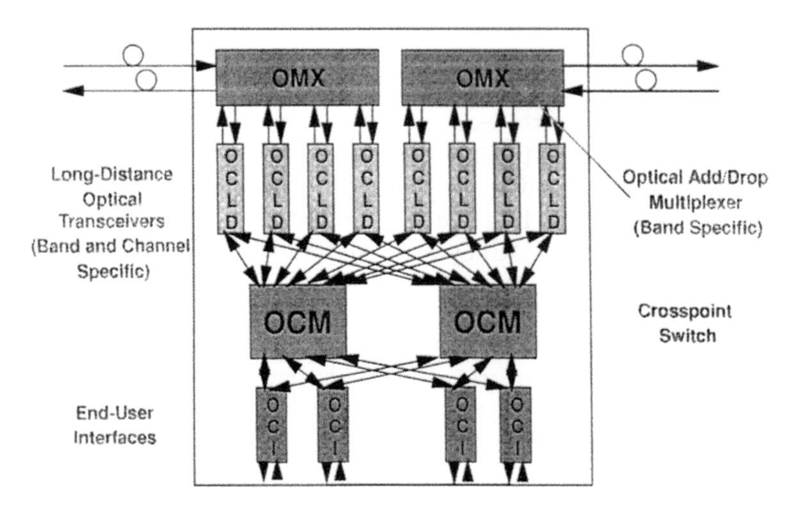

Fig. 5.4.14 2029 shelf traffic flow in high availability or protected mode.

Optical signal multiplexing is performed using passive thin film interference filters in the OMX cards. The daisy chain connections between OMX cards for both point-to-point and hubbed ring configurations are shown in Fig. 5.4.16 and Fig. 5.4.17 as an example. Other topologies, including ring mesh, are also available. The maximum distance for point-to-point links is 50 km/15 dB, while the maximum distance between any 2 nodes on a hubbed or dual hubbed ring is 35 km/10.5 dB. The difference is due to the inclusion of additional mux/demux stages in ring topologies to allow for passive add/drop of individual wavelength channels at any point on the ring, and optical pass-through for wavelengths destined for other nodes (shelves). For example, a point-to-point topology with 3 shelves is shown in Fig. 5.4.18, giving the resultant mapping of logical and physical connectivity; the flow of traffic in base and high availability modes is shown in Fig. 5.4.19 and Fig. 5.4.20. Note that the shelves are connected in reverse order at the hub and remote sites, so that all wavelengths experience an equal amount of delay propagating through the network; this principle is known as first add/last drop or last add/first drop. The corresponding diagram for a hubbed ring is given in Fig. 5.4.21; each 2029 shelf can act as a node, meaning that a hubbed ring will consist of up to 8 remote locations which logically communicate with a central hub site. High availability channels work in the same manner between the hub and remote locations on a ring. A meshed ring supports re-use of wavelength channels, so that any 2 nodes may communicate with each other; the number of nodes in this case is limited only by the network configuration, not the DWDM technology. The multiplexers for each wavelength band are configured as either an optical seam or optical bypass, depending on their location in the ring, to facilitate the largest possible link budgets. For example, in Fig. 5.4.17 the hub site is configured as an optical seam, while the remote site is configured as an optical bypass. The dual counter-rotating rings are also self-healing (a break in the ring does not disrupt traffic to other points on the ring) and fault tolerant (an electrical power failure at any shelf does not impact other bands that pass through that shelf, since the OMX is passive), while the shelf supports concurrent maintenance (replacement of cards without affecting other channels).

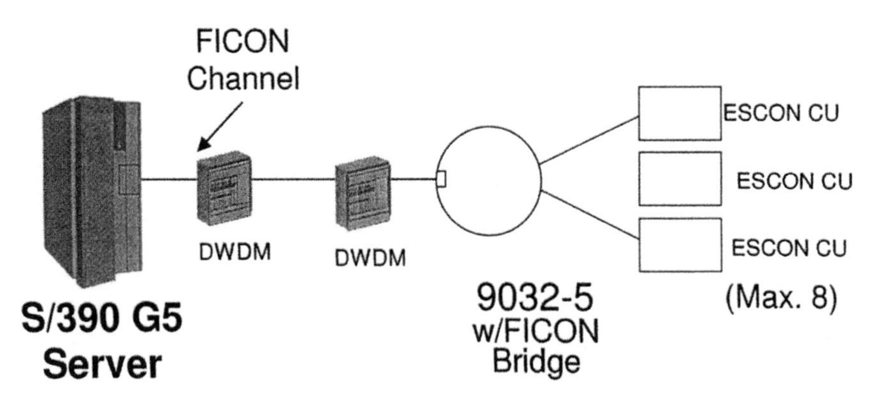

Fig. 5.4.15 DWDM used with the 9032-5 ESCON Director (FICON bridge feature) and S/390 model G5 enterprise server to achieve an effective 8 to 1 TDM of ESCON channels over WDM.

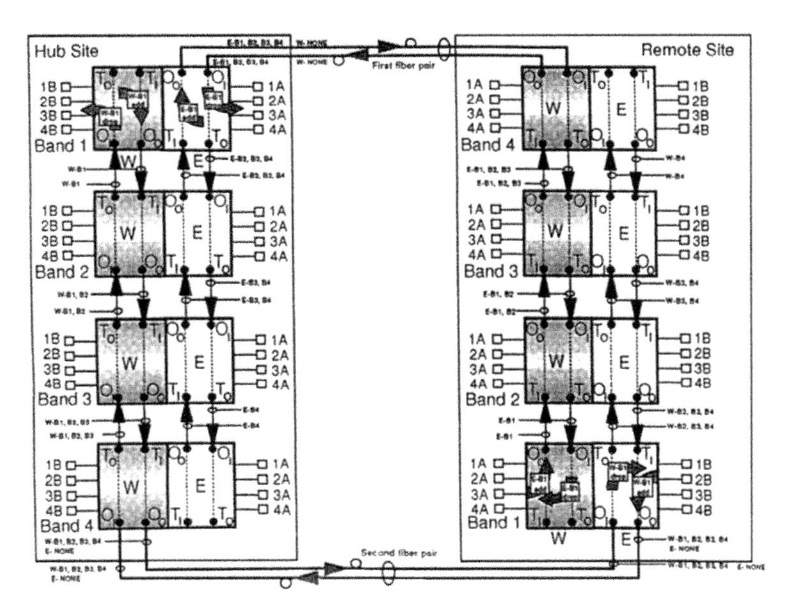

Fig. 5.4.16 2029 OMX connectivity, point-to-point configuration (4 shelves shown as an example).

Because all data channels are 3R retimed, the 2029 also supports cascading of up to 4 networks with 32 channels each in series, which increases the total distance to 200 km in a cascaded point-to-point configuration.

Both point-to-point and ring networks may be cascaded, which can result in many different topologies; a few of these were illustrated earlier. Optical amplifiers may also be used to increase the working distance to over 100 km point-to-point, or over 400 km in a 4 system cascade; the amplifiers occupy a separate shelf, and may be used either as pre-amps, post-amps, or in-line amps with various tradeoffs in the achievable distance and link budgets; appropriate equalization must also be used.

Unidirectional switching is employed in high-availability mode on the transmit and receive fiber separately to ensure that there are no single points of failure in a protected 2029 channel. With a worst case switching time of 50 ms, this is a significant improvement over the 9729 and allows uninterrupted operation of many protocols; some interfaces, such as HiPerLinks and sysplex timer, should still rely on link redundancy for continuous application availability. There is a third protection option available on the 2029, known as switched base mode; this uses a dual fiber optical switch to detect fiber breaks and switch all traffic to a redundant backup path. The switch is intended as a lower cost option than high availability

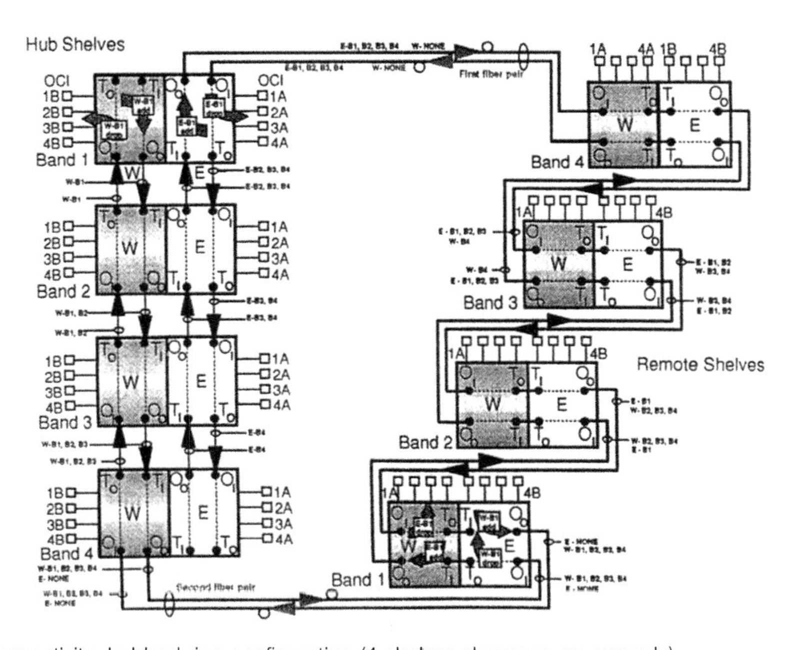

Fig. 5.4.17 2029 OMX connectivity, hubbed ring configuration (4 shelves shown as an example).

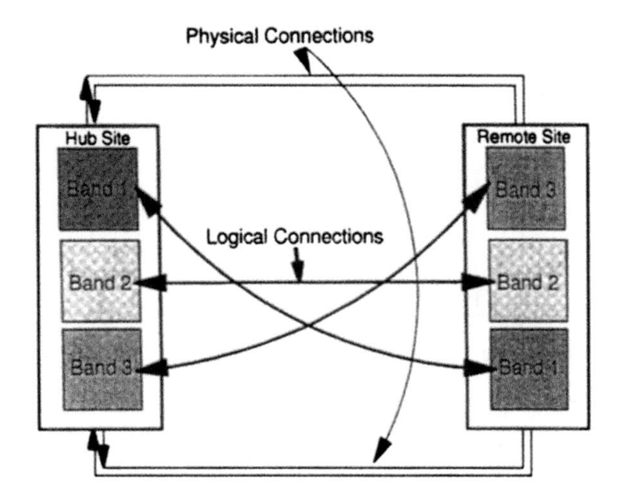
Physical Connections
Hub Site
Remote Site
Band 1
Band 2
Band 3
Logical Connections
Band 3
Band 2
Band 1
• Consists of only two sites
• Both sites have the same number of shelves (max. 8)
• The number of shelves between a shelf pair is the same

Fig. 5.4.18 Example of 3 shelf point-to-point configuration showing physical and logical connections.

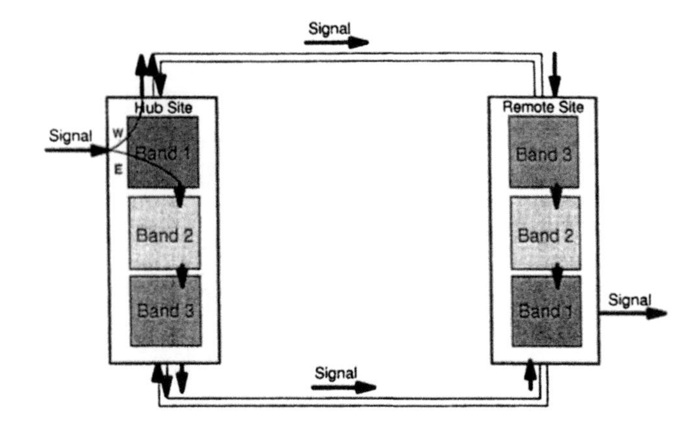
Signal
Hub Site
Remote Site
Signal
W
E
Band 1
Band 2
Band 3
Band 3
Band 2
Band 1
Signal
Signal
• A signal from a channel on either side of the shelf travels over both fibers
• The stronger signal will be chosen at the other end (this is determined at configuration time)
• The path can be switched manually or automatically (50 milliseconds)
• Max. number of channels is 32 with 8 shelves at each site

Fig. 5.4.20 Example of 3 shelf point-to-point traffic flow in high-availability mode.

for environments in which fiber breaks are more common than equipment failures, since the dual fiber switch protects only the fiber path and not the equipment cards. The backup fiber path is monitored with a heartbeat function to ensure that it remains available at all times. The switch also implements bidirectional switching, meaning that it can support some protocols, such as Sysplex Timer links, which may not function properly in high availability mode. There are 2 switches, one each for the east and west side of the network; this is shown schematically in Fig. 5.4.22.

This platform provides for protection and restoration of communication services within the optical (physical) layer of the network, without requiring protection at higher level network protocols. Conventional MANs have deployed SONET-based networking to support a variety

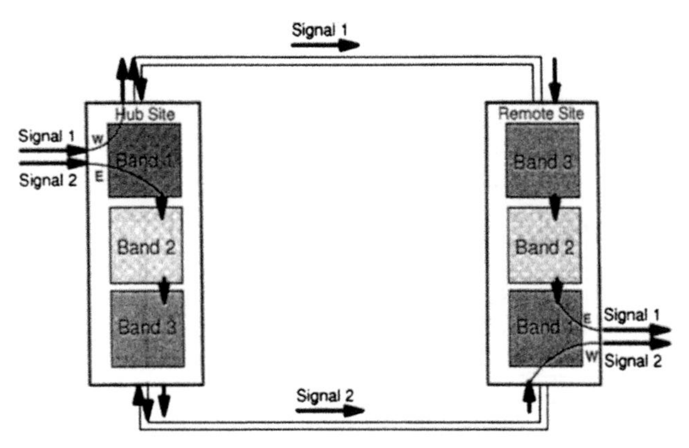
Signal 1
Hub Site
Remote Site
Signal 1
Signal 2
W
E
Band 1
Band 2
Band 3
Band 3
Band 2
Band 1
E
W
Signal 1
Signal 2
Signal 2
• Signal 1 coming from a channel on the West side of the shelf travels over the West fiber (max. 4 channels)
• Signal 2 coming from a channel on the East side of the shelf travels over the East fiber (max. 4 channels)
• If a fiber cut occurs on the West fiber, only West side traffic is affected
• The max. number of channels is 64 with 8 shelves at each site

Fig. 5.4.19 Example of 3 shelf point-to-point traffic flow in base mode.

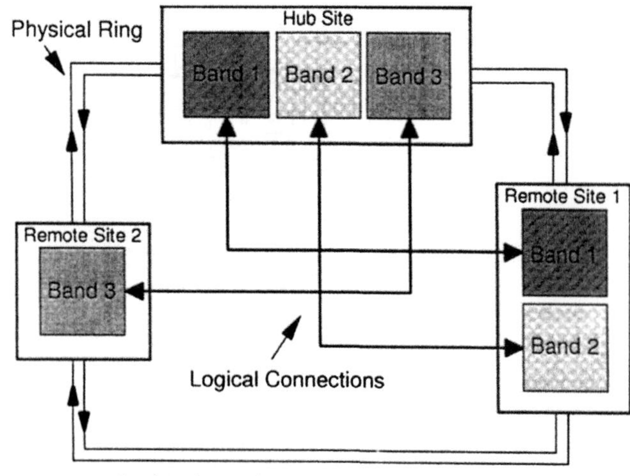
Physical Ring
Hub Site
Band 1
Band 2
Band 3
Remote Site 1
Band 1
Band 2
Remote Site 2
Band 3
Logical Connections
• Consists of up to nine sites
• Multiple shelves are at the Hub site and one or more at the remote sites
• The number of shelves between a shelf pair is the same

Fig. 5.4.21 Example of 3 shelf hubbed ring configuration showing physical and logical connections.

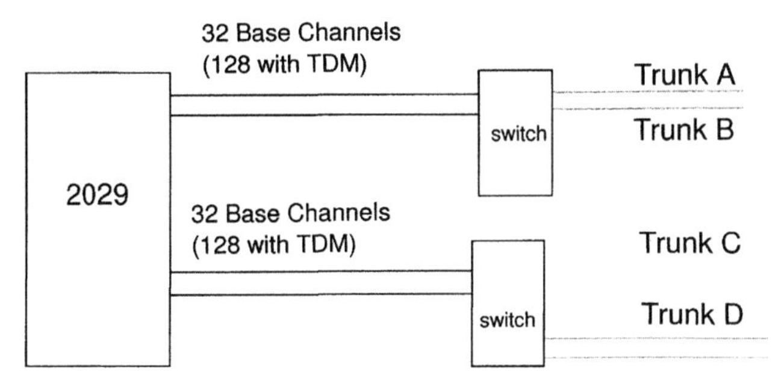

Recommended Configuration: Trunks A & C take same physical path and Trunks B & D take same physical path - assumes a break in the fiber path affects all fibers on that path

Fig. 5.4.22 Example of dual fiber switch implementation for east and west side of a 2029 system, switched base channels, point-to-point configurations only.

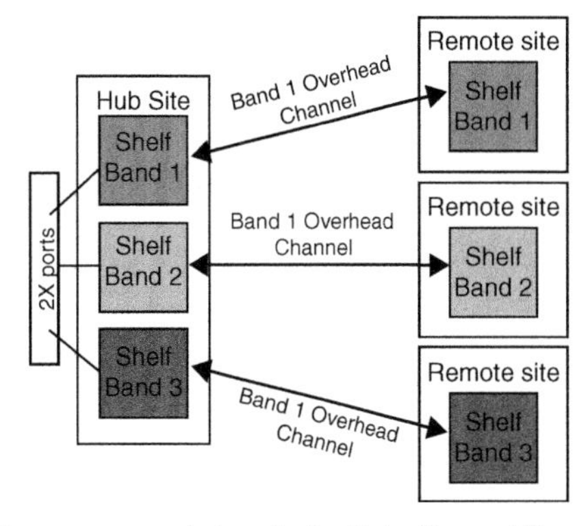

- With three or more shelves in the Hub site, an Ethernet hub is required
- All Shelves are connected to the Ethernet hub for the 2X ports with STP crossover cables

Fig. 5.4.23 Cross-band network management using per wavelength overhead channels and Ethernet hub interconnections at the hub site in a 2029 network.

of services including traditional asynchronous networks (OC-3, DS-1, etc.) as well as new services such as ATM, IP, compressed video, circuit switching, and others. In particular, ATM over SONETallows conventional "1 + 1" protection switching to be implemented over the network, providing redundancy in the event of a fiber break or equipment failure. Because of the proprietary nature of many new protocol signals, overlay networks must be implemented to support the full range of services; mapping all of the desired protocols to SONET is not always possible because many signal types are not compatible with SONET (for example, FICON and Parallel Sysplex links). New services offered in this manner cannot necessarily take advantage of protection switching and other SONET features, although these features are highly desirable in the design of fault-tolerant computing systems, which require high reliability and availability. Using the 2029, these multiple infrastructures are combined into a single, protocol-independent backbone implemented at the optical layer. The 2029 supports native attachment of all industry-standard protocols up to a maximum data rate of 1.25 Gbit/s, including ESCON, FICON, Fibre Channel, Parallel Sysplex links (HiPerLinks, ETR, and CLO), ATM 155 and 622 (OC-3 and OC-12), FDDI, Fast Ethernet, and Gigabit Ethernet (LX and SX). Propagation of OFC protocols on HiPerLinks is supported using proprietary signaling between 2029 devices, similar to the 9729; because GDPS protocols remain IBM proprietary as of this writing, the 2029 is the only currently available product tested and supported by IBM for Parallel Sysplex and GDPS applications. Many of these protocols use common hardware; the 2029 includes the necessary optical attenuators, patch cables, mode conditioning, and optical adapters as required for ESCON, ETR, CLO, and other links. Future work is planned to be compatible with other optical WAN products, including Nortel's recently announced 1.6 Tbit/s multiplexers, currently the largest in the world.

As the DWDM infrastructure scales, network management and security become increasingly important. Each data channel on the 2029 includes an overhead service channel multiplexed on the same path as the data at much lower speed; this allows individual channels to communicate in-band across the 2029 network. This approach avoids any single points of failure in the network monitoring systems, which is a potential concern with telecom standard systems that carry all management information over a separate outband wavelength channel. External management in a 2029 network is accomplished using Ethernet routing; all network shelves are connected to an Ethernet hub in the 2029 model 1 using a 2X crossover cable as shown in Fig. 5.4.23. The 2029 network management, commissioning, and provisioning is accomplished by Java-based applets that run System Manager software on an attached personal computer (PC); this provides a graphic user interface running under a Web browser. Either the Windows 95 or Windows NT operating system is currently supported for the attached PC. Using the System Manager software running on an attached PC, protection switching can be turned on or off for individual channels, and the user can always monitor which channels are being used to transport data. As shown in Fig. 5.4.24, the entire 2029 system can be viewed and managed from any point on the network; in practice, multiple PCs are used for redundancy and to provide information to users at different locations. The PC attaches to the 2029 via an Ethernet interface, and can be located anywhere on an

Ethernet LAN. The user provides a set of sub-netted IP addresses for the 2029 shelves; internally, the 2029 implements proxy ARP serving through one or more gateway interfaces, so that it can be treated in the same

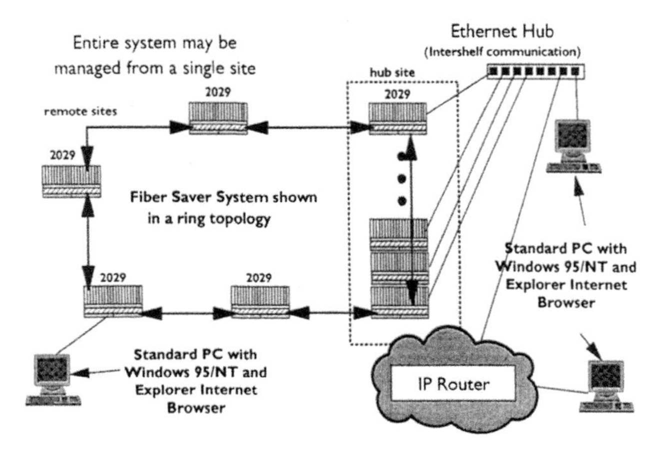

Fig. 5.4.24 Single point of control management of a 2029 network.

manner as any other network attached device. A typical IP network management system is illustrated in Fig. 5.4.25, which employs multiple points of control for both the end user and the fiber service provider. Security is provided by user-defined passwords, which permit logging into the 2029 as either an observer (able to view the configuration but not change it) or an administrator (able to both view and change the configuration). Ease of use features include a graphic display with user-defined names for each channel and system, and color-coded alarm banners (for example, green for a working channel, red for a failure). The management is based on Standard Network Management Protocol (SNMP) version 1.0, and the 2029 can be remotely managed under a variety of software applications such as Net View; alternately, TL-1 management is also supported. The 2029 is configured to send alarms and alerts to any other user-defined IP address on the Ethernet LAN; multiple gateway devices, such as routers, are supported for redundancy at different points on the 2029 network using OSPF protocols. The 2029 also provides integrated management from an IBM System/390 Enterprise Server Hardware Management Console (HMC), which implements an automatic call-home feature to proactively report equipment problems to a service center; this can be configured, for example, as part of the System Integrator software in a GDPS installation.

As part of product testing and qualification, a large data system using the 2029 network at the IBM TeraPlex Center is on the S/390 test floor in Poughkeepsie, New York, to evaluate its performance. Results of this work have been published elsewhere [32,33]; we will provide a brief summary of the results here. A fully protected 32-channel 2029 system was used with a variety of input/output devices. Two IBM G5 Enterprise Servers (air-cooled CMOS) were configured in a 40-km GDPS using two sysplex timers. Four ESCON channels were routed through the 2029, optically looped back at the far end, and returned to the original processor to evaluate the effect of multiple passes through the 2029 network. Four FICON channels, each carrying a time division multiplex of 4 ESCON channels, were also run from the processor through the 2029s and into a 9032-5 ESCON Director where they were broken out into single channels and used to drive various storage devices,

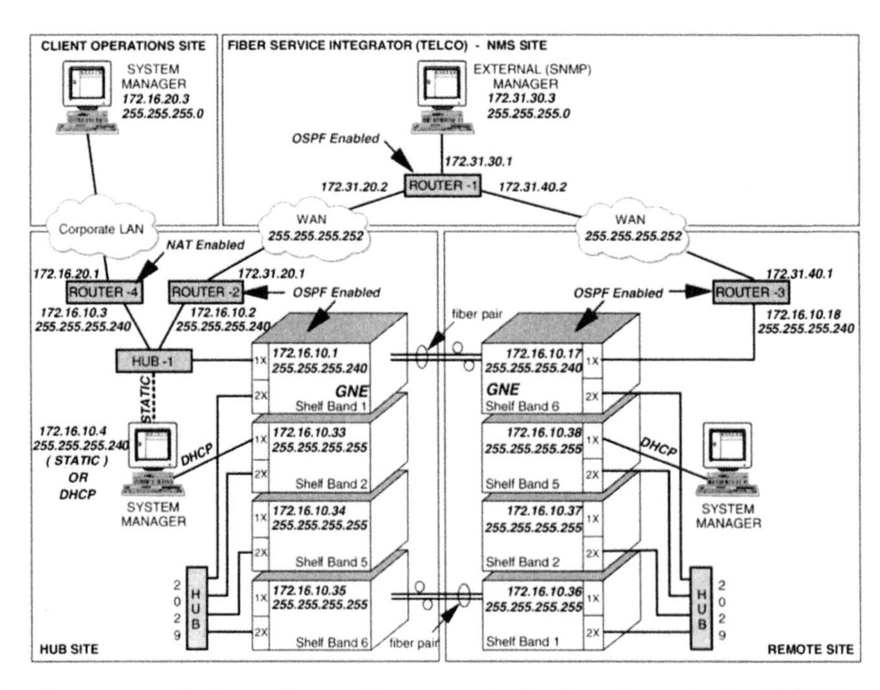

Fig. 5.4.25 2029 network management using service provider managed network service and external TCP/IP network to both the end user and service provider locations.

including an IBM 3945 automated tape library. One channel each of FDDI, ATM 155 over SONET, and Gigabit Ethernet LX was run from the G5 Open System Adapter (OSA) interface cards through the 2029 network to LAN connections; for example, the Gigabit Ethernet link was connected to a Cisco Catalyst 5000 router. Spools of fiber were used to simulate inter-site links, with the maximum 15 dB link budget and at least eight ST-type optical connectors in the link to verify that there was no effect from connector return loss or modal noise. The processors were logically partitioned into 15 processing zones, the maximum allowed for this machine type. Various applications were run on the sysplex to simulate stressful traffic under the OS/390 and MVS operating systems, including Lotus Domino and Notes servers, a UNIX-branded operating system partition (S/390 is officially branded as UNIX compatible) and transaction processing using secure encryption methods (IBM offers the only cryptography coprocessor certified as Level 4, the highest achievable level, by the National Institute of Standards and Technology). System code on the processors was used to log bit errors on all links as well as any other conditions indicating either failure or degraded performance; the 2029 system manager was also monitored during this testing. All fiber optic links operated error free (extrapolated to 10^{-15} bit error rate) over a 72-hour test run. By opening individual fiber sections throughout the system, pulling and reseating cards with the system powered on, and failing power supply components, we verified that there were no single points of failure in the system and no cross-talk effects between adjacent channels, either electrical or optical.

5.4.4 Intelligent optical internetworking

In recent years, the growing demand for bandwidth in MANs has driven the acceptance of DWDM technology. The nature of the metropolitan network is changing; many current networks are service specific, providing multiple overlay networks for different communication protocols and applications. This approach is well understood and has the advantage of a large installed base and simple network design rules. Such systems require adaptation at the network layer in order to accommodate both data and voice protocols; as the balance of traffic shifts from predominantly voice to data, the limited scalability of this approach makes it difficult to offer new services while maintaining support for legacy systems. For this reason, MANs are evolving into a more service-transparent structure based on DWDM. This offers the advantages of a highly scaleable, low-cost network infrastructure that is both bit rate and protocol transparent. Eventually, this roadmap may also lead to transparent all-optical networks with add/drop capability and over 10 Gbit/s serial line rates per channel, although it appears that service transparency alone delivers most of the value in an optically transparent network with lower cost and less complexity. In the near term, a combination of circuit and packet switching is likely to prevail in the DWDM MAN or WAN. In this section, we will discuss some emerging trends and directions in technology and services that are expected to play an important role in the rapidly developing WDM market.

5.4.4.1 IP over WDM: Digital wrappers

The ability to transmit multiple data channels over a common physical media has helped to alleviate fiber exhaust in densely populated areas, where the cost and availability of optical fiber have become obstacles to the growth of new applications. However, it has become apparent that DWDM technology alone is not sufficient to address the requirements of growing MAN environments; some level of electronic signal processing is also required of the data in order to assure the necessary data integrity, quality of service, reliability, security, and manageability of the network infrastructure. Previously, these features have been provided over the telecommunications infrastructure using a combination of asynchronous transfer mode (ATM) over synchronous optical networks (SONET). Indeed, SONET-based traffic is already carried over a physical layer that makes extensive use of DWDM optical fiber interfaces. However, the growth of IP traffic has led to a complicated arrangement with up to 4 separate transport layers (IP over ATM over SONET over DWDM). In an effort to simplify this approach, there is a clearly emerging trend toward elimination of the ATM and SONET layers, and toward the direct transmission of IP over DWDM [34].

The combination of the most widely used networking protocol (IP) and the almost unlimited bandwidth of DWDM offers the potential for many new networking architectures. In practice, this term is a bit misleading: "IP over WDM" implies one of a multiplicity of mappings of IP onto fiber (or wavelengths) as illustrated in Fig. 5.4.26. To state that any one particular mapping represents IP over WDM is very disingenuous, and ignores the fact that every data network is unique in a marketplace governed by differentiation. For example, some of the mappings shown in this figure continue to exploit the advantages of the installed ATM infrastructure in the WAN, commonly known as "everything over SONET" (EOS). Many of these models can also be extended to encapsulation of other protocols such as ESCON or Fibre Channel. Commensurate with the emergence of IP over WDM protocol mappings is the

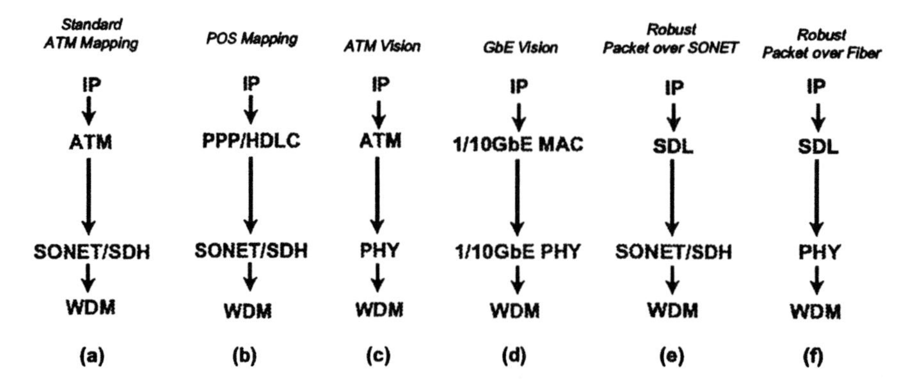

Fig. 5.4.26 Different types of IP over WDM (a) standard ATM mapping (b) point-of-service mapping (c) ATM future mapping (envisioned) (d) Gigabit Ethernet mapping (e) robust packet over SONET mapping (f) robust packet over fiber mapping.

demand for transport networking at unprecedented levels of granularity — on the order of Gigabits to tens of Gigabits per second — and the evolution from exclusive use of SONET or SDH-based time division multiplexing to Optical Transport Networking (OTN) via WDM. This may be the next step in the evolution of the transport network, which will be predicated upon the support of operations, administration, maintenance, and provisioning (OAM&P) functions at the optical layer.

Current deployments of IP networks presuppose that data delivery will be on a "best effort" basis. However, many network operators want to provide new services that require traffic management of IP flows. One of the most significant of these is virtual private networking (VPN), where one customer's traffic is separated from another's in a single IP network with well-defined quality of service and security. While this is not possible with today's best effort IP network, it can be achieved in the future by means of a protocol that allows IP packets to be partitioned into traffic flows that can be manipulated in their own right. One example of such a protocol is MPLS. The partitioned flows must then be encapsulated to be carried over WDM (from the perspective of IP, it is of no consequence if IP is partitioned or not, only that it is encapsulated).

The management of OTNs using digital wrappers such as those in Fig. 5.4.26 will also undergo a profound evolution. Digital wrappers could enable protocol-independent OTNs with similar management features to those existing today only in SONET and SDH networks, while at the same time opening the optical layer to a wide variety of client traffic. This could allow service providers to accommodate different protocol stacks and network architectures for their next-generation IP backbones. This is known as service transparency; together with the required technologies for optical switching, protection, and restoration, this is critical for future WDM networks that will be governed by service differentiation and rapid service provisioning. This new model requires the DWDM layer to assume many of the traditional functions associated with ATM over SONET, such as protection switching. In general, the more intelligence that can be built into the product, the more value to the end users; but where the intelligence is placed in the network may ultimately determine what the final applications will be. We will discuss alternate approaches to design of next-generation DWDM devices and networks that use electronic signal processing to enhance the performance of optical interconnects.

5.4.4.2 WAN traffic engineering

Optical Internetworking is a concept introduced in 1998, when equipment was introduced that allowed the IP switching/routing layer to operate at the same data rates as the DWDM equipment that provides the physical layer interface. Typically this is realized by providing an OC-48c interface (2.4 Gigabit/second) on IP switches offered by vendors such as Cicso, Ascend, and others; in this model, the IP switch/router serves as the central location for all network intelligence. This data rate matching removes the need for SONET aggregation equipment in the core of a data network, which is seen as the first step in an evolutionary path toward full transmission of IP over DWDM. IP router interfaces may also be directly attached to optical ADM ports or OXCs. This forms an overlay network in which DWDM acts as the server layer and IP behaves like the client layer. Of course, there are many alternatives to forming a virtual topology between IP routers and DWDM equipment. Intelligence required to manage resources of both IP and WDM layers could reside in the IP router, or the router could simply act as a "black box" with optical interfaces.

There is vigorous ongoing discussion in the technical community concerning the best way to implement intelligence in the network topology, and thereby exploit the strengths of both optics and electronics. For example, design rules for WDM LAN and MAN routing have been proposed [35–41]. In electronics, operations such as buffering, adding and dropping packets, or merging packet streams, are done with ease. In optics, however, buffering is onerous. Operations such as retrieving a packet from a traffic stream affect the whole stream. Therefore buffering in the network, with the attendant issues of cost and possibility of overflow, is to be avoided, and consequently so is traffic merging. Traffic splitting, on the other hand, doesn't require buffering. While traffic to remote locations should be handled by switches, local traffic in the WAN need not be handled by these switches. In fact, it can be desirable to isolate local traffic from the vagaries of congestion associated with external traffic. Also, switches are expensive and make extendibility of the network more difficult; switch upgrades or outages may also disrupt the network. For these reasons, the network should allow traffic that does not need to be routed through switches to bypass them. Finally, routing should be well suited to packet-switched traffic but not rely on a specific protocol such as IP or ATM; instead, it should operate at the physical layer, below such protocols. This includes recovery of failed components at the optical layer, in a manner that is sufficiently simple, robust, and rapid that it does not trigger recovery attempts by the higher order protocols. Such an approach is well suited to overlay network engineering, where operations in one layer are pursued independently from those in other layers.

Traffic engineering solutions developed for either IP networks or WDM networks can be applied directly to their respective layers in this model, with little or no inter-layer coordination. In an IP over WDM approach, traffic engineering in the IP layer can theoretically be effected via IP routing algorithms that can adapt the IP packet routes and perform load balancing. Likewise, traffic engineering in the WDM layer can be effected through light path reconfiguration that adapts the IP network's virtual topology to the evolving traffic pattern. In the case of switched WDM, the optical layer can also adjust the number of wavelengths allocated to each virtual link, thereby affecting the bandwidth and contention statistics of the virtual link. In practice, because traditional IP routing algorithms are all oblivious to traffic loading, traffic engineering in current IP networks often relies entirely on link layer adjustments. For example, an IP over ATM network typically maintains a full IP mesh connectivity, and relies on the ATM layer to adjust the bandwidths provisioned for the ATM virtual circuits that support the IP links. These ATM layer actions are completely transparent to the IP layer routing algorithm. Overlay traffic engineering is most likely to be adopted in near-term deployment of overlay IP/WDM networks. Network engineering rules for IP over WDM will likely consist of a number of basic functional components, including the following [42, 43]:

- Traffic monitoring, analysis, and aggregation is responsible for collecting data traffic statistics from the network elements (IP routers and OXCs). These statistics are then analyzed and/or aggregated to prepare for the traffic engineering and network reconfiguration related changes; depending on the implementation, this may be either local or end-to-end performance monitoring. In addition, the optical channels may be monitored for other statistics, such as average optical power (to predict laser end of life and estimate bit error rates), optical signal-to-noise ratio (OSNR), or wavelength stability. Various types of traffic monitoring are possible in a WDM system, including schemes based on the power spectral density of the WDM signals.
- Bandwidth demand projection uses past and present measurement of the network and characteristics of the traffic arrival process to forecast bandwidth requirements in the near future. These projections are used for planning subsequent bandwidth allocations.
- Reconfiguration triggers consist of a set of network management policies that decide when a network level reconfiguration is to be performed. This can be based on traffic measurements, bandwidth predictions, and operational issues (such as allowing adequate time for the new network to converge and stabilize while minimizing transient effects).
- Topology design provides reconfiguration of the network topology based on traffic measurements and predictions. Conceptually, this can be viewed as optimizing some performance function (number of IP routers connected by optical paths in the WDM layer) to meet a specific objective (maximizing throughput) subject to certain constraints (interface capacity or nodal degree of the network) for a given set of load conditions applied to the network. This is generally a very complex problem. Because reconfiguration is regularly triggered by the continually changing traffic patterns, an optimized solution may not be stable. It may be more practical to develop a metric that emphasizes factors such as fast convergence and minimal impact to ongoing traffic rather than optimizing over the whole network.
- Topology migration refers to a set of algorithms or policies to coordinate the network migration from an old topology to a new one. As WDM reconfiguration deals with large capacity channels, changing allocation of channel resources with the resulting coarse level of granularity has a significant impact on many end user communication links at the same time. Traffic flows need to adapt to the light path changes during and after each migration step; the effects can potentially spread over the routing pattern of the whole network and impact many user data channels.

In future networks, protocols such as MPLS, MPlambdaS, and related types of optical label switching (OLS) may provide a unified control plane across both IP and WDM layers. MPLS would effectively serve as the intermediate layer between IP and WDM. This could make it possible to optimize performance across both layers together with some form of integrated traffic engineering. All of the IP and WDM network traffic management discussed above still exists, but could now be coordinated together. Specific traffic management methods can be applied to different layers, subject to the prevailing network traffic load, granularity, and time scale considerations. Note that another consequence of the emerging converged IP/WDM environment in the WAN is a change in the network traffic patterns, the provisioning of transport resources, and the sources of network traffic. Traditionally, nationwide transport networks were operated by a single telecommunications provider to meet the needs of their subscribers. A new model of dynamically reconfigurable networks is emerging, in which the OTN operator provides capacity on demand in the form of lightpath connections (wavelengths) to independent users such as independent service providers (ISPs), storage solution providers (SSPs), or even telephone companies. Because the light paths are service transparent, a leased wavelength could be managed according to the needs of a specific application. Today these are primarily narrowcast (NC) services such as telephony or video on demand, but they will expand to include many other features as well. This also implies that traditional methods of forecasting average traffic demand in the WAN based on past history (so-called static traffic capacity planning [44]) becomes very difficult to apply. Not only do emerging networks offer new services for which no prior history exists, they also ignore both the number of clients in use at any given time in the future and what kind of service will be required for these clients. This has led to ongoing research into dynamic traffic modeling, which means that lightpath requests can arrive at random intervals while other connections are already active. If the new request can be accommodated without disrupting existing connections, a new lightpath is created and maintained for the duration of the service; otherwise the request is blocked. Dynamic network performance is usually measured by the blocking probability. As one might expect, this is a multi-variable optimization problem; for example, in some cases blocking probability decreases by increasing the nodal degree (number of add/drop points in the network) which is a network topology consideration.

Another measure of network performance compares the blocking probability of connections between adjacent nodes and widely separated nodes; this is called the fairness of the network. For example, it has been suggested [45] that voice and video traffic follow a Poisson distribution, while IP data is a self-similar profile. Various models incorporating link utilization metrics, packet loss rates, and different loading considerations have been proposed [45], including migration plans from existing legacy networks. For example, it is now possible to construct separate ATM and IP overlay networks that attach directly to DWDM equipment; then, as the mix of traffic in the MAN tends toward more data than voice, especially IP data, the two networks can gradually be converged into a single backbone. Previously, network service providers have deployed dedicated networks for each type of traffic; these overlay networks do not scale well, due to unplanned and rapid growth in many different types of data protocols (Fiber channel, Gigabit Ethernet, ATM, etc.). Basic networks have been optimized for time division multiplexed circuit-switched voice traffic; the new push is toward packet switched, data centric networks.

5.4.4.3 Emerging standards

Mapping all of the desired protocols into one network is not always possible because many signal types are not compatible with SONET (for example, Fibre Channel and Parallel Sysplex links). New services offered in this manner cannot necessarily take advantage of protection switching and other SONET features, although these features are highly desirable in the design of fault-tolerant computing systems, which require high reliability and availability. For these reasons, there has been a need for protocol-independent DWDM solutions that combine these multiple infrastructures into a single backbone at the optical layer. Some more aggressive regional telecom service providers have chosen to skip the intermediate step altogether, and jump directly to IP over DWDM. The IP switch must also provide protection switching, quality of service, and related functions.

An alternate viewpoint is known as the "intelligent optical network"; this maintains that IP switches/routers can't provide the wavelength switching, service provisioning, and rapid path restoration functions required by future networks, and these functions best reside in the optical layer. Such devices would either be integrated with or interface with existing DWDM equipment, and would provide the above functions in addition to handling the details of signal transport. Generally, these two are viewed as complimentary technologies; relatively small, purely IP networks will be built following the optical internetworking vision, but larger more complex networks will require intelligent optical networking to either supplement or replace these functions. Optical internetworking is better suited to pure data environments, especially IP data; multiservice provisioning is better suited to an intelligent optical network. Intelligent networking offers the advantage of bandwidth management, including integrated TDM functions at all data rates. Also, it can respond to protection switch events within the maximum SONET switch time of 50 ms, with typical switch times of 25 ms or better; this helps prevent conflict between the two versions, as intelligent optical networks will respond faster to remedy service problems before switch/routers can respond. However, standards are not yet defined to negotiate the division of responsibility in networks that contain both elements.

Generally, DWDM can best respond to physical path problems, while switch/router can better respond to application and quality-of-service issues. Equipment must also be able to interface with many different kinds of management systems, both legacy telecom carrier (TL-1) and modern datacom (simple network management protocol or SNMP-based) management, and there are distinct advantages to integrating both element and network management systems without requiring outband management. Future network devices such as switches and routers based on Fiber Channel or similar standards will likely be able to automatically recognize attached devices and perform a configuration map of the network, as well as respond to in-band management commands. Currently, DWDM devices typically employ their own network management interface which is transparent to the attached devices; this can create conflicts, for example DWDM implementation of ring topologies that are not necessarily compliant with Fiber Channel arbitrated loop or other topologies. Existing industry standards do not provide for the management of transparent, out-of-band devices such as repeaters, protocol converters, or multiplexers; thus, some level of intelligent DWDM is needed to assist with network management of these devices. Wavelength switching and routing are also key advantages of intelligent networks; this enables the creation of virtual optical data pipes with bandwidth management and simultaneously provisioning IP services. It is an advantage if current DWDM designs offer fully cross-connected backplanes, as this should allow the architecture to scale into wavelength switching and routing in the future or integrate with WAN devices using high-speed parallel optical backplane bus extensions. An advantage of integrating the intelligence with the multiplexer is that DWDM is a modular platform, providing building blocks with individual channel control; this leads to improved reliability, availability, serviceability (concurrent maintenance), and scalability between MAN and WAN networks. There is ongoing debate whether this intelligence should reside at the edge of the network (into the end-user premises or private networks) vs. in the central office of telcom or service providers who may wish to offer line-drop services.

As SONET-based termination of circuit-switched, TDM environments is being embedded in other packet-based network layers, future network designs will require new ways of allocating network bandwidth to IP traffic. With the advent of "always-on" network services and the growing need to link high-speed optical backbones with slower legacy equipment, a coalition of optical networking experts and IP specialists would seem to be a natural step for the next generation of internetworking standards. Indeed, there are a number of promising efforts underway in the telecommunications industry to more closely integrate the electrical and optical signaling domains. In particular, two standards efforts are currently underway to link IP data packets directly to DWDM optical wavelengths, so that the optical network can take some advantage of the intelligence embedded in IP traffic. This so-called "optical IP" effort could eventually allow customers to dynamically request portions of a fiber cable's bandwidth for a particular time or service. It also represents a radical change for service providers, who would be able to build intelligent optical networks with a seamless interface between switched packet and optical services. One standards effort that held its first meeting in Boston this past January is the Optical Domain Service Interconnect (ODSI) coalition [46], a loose connection of vendors providing optical transmission equipment, access services, terabit routers, switches, and network provisioning software. ODSI seeks to define common control interfaces between optical or electrical physical layers and IP media access layers of the Open Systems Interconnect (OSI) model. This work may ultimately rely on derivatives of the Internet Engineering Task Force (IETF) MPLS standard, which basically provides a means of defining IP flows and is already being used in the electrical signaling domain. ODSI will propose low-level (below Layer 3) control plane standards that must be met by vendors of both optical transmission equipment and broadband IP switches/routers; these standards would also be offered to the IETF, the OIF, and other standardization groups.

A separate but complementary effort, also based on MPLS, is being drafted within the IETF itself. Although this effort is somewhat smaller at present than ODSI, it has proposed methods to link optical cross-connects and gigabit routers through a Layer 3 switching methodology called Multi-Protocol Lambda Switching (MPLmS; lambda refers to switching by native wavelengths). Operating at a higher level than ODSI, this proposal has the advantage of not requiring a new set of protocols for quality-of-service and bandwidth control. This work may have implications to many other OSI layers as well; for example, it has been suggested that MPLmS could be used as a control mechanism for optoelectronic interfaces. Other proposals for the control plane of OTNs are also under consideration by the IETF, ODSI, ANSI Tl, and ITU standards bodies [47].

These efforts are still in the early stages of development, providing opportunities for technical input and new services to evolve. First interoperability demonstrations of ODSI are planned for the latter half of 2001, but there have already been product demonstrations of an MPLS-based router as a conduit between IP and optical networks, which associates optical wavelengths directly with IP services. It has also been suggested that these efforts could lead to a "collapsed central office" architecture, in which optical assignment switching is performed at locations far from a telecom carrier central office and the typical distinctions between the MAN and WAN environments begin to blur together.

5.4.4.4 Converged and hierarchical networks

Within the datacom industry, there are a number of ad hoc efforts to standardize on the next-generation input/output (NGIO) interfaces [48] for computer equipment. The recent emergence of SANs, or switch-based private networks interconnecting computer processors with remote storage devices, has helped accelerate this trend. One promising effort is centered around the Infiniband consortium [49], which seeks to define the next generation of parallel copper and parallel optical interfaces for I/O subsystems. Although most efforts to date have concentrated on small form factor parallel copper interfaces, there has also been a recent draft proposal for multi-channel parallel optical interconnects. This standards effort is expected to drive a need for higher density optoelectronic packaging and closer integration between the parallel optical transceivers and associated electronics for data manipulation.

There is also a strong push toward terabit networking in both the datacom and telecom industries. As discussed earlier, DWDM has emerged as a solution to bandwidth constraints at the physical layer; survivability has been integrated into the physical, or transport, layer (extending traditional $1+1$ SONET protection to all protocols) and bandwidth optimization has been provided by sub-rate TDM over DWDM; future systems may also take advantage of code division multiple access or other bandwidth provisioning techniques. However, this approach will require the service layer (which provides IP, ATM, and other protocols) to upgrade more than twice as fast as the transport layer, or roughly double capacity every six months, in order to keep pace with the demand for bandwidth. Some estimates have shown the service layer growing over 70 times by 2003. A more realistic approach is to have the transport and service layers evolve together, although this still requires the service layer capacity to double on a yearly basis.

This trend is likely to be accompanied by a blurring of the conventional boundaries between the LAN (traditionally 0–10 km), MAN (10–100 km), and long haul or WAN (100 km +). Convergence of the datacom and telecom environments may drive traditional carriers and service providers to deploy networks with hierarchical logical structures [50]. In a hierarchical network, each layer is designed to hide the details of its operations from the layers above, thus enabling cost-effective and scalable network growth. At each layer, the appropriate traffic granularity and switching are selected, as well as suitable traffic "grooming" (sorting services by type and destination to wide bandwidth optical paths); both physical and logical paths are then selected. Significant cost benefits can be realized with this approach, including both equipment capital cost and operational expenses. Multi-tiered architectures can take advantage of cost-effective banded switching technologies, enable a large amount of optical bypass (which reduces the amount of required electronic termination equipment), and reduce the number of elements that need to be managed. A hierarchical scheme also provides efficient use of restoration capacity, reduces the amount of electronic access equipment required in order to effectively share this capacity, and simplifies the restoration policy algorithms [51].

One example of a hierarchical architecture is that of the two-tiered scheme shown in Figs. 5.4.27 and 5.4.28 [50]. In this architecture, the upper layer is comprised of localized collector rings, operating at granularities of DS-3 or OC-3, whereas the lower layer consists of a nationwide "express mesh" operating at granularities of OC-48, OC-192, or higher. Each collector ring is typically populated with ADMs, with all nodes in the ring being logically connected (although not necessarily at all wavelengths). The functions of the collector ring are to route traffic within the ring and to collect inter-ring traffic for delivery to the large mesh nodes located on that ring. The ring topology is suitable for this layer due

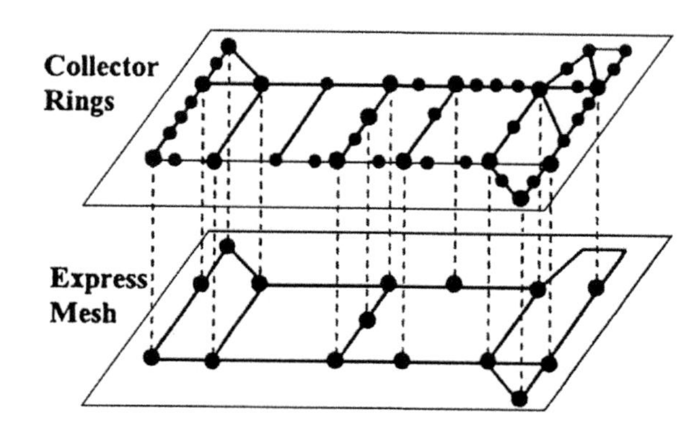

Fig. 5.4.27 Example of a two-level hierarchical network in the WAN, where a group of collector rings aggregates and grooms fine granularity traffic before delivering it to a streamlined, coarse granularity express mesh.

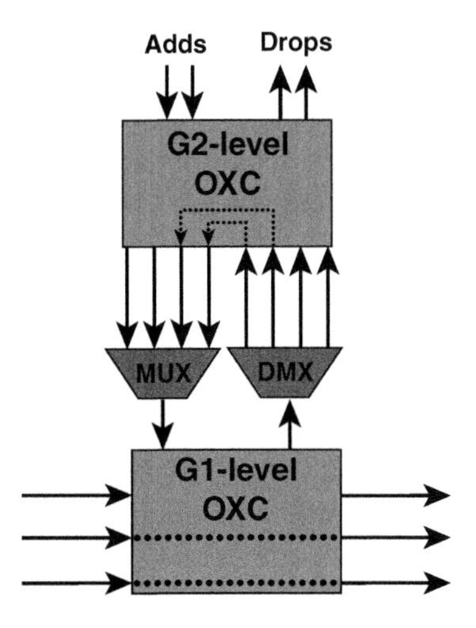

Fig. 5.4.28 Advantages of hierarchical switching; at the lowest level, an optical cross-connect (OXC) switches traffic at a coarse granularity 61, and only a small percentage of the traffic needs to be switched at a finer granularity (62).

to the limited geographic extent of the collector rings and the simple protection properties of rings. Furthermore, many carriers and ISPs already have a large legacy investment in SONET ring-based networks, whose capacity is presently being exhausted by the tremendous growth of traffic, exacerbated by the inefficiencies of inter-ring routings. These legacy rings can be more effectively used if their role is restricted to serving as the collector network; alternately, this network could be implemented as a mesh architecture with grooming cross-connects.

The collector network is designed with fine granularity and limited distances; by contrast, the lower-layer express mesh of Fig. 5.4.27 must carry large amounts of traffic for longer distances. These requirements could be met with emerging high-capacity, ultra-long-reach transport networks, possibly using all-optical switches. Because the collector rings perform an aggregation function, the express mesh could carry traffic at coarser granularities (OC-38 and above). The logical mesh does not include all network nodes, only those that generate traffic at levels comparable to the granularity of this level, or those that are strategically located. As a result, the mesh may completely bypass many of the nodes, depending on network traffic patterns; the express mesh may consist of only about 20 nodes nationwide. To take full advantage of this approach, traffic needs to be routed between and through lower-layer nodes without any intermediary regeneration; this requires the development of very long reach amplified optical networks. The combination of a sparsely populated express mesh with nationwide transport distances eliminates a large part of the electronic terminations in this proposed network [51]. In addition to limiting the nodes that are included in the lower layer, it is also important to create a streamlined physical topology; thus, fiber links that do not pass by major nodes may not be included in this layer. A mesh topology is most appropriate for the lower level since it can use switches to create virtual topologies that can be optimized for traffic routing and sharing of protection bandwidth. A significant benefit of the hierarchical approach is that it enables deployment of a hierarchical switch architecture, which has advantages in terms of cost and scalability. The streamlined design of an express mesh creates large bundles of traffic that can be switched as a single unit. For example, increasing the switching granularity by a factor of B decreases the switch fabric size by a factor between $B \log B$ and B^2, depending on the switch topology [50]. This translates into a savings in switch cost, size, power, number of physical interfaces, and states that must be managed.

The preceding example has discussed a two-tiered approach to network design. Ultimately, there may be a multi-layered hierarichal approach, where the number and granularity of layers is determined by the available technology and relative cost. For example, assume that the cross-sectional traffic on any link in a particular layer of the network is T, and that the finest level of granularity required at that layer is G. Then the number of elements per link that are potentially switched is T/G. At a node of degree N, this translates to a switch size of about $(NT/G)^2$ [50]. Due to limits of technology and physical space, the switch size may be limited, resulting in multiple tiers, where the factor NT/G is chosen for each tier to enable the network to scale easily. This approach may also lead to a compromise between the two competing views of the WAN as an opaque network (regeneration provided on every channel at every configuration point) vs. the construction of smaller, transparent sub-networks interconnected by regeneration sites (the so-called "islands of transparency" model) [51].

5.4.5 Future directions and conclusions

DWDM is being adopted in large data communication systems as a practical solution to problems of fiber exhaust, and a way to offer new high-availability services with fault recovery at the physical layer that do not need to be based on a SONET infrastructure. DWDM has enabled new computer architectures such as Parallel Sysplex, and made it practical to implement real-time disaster recovery and data mirroring at extended distances. New datacom network topologies are also enabled by DWDM coupled with flexible network management and security. Efficient bandwidth utilization is possible by using a TDM front end in combination with DWDM. Future systems are expected to continue this trend toward higher data rates, extended distances, more complex topologies, and new applications or service offerings; emerging technologies including multi-terabit routing and space-time conversion are expected to play an increasingly important role in these systems, which may be the first step toward all-optical networks.

There has been an explosion of research and development activity in WDM recently, including many new product announcements and developing industry standards. Given the rapid pace of growth and change, it is difficult to predict future trends in this area. Generally speaking, we can expect WDM systems to increase their maximum data rate per channel from OC-48 to 10 Gbit/s or 40 Gbit/s and beyond, with some form of digital wrapper or sub-rate TDM scheme to provide efficient bandwidth management. The number of wavelengths is expected to grow as well, from the current 32 up to 128 or more. With the combined aggregate bandwidth of WDM networks approaching multi-terabit rates, even in the MAN environment, significant challenges will emerge for network management, restoration, monitoring, topology configuration, and intelligence. The convergence of MAN and WAN applications will continue to drive use of optical amplifiers for extended distance, as part of either 2R or 3R repeater designs. Use of new topologies including ring mesh in the MAN or hierarchical networks at the nationwide level, combined with optical switching and cross-connects, promises to open up even more applications.

Because there are no ultra-dense WDM systems commercially available yet, a combination of TDM and DWDM is currently used to maximize use of existing installed fibers, enable new applications such as Parallel Sysplex, and drive down the per channel cost of datacom systems. This approach offers some advantages; because TDM is an established technology, it can be implemented immediately with good reliability and reasonable cost; it also does not require changing the DWDM wavelengths currently in use, so that it can be made backward compatible with existing systems. There is also a well-established roadmap for incorporating TDM into datacom networks; care must be taken in extending the SONET TDM approach to protocol-independent TDM systems, because the timing jitter and other factors may vary significantly as a function of data rate. Traditionally, optical communication solutions have been successful in those areas that require higher data rates and longer unrepeated distances than can be provided by copper solutions. In the 1960s most computer interconnections were performed by using parallel copper cables with 8 to 16 wires in a duplex link; these attachments were limited in distance to about 100 meters, and to data rates of about 3.5 Mbyte/s. As the demand for higher data rates became apparent and fiber optics became affordable, many low-speed parallel copper links were combined into a single high-speed serial data link. The classic example of this approach is the ESCON protocol, which provides a 17 Mbyte/s serial data stream and performs serial-to-parallel conversion at either end of the link to recover the low-speed data links. Over time, this approach has been applied again and again to increase the effective data rate. Currently, FICON links are capable of time division multiplexing up to 8 ESCON links over one gigabyte link using the FICON Bridge feature in the 9032-5 ESCON Director. In the near future, we can expect this trend to continue as the most cost-effective way to implement higher bandwidth connectivity.

Recently, TDM has been supplemented by DWDM techniques; for example, combining the FICON Bridge with the 2029 allows up to 128 ESCON channels to be multiplexed over a single fiber pair. Currently, the 2029 provides 40 Gbytes total bandwidth over a fiber pair; as the per channel data rate of DWDM is increased, future plans include a protocol-independent TDM sub-rate multiplexing (SRM) function integrated with the DWDM products. For example, a 4:1 TDM of ESCON channels (at 100% utilization) could be accommodated in a single 1.25 Gbit/s DWDM channel; if the 4:1 TDM card occupies only a single slot in the DWDM card cage, then the total capacity of the 2029 can be increased from 32 to 128 protected channels or 256 unprotected channels (the total capacity of an S/390 Enterprise Server). The use of TDM as a front-end for DWDM in this manner offers significantly improved bandwidth utilization and lower cost per channel.

Next, we will discuss other approaches that have the potential to be used in future applications. Before we consider these future systems, let us consider the fundamental building blocks of time-bandwidth allocation or space-spatial bandwidth allocation in a communications network. If we consider time-based systems only we can implement many well-known operations such as TDM, which is also known in wireless communications as time division multiple access (TDMA), frequency multiplexing or frequency division multiple access (FDMA), and code division multiple access (CDMA) networks. The consideration of space-based systems leads us to examples such as parallel fiber ribbons and linear arrays of optical transmitters and receivers. Among the time-based systems, TDMA and FDMA have been the most widely available technologies; however, CDMA has received increasing attention lately as the technology has become more advanced. Wireless systems using CDMA with multiple antenna arrays are an example of combining time-based and space-based bandwidth considerations into a single system. By contrast, although fiber optic networks are quite advanced they still do not possess the necessary building blocks to combine time-based and space-based approaches. As an example, consider that the bandwidth of a single-mode fiber is on the order of 20 THz; thus, we should be able to work with optical pulses as small as 50 femtoseconds (actually, in the future we can envision a parallel array of semiconductor femtosecond pulses lasers, just as we currently use nanosecond radio frequency (R.F.) pulses for electronic communication). We can then consider a simple TDMA system with M channels; as a numerical example, we will consider $M = 32$. Assume further that each channel carries a signal of 10 Gbit/s; this could represent advanced channels in a future Parallel Sysplex, or perhaps the emerging 10 Gigabit Ethernet standard. The total system throughput is thus 320 Gbit/s, with each input channel modulating a series of 50 femtosecond pulses, and the TDMA circuitry running at 320 GHz. In this example, channel one modulates pulses 0, M, $2M$, etc. while channel two modulates pulses 1, $M+1$, $2M+1$, and so on. The modulated pulses are combined and transmitted along a single fiber; the combiner is really acting as a parallel-to-serial converter or a multiplexer. At the receiver we need to perform demultiplexing or serial-to-parallel conversion capable of handling 50 femtosecond pulses. Although elements of this system have already been demonstrated, including the 50 femtosecond lasers, the technology required to produce the serial/parallel and parallel/serial conversions is very difficult to realize and has been the bottleneck to producing systems like the one in our example. Before we discuss a possible solution to this problem, we point out that demultiplexing from a series of fast pulses to a parallel group of slower pulses is a well-known, fundamental problem that arises whenever high bandwidth optical fiber is coupled to comparatively low bandwidth electronics. Sometimes this serial-to-parallel conversion is called a time-space convolver, N-path convolver, or channelizer; its equivalent model uses z-transforms and multirate filters [34]. In the history of networking, such devices are often the bottleneck whenever a new, faster technology is introduced and it is desirable to interoperate with legacy, slower technology. There are many examples of this, including the replacement of parallel copper cables with ESCON fiber optic channels in many IBM enterprise servers.

Although there are many design proposals and laboratory demonstrations of methods to deal with this problem, we will note two possible variations of the so-called time-space multiplexer that are promising for long-term implementation. Of these, the option that uses DWDM as part of its functionality is probably more viable. In the optics community, the serial-to-parallel converter is generally referred to as a time-space converter because the input time signal is spread into M channels, which are distributed in space. One example of a time-space converter based on a nonlinear crystal or holographic element has been demonstrated [52]. This device has the ability to manipulate the amplitude and phase of each channel. A second approach uses DWDM filters to replace the nonlinear optical crystal in this implementation. As in the previous case, input signal pulses are multiplied with a time chirp signal, which basically maps the input pulses into different wavelengths (or frequencies). This can be done using a chirped fiber grating or dispersive filter, for example. The output is then applied to an M-element DWDM filter, which separates the pulses. Note that in this case, the phase information is lost because we are using a chirp signal. There are many technology and engineering issues to be resolved for these two serial-to-parallel converter designs. However, it appears that there is no fundamental limitation that prevents their realization; in particular, the WDM filters required for the second approach are already commercially available.

This approach leads naturally to the capability of implementing CDMA using femtosecond pulses, similar to the proliferation of this technology in wireless communications. The same time-space convolver discussed earlier can be easily modified to implement spread spectrum coding, by placing a transmission mask at the input spatial signal plane [52]. As the signal is multiplied by a chirp, it can be shown that the spatial distribution of light in this plane of the figure is the equivalent of a spatial Fourier transform of the input signal. By placing a mask in this plane whose transmittance function contains the Fourier transform of a spread spectrum code, the desired encoded pulse is obtained at the output plane. This is possible because the last half of the optical system shown is equivalent to performing the inverse Fourier transform. Although there are still many technology challenges involved in adopting R.F. solutions to the optical domain, once devices such as this come out of the laboratory and into commercial use, it is possible that optical networking will follow the history of wireless communications in the adoption of CDMA.

Acknowledgments

The author would like to thank the IBM TeraPlex Center support staff (Rich Hamilton, Mario Borelli, and Jack Myers) for support during the 2029 testing, the IBM 2029 product development team and GDPS support team (Ray Swift, Gary Vitullo, Simon Yee, JoAnn Transue, John Matcham, Ernie Swanson, John Torok, Jim Keller, John Deatcher, Steve Rohersen, Ken Knipple, Juan Parrilla, Dave Petersen, and Noshir Dhondy) for their months of dedicated work on this program.

References

1. C. DeCusatis, E. Maass, D. Clement, and R. Lasky, eds. *Handbook of Fiber Optic Data Communication.* New York: Academic Press (2001).
2. *IBM Journal of Research & Development.* Special issue on "IBM System/390: Architecture and Design," vol. 36, no. 4. (1992).
3. C. DeCusatis, "Optical data communication: fundamentals and future directions." *Opt. Eng.* **37**(12): 3082–3099 (December 1998).
4. "Market Report on Future Fiber Optic Technologies." Available from Frost and Sullivan, New York, N.Y. (May 1999). For details on 10 Gbit/s Ethernet, see also the homepage of the Optical Internetworking Forum (OIF) at http://www.oiforum.com
5. "Optical interface for multichannel systems with optical amplifiers." Draft standard G.MCS, annex A4 of standard COM15-R-67-E, available from the International Telecommunication Union (1999).
6. M. Ferries, "Recent developments in passive components and modules for future optical communication systems. " Paper MIl, *Proc.* OSA Annual Meeting, Santa Clara, Calif., p. 61 (1999).
7. C. G., Gyaneshwar, et al. Paper TuJ7-2, *Proc. OFC 2000,* Baltimore, Md (2000).
8. J. Simon, C. L. Billes, and L. Bramerie. "All optical regeneration." Paper FC4.1, *Proc. IEEE Summer Topical Meeting,* Boca Raton, Fla (2000).
9. B. Meagher, X. Yang, J. Perreault, and R. MacFarland. "Burst mode optical data switching in WDM networks." Paper FC3.2, *Proc. IEEE Summer Topical Meeting,* Boca Raton, Fla (2000).
10. J. Wei, "The role of DCN in Optical WDM Networks." *Proc. OFC 2000,* Baltimore, Md (2000).
11. J. S. Turnet, "WDM burst switching for petabit data networks." *Proc. OFC 2000,* Baltimore, Md (2000).
12. G. K. Chang, et al. "A proof-of-concept, ultra-low latency optical label switching testbed demonstration for next generation internet networks." *Proc. OFC 2000,* Baltimore, Md (2000).
13. N. Henmi, et al. "OADM workshop." *EURESCOM P615,* p. 36 (1998).
14. N. Henmi, "Beyond terabit per second capacity optical core networks." Paper FC2.5, *Proc. IEEE Summer Topical Meeting,* Boca Raton, Fla (2000).
15. Telcordia Technologies standard GR-485-CORE. "Common language equipment coding procedures and guidelines: generic requirements," issue 3 (May 1999).
16. H. Dutton, *Optical Communications.* San Diego, Calif.: Academic Press (1999).
17. C. DeCusatis, and P. Das. "Spread spectrum techniques in optical communication using transform domain processing." *IEEE Trans. Selected Areas in Communications* 8(8): 1608–1616 (1990).
18. T. Brown, "Optical fibers and fiber optic communications." Chapter 1 in *Handbook of Optics* vol. IV, OSA Press (2000).
19. K. Hill, 2000. "Fiber Bragg Gratings." Chapter 9 in *Handbook of Optics* vol. IV, OSA Press (2000).
20. B. Poumellec, P. Niay, M. Douay et al. "The UV induced refractive index grating in Ge:SiO_2 Preforms: Additional CW experiments and the macroscopic origins of the change in index." *Journ. Of Physics D, App. Phys.* 29:1842–1856 (1996).
21. K. O. Hill, B. Malo, F. Bilodeau, et al. "Photosensitivity in optical fibers." *Ann. Review of Material Science* 23: 125–157 (1993).
22. B. S. Kawasaki, K. O. Hill, D. C. Johnson, et al. "Narrowband Bragg reflectors in optical fibers." *Opt. Lett.* 3: 66–68 (1978).
23. K. Hill, and G. Meltz. "Fiber Bragg grating technology: fundamentals and overview." *Journ. of Lightwave Tech.* 15: 1263–1276 (1997).
24. R. Stolen, "Nonlinear properties of optical fiber." Chapter 5 in *Optical Fiber Communications.* S.E. Miller and A.G. Chynoweth, eds. New York: Academic Press (1979).
25. C. S. Li, and F. Tong. "Emerging technology for fiber optic data communication." Chapter 21 in *Handbook of Fiber Optic Data Communication,* pp. 759–783 (1998).
26. W. Xin, G. K. Chang, and T. T. Gibbons. "Transport of gigabit Ethernet directly over WDM for 1062 km in the MONET Washington, D.C. network." Paper WC1.4, *Proc. IEEE Summer Topical Meeting,* Boca Raton, Fla (2000).
27. C. Bouchat, C. Martin, E. Ringoot, et al. "Evaluation of Super PON demonstrator." Paper ThC2.3, *Proc. IEEE Summer Topical Meeting,* Boca Raton, Fla (2000).
28. A. Manzalini, A. Gladisch, and G. Lehr. "Management optical networks: view of the EURESCOM project P918." Paper ThC1.2, *Proc. IEEE Summer Topical Meeting,* Boca Raton, Fla (2000).
29. C. DeCusatis, D. Petersen, E. Hall, F. Janniello. "Geographically distributed parallel sysplex architecture using optical wavelength division multiplexing." *Optical Engineering,* special issue on Optical Data Communication **37**(12): 3229–3236 (Dec 1998).
30. C. DeCusatis, "Wavelength and channel multiplexing for data communications." *Proc. OSA Annual Meeting,* Portland, Ore., p. 115 (1995).
31. C. DeCusatis, D. Stigliani Jr., W. Mostowy, M. Lewis, D. Petersen, and N. Dhondy. "Fiber optic interconnects for the IBM S/390 Parallel Enterprise Server G5." *IBM Journal of Research and Development* **43**(5/6): 807–828 (Sept/Nov 1999).
32. C. DeCusatis, and P. Das. "Subrate multiplexing using time and code division multiple access in dense wavelength division multiplexing networks." *Proc. SPIE Workshop on Optical Networks,* Dallas, Tex., pp. 1–8 (2000).
33. C. DeCusatis, and D. Priest. "Dense wavelength division multiplexing devices for metropolitan area datacom and telecom networks." *International Conference on Applications of Photonics Technology (Photonics North),* Quebec, Canada, pp. 8–9 (June 2000).
34. J. Wei, C. Liu, and K. Liu. "IP over WDM traffic engineering." Paper WC1. 3, *Proc. IEEE Summer Topical Meeting,* Boca Raton, Fla (2000).
35. D. O. Awduce, Y. Rckhter, J. Drake, et al. "Multiprotocol lambda switching: combining MPLS traffic engineering control with optical crossconnects," IETF draft proposal at draft-awduche-mpls-te-optical-01.txt
36. S. Chaudhuri, G. Hjalmtysson, and J. Yates. "Control of lightpaths in an optical network." IETF draft proposal: draft-chaudhuri-ip-olx-control-00.txt
37. D. Basak, D. O. Awduche, J. Drake, et al. "Multiprotocol lambda switching: combining MPLS traffic engineering control with optical crossconnects."

IETF draft proposal: draft-basak-mpls-oxc-issues-01.txt

38. M. Krishnaswamy, G. Newsome, J. Gajewski, et al. "MPLS control plane for switched optical networks." IETF draft proposal: draft-krishnaswamy-mpls-son-00.txt
39. B. Rajagopalan, D. Saha, B. Tang, et al. "Signalling framework for automated provisioning and restoration of paths in optical mesh networks." IETF draft proposal: draft-rstb-optical-signaling-framework-00.txt
40. G. Wang, D. Fedyk, V. Sharma, et al. "Extensions to OSPF/IS-IS for optical routing." IETF draft proposal: draft-wang-ospf-isis-lambda-te -routing-00.txt
41. Y. Fan, P. A. Smith, V. Sharma, et al. "Extensions to CR-LDP and RSVP-TE for optical path set-up." IETF draft: draft-fan-mpls-lambda-signaling-00.txt
42. J. Wei, C. Liu, and K. Liu. "IP over WDM traffic engineering." Paper WC1. 3, *Proc. IEEE Summer Topical Meeting,* Boca Raton, Fla (2000).
43. P. Bonenfant, A. Moral. "Digital wrappers for IP over WDM systems." Paper WC1.2, *Proc. IEEE Summer Topical Meeting,* Boca Raton, Fla (2000).
44. M. Medard, S. Lumetta. "Robust routing for local area optical access networks." Paper FC1.4, *Proc. IEEE Summer Topical Meeting,* Boca Raton, Fla (2000).
45. Chung, Y. C. "Optical monitoring techniques for WDM networks." Paper FC2.2, *Proc. IEEE Summer Topical Meeting,* Boca Raton, Fla (2000).
46. Information on ODSI is available in Electrical Engineering Times issue 1096 p. 1 available at http://www.eet.com.
47. See, for example, ANSI X3.230–1994 rev. 4.3, Fibre channel —physical and signaling interface (FC-PH) May 30, 1995; ANSI X3.272-199x, rev. 4.5, Fibre channel — arbitrated loop (FC-AL), June 1995; ANSI X3. 269-199x, rev. 012, Fiber channel protocol for SCSI (FCP), May 30, 1995.
48. For the most current information on NGIO, see http://www.intercast.de/design/servers/future_server io/documents/071 cameron/slide014.htm.
49. For the most current information on Infiniband, see Infiniband (SM) trade association at http://www.connectedpc.com/design/servers/future_server_io/link_spec.htm.
50. J. Simmons, "Economic and architectural benefits of hierarchical backbone networks." Paper WC1.1, *Proc. IEEE Summer Topical Meeting,* Boca Raton, Fla (2000).
51. J. Simmons, "Hierarchical restoration in a backbone network," *Proc. OFC,* San Diego, Calif (1999).
52. P. C. Sun, Y. T. Mazurenko, W. S. C. Chang, P. K. L. Yu, and Y. Fainman. "All optical parallel-to-serial conversion by holographic spatial-to-temporal frequency encoding." *Optics Letters* 20: 1728–1730 (1995).

Section Six

Optical networks

Chapter 6.1

6.1

Passive optical network architectures

Lam

6.1.1 FTTx overview

The general structure of a modern telecommunication network consists of three main portions: backbone (or core) network, metro/regional network, and access network (Fig. 6.1.1).

On a very high level, core backbone networks are used for long-distance transport and metro/regional networks are responsible for traffic grooming and multiplexing functions. Structures of backbone and metro networks are usually more uniform than access networks and their costs are shared among large numbers of users. These networks are built with state-of-the-art fiber optics and wavelength division multiplexing (WDM) technologies to provide high-capacity connections.

Access networks provide end-user connectivity. They are placed in close proximity to end users and deployed in large volumes. As can be seen from Fig. 6.1.1, access networks exist in many different forms for various practical reasons. In an environment where legacy

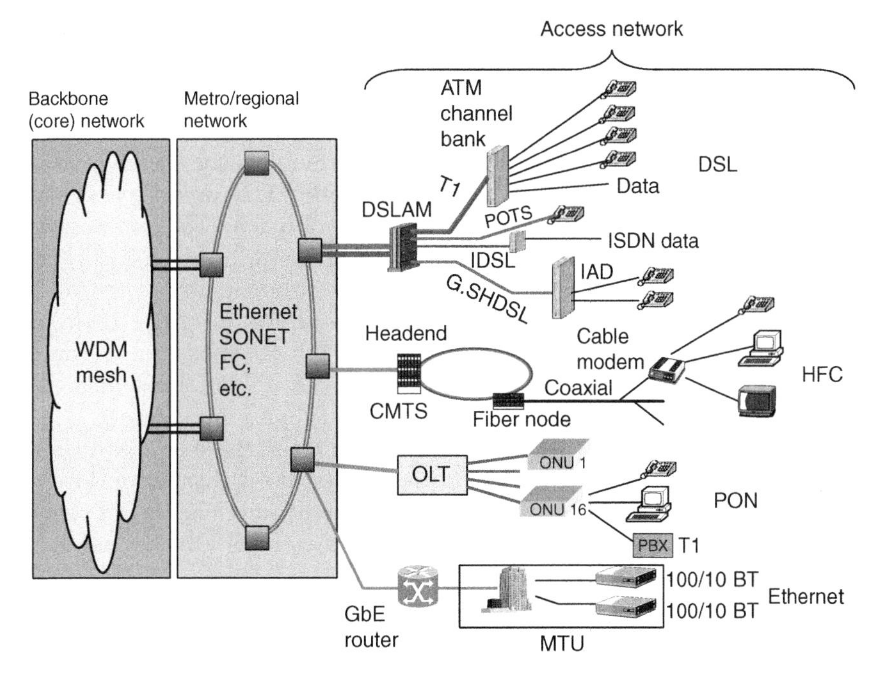

Fig. 6.1.1 Generic structure of a modern telecommunication network.

Passive Optical Networks; ISBN: 9780123738530

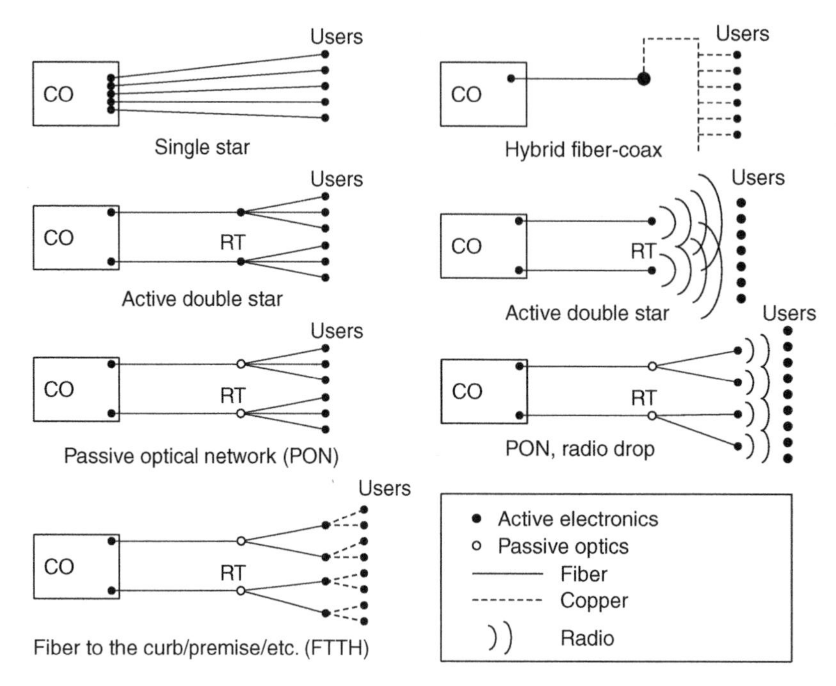

Fig. 6.1.2 FTTx alternatives (from [1] copyright [2004] by IEEE, reprinted with permission).

systems already exist, carriers tend to minimize their capital investment by retrofitting existing infrastructure with incremental changes, whereas in a green-field environment, it often makes more sense to deploy future-proof new technologies which might be revolutionary and disruptive.

Compared to traditional copper-based access loops, optical fiber has virtually unlimited bandwidth (in the range of tera-hertz or THz of usable bandwidth). Deploying fiber all the way to the home therefore serves the purpose of future-proofing capital investment. A passive optical network (PON) is a form of fiberoptic access network. Most people nowadays use PON as a synonym of FTTx, despite the fact that the latter carries a much broader sense.

Figure 6.1.2 shows the alternatives of FTTx [1]. As seen from the figure, in the simplest case, individual optical fibers can be run directly from the central office (CO) to end users in a single star architecture. Alternatively, an active or passive remote terminal[1] (RT) with multiplexing functions may be placed in the field to reduce the total fiber mileage in the field. A PON network is characterized by a passive RT.

In an optical access network, the final drop to customers can be fiber (FTTH), coaxial cable (as in an HFC system), twisted pairs or radio (FTTC). In fact, a PON system can be used for FTTH or FTTC/FTTP depending on whether the optical fiber termination (or the ONU location) is at the user, or in a neighborhood and extended through copper or radio links to the user. In this book, we do not make a distinction between FTTH and FTTC/FTTP.

6.1.2 TDM-PON Vs WDM-PON

Figure 6.1.3 shows the architecture of a time division multiplexing PON (TDM-PON) and a wavelength division multiplexing (WDM) PON [2]. In both structures, the fiber plant from the optical line terminal (OLT) at a CO to the optical network units (ONUs) at customer sites is completely passive.

A TDM-PON uses a passive power splitter as the remote terminal. The same signal from the OLT is broadcast to different ONUs by the power splitter. Signals for different ONUs are multiplexed in the time domain. ONUs recognize their own data through the address labels embedded in the signal. Most of the commercial PONs (including BPON, G-PON, and EPON) fall into this category.

A WDM-PON uses a passive WDM coupler as the remote terminal. Signals for different ONUs are carried on different wavelengths and routed by the WDM coupler to the proper ONU. Since each ONU only receives its own wavelength, WDM-PON has better privacy and better scalability. However, WDM devices are

[1] Another term for RT is remote node (RN). In this book, these two terminologies will be used interchangeably from place to place.

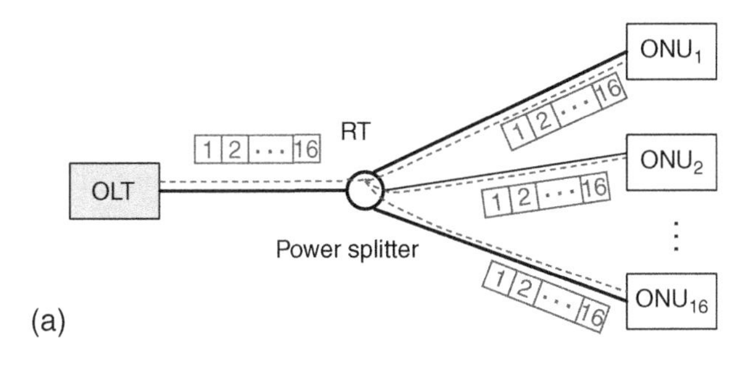

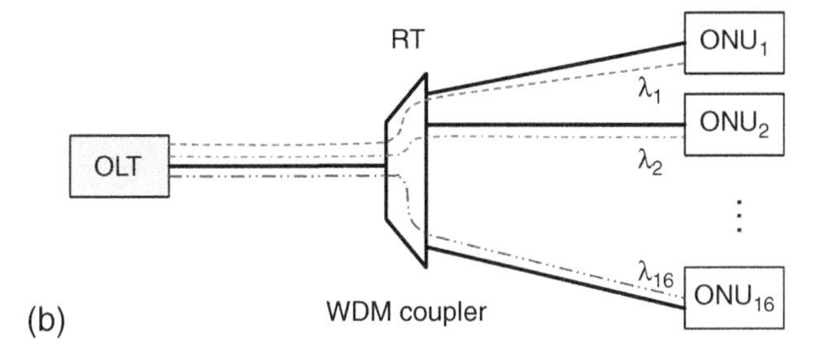

Fig. 6.1.3 Architecture of (a) TDM-PON and (b) WDM-PON.

significantly more expensive, which makes WDM-PONs economically less attractive at this moment.

6.1.3 Optical transmission system

6.1.3.1 Optical fiber

The low loss, low noise, and exceptionally large bandwidth of optical fiber makes it ideal for long-distance backbone network transmission. Recently, the field of fiber-optic communications has experienced tremendous growth, thanks to the development of WDM technologies. As a result of this development, costs of fiber-optic components have dramatically decreased to the point that it is now commercially viable to apply fiber-optic technologies in access networks.

Optical fibers are waveguides made of high-purity glasses [3]. The cylindrical core of a fiber has a slightly lower refractive index than the cladding surrounding it. Optical fibers can be classified as single mode or multimode. Standard single-mode fiber (SMF) has a small core diameter of about 10 μm and requires high mechanical precision for signal coupling. On the other hand, multimode fibers (MMFs) have large core diameters for easy alignment and coupling. There are two commonly seen MMFs with core diameters of 50 μm and 62.5 μm respectively.

As shown in Fig. 6.1.4, light can only propagate in one mode in an SMF whereas there are multiple modes that light signals can propagate in an MMF because of the large core size [3]. These modes propagate at different speeds and result in modal dispersion in MMF. Modal dispersion causes signal pulses to broaden, thus limiting the signal bandwidth and transmission distance. The bandwidth-distance product of MMF is measured in mega-hertz kilometer (MHz km).

Traditionally, SMF has been used for long-distance backbone transmissions and MMF for local building connections. With the existing twisted copper pair infrastructure, one can already achieve tens of megabit per second data rate. Therefore, in order for PON, or FTTx system to make economic sense and future-proof the investment, it has to offer unparalleled capabilities that traditional copper plant cannot provide, i.e. gigabit per second data rate at kilometer distances. For this reason, high-speed optical access networks use SMF as opposed to MMF. In this book, we mainly deal with SMF.

6.1.3.2 Chromatic dispersion

Modal dispersion is not an issue in SMF. However, chromatic dispersion exists in SMF. It is caused by the different propagation speeds of light signals of different wavelengths or frequencies. As data symbols occupy finite frequency spans, chromatic dispersion broadens optical signal pulses as they propagate through optical fibers and produces power penalties at the receiver.

Chromatic dispersion is characterized by the dispersion parameter D, which is measured in units of ps/nm/km. It gives the broadening ΔT (in pico-seconds) of a pulse with bandwidth $\delta\lambda$ of 1 nm on the optical spectrum, after the pulse propagates through 1 km distance of fiber. For an arbitrary optical pulse propagating in a fiber network, the total broadening is given by:

$$\Delta T = \mathrm{D} \cdot \delta\lambda \cdot L \qquad (6.1.1)$$

where L is the transmission distance given in kilometers.

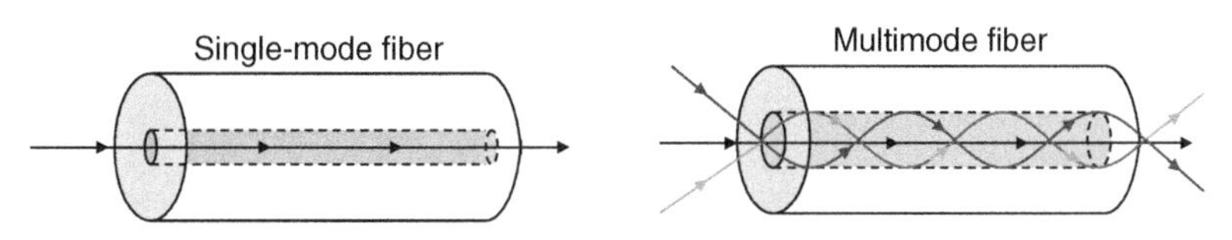

Fig. 6.1.4 Single-mode fiber (SMF) vs multimode fiber (MMF).

The modulation bandwidth $\delta\lambda$ of a transform limited optical signal with NRZ (nonreturn zero) modulation, is roughly related to the data rate R as:

$$\delta\lambda\cdot = (\lambda^2/c)\cdot R \qquad (6.1.2)$$

where λ is the wavelength of the optical carrier. Such a signal is usually produced with a high-quality single-wavelength laser through external modulation. In many cases, the output from the optical source spreads over a finite spectral region much wider than the modulation bandwidth of the data, which dominates $\delta\lambda$ and hence the chromatic dispersion.

The dispersion coefficient is a function of optical wavelength [3]. Figure 6.1.5 shows the dispersion coefficients as a function of wavelengths for different types of optical fibers. Standard single-mode fiber is usually the only type of fiber used in a PON system. As can be seen from Fig. 6.1.5, standard SMF has nearly 0 dispersion around the 1.3 μm wavelength region and 17 ps/nm/km dispersion around 1.55 μm, which is where erbium doped fiber amplifier (EDFA)—the most mature optical amplification technology works the best. Chromatic dispersion affects the choice of optical wavelength and transmitter technologies for upstream and downstream connections, which will be discussed later.

6.1.3.3 Fiber loss

Optical fiber loss affects the power budget, which limits the physical distance and splitting ratio that can be achieved in a PON system. Standard fibers are made of Silica (SiO_2), the very same material of glass and sand. Figure 6.1.6 shows the loss in a Silica optical fiber at different wavelengths.

In the short-wavelength region, optical signal loss is limited by Rayleigh scattering whereas in the long-wavelength region, optical scattering due to lattice vibrations limits signal loss. Figure 6.1.6 also shows the names of different optical bands and their spectral locations. It indicates that the C-band (stands for conventional band) wavelengths experience the minimum signal loss. As mentioned earlier, this spectral region is the region where optical amplification can be easily achieved with EDFAs [4]. This makes it the most suitable wavelength band of choice for long-haul WDM transmissions and for analog CATV transmissions [5] where high power is required to achieve the stringent carrier-to-noise ratio (CNR) requirement.

In conventional fibers, the attenuation peaks locally around the 1.38 μm wavelength (inside the E-band). This local peak (also called the water peak) is due to the absorption by the OH^- impurities left from fiber manufacturing. Better purification techniques have removed the water peak in new fibers such as the All-Wave fiber from OFS-Fitel Corporation and the SMF-28e fiber from Corning Inc. This makes the E-band spectrum available for coarse WDM applications.

PON systems are mostly built with 1.3 μm (O-band) for upstream signal transmission, 1.49 μm (S-band) for downstream transmission, and 1.55 μm (C-band) wavelengths for an optional analog CATV signal overlay. The use of these bands will be explained in more details in a latter section. The loss of new fibers around the 1.55 μm wavelength region can be as low as 0.19 dB/km [6]. It should also be noticed that around the 1.3 μm wavelength region used for upstream transmission, optical fiber loss is around 0.33-0.35 dB/km, significantly higher than that in the 1.5 μm wavelength region.

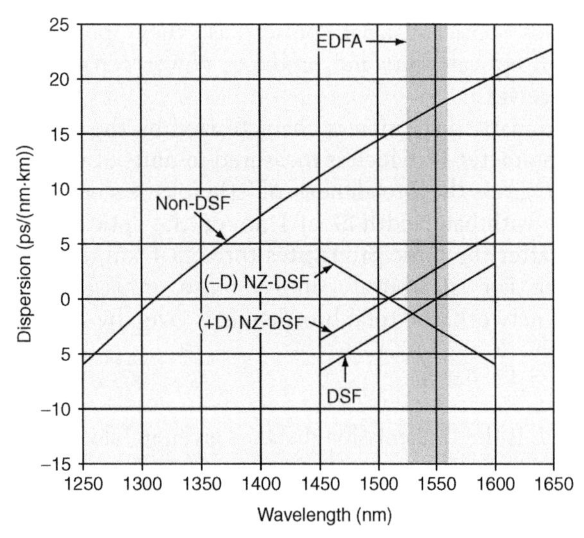

Non-DSF: Nondispersion shifted fiber, which is also called standard single-mode fiber or ITU-T G.652 fiber
(–D) NZ-DSF: negative nonzero dispersion shifted fiber
(+D) NZ-DSF: positive nonzero dispersion shifted fiber
DSF: dispersion shifted fiber with nearly zero dispersion in C-band

Fig. 6.1.5 Dispersion coefficient as a function of wavelengths for various types of optical fibers.

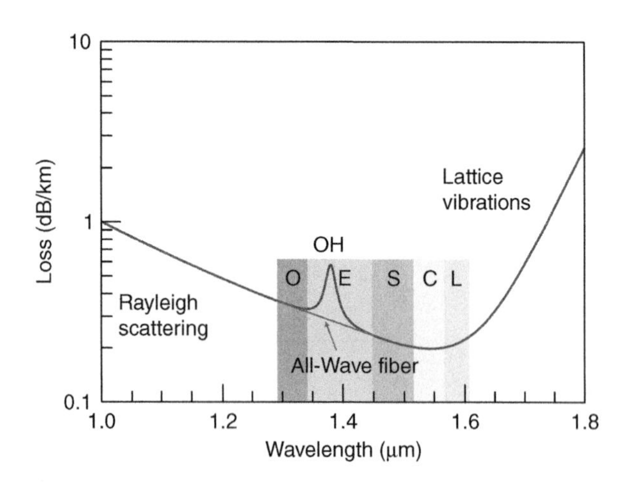

Fig. 6.1.6 Loss in an optical fiber at different wavelengths.

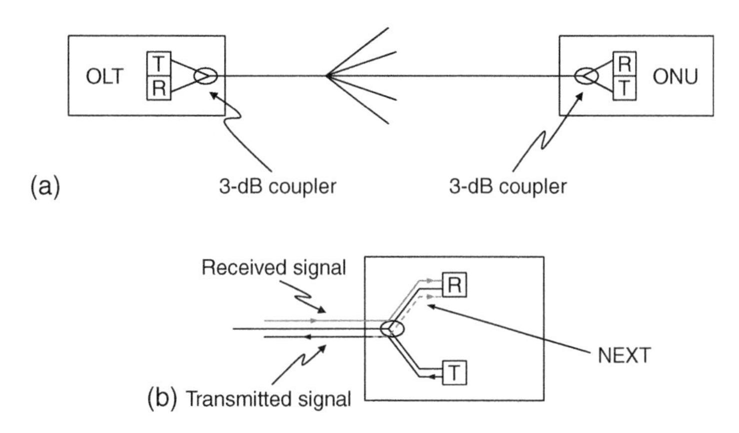

Fig. 6.1.7 (a) One-fiber single-wavelength bidirectional transmission. (b) Near-end cross talk (NEXT).

6.1.3.4 Bidirectional transmission

6.1.3.4.1 Two-Fiber vs One-Fiber

Conventional fiber-optic communication systems use two separate fibers for bidirectional communications. This is also called space division duplex, or two-fiber approach. This straightforward method does not require separation of the upstream (ONU to OLT) and downstream (OLT to ONU) signals in time, frequency, or wavelength domains, and is simple to implement. In a TDM-PON system, the two-fiber approach requires two optical power splitters in field whereas in a WDM-PON system, one or two WDM multiplexers may be used in the field (Chap. 3). Our main focus in this section will be power-splitting-based TDM-PON. WDM-PONs will be discussed in a separate section.

Usually, for power-splitting PONs with two fibers, the 1.3 μm wavelength is used for both upstream and downstream transmissions because low-cost Fabry-Perot (FP) lasers are readily available at this wavelength, and they can be used without much worry about fiber dispersion effects.

Despite its simplicity, because of the extra fiber required and the necessity to terminate and manage a second splitter, the two-fiber solution is more costly than one-fiber solutions from both the capital and operational standpoints.

6.1.3.4.2 One-Fiber single-wavelength full duplex

In this approach, only one optical fiber is used for both upstream and downstream connections. A simple 3-dB 1:2 directional coupler is used at the OLT and ONU to separate the upstream and downstream optical signal. Such a system is illustrated in Fig. 6.1.7 (a).

The problem of this approach is that the 3 dB couplers introduce about 3.5 dB signal loss on each end of the transmission link and hurt the system power budget. Moreover, the transmitted signal can be scattered into the local receiver as near-end cross talk (NEXT), which is illustrated in Fig. 6.1.7 (b). Given that the received downstream signal strength is reduced by the optical coupler at the RT and local transmitter power is usually high, NEXT puts a stringent requirement on reflection controls [7]. Furthermore, if the ONU and OLT wavelengths happen to be very close, NEXT will produce coherent cross talk, which is even more detrimental.

6.1.3.4.3 Time division duplex

In the time division duplex approach, the OLT and ONU take turns to use the fiber in a ping-pong fashion for upstream and downstream transmissions. Similar to the one-fiber single-wavelength full duplex approach, directional couplers are used at OLT and ONUs to separate upstream and downstream optical signals. The NEXT effect is avoided by separating upstream and downstream signals in the time domain, at the cost of reducing the overall system throughput by about 50%. The OLT coordinates the time slots assigned for upstream and downstream transmissions. Burst mode receivers are required at both the OLT and ONU.

6.1.3.5 Wavelength division duplex

The wavelength division duplex method separates upstream and downstream transmission signals using different wavelengths. To ease wavelength control, a coarse 1.3/1.5 μm wavelength duplexing scheme[2] is chosen to separate upstream and downstream signals (Fig. 6.1.8) [8]. The window at each wavelength is made sufficiently

[2] As mentioned earlier, the exact downstream wavelength used in industry standard PONs is actually 1.49 μm.

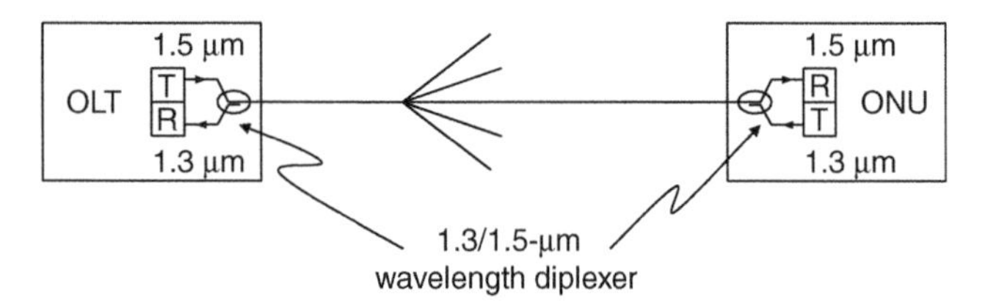

Fig. 6.1.8 Wavelength division duplex uses 1.3/1.5-μmm coarse WDM coupler (diplexer) to separate upstream and downstream signals.

large so that no temperature control is needed to stabilize laser output wavelengths.

To lower the overall system cost, it is important to employ low-cost optical components. F-P lasers are simple to manufacture and have good output power and reliability. However, these lasers emit more than one wavelength defined by repetition frequency of the cavity (Chap. 3). As a result of chromatic dispersions, these different longitudinal modes will propagate at different speeds, leading to pulse-broadening and intersymbol interference.

At high speed, dispersion effects due to the multiple F-P laser emission modes can limit the transmission distance. As mentioned earlier, such effect is minimum for standard SMF near the 1.3 μm wavelength region where the dispersion coefficient is nearly zero. F-P lasers at 1.5 μm wavelength region ($D = 17$ ps/nm/km) has a typical spectral width of $\delta\lambda = 2.5$ nm. Therefore, for a transmission distance of 10 km, the pulse-broadening $\Delta T = 425$ ps according to Eq. (6.1.1). As an example, in the IEEE802.3ah EPON standard [9, Clause 60], the symbol rate is 1.25 Gbaud/s after 8B10B physical layer encoding. The pulse-broadening due to typical 1.5 μm F-P laser at 10 km transmission distance is more than half of the symbol period (800 ps) (assuming NRZ modulation). Therefore, without dispersion compensation, F-P lasers are unsuitable at 1.5 μm transmission wavelength.

The more expensive single-mode DFB (distributed feedback) lasers [10] with narrow-output spectrum are needed for 1.5 μm transmission. Therefore, low-cost 1.3 μm F-P lasers are used at ONUs for upstream transmission and 1.5 μm DFB lasers are used at the OLT for downstream transmission where its cost can be shared by the multiple ONUs connected to the OLT.

One of the advantages of wavelength division duplex is that the reflected downstream light from unterminated splitter ports is also flittered by the diplexer at the OLT. This reduces the connector reflectivity requirements at the remote node power splitter. During the initial deployment stage of a PON system, the take-rate would be low and most of the remote node splitter fan-out ports may be unused. Without the wavelength diplexer, this could produce a lot of unwanted interference to the upstream-received signal at the OLT if those connectors are not properly terminated. Another source of unwanted reflected light is from breaks of distribution fibers in the field.

6.1.4 Power-splitting strategies in a TDM-PON

6.1.4.1 Splitting architectures

The purposes of power splitting include: (1) sharing the cost and bandwidth of OLT among ONUs and (2) reducing the fiber mileage in the field. Apart from the simple one-stage splitting strategy (Fig 6.1.9 (a)),

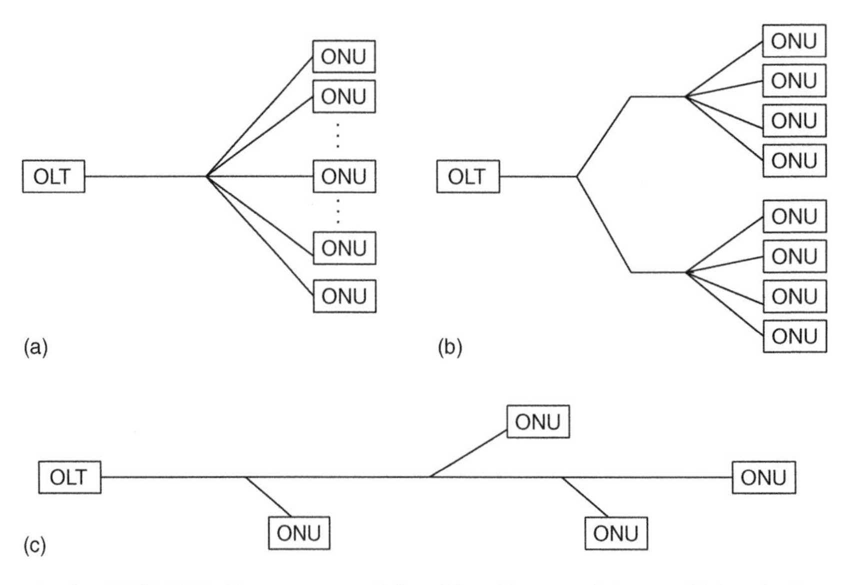

Fig. 6.1.9 Splitting strategies in a TDM-PON: (a) one-stage splitting, (b) multistage splitting, and (c) optical bus.

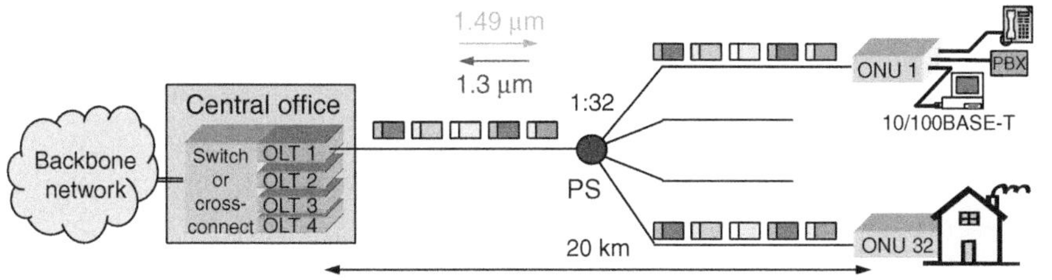

Fig. 6.1.10 Standard commercial TMD-PON architecture.

splitters may also be cascaded in the field as shown in Fig. 6.1.9 (b). In the most extreme case, the feeder fiber forms an optical bus and ONUs are connected to it at various locations along its path through 1:2 optical tap splitters as shown in Fig. 6.1.9 (c).

The actual splitting architecture depends on the demography of users and the cost to manage multiple splitters. From a management point of view, it is usually simpler to have a single splitter for distribution in the field, which makes splicing easier and minimizes connector and splicing losses.

In a bus or tree architecture like Fig. 6.1.9 (c), if all the splitters have the same power splitting ratio, the furthest ONU will suffer the most transmission and splitting loss and become the system bottleneck. Splitters with uneven splitting ratios may be used to improve the overall power margin. However, such optimization requires stocking nonuniform splitters and is hence difficult to manage.

6.1.4.2 Splitting ratio

Most of the commercial PON systems have a splitting ratio of 1:16 or 1:32. A higher splitting ratio means that the cost of the PON OLT is better shared among ONUs. However, the splitting ratio directly affects the system power budget and transmission loss. The ideal splitting loss for a 1:N splitter is 10 x log(N) dB. To support large splitting ratio, high-power transmitters, high-sensitivity receivers, and low-loss optical components are required. Higher splitting ratio also means less power left for transmission fiber loss and smaller margin reserved for other system degradations and variations. Therefore, up to a certain point, higher splitting ratio will create diminishing returns. Studies showed that economically the most optimal splitting ratio is somewhere around 1:40 [11].

A high splitting ratio also means the OLT bandwidth is shared among more ONUs and will lead to less bandwidth per user. To achieve a certain bit error rate (BER) performance, a minimum energy per bit is required to overcome the system noise. Therefore, increasing the bit rate at the OLTwill also increase the power (which is the product of bit rate and bit energy) required for transmission. The transmission power is constrained by available laser technology (communication lasers normally have about 0–10 dBm output power) and safety requirements issued by regulatory authorities [12].

6.1.5 Standard commercial TMD-PON infrastructure

Figure 6.1.10 shows the architecture of a standard commercial TDM-PON structure. This general architecture applies to APON, EPON, and G-PON.

In this architecture, an OLT is connected to the ONUs via a 1:32 splitter.[3] The maximum transmission distance covered is usually 10–20 km. Upstream direction traffic from ONUs is carried on 1.3 μm wavelength and downstream traffic from OLTs on 1.49 μm wavelength.

An ONU offers one or more ports for voice connection and client data connections (10/100BASE-T Ethernets). The voice connections can be T1/E1 ports for commercial users and plain old telephone service (POTS) for residential users.

Multiple OLTs in the CO are interconnected with a backbone switch or cross-connect, which also connects them to the backbone network. In a carrier environment, OLTs are usually constructed as line cards that are inserted into a chassis. The chassis can also host the backbone switch/cross-connect and provide the interconnect to the OLTs through a high-speed back plane.

The connection between the OLT and ONU is called the PON section. The signals transported in this section can be encoded and multiplexed in different formats and schemes depending on the PON standard implemented. Nevertheless, beyond the PON section, standard format signals are used for client interface hand-off, switching, and cross-connect. As mentioned in Chap. 1, the most common standard interface used today is the Ethernet interface.

[3] In some cases, smaller splitting ratios such as 1:8 or 1:16 may be used.

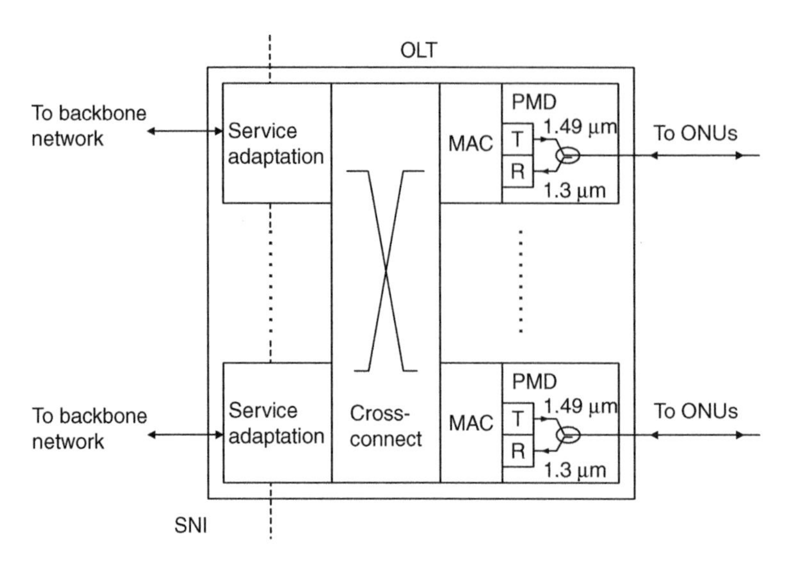

Fig. 6.1.11 Generic structure of a standard TDM-PON OLT. This diagram represents a chassis with multiple OLT cards, which are interconnected through a back plane switch. Each OLT card with its own MAC and PMD layer serves a separate PON.

In the PON section, signals from and to different ONUs are frame interleaved. Each frame is identified with a unique ONU ID in the frame header. The directional property of 1:N optical splitter made the downstream link a one-to-many broadcast connection. On the other hand, the upstream direction is a many-to-one connection, i.e. frames sent from all ONUs will arrive at the OLT, but no two ONUs can directly send signals toward each other on the optical layer. This implies that communications between ONUs need to be forwarded to the CO and relayed with the help of the OLT.

6.1.5.1 OLT and ONU structures

Figures 6.1.11 and 6.1.12 show the generic structure of a TDM-PON OLT and ONU respectively [8]. The PMD (physical layer dependent) layer defines the optical transceiver and the wavelength diplexer at an OLT or ONU.

The medium access control (MAC) layer schedules the right to use the physical medium so that contention for the shared fiber link is avoided among different ONUs. In a PON system, the MAC layer at the OLT serves as the master and the MAC layer at an ONU as a client. The OLT specifies the starting and ending time that a particular ONU is allowed to transmit.

As shown in Fig. 6.1.11, an OLT may contain multiple MAC and PMD instances so that it may be connected to multiple PON systems. A cross-connect at the OLT provides the interconnection and switching among different PON systems, ONUs, and the backbone network. The service adaptation layer in an OLT provides the translation between the backbone signal formats and PON section signals. The interface from an OLT to the backbone network is called service network interface (SNI).

An ONU provides the connection to the OLT in the PON section through the ONU MAC and PMD (Fig. 6.1.12). The service adaptation layer in the ONU provides the translation between the signal format

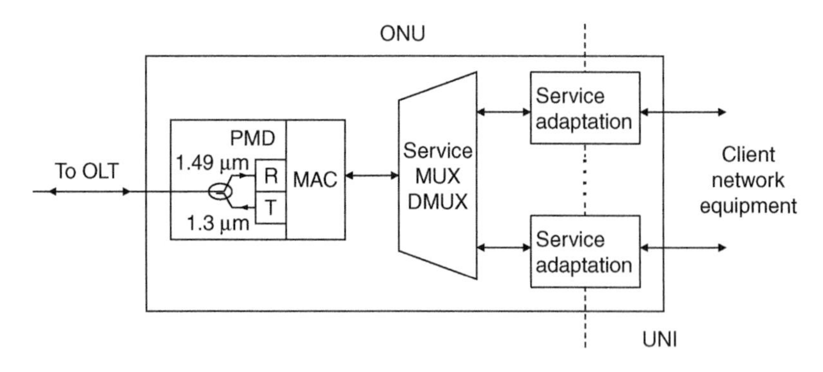

Fig. 6.1.12 Generic structure of a standard TDM-PON ONU.

required for client equipment connection and the PON signal format. The interface from an ONU to client network equipment is called user network interface (UNI). The Service MUX/ DMUX section provides multiplexing function for different client interfaces. Usually, multiple UNIs are available from an ONU for different type of services (e.g., data, voice). Each UNI may support a different signal format and require its own different adaptation service.

6.1.5.2 Burst mode operation and ranging

In the downstream direction, the OLT interleaves frames destined for different ONUs as a continuous stream and broadcast to all ONUs. Each ONU extracts its own frame based on the header address.

In the upstream direction, each ONU has its own optical transmitter to communicate with the OLT. Since there is only one optical receiver at the OLT, ONUs need to take turns to send their data to the OLT. In a conventional transmission system, the transmitter and its receiver always maintain synchronization by transmitting an idle pattern when there is no data to send. In a PON system, when an ONU is not sending upstream data, it has to turn off its transmitter to avoid interfering with other ONUs' upstream transmission. Therefore, in a TDM-PON system, burst mode transmission is used in the upstream direction. Burst mode transmission was also used in early generation coaxial-cable Ethernet LAN systems such as 10BASE-2 and 10BASE-5.

Every time an ONU transmits a signal burst to the OLT, it needs to first send a preamble sequence to the OLT. The OLT uses the preamble as a training sequence to adjust its decision threshold and perform synchronization with the ONU. Moreover, a guard time is reserved between bursts from different ONUs to allow the OLT receiver to recover to its initial state before the next burst. The guard time between bursts adds overheads to bandwidth efficiency and should be minimized in system designs.

To avoid collision between bursts from different ONUs, it is necessary to coordinate or schedule upstream transmissions. Early generations of Ethernet uses a distributed control mechanism called carrier sense multiple access with collision detection (CSMA/CD) to avoid scheduling [9]. However, the CSMA/ CD mechanism cannot be used in a PON system for the following reasons: (1) directional power splitters makes carrier sense and collision detection impossible because without using special tricks no ONU can monitor the optical transmission from other ONUs on the same PON; and (2) the data rate and distance covered by a PON system greatly exceeds the limits imposed by the CSMA/CD protocol. The CSMA/CD protocol becomes very inefficient under high bandwidth and long transmission distance [13, Chap. 4].

Scheduling is performed by the OLT MAC layer in a PON system. ONUs are located at different distances from the OLT. Therefore, signals from different ONUs will experience different delays before reaching the OLT. It is therefore important to establish a timing reference between the OLT and an ONU so that after accounting for the fiber delay, when the ONU signal arrives at the OLT, it arrives at precisely the same moment that the OLT intends for the ONU to transmit. The timing reference between the OLT and ONUs is established through the *ranging* process.

Ranging measures the round-trip delay between an ONU and OLT. In order to perform ranging, the OLT sends out a *ranging request* to ONU(s) to be ranged. An ONU participating in ranging then replies with a *ranging response* to the OLT. The OLT measures the round-trip time (RTT) from the ranging response and updates the ONU with this delay. The RTT is stored in the OLT or ONU which uses it to adjust the time that data frames from the ONU should be transmitted. All the ONUs are aligned to a common logical time reference after ranging so that collision does not occur in a PON system.

When ranging request is sent out, the OLT must reserve a time period called ranging window for unranged ONUs to respond. The size of the ranging window depends on the maximum differential delays between the closest ONU and the furthest ONU. Optical signal delay in 1 km of fiber is 5 μs. Therefore, for 20 km of differential distances between ONUs, an RTT difference of 200 μs needs to be reserved in the ranging window. Most PON standards specify a maximum physical distance of 20 km from an ONU to the OLT [8, 9, Clause 60, 14]. It should be realized that if the upper bound and lower bound of ONU separations are known to the OLT (e.g., through management provision), then instead of reserving a ranging window covering the maximum allowed separation between an ONU and an OLT, the size of ranging window can be reduced to cover only the maximum differential distance among ONUs. This reduces the ranging overhead and improves the overall bandwidth efficiency.

Ranging is usually done at the time an ONU joins a PON. The OLT periodically broadcasts ranging requests for ONU discovery. A new ONU detects the ranging request and responses to the OLT within the reserved ranging window after the ranging request. If multiple ONUs attempt to join the PON at the same time, collision may occur. Collisions in discovery are resolved by ONUs backing off with a random delay.

During operation, the ONU and OLT may continuously monitor the fluctuation of RTT due to changes such as temperature fluctuation, and perform fine adjustment by updating the RTT register value.

6.1.5.3 C-Band analog CATV signal overlay

In addition to broadband data delivery, TV program distribution was envisioned as a primary PON service in the very beginning. As pointed out in Chap. 1, analog TV services have been distributed using the 1550 μm wavelength in an HFC system for a long time [5]. Despite its limitations, one way analog broadcast TV is a simple and efficient method to stream video services to a large group of users.

A quick and easy method to offer TV services on a TDM-PON is to directly broadcast analog TV signal to end users on the 1.55 μm wavelength (Fig. 6.1.13) using a wavelength coupler [16]. As indicated in Fig. 6.1.13, the 1.55 μm wavelength can be amplified by an EDFA at the CO for broadcasting to multiple PONs to achieve better sharing of the analog TV resources.

The analog TV signal is peeled off at the ONU and converted into the regular RF signal on a coaxial cable using a set-top box (STB), which provides a simple OE (Optical to Electrical) conversion function. In reality, the analog conversion function is built into the ONU which offers a standard 75 Ω coaxial cable client interface. Furthermore, a 1.3/1.49/1.55 μm wavelength triplexer is used instead of two cascaded diplexers as shown in Fig. 6.1.13.

One potential issue in such configuration is the degradation of the subcarrier analog TV signals on the 1.55 μm wavelength by the 1.49 μm wavelength downstream digital signal, which acts a Raman pump to the 1.55 μm analog signal [17]. Since the downstream wavelength is modulated, it degrades the carrier-to-noise ratio (CNR) of the analog signals especially for low-frequency subcarrier channels.

In the past few years, digital and on-demand IPTV services are fast growing and replacing traditional analog TV services. The current trend is to take the advantage of a converged IP network and low-cost Ethernet technology to encode TV signals as MPEG/IP/Ethernet packets. Such a trend speeds up the unification of TV distribution network with data networks, which not only reduces the total capital expenditure (CAPEX) but also simplifies network management.

In the United States, MSOs are busy retiring analog services. With proper QoS (quality of service) control, digital IPTV services can be more cost-effectively delivered through the data channel. The author expects the 1.55 μm analog overlay technique to be less and less popular because of the following reasons:

1. The triplexer and analog RF receiver increase the cost of ONU.
2. The overlaid analog service creates additional management complexities. MSOs may be able to take the advantage of their existing legacy analog HFC system and the know-how they have developed for that. However, for conventional telecom operators entering the triple-play market, it is much better to leapfrog with digital IPTV technology as they do not have the legacy burden of analog TV distribution systems.

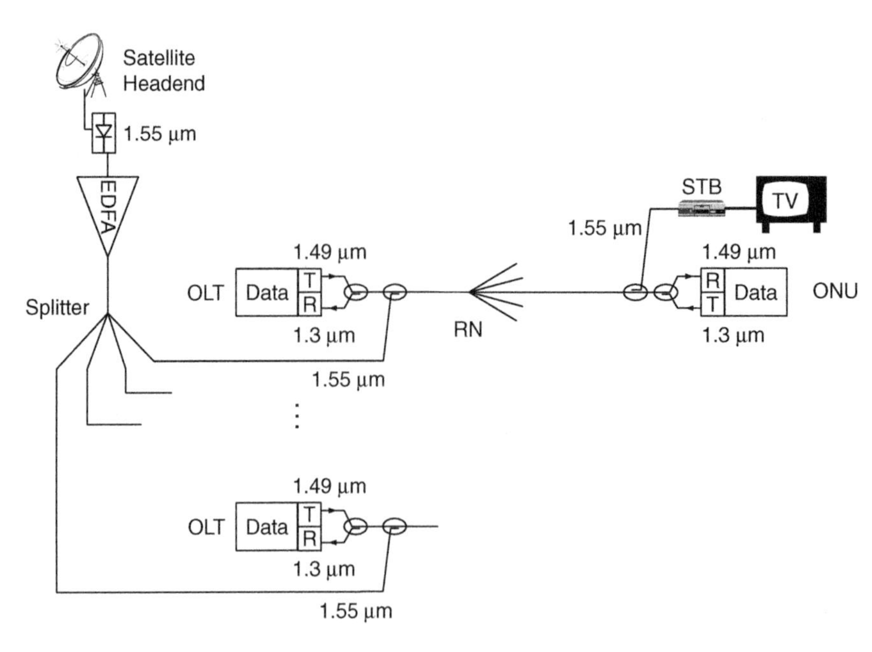

Fig. 6.1.13 Overlaying analog broadcast TV services on a TDM-PON using the 1.55 μm wavelength.

3. Regulatory bodies around the world are now phasing out analog TV broad cast and replacing them with digital services.

6.1.5.4 Security concerns in power-splitting PON

One important concern of power-splitting PON is the broadcast nature of the downstream channel, which makes it easy to eavesdrop downstream communication signals by malicious users. It should be realized that because of the directional nature of the optical star coupler at the remote node, an ONU cannot listen to the upstream transmission from another ONU.

The biggest security exposure is in the ranging process when the OLT broadcasts the serial number and ID of the ranged ONU. A malicious user can make use of this information for spoofing. This problem can be avoided through an authentication process during which the ONU is verified by a password known only to the OLT (e.g., through management provision).

To improve security, the ITU-T G.983.1 standard (Sect. 8.3.5.6, [8]) defines a churning procedure to scramble the data for downstream connections with a key established between the ONU and OLT. The encryption key is sent from an ONU to the OLT with a defined protocol. (Note that the encryption key is only sent in the secure upstream link.) To further enhance the security, the encryption key can be periodically updated. Churning provides some level of security in the physical layer. When security concern becomes very important, encryption should be used at the application layer. The IEEE 802.3ah EPON standard does not include transmission encryption. Nevertheless, many ASIC vendors provide their own encryption mechanisms on EPON chip sets. Depending on the security and interoperability requirements, users can decide to enable or disable these vendor-specific encryption mechanisms.

6.1.6 APON/BPON and G-PON

APON/BPON and G-PON are standardized by the ITU-T Study Group 15 (SG15). APON (ATM-PON) and BPON (Broadband PON) are different aliases of the TDM-PON architecture based on the ITU-T G.983 series standards [8, 16, 18]. While the name BPON serves its marketing purpose, APON clearly conveys that ATM frames are used for transport in the ITU-T G.983 standards. For this reason, we will simply use APON to refer to this class of PON designs. G-PON stands for gigabit-capable PON and is covered by the ITU-T G.984 series standards [14, 15]. It is the next generation PON technology developed by ITU-T after APON.

Both APON and G-PON defined line rates as multiples of 8 kHz [8, 14], the basic SONET/SDH frame repetition rate. As a matter of fact, the OLT distribute the 8 kHz clock timing from OLT to ONUs. This makes it easier to support TDM services on APON and GPON.

6.1.6.1 ATM-PON and ITU-T G.983

6.1.6.1.1 APON system description

The work on APON was started by the full service access network (FSAN) consortium [19] and later transferred to ITU-T SG15 [20] as the G.983 standards. APON systems were mostly deployed in North America by RBOCs for their FTTP projects. Many ideas covered in the G.983 standards were carried over to the G.984 G-PON standards.

The original G.983.1 standard published in 1998 defined 155.52 Mbps and 622.08 Mbps data rates. A newer version of the standard published in 2005 [8] added 1244.16 Mbps downstream transmission rate. APON vendors can choose to implement symmetric or asymmetric downstream and upstream transmission rates. Table 6.1.1 shows the possible combinations of downstream and upstream data rates for an APON system.

In addition to the one-fiber wavelength diplex solution explained earlier, both the G.983.1 and G.984.1 standards specify a two-fiber solution with dedicated upstream and downstream transmission fibers. The 1.3 μm wavelength is used in both directions in the two-fiber solution. However, to the knowledge of the author, no system has been deployed with the two-fiber solution.

All ITU PON standards feature three classes of optical transmission layer designs with different ODN (Optical Distribution Network) attenuations between ONU and OLT. The three classes are specified in ITU-T G.982 [21] as:

- Class A: 5-20 dB
- Class B: 10-25 dB
- Class C: 15-30 dB

Table 6.1.1 APON downstream/upstream bit-rate combinations.

Downstream	Upstream
1. 155.52 Mbps	155.52 Mbps
2. 622.08 Mbps	155.52 Mbps
3. 622.08 Mbps	622.08 Mbps
4. 1,244.16 Mbps	155.52 Mbps
5. 1,244.16 Mbps	622.08 Mbps

Class C design is a very demanding power budget requirement for a passive fiber plant. For practical implementation yield and cost reasons, Class B+ with 28-dB attenuation was later introduced by most PON transceiver vendors.

ITU-T G.983.1 specifies the reference architecture, transceiver characteristics, transport frame structures, and ranging functions in APON [8]. APON signals are transported in time slots. Each time slot contains either an ATM cell or a PLOAM (physical layer OAM) cell. PLOAM cells are used to carry physical layer management information such as protocol messages for ranging, churning key request and update, time slot requests from ONUs to OLT, assignments of upstream time slots to various ONUs by the OLT, system error, and performance monitoring reports, etc. The definition of PLOAM cell contents can be found in Sect. 8 of G.983.1.

The downstream time slots are exact 53-octet (byte) long ATM or PLOAM cells. A PLOAM cell is inserted every 28 time slots (or 27 ATM cells). At 155.52 Mbps speed, APON designates 56 time slots (with 54 ATM cells and 2 PLOAM cells) as a downstream frame. Each 155.52 Mbps upstream frame contains 53 time slots of 56 octets. An upstream time slots contains a 3-octet overhead in addition to an ATM or PLOAM cell. Besides, each upstream time slot may be optionally divided into multiple mini-slots at the request of OLT if necessary. Details of mini-slot definitions can be found in ITU-T G.983.4 [22], which covers the dynamic bandwidth allocation (DBA) mechanisms for APON systems. Figure 6.1.14 illustrates APON frames at 155.52 Mbps.

Frame structures at 622.08 Mbps and 1244.16 Mbps are similar except that the numbers of time slots per frame are multiplied by 4 and 8 respectively. Transmission of an ATM cell, PLOAM cell, or divided slot in the upstream direction is controlled by OLT using downstream PLOAM cells. G.983 requires a minimum of one PLOAM cell per ONU in every 100 ms [8].

The 3-octet upstream overhead contains the guard time, preamble, and the cell-start delimiter. APON specifies a minimum guard time of 4 bits. The contents of these fields are programmable and defined by OLT in the "Upstream_overhead" message, which is broadcast through downstream PLOAM cells.

6.1.6.1.2 Services in APON

Connections between ONU and OLT are established as ATM virtual circuits in an APON system. ATM services are connection-oriented. Each virtual circuit is identified by a virtual path identifier (VPI) and a virtual channel identifier (VCI) which are embedded in the cells comprising its data flow. VPI and VCI are indices providing different levels of ATM signal multiplexing and switching granularity. Multiple virtual circuits (VCs) can exist within a single virtual path (VP). An ATM connection is identified by its VPI/VCI pair. Figure 6.1.15 illustrates the idea of ATM signal switching.

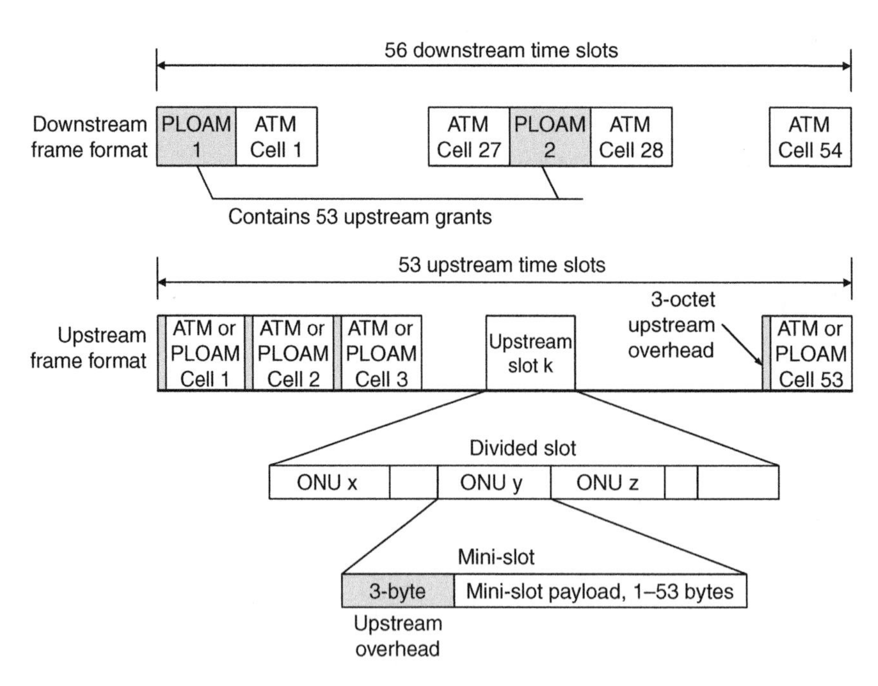

Fig. 6.1.14 Downstream and upstream APON frame formats at 155.52 Mbps speed. For 622.08 Mbps and 1244.16 Mbps speed, the numbers of time slots are simply multiplied by 4 and 8 to the numbers shown in the above diagrams.

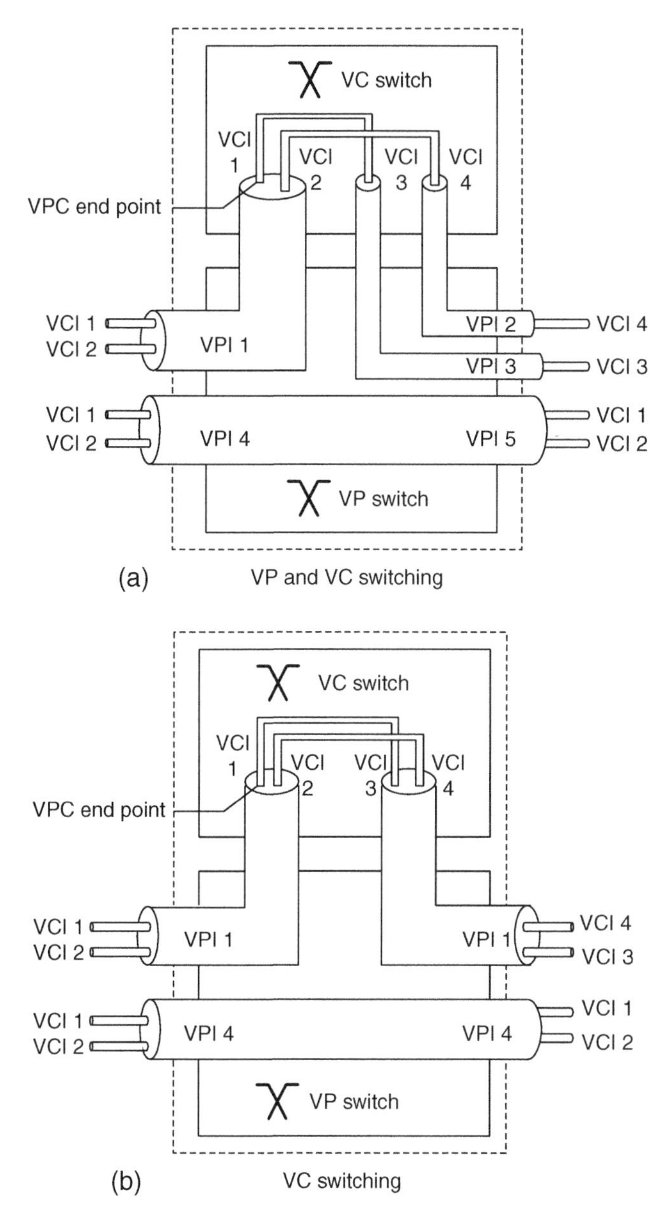

Fig. 6.1.15 ATM switching examples: (a) VP and VC switching and (b) VC switching.

Services in APON are mapped to ATM virtual circuits through the ATM adaptation layer (AAL). ATM cells include information cells, signaling cells, OAM cells, unassigned cells, and cells used for cell-rate decoupling [23]. AAL implements different levels of quality of service (QoS). ITU-T G.983.2 includes three different ATM adaptations for APON [18]: AAL-1, AAL-2, and AAL-5. AAL-1 provides adaptation functions for time-sensitive, constant bit-rate, and connection-oriented services such as T1 and E1 circuits. AAL-2 is used for variable-bit-rate connection-oriented services such as streaming audio and video signals and AAL-5 is used for connectionless data services such as TCP/ IP applications.

The OLT and ONU[4] as a whole can function as a VP or VC switch. Depending on the implementation, an ONU can cross-connect traffic at the VP level or VC level.

6.1.6.1.3 APON control and management

APON control and management functions are specified in the ITU-T G.983.2 standard [18]. An OLT manages its ONUs through a master-slave interface called ONT management and control interface (OMCI), with the OLT function as the master and ONU as slaves.

An ATM connection called ONT management and control channel (OMCC) provides the communication channel for OMCI. An OMCC is established by activating a pair of VPI/VCI using the particular PLOAM messages specified in ITU-T G.983.1. ITU-T G.983.2 requires each ONU to use a different VPI for the OMCC channel.

The G.983.2 standard specifies the management information bases (MIBs) with different parameters required for managing various APON objects and their behaviors. The OAM functions and parameters described in this standard are also applicable to the APON successor G-PON, except that different implementations of OMCI and OMCC are adopted.

6.1.6.2 Collision resolution in APON/ G-PON

In normal operation, a TDM-PON performs scheduled upstream transmission coordinated by the OLT and is therefore collision-free. The only exception is during the ranging process when multiple ONUs may respond to the broadcast ranging request at the same time. A collision will occur.

To resolve this contention, an APON system uses a binary tree mechanism specified in G.983.1 (Sect. 8.4.4, [18]). After detecting a ranging cell collision, an OLT will send a *Serial_number_mask* message followed by a ranging grant to allow ONUs whose serial number matches the mask to transmit a ranging cell. The size of the *Serial_number _mask* is increased by one bit at a time until only one ONU is transmitting a ranging cell. This resolves the contention and allows ONUs to be ranged one at a time. This mechanism may also help to avoid overloading the optical input of the OLT during ONU power setup.

[4] The G.983.2 standard refers to ONU as optical network terminator (ONT). The G.983.1 standard defines the ONU as the PON optical termination unit at the customer end in an FTTH system and ONT as the remote termination unit for FTTB systems. However, the use of these two terms seems to be quite arbitrary and interchangeable in the G.983 series standards. In this book, we do not distinguish between ONU and ONT.

In a G-PON system, ONUs use a random delay method to avoid collision during ONU initial registration process. The OLT may or may not enable the *Serial_-number _mask* mechanism in G-PON systems as specified in G.984.3 (Sect. 10, [24]).

6.1.6.3 Wavelength overlay in APON/ G-PON

The original APON system only specified 1.3 μm wavelength for upstream and 1.5 μm wavelength for downstream transmissions. To leverage the vast bandwidth in optical fiber and make the value out of capital-intensive fiber plants, ITU-T G.983.3 [16] specified coarse WDM overlay on the original APON system. This wavelength plan has been adopted by most PON manufactures to transport different services on the same physical infrastructure.

Figure 6.1.16 illustrates the wavelength allocation plan specified in ITU-T G.983.3. This ITU recommendation specifies the C-band as enhanced band for carrying either analog TV signals or additional digital services such as SONET/ SDH links. Two more bands, (1) 1360–1480 nm and (2) 1565 nm and beyond, are reserved for future studies. Additional wavelength multiplexers are required at ONU and OLT in order to make use of the enhancement bands specified in G.983.3. These additional wavelength multiplexers add losses and cross talks to the system. The G.983.3 standard also defines the loss and isolation requirements of these additional wavelength multiplexers.

6.1.6.4 G-PON and ITU-T G.984

As explained before, ITU-T G.983 was based on the ATM technology. Unfortunately, ATM did not live up to the expectation of becoming the universal network protocol to carry different applications. Instead, Ethernet and IP successfully evolved into that role. In ITU-T G.983, the OLT and ONU as a whole function as VP and VC switches. This requires APON ONU and OLT to implement ATM switching capabilities. The complicated adaptation model and QoS support made ATM switches costly. Moreover, it is necessary to implement translations between ATM and the protocols used at UNI and SNI. These requirements increase the system cost and complexity, and hinder the growth of APON systems in a fast-evolving broadband communication world.

To better cope with the changes in communication technologies and meet fast-growing demand, ITU-T created the G.984 series standards for PONs with Gigabit capabilities, or G-PON [14, 15, 24].

6.1.6.4.1 G-PON architecture

ITU-T G.984.1 gives a high-level overview of G-PON components and reference structure. G-PON PMD layer or transceiver requirements are covered by the ITU-TG.984.2 standard. Similar to APON, G-PON also defined single-fiber and dual-fiber PMDs. The bit rates defined in G.984 are:

- Downstream: 1244.16 Mbps/2488.32 Mbps
- Upstream: 155.52 Mbps/622.08 Mbps/1244.16 Mbps/2488.32 Mbps

Appendix I lists the optical layer characteristics for gigabit speed interfaces specified in G.984.2. At the time of writing, characteristics of 2488.32 Mbps upstream transmission link are yet to be studied and finalized.

As the bit rate advances into the gigabit regime, PON optical layer starts to become challenging. First, to cover the full 20 km transmission distance, multi-longitudinal-mode (MLM) lasers cannot be used at ONU any more in order to avoid excessive dispersion penalty. Second, to cover the loss budget requirements for Class B (10-25 dB) and Class C (15-30 dB) fiber plants, more sensitive avalanche photo-diodes (APDs) are required instead of the lower cost PIN receivers [25]. Without proper protection circuits, APDs are susceptible to damages due to avalanche breakdown if the inputs optical power becomes too high.

With the same fiber plant loss budget as in APON, to support the high bit rates, higher power transmitters are used in G-PON to meet the power budget requirements. This also implies that G-PON receivers need to handle higher receiver overload powers and therefore larger dynamic ranges. To ease the requirements and

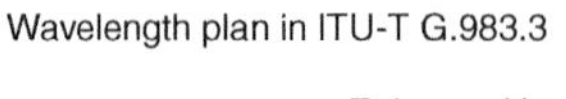

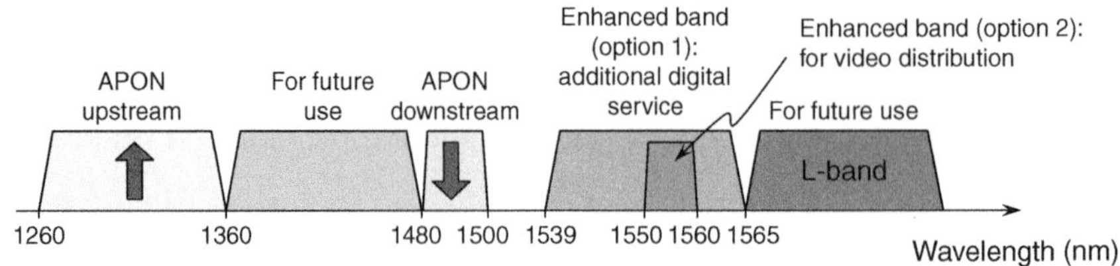

Fig. 6.1.16 Wavelength allocation plan in ITU-T G.983.3.

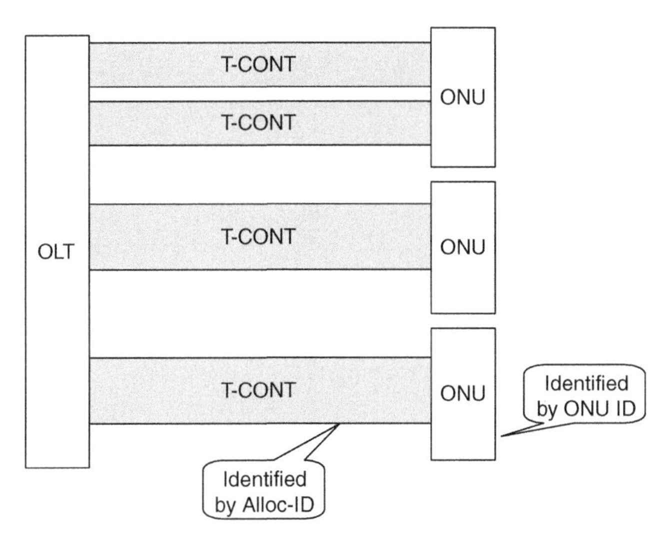

Fig. 6.1.17 A T-CONT represents a logical link between the OLT and an ONU.

implementation of the upstream OLT burst mode receiver, G-PON has specified a power-leveling mechanism for "dynamic" power control (Sect. 8.3, [15]).

In the power-leveling mechanism, the OLT tries to balance the power it received from different ONUs by instructing ONUs to increase or decrease the launched power. Consequently, an ONU which is closer to the OLT and seeing less loss, will launch at a smaller power than an ONU which is further apart and experiencing more loss. Such concepts of power-leveling or power control have long existed in cellular networks to deal with the near-far cross talk effect and save cellular device battery power.

6.1.6.4.2 G-PON transmission convergence layer

The main function of the G-PON transmission convergence (GTC) layer is to provide transport multiplexing between the OLT and ONUs [24]. Other functions provided by the GTC layer include:

- Adaptation of client layer signal protocols
- Physical layer OAM (PLOAM) functions
- Interface for dynamic bandwidth allocation (DBA)
- ONU ranging and registration
- Forward error correction (optional)
- Downstream data encryption (optional)
- Communication channel for the OMCI

GTC functions are realized through transmission containers or T-CONTs. Each T-CONT, which is identified by an allocation ID (Alloc-ID) assigned by the OLT, represents a logical communication link between the OLT and an ONU. A single ONU may be assigned with one or more T-CONTs as shown in Fig. 6.1.17. Five different types of T-CONTs with different QoS attributes have been defined in the ITU-T G.983.4 standards. G-PON standards defined two different operation modes, ATM and GEM (G-PON encapsulation mode). The GEM mode encapsulation is similar to generic framing procedure (GFP) [26]. A T-CONT can be either ATM- or GEM-based. An ATM-based T-CONT multiplexes virtual circuits identified by VPI and VCI, whereas a GEM-based T-CONT contains connections identified by 12-bit port numbers. These are illustrated in Fig. 6.1.18.

The protocol stack of GTC is shown in Fig. 6.1.19. Although the physical reach and splitting ratio are defined as 20 km and 1:32 respectively in G.984.2, the GTC layer has specified a maximum logical reach of 60 km and splitting ratio support of 128, in anticipation for future technology developments. Such high splitting ratio and longdistance coverage are being studied under a series of super PON projects around the world. ONUs and OLTs may support either ATM-based or GEM-based T-CONT or a mix of both (dual-mode). A G-PON using ATM-based T-CONT is however not backward compatible with APONs. Therefore, in reality, most of the G-PONs only implement GEM-based T-CONT. In the rest of the discussion, we will mainly focus on GEM.

The GTC layer supports the transports of 8 kHz clock from the OLT to ONUs and an additional 1 kHz reference signal from the OLT to ONUs using a control channel.

6.1.6.4.2.1 GTC framing

The GTC framing sublayer offers multiplexing capabilities and embedded OAM functions for upstream time slot grants and dynamic bandwidth allocation (DBA). Embedded OAM is implemented in the GTC frame header.

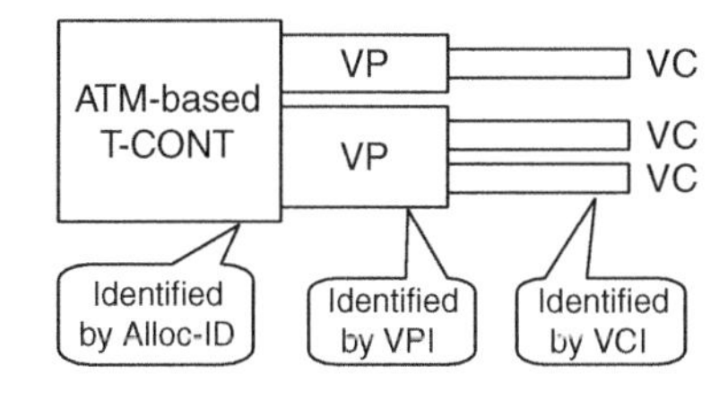

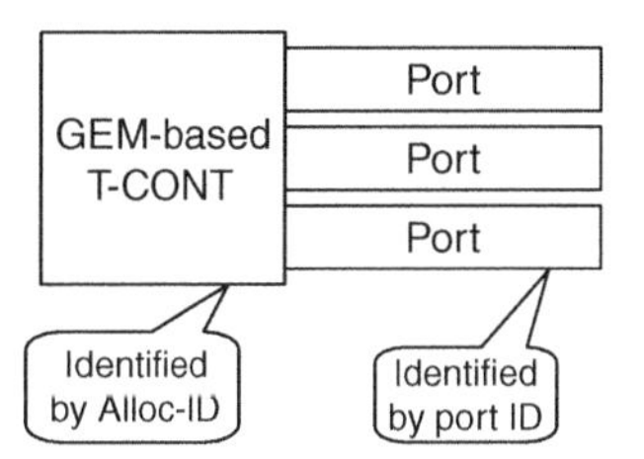

Fig. 6.1.18 ATM-based T-CONT vs GEM-based T-CONT.

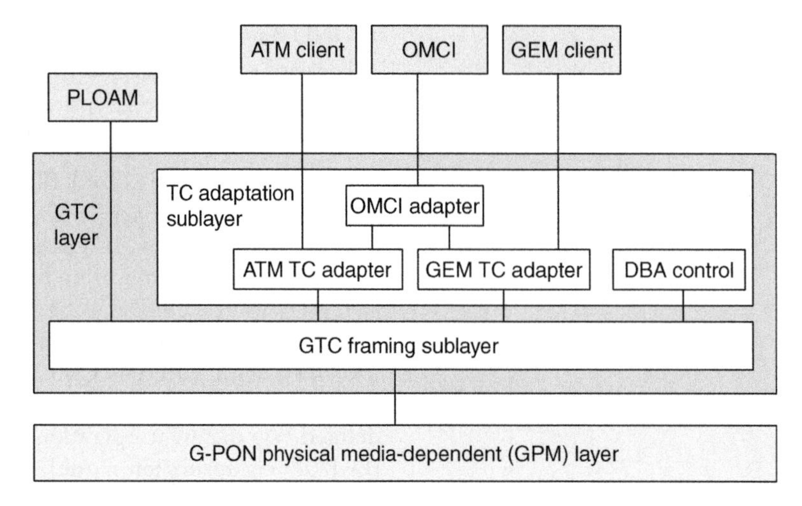

Fig. 6.1.19 Protocol stack of the G-PON transmission convergence (GTC) layer (from [24], reproduced with kind permission from ITU).

6.1.6.4.2.1.1 GTC downstream framing

Each GTC downstream frame is 125 μs long, which contains a downstream physical control block (PCBd) and a payload section as shown in Fig. 6.1.20.

The concepts of G-PON media access control is depicted in Fig. 6.1.21. The PCBd contains media access control information in the *upstream* (US) *bandwidth* (BW) *map*. As illustrated in the figure, the OLT specifies the start time and end time that each T-CONT can use to transmit upstream data using pointers in the US BW map. The pointers are given in units of bytes. This allows the upstream bandwidth to be controlled with 64 kb/s granularity. However, the standard allows vendors to implement larger granularities.

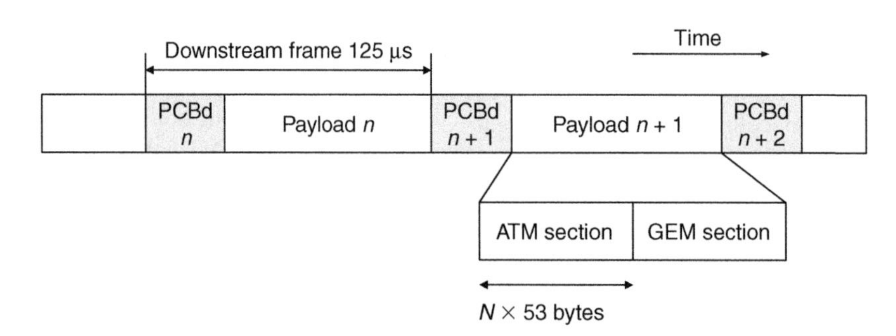

Fig. 6.1.20 GTC downstream signal consists of 125 μs frames with a PCBd header and a payload section.

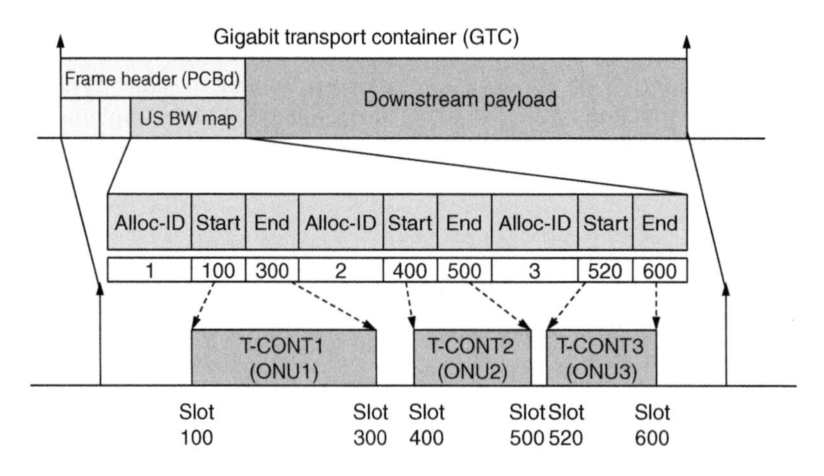

Fig. 6.1.21 GTC downstream frame and media access control concept (from [24], reproduced with kind permission from ITU).

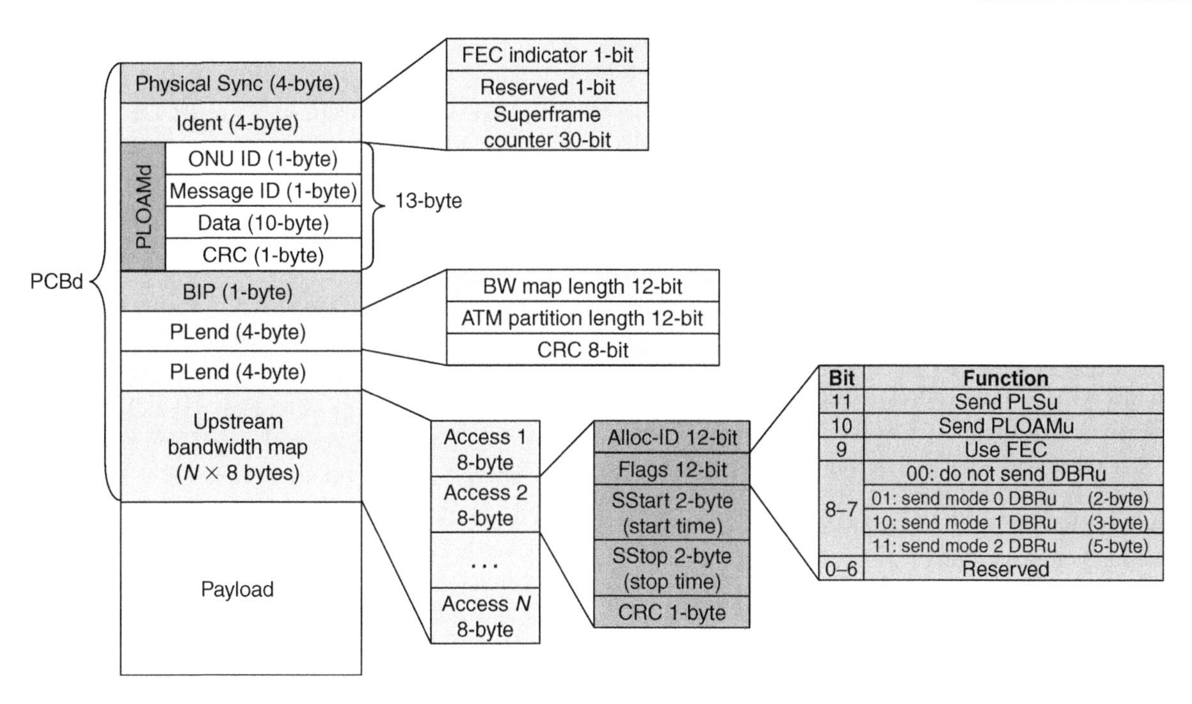

Fig. 6.1.22 GTC downstream frame formats.

Figure 6.1.22 shows the details of downstream frame formats. The PCBd header consists of a fixed part and a variable part. The fixed part contains the Physical Sync field, Ident field, and a PLOAM field. These fields are protected by a 1-byte bit-interleaved parity check. The 4-byte unscrambled physical synchronization pattern indicates the beginning of a downstream frame. The 4-byte Ident field indicates whether FEC is used. In addition, it also implements a 30-bit wraparound superframe counter, which can be used to provide a low-rate synchronization reference signal. The 13-byte PLOAM field in PCBd is used to communicate the physical layer OAM messages to ONUs. PLOAM functions include ONU registration and deregistration, ranging, power leveling, cryptographic key update, physical layer error reports, etc.

The variable part of the PCBd header contains duplicated payload length downstream (PLend) descriptor, which specifies the length of the upstream bandwidth map and that of the ATM partition in the T-CONT. As mentioned before, each ONU may be configured with multiple T-CONTs. The US BW map specifies the upstream bandwidth allocation with access entries. Each 8-byte access entry in the US BW map contains the Alloc-ID of a T-CONT, the start time and end time to transmit that T-CONT in the upstream direction and a 12-bit flag indicating how the allocation should be used. The rest of the downstream frame is downstream payload. Since each downstream frame has a duration of 125 μs, the length of downstream frames are different for 1.24416 Gb/s and 2.48832 Gb/s speeds, which are 19,440 and 38,880 bytes respectively. The PCBd block is the same for both speeds.

6.1.6.4.2.1.2 *GTC upstream framing*

The GTC upstream transmission consists of upstream virtual frames of 125 μs duration as shown in Fig. 6.1.23. Upstream virtual frames are made up of bursts from

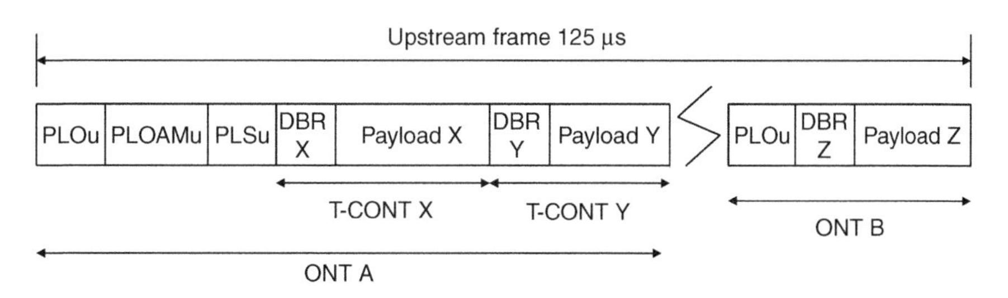

Fig. 6.1.23 GTC upstream framing: each ONT starts the upstream transmission with PLOu. An ONT assigned with two Alloc-ID in two consecutive upstream allocations only needs to transmit PLOu once.

different ONTs. Each burst starts with a physical layer overhead upstream (PLOu).

The PLOu begins with a preamble to help the burst mode receiver at the OLT to synchronize with the ONT (Fig. 6.1.24) transmitter. A delimiter following the preamble signifies the beginning of an upstream burst. As in APON, the length and format of preamble and delimiter are specified by the OLT using downstream PLOAM messages. An indication field (Ind) in PLOu provides real-time ONU status reports to the OLT.

As mentioned earlier, a single ONU may be assigned multiple T-CONTs. If one ONU is allocated contiguous time slots for its multiple T-CONTs with different Alloc-IDs, PLOu only needs to be transmitted once. This is shown in Fig. 6.1.23 for ONT A.

Following the PLOu field, there are three optional overhead fields in each burst:

1. Physical layer operation, administration, and management upstream (PLOAMu)
2. Power leveling sequence upstream (PLSu)
3. Dynamic bandwidth report upstream (DBRu)

Transmissions of these fields are dictated by the OLT through flags in the US BW map in PCBd. When requested by the OLT, the 120-byte PLSu field is sent by ONU for power measurement purposes.

A DBRu field is tied to each T-CONT for reporting the upstream traffic status of the associated T-CONT. The DBRu field contains the DBA report which is used to indicate the upstream queue length for dynamic bandwidth allocation. G-PON supports status-reporting (SR) and nonstatus-reporting (nSR) DBA mechanisms. In nSR DBA, the OLT monitors the upstream traffic volume and there is no protocol required for ONU to communicate queue status to the OLT. In SR DBA, an ONU informs OLT the status of traffic waiting in the queuing buffer.

There are three ways that ONUs may communicate the status of pending traffic to the OLT:

1. Status indication bits in PLOu (as shown in Fig. 6.1.24). Status indication gives OLT a quick and simple way of knowing the type of upstream traffic waiting, without much details.
2. Piggyback DBA report in the DBRu. The DBA report have three modes (Mode 0, Mode 1, and Mode 2) corresponding to DBA field lengths of 1, 2, or 4 bytes (i.e. 2-, 3-, or 5-byte DBRu) in Fig. 6.1.24. Transmission of DBRu and its format are specified by the OLTusing bits 8 and 7 of the flags in the US BW map (Fig. 6.1.22). Piggyback DBA in DBRu allows ONUs to continuously update the traffic of a specific T-CONT. The DBA report contains the number of ATM cells or GEM blocks (which are 48-byte in length) waiting in the upstream buffer.
3. DBA upstream payload. An ONU can send a dedicated whole report of traffic status on any or all its T-CONT in the upstream payload. Details of DBA upstream payload can be found in ITU-T G.984.3.

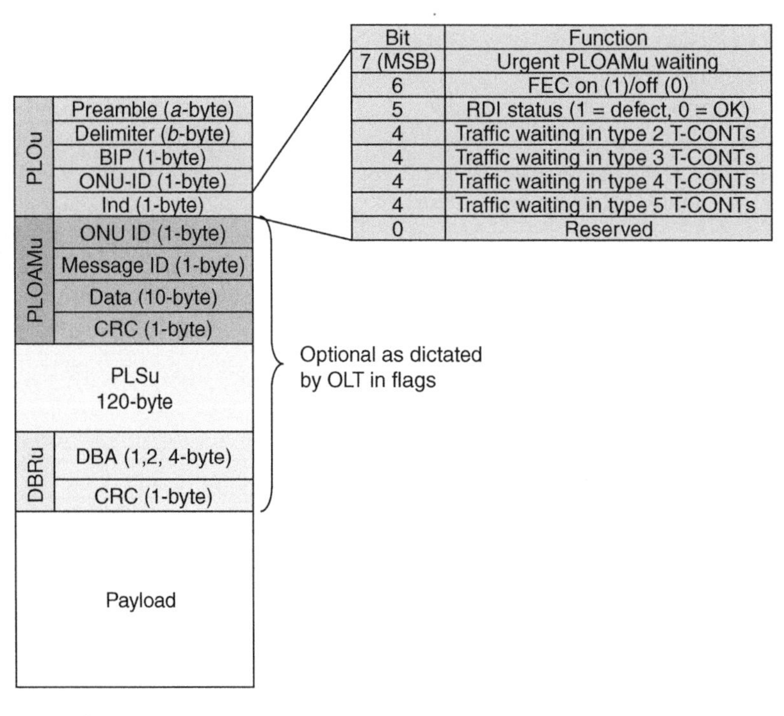

Fig. 6.1.24 GTC upstream frame format.

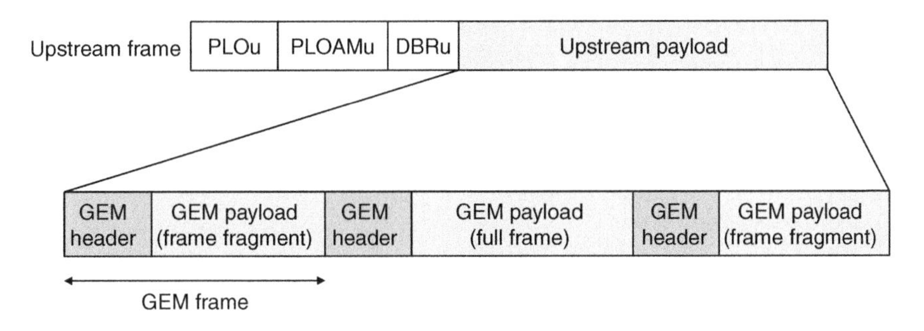

Fig. 6.1.25 GEM frames in upstream payload (from [24], reproduced with kind permission from ITU).

A G-PON system does not have to support all the reporting modes. Reporting capabilities are negotiated and agreed upon between ONU and OLTs using OMCI handshakes.

6.1.6.4.2.1.3 GEM encapsulation

In the downstream direction, GEM frames are carried in the GEM section of downstream payloads (Fig. 6.1.20). In the upstream direction, GEM frames are carried in upstream payloads as shown in Fig. 6.1.25.

GEM frame encapsulation serves two functions: (1) multiplexing of GEM ports and (2) payload data fragmentation. The format of GEM encapsulation is shown in Fig. 6.1.26. The GEM header contains a 12-bit payload length indicator (PLI) which specifies the GEM payload length in bytes. This allows a maximum payload length to be 4095 bytes. Any user-payload structure longer than 4095 bytes must be fragmented. The 3-bit payload type indicator (PTI) indicates whether the payload contains user data frames or GEM OAM frames. It also indicates whether the user data frame in the payload is the last fragment. The 12-bit port ID allows 4096 port numbers for traffic multiplexing.

The header error control (HEC) field uses a BCH code for header signal integrity protection. This field is also used to maintain GEM frame synchronization.

Figure 6.1.27 shows the fragmentation process in GEM encapsulation. Fragments of a user data frame must be transmitted contiguously. An advantage of the GEM fragmentation process is the convenience to support high-priority traffic. Urgent frames with high priorities can be inserted at the beginning of each GTC frame (Fig. 6.1.28) with 125 μs latency corresponding to the frame period. This makes it very easy to support TDM services with deterministic data rate in a G-PON system.

To decouple user data rate, an idle GEM with all zero payload has also been defined. When there is no user data to send to, idle GEM frames are sent to fill up empty slots and maintain receiver synchronization.

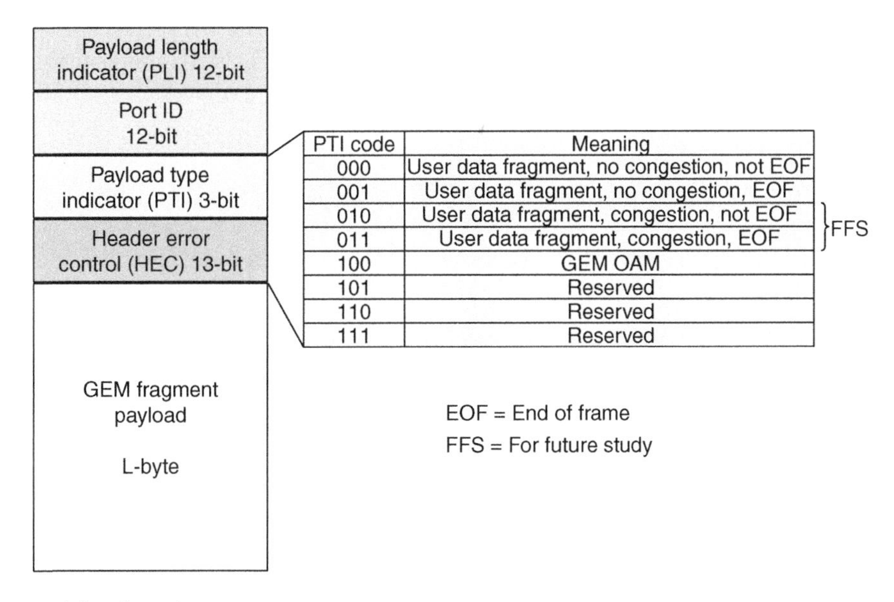

PTI code	Meaning
000	User data fragment, no congestion, not EOF
001	User data fragment, no congestion, EOF
010	User data fragment, congestion, not EOF
011	User data fragment, congestion, EOF
100	GEM OAM
101	Reserved
110	Reserved
111	Reserved

Fig. 6.1.26 GEM encapsulation formats.

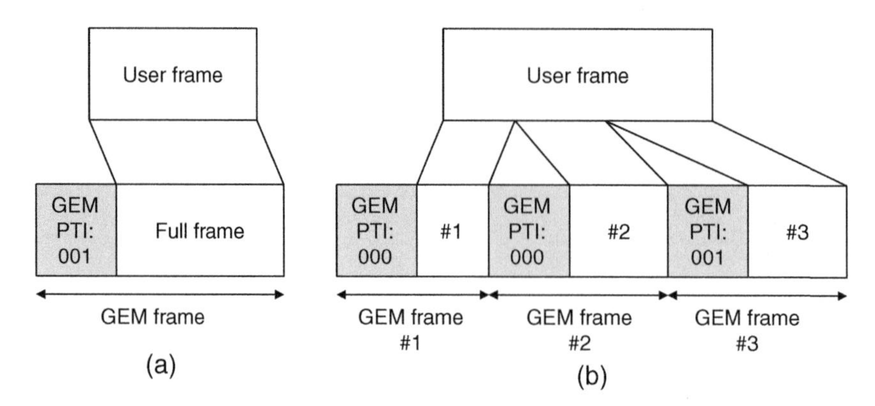

Fig. 6.1.27 Mapping and fragmentation of user data frames into GEM payload (from [24], reprinted with kind permission from ITU).

6.1.6.4.3 Forward Error Correction (FEC) in G-PON

To improve optical link budgets in G-PON, ITU-T G.984.3 defines an optional FEC capability using RS(255, 239) code in GTC framing. The same code is also used in the ITU-T G.709 optical transport network (OTN) standard [27]. FEC gives about 3–4 dB extra margin in optical budget but reduces the user data throughput by 6% because the symbol rate and GTC frame length are both kept unchanged when FEC is turned on. Therefore, the effective payload data rate has to be reduced in order to accommodate FEC overheads.

6.1.6.5 APON/G-PON protection switching

The reference models for APON/G-PON resiliency and redundancy are specified in the ITU-T G.983.5 standard [28]. ITU-T G.983.6 [29] covers the control and management interface to support APON/G-PON with protection features.

6.1.7 EPON

EPON is a new addition to the Ethernet family. The work of EPON was started in March 2001 by the IEEE 802.3ah study group and finished in June 2004 [30].

6.1.7.1 Ethernet layering architecture and EPON

Ethernet covers the physical layer and data link layer of the open system interconnect (OSI) reference model. Figure 6.1.29 shows a comparison of the layering model of the traditional point-to-point (P2P) Ethernet and the point-to-multipoint (P2MP) EPON architecture [9].

It can be seen from the figure that EPON layering is very similar to that of traditional P2P Ethernet. The Ethernet standard further divides the OSI physical layer and data link layer into multiple sublayers. The physical layer (PHY) is connected to the data link layer using the media-independent interface (MII) or gigabit media-independent interface (GMII).

The optional MAC sublayer in P2P Ethernet is replaced with a mandatory multipoint media access

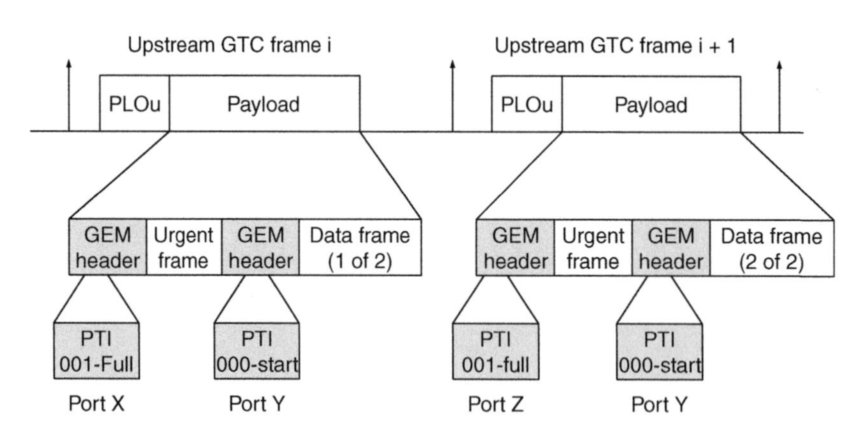

Fig. 6.1.28 Multiplexing of urgent data using GEM fragmentation process (from [24], reprinted with kind permission from ITU).

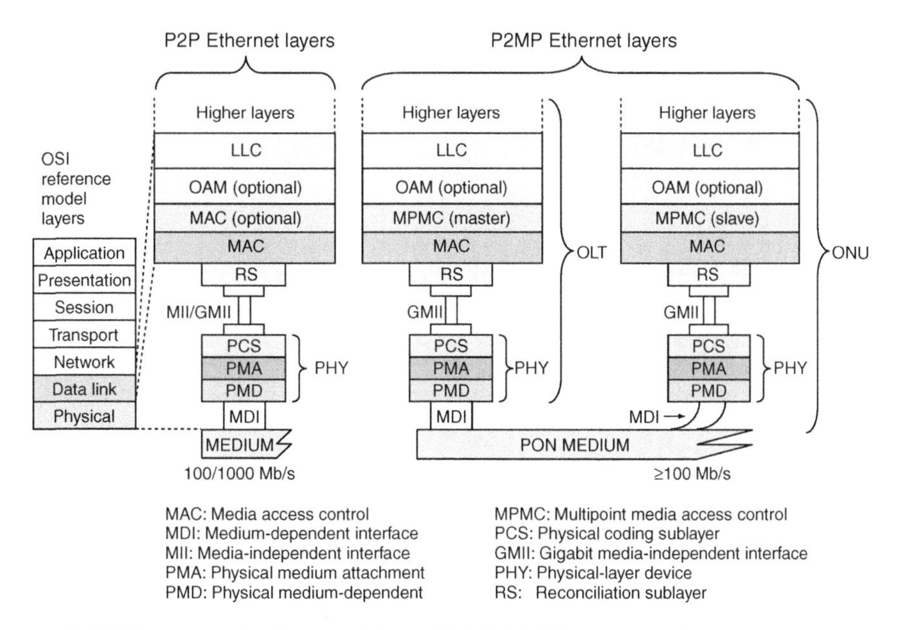

Fig. 6.1.29 Point-to-point (P2P) Ethernet and point-to-multipoint (P2MP) EPON layering architecture.

control (MPMC) layer in EPON [9, Clause 64]. The MPMC layer runs the multipoint control protocol (MPCP) to coordinate the access to the shared PON medium among EPON ONUs. Although the OLT and ONU stacks look nearly identical (Fig. 6.1.29), the MPCP entity in an OLT functions as the master and the MPCP entity in ONU as the slave.

It will be explained later that the reconciliation sublayer has also been extended in EPON to handle a P2P emulation function so that IEEE 802.1-based bridging protocols [31] will continue to function with EPON.

6.1.7.2 EPON PMD layer

The PMD layer specifies the physical characteristics of the optical transceivers. Ethernet has the tradition of adopting mature low-cost designs to promote mass deployments. This philosophy has been the key to the tremendous commercial successes of Ethernet.

As opposed to the 1:32 splitting ratio used in G-PON, the IEEE 802.3ah standard specified a minimum splitting ratio of 1:16. It should be realized that the reconciliation sublayer can support up to 32,768 different logical ONUs through a 15-bit logical link identifier (LLID).

Two different reaches between the OLT and ONU, 10 km and 20 km, have been defined in EPON standards [9, Clause 60]. The 1000BASE-PX10-D PMD and 1000BASE-PX10-U PMD define the OLT and ONU transceiver characteristics for a 10 km reach. The 20 km reach OLT and ONU PMDs are defined as 1000BASE-PX-20-D and 1000BASE-PX20-U. EPON basically inherited the ITU-G.983.3 wavelength allocation plan. Tables 6.1.2 and 6.1.3 present selected properties of 1000BASE-PX10 and 1000BASE-PX20 PMDs from the IEEE802.3ah standards. One can notice that the ONU properties for 10 km and 20 km transmission distances are almost identical. Most of the changes are made at the OLT when the transmission distance is doubled from 10 km to 20 km. This helps to improve the ONU device cost by creating a large economy scale. It also improves the chance that end users can continue to use the same ONU when the transmission distance is increased.

6.1.7.3 Burst mode operation and loop timing in EPON

Ethernet protocol is a burst mode protocol. However, modern P2P Ethernet uses dedicated transmitting and receiving channels between a hub and Ethernet workstations. Such a system maintains the clock synchronization between the receiver and transmitter by transmitting idle symbols when there is no data to be sent. Therefore, even though the Ethernet protocol itself is bursty, the physical layer of modern P2P Ethernets is no longer bursty. Although the preamble has been preserved in modern P2P Ethernet, they have no practical significance except for backward compatibility with first-generation Ethernet devices.

Since EPON upstream physical connectivity is bursty, preambles are needed again to help the OLT burst mode receiver to synchronize with the ONU. Moreover, preambles are modified in EPON to carry the logical link ID (LLID) used in P2P emulation [9, Clause 65].

Table 6.1.2 Selected properties of IEEE802.3ah EPON transmitters.

	1000BASE-PX10-D (OLT)	1000BASE-PXIO-U (ONU)	1000BASE-PX20-D (OLT)	1000BASE-PX20-U (ONU)	Unit
Transmitter type	Long-wave Laser	Long-wave Laser	Long-wave laser	Long-wave Laser	
Signaling speed (range)	1.25 ± 100 ppm	1.25 ± 100 ppm	1.25 ± 100 ppm	1.25 ± 100 ppm	GBd
Wavelength (range) RMS spectral width (max)	1,480–1,500 Depends on wavelengths	1,260–1,360	1,480–1,500	1,260–1,360	nm nm
Average launch power (max)	+2	+4	+7	+4	dBm
Average launch power (min)	–3	–1	+2	–1	dBm
Average launch power of OFF transmitter	−39	−45	−39	−45	dBm
Extinction ratio (min)	6	6	6	6	dB
RIN (max)	–118	–113	–115	–115	dB/Hz
Launch OMA (min)	0.6	0.95	2.8	0.95	mW
Ton (max)	N.A.	512	N.A.	512	ns
Toff (max)	N.A.	512	N.A.	512	ns
Optical return loss tolerance	15	15	15	15	dB
Transmitter dispersion penalty	1.3	2.8	2.3	1.8	dB

Table 6.1.3 Selected properties of IEEE 802.3ah receiver characteristics.

	1000BASE-PX10-D (OLT)	1000BASE-PX10-U (ONU)	1000BASE-PX20-D (OLT)	1000BASE-PX20-U (ONU)	Unit
Signaling speed (range)	1.25 ± 100 ppm	1.25 ± 100 ppm	1.25 ± 100 ppm	1.25 ± 100 ppm	GBd
Wavelength (range)	1,260–1,360	1,480–1,500	1,260–1,360	1,480–1,500	nm
Average receive power (max)	-1	-3	-6	-3	dBm
Receive sensitivity	-24	-24	-27	-24	dBm
Receiver reflectance (min)	-12	-12	-12	-12	dB
Stressed receive sensitivity	-22.3	-21.4	-24.4	-22.1	dBm
Vertical eye closure penalty (min)	1.2	2.2	2.2	1.5	dB
T_receiver-settling (max)	400	N.A	400	N.A	ns

To maintain low cost, traditionally all Ethernet transmitters are running asynchronously on their own local clock domains. There is no global synchronization. A receiver derives the clock signal for gating the received data from its received digital symbols. Mismatches between clock sources are accounted for by adjusting the interframe gap (IFG) between Ethernet frames.

In an EPON system, the downstream physical link maintains continuous signal stream and clock synchronization. In the upstream direction, in order to maintain a common timing reference with the OLT, ONUs use loop timing for the upstream burst mode transmission, i.e. the clock for upstream signal transmission is derived from the downstream received signal.

6.1.7.4 PCS layer and forward error correction

The physical coding sublayer (PCS) is the layer that deals with line-coding in Ethernet physical layer devices. EPON defines a symmetric throughput of 1.0 Gbps both in the upstream and downstream directions, and adopted the 8B/10B line coding used in the IEEE802.3z gigabit Ethernet standard [9, Clause 36]. The 8B/10B code adds an overhead of 25% to limit the running disparity[5] to 1. Besides conveying physical layer control sequences, the 8B/10B code produces a DC balanced output and enough transitions for easy clock recovery. Nonetheless, it increases the symbol rate to 1250 Mbaud/s.[6]

The use of FEC is optional in EPON. The IEEE 802.3ah standard defines RS(255, 239) block codes in the EPON PCS layer [9, Clause 65]. This is the same code used in G-PON. Parity bits are appended at the end of each frame. Similar to G-PON, since the clock rate does not change when FEC parities are appended, the data throughput is decreased when FEC is used. The RS(255, 239) block code does not change the information bits. This allows ONUs which do not support FEC to coexist with ONUs supporting FEC coded frames. An ONU with no FEC support will simply ignore the parity bits albeit running at a higher bit error rate (BER).

6.1.7.5 Ethernet framing

EPON transmits data as native Ethernet frames in the PON section. Ethernet frames are variable size frames. Standard Ethernet frame format [9] is shown in Fig. 6.1.30. It starts with a preamble and a 1-octet start frame delimiter (SFD) to signify the beginning of a frame.[7] The EPON standard makes use of the preamble field with modifications to carry LLIDs for ONUs [9, Clause 65].

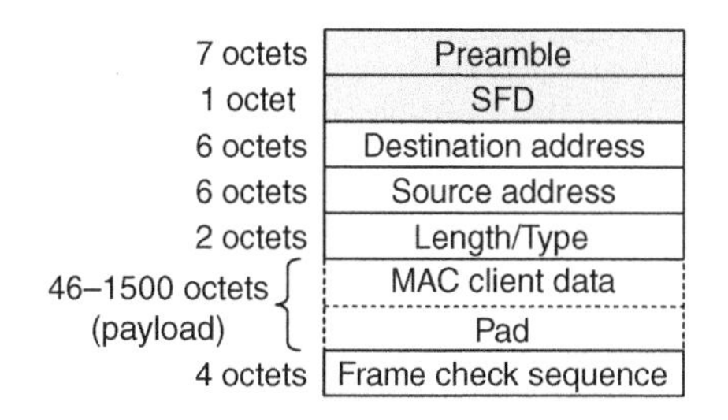

Fig. 6.1.30 Standard Ethernet frame format.

Each frame carries the destination and source MAC addresses, which are both 6-octet fields. A 2-octet length/type field is used to represent the length of the payload when its value is between 0 and 1500 (the maximum payload length). When its value is between 1536 and 65,535, it is used to represent the type of Ethernet frames. The use of this field to represent length and type is therefore mutually exclusive. Ethernet frames have a variable payload size from 46 to 1500 octets. Following the payload is a 4-octet frame check sequence (FCS) using cyclic redundancy check (CRC).

It can be seen from Fig. 6.1.30 that Ethernet frames carry minimum overhead bytes to convey management and protocol information. In the Ethernet field, management and OAM information are carried using protocol data units (PDUs) and OAM frames, which are standard Ethernet frames identified by special length/type values. Protocol and OAM information is carried in the payload field of PDUs and OAM frames. These frames are multiplexed in-band with other Ethernet frames carrying actual user data payloads.

6.1.7.6 Multipoint Control Protocol (MPCP)

The MPCP [9, Clause 64] in the MPMC sublayer uses multipoint control protocol data units (MPCPDUs) to perform ONU discovery and ranging functions. It also provides the arbitration mechanism for upstream medium access control among multiple ONUs.

6.1.7.6.1 Ranging in EPON

The ranging process measures the RTT between the OLT and ONUs so that the OLT can appropriately offset the upstream time slots granted to ONUs to avoid upstream

[5] The running disparity represents the difference in the number of 0s and 1s in the transmitted symbols.
[6] Many EPON marketing materials fraudulently claim EPON data rate as 1.25 Gb/s.
[7] Ethernet frame lengths are counted without including the preamble and SFD fields.

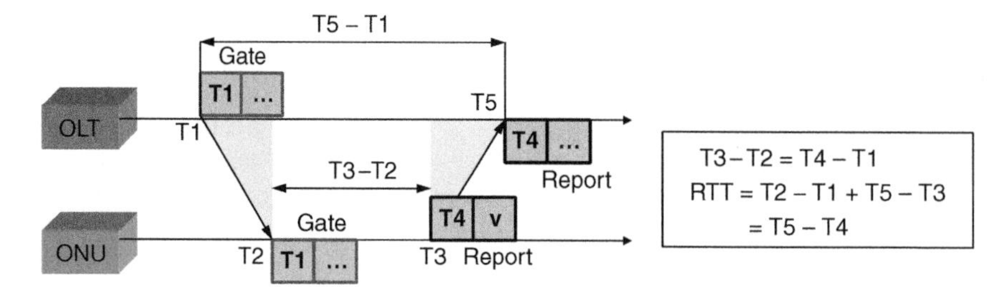

Fig. 6.1.31 EPON ranging process.

collision. It is achieved through *Gate* and *Report* MPCPDUs as shown in Fig. 6.1.31. Both the OLT and ONU maintain their own local 32-bit counters that increment in 16-ns time quantum. In the ranging process, the OLT sends a *Gate* message with time stamp T1 (Fig. 6.1.31) which represents the absolute time. The ONU receives the Gate message at T2 after the transmission delay and resets its timer to T1. After some processing delay, the ONU sends a Report message at time T3, with time stamp T4 = T1 + (T3–T2). The OLT receives the Report message with time stamp T4 at absolute time T5 as indicated in Fig. 6.1.31. It can be seen that the RTT is simply T5–T4.

6.1.7.6.2 Gate and report operation

The Gate operation provides the mechanism for the OLT to specify the time slots that ONUs can transmit. Unlike in G-PON where the OLT specifies the start and stop time in increments of 1-byte, the Gate operation specifies the start time and length of time slots in increments of 16 ns (equivalent to 2-byte in gigabit EPON) time quanta [9, Clause 64].

Figure 6.1.32 shows the EPON Gate operation. The *Gate* MPCPDU contains time stamp, the start time, and length of upstream transmission. Upon receiving the Gate frame, the ONU updates its time stamp register, slot-start register, and slot-length register.

In the report operation, an ONU informs the OLT with its queue lengths and provides the timing information to calculate RTT by sending the Report MPCPDU. Upon receiving the Report MPCPDU, the OLT updates ONU queue length registers and RTT register as shown in Fig. 6.1.33.

6.1.7.6.3 Dynamic bandwidth allocation

The EPON standard does not specify the implementation details of DBA. However, the Gate and Report operation provides the necessary interface and mechanism for controlling ONU bandwidth. It is up to vendors to design the DBA strategies and algorithms.

6.1.7.6.4 Multipoint Control Protocol Data Unit (MPCPDU)

MPCPDUs are 64-byte-long MAC frames with no VLAN tagging. The generic format of MPCPDU is shown in Fig. 6.1.34. MPCPDUs are recognized by the

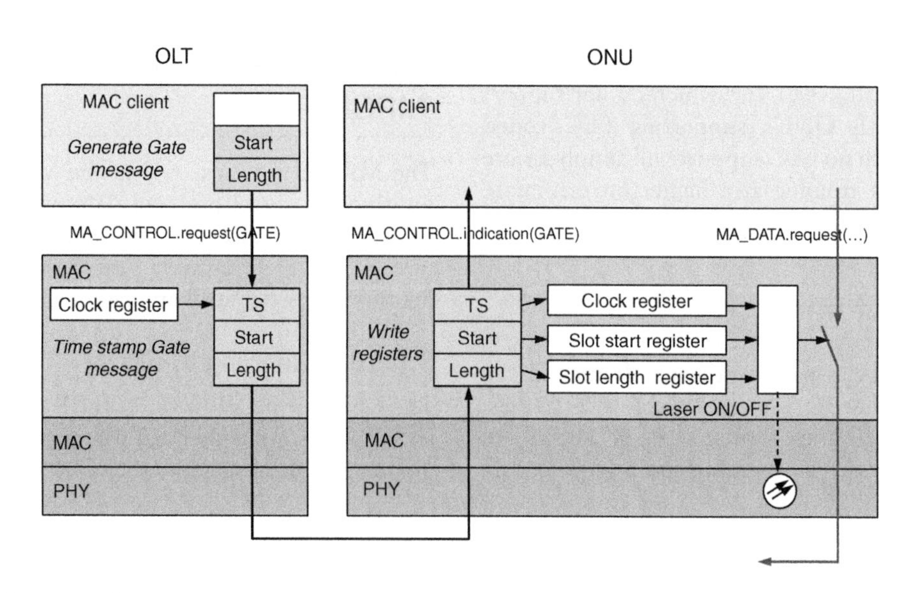

Fig. 6.1.32 EPON Gate operation.

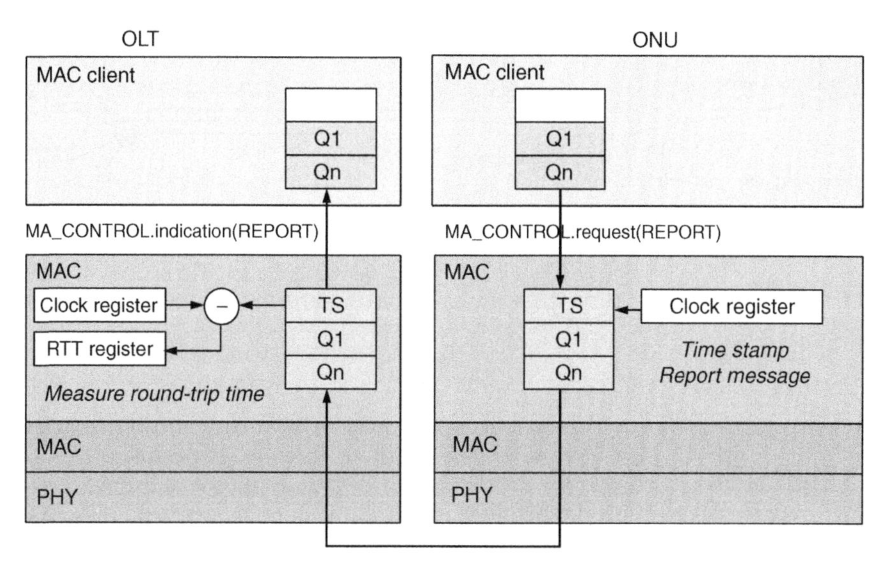

Fig. 6.1.33 EPON Report operation.

MAC frame type 0x88–08 in the length/type field. The 2-octet op-code field identifies the type of MPCPDU message. Figure 6.1.34 also shows the op-codes for all the MPCPDU types defined in EPON. Every MPCPDU message contains a 4-octet time stamp field so that OLT and ONUs can continuously update each other to correct for timing drifts (e.g. due to changes in fiber temperatures and stresses).

The data/pad portion of the MPCPDU contains the MAC parameters used in MPCP and the necessary zero padding to maintain the 64-byte frame size. More details of MPCPDU are given in Appendix II. Interested readers should refer to Clause 64 of the IEEE802.3 standard for the exact details of MPCPDU.

6.1.7.6.5 Autodiscovery of EPON ONU

The autodiscovery process allows EPON ONUs to register and join the system after powering up. The OLT allocates the virtual MAC and assigns LLIDs to ONUs for point-to-point emulation (P2PE).

During autodiscovery, ONU and OLTs exchange each other's capabilities such as the *synchronization time* of the OLT burst mode receiver. *Synchronization time* is the time required by the OLT, after receiving a burst of data, to lock itself to the ONU transmitter clock and adjust its decision threshold to account for the differences in received power levels from different ONUs.

To perform autodiscovery, the OLT broadcasts discovery Gate frames periodically. As shown in Fig. 6.1.35, a reserved discovery window is granted by the discovery Gate. The OLT also transmit its burst mode receiver synchronization time to ONUs in the discovery gate so that ONUs know to format the upstream signal with idle symbols during the initial burst synchronization time. An ONU intending to register receives the discovery gate and sends a Register Request after waiting for a random delay. The OLT receives the Register Request and

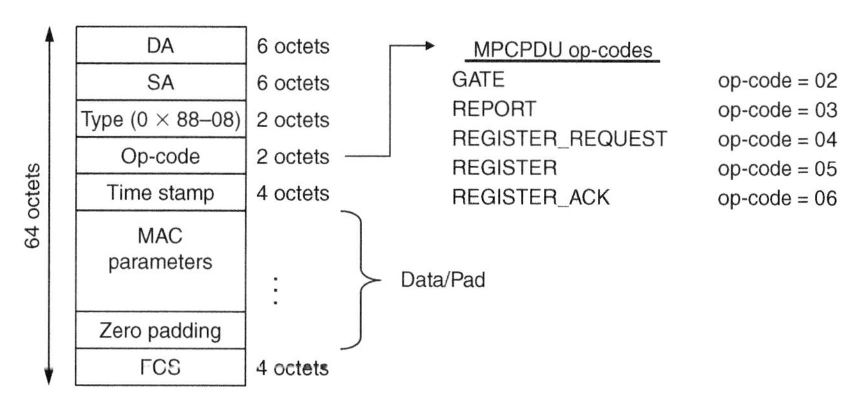

Fig. 6.1.34 Generic format of MPCPDU.

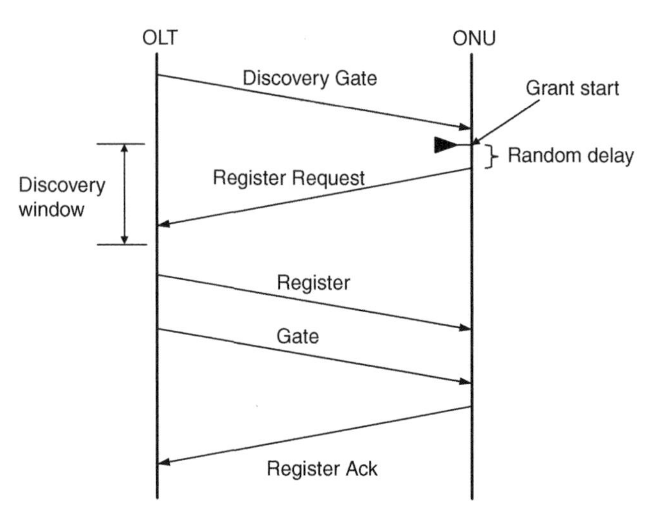

Fig. 6.1.35 Autodiscovery process (from [9], reprinted with permission from IEEE Std. IEEE Std 802.3, 2005, carrier sense multiple access with collision detection (CSMA/CD) access method and physical layer specifications, copyright [2005], by IEEE).

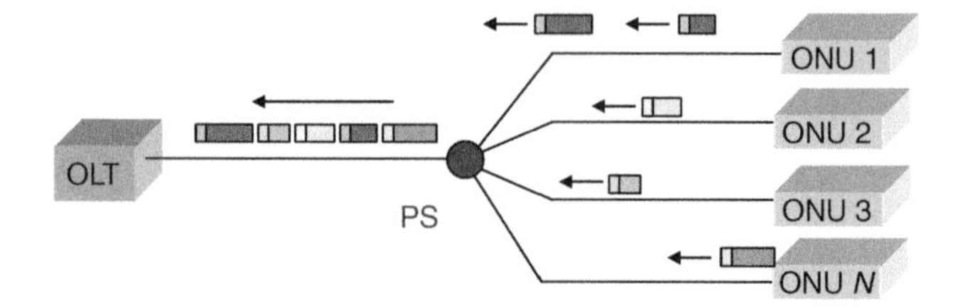

Fig. 6.1.36 Although all the ONU traffic arrives at the same physical port at the OLT, because of the directional power-splitting coupler used at the remote node, ONUs cannot see each other's traffic without the forwarding aid of OLT.

allocates the ONU with the **LLID.** The OLT then sends another Gate frame to grant the upstream time slot for the ONU with the newly allocated **LLID** to transmit a Register Acknowledgment frame, which signifies the finish of the registration process.

During the discovery window, multiple ONUs may attempt to register at the same time and cause collisions at the OLT. Contention in the discovery window is resolved by adding a random delay before each ONU transmits the Register frame. It is also possible for an OLT to receive multiple valid Register frames in a discovery window.

6.1.7.7 Point-to-Point Emulation (P2PE) in EPON

Ethernet MAC interfaces are interconnected with Layer-2 (L2) switches or Layer-3 (L3) routers. L2 switches directly interconnect network devices on the Ethernet layer and are therefore more efficient, less costly, and easier to manage in a LAN environment. On the other hand, routing is used in a wide area network to interconnect multiple Layer-2 networks [13]. In an Ethernet environment, L2 connection is achieved using IEEE802.1 based bridges[8] or switches [31, 32]. An Ethernet switch forwards the traffic among its multiple ports. Each port is connected to a different broadcast domain containing one or more MAC devices. A switch assumes that within the same broadcast domain MAC devices can forward traffic directly toward each other without its help. It performs the L2 switching function by examining the SA and DA of each received frame. If both of them belong to the same domain (i.e. connected to the same port), it filters out the packet without forwarding it. This helps to preserve the bandwidth in other parts of the network and improves the network performance.

In an EPON system, the P2P symmetric Ethernet connectivity is replaced by the asymmetric P2MP connectivity. Because of the directional nature of the remote node, ONUs cannot see each other's upstream traffic directly (Fig. 6.1.36). In a subscriber network, this directional property provides an inherent security advantage. Nevertheless, it also requires the OLT to help forwarding inter-ONU transmissions.

Without any treatment, an IEEE 802.1 switch connected to the OLT would see all the inter-ONU frames with SA and DA belonging to MAC entities connected to the same switch port and thus they would be within the same domain. As a result, the switch would not forward the traffic between different ONUs connected to the same OLT.

To resolve this issue, a P2PE function has been created in the RS layer. The P2PE function maps EPON frames from each ONU to a different MAC in the OLT, which is then connected to a higher-layer entity such as L2 switch (Fig. 6.1.37).

The P2PE function is achieved by modifying the preamble in front of the MAC frame to include an LLID. The modified preamble with the LLID is used in the PON section. The format of the modified EPON preamble is shown in Fig. 6.1.38. It starts with a start LLID delimiter (SLD) field, followed by a 2-byte offset and a 2-byte LLID. A 1-byte CRC field protects the data from the SLD to the LLID. The first bit of the LLID is a mode bit indicating broadcast or unicast traffic. The rest of the 15 bits are capable of supporting 32,768 different logical ONUs.

[8] Bridge and switch are interchangeable terms used to describe L2 frame forwarding devices. Bridge is the original term. Early bridges were implemented using software. Switch is more commonly used to describe bridges implemented in Silicon chips.

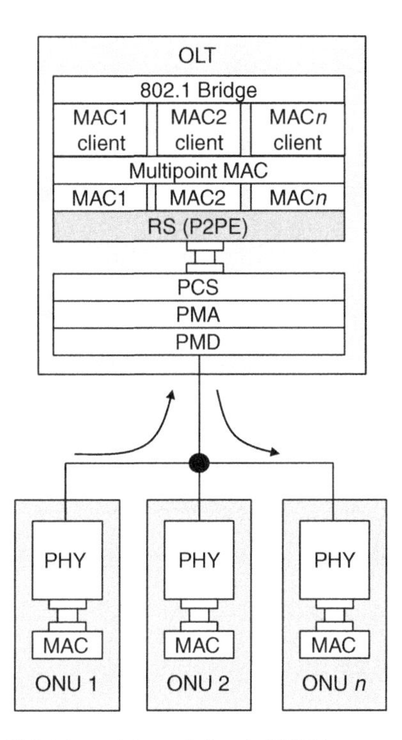

Fig. 6.1.37 Point-to-point emulation in EPON.

The mode bit is set to 0 for P2PE operation. Figure 6.1.39 shows principle of EPON P2PE. When the mode bit is set to 1, the OLT uses the so-called single-copy broadcast (SCB) MAC to broadcast traffic to all ONUs. It takes the advantage of native EPON downstream broadcast operation. To prevent broadcast storm in L2 switches, EPON standard recommends avoiding the connection of the SCB port to 802.1 switches, and using it only to connect to L3 routers[9] or servers for disseminating information. Figure 6.1.40 illustrates the SCB MAC and emulated P2P MACs in an EPON model.

One may have noticed that the concept of LLID in EPON and that of Alloc-ID in G-PON are similar. In a G-PON, single ONU can be allocated with multiple T-CONTs with different Alloc-IDs. As a matter of fact, an EPON ONU may also be allocated with multiple LLIDs. Certain implementations of EPON make use of the multiple LLIDs to implement different quality of service classes [36].

6.1.7.8 EPON encryption and protection

The IEEE802.3ah standard does not specify encryption and protection mechanisms for EPON. Encryption is important to ensure privacy when ONUs are directly connected to users as in FTTH applications. Protection is important when ONUs are shared among groups of users as in FTTB/FTTC applications.

Many implementations of EPON chips include vendor-specific encryption mechanisms which can be enabled by service providers if necessary.

6.1.7.9 Ethernet OAM sublayer

Besides standardizing the EPON technology, another charter of the IEEE 802.3ah task force was to develop the OAM sublayer for Ethernet. The OAM sublayer [9, Clause 57] is an optional Ethernet MAC sublayer (Fig. 6.1.29) which not only can be built in EPON but also in other Ethernet devices. In fact, EPON only consists of a subset of the objects manageable by the OAM sublayer. Most of the EPON ASICs are built with the OAM sublayer which can be used for managing OLT and ONU.

6.1.8 G-PON and EPON comparison

A high-level comparison of EPON and G-PON is given in Table 6.1.4. From a technical viewpoint, one major difference between G-PON and EPON is the support for TDM circuits. G-PON divides upstream and downstream signals into 125 μs frames. Data frames are encapsulated using the GEM encapsulation with segmentation capability, which allows TDM circuits with guaranteed bandwidth and granularities of 64 kb/s to be created between OLT and ONUs.

On the other hand, in an EPON system, variable length native Ethernet frames are used in the transport layer. Circuit emulation [33–35] is needed to implement fixed-bandwidth TDM circuits.

While EPON has been optimized for native Ethernet packet transport, the GEM encapsulation allows easier adaptation of other format signals. In an EPON system, bandwidth report and grant functions are implemented using MPCPDUs. Each logical ONU port with a different LLID requires its own separate Gate and Report frames, which are full Ethernet frames. The more logical IDs are allocated, the higher the overhead of the MPCP [9, Clause 64]. On the contrary, G-PON bandwidth report and grant functions can be piggybacked into the PCB overhead of the GTC frame [24]. Each PCBd overhead contains the bandwidth allocation information

[9] Broadcast storm is caused by loops in a network, allowing packets to replicate and circulate which eventually use up the network bandwidth. L2 switches uses spanning tree protocol (STP) to ensure no multiple paths exist between two nodes and thus avoid the possibility of forming loops. When both the emulated P2P link and SBC link exist between the OLT and ONU, STP will get confused. Unlike L2 switches, L3 routing protocols can make use of multiple signal paths for load balancing. They can also use the time-to-live (TTL) field to avoid loops.

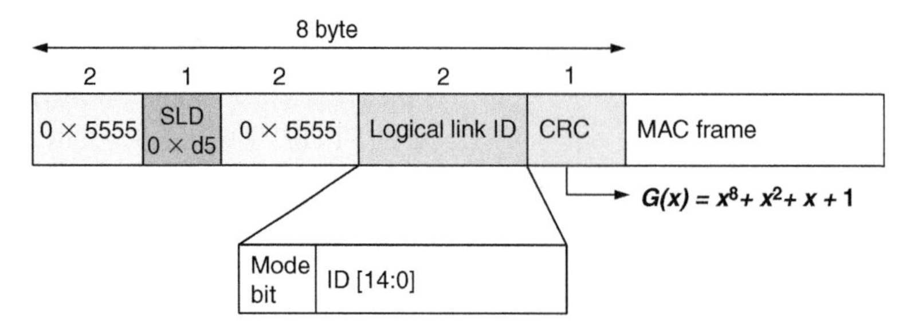

Fig. 6.1.38 Modified preamble with LLID for point-to-point emulation in EPON.

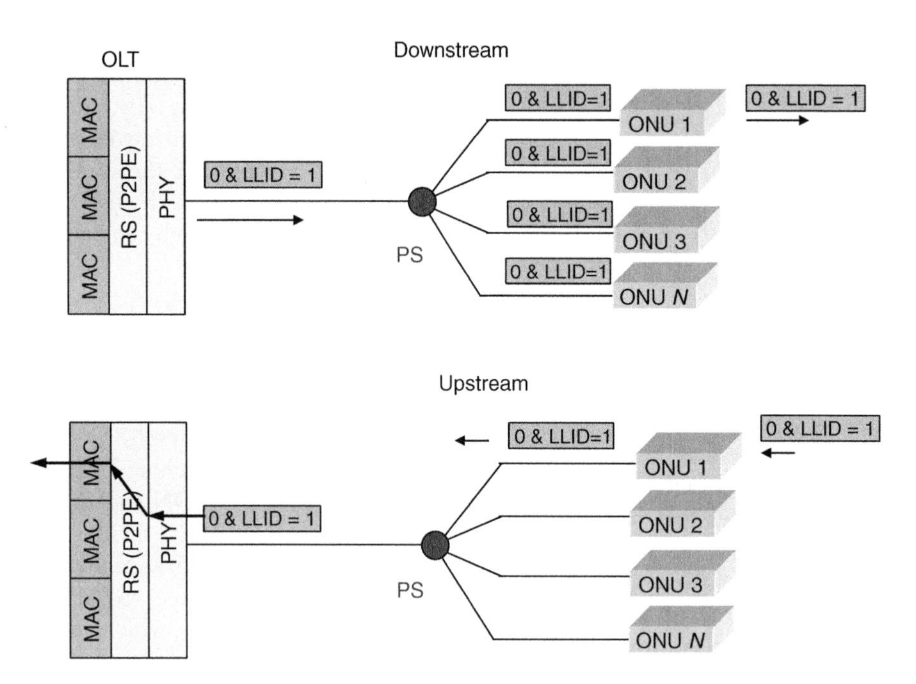

Fig. 6.1.39 EPON point-to-point emulation operation.

for all the allocated T-CONTs. This makes G-PON overhead very efficient compared to EPON. A very good analysis of the APON, G-PON, and EPON overhead efficiency can be found in [36].

Assuming 32-way[10] split for both EPON and G-PON, the EPON system offers an average bandwidth of 31.25 Mbps per ONU in both downstream and upstream directions. A G-PON with 2488 Mbps symmetric bandwidth will offer 77.75 Mbps for each ONU. The RBOCs in the United States do not think the EPON capacity meets the requirements in a modern data-centric society. This is one of the reasons that G-PON has been selected as the technology of choice for the joint request for proposal (RFP) issued by Verizon, SBC, and Bell South.[11] To address the growing demand, at the time of writing,

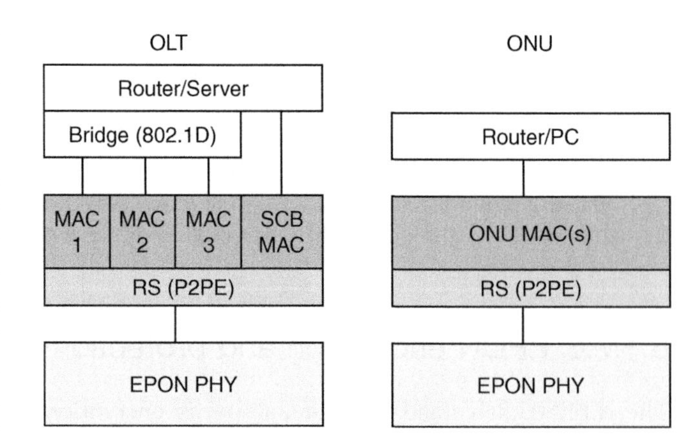

Fig. 6.1.40 Point-to-point and single-copy broadcast (SCB) MACs in an EPON model.

[10] The IEEE 802.3ah standard only specified 1:16 splitting ratio. Most commercial EPON PHY do support the 1:32 splitting ratio.
[11] SBC, AT&T, and Bell South have merged into one single company, which became the new "at&t."

Table 6.1.4 G-PON and EPON comparison.

	EPON	G-PON
Downstream data rate (Mbps)	1,000	1,244 or 2,488
Upstream data rate (Mbps)	1,000	155, 622, 1,244, or 2,488
Payload encapsulation	Native ethernet	GEM
TDM support	Circuit emulation	Yes
Laser on/off	512 ns	≈13 ns
AGC	≤400 ns	44 ns
CDR	≤400 ns	

IEEE has started a new task force 802.3av to charter the development of 10 Gb/s EPON [38].

6.1.9 Super PON

Super PON was proposed to achieve better economy by increasing the reach of PON systems beyond 20 km and supporting higher splitting ratios of 1:64 or even 1:128. Most of the super PON studies were performed within the European Community [39-41].

However, super PON faces the following challenges:

1. The bit rate in PON systems will keep increasing as demand for bandwidth grows. A minimum energy per bit is required to keep the BER. This means that by increasing the bit rate, the power required, which equals the product of bit rate and energy per bit, will increase proportionally. To support a bigger share group size with the same average user bandwidth requires increasing the physical layer bit rate. Increasing the splitting ratio and transmission distance will increase the loss between the OLT and ONU. As a result, a significant increase in power budget between the OLT and ONU is necessary for super PONs. In fact, super PON requires amplifiers in the field to provide the needed power budget.
2. Economical fiber amplifiers are not available at the 1.3=1.49 μm upstream and downstream wavelengths. The most mature optical amplification technology, erbium-doped fiber amplifier (EDFA) works in the C-band, which is also the spectral range with minimum fiber loss. The fiber loss at 1.3 μm almost doubles that at 1.55 μm. Although other amplification technologies such as Semiconductor Optical Amplifier (SOA) [42] or Raman Amplifiers [43] can be used, these technologies are still very expensive. Raman amplification requires very high pump power and poses safety issues, which is tricky in an access network environment. SOAs generally have very high noise figure.
3. Downstream optical amplifications can be easily shared amongst multiple ONUs. However, to share an upstream amplifier could be challenging because ONUs are located at different distance with different losses to the optical amplifier. In addition, upstream data are sent in burst mode, requiring optical amplifier to have fast transient control capabilities.
4. Dispersion effect around 1.5 μm wavelength is nonnegligible. Dispersion effect increases as the square of bit rate [3]. So without using dispersion-tolerant transceivers, when both bit rate and distance increases, dispersion penalty will become significant. In a cost-sensitive access system, this may not make economic sense, at least for the foreseeable future.
5. RTT increases between the OLT and ONU. For a 20 km transmission distance, the RTT is 200 ms. For 60 km, this time increases to 600 μs. At 1 Gb/s transmission speed, this is equivalent to 750,000 bytes of data. Our previous discussion showed that the size of the discovery window reserved for new ONU registration and ranging needs to cover the maximum differential distances between ONUs. Increasing the PON size means that more bandwidth needs to be reserved for autodiscovery of ONUs, unless the frequency of autodiscovery windows is decreased. Also, a bigger share group will increase the chances of collision due to ONU registration and increase the time required for ONU registration. The increased RTT will also increase the latency from bandwidth request to bandwidth grant, which might be problematic for real-time applications.
6. In an EPON system, separate Gate and Report frames are required for each LLID. Increasing the share group size will increase the bandwidth overhead required for the MPCP used in a super EPON system [37].
7. Better protection mechanisms are required as the share group size increases because a failure will have a more significant business impact than that with a small share group size.

These challenges of super PON in both the physical and MAC layers may demand advanced technologies which eventually outweigh the economic advantages achieved through longer transmission distances and bigger share groups.

6.1.10 WDM-PON

An alternative to expand PON capabilities besides super PON is WDM-PON. A WDM-PON is characterized by a WDM coupler, which replaces the power splitter at the remote terminal [44].

6.1.10.1 Advantages and challenges of WDM-PON

WDM-PON offers the following advantages:

1. The optical distribution plant is still passive and therefore has the same low-maintenance and high-reliability properties of PS-PON.
2. Each user receives its own wavelength. Therefore WDM-PON offers excellent privacy.
3. P2P connections between OLT and ONUs are realized in wavelength domain. There is no P2MP media access control required. This greatly simplifies the MAC layer. There will not be distance limitation imposed by ranging and DBA protocols.
4. Easy pay-as-you-grow upgrade. In a PS-PON, if the OLT speed is increased, all ONUs need to be upgraded at the same time. Such a problem does not exist with WDM-PONs. Each wavelength in a WDM-PON can run at a different speed as well as with a different protocol. Individual user pays for his or her own upgrade.

The challenges of WDM-PON include:

1. High costs of WDM components. However, the costs of WDM components have dropped tremendously in recent years which made WDM-PONs economically more viable. For instance, Korean Telecom has already begun its WDM-PON trial. Interests in WDM-PON standardization has also been generated in ITU-T SG15 recently.
2. Temperature control. WDM components' wavelengths tend to drift with environmental temperatures. Temperature control consumes power and requires active electronic parts in the optical distribution network. To remove the need for temperature control, tremendous component and architectural progresses have been made in producing athermal WDM components and systems which are temperature-agnostic.
3. Colorless ONU operation. In a WDM-PON, each ONU needs a different wavelength for upstream connection. This presents a rather serious operational and economical issue. Wavelength specific ONU introduces significant challenges in managing production lines, inventory stocks, sparing, and maintenance. A lot of the solutions have been invented to realize colorless ONUs in the last 20 years.

Technological progresses to materialize economical WDM-PON system will be reviewed in detail in Chap. 3.

6.1.10.2 Arrayed Waveguide Grating (AWG) router

The AWG router is a key element in many WDM-PON architectures. A conventional N-wavelength WDM coupler is a 1 x N device as shown in Fig. 6.1.41 (a).

Components of an input optical signal are directed to specific output ports according to their wavelengths as indicated in Fig. 6.1.41 (a). A general AWG router consists of two star couplers joined together with arms of waveguides of unequal lengths as shown in Fig. 6.1.41 (b) [45]. Each arm is related to the adjacent arm by a constant length difference. These waveguides function as an optical grating to disperse signals of different wavelengths.

Another name of AWG in literature is wavelength grating router (WGR). A very important and useful characteristic of the AWG is its cyclical wavelength routing property illustrated by the table in Fig. 6.1.41 (b) [2]. With a normal WDM multiplexer in Fig. 6.1.41 (a), if an "out-of-range" wavelength, (e.g., λ_{-1} or λ_4,λ_5) is sent to the input port, that wavelength is simply lost or "blocked" from reaching any output port. An AWG device can be designed so that its wavelength demultiplexing property repeats over periods of optical spectral ranges called free spectral ranges (FSR). Moreover, if the multi-wavelength input is shifted to the next input port, the demultiplexed output wavelengths also shift to the next output ports accordingly. Cyclical AWGs are also referred to as colorless AWGs, although the author does not like this term personally.

The wavelength-cyclic property of AWG can be exploited to enable many clever architectural innovations. For example, by using two adjacent ports for upstream and downstream connections, the same wavelength can be "reused" for transmission and reception at an ONU.

6.1.10.3 Broadcast emulation and point-to-point operation

Broadcast is a very efficient way to distribute a large amount of information to multiple users. Figure 6.1.42 illustrates a method to emulate a broadcast service in

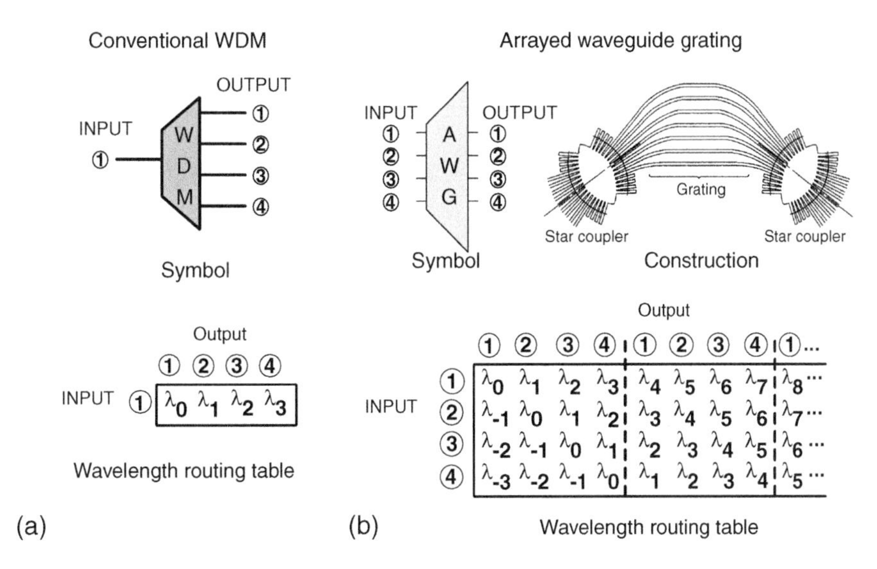

Fig. 6.1.41 Conventional WDM coupler vs arrayed waveguide grating.

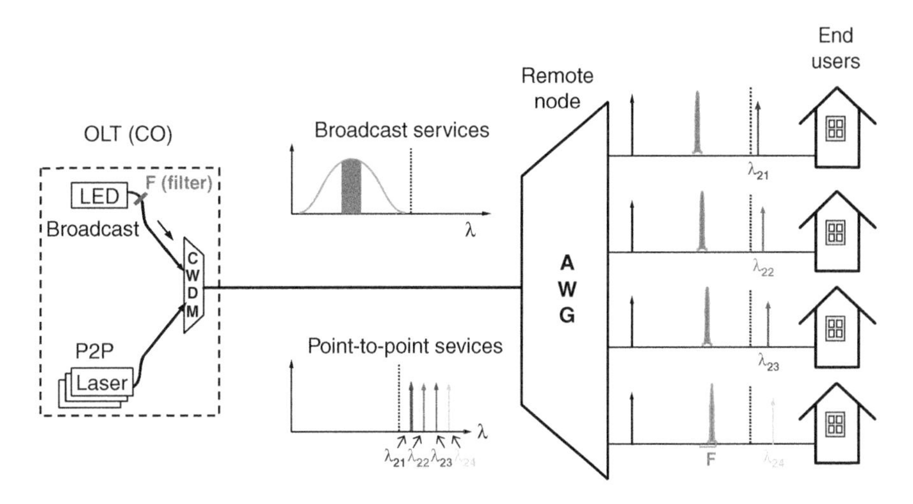

Fig. 6.1.42 Emulation of broadcast services on a WDM-PON with a broadband source.

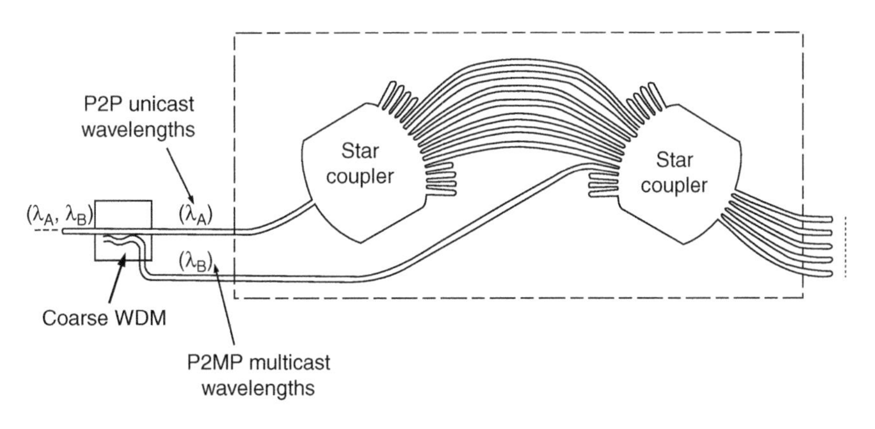

Fig. 6.1.43 A modified AWG for 2-PONs-in-1.

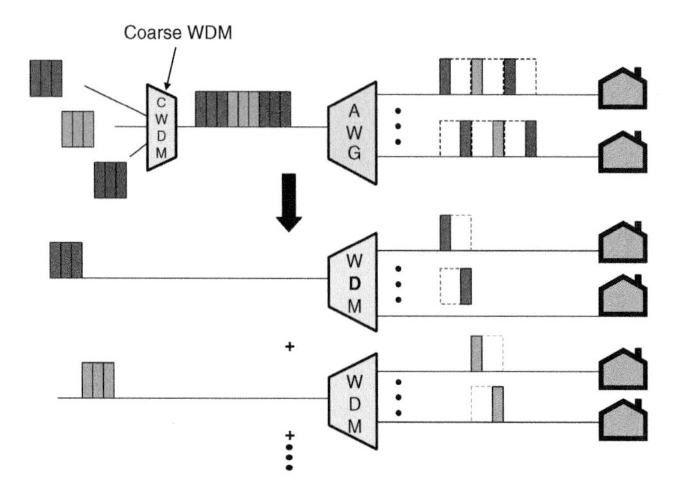

Fig. 6.1.44 By using CWDM devices to combine and separate optical signals in multiple FSRs of an AWG device, a highly flexible WDM-on-WDM system can be achieved.

WDM-PON with a broadband optical source such as LED [46, 47].

In Fig. 6.1.42, a broadband amplified spontaneous emission (ASE) light source such as an LED is first modulated with the broadcast data. The output of the LED is filtered (by filter F) to one FSR of the AWG before being delivered to the end users. After going through the AWG, each user receives a slice of the modulated broadband source with the same data modulation, albeit at different wavelengths. The same figure also shows P2P services being delivered in a different FSR using laser diodes which are line sources. The broadcast and P2P services are multiplexed at the CO using a coarse WDM (CWDM) device.

6.1.10.4 2-PONs-In-1

A problem with spectral slicing is the limitation due to spontaneous-spontaneous emission beat noise in ASE sources, which is proportional to the ratio of the receiver's electrical bandwidth to the received optical spectral bandwidth [48]. As the bit rate increases and channel spacing decreases, the BER is eventually limited by the spontaneous-spontaneous emission beat noise.

An alternative to support broadcast operation on a WDM-PON is to use the 2-PONs-in-1 architecture. A way to realize this is to use the device shown in Fig. 6.1.43 [49]. In this arrangement, the broadcast wavelength (for PS-PON) is separated from the routed wavelengths using a coarse WDM multiplexer. The broadcast wavelength only goes into the second start coupler and is broadcast to all the output ports while the P2P WDM wavelengths are routed to their respective output port.

6.1.10.5 WDM-on-WDM

By using multiple FSRs of an AWG, one can deliver multiple wavelength services to the same ONU. A coarse

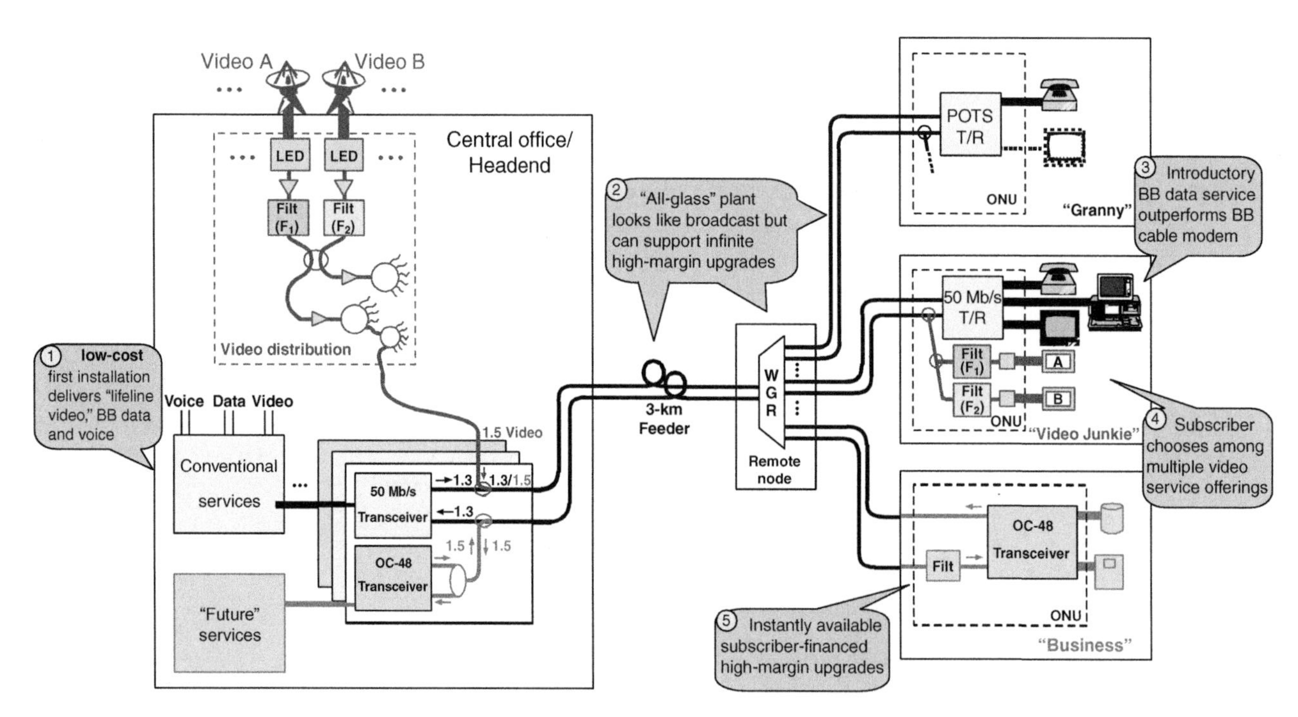

Fig. 6.1.45 Split-net test-bed demonstrating the WDM-on-WDM concept (from [50], copyright [1997] by IEEE, reprinted with permission).

wavelength division multiplexer (CWDM) is used at the OLT to combine services occupying different FSRs (Fig. 6.1.44). A similar CWDM at an ONU separates the individual services. In effect, on one physical fiber distribution infrastructure, one has spawned multiple WDM-PONs segregated by different spectral domains defined by the AWG FSRs. This WDM-on-WDM architecture offers the path to achieve unparalleled flexibility and growth ability [50, 51].

The concept of WDM-on-WDM has been demonstrated on the split-net test-bed by AT&T Labs in 1997 [50]. Figure 6.1.45 describes the test-bed setup. An 8 × 8 AWG with two-fiber configuration was used in this pioneer demonstration. The power budget has been designed to simulate a 16-way split.

Three services were "offered" in this early-day setup. (1) A 50 Mbps low-cost baseband data, voice and video service carried in one FSR near the 1.3 μm wavelength. Uncooled LEDs and spectral slicing were used to carry this service. (2) Two sets of 80-channel digital broadcast TV service carried on two FSRs centered at 1545 nm and 1565 nm wavelengths. These services used uncooled LEDs which were directly modulated with QPSK RF subcarriers. The LED outputs were amplified with enough power to serve over 1000 people, demonstrating the sharing of optical backbone equipment. (3) 2.5 Gbps OC-48 highspeed "future" services carried by DFB lasers with wavelengths in the 1.5 μm region.

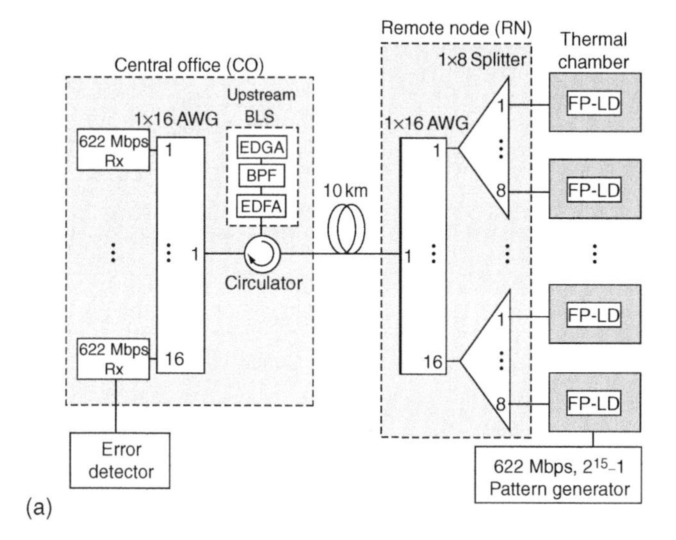

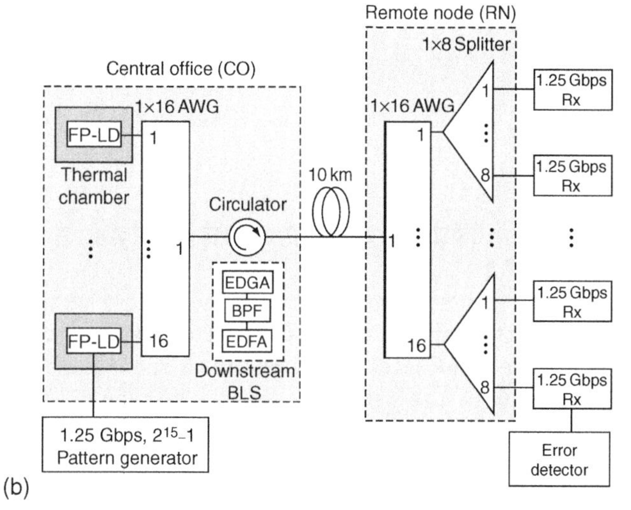

Fig. 6.1.46 Demonstration of a hybrid WDM/TDM-PON with wavelength-selection-free transmitters: (a) downstream link and (b) upstream link (from [52], copyright [2005] by IEEE, reprinted with permission).

6.1.10.6 Hybrid WDM/TDM-PON

Another way to increase PON-system scalability besides the brute-force super-PON approach is to use a hybrid WDM/TDM-PON architecture. A demonstration of such a system is given in [52] and shown in Fig. 6.1.46. Using a 16-wavelength AWG with 1 × 8 power splitter, operation of 128 ONUs was demonstrated on the same feeder plant. Colorless ONU and OLT transmitters are realized by Fabry-Perot lasers whose output wavelengths are injection-locked to spectrally sliced ASE sources and therefore automatically aligned with the AWG wavelength grid. Burst mode operations over a temperature range of 0-60 °C were demonstrated in this setup.

The WDM Ethernet-PON (WE-PON) project jointly developed by Electronics and Telecommunications Research Institute, Korea (ETRI) and Korea Telecom [53] combines EPON and WDM-PON technologies together. Using 32 wavelengths and 1 × 32 splitters, 1000 users per WE-PON is possible.

In the WE-PON test-bed, two types of upstream transmitters were used. The optical layer block diagrams of these two designs are shown in Fig. 6.1.47. The first one uses a gain-saturated Reflective SOA (RSOA) (Fig. 6.1.48). The partially modulated downstream optical signal is reverse-modulated by a feed-forward current signal in the RSOA to remove the downstream modulation (Fig. 6.1.49). This same-wavelength signal is then modulated with upstream data and transmitted back to the OLT. Two fibers are used at the OLT to separate the downstream and upstream signal as shown in Fig. 6.1.47 (a).

The second type of upstream stream transmitter uses an integrated PLC-ECL planner lightwave circuit external cavity tunable laser (Fig. 6.1.50). The downstream and upstream signals are separated into the L-band and C-band respectively, making use of the multiple FSRs of a cyclical AWG. A single-feeder fiber is used to connect the OLT to the RT (Fig. 6.1.47 (b)).

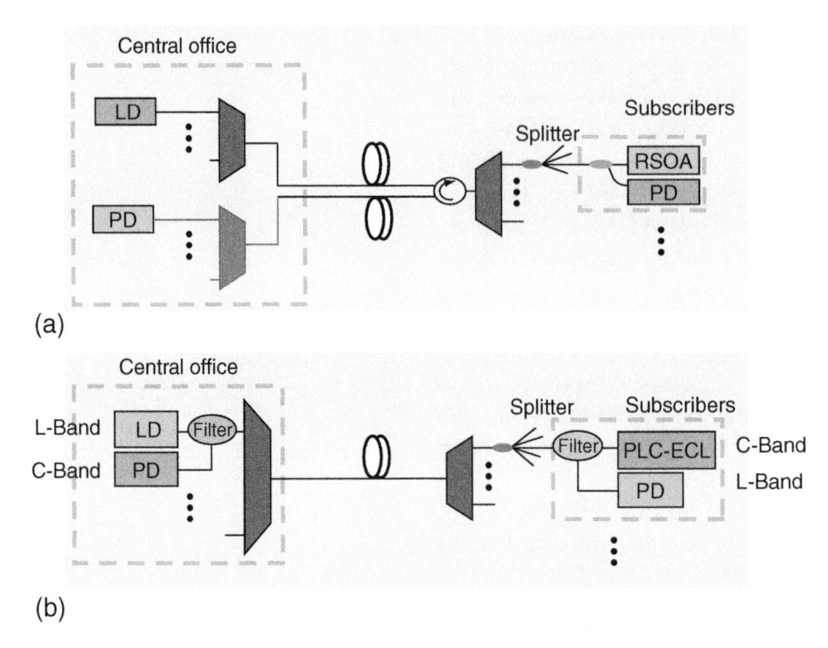

Fig. 6.1.47 WE-PON optical layer block diagram (Courtesy of ETRI and Korea Telecom.).

6.1.10.7 Multiply fiber plant utility by WDM

Fiber plant comprises the most important infrastructure of a fiber-optic access network. Constructing a fiber plant requires extensive investments in capital, time, labor, license approval, and complicated right-of-way negotiations. Techniques to improve fiber plant utilization are therefore very valuable to infrastructure owners.

WDM provides a path to multiply the utility of a fiber plant by spectral segregation [54]. For example, one can imagine a broadband WDM wavelength interleaver which we call a "bandwidth sorter" here. The bandwidth sorter segregates the optical spectrum into interleaved bands as shown in Fig. 6.1.51. Such band sorters may be economically realized using optical thin-film filters, fiber Bragg gratings, or planner light waveguide technologies.

Different PON technologies can then be deployed on the same physical fiber plant using interleaved bands. For example, the "–" band in Fig. 6.1.51 can be occupied by an EPON while the "+" band, by an APON or G-PON, which is shown in Fig. 6.1.52. Thus, by a simple modification of the transceiver wavelengths and using low-cost

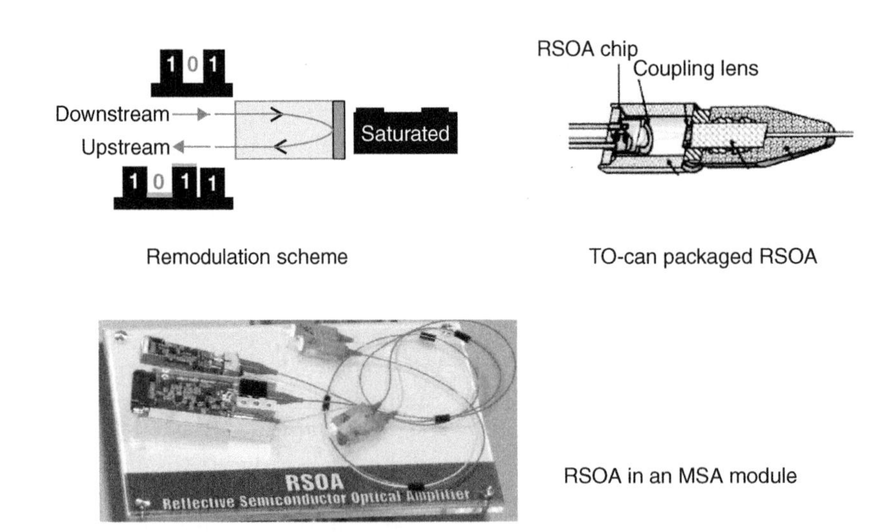

Fig. 6.1.48 RSOA module used in ETRIWE-PON prototype (Courtesy of ETRI and Korea Telecom.).

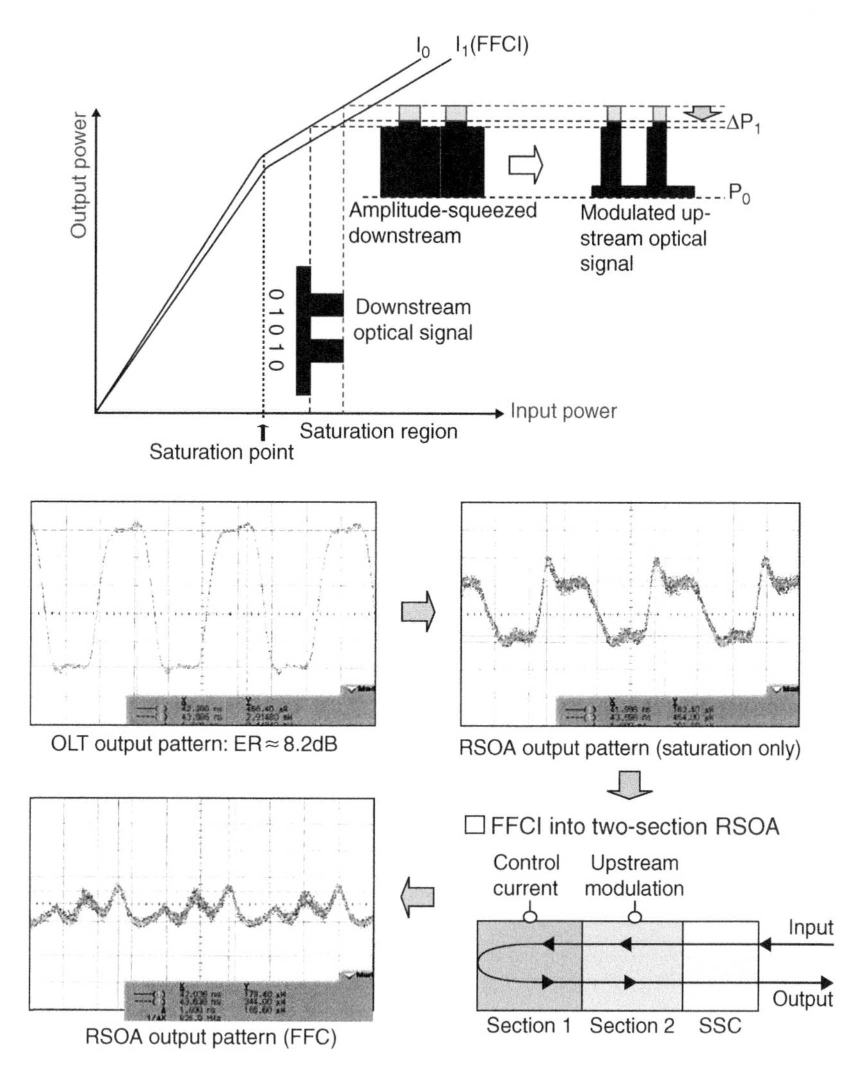

Fig. 6.1.49 Feed-forward current injection to RSOA in WE-PON (Courtesy of ETRI and Korea Telecom.).

band sorters, one can realize two PONs for one infrastructure.

The ITU-T G.694.2 standard [55] defines optical bands on a 20 nm CWDM (coarse WDM) grid, which are wide enough to allow uncooled DFB laser operation in each band. These CWDM bands could form the basis for band-sorter designs.

6.1.11 Summary

In summary, we have reviewed various PON architectures in this chapter. PON has become an important access technology to offer next generation broadband services. It offers long-reach and high-bandwidth passive local loops using single-mode optical fiber. There are two major ways to realize a passive distribution plant: power splitting (TDM-PON) and wavelength multiplexing (WDM-PON).

Time division multiplexing is used for multiplexing data from different ONUs on a power-splitting PON. A ranging process is required to set up ONUs at different transmission distances from the OLT with a common logical timing reference, which is used in scheduling ONU transmissions. In a power-splitting PON, the PON bandwidth is shared among ONUs. Dynamic bandwidth allocation can be used to improve the bandwidth efficiency and user experience.

Power budget eventually limits the speed, distance, and ONU count in a power-splitting PON. A WDM-PON provides an upgrade path to overcome the limitations of a power-splitting PON. In a WDM-PON, a WDM multiplexer is used at the RT to combine the signals from different ONUs. Since each user is allocated with its own wavelengths, WDM-PON offers better capacity, security, privacy, as well as easier upgrade because each user can be individually upgraded.

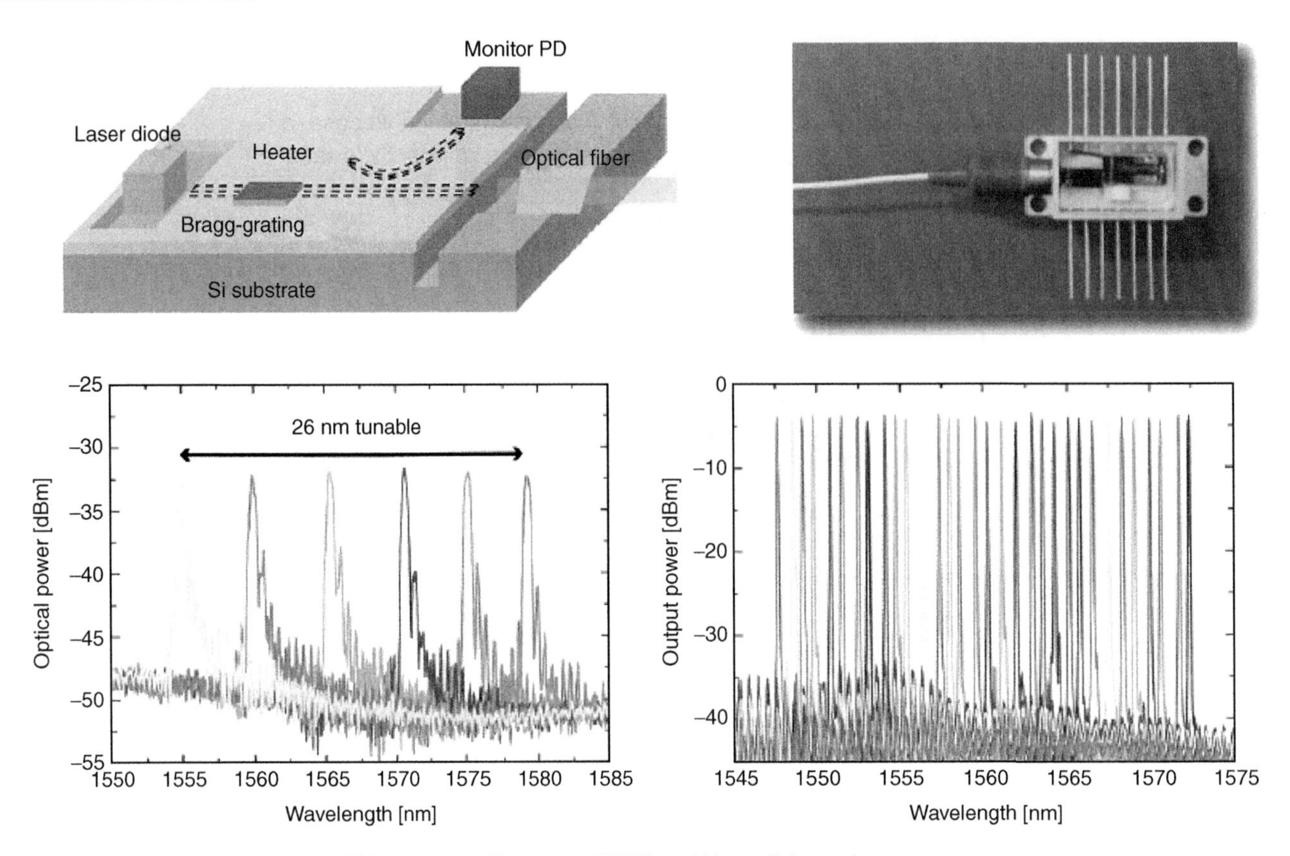

Fig. 6.1.50 PLC-ECL used in WE-PON prototype (Courtesy of ETRI and Korea Telecom.).

The WGR is an important optical element for WDM-PON remote node. It enables many scalable and flexible WDM-PON designs. By using CWDM technology on a PON infrastructure, one can reuse the fiber plant infrastructure through different FSRs in a WDM-PON or overlaying multiple PON technologies.

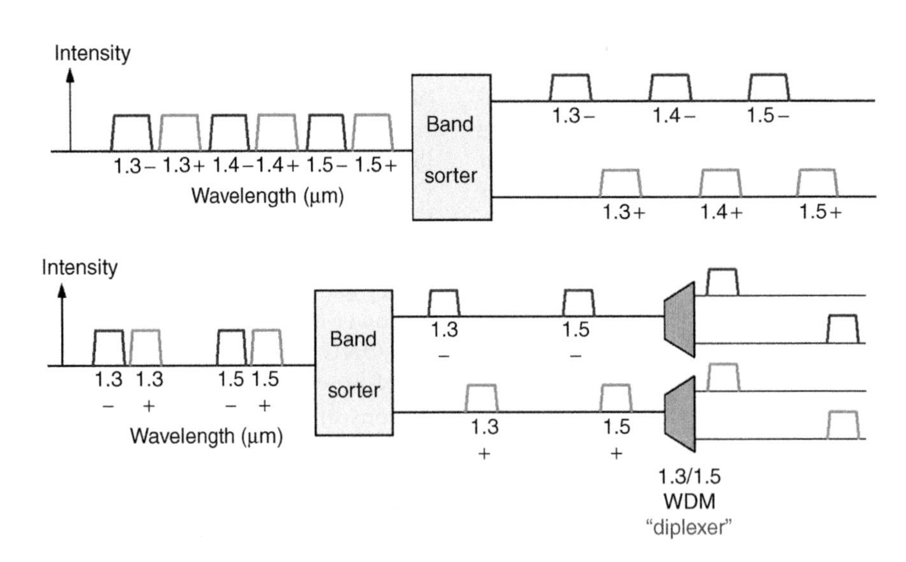

Fig. 6.1.51 A band sorter interleaves the optical spectrum into bands (top) which can be further separated with diplexers (bottom) into bands for upstream and downstream PON connections.

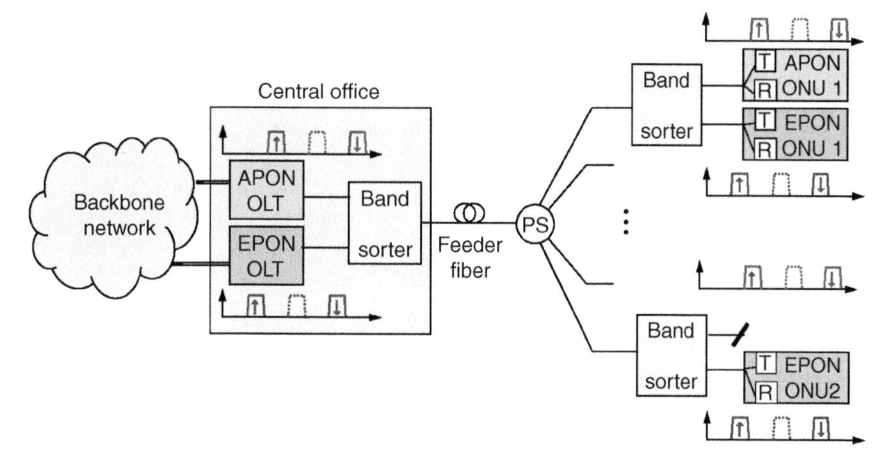

Fig. 6.1.52 Overlaying APON and EPON on the same fiber plant using "band sorters."

References

[1] N.J. Frigo, P.P. Iannone, and K.C. Reichmann, "A view of fiber to the home economics," *IEEE Opt. Commun.*, vol.2, no.3, ppS16-S23, Aug. (2004).

[2] N.J. Frigo, "A Survey of Fiber Optics in Local Access Architectures," in *Optical Fiber Telecommunications, IIIA*, edited by I.P. Kaminow and T.L. Koch, Academic Press, pp461-522 (1997).

[3] G.P. Agrawal, *Fiber-Optic Communication Systems*, 2/e, John Wiley & Sons (1997).

[4] E. Desurvire, *Erbium-Doped Fiber Amplifiers*, John Wiley & Sons (1994).

[5] W.I. Way, *Broadband Hybrid Fiber/Coax Access System Technologies*, Academic Press (1999).

[6] Corning SMF-28e Fiber Data Sheet, available from http://doclib.corning.com/Optical-Fiber/pdf/pi1463.pdf

[7] P. Bohn, S. Das, "Return loss requirements for optical duplex transmission," *J. Lightwave Tech.*, vol. 5, pp234-254 (1987).

[8] ITU-T Recommendation G.983.1, Broadband Optical Access Systems Based on Passive Optical Networks (PON) (2005).

[9] IEEE Standard 802.3, 2005 Edition, Carrier Sense Multiple Access with Collision Detection (CSMA/CD) access method and physical layer specifications (2005).

[10] L.A. Coldren and S.W. Corzine, *Diode Lasers and Photonic Integrated Circuits*, John Wiley & Sons (1995).

[11] J. Wellen, "High-speed FTTH technologies in an open access platform—the European MUSE project," in *Broadband Optical Access Networks and Fiber-to-the-Home*, edited by C. Lin, pp139-166, John Wiley & Sons (2006).

[12] IEC 60825-1, Safety of laser products—Part 1: Equipment classification, requirements and user's guide (2001).

[13] S. Tanenbaum, *Computer Networks*, 3/e, Prentice Hall (1996).

[14] ITU-T G.984.1, Gigabit-capable Passive Optical Networks (GPON): General Characteristics (2003).

[15] ITU-T G.984.2, Gigabit-capable Passive Optical Networks (GPON): Physical Media Dependent (PMD) layer specification (2003).

[16] ITU-T G.983.3, A broadband optical access system with increased service capability by wavelength allocation (2001).

[17] F. Coppinger, L. Chen, and D. Piehler, "Nonlinear Raman cross-talk in a video overlay passive optical network," paper TuR5, *OFC 2003 Tech. Digest*, vol.1, pp285-286 (2003).

[18] ITU-T G.983.2, ONT management and control interface specification for B-PON (2005).

[19] FSAN website: http://www.fsanweb.org/

[20] ITU-T SG1 5 website: http://www.itu.int/ITU-T/studygroups/com15/index.asp

[21] ITU-T G.982, Optical access networks to support services up to the ISDN primary rate or equivalent bit rates (1996).

[22] ITU-T G.983.4, A broadband optical access system with increased service capability using dynamic bandwidth assignment (2001).

[23] W.J. Goralski, *Introduction to ATM Networking*, McGraw-Hill (1995).

[24] ITU-T G.984.3, Gigabit-capable Passive Optical Networks (G-PON): Transmission convergence layer specification (2004).

[25] S.M. Sze, *Physics of Semiconductor Devices*, 2/e, John Wiley & Sons (1981).

[26] ITU-T Recommendation G.7041/Y. 1303 *Generic framing procedure (GFP) (2003)*.

[27] ITU-T G.709/Y.1331, *Interfaces for the Optical Transport Network (OTN)* (2003).

[28] ITU-T G.983.5, A broadband optical access system with enhanced survivability (2002).

[29] ITU-T G.983.6, ONT management and control interface specifications for B-PON system with protection features (2002).

[30] Ethernet for the first mile (EFM) website: http://www.ieee802.org/3/efm/

[31] IEEE Standard 802.1d, *Media Access Control Bridge* (1993).

[32] J.J. Roese, *Switched LANs*, McGraw-Hill (1999).

[33] Metro Ethernet Forum White Paper, "Introduction to circuit emulations services over Ethernet," available from http://www.metroethernetforum.org/PDFs/WhitePapers/Introduction-to-CESoE.pdf

[34] IETF draft "Emulation of structure agnostic (unstructured) PDH services (T1/E1/T3/ E3)," available from http://www.ietf/org/internet-drafts/draft-ietf-pwe3-satop-01.txt

[35] IETF PW3 working group, available from http://www.ietf.org/html.charters/pwe3-charter.html

[36] M. Hajduczenia, H.J. da Silva, and P.P. Monteiro, "Flexible logical-link-identifier assignment policy for Ethernet passive optical networks based on extended multipoint-control-protocol DU flow control," J. *Optical Networking*, vol.5, no.9, pp681-698, (2006).

[37] M. Hajduczenia, H.J. da Silva, and P.P. Monteiro, "EPON versus APON and G-PON: a detailed performance comparison," *J. Optical Networking*, vol.5, no.4, pp298-319, (2006).

[38] IEEE 802.3av 10GEPON task force website: http://www.ieee802.org/3/av/index.html

[39] Van de Voorde, C. Martin, J. Vandewege, and X.Z. Qiu, "The SuperPON demonstrator: an exploration of possible evolution paths for optical access networks," *IEEE Commun. Mag.*, vol.38, no.2, pp74-82, (2000).

[40] C. Bouchat, C. Martin, E. Ringoot, M. Tassent, I. Van de Voorde, B. Stubbe, P. Vaes, X.Z. Qiu, and J. Vandewege, "Evaluation of superPON demonstrator," 2000 IEEE/LEOS Summer Topical Meetings, Broadband Access Technologies, pp25-26, July 24-28, Aventura, Florida (2000).

[41] J.M.P. Delavaux, G.C. Wilson, C. Hullin, B. Beyret, and C. Bethea, "QAM-PON and Super PON for Access Distribution Networks," paper WN2, *OFC 2001 Technical Digest* (2001).

[42] L. Spiekman, "Applications of Semiconductor Optical Amplifiers," Tutorial We 1.3, Proceedings of ECOC 2003, vol.5, pp268-297, Remini, Italy (2003).

[43] C. Headley and G.P. Agrawal, *Raman Amplification in Fiber Optical Communication Systems*, Academic Press (2005).

[44] A. Banerjee, Y. Park, F. Clark, H. Song, S. Yang, G. Kramer, K. Kim, and B. Mukerjee, "Wavelength-division-multiplexed passive optical network (WDM-PON) technologies for broadband access: a review," *J. Optical Networking*, vol.4, no.11, pp737-756, (2005).

[45] Dragone, "A NxN optical multiplexer using a planar arrangement of two star couplers," *IEEE Photon. Technol. Lett.*, vol.3, pp812-815 (1991).

[46] P.P. Iannone, K.C. Reichmann, and N. J. Frigo, "Broadcast Digital Video Delivered over WDM Passive Optical Networks," *IEEE Phot. Technol. Lett.*, vol.8, pp930-932 (1996).

[47] C.F. Lam, K.C. Reichmann, and P.P. Iannone, "Cascadable Modular Transmitter and Receiver for Delivering Multiple Broadcast Services on WDM Passive Optical Networks," *Proceedings of 26th European Conference on Optical Communication*, vol.1, pp109-110, Munich, Germany, (2000).

[48] C.F. Lam, M.D. Feuer, and N.J. Frigo, "Performance of PIN and APD receivers in highspeed WDM data transmission systems employing spectrally sliced spontaneous emission sources," *Elect. Lett.*, vol.36, no.8, ppl 572-1574, (2000).

[49] Y. Li, US Patent 5,926,298, Optical multiplexer/demultiplexer having a broadcast port, July (1999).

[50] N.J. Frigo, K.C. Reichmann, P.P. Iannone, J.L. Zyskind, J.W. Sulhoff, and C. Wolf, "A WDM PON architecture delivering point-to-point and multiple broadcast services using periodic properties of a WDM router," post-deadline paper PD24, *OFC 97*.

[51] M.Birk, P.P. Iannone, K.C. Riechmann, N.J. Frigo, and C.F. Lam, US Patent 7,085,495, "System for flexible multiple broadcast service delivery over a WDM passive optical network based on RF block-conversion of RF service bands within wavelengths bands," (Aug., 2006).

[52] D.J. Shin, D.K. Jung, H.S. Shin, J.W. Kwon, S. Hwang, Y. Oh, and C. Shim, "Hybrid WDM/TDM PON with wavelength-selection-free transmitters," *IEEE J. Lightwave Tech.*, vol.23, no.1 (Jan., 2005).

[53] J. Yoo, H. Yun, T. Kim, K. Lee, M. Park, B. Kim, and B. Kim, "A WDM-Etherent hybrid passive optical network architecture," *8th International Conference on Advanced Communication Technology (ICACT 2006)* (Feb., 2006).

[54] N. Frigo and C.F. Lam, "WDM Overlay of APON with EPON—a carrier's perspective," http://grouper.ieee.org/groups/802/3/efm/public/sep01/lam_1_0901.pdf

[55] ITU-T G.694.2, Spectral grids for WDM applications: CWDM wavelength grid (2002).

Chapter 6.2

6.2

Fiber optic transceivers

DeCusatis

6.2.1 Introduction

Fiber-optic transceivers (TRX), each a combination of a transmitter (Tx) and a receiver (Rx) in a single common housing, have already replaced discrete solutions in most of the datacom applications with two-fiber, bidirectional transmission in the last 12 years. The transmitter unit, with a light-emitting diode (LED) or an edge- or vertical-emitting laser diode (LD) as its radiation source, and the receiver unit, with in most cases a positive-intrinsic-negative (PIN) photodiode, are connected together to the transmission medium, either multimode or single-mode fiber. Thus, contrary to typical telecom applications with single-fiber-optic connectors at the ends of fiber pigtails, transceivers characteristically have duplex optical receptacles at one side of the housing, which fit to the corresponding duplex connectors. Depending on the application, more or less electronics with integrated circuits and passive components are implemented into the Tx and Rx units.

In this chapter we discuss the operation and application of these fiber-optic transceivers in the physical layer (PHY), along with some innovative tendencies for integration of more data transport functions into these components or for a significant reduction of size and power consumption, respectively.

The typical datacom transceiver adheres to one or more national or international standards and manufacturers may create multisource agreements (MSAs) to standardize form factor, pinout, power consumption, and so on. An example standard is the ANSI Fibre Channel (FC-PI-2) for serial transceivers with data rates of 1.0625 Gbps, 2.125 Gbps, and 4.250 Gbps. The GbE standard has a data rate of 1.0 Gbps. When data rates are close, such as GbE and 1G FC, a transceiver can be designed to handle multiple standards because it is protocol agnostics (see Fig. 6.2.1). An example MSA is small form factor (SFF), which defines a pin-through-hole device shown in Fig. 6.2.2.

6.2.1.1 Physical layer interface

Transceivers are typically implemented as part of a physical link structure that incorporates data serialization.

Fig. 6.2.1 Multistandard transceiver.

Handbook of Fiber Optic Data Communication; ISBN: 9780123742162

Fig. 6.2.2 SFF GbE transceiver.

The outgoing bytes undergo 8B/10B encoding, to ensure a balanced pattern, followed by a parallel to serial conversion.

In this manner an 8-bit wide 100 MBps datastream would be converted to a 1-bite-wide 1 Gbps datastream to be delivered to the serial fiber-optic transceiver. After passing through the transmission medium, the reverse of this process is carried out.

6.2.1.1.1 Clock oscillator and regenerator

Clock generation multiplication is mainly realized using a quartz oscillator for very stable clock sources. For example, the ANSI Fibre Channel requires a clock frequency accuracy of 100 ppm. Clock regeneration/synchronization can be realized by either surface wave filter (SWF) or phase-locked loop (PLL), the latter being more common in today's applications.

6.2.1.1.2 Serializer and deserializer (or multiplexer and demultiplexer)

The serializer/multiplexer accepts parallel data from the encoder once per byte timeframe and shifts it into the serial interface output buffers using a PLL-multiplied bit clock. The deserializer/demultiplexer accepts serial bit-by-bit data from the serial receiver output, as clocked by the recovered receiver sync clock, and shifts it back into a parallel datastream.

6.2.1.1.3 Encoding and decoding

The dominant intention of coding for transmission of high-speed data is to maintain the DC balance by bounding the maximum run length of the code. Typical kinds of coding in data communication include 4B/5B coding (e.g., in FDDI) or 8B/10B coding (e.g., in ESCON/SBCON and Fibre Channel). This means high-speed receiver designs in fiber-optic transceivers are normally AC-coupled so that each DC component inside the datastream reduces the signal-to-noise ratio in the preamplifier stages.

During clock recovery, the PLLs used in today's applications require a certain edge density to ensure that the receiving PLL remains synchronized to the incoming data. In addition, word alignment can be provided by special transmission characters (e.g., a K28.5 pattern). In general, two types of characters are defined: data characters and special characters.

6.2.2 Technical description of fiber-optic transceivers

When selecting a transceiver, the key parameters are optical output power, operating data rate, receiver saturation level, receiver sensitivity, transmit and receive pulse quality (defined by rise/fall times and over/undershoot), jitter, and extinction ratio. For transceivers, an optimal interoperation with the associated circuitry is the key to success of the application. Thus, the high-speed characteristics and immunity against external influences have to be carefully evaluated during the design phase.

6.2.2.1 Serial transceivers

Serial fiber-optic transceivers are the interface between the serial electrical signal and the transmission medium, the optical fiber. They are composed of the transmitter and the receiver functions, which operate in one common housing, but are electrically independent of each other. The implementation of serial transceivers requires careful design of filter circuitry as well as sophisticated transmission line considerations for data input and output tracks to connect to the internal or additional external circuitry.

Figure 6.2.3 shows a simplified block diagram of a fiber-optic transceiver containing a transmitter side with a laser driver circuit and a laser diode, together with its monitor diode, for tracking the laser output. Also shown is an alarm circuitry that activates if the laser crosses a preset upper or lower optical power threshold.

The receiver side shows a PIN photodiode as opto-electric converter, a transimpedance preamplifier that converts extremely low AC currents (in the range of nA) into differential voltage signals with some mV amplitude, and the buffer stage with integrated postamplifier for ECLPECL line driving.

In most of the standards mentioned in Section 6.2.1, such as Ethernet, FDDI, ATM, and Fibre Channel (FC) including ESCON/SBCON, pure serial fiber-optical transceiver solutions are described, all including so-called multistandard transceivers. In order to minimize efforts

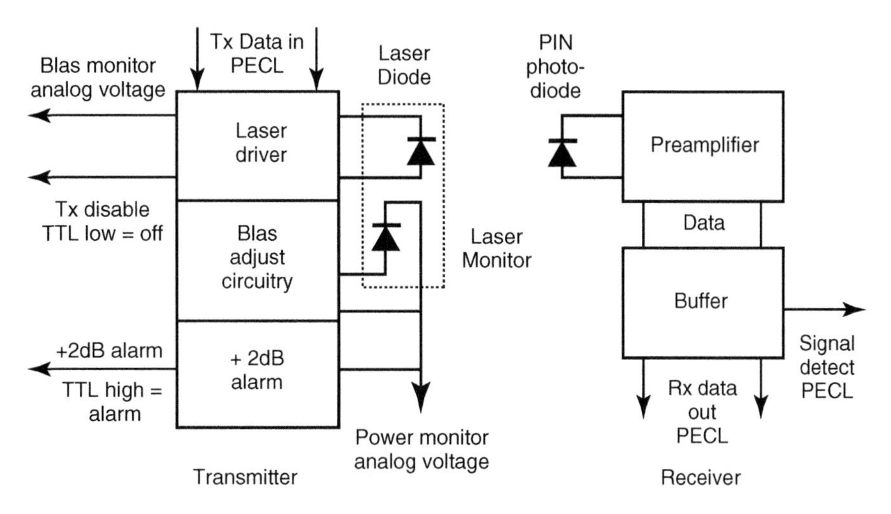

Fig. 6.2.3 Simplified block diagram of a serial laser transceiver.

in design, manufacturing, and testing of boards, a couple of parallel transceiver designs have also arisen in the interim.

6.2.2.2 Parallel transceivers

Transceivers (TRX) with parallel electrical interfaces have become increasingly popular in fiber-optic link applications. Parallel TRXs provide high potential for cost saving due to

- significant board space savings by higher functional integration
- simplified board design by avoiding higher frequencies on the PC board
- low risk for development effort and design-in
- much reduced overall power consumption
- logistical benefits on all levels.

6.2.2.2.1 Transponders

The electro-optical functions of transponders are basically very similar to parallel transceivers. But contrary to the typical design of transceivers with fiber-optic connector receptacles, transponders have connectorized fiber pigtails. They are also plugged into the board rather than soldered or surface mounted, whereas transceivers are typically directly soldered or mated to a board-mounted connector.

Figure 6.2.4 shows as an example the OC-48 transponder, which supports ATM/SDH/SONET applications by converting four parallel LVDS lines of 622.08 Mbit/s input/output datastream (equal to the STM-4/OC-12 hierarchy) into 2.48832 Gbit/s serial optical output/input data stream (equal to the STM-16/OC-48 hierarchy) [10 12]. This transponder is also compliant with Proposal 99.102 of the optical internetworking forum OIF.

Due to high functional integration with serializing/deserializing and, in addition, 8B/10B coding/decoding, the power consumption is the order of 3 W. Thus, heat dissipation is a serious challenge that can only be reliably managed by numerous cooling studs on top of the all-metal housing and by forced airflow inside the system device. A block diagram of the circuitry of the 4:1 OC-48 transponder is shown in Fig. 6.2.5.

6.2.3 The optical interface

An interface is the meeting of two objects. In the case of the optical interface, the fiber comes together with the optical transceiver elements: the transmitter diode and the receiver diode receptacle unit. The operating link demands that the mechanical and optical properties of these two objects match. These attributes are defined in various standards to maintain a reliable interoperability of products of different manufacturers [5–8].

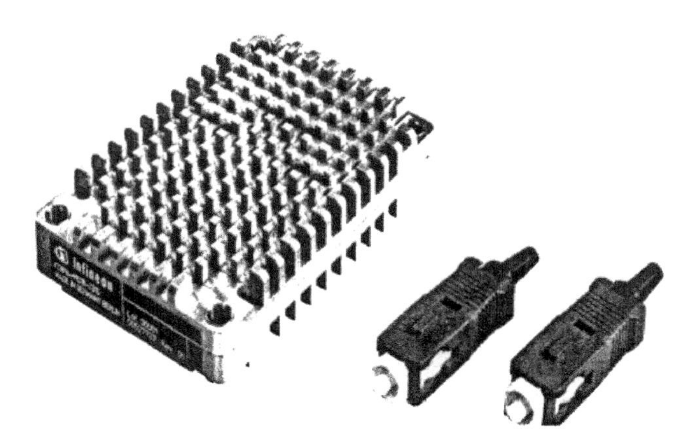

Fig. 6.2.4 The 4:1 OC-48 transponder.

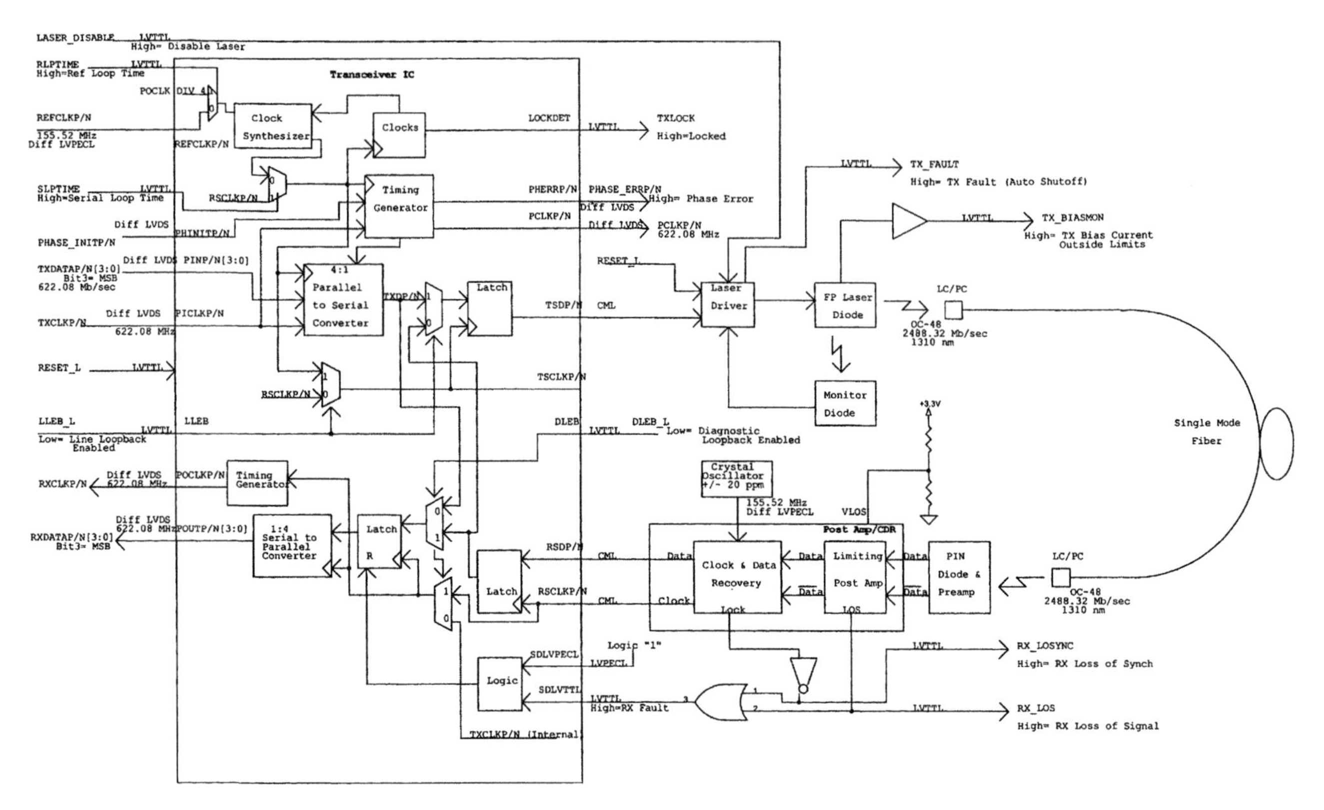

Fig. 6.2.5 Block diagram of the 4:1 OC-48 transponder.

6.2.3.1 Fiber-optic connectors and active device receptacles

A brief, closer look shows that a transceiver has not one but two optical interfaces: first, the connection of the outgoing fiber at the output of the transmitter, and second, the connection of the incoming fiber at the input of the receiver. The optical subassembly of the transmitter launches the radiation into the fiber, which will be taken up by the optical subassembly of the receiver at the other side of the link.

For a flexible configuration or a reconfigurable network, it is necessary to have repeatedly connectable and disconnectable devices. This function is realized by a single, duplex, or multifiber connector at the end of the fiber-optical cable and a port at the transceiver that accepts the fiber connector. There are various different design families on the market, such as the EC, ESCON/SBCON, FC/PC, FDDI MIC, LSA (screwed), LSG (push-pull version of LSA), LSH (E2000), SC, SMA, ST/BFOC, LC, MF, MLT (mini-SC), MT, MT-RJ, and VF-45 international standardized connector families[8]. Most of the single-fiber connectors suitable for single-mode fiber application are available in either a version with physical contact (PC) or angled polished contact (APC). Nearly all push-pull or latch-type single-fiber connectors are also available in a duplex version. ESCON/SBCON and FDDI MIC were originally duplex connectors.

Most connector systems use a ferrule-a cylindrical tube containing the fiber end that fits within a precise mechanical tolerance into the transceiver's optical receptacle. The VF-45 connector is the only one with a totally different concept based on V-grooves instead of ferrules.

For ferrule-based systems, the most popular diameter of the cylindrical ferrule is 2.5 mm, while 1.25 mm has established itself for small form factor (SFF) designs such as MU- and LC-type connector families. The SMA connector is the only one with a 3.125 mm ferrule diameter. However, a ferrule diameter tolerance within a few microns must be maintained in all cases in order to achieve a satisfactory coupling of light and to ensure intermateability to receptacles of different suppliers.

Besides these single and duplex connectors, it is also possible to integrate 12 fibers into one connector to be used for parallel optical links (see Section 6.2.8). The most popular multifiber connectors are based on the MT design with its rectangular-shaped plastic ferrule. In general, any fiber connector has to be locked to the receptacle, for example, by screwing on a coupling ring (e.g., LSA) or snapping on a bayonet (ST) or a latch mechanism. Moreover, some connectors can be provided with a key to distinguish multimode and single-mode connectors.

For bidirectional links in datacom applications, cross-plugging must be avoided by using asymmetric-shaped connector or corresponding key structures both inside the receptacle and at the connector. In the case of parallel links using MPO connectors for a 12-lane transmitter and separate 12-lane receiver, the two 12x cables are often bundled together and the ends are labeled TX and RX.

Pigtail solutions are preferred for telecom applications. These modules do not have an optical port, but instead have one or more permanently fixed fibers of short length (i.e., on the order of 1 meter). In this case, the optical interface of the transceiver/transponder is situated at the connectorized end of the pigtail (see also Section 6.2.2.2.1. and Fig. 6.2.4).

6.2.3.2 The optical fiber

When considering the optical aspects of the interface, it is useful to look at the parameters of the fiber, the transmission medium between transmitter and receiver. An optical fiber can be described by the properties of core diameter, refractive index profile and numerical aperture, bandwidth distance product, and attenuation. These interdependent parameters more or less directly influence the requirements for the output of the transmitter and the input of the receiver.

The core diameter of the fiber defines the maximum waist of the output beam at the transmitter, because only light inside this waist will be guided by the fiber. This is why the numerical aperture, which is determined by the index profile of the fiber, limits the maximum divergence of the input beam. Launched radiation exceeding this maximum divergence will be lost. An efficient optical coupling between transmitter and fiber can only be achieved if the core diameter matches the beam waist and the numerical aperture matches the divergence. At the receiver side, the optical coupling is mainly influenced by the waist and divergence of the beam leaving the fiber. The optical subassembly of the receiver must focus this beam onto the sensitive area of the photodiode.

In some cases, it is necessary to consider not only the coupling efficiencies at the optical interfaces of the transceiver but also the fraction of light that is reflected back to the light source. Lasers can be very sensitive to backreflections that can disturb the laser emission and cause some noise on the optical signal. To prevent these effects, the return loss at the optical interfaces has to be high. The backreflections can be kept low by using angled polished physical contacts at the connectors or by inserting an optical isolator in front of the laser. Another method uses a design wherein the axis of the laser beam is slightly tilted to the axis of the optical fiber.

The fiber bandwidth is determined by the modal dispersion and the chromatic dispersion. Modal dispersion occurs in multimode fibers and causes a pulse broadening due to the different propagation speeds of the different paths (modes) taken by light traveling down the fiber. The excitation of fiber modes depends strongly on the optical coupling at the transceiver. Due to its index profile, single-mode fiber only supports the fundamental mode of propagation and therefore does not suffer from modal dispersion.

Pulse broadening caused by chromatic dispersion requires control of the spectral bandwidth of the transmitter. For low bit rates and short distances, LEDs with a typical full width at half maximum (FWHM) of 40 to 60 nm are sufficient. However, higher bit rates and longer distances require the application of lasers with a typical spectral width of a few nanometers down to the subnanometer region, which is achieved by distributed-feedback (DFB) lasers with only one spectral mode.

The transparency of the fiber material at the different wavelengths of radiation determines the spectral parameters of transmitter and receiver. In the past, three "classical" optical windows at 850 nm (first window), 1310 nm (second), and 1550 nm (third) of fused silica fibers were mostly used with only about ±30 nm wavelength deviation from the window's center (Fig. 6.2.6).

Today, these windows are broader due to the reduced OH concentration in fused silica fibers. With the availability of new types of emitters—dense wavelength division multiplexers/demultiplexers (DWDM) and different optical amplifiers (0A)—the new wide window from 1440 to 1625 nm is now applicable especially for high-speed long-haul telecom systems. The wavelength region between the second window and the fifth window, named 2e-window, may also be used in the near future by the application of special "zero-OH" fibers. Moreover, other materials—for example, polymethylmethacrylate (PMMA) plastic optical fibers (POF)—open new windows in the visible region.

Basically, the wavelength of the transmitter and the spectral sensitivity of the receiver have to fit with these windows of low attenuation. The attenuation of the fiber (as well as any intermediate connectors), together with the sensitivity of the receiver, determines the minimum launched power in the fiber. This defines the minimum optical power of the transmitter at the optical interface if the coupling efficiency of the transmitter is provided.

6.2.4 Noise testing of transceivers

In everyday life, the dramatic increase in the use of electronic devices in general and mobile telecommunication components in particular has evoked a strong demand for electromagnetic compatibility (EMC) and

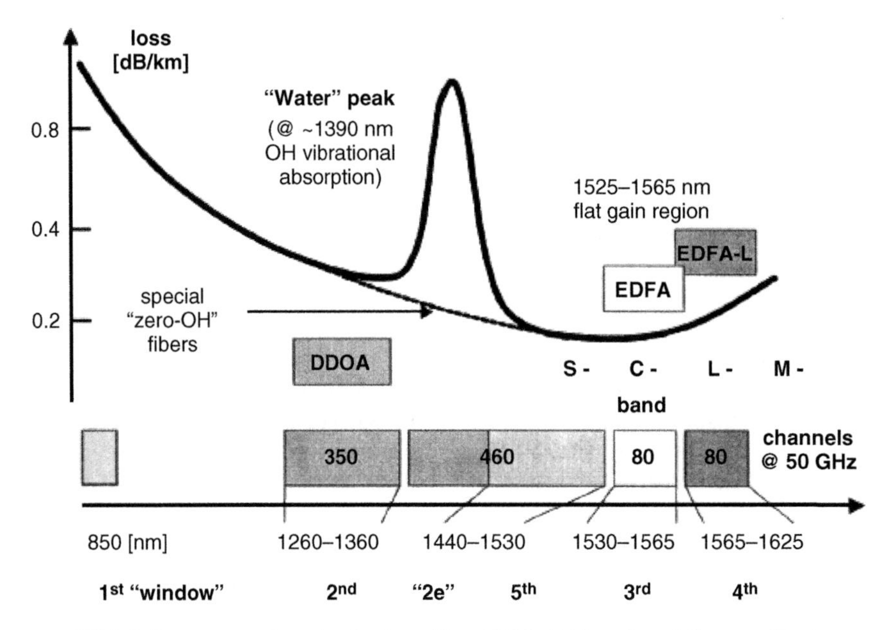

Fig. 6.2.6 Typical spectral attenuation of silica optical fibers.

electromagnetic immunity (EMI) of electronic devices. With growing efforts in the field of high-speed data transmission along with ever-increasing computer or processor integration and speed, this demand has also gained major influence on design, production, and testing of today's fiber-optic components and transmission equipment.

In close cooperation with computer, telecom, and datacom equipment manufacturers, active fiber-optic components of high EMCEMI performance have been designed. As an important foundation for success, a common understanding of the really relevant phenomena had to be reached, and reliable and repeatable test methods had to be evaluated. This gave the basis necessary to define, then to test and to meet standardized EMCEMI requirements for system application of fiber-optic transceivers, and finally to guarantee reliable operation of these components.

This section describes established measurement procedures for the EMC/EMI performance of active fiber-optic components. The methods presented here deal with noise on supply voltage, with emission of electromagnetic noise into the surrounding environment, and with immunity against electromagnetic noise from outside.

6.2.4.1 Description of the device under test

For all described measurements, the device under test (DUT) is a fiber-optic transmission module, for example, a transceiver or a combination of a transmitter and a receiver. The DUT is connected to a printed circuit board (PCB), which is populated with power supply filtering elements. The electrical data output lines of the receiver are connected directly to the electrical data input lines of the nearby transmitter. The top and bottom layers of the PCB are connected to ground. Connections to the measurement setup are fiber-optic jumper cables and power supply lines.

6.2.4.2 Noise on Vcc

Fiber-optic transceivers usually operate in an environment with fast-switching circuitry. This results in high-frequency noise on the DC power supply Vcc. High-frequency noise on Vcc cannot always be sufficiently suppressed by capacitors or inductors because these components show resonant behavior of their own, with resonant frequencies lying close to the noise frequencies. Therefore, it is essential for fiber-optic transceivers to be able to withstand high-frequency noise on Vcc. Another noise effect, commonly known as ripple, is less harmful to fiber-optic modules because ripple frequencies are lower than frequencies of noise caused by fast switching.

To perform a test on a DUT's behavior concerning noise on Vcc, a noise signal has to be coupled into the Vcc line using a bias tee. When transceivers are tested, the receiver part should act directly as a driver for the transmitter part.

If the DUTcontains internal decoupling elements, noise of certain frequencies may be sufficiently suppressed.

In these cases, noise voltages at Vcc device pins are close to zero. At frequencies at which noise suppression is insufficient, noise voltages can be measured at the pins.

At frequencies that are sufficiently suppressed, high noise generator amplitudes would be necessary to keep the noise voltage at the pins at a constant level. Testing a DUT under such rather artificial conditions is overly demanding. It is therefore recommended to merely keep the noise generator amplitude at a constant level for all frequencies.

For the measurement, regular data transmission is established and then optical power is attenuated to a level that causes a certain bit error rate (BER) value. When noise is applied on Vcc, the BER increases. This increase in BER is a measure of the DUT's sensitivity to noise. The optical power is then increased until the DUT reaches the former BER. The device meets the requirements if the difference in optical power does not exceed a defined level. The measurement is repeated for different noise frequencies. It is always important to use identical test setups to ensure that measurements are consistent.

6.2.4.3 Electromagnetic compatibility

6.2.4.3.1 Emission

The emission of electromagnetic radiation may cause improper operation of components located near a radiating device. Therefore, maximum allowable electromagnetic noise levels (depending on noise frequency) are defined. For information technology equipment the emission limits are defined in standard EN 55022, derived from IEC CISPR 22, and in FCC CFR 47, part 15, classes A and B.

The measurements are done in an anechoic chamber up to 40 GHz for any defined orientation of the DUT, or for continuous 360-degree rotation of the DUT on a wooden turntable. During testing the DUT is battery powered, and the data pattern used is a Fibre Channel test pattern at nominal data rate. The DUT's emission is measured in a range of 250 MHz up to 18 GHz and, if necessary, up to 40 GHz. The DUT's emissions are tolerated if noise remains 6 dB below these limits in the appropriate range.

6.2.4.3.2 Immunity

When external electromagnetic fields are applied to the DUT, bit errors can occur. The most sensitive part of a DUT is the receiver section around the trans-impedance amplifier. In this section, very-low-input currents (in the order of nA) are converted into low-voltage signals. In typical transceiver designs, this part of the receiver circuitry is located near the fiber-optic connector receptacle.

Two measurement methods are established to determine the DUT's immunity: radio frequency (RF) immunity and electrostatic discharge (ESD) immunity.

6.2.4.3.2.1 Noise immunity against radio frequency electromagnetic fields

This immunity test can be performed based on IEC 61000-4-3 standard. At the start of the measurement, no noise is applied. Regular data transmission is established (27-1 pseudorandom bit sequence—PRBS—at nominal data rate), and then the optical attenuator is set to a value at which a bit error rate of 10-6 is detected. At this operating point the DUT is more sensitive to external noise than during normal operation. The sine wave noise carrier source is switched on and regulated to generate a field strength of 3 or 10 V/m root mean square (RMS) at the DUT's location. Then amplitude modulation is applied to the carrier signal (modulation factor 80%, sine wave modulation 1 kHz). The noise carrier frequency starts at 80 MHz and is increased in steps of 1 or 10 MHz up to 2 GHz. The DUT is tested at its worst case position (evaluated experimentally) regarding its orientation and polarization of the noise field. Bit error rate (BER) values are measured for the different noise carrier frequencies.

Test results show that BER values depend on the noise frequency applied. Certain values of BER can be achieved by reducing optical power, so there is a corresponding optical power level for each BER value. Using this correspondence, test results can finally be presented as the minimum optical power necessary to achieve a certain constant BER value depending on the noise frequency. The DUT meets the requirements if the sensitivity values given in the DUT's specification are reached for all noise frequencies.

For a more detailed investigation, the difference between the minimum optical input power in the undisturbed case and the worst measurement result with noise applied can be required to be lower than a specified value. This measurement can be used in the development process to indicate improvement or degradation of a certain design.

6.2.4.3.2.2 Immunity against electrostatic discharge (ESD)

As with most electronic devices, transceivers are sensitive to ESD [14]. ESD immunity tests deal with noise caused by fast transients from electrostatic discharges. A transient electromagnetic field based on the human body model is generated and influences the module behavior with respect to bit errors. Two tests are proposed: The first one is based on EN 61000-4-2 = IEC 61000-4-2, and the second one is suitable for a more detailed analysis. In both cases, the source of the discharges is a so-called ESD gun.

The DUT is operated with a 27-1 PRBS at nominal data rate. The optical input power is set to the lower specification limit. The DUT's receptacle is fed through a grounded plate simulating a rack panel in a possible application environment. The following are the two steps involved in this test:

1. *Without vertical coupling plate:* Electrostatic discharges (both polarities) at any point on the rack panel or at the receptacle of the DUT are applied to induce bit errors during a defined time interval of discharge. After the discharge, the DUT continues error-free operation.
2. *With vertical coupling plate:* Electrostatic discharges are applied at any point on the vertical coupling plate and thereafter to the horizontal coupling plate to induce bit errors.

The discharge voltage and the number of bit errors in the time interval observed describe the ESD immunity of the DUT.

6.2.5 Packaging of transceivers (TRX)

6.2.5.1 Basic considerations

As also discussed in Chapter 5, the most serious task of any packaging is to maintain the functional integrity of a device over its total lifetime under all specified conditions. One of these tasks is to protect the electronic circuitry inside the transceiver from electromagnetic influences radiated or conducted into the module as well as to prevent electronics outside the module from being influenced by the module itself. This EMI shielding may be performed by means of a conductive metallic housing, or a conductive plating on a nonconductive housing, or even separate shielding structures inside the housing. Naturally, the electronic circuitry itself should also be carefully decoupled by properly selected capacitors and other filter components where needed (see also Section 6.2.4.2).

The next task is to avoid possible damage from corroding chemical agents and/or humidity to sensitive devices and surfaces during the card assembly process or during normal operation. Otherwise, for example, corrosion effects in the presence of ionic impurities and humidity will occur. These effects are particularly pronounced with electronic circuitry where electromigration in the presence of DC power occurs. Therefore, the appearance of intermittent short circuits or even other catastrophic failures due to complete destruction of joints or components is almost inevitable.

Another task of the TRX housing is to avoid the influences of external mating or withdrawal forces applied by the fiber-optic connector or the cable. This is a question of mechanical stability based on design, materials, and techniques used for assembling the transceiver itself and the transceiver to a motherboard. Some typical features may be explained in more detail by the following examples.

Figure 6.2.7 shows the second design generation of the ESCON/SBCON transceiver, with a housing containing a transmitter optical subassembly (OSA) and a receiver OSA. Also included is a chip-on-board (COB) electronic circuit with a driver IC for the infrared radiation emitting diode (IRED) on the transmitter side and a photodiode preamplifier IC and a separated buffer IC on the receiver side. The plastic housing has ground pins that are directly molded in during injection molding. After molding, the housing is plated with copper and nickel and a flash of gold, including the ground pins. Therefore, in combination with the metallized plastic cover, a completely closed metallic shielding and an extremely low ohmic-resistance connection to ground pins is achieved. Also, the Au-flashed nickel avoids oxidation or corrosion effects on any surface of the housing and the pins, which guarantees proper solderability.

Functional pins are press-fitted into the printed circuit board (PCB) and penetrate corresponding holes in the module housing's bottom. Also, the OSAs are press-fitted into noses on the front side of the housing. The PCB and cover are attached to the housing by conductive epoxy. To achieve washable seal, the housing's underside, as well as the nose area, are sealed by a suitable epoxy potting material.

6.2.5.2 Techniques for assembling transceivers to motherboard

The transceiver shown in Fig. 6.2.7 is a typical example for a module under assembly by classic through-hole technique to a motherboard. After insertion of the module pins into the corresponding holes in the motherboard, the connector shell or the transceiver housing is screwed to the board. The transceiver is then connected to the board by a common wave soldering or fountain soldering technique followed by water-jet washing, rinsing, and hot-air blower drying. During this procedure the optical ports in the connector receptacle are protected from processing chemicals and water by a process plug.

The surface mount technique (SMT) is especially used for transceivers with higher pin counts. Also, the possibility of having components on both sides of a motherboard is a basic advantage of this technique.

Gull-wing leads in principle allow partial soldering techniques such as hot ram (thermode) or hot gas or even laser soldering. The important advantage of these

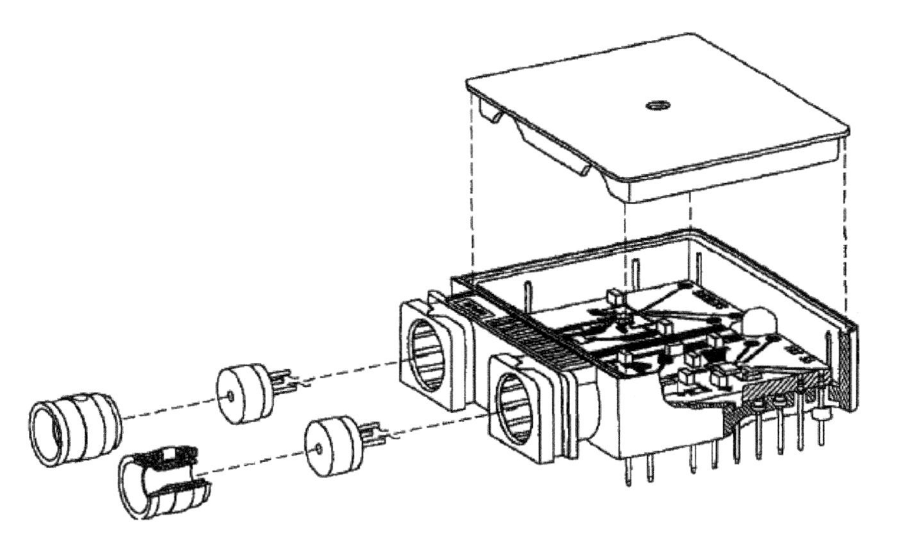

Fig. 6.2.7 Exploded view of serial ESCON transceiver (connector shell not shown).

processes is that only the leads are heated up to the liquid temperature of the solder and not the transceiver itself. Also, no washing procedure after soldering is needed if less aggressive fluxes are used. Moreover, a visual inspection of the solder joint quality is easily possible.

The main reason J-leaded transceivers are not widely established is their basic disadvantage with respect to partial soldering techniques and inspection. This is also a principal concern for the ball grid array (BGA) technique for transceivers with high external lead counts due to very high functional integration.

In recent years, however, the market has clearly tended toward pluggable connection of transceivers to motherboards or backplanes. An obvious advantage is the possibility to change a failed transceiver quickly and easily without any special tools. The transponder mentioned in Section 6.2.2.2.1 is an example of such a solution. Other examples are described in Sections 6.2.5.4 and 6.2.5.5.

6.2.5.3 Washable sealed module design

The previously mentioned soldering processes with washing and rinsing suggest a module that should be designed for hermeticity or a washable seal, which protects against intrusion of aqueous liquids during washing/rinsing. That also means that, on the one hand, no additional care must be taken to protect single electronic devices inside the module, which certainly will reduce running cost. On the other hand, the total design and nearly all assembling agents and processes have to be adapted to the demands of the soldering process. For example, the application of a standard soldering process with aggressive fluxes may be strictly forbidden inside a sealed package. In addition, during manufacturing, a suitable and cost-saving procedure for testing the seal must be implemented in a production line. The transceiver shown in Fig. 6.2.7 is an example of a washable sealed design.

6.2.5.4 Open module design

The alternative to the sealed module is an open design. Such designs are well established worldwide and are becoming increasingly popular. This design is based on the simple consideration that any possibly damaging gaseous, liquid, or dusty agent going into an open housing may also be washed out or evaporate. This assumption may in general be correct, but there are design complications and challenges to open packages.

In open packages, individual protection of single electronic devices such as ICs by separate housing or encapsulation by, for example, globe top is necessary in the case of a chip-on-board (COB) technique. In addition, special care must be taken regarding the circuit layout and the circuit assembly and protection in order to prevent electromigration effects. Finally, the EMI shielding must be managed individually by separate structures such as metallic sheets over sensitive areas of the circuitry.

With open module design, the transceiver housing itself may now be a simple injection-molded plastic part, and the final assembly of the TRX could be done by a snap-together technique. No additional sealing or corresponding test procedures are necessary.

6.2.5.5 Open card module design

As mentioned previously, open card module designs are taking their place in fiber-optic datacom. The mechanical basis of a card module typically is a PCB combined with a peripheral plastic frame. This frame also contains the optical subassemblies (OSAs) (see also Chapter 5) and the connector receptacle on its front side. For the design of such open cards, the considerations for the previously mentioned open module designs generally apply.

Figure 6.2.8 shows as an example the Gigalink Card (GLC), which is a laser-based transceiver card compliant to ANSI FC-0 100-SM-LL-I standard [5] with 1.0635 GBd at 1300 nm for single-mode fiber application and 20-bit parallel electrical TTL idout. Therefore, the GLC works as a transponder. The GLC is connected to a motherboard by an 80-pin connector on the rear area of the card's bottom side and secured by a pair of snapping legs. Total insertion and withdrawal forces for that connector are on the order of nearly 100 N. Therefore, a special withdrawal lever is designed as an integral part of the frame in order to prevent damage of the module card and the motherboard during withdrawal apply.

The GLC developed during the first half of the 1990s for storage area network (SAN) applications in mainframe computer systems represents the first product worldwide with the described properties.

Another pluggable module card design family is the Gigabit Interface Card (GBIC), which is becoming increasingly popular. Characteristic for the GBIC design is that the insertion/withdrawal direction is identical for both the optical and the electrical connection.

6.2.5.6 Small form factor pluggable design

The clear tendency toward pluggable transceivers mentioned in Section 6.2.5.2, combined with the other clear tendency toward increased port densities mentioned in Section 6.2.1.1.1, leads to the small form factor pluggable (SFP) design family. The SFP transceivers have at the bottom of the rear end an electrical connector specially designed to support hot pluggability. This means that during insertion in the motherboard, first the signal ground, then the power lines, and finally the data lines are connected. Disconnecting works in inverse order. Therefore, replacement of failed devices is possible without withdrawal of the motherboard or shutdown of the datastream of other links in operation on this motherboard.

Fig. 6.2.8 The Gigalink Card (left: bottom view; right: top view).

6.2.6 Series production of transceivers

6.2.6.1 Basic considerations for production processes and their reliability

For every series production, a basic precondition for cost savings and high-quality output is a continuously running production with qualified equipment as well as qualified and reliable processes. Correct equipment and processes mean that within a defined variation of process parameters, the so-called process window, a uniform high-quality result is achieved with a high yield. Naturally, a process is safer as its window is made wider. Therefore, all equipment and processes must be qualified before the start of series production. This will typically be performed by careful and adequate testing of a statistically significant number of parts (i.e., 40 pieces at the absolute minimum) that had previously been produced on the related equipment by the corresponding process. The test results must be evaluated in a statistical manner in which the number of tested parts defines the confidence level of the result.

6.2.6.1.1 Qualification of processes, the process capability

A brief example is provided as an explanation: A mechanical part A shall be attached to another part B by an epoxy adhesive. The key parameter for the process may be the mechanical strength characterized by, for example, the pull force needed for destruction of the epoxy attachment between parts A and B. During tests the values for the forces are monitored, and then their mean value, F_{mean}, and the standard deviation, σ, are calculated. These values are now put into a simple equation for calculating the so-called process capability:

$$C_{pk} = \frac{F_{mean} - F_{min}}{n\sigma} \geq S$$

where S is the capability index, F_{min} is the minimum force specified for the attachment, and $1 \leq n \leq 6$ is the value for the statistical sharpness. For $n = 3$, for instance,

S is 1.33 in order to reach a statistical safety of $\pm 3\sigma$ or 99.73%, or a failure probability of 2700 ppm. In the case of $n = 4$ and $C_{pk} \geq S = 1.66$, the statistical safety will be $\pm 4\sigma$ or 99.9937%, or the failure probability will be 63 ppm. Therefore, an improvement of a process is simply determined by increasing its C_{pk}.

The statistical sharpness indicated by n and the correlated value of S are defined indirectly by the requested product reliability or the permitted maximum number of failures in time (FIT) in the product specification, respectively.

6.2.6.1.2 Correlated environmental tests on subcomponents

Here it makes sense to combine the evaluation of process capability with environmental tests that are also typically defined in a product specification. These tests normally assess long-term temperature and/or humidity stress, temperature cycling, shock stress, mechanical shock, vibration stresses, and also some additional stresses determined by the customer's special application. Upper and lower stress limits or values are mostly defined by commonly known MIL or Telcordia (formerly Bellcore) or IEC generic standards named in the product specification.

The major advantage in performing tests as early as possible during the development phase and on the subcomponent level is clear: Any weakness of a construction detail, a subcomponent, or a manufacturing process will be detected earlier, resulting in time and cost savings, and improvement or general change may be more easily implemented in the design itself and in the production flow.

Naturally, all tests will also be performed at the end of product development on the completed and final product. Normally, no major weakness will occur if all pretesting has been performed carefully during the design phase.

6.2.6.1.3 Equipment qualification, the machine capability

A similar procedure to C_{pk} may be used for calculating the so-called machine capability, C_{mk}, which is defined as the statistical safety for processes performed by a tool or a machine. For example, the uniformity of a dispensed epoxy volume will be qualified by evaluation of C_{mk} of the automated dispenser. If C_{mk} exceeds the indicated value of capability index S, then the equipment will be safe and capable of its task. A combination with environmental stress for this evaluation does not make sense.

6.2.6.1.4 Frozen process

After qualification and release of equipment and a corresponding process, this production step generally should be frozen. Frozen means that in case of major changes of equipment—type or location of the equipment, process parameters, agents, or even parts—partial or total requalification of the production step must be performed. The meaning of major change or even minor change should be defined in close communication with the customer. Also, the release of a changed production step for series production typically should be with customer agreement.

6.2.6.2 Statistical process control and random sampling in a running series production

During running production, some evident quality parameters should be fully monitored and some others checked by statistical random sampling in order to continuously assess the uniformity of processes and the quality of their output. Any deviation from the expected output quality or creeping degradation of process parameters can be detected in time for corrective action to be taken. For key process steps, a higher percentage of checks should be performed or a higher number of samples tested than would be for proven stable and uniform processes.

A commonly known tool for monitoring the quality level of any production process is the so-called statistical process control (SPC) card. Here, the actual level of a parameter is compared statistically to previously defined lower and upper limits for this most characteristic parameter that determines the quality of a component or a process. In addition, the actual level may also be compared to the so-called warning limits that typically are defined as 1σ closer than the lower and upper limit. If during production an actual value goes close to a warning limit or even exceeds it, then an early corrective action may happen immediately. Therefore, components will almost never be produced out of specification. Figure 6.2.9 shows an example of a SPC card.

6.2.6.3 Zero failure quality burn-in, final outgoing inspection, and ship to stock

Any technical system or component behaves with respect to its failure probability according to a well-known time-dependent failure function, the so-called bathtub life curve. This curve describes the fact that a component will most probably fail at a higher rate at the early beginning of its lifetime (BOL) and very late at the end of its lifetime (EOL). In between, the probability of failure is low and constant.

A well-established method for identifying early failing parts, especially in electronics, is to perform a burn-in. During burn-in, the fiber-optic transceivers are operated at an elevated temperature level over a defined

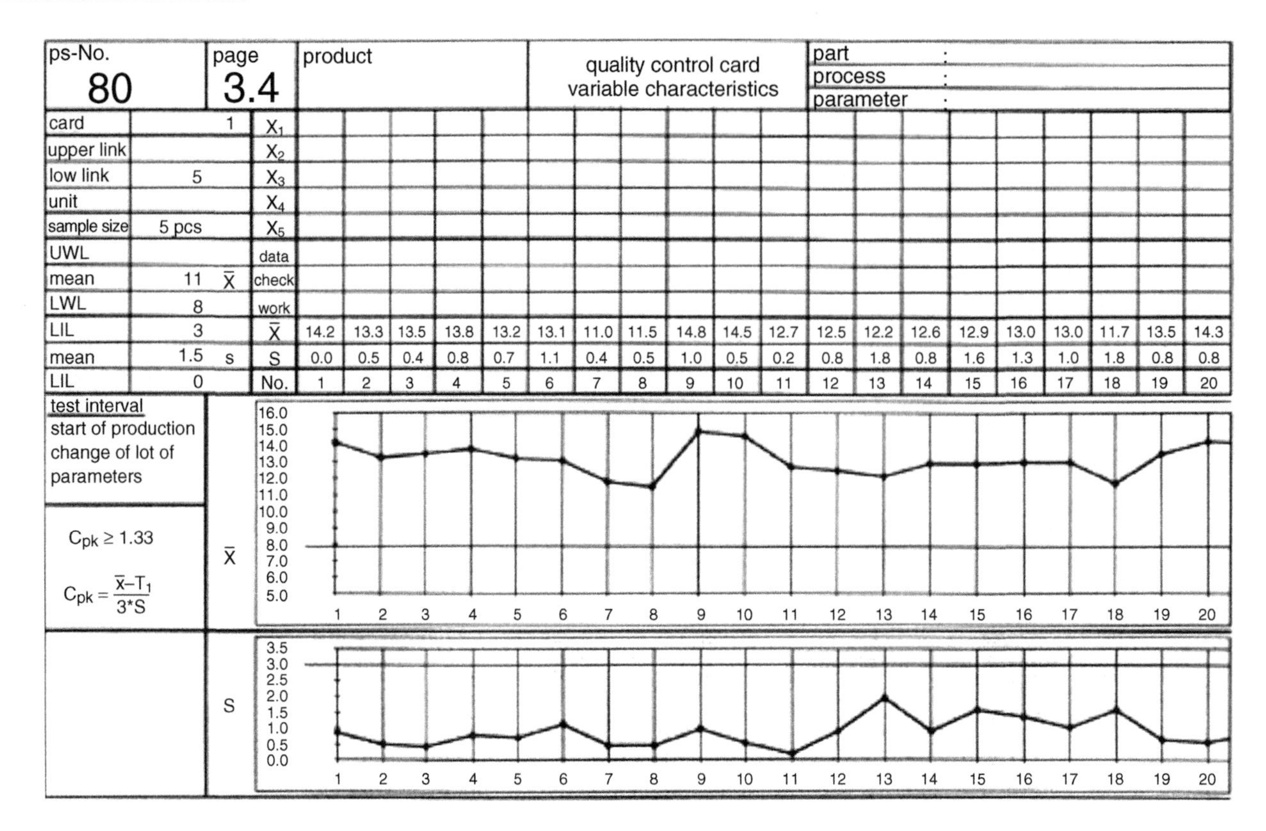

Fig. 6.2.9 Example of an SPC card.

period of time. The time, temperature, and possibly some DC power overload will define the confidence level for effective screening. Before and after burn-in, the transceivers will run their normal complete inspection of all relevant electro-optical parameters, and the measured values will be compared. Individual transceivers will be rejected if there is a delta in any parameter exceeding a defined maximum value. Therefore, the failure rate for field-installed components can be reduced dramatically, and the goal of a real zero-failure quality is approached.

6.2.7 Transceivers today and tomorrow

6.2.7.1 Transceivers today

Fiber-optic transceivers for applications in the field of datacom are mostly characterized by a couple of established international standards. These standards define the electro-optical performance of a transceiver/transponder as well as its pinout and its physical outline and package, including the corresponding fiber-optic connector interfaces [6, 7, 8].

Fiber-optic transceivers meeting these standards are operating worldwide in numerous applications in mainframes, server clusters, storage area networks, wide area networks, and local area networks, and currently around 20 to 30 worldwide competing suppliers have been established. The number of partners involved in some important multisourcing agreements has seen an increase since 1989. This is also indicative of the increasing importance of industrial associations where both suppliers and applicators are represented. This speeds up the market penetration of novel components, systems, and applications. Nowadays, this does not seem to generate conflicts with the commonly agreed normative power of international standardization organizations such as the International Organization for Standardization (ISO), International Electrotechnical Commission (IEC), and International Telecommunication Union (ITU).

The demand for these transceivers has continuously increased during the past 10 years, and the prices have shown dramatic decreases of the order of 25% per year. Consequently, the goal of all manufacturers is to offer a high level of performance, reliability, quality, and serviceability while maintaining cost-effective production in the face of drastically increased volumes to meet the market pricing.

6.2.7.2 Some aspects of tomorrow's transceivers

The bit rates of fiber-optic transceivers are continuously increasing in order to meet the worldwide demand for ever higher bandwidths. These bandwidth increases are called for by both existing storage and networking markets, as well as the parallel computing industry and high-end server design.

6.2.7.2.1 Geometrical outline of transceivers

In the past 10 years, a significant reduction of module/transceiver size was possible due to significant progress in the downsizing of optical subassemblies (see Chapter 5) and associated passive and active electronic components and circuitry. Figure 6.2.10 shows an in-scale comparison of the ESCON/SBCON outline (left), multistandard, small form factor (SFF), and parallel SNAP-12 transceivers (right). The function of the transceivers shown is described in detail in Section 6.2.1.1.

If one combines the increase of bit rate with the reduction of size, the success of the development efforts of the past 10 years is obvious. Figure 6.2.11 shows a graph for the bit rate per square millimeter, named "rate-density," versus the years of introduction of the products to the market. The dots represent, from left to right: ESCON/SBCON, MS 155 Mbit/s, MS 622 Mbit/s, SFF 1 Gbit/s, and SFF 2.5 Gbit/s. The first dot differs from the last dot by a factor of 100. There is no obvious reason why this trend should change in the near future.

6.2.7.2.2 Functional integration

Another direction for the next generation of transceivers is the inclusion of additional electronic functions in a common module housing, such as

- Serialization and deserialization of parallel digital bitstreams
- Encoding and decoding of serial bit streams

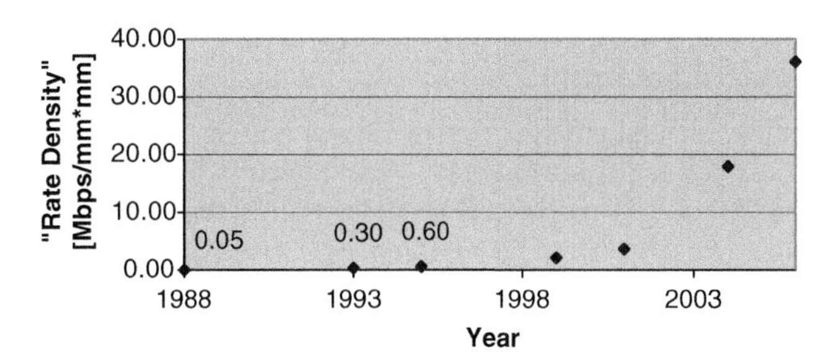

Fig. 6.2.11 "Rate-density" of transceivers.

- Clock synchronization/regeneration on the receiver side
- Laser control and laser safety functions in laser-based transceivers/transponders as previously discussed in Section 6.2.2.2.

The main advantage for the user of such higher levels of integration is that no high-speed signals are on the system board, with the related cost savings. One challenge developers need to solve is the issue of heat dissipation caused by such increased integration of a large number of high-speed digital electronic functions within the package. The only reasonable solution is to reduce the power consumption by application of low-power IC technologies with a supply voltage of 3.3 V or less. Such ICs have been already introduced and will be continuously improved for reduced power consumption.

An additional complication arises when some of the ICs in transceivers have to operate mixed signals, which means pure digital signals combined with analog signals, and in addition DC bias voltages and control functions. In the case of laser-driver ICs, the bandgap of the laser's active radiating material defines the absolute minimum of the supply voltage, given by fundamental physical laws. The lower the wavelength emitted by the radiation source, the higher the bandgap energy and, consequently, the higher the required bias voltage. Therefore, the supply voltages of fiber-optical transceivers may not completely follow the general tendency in digital electronics toward continuously decreasing supply voltages.

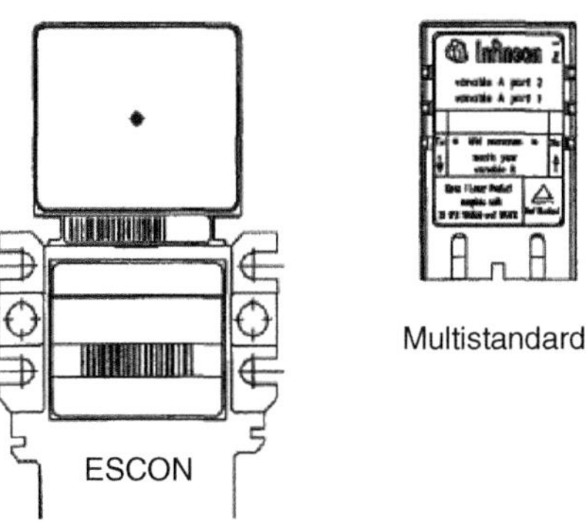

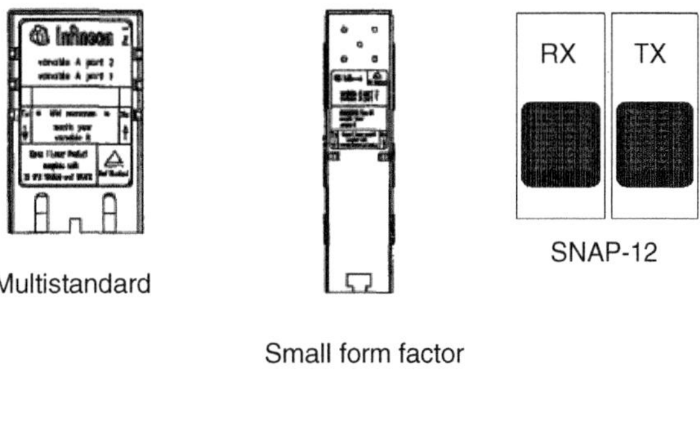

Fig. 6.2.10 Comparison of the outlines of different transceiver generations.

6.2.7.2.3 Edge-emitting lasers and VCSELs as optical sources

If one exceeds the bit rate of approximately 300 Mbit/s in a fiber-optic intermediate-range multimode fiber link, the commonly used IRED on the transmitter side will be too slow and must be replaced by a laser diode (LD). However, the very fast (up to 10 Gbit/s with direct modulation) conventional edge-emitting laser diode (EELD) is more complicated in application than an IRED because of the following:

1. An EELD needs a control circuit that monitors the output optical power and compensates for temperature and aging effects.
2. Accurate and reliable optical coupling of an EELD into a single-mode fiber is much more difficult and therefore more expensive compared to coupling an IRED into a multimode fiber. The position accuracy and stability needed for EELD in the range of 0.1 micrometer is approximately an order of magnitude higher than for coupling an IRED.
3. Laser safety due to potentially high optical output power has to be taken into account. The limitation of radiated optical power can be achieved by optical means or by electrical limitation of LD power output. Nevertheless, products for most datacom applications are unlikely to be successful in the market without certification as laser class 1 safe according to IEC 60825-1 or corresponding regulations such as those of the FDA (see also Section 6.2.3.3).

Currently, a new type of laser is becoming dominant in some specific applications, the vertical cavity surface-emitting laser (VCSEL). This source was originally developed as 980-nm pumping of erbium doped fiber amplifiers for long-haul telecom transmission lines.

One of the key advantages of a VCSEL compared to an EELD is the IRED-like technology. This allows one to produce VCSELs with all processing steps, including burn-in and final testing, completely at a wafer level. Some additional advantages of VCSELs with respect to the EELD are listed in Table 6.2.1. A disadvantage of VCSELs is that not all of the wavelength bands covered by EELDs are available with VCSEL technology. Currently, only VCSELs for the 850-nm band are available for volume production with proven reliability and lifetime. VCSELs for the 1300-nm and the 1550-nm bands are still under basic research and design development. The experts estimate that possibly in the next four years 1300-nm VCSELs will also be available in small volumes with acceptable yield.

Table 6.2.1 Comparison of Features: Edge-emitting laser diode (EELD) vs. vertical cavity surface-emitting laser (VCSEL).

Feature	EELD	VCSEL
Wavelength bands	650, 850, 1300 to 1660 nm	(650), 850, (1300, 1550) nm
Spectral bandwidth	Very narrow	Narrow
Size of active area	Typically 0.5–1 × 2–10 μm	Variable, 5–50 μm diameter
Beam geometry	Strong elliptic	Circular
Beam divergence	High, up to 60° × 20°	Low, ca, 5°
Number of modes	Typically 1 or few	1 or even up to many 10 s
Coupling to fiber	Difficult and sensitive	Easy
Coupling efficiency	Moderate	High
Threshold current	Approximately 10 mA	Some mA
Direct modulation bandwidth	High, up to 10 Gbit/s	High, up to 10 Gbit/s
Temperature drift of P_{opt}	Fairly high	Tendentially low
Environmental sensitivity	Extremely high	Moderate
Processing of chip	Very specific	Similar to LED
Final processing	Single bar	On wafer
Burn-in and functional test	Single on heatsink	On wafer

6.2.7.2.4 Laser diodes for multimode fibers, mode underfill

Worldwide there are many miles of graded-index (GI) multimode fibers installed in buildings and campuses. However, the speed and transmission field length of fiber-optic links with GI multimode fibers combined with IREDs is limited due to power budget and bandwidth-length limits.

In order to safeguard this investment and use this current cabling even for higher speed transmission over distances of more than 100 m, the concept emerges of using laser diodes as sources for GI multimode fibers. There are groups studying the idea of extending the limits of GI multimode fibers by means of the so-called mode underfill launch condition. That would mean that coupling of optical power from a LD with limited focal diameter and numerical aperture into a GI multimode fiber would establish only a few low-order propagation modes near the center of the fiber core. The result would be a significant increase of the fiber's bandwidth-length limit.

Experimental investigations have confirmed the theoretical assumptions. Therefore, transmission of up to 1 Gbps with 1300 nm wavelength over more than 500 m of standard graded-index multimode fibers would work well. This direction is still receiving intensive discussion in the related standardization groups.

However, this technique will establish itself only if the price and performance for laser-based products are drastically improved. One key component would be an inexpensive laser optical subassembly (see also Chapter 5) with a laser diode that operates uncooled over the temperature range of category C, controlled environment (−10°C to +60°C), according to IEC 61300-2-22, a typical office or building environment.

6.2.8 Parallel optical links

6.2.8.1 High-density point-to-point communications

Fiber-optic transceivers have become well established for applications requiring high-bandwidth transmission of data. Such applications include backbone switching for telecommunications, high-end routers, storage area networks (SANs), cross data center communications (Ethernet), and data flow for disk clusters.

Point-to-point communications are often configured as "patch panels," in which the fiber-optic transceivers are mounted onto a front panel, with the fiber sockets accessed through holes in the front panel. A duplex fiber cord is routed from one transceiver in one rack to the next desired transceiver in an adjacent rack. The number of fiber cords that a given panel can support, and hence the total aggregate bandwidth available from standard fiber-optic transceivers, is practically limited by the number of fiber sockets that can be installed on the panel. Consider, for example, a small form factor transceiver with a width of approximately 14 mm, operating at a data rate of 2.5 Gbit/s. Such a transceiver offers a bandwidth per front panel width of almost 200 Mbit/s per millimeter; let us call this the bandwidth density.

As bandwidth density requirements increase, the density limit imposed by single-channel transceivers becomes increasingly burdensome. This density constraint can be significantly relaxed by using a combination of multiple-channel fiber-optic modules and multifiber ribbon cable.

SANs and cross data center communication applications are stressed with more and more data every year. The high-volume serial protocols respond to this with regular increases in data rate; Ethernet has recently increased from 1 Gbps to 10 Gbps, and FICON/FC has gone from 1 Gbps/2 Gbps multirate transceivers to 1 G/2 G/4 Gbps multirate parts, with 8 Gbps coming soon. These serial transceivers allow host bus adapter (HBA) cards to be designed with several high-bandwidth ports, and directors and switches to be designed with tens of ports brickwalled on both sides of the client cards.

The parallel computing industry supplies products to meet the demands of most complex simulation and modeling problems, such as global climate modeling and protein folding. These problems are split into thousands of small chunks that are computed by individual processors. The results from, and new inputs to, the processors must be communicated through a switching fabric to keep the program moving forward, which results in very high IO bandwidth requirements from the card edge. Parallel transceivers operating are available today with single-lane bandwidths from 2 to 6 Gbps (with Double Date Rate Infiniband, DDR-IB, at 5 Gbps as one standard example) and individual transmitters and receivers housing from 4 to 12 lanes. One MSA related to such parallel transceivers is the SNAP-12 standard. Using a typical 20 mm center-to-center spacing, one can fit a transmitter/receiver pair in 40 mm of card edge with an aggregate bandwidth of 60 Gbps at DDR-IB.

6.2.8.2 Common parallel optic module configurations

Just as multifiber cables improve the bandwidth density of the front panel, 12-channel fiber-optic modules dramatically improve the area utilization of the printed circuit board (bandwidth per unit board area). A transmitter module consists of a linear array of 12 lasers plus associated drive electronics; a receiver module consists of a linear array of 12 PIN diodes plus associated transimpedance amplifiers. The operation of each channel is independent of that of the next adjacent channel.

VCSELs are by far the most common choice for laser in the transmitter modules because of low cost and ease of launching laser light into the optical fiber. Currently, the state of the art is 850 nm emission of multimode light; thus the fiber cable should also be multimode. VCSEL arrays operating at 1310 nm are available from a number of manufacturers.

An alternative configuration to the 12-channel parallel optics combines four transmitters and four receivers into a single package. A 12-fiber ribbon cable is typically used, with the center four fibers "dark."

6.2.8.3 Link reach

One of the most critical questions about a parallel optical link is, "What is a reasonable link reach?' This means, "What fiber cable length can be supported while still obtaining acceptable link performance in a low-cost installation?" This is a complex issue that prompts at least three distinct questions.

The first question concerns technical feasibility: What link reaches can be demonstrated in a laboratory? The current state of the art is for a per-channel bandwidth of 2.5 Gbit/s at an operating wavelength of 850 nm. Such an optical signal propagating through multimode becomes degraded through one of three mechanisms: optical absorption (which is significantly higher at 850 nm than at 1310 nm), chromatic dispersion (which is much more severe at 850 nm than at 1310 nm), and modal dispersion. Optical attenuation and chromatic dispersion performances are largely defined by the glass materials system, and hence are not likely to exhibit major improvements. However, the third mechanism, modal dispersion, is highly sensitive to the fiber manufacturing process. A number of fiber manufacturers are optimizing their processes to produce very low modal dispersion fiber for operation at 850 nm. Lucent's LazrSpeed and Coming's NGMM fiber are examples of this effort. A number of companies have demonstrated excellent link performance over a reach of at least a kilometer at 2.5 Gbit/s using such fiber.

The second question concerns prudent system design: What is a reasonable link budget that gives high assurance of successful operation under essentially all circumstances? Just because a particular link reach can be (routinely) demonstrated in the laboratory does not make this a prudent choice for system design. Typically, a desired system design is expressed in terms of a link budget. The lowest expected laser power (over the life of the laser, for all allowed performance limits of temperature and voltage), minus the worst-case receiver sensitivity, defines a possible operating range. From this range must be subtracted expected losses such as worst-case fiber connector losses and fiber attenuation. Additional "penalties" are deducted to account for degradation mechanism such as laser residual intensity noise (RIN).

Parallel optics modules are at a disadvantage compared with single-channel transceivers for achieving link budgets. Specifications on expected transmitter optical power need to be wider for parallel optics module to account for expected channel-to-channel power variations. Furthermore, laser safety limits are more restrictive for a parallel optics module because of the multiple channels. As performance goals become ever more aggressive, finding an optimum balance between laser safety constraints and a prudent link budget becomes an ever more difficult challenge for parallel optics.

The third question concerns the cost of the link, as fiber cable costs (especially for high-performance multimode ribbon fiber) can be a substantial fraction of the total link cost.

6.2.8.4 A look to the future

Cost and performance analysis changes dramatically if parallel links are configured with single-mode fiber operating at 1310 nm wavelength. Mode dispersion is eliminated because of the single-mode fiber. Optical attenuation is significantly reduced, as is chromatic dispersion. Laser safety is much less restrictive at the longer wavelength, so higher optical powers can be considered. Thus significantly longer link reach can be realized at 1310 nm. But the most dramatic change is the cable cost. Single-mode multifiber cables should have the lowest cost of all multifiber possibilities.

Such a link would require two key changes, however. One is the use of a long-wavelength laser. For reasons of cost and ease of light launch, the laser array is preferably a VCSEL. The second major change is a dramatic tightening of optomechanical tolerances. Single-mode fiber has a core diameter that is almost an order of magnitude smaller than standard multimode fiber. This will make the manufacture of such a parallel optics module much more challenging. While 1310 nm VCSELs have been available now for a couple of years, their adoption has been slow. Thus, single-mode 1310 nm parallel optics modules are not yet available.

Future transceiver design is likely to focus on power consumption, electromagnetic compatibility and immunity, and density. As data rates continue to increase, we will start to see transceivers used closer to the ICs on the board and not just at the card edge. It has also been demonstrated that it is possible to incorporate optical components onto a chip, completely avoiding the deficiencies of high-speed signals on copper board traces. While these advancements may take their place in high-end computing systems, classical card edge transceivers are likely to continue to play their role into the foreseeable future to allow fiber cable connection for SANs and networking.

Acknowledgments

Many thanks to all colleagues for their help in giving hints for corrections and updates, in particular:

- Thomas Murphy for careful check of grammar and wording in this chapter
- Herwig Stange for the update of currently valid laser safety limits
- Mario Festag for checking and updating Section 6.2.4.3
- Ursula Annbrust and Renate Lindner for their help in preparing the figures, graphs, and photos.

References

1. Govind P. Agrawal *Fiber-optic communication systems*, 2nd ed. New York: Wiley (1997).
2. B. E. A. Saleh, and M. C. Teich. *Fundamentals of photonics*. New York: Wiley (1991).
3. Proceedings of 26th ECOC. September 3–7, 2000. Munich, Germany: VDE-Verlag (2000).
4. IEC CA/1727/QP, 2000, March. SB4 FWG: Survey of future telecommunications scenario.
5. ANSI X3T9.x and T1l.x. Fibre Channel (FC) Standards incl. FDDI, SBCON and HIPPI-6400, URLs: http://web.ansi.org/default.htm and http://www.fibrechannel.com.
6. IEC SC86C Drafts, released or midterm to be released IEC Standards, Group 62 148-xx, Discrete/integrated optoelectronic semiconductor devices for fiber optic communication—Interface Standards, URL: http://www.iec.ch.
7. IEC SC86C Drafts, released or midterm to be released IEC Standards, Group 62 149–xx, Discrete/integrated optoelectronic semiconductor devices for fiber optic communication including hybrid devices—Package interface standards.
8. IEC SC86B Drafts, released or midterm to be released IEC Standards, Group 61 754–xx, Fibre Optic Connector Interfaces.
9. IEEE Projects 802.x, LAN/MAN Standards and Drafts URL: http://standards.ieee.org.
10. Telcordia Technologies (formerly BELLCORE) GR-253-CORE. 2000, September. Issue 3. Synchronous Optical Network (SONET) Transport Systems: Common Generic Criteria.
11. ITU-T G.957. Optical interfaces for equipment and systems relating to the synchronous digital hierarchy (SDH) (June, 1999).
12. ITU-T G.958. Digital line systems based on the synchronous digital hierarchy (SDH) for use on optical fiber cables (Nov., 1994).
13. International Standard IEC 60825-1,1993 incl. Amendment 2, January 2001, ISBN 2-8318- 5589-6, Safety of laser products—Part 1: Equipment classification, requirements and user's guide.
14. R., Atkins, and C. DeCusatis. March 27–28. Latent electro-static damage in vertical cavity surface emitting semiconductor laser arrays. Proc. 2006, IEEE Sarnoff Symposium, Princeton, NJ (2000).

Chapter 6.3

Optical link budgets and design rules

DeCusatis

6.3.1 Fiber-optic communication links (telecom, datacom, and analog)

There are many different applications for fiber-optic communication systems, each with its own unique performance requirements. For example, analog communication systems may be subject to different types of noise and interference than digital systems, and consequently require different figures of merit to characterize their behavior. At first glance, telecommunication and data communication systems appear to have much in common, as both use digital encoding of datastreams. In fact, both types can share a common network infrastructure. Upon closer examination, however, we find important differences between them. First, datacom systems must maintain a much lower bit error rate (BER), defined as the number of transmission errors per second in the communication link (we will discuss BER in more detail in the following sections). For telecom (voice) communications, the ultimate receiver is the human ear, and voice signals have a bandwidth of only about 4 kHz. Transmission errors often manifest as excessive static noise such as encountered on a mobile phone, and most users can tolerate this level of fidelity. In contrast, the consequences of even a single bit error to a datacom system can be very serious; critical data such as medical or financial records could be corrupted, or large computer systems could be shut down. Typical telecom systems operate at a BER of about 10^{-9}, compared with about 10^{-12} to 10^{-15} for datacom systems.

Another unique requirement of datacom systems is eye safety vs. distance tradeoffs. Most telecommunications equipment is maintained in a restricted environment and is accessible only to personnel trained in the proper handling of high-power optical sources. Datacom equipment is maintained in a computer center and must comply with international regulations for inherent eye safety; this limits the amount of optical power that can safely be launched into the fiber, and consequently limits the maximum distances that can be achieved without using repeaters or regenerators. For the same reason, datacom equipment must be rugged enough to withstand casual use, while telecom equipment is more often handled by specially trained service personnel. Telecom systems also tend to make more extensive use of multiplexing techniques, which are only now being introduced into the data center, and more extensive use of optical repeaters.

6.3.2 Figures of merit: SNR, BER, and MER

Several possible figures of merit may be used to characterize the performance of an optical communication system. Furthermore, different figures of merit may be more suitable for different applications, such as analog or digital transmission. In this section, we will describe some of the measurements used to characterize the performance of optical communication systems. Even if we ignore the practical considerations of laser eye safety standards, an optical transmitter is capable of launching

Handbook of Fiber Optic Data Communication; ISBN: 9780123742162

a limited amount of optical power into a fiber. Similarly, there is a limit as to how weak a signal can be detected by the receiver in the presence of noise and interference. Thus, a fundamental consideration in optical communication systems design is the optical link power budget, or the difference between the transmitted and received optical power levels. Some power will be lost due to connections, splices, and bulk attenuation in the fiber. There may also be optical power penalties due to dispersion, modal noise, or other effects in the fiber and electronics. The optical power levels define the signal-to-noise ratio (SNR) at the receiver, which is often used to characterize the performance of analog communication systems. For digital transmission, the most common figure of merit is the bit error rate (BER), defined as the ratio of received bit errors to the total number of transmitted bits. Signal-to-noise ratio is related to the bit error rate by the Gaussian integral

$$\mathrm{BER} = \frac{1}{\sqrt{2\pi}}\int_{Q}^{\infty} e^{-\frac{Q^2}{2}} dQ \cong ; \frac{1}{Q\sqrt{2\pi}} e^{-\frac{Q^2}{2}} \tag{6.3.1}$$

where Q represents the SNR for simplicity of notation [1–4]. From Eq. (6.3.1), we see that a plot of BER vs. received optical power yields a straight line on a semilog scale, as illustrated in Fig. 6.3.1. Nominally, the slope is about 1.8 dB/ decade; deviations from a straight line may indicate the presence of nonlinear or non-Gaussian noise sources. Some effects, such as fiber attenuation, are linear noise sources; they can be overcome by increasing the received optical power, as seen from Fig. 6.3.1, subject to constraints on maximum optical power (laser safety) and the limits of receiver sensitivity. There are other types of noise sources, such as mode partition noise or relative intensity noise (RIN), which are independent of signal strength. When such noise is present, no amount of increase in transmitted signal strength will affect the BER; a noise floor is produced, as shown by curve B in Fig. 6.3.1. This type of noise can be a serious limitation on link performance. If we plot BER vs. receiver sensitivity for increasing optical power, we obtain a curve similar to Fig. 6.3.2, which shows that for very-high-power levels, the receiver will go into saturation. The characteristic "bathtub"-shaped curve illustrates a window of operation with both upper and lower limits on the received power. There may also be an upper limit on optical power due to eye safety considerations.

We can see from Fig. 6.3.1 that receiver sensitivity is specified at a given BER, which is often too low to measure directly in a reasonable amount of time (for example, a 200 Mbit/s link operating at a BER of 10^{-15} will only take one error every 57 days on average, and several hundred errors are recommended for a reasonable BER measurement). For practical reasons, the BER is

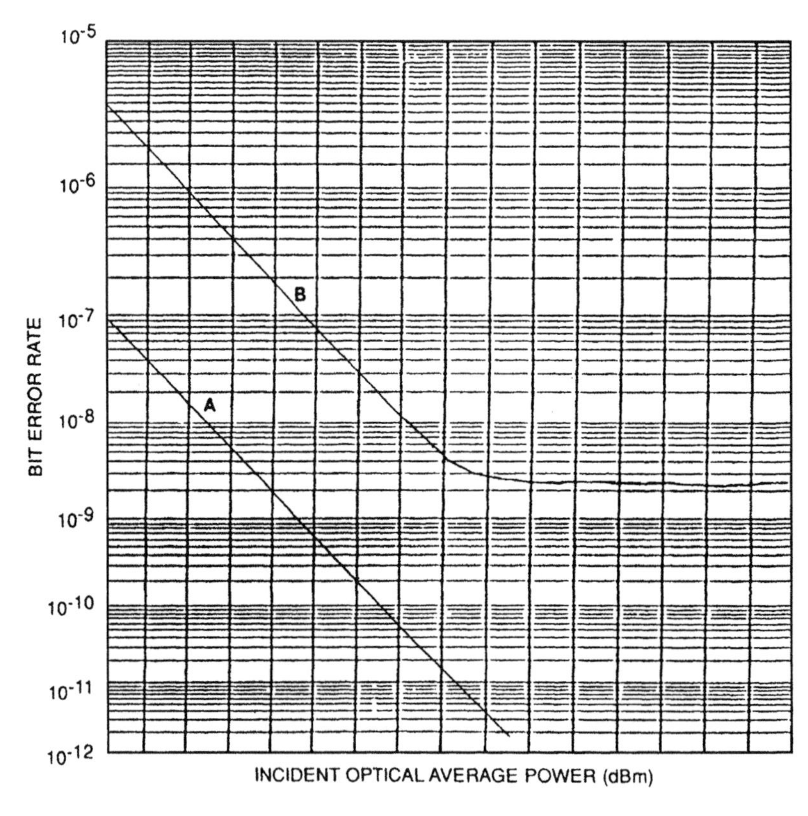

Fig. 6.3.1 Bit error rate as a function of received optical power. Curve A shows typical performance, whereas curve B shows a BER floor.

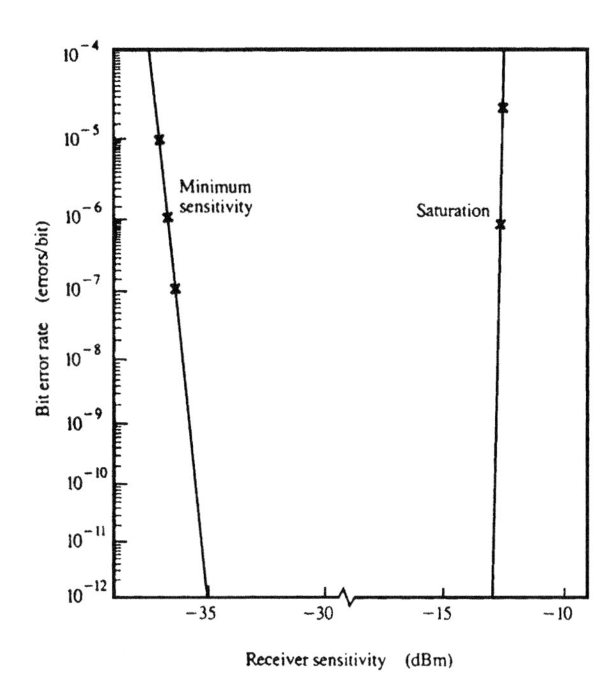

Fig. 6.3.2 Bit error rate as a function of received optical power illustrating range *c*, operation from minimum sensitivity to saturation.

typically measured at much higher error rates, where the data can be collected more quickly (such as 10^{-4} to 10^{-8}) and then extrapolated to find the sensitivity at low BER. This assumes the absence of nonlinear noise floors, as cautioned previously. The relationship between optical input power, in watts, and the BER is the complementary Gaussian error function

$$\mathrm{BER} = 1/2\ \mathrm{erfc}\ (\ P_{\mathrm{out}} - P_{\mathrm{signal}}/\mathrm{RMS\ noise}), \tag{6.3.2}$$

where the error function (erfc) is an open integral that cannot be solved directly. Several approximations have been developed for this integral, which can be developed into transformation functions that yield a linear least squares fit to the data [1]. The same curve-fitting equations can also be used to characterize the eye window performance of optical receivers. Clock position/phase vs. BER data are collected for each edge of the eye window; these data sets are then curve fitted with the above expressions to determine the clock position at the desired BER. The difference in the two resulting clock positions on either side of the window gives the clear eye opening [1–4].

In describing Figs. 6.3.1 and 6.3.2, we have also made some assumptions about the receiver circuit. Most data links are asynchronous and do not transmit a clock pulse along with the data; instead, a clock is extracted from the incoming data and used to retime the received datastream. We have made the assumption that the BER is measured with the clock at the center of the received data bit; ideally, this is when we compare the signal with a preset threshold to determine if a logical "1" or "0" was sent. When the clock is recovered from a receiver circuit such as a phase-locked loop, there is always some uncertainty about the clock position; even if it is centered on the data bit, the relative clock position may drift over time. The region of the bit interval in the time domain where the BER is acceptable is called the eyewidth; if the clock timing is swept over the data bit using a delay generator, the BER will degrade near the edges of the eye window. Eyewidth measurements are an important parameter in link design, which will be discussed further in the section on jitter and link budget modeling.

In the design of some analog optical communication systems, as well as some digital television systems (for example, those based on 64-bit quadrature amplitude modulation), another possible figure of merit is the modulation error ratio (MER). To understand this metric, we will consider the standard definition of the Digital Video Broadcasters (DVB) Measurements Group [5]. First, the video receiver captures a time record of N received signal coordinate pairs, representing the position of information on a two-dimensional screen. The ideal position coordinates are given by the vector (X_j, Y_j). For each received symbol, a decision is made as to which symbol was transmitted, and an error vector $(\Delta X_j, \Delta Y_j)$ is defined as the distance from the ideal position to the actual position of the received symbol. The MER is then defined as the sum of the squares of the magnitudes of the ideal symbol vector divided by the sum of the squares of the magnitudes of the symbol error vectors:

$$\mathrm{MER} = 10 \log \frac{\sum_{j=1}^{N} (X_j^2 + Y_j^2)}{\sum_{j=1}^{N} (\Delta X_j^2 + \Delta Y_j^2)\ \mathrm{dB}}. \tag{6.3.3}$$

When the signal vectors are corrupted by noise, they can be treated as random variables. The denominator in Eq. (6.3.3) becomes an estimate of the average power of the error vector (in other words, its second moment) and contains all signal degradation due to noise, reflections, transmitter quadrature errors, and so on. If the only significant source of signal degradation is additive white Gaussian noise, then MER and SNR are equivalent. For communication systems that contain other noise sources, MER offers some advantages; in particular, for some digital transmission systems there may be a very sharp change in BER as a function of SNR (a so-called cliff effect), which means that BER alone cannot be used as an early predictor of system failures. MER, on the other hand, can be used to measure signal-to-interference ratios accurately for such systems. Because MER is a statistical measurement, its accuracy is directly related to the number of vectors, N, used in the computation. An accuracy of 0.14 dB can be obtained with N = 10,000, which would require about 2 ms to accumulate at the industry standard digital video rate of 5.057 Msymbols/s.

In order to design a proper optical data link, the contribution of different types of noise sources should be assessed when developing a link budget. There are two basic approaches to link budget modeling. One method is to design the link to operate at the desired BER when all the individual link components assume their worst case performance. This conservative approach is desirable when very high performance is required, or when it is difficult or inconvenient to replace failing components near the end of their useful lifetimes. The resulting design has a high safety margin; in some cases, it may be overdesigned for the required level of performance. Since it is very unlikely that all the elements of the link will assume their worst case performance at the same time, an alternative is to model the link budget statistically. For this method, distributions of transmitter power output, receiver sensitivity, and other parameters are either measured or estimated. They are then combined statistically using an approach such as the Monte Carlo method, in which many possible link combinations are simulated to generate an overall distribution of the available link optical power. A typical approach is the 3-sigma design, in which the combined variations of all link components are not allowed to extend more than 3 standard deviations from the average performance target in either direction. The statistical approach results in greater design flexibility and in generally increased distance compared with a worst case model at the same BER.

6.3.3 Link budget analysis: installation loss

It is convenient to break down the link budget into two areas: installation loss and available power. Installation or DC loss refers to optical losses associated with the fiber cable plant, such as connector loss, splice loss, and bandwidth considerations. Available optical power is the difference between the transmitter output and receiver input powers, minus additional losses due to optical noise sources on the link (also known as AC losses). With this approach, the installation loss budget may be treated statistically and the available power budget, as worst case. First, we consider the installation loss budget, which can be broken down into three areas: transmission loss, fiber attenuation as a function of wavelength, and connector or splice losses.

6.3.3.1 Transmission loss

Transmission loss is perhaps the most important property of an optical fiber; it affects the link budget and maximum unrepeated distance. Since the maximum optical power launched into an optical fiber is determined by international laser eye safety standards [8], the number and separation between optical repeaters and regenerators are largely determined by this loss. The mechanisms responsible for this loss include material absorption as well as both linear and nonlinear scattering of light from impurities in the fiber [1–5]. Typical loss for single-mode optical fibers is about 2 to 3 dB/km near 800 nm wavelength, 0.5 dB/km near 1300 nm, and 0.25 dB/km near 1550 nm. Multimode fiber loss is slightly higher, and bending loss will only increase the link attenuation further.

6.3.3.2 Attenuation vs. wavelength

Since fiber loss varies with wavelength, changes in the source wavelength or use of sources with a spectrum of wavelengths will produce additional loss. Transmission loss is minimized near the 1550 nm wavelength band, which unfortunately does not correspond with the dispersion minimum at around 1310 nm. An accurate model for fiber loss as a function of wavelength has been developed by Walker [9]; this model accounts for the effects of linear scattering, macrobending, and material absorption due to ultraviolet and infrared band edges, hydroxide [OH] absorption, and absorption from common impurities such as phosphorous. Using this model, it is possible to calculate the fiber loss as a function of wavelength for different impurity levels; the fiber properties can be specified along with the acceptable wavelength limits of the source to limit the fiber loss over the entire operating wavelength range. Design tradeoffs are possible between center wavelength and fiber composition to achieve the desired result. Typical loss due to wavelength dependent attenuation for laser sources on single-mode fiber can be held below 0.1 dB/km.

6.3.3.3 Connector and splice losses

There are also installation losses associated with fiber-optic connectors and splices; both of these are inherently statistical in nature and can be characterized by a Gaussian distribution. There are many different kinds of standardized optical connectors, some of which have been discussed previously. Some industry standards also specify the type of optical fiber and connectors suitable for a given application [10]. There are also different models which have been published for estimating connection loss due to fiber misalignment [11, 12]. Most of these models treat loss due to misalignment of fiber cores, offset of fibers on either side of the connector, and angular misalignment of fibers. The loss due to these effects is then combined into an overall estimate of the connector performance. No general model is available to treat all types of connectors, but typical connector loss values average about 0.5 dB worst case for multimode, slightly higher for single mode (see Table 6.3.1).

Optical splices are required for longer links, since fiber is usually available in spools of 1 to 5 km, or to repair broken fibers. There are two basic types: mechanical splices (which involve placing the two fiber ends in a receptacle that holds them close together, usually with epoxy) and the more commonly used fusion splices (in which the fibers are aligned and then heated sufficiently to fuse the two ends together).

If the optical fiber is improperly cabled or installed, additional loss can be experienced due to bending of the fibers. This falls into two categories: microbending (due to nanometer-scale variations in the fiber) and macrobending (due to much larger, visible bends in the fiber). Because both types of bending loss may contribute to the attenuation of a single-mode or multimode fiber, it is obviously desirable to minimize bending in the application whenever possible. Most qualified optical fiber cables specify a maximum bend radius to limit macrobending effects, typically around 10–15 mm, although this varies with the cable type and manufacturer's recommendations.

6.3.4 Link budget analysis: optical power penalties

Next, we will consider the assembly loss budget, which is the difference between the transmitter output and receiver input powers, allowing for optical power penalties due to noise sources in the link. We will follow the standard convention in the literature of assuming a digital optical communication link that is best characterized by its BER. Contributing factors to link performance include the following:

- Dispersion (modal and chromatic) or intersymbol interference
- Mode partition noise
- Mode hopping
- Extinction ratio
- Multipath interference
- Relative intensity noise (RIN)
- Timing jitter
- Radiation-induced darkening
- Modal noise.

Higher order, nonlinear effects, including Stimulated Raman and Brillouin scattering and frequency chirping, will also be discussed.

Table 6.3.1 Datacom vs Telecom Requirements.

	Datacom	Telecom
BER	10e–12 to 10e–15	10e–9e
Distance	20–50 km	Varies with repeaters
No. transceivers/km	Large	Small
Signal bandwidth	00Mb–1Gb	3–5 Kb
Field service	Untrained users	Trained staff
No. fiber replugs	250–500	<100 over lifetime

6.3.4.1 Dispersion

The most important fiber characteristic after transmission loss is dispersion, or intersymbol interference. This refers to the broadening of optical pulses as they propagate along the fiber. As pulses broaden, they tend to interfere with adjacent pulses; this limits the maximum achievable data rate. In multimode fibers, there are two dominant kinds of dispersion: modal and chromatic. Modal dispersion refers to the fact that different modes will travel at different velocities and cause pulse broadening. The fiber's modal bandwidth in units of MHz-km is specified according to the expression

$$BW_{modal} = BW_1/L^{\gamma}, \qquad (6.3.4)$$

where BW_{modal} is the modal bandwidth for a length L of fiber, BW_1 is the manufacturer-specified modal bandwidth of a 1 km section of fiber, and γ is a constant known as the modal bandwidth concatenation length scaling factor. The term γ usually assumes a value between 0.5 and 1, depending on details of the fiber manufacturing and design as well as the operating wavelength; it is conservative to take $\gamma = 1.0$. Modal bandwidth can be increased by mode mixing, which promotes the interchange of energy between modes to average out the effects of modal dispersion. Fiber splices tend to increase the modal bandwidth, although it is conservative to discard this effect when designing a link.

The other major contribution is chromatic dispersion, BW_{chrom}, which occurs because different wavelengths of light propagate at different velocities in the fiber. For multimode fiber, this is given by an empirical model of the form

$$BW_{chrom} = \frac{L^{\gamma c}}{\sqrt{\lambda_W}\,(a_0 + a_1|\lambda_c - \lambda_{eff}|)}, \qquad (6.3.5)$$

where L is the fiber length in km; λ_c is the center wavelength of the source in nm; λ_w is the source FWHM spectral width in nm; λ_c is the chromatic bandwidth length scaling coefficient, a constant; λ_{eff} is the effective wavelength, which combines the effects of the fiber zero-dispersion wavelength and spectral loss signature; and the constants a_1 and a_o are determined by a regression fit of measured data. From Ref. [13], the chromatic bandwidth for 62.5/125 micron fiber is empirically given by

$$BW_{chrom} = \frac{10^4 \, L^{-0.69}}{\sqrt{\lambda_W}\,(1.1 + 0.0189|\lambda_c - 1370|)}. \tag{6.3.6}$$

For this expression, the center wavelength was 1335 nm, and λ_{eff} was chosen midway between λ_c and the water absorption peak at 1390 nm. Although λ_{eff} was estimated in this case, the expression still provides a good fit to the data. For 50/125 micron fiber, the expression becomes

$$BW_{chrom} = \frac{10^4 \, L^{-0.65}}{\sqrt{\lambda_W}\,(1.01 + 0.0177|\lambda_c - 1330|)}. \tag{6.3.7}$$

For this case, λ_c was 1313 nm, and the chromatic bandwidth peaked at $\lambda_{eff} = 1330$ nm. Recall that this is only one possible model for fiber bandwidth [1]. The total bandwidth capacity of multimode fiber BW_t is obtained by combining the modal and chromatic dispersion contributions, according to

$$\frac{1}{BW_t^2} = \frac{1}{BW_{chrom}^2} + \frac{1}{BW_{model}^2}. \tag{6.3.8}$$

Once the total bandwidth is known, the dispersion penalty can be calculated for a given data rate. One expression for the dispersion penalty in dB is

$$P_d = 1.22\left[\frac{\text{Bit Rate (Mb/s)}}{BW_t(\text{MHz})}\right]^2. \tag{6.3.9}$$

For typical telecommunication grade fiber, the dispersion penalty for a 20 km link is about 0.5 dB.

Dispersion is usually minimized at wavelengths near 1310 nm; special types of fiber have been developed that manipulate the index profile across the core to achieve minimal dispersion near 1550 nm, which is also the wavelength region of minimal transmission loss. Unfortunately, this dispersion-shifted fiber suffers from some practial drawbacks, including susceptibility to certain kinds of nonlinear noise and increased interference between adjacent channels in a wavelength multiplexing environment. There is a new type of fiber that minimizes dispersion while reducing the unwanted crosstalk effects, called dispersion optimized fiber. By using a very sophisticated fiber profile, it is possible to minimize dispersion over the entire wavelength range from 1300 to 1550 nm, at the expense of very high loss (around 2 dB/km); this is known as dispersion flattened fiber. Yet another approach is called dispersion conpensating fiber; this fiber is designed with negative dispersion characteristics, so that when used in series with conventional fiber it will offset the normal fiber dispersion. Dispersion compensating fiber has a much narrower core than standard single-mode fiber, which makes it susceptible to nonlinear effects; it is also birefringent and suffers from polarization mode dispersion, in which different states of polarized light propagate with very different group velocities. Note that standard single-mode fiber does not preserve the polarization state of the incident light. There is yet another type of specialty fiber, with asymmetric core profiles, capable of preserving the polarization of incident light over long distances.

By definition, single-mode fiber does not suffer modal dispersion. Chromatic dispersion is an important effect, though, even given the relatively narrow spectral width of most laser diodes. The dispersion of single-mode fiber corresponds to the first derivative of group velocity τ_g with respect to wavelength and is given by

$$D = \frac{d\tau_g}{d\lambda} = \frac{S_o}{4}\left(\lambda_c - \frac{\lambda_o^4}{\lambda_c^3}\right), \tag{6.3.10}$$

where D is the dispersion in ps/(km-nm) and λ_c is the laser center wavelength. The fiber is characterized by its zero-dispersion wavelength, λ_o, and zero-dispersion slope, S_o. Usually, both center wavelength and zero-dispersion wavelength are specified over a range of values; it is necessary to consider both upper and lower bounds in order to determine the worst case dispersion penalty. This can be seen from Fig. 6.3.3, which plots D vs. wavelength for some typical values of λ_o and λ_c; the largest absolute value of D occurs at the extremes of this region. Once the dispersion is determined, the intersymbol interference penalty as a function of link length,

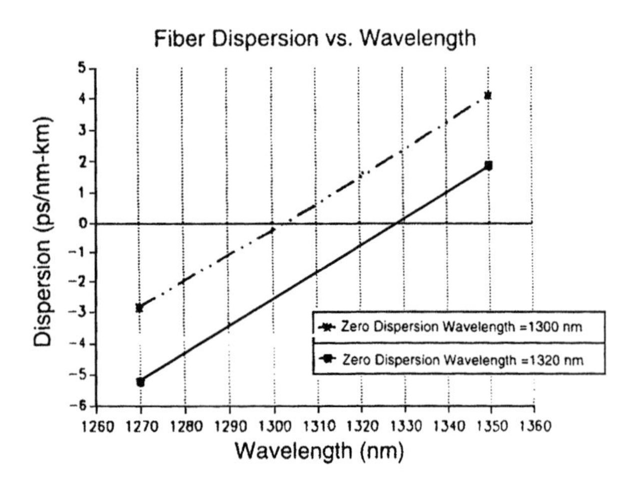

Fig. 6.3.3 Single-mode fiber dispersion as a function of wavelength.

L, can be determined to a good approximation from a model proposed by Agrawal [14]:

$$P_d = 5\log(1 + 2\pi(BD\,\Delta\lambda)^2 L^2), \qquad (6.3.11)$$

where B is the bit rate and $\Delta\lambda$ is the root mean square (RMS) spectral width of the source. By maintaining a close match between the operating and zero-dispersion wavelengths, this penalty can be kept to a tolerable 0.5–1.0 dB in most cases.

6.3.4.2 Mode partition noise

Group velocity dispersion contributes to another optical penalty that remains the subject of continuing research, mode partition noise and mode hopping. This penalty is related to the properties of a Fabry-Perot type laser diode cavity; although the total optical power output from the laser may remain constant, the optical power distribution among the laser's longitudinal modes will fluctuate. This is illustrated by the model depicted in Fig. 6.3.4. When a laser diode is directly modulated with injection current, the total output power stays constant from pulse to pulse; however, the power distribution among several longitudinal modes will vary between pulses. We must be careful to distinguish this behavior of the instantaneous laser spectrum, which varies with time, from the time-averaged spectrum that is normally observed experimentally. The light propagates through a fiber with wavelength-dependent dispersion or attenuation, which deforms the pulse shape. Each mode is delayed by a different amount due to group velocity dispersion in the fiber; this leads to additional signal degradation at the receiver, in addition to the intersymbol interference caused by chromatic dispersion alone, discussed earlier. This is known as mode partition noise; it is capable of generating bit error rate floors, such that additional optical power into the receiver will not improve the link BER. This is because mode partition noise is a function of the laser spectral fluctuations and wavelength-dependent dispersion of the fiber, so the signal-to-noise ratio due to this effect is independent of the signal power. The power penalty due to mode partition noise was first calculated by Ogawa [15] as

$$P_{mp} = 5\log\frac{1}{(1 - Q^2\sigma_{mp}^2)}, \qquad (6.3.12)$$

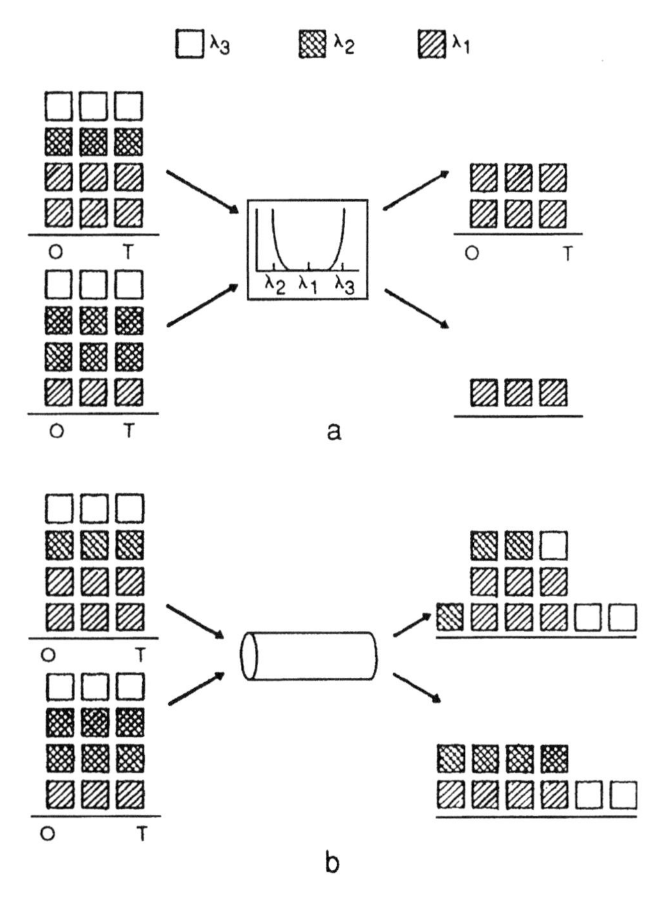

Fig. 6.3.4 Model for mode partition noise; an optical source emits a combination of wavelengths, illustrated by different color blocks: (a) wavelength-dependent loss; (b) chromatic dispersion.

where

$$\sigma^2_{mp} = \frac{1}{2}K^2(\pi B)^4[A_1^4\Delta\lambda^4 + 42A_1^2A_2^2\Delta\lambda^6 + 48A_2^4\Delta\lambda^8], \quad (6.3.13)$$

$$A_1 = DL, \quad (6.3.14)$$

and

$$A_2 = \frac{A_1}{2(\lambda_c - \lambda_o)}. \quad (6.3.15)$$

The mode partition coefficient, *k*, is a number between 0 and 1, which describes how much of the optical power is randomly shared between modes; it summarizes the statistical nature of mode partition noise. According to Ogawa, *k* depends on the number of interacting modes and rms spectral width of the source, the exact dependence being complex. However, subsequent work has shown [16] that Ogawa's model tends to underestimate the power penalty due to mode partition noise because it does not consider the variation of longitudinal mode power between successive baud periods, and because it assumes a linear model of chromatic dispersion rather than the nonlinear model given in the above equation. A more detailed model has been proposed by Campbell [17], which is general enough to include effects of the laser diode spectrum, pulse shaping, transmitter extinction ratio, and statistics of the datastream. While Ogawa's model assumed an equiprobable distribution of zeros and ones in the data stream, Campbell showed that mode partition noise is data dependent as well. Recent work based on this model [18] has re-derived the signal variance:

$$\sigma^2_{mp} = E_{av}(\sigma_0^2 + \sigma_{+1}^2 + \sigma_{-1}^2), \quad (6.3.16)$$

where the mode partition noise contributed by adjacent baud periods is defined by

$$\sigma^2_{+1} + \sigma^2_{-1} = \frac{1}{2}K^2(\pi B)^4[1.25A_1^4\Delta\lambda^4 + 40.95A_1^2A_2^2\Delta\lambda^6 + 50.25A_2^4\Delta\lambda^8], \quad (6.3.17)$$

and the time-average extinction ratio $E_{av} = 10\log(P_1/P_0)$, where P_1, P_0 represent the optical power by a "1" and "0", respectively. If the operating wavelength is far away from the zero-dispersion wavelength, the noise variance simplifies to

$$\sigma^2_{mp} = 2.25\frac{k^2}{2}E_{av}(1 - e^{-\beta L^2})^2, \quad (6.3.18)$$

which is valid provided that

$$\beta = (\pi BD\Delta\lambda)^2 \ll 1. \quad (6.3.19)$$

Many diode lasers exhibit mode hopping or mode splitting in which the spectrum appears to split optical power between two or three modes for brief periods of time. The exact mechanism is not fully understood, but stable Gaussian spectra are generally only observed for CW operation and temperature stabilized lasers. During these mode hops, the above theory does not apply since the spectrum is non-Gaussian, and the model will over-predict the power penalty. Hence, it is not possible to model mode hops as mode partitioning with $k = 1$. There is no currently published model describing a treatment of mode hopping noise, although recent papers [19] suggest approximate calculations based on the statistical properties of the laser cavity. In a practical link, some amount of mode hopping is probably unavoidable as a contributor to burst noise; empirical testing of link hardware remains the only reliable way to reduce this effect. A practical rule of thumb is to keep the mode partition noise penalty less than 1.0 dB maximum, provided that this penalty is far away from any noise floors.

6.3.4.3 Extinction ratio

The receiver extinction ratio also contributes directly to the link penalties. The receiver BER is a function of the modulated AC signal power; if the laser transmitter has a small extinction ratio, the DC component of total optical power is significant. Gain or loss can be introduced in the link budget if the extinction ratio at which the receiver sensitivity is measured differs from the worst case transmitter extinction ratio. If the extinction ratio E_t at the transmitter is defined as the ratio of optical power when a one is transmitted vs. when a zero is transmitted,

$$E_t = \frac{\text{Power}(1)}{\text{Power}(0)}, \quad (6.3.20)$$

then we can define a modulation index at the transmitter M_t according to

$$M_t = \frac{E_t - 1}{E_t + 1}. \quad (6.3.21)$$

Similarly, we can measure the linear extinction ratio at the optical receiver input and define a modulation index M_r. The extinction ratio penalty is given by

$$P_{er} = -10\log\left(\frac{M_t}{M_r}\right), \quad (6.3.22)$$

where the subscripts *T* and *R* refer to specifications for the transmitter and receiver, respectively. Usually, the extinction ratio is specified to be the same at the transmitter and receiver and is large enough so that there is no power penalty due to extinction ratio effects.

Furthermore, the extinction ratio is used to calculate the optical modulation amplitude (OMA), which is sometimes specified in place of receiver sensitivity (for example, in the ANSI Fibre Channel Standard recent revisions). OMA is defined as the difference in optical power between logic levels of 1 and 0; in terms of average optical power (in microwatts) and extinction ratio, it is given by

$$\mathrm{OMA} = 2P_{av}\left(\frac{E-1}{E+1}\right), \tag{6.3.23}$$

where the extinction ratio in this case is the absolute (unitless linear) ratio of average optical power (in microwatts) between a logic level 1 and 0, measured under fully modulated conditions in the presence of worst case reflections. In the Fibre Channel Standard, for example, the OMA specified at 1.0625 Gbit/s for short-wavelength (850 nm) laser sources is between 31 and 2000 microwatts (peak-to-peak), which is equivalent to an average power of −17 dBm and an extinction ratio of 9 dB. Similarly, the OMA specified at 2.125 Gbit/s for short-wavelength (850 nm) laser sources is between 49 and 2000 microwatts (peak-to-peak), which is equivalent to an average power of −15 dBm and an extinction ratio of 9 dB.

6.3.4.4 Multipath interference

Another important property of the optical link is the amount of reflected light from the fiber end faces that return up the link back into the transmitter. Whenever there is a connection or splice in the link, some fraction of the light is reflected back; each connection is thus a potential noise generator, since the reflected fields can interfere with one another to create noise in the detected optical signal. The phenomenon is analogous to the noise caused by multiple atmospheric reflections of radio waves and is known as multipath interference noise. To limit this noise, connectors and splices are specified with a minimum return loss. If there are a total of *N* reflection points in a link and the geometric mean of the connector reflections is alpha, then based on the model of Duff et al. [20] the power penalty due to multipath interference (adjusted for bit error rate and bandwidth) is closely approximated by

$$P_{mpi} = 10\log(1 - 0.7Na). \tag{6.3.24}$$

Multipath noise can usually be reduced well below 0.5 dB with available connectors, whose return loss is often better than 25 dB.

6.3.4.5 Relative intensity noise (RIN)

Stray light reflected back into a Fabry-Perot type laser diode gives rise to intensity fluctuations in the laser output. This is a complicated phenomenon, strongly dependent on the type of laser; it is called either reflection-induced intensity noise or relative intensity noise (RIN). This effect is important because it can also generate BER floors. The power penalty due to RIN is the subject of ongoing research; since the reflected light is measured at a specified signal level, RIN is data dependent, although it is independent of link length. Since many laser diodes are packaged in windowed containers, it is difficult to correlate the RIN measurements on an unpackaged laser with those of a commercial product. Several detailed attempts have been made to characterize RIN [21, 22]. Typically, the RIN noise is assumed Gaussian in amplitude and uniform in frequency over the receiver bandwidth of interest. The RIN value is specified for a given laser by measuring changes in the optical power when a controlled amount of light is fed back into the laser. It is signal dependent, and it is also influenced by temperature, bias voltage, laser structure, and other factors that typically influence laser output power [22]. If we assume that the effect of RIN is to produce an equivalent noise current at the receiver, then the additional receiver noise σ_r may be modeled as

$$\sigma_r = \gamma^2 S^{2g} B, \tag{6.3.25}$$

where *S* is the signal level during a bit period, *B* is the bit rate, and *g* is a noise exponent that defines the amount of signal-dependent noise. If $g=0$, noise power is independent of the signal, while for $g=1$, noise power is proportional to the square of the signal strength. The coefficient γ is given by

$$\gamma^2 = S_i^{2(1-g)} 10^{(RIN_i/10)}, \tag{6.3.26}$$

where RIN_i is the measured RIN value at the average signal level S_i, including worst case backreflection conditions and operating temperatures. The Gaussian BER probability due to the additional RIN noise current is given by

$$P_{error} = \frac{1}{2}\left[P_e^1\left(\frac{S_1 - S_o}{2\sigma_1}\right) + P_e^O\left(\frac{S_1 - S_o}{2\sigma_o}\right)\right] \tag{6.3.27}$$

where σ_1, σ_o represent the total noise current during transmission of a digital 1 and 0, respectively, and P_e^1, P_e^o are the probabilities of error during transmission of a 1 or 0, respectively. The power penalty due to RIN may then be calculated by determining the additional signal power required to achieve the same BER with RIN noise present as without the RIN contribution. One approximation for the RIN power penalty is given by

$$P_{rin} = -5\log\left[1 - Q^2(BW)(1+M_r)^{2g} \times (10^{RIN/10})\left(\frac{1}{M_r}\right)^2\right], \qquad (6.3.28)$$

where the RIN value is specified in dB/Hz, BW is the receiver bandwidth, M_r is the receiver modulation index, and the exponent g is a constant varying between 0 and 1, which relates the magnitude of RIN noise to the optical power level. The maximum RIN noise penalty in a link can usually be kept to below 0.5 dB.

6.3.4.6 Jitter

Although it is not strictly an optical phenomenon, another important area in link design deals with the effects of timing jitter on the optical signal. In a typical optical link, a clock is extracted from the incoming data signal, which is used to retime and reshape the received digital pulse. The received pulse is then compared with a threshold to determine if a digital 1 or 0 was transmitted. So far, we have discussed BER testing with the implicit assumption that the measurement was made in the center of the received data bit; to achieve this, a clock transition at the center of the bit is required. When the clock is generated from a receiver timing recovery circuit, it will have some variation in time and the exact location of the clock edge will be uncertain. Even if the clock is positioned at the center of the bit, its position may drift over time.

There will be a region of the bit interval, or eye, in the time domain where the BER is acceptable; this region is defined as the eyewidth [1–3]. Eyewidth measurements are an important parameter for evaluation of fiber-optic links; they are intimately related to the BER, as well as the acceptable clock drift, pulsewidth distortion, and optical power. At low optical power levels, the receiver signal-to-noise ratio is reduced; increased noise causes amplitude variations in the received signal. These amplitude variations are translated into time domain variations in the receiver decision circuitry, which narrows the eyewidth. At the other extreme, an optical receiver may become saturated at high optical power, reducing the eyewidth and making the system more sensitive to timing jitter. This behavior results in the typical "bathtub" curve shown in Fig. 6.3.2. For this measurement, the clock is delayed from one end of the bit cell to the other, with the BER calculated at each position. Near the ends of the cell, a large number of errors occur; toward the center of the cell, the BER decreases to its true value. The eye opening may be defined as the portion of the eye for which the BER remains constant; pulsewidth distortion occurs near the edges of the eye, which denotes the limits of the valid clock timing. Uncertainty in the data pulse arrival times causes errors to occur by closing the eye window and causing the eye pattern to be sampled away from the center. This is one of the fundamental problems of optical and digital signal processing, and a large body of work has been done in this area [23, 24]. In general, multiple jitter sources will be present in a link; these will tend to be uncorrelated. However, jitter on digital signals, especially resulting from a cascade of repeaters, may be coherent.

International standards on jitter were first published by the CCITT (Central Commission for International Telephony and Telegraphy, now known as the International Telecommunications Union, or ITU). This standards body has adopted a definition of jitter [24] as short-term variations of the significant instants (rising or falling edges) of a digital signal from their ideal position in time. Longer-term variations are described as wander; in terms of frequency, the distinction between jitter and wander is somewhat unclear. The predominant sources of jitter include the following:

- Phase noise in receiver clock recovery circuits, particularly crystal-controlled oscillator circuits; this may be aggravated by filters or other components that do not have a linear phase response. Noise in digital logic resulting from restricted rise and fall times may also contribute to jitter.
- Imperfect timing recovery in digital regenerative repeaters, which is usually dependent on the data pattern.
- Different data patterns, which may contribute to jitter when the clock recovery circuit of a repeater attempts to recover the receive clock from inbound data. Data pattern sensitivity can produce as much as 0.5 dB penalty in receiver sensitivity. Higher data rates are more susceptible (>1Gbit/s); data patterns with long run lengths of 1's or 0's, or with abrupt phase transi tions between consecutive blocks of 1's and 0's, tend to produce worst case jitter.
- At low optical power levels, the receiver signal-to-noise ratio, Q, is reduced; increased noise causes amplitude variations in the signal, which may be translated into time domain variations by the receiver circuitry.
- Low frequency jitter, also called wander, resulting from instabilities in clock sources and modulation of transmitters.
- Very-low-frequency jitter caused by variations in the propagation delay of fibers, connectors, and the like, typically resulting from small temperature variations (making it especially difficult to perform long-term jitter measurements).

In general, jitter from each of these sources will be uncorrelated; jitter related to modulation components of the digital signal may be coherent, and cumulative jitter from a series of repeaters or regenerators may also contain some well-correlated components.

There are several parameters of interest in characterizing jitter performance. Jitter may be classified as either random or deterministic, depending on whether it is associated with pattern dependent effects. These are distinct from the duty cycle distortion that often accompanies imperfect signal timing. Each component of the optical link (data source, serializer, transmitter, encoder, fiber, receiver, retiming/clock recovery/deserialization, decision circuit) will contribute some fraction of the total system jitter. If we consider the link to be a "black box" (but not necessarily a linear system), then we can measure the level of output jitter in the absence of input jitter; this is known as the "intrinsic jitter" of the link. The relative importance of jitter from different sources may be evaluated by measuring the spectral density of the jitter. Another approach is the maximum tolerable input jitter (MTIJ) for the link. Finally, since jitter is essentially a stochastic process, we may attempt to characterize the jitter transfer function (JTF) of the link, or estimate the probability density function of the jitter. When multiple traces occur at the edges of the eye, this can indicate the presence of data-dependent jitter or duty cycle distortion; a histogram of the edge location will show several distinct peaks. This type of jitter can indicate a design flaw in the transmitter or receiver. By contrast, random jitter typically has a more Gaussian profile and is present to some degree in all data links.

The problem of jitter accumulation in a chain of repeaters becomes increasingly complex; however, we can state some general rules of thumb. It has been shown [25] that jitter can be generally divided into two components, one due to repetitive patterns and one due to random data. In receivers with phase-locked loop timing recovery circuits, repetitive data patterns will tend to cause jitter accumulation, especially for long run lengths. This effect is commonly modeled as a second-order receiver transfer function. Jitter will also accumulate when the link is transferring random data. Jitter due to random data is of two types: systematic and random. The classic model for systematic jitter accumulation in cascaded repeaters was published by Byrne [26]. The Byrne model assumes cascaded identical timing recovery circuits, and then the systematic and random jitter can be combined as rms quantities so that total jitter due to random jitter may be obtained. This model has been generalized to networks consisting of different components [27] and to nonidentical repeaters [28]. Despite these considerations, for well designed practical networks the basic results of the Byrne model remain valid for N nominally identical repeaters transmitting random data; systematic jitter accumulates in proportion to $N^{1/2}$ and random jitter accumulates in proportion to $N^{1/4}$. For most applications, the maximum timing jitter should be kept below about 30% of the maximum receiver eye opening.

6.3.4.7 Modal noise

An additional effect of lossy connectors and splices is modal noise. Because high-capacity optical links tend to use highly coherent laser transmitters, random coupling between fiber modes causes fluctuations in the optical power coupled through splices and connectors; this phenomenon is known as modal noise [29]. As one might expect, modal noise is worst when using laser sources in conjunction with multimode fiber; recent industry standards have allowed the use of short-wave lasers (750–850 nm) on 50-micron fiber that may experience this problem. Modal noise is usually considered to be nonexistent in single-mode systems. However, modal noise in single-mode fibers can arise when higher order modes are generated at imperfect connections or splices. If the lossy mode is not completely attenuated before it reaches the next connection, interference with the dominant mode may occur. The effects of modal noise have been modeled previously [29], assuming that the only significant interaction occurs between the LP01 and LP11 modes for a sufficiently coherent laser. For N sections of fiber, each of length L in a single-mode link, the worst case sigma for modal noise can be given by

$$\sigma_{\mathrm{m}} = \sqrt{2}\mathrm{N}\eta(1-\eta)\mathrm{e}^{-\mathrm{aL}}, \tag{6.3.29}$$

where α is the attenuation coefficient of the LP_{11} mode, and η is the splice transmission efficiency, given by

$$\eta = 10^{-(\eta_o/10)}, \tag{6.3.30}$$

where η_o is the mean splice loss (typically, splice transmission efficiency will exceed 90%). The corresponding optical power penalty due to modal noise is given by

$$\mathrm{P} = -5\log\left(1-\mathrm{Q}^2\sigma_{\mathrm{m}}^2\right), \tag{6.3.31}$$

where Q corresponds to the desired BER. This power penalty should be kept to less than 0.5 dB.

6.3.4.8 Radiation-induced loss

Another important environmental factor, as mentioned earlier, is exposure of the fiber to ionizing radiation damage. There is a large body of literature concerning the effects of ionizing radiation on fiber links [30, 31]. Many factors can affect the radiation susceptibility of optical fiber, including the type of fiber, type of radiation (gamma radiation is usually assumed to be representative), total dose, dose rate (important only for higher exposure levels), prior irradiation history of the fiber, temperature, wavelength, and data rate. Optical fiber with a pure silica core is least susceptible to radiation damage. However, almost all commercial fiber is intentionally doped to control the refractive index of the core and cladding, as well as dispersion properties. Trace impurities are also introduced, which become important only under irradiation. Among the most important are Ge dopants in the core of graded-index (GRIN) fibers, in addition to F, Cl, P, B, OH content, and the alkali metals. In general, radiation sensitivity is worst at lower temperatures and is also made worse by hydrogen diffusion from materials in the fiber cladding. Because of the many factors involved, there does not exist a comprehensive theory to model radiation damage in optical fibers.

The basic physics of the interaction has been described [30, 31]; there are two dominant mechanisms, radiation-induced darkening and scintillation. First, high-energy radiation can interact with dopants, impurities, or defects in the glass structure to produce color centers that absorb strongly at the operating wavelength. Carriers can also be freed by radiolytic or photochemical processes. Some of these become trapped at defect sites, which modifies the band structure of the fiber and causes strong absorption at infrared wavelengths. This radiation-induced darkening increases the fiber attenuation. In some cases, it is partially reversible when the radiation is removed, although high levels or prolonged exposure will permanently damage the fiber. A second effect is caused if the radiation interacts with impurities to produce stray light, or scintillation. This light is generally broadband, but will tend to degrade the BER at the receiver; scintillation is a weaker effect than radiation-induced darkening. These effects will degrade the BER of a link; they can be prevented by shielding the fiber, or can be partially overcome by a third mechanism, photobleaching. The presence of intense light at the proper wavelength can partially reverse the effects of darkening in a fiber. It is also possible to treat silica core fibers by briefly exposing them to controlled levels of radiation at controlled temperatures; this increases the fiber loss, but makes the fiber less susceptible to future irradiation. These so-called radiation hardened fibers are often used in environments where radiation is anticipated to play an important role. Recently, several models have been advanced [31] for the performance of fiber under moderate radiation levels; the effect on BER is a power law model of the form

$$\mathrm{BER} = \mathrm{BER}_o + A(\mathrm{dose})^b, \qquad (6.3.32)$$

where BER_0 is the link BER prior to irradiation, the dose is given in rads, and the constants A and *b* are empirically fitted. The loss due to normal background radiation exposure over a typical link lifetime can be held below about 0.5 dB.

6.3.4.9 Nonlinear noise effects

Except for those penalties that produce BER floors such as mode partitioning and RIN, most penalties can be reduced by increasing the transmitted or received optical power. This brute-force approach is subject to limitations such as maintaining class 1 laser safety; even if it were possible to increase optical power significantly (as will be discussed later using optical amplifiers), a new class of nonlinear optical penalties becomes important at high-power levels. Class 1 laser systems typically do not experience these nonlinear effects, although they may be important for optical fiber amplifiers or systems using open fiber control (OFC), for which significantly higher power levels may be present in the fiber.

At higher-optical power levels, nonlinear scattering may limit the behavior of a fiber-optic link. The dominant effects are stimulated Raman and Brillouin scattering. When incident optical power exceeds a threshold value, significant amounts of light may be scattered from small imperfections in the fiber core or by mechanical (acoustic) vibrations in the transmission media. These vibrations can be caused by the high-intensity electromagnetic fields of light concentrated in the core of a single-mode fiber. Because the scattering process also involves the generation of photons, the scattered light can be frequency shifted. Put another way, we can think of the high-intensity light as generating a regular pattern of very slight differences in the fiber refractive index. This creates a moving diffraction grating in the fiber core, and the scattered light from this grating is Doppler shifted in frequency by about 11 GHz. This effect is known as stimulated Brillouin scattering (SBS); under these conditions, the output light intensity becomes nonlinear as well. Stimulated Brillouin scattering will not occur below the optical power threshold defined by

$$\mathrm{Pc} = 21A/G_b L_c \text{ watts}, \qquad (6.3.33)$$

where L_c is the effective interaction length, A is the cross-sectional area of the guided mode, and G_b is the Brillouin gain coefficient. Brillouin scattering has been observed in single-mode fibers at wavelengths greater than cutoff with optical power as low as 5mW. It can be a serious problem in long-distance communication systems when the span between amplifiers is low and the bit rate is less than about 2 Gbit/s, in WDM systems up to about 10 Gbit/s when the spectral width of the signal is very narrow, or in remote pumping of some types of optical amplifiers. In general, SBS is worse for narrow laser linewidths (and is generally not a problem for channel bandwidth greater than 100 MHz), wavelength (SBS is worse near 1550 nm than near 1300 nm), and signal power per unit area in the fiber core. All of these factors are summarized in the above expression.

SBS can be a concern in long-distance communication systems, when the span between amplifiers is large and the bit rate is below about 2.5 Gbit/s, in WDM systems where the spectral width of the signal is very narrow, and in remote pumping of optical amplifiers using narrow linewidth sources. In cases where SBS could be a problem, the source linewidth can be intentionally broadened by using an external modulator or additional RF modulation on the laser injection current. However, this is a tradeoff against long-distance transmission, as broadening the linewidth also increases the effects of chromatic dispersion. When the scattered light experiences frequency shifts outside the acoustic phonon range, due instead to modulation by impurities or molecular vibrations in the fiber core, the effect is known as stimulated Raman scattering (SRS). The mechanism is similar to SBS, and scattered light can occur in both forward and backward directions along the fiber. The threshold below which Raman scattering will not occur is given by

$$P_t = 16A/G_R L_c \text{ watts}, \qquad (6.3.34)$$

where G_R is the Raman gain coefficient [54]. Note that SRS is also influenced by fiber dispersion and that standard fiber reduces the effect of SRS by half (3 dB) compared with dispersion-shifted fiber.

As a rule of thumb, the optical power threshold for Raman scattering is about three times larger than for Brillouin scattering. Another good rule of thumb is that SRS can be kept to acceptable levels if the product of total power and total optical bandwidth is less than 500 GHz-W. This is quite a lot; for example, consider a 10-channel DWDM system with standard wavelength spacing of 1.6 nm (200 GHz). The bandwidth becomes $200 \times 10 = 2000$ GHz, so the total power in all 10 channels would be limited to 250 mW in this case (in most DWDM systems, each channel will be well below 10 mW for other reasons such as laser safety considerations). In single-mode fiber, typical thresholds for Brillouin scattering are about 10 mW and for Raman scattering about 35 mW. These effects rarely occur in multimode fiber, where the thresholds are about 150 mW and 450 mW, respectively. In general, the effect of SRS becomes greater as the signals are moved further apart in wavelength (within some limits); this introduces a tradeoff with FWM, which is reduced as the signal spacing increases. Optical amplifiers can be constructed using the principle of SRS. If a pump signal with relatively high power (half a watt or more) and a frequency 13.2 THz higher than the signal frequency is coupled into a sufficiently long length of fiber (about 1 km), then amplification of the signal will occur. Unfortunately, more efficient amplifiers require that the signal and pump wavelengths be spaced by almost exactly the Raman shift of 13.2 THz. Otherwise the amplification effect is greatly reduced.

It is not possible to build high-power lasers at arbitrary signal wavelengths; one possible solution is to build a pump laser at a convenient wavelength, and then wavelength shift the signal by the desired amount. However, another good alternative to SRS amplifiers are the widely used erbium doped fiber amplifiers (EDFAs). These allow the amplification of optical signals along their direction of travel in a fiber, without the need to convert back and forth from the electrical domain. While there are other types of optical amplifiers based on other rare-earth elements such as praseodymium (Pd) or neodymium (Nd), and even some optical amplifiers based on semiconductor devices, the erbium doped amplifiers are the most widely used because of their maturity and good performance at wavelengths of interest near 1550 nm. Note that there is still a good deal of research going on in this area, most recently including thulium-doped fiber amplifiers (TDFAs) based on flurozirconiate (ZBLAN) fiber for use at wavelengths between 1450 and 1510 nm(theS-band).

The final nonlinear effect we will consider is frequency chirping of the optical signal. Chirping refers to a change in frequency with time. It takes its name from the sound of an acoustic signal whose frequency increases or decreases linearly with time. There are three ways in which chirping can affect a fiber-optic link. First, the laser transmitter can be chirped as a result of physical processes within the laser the effect has its origin in carrier-induced refractive index changes, making it an inevitable consequence of high-power direct modulation of semiconductor lasers. For lasers with low levels of relaxation oscillation damping, a model has been proposed for the chirped power penalty:

$$P = 10 \log\left(1 + \frac{\pi B^2 \lambda^2 L D \alpha}{4c}\right), \qquad (6.3.35)$$

where c is the speed of light, B is the fiber bandwidth, λ is the wavelength of light, L is the length of the fiber, D is the dispersion, and α is the linewidth enhancement factor (a typical value is -4.5). This model is only a first approximation because it neglects the dependence of chirp on extinction ratio and nonlinear effects such as spectral hole burning. Second, a sufficiently intense light pulse will be chirped by the nonlinear process of self-phase modulation in an optical fiber. The effect arises from the interaction of the light and the intensity-dependent portion of the fiber's refractive index. It is thus dependent on the material and structure of the fiber, polarization of the light, and shape of the incident optical pulse. Based on a model from Ref, the maximum optical power level before the spectral width increases by 2 nm is given by

$$P = \frac{n^2 A}{377 n_2 \kappa L_e} \text{ watts,} \tag{6.3.36a}$$

where

$$L_e = (1/a_0)(1 - \exp(-a_0 L)), \tag{6.3.36b}$$

where n is the fiber's refractive index, κ is the propagation constant (a typical value is 7×10^4 at a wavelength of 1.3 μm) a_0 is the fiber attenuation coefficient, A is the fiber core cross-sectional area, and n_2 is the nonlinear coefficient of the fiber's refractive index (a typical value is 6.1×10^{-19} and L_e is the effective interaction length for the nonlinear interaction, which is related to the actual length of the fiber, L, by Eq. (6.3.36). This expression should be multipled by 2 if the fiber is not polarization preserving. Typically, this effect is not significant for optical power levels less than 950 mW in a single-mode fiber at 1.3 μm wavelength. Finally, there is a power penalty arising from the propagation of a chirped optical pulse in a dispersive fiber because the new frequency components propagate at different group velocities. This may be treated as simply a much worse case of the conventional dispersion penalty, provided that one of the first two effects exists to chirp the optical signal.

6.3.5 Gigabit ethernet link budget model

When the IEEE 802.3z committee began investigating high-data-rate Ethernet links in the late 1990s, a model was developed to predict the performance of laser-based multimode optical fiber links and potential tradeoffs between the various link penalties as part of preparing the link specifications [32, 33]. This model was first released to the public in 2001, has been enhanced periodically over the years, and now forms the basis for the gigabit Ethernet and 10 gigabit Ethernet specifications, as well as other multimode fiber standards including InfiniBand and several types of parallel optical links. (Note that this is not a transceiver design tool; rather it provides a framework for comparing different link design options.)

As this model has evolved over the years, there have been many updates, including the addition of single-mode specifications and polarization mode dispersion penalties, deterministic jitter, baseline wander, and other features. For the purposes of this chapter, we will review only the basic model, following the description in Ref. [32] (a spreadsheet implementation of the latest model with a change history is available from the IEEE [33]). The basic model is an extension of previously reported work on LED-based links [34, 35]. In addition to considering loss due to fiber attenuation, connectors, and splices, power penalties are calculated to account for the effects of intersymbol interference (ISI) [36], mode partition noise [37], extinction ratio, modal noise [38], and relative intensity noise (RIN). Since all power penalties except ISI are assumed to be independent of the link length, the ISI penalty dominates at longer distances and, along with link attenuation, determines the maximum link length for Ethernet applications.

This model assumes that the laser and multimode fiber impulse responses are Gaussian and that the optical receiver is nonequalized, with a single-pole filter having a specified 3 dB electrical bandwidth. Variations on this model have been published with results for the case of a nonequalized receiver having a raised cosine response [35] and for the case where the receiver has an exponential impulse response [34, 35].

The model includes expressions that convert the root mean square (rms) impulse response of the laser, fiber, and optical receiver to rise times, fall times, and bandwidths, which are used to determine the fiber and composite channel output impulse response and the ISI penalty. It is assumed that rise and fall times are equal; to be conservative when modeling real components, the larger of the experimentally measured rise or fall time should be used.

First, we will consider the effects of rise/fall times, pulsewidth, and bandwidth; by calculating these factors, we can estimate the resulting eye diagrams for various link designs. Note that the Gigabit and 10 Gigabit Ethernet standards define two types of receiver sensitivity, namely, "normal" sensitivity (measured with a typical or good performance transmitter) and "stressed" sensitivity (measured with a transmitter having worst case jitter performance). While the stressed sensitivity is intended to prove interoperability, it is not required in order to use this link model. It has been shown [39] that for two positive signal pulses h1(t) and h2(t), if their convolution is represented by a signal h3(t) such that

$$h3(t) = h1(t)^{*}\, h2(t), \tag{6.3.37}$$

then

$$\vartheta_3 = \sqrt{\vartheta_1^2 + \vartheta_2^2}, \tag{6.3.38}$$

where σi is the rms pulsewidth of the individual components. The 10% to 90% rise time, T_i and bandwidth of individual components, BW_i, are related by constant conversion factors, a_i and b_i such that:

$$\vartheta_3\ (\mathrm{BW}) = \frac{a_i}{BW_i}, \qquad (6.3.39)$$

and

$$\vartheta_i\ (T) = \frac{T_i}{b_i}, \qquad (6.3.40)$$

therefore

$$T_i = \frac{a_i b_i}{BW_i(6dB)}, \qquad (6.3.41)$$

where BW(6dB) is the 6 dB electrical bandwidth (equivalent to the 3 dB optical bandwidth). Since equation 6.3.40 can be generalized to include the sum of an arbitrary number of signals, the rms pulsewidths of the individual components may be used to calculate the bandwidth or the 10% to 90% rise time of the composite system if the appropriate conversion factors for each individual component are known [35]. For example, the overall system rise time, T_s, may be calculated using:

$$T_s^2 = \sum_i \left(\frac{b_s T_i}{b_i}\right)^2 = \sum_i \left(\frac{a_i b_i}{BW_i(6dB)}\right)^2 = \sum_i \left(\frac{C_i}{BW_i(6dB)}\right)^2. \qquad (6.3.42)$$

Using this approach and the central limit theorem, it can be shown that the composite impulse response of multimode fiber-optic links tend to a Gaussian impulse response [2]. For these systems, it can be shown that the conversion factors a and b are equal to 0.187 and 2.563, respectively, so that C = 0.48 [35]. These conversion factors can be applied to the laser and fiber.

The simplest form of optical receiver is a nonequalized receiver with a single-pole filter. This type of receiver can be modeled by an exponential impulse response of the form [35]:

$$hr(t) = \frac{1}{\tau}\exp(-t/\tau), \qquad (6.3.43)$$

for t greater than or equal to zero; $hr(t)$ is zero otherwise. Here, τ is called the rise time constant; this impulse response has an rms width of τ. If the receiver is excited by a step function, then the 10% to 90% rise time of the source is [35]:

$$tr = \mathrm{In}(9).\tau, \qquad (6.3.44)$$

and the 3 dB bandwidth is [35]:

$$BW_r\ (3dB) = 0.1588/\tau = 01588/\vartheta. \qquad (6.3.45)$$

By substitution we have

$$tr = \mathrm{In}(9)\left(\frac{0.1588}{BW_r(dB)}\right). \qquad (6.3.46)$$

Therefore, a = 0.1588 and b = In(9), for a component or system with an exponential impulse response. With the assumption that the fiber exit impulse response is Gaussian, we can calculate the fiber 10% to 90% exit response time (Te):

$$T_{exit} = \sqrt{\left(\frac{C_1}{BW_m}\right)^2 + \left(\frac{C_1}{BW_{ch}}\right)^2 + T_s^2}, \qquad (6.3.47)$$

where BW_m is the 3 dB optical modal bandwidth of the fiber link, BW_{ch} is the 3 dB chromatic bandwidth of the fiber link, and T_s is the 10% to 90% laser rise time. If we assume that the fiber has a Gaussian response, $C_1 = 0.48$. The approximate 10% to 90% composite channel exit response time (T_c) is then:

$$T_{exit} = \sqrt{\left(\frac{C_1}{BW_m}\right)^2 + \left(\frac{C_1}{BW_{ch}}\right)^2 + T_s^2\left(\frac{0.4}{BW_r}\right)^2}. \qquad (6.3.48)$$

Note that if a raised cosine receiver is used, the last term in Eq. (6.3.48) would be (0.35/BWr). It can be shown [32, 33] that for a channel having a Gaussian impulse response, the ISI penalty is approximated by

$$P_{ISI} = \frac{1}{1 - 1.425\exp(-1.28T/T_c)^2}, \qquad (6.3.49)$$

where T is the bit period. This equation is used in the spreadsheet implementations of the model [33]. Experimental results have shown [32] that this model predicts worst case ISI power penalties up to 5 dB within 0.3 dB of the exact solution (the maximum allocation for the original optical Ethernet link budget is 3 dB). For higher power penalties (less than 20 dB), the model is accurate to within about 1 dB, with an uncertainty of approximately 10% in link length. For these power penalties, the uncertainty of the measured eye center power penalty increases significantly due to increased timing jitter.

The various wavelength components of a laser output will travel at slightly different velocities through a fiber. If the power in each laser mode remained constant, then BW_{ch}, duc to thc lascr time averaged spectrum, would accurately account for chromatic dispersion-induced ISI. However, in a multimode laser, although the total output power is constant, the power in each laser mode is not constant. As a result, power fluctuations between laser modes leads to an additional ISI component. This is usually referred to as mode partition noise [37]; the Gigabit Ethernet link model uses the same expression for MPN power penalty as discussed in the previous section.

The chromatic dispersion of the multimode fiber, in MHz, is given in this model by [34, 35]:

$$BW_{ch} = \frac{0.187}{L\vartheta\lambda}\frac{1}{\sqrt{D_1^2 + D_2^2}}, \quad (6.3.50)$$

where D_1 is the worst case chromatic dispersion, and

$$D_2 = 0.7S_0\vartheta\lambda. \quad (6.3.51)$$

The extinction ratio power penalty (associated with transmitting a nonzero power level for a zero) is given in this model by [39]:

$$P_E = \frac{1+E}{1-E}, \quad (6.3.52)$$

where E is defined as the laser extinction ratio (i.e., the ratio of the power when a zero is sent to the power when a one is sent).

The worst case noise variance, σ_{rin}^2, due to laser RIN can be calculated using the following equation:

$$\vartheta_{RIN}^2 = 4BW_r(3dB)10^{RIN/10}, \quad (6.3.53)$$

where RIN is the laser RIN in dB/Hz and it has been assumed that the RIN is worst during transmission of a one, so that the peak laser power is used to calculate the noise variance. Assuming equiprobable symbols, zero-transmitted power on a logical "zero," and unity photodiode responsivity, the peak detected electrical power will be four times the average detected electrical power due to square law detection; hence the factor of four in Eq. (6.3.55) for the RIN variance. The worst case RIN-induced power penalty is then given by the same form described previously in this chapter, where the RIN noise variance replaces the mode partition noise variance.

Since the attenuation of optical fiber decreases as a power of the wavelength, this model calculates attenuation according to the expression [34]:

$$\text{Attenuation} = \frac{A_{ref}L}{\lambda_c^{3.2}}, \quad (6.3.54)$$

where A_{ref} is the fiber attenuation (in dB) at the reference wavelength λ_{ref}, λ_c is the laser center wavelength (in nm), and L is the link length (in km).

The worst case power budget is the difference between the minimum allowed laser launch power and the maximum allowed receiver sensitivity at the specified BER. The power budget can be calculated and plotted as a function of link length. If the summation of the worst case power losses and penalties is less than the power budget, then the link will remain within specification. Since some of the power penalties and losses vary with link length, there will be a maximum link length that can be supported when all penalties and losses are set to their worst case values. As this model is extended to 10G Ethernet and other types of highspeed links, ongoing enhancements are made. These include improving the accuracy of MPN noise calculations.

6.3.6 Link budgets with optical amplification

The principles we have used thus far apply to the design of point-to-point optical communication links, which may be either loss-limited or dispersion-limited. For loss-limited systems, the maximum transmitted optical power places a fundamental limit on the BER or Optical Signal to Noise Ratio (OSNR) that can be achieved at a given distance. It may not be practical to increase the transmitter optical power beyond certain limits (for example, due to laser eye safety considerations or generation of nonlinear effects within the fiber). One approach to achieving longer links is the placement of regenerating equipment (switches, routers, or other devices) that performs optical to electrical conversion. This amounts to breaking a long link into several shorter links, each with a more manageable link budget. However, for some types of systems it is possible to avoid this constraint and directly amplify the optical signal. Although optical amplifiers can be designed to operate at various wavelengths (see Chapter 15), a common example is the use of erbium doped fiber amplifiers (EDFAs) in long-haul communication links and wavelength multiplexing systems. Optically amplified systems can be used for data communication applications such as disaster recovery and grid computing; optical amplifiers may also be required on shorter distance links with high attenuation. We will briefly describe a common link budget design approach for these systems.

Consider a long-distance optical link (typically >100 km), with optical amplifiers placed periodically along its length to boost the signal power. We assume that the amplifiers are equally spaced, dividing the link into segments of equal length. However, both the signal and noise are amplified at each link segment. Furthermore, each amplifier adds its own component of noise (called amplified spontaneous emission, or ASE), which further degrades the OSNR. It is possible to design the system to produce a desired OSNR at the end of the final link segment. It can be shown [40–42] that the OSNR for each link segment is given by

$$\mathrm{OSNR} = \frac{P_{in}}{(NF)\, h\nu(\Delta f)\Gamma N}, \tag{6.3.55}$$

where (NF) is the noise figure of the amplifier (i.e., the amplifier output when there is no input), h is Planck's constant, ν is the optical frequency, N is the number of amplifiers, Γ is the loss of one link segment, or span loss, and Δf is the bandwidth used to measure the noise factor (typically, 0.1 nm or 12.5 GHz). Since each span is of equal length, we assume that all spans have the same loss; furthermore, we assume that all amplifiers have the same noise factor. These are reasonably good assumptions based on the uniformity of fiber and components available; either assumption can be changed in order to improve the design accuracy. The total OSNR for the system is given by a reciprocal sum of the OSNR for each link segment:

$$\frac{1}{\mathrm{OSNR}_{final}} = \sum_{i=1}^{N} \frac{1}{\mathrm{OSNR}_i}. \tag{6.3.56}$$

Taking the common log (base 10) of both sides of Eq. (6.3.58) to convert into dB, and assuming the typical value for Δf, yields

$$\mathrm{OSNR}_{dB} = 58 + P_{in} - \Gamma_{dB} - NF_{dB} - 10 \log N. \tag{6.3.57}$$

The expression is typically written in this form because both the span loss and noise factor are specified in dB, rather than in linear form, so they do not need to be converted. Equation (6.3.57) provides a useful approximation to the system OSNR. Additional loss can be subtracted from the right-hand side of this expression to account for other factors, such as gain flatness and gain tilt of the amplifiers, and polarization dependent noise. In a multiwavelength system, the design should be based on the OSNR for the worst wavelength in the system (this is sometimes assumed to be the first or last wavelength). Alternatively, some designs assume that optical power and noise are uniformly divided across all wavelengths of the system, and either divide the total power by the number of wavelengths or multiply the total noise by the number of wavelengths. Various forms of this expression are used in the literature, depending on the assumptions made in the link design [40–42].

Case study WDM link budget design

There has always been a need for long-distance, disaster recovery networks in the data communications industry. These may be used for extending storage area networks, enabling grid computing applications, or as part of the ongoing convergence between data communication and telecommunications. A common design problem involves determining whether a wavelength-division multiplexing (WDM) system, used as a protocol independent channel extension, can operate with sufficient fidelity (low enough bit error rate). Since these links are commonly loss limited rather than dispersion limited, we can estimate the requirements from an OSNR.

Consider as an example a WDM link 80 km long, originally designed for use with telecommunication systems, which is being repurposed for an extended distance datacom link. Since the link loss is too great for a point-to-point WDM system, it has been divided into four equal segments, each having 25 dB loss, using optical amplifiers between each span. Each amplifier has a fixed gain of 22 dB and a noise factor of 5 dB. We further assume that the wavelengths are closely enough spaced that their behavior can be approximated by an average wavelength of around 1550 nm with a small spectral width (25 kHz). Using this information, we can determine that the output OSNR for this link is 27 dB. If we attempt to connect a piece of datacom equipment with a lower receiver sensitivity (say, 25 dB), then the span will not function despite the use of several optical amplifiers in the path. This illustrates the importance of not only computing the OSNR, but determining if it is adequate for the system to be used.

REFERENCES

1. S. E. Miller, and A. G. Chynoweth, eds. *Optical fiber telecommunications*. New York: Academic Press (1979).
2. J. Gowar, *Optical communication systems*. Englewood Cliffs, N.J.: Prentice Hall (1984).
3. C. DeCusatis, ed. 1998, December. *Handbook of fiber optic data communication*. New York: Elsevier/ Academic Press, (second edition 2002); see also *Optical Engineering* special issue on Optical Data Communication.
4. R. Lasky, U. Osterberg, and D. Stigliani, eds. *Optoelectronics for data communication*. New York: Academic Press (1995).
5. Digital video broadcasting (DVB) Measurement Guidelines for DVB systems, European Tele-communications Standards Institute ETSI Technical Report ETR 290, May 1997; Digital Multi-Programme Systems for Television Sound and Data Services fo rCable Distribution, International Telecommunications Union ITU-T Recommendation J.83,1995; Digital Broadcasting System for Television, Sound and Data Services; Framing Structure, Channel Coding and

Modulation for Cable Systems, European Telecommunications Standards Institute ETSI 300 429, 1994.
6. W. E. Stephens, and T. R. Hoseph. System characteristics of direct modulated and externally modulated RF fiber-optic Links. *IEEE J. Lightwave Technol.* LT-5(3):380–387(1987).
7. C. H. Cox, III and E. I. Ackerman. Some limits on the performance of an analog optical link. *Proceedings of the SPIE—The International Society for Optical Engineering.* 3463:2–7 (1999).
8. United States laser safety standards are regulated by the Department of Health and Human Services (DHHS), Occupational Safety and Health Administration (OSHA), and Food and Drug Administration (FDA) Code of Radiological Health (CDRH) 21 Code of Federal Regulations (CFR) subchapter J; the relevant standards are ANSI Z1 36.1, "Standard for the safe use of lasers" (1993 revision) and ANSI Z136.2, "Standard for the safe use of optical fiber communication systems utilizing laser diodes and LED sources" (1996-1997 revision); elsewhere in the world, the relevant standard is International Electrotechnical Commission (IEC/CEI) 825 (1993 revision).
9. S. S. Walker, Rapid modeling and estimation of total spectral loss in optical fibers. *IEEE Journ. Lightwave Tech.* 4:1125–1132 (1996).
10. Electronics Industry Association/ Telecommunications Industry Association (EIA/TIA) commercial building telecommunications cabling standard (EIA/TIA-568-A), Electronics Industry Association/ Telecommunications Industry Association (EIA/TIA) detail specification for 62.5 micron core diameter/125 micron cladding diameter class 1a multimode graded index optical waveguide fibers (EIA/ TIA-492AAAA), Electronics Industry Association/Telecommunications Industry Association (EIA/TIA) detail specification for class IV-a dispersion unshifted single-mode optical waveguide fibers used in communication s systems (EIA/ TIA-492BAAA), Electronics Industry Association, New York, N.Y.
11. D. Gloge, Propagation effects in optical fibers. *IEEE Trans. Microwave Theory and Tech.* MTT-23:106–120 (1975).
12. P. M. Shanker, Effect of modal noise on single-mode fiber-optic network. *Opt. Comm.* 64:347–350 (1988).
13. J. J. Refi, LED bandwidth of multimode fiber as a function of source bandwidth and LED spectral characteristics. *IEEE Journ. of Lightwave Tech.* LT-14:265–272 (1986).
14. G. P. Agrawal, et al. Dispersion penalty for 1.3 micron lightwave systems with multimode semiconductor lasers. *IEEE Journ. Lightwave Tech.* 6:620–625 (1988).
15. K. Ogawa, Analysis of mode partition noise in laser transmission systems. *IEEE Journ. Quantum Elec.* QE-18: 849–9855 (1982).
16. K. Ogawa, Semiconductor laser noise; mode partition noise, in *Semiconductors and Semi-metals*, Vol. 22C, R.K. Willardson and A. C. Beer, (eds.). New York: Academic Press (1985).
17. J. C. Campbell, Calculation of the dispersion penalty of the route design of single-mode systems. *IEEE Journ. Lightwave Tech.* 6:564–573 (1988).
18. M. Ohtsu, et al. Mode stability analysis of nearly single-mode semiconductor laser. *IEEE Journ. Quantum Elec.* 24: 716–723 (1988).
19. M. Ohtsu, and Y. Teramachi, Analysis of mode partition and mode hopping in semiconductor lasers. *IEEE Quantum Elec.* 25:31–38 (1989).
20. D. Duff, et al. Measurements and simulations of multipath interference for 1.7 Gbit/s lightwave systems utilizing single and multifrequency lasers. *Proc. OFC:* 128 (1989).
21. J. Radcliffe, Fiber optic link performance in the presence of internal noise sources. *IBM Technical Report.* Endicott, N.Y.: Glendale Labs (1989).
22. L. L. Xiao, C. B. Su, and R. B. Lauer. Increae in laser RIN due to asymmetric nonlinear gain, fiber dispersion, and modulation. *IEEE Photon. Tech. Lett.* 4: 774–777 (1992).
23. P. Trischitta, and P. Sannuti. The accumulation of pattern dependent jitter for a chain of fiber optic regenerators. *IEEE Trans. Comm.* 36: 761–765 (1988).
24. CCITT Recommendations G.824, G. 823,O.171, and G.703 on timing jitter in digital systems (1984).
25. R. J. S. Bates, A model for jitter accumulation in digital networks. *IEEE Globecom Proc.:* 145–149 (1983).
26. C. J. Byrne, B. J. Karafin, and D. B. Robinson Jr. Systematic jitter in a chain of digital regenerators. *Bell System Tech. Journal.* 43:2679–2714 (1963).
27. R. J. S. Bates, and L. A. Sauer. Jitter accumulation in token passing ring LANs. *IBM Journal Research and Development* 29:580–587 (1985).
28. C. Chamzas, Accumulation of jitter: a stochastic model. *AT&T Tech. Journal:* 64 (1985).
29. D. Marcuse, and H. M. Presby. Mode coupling in an optical fiber with core distortion. *Bell Sys. Tech. Journal.* 1:3 (1975).
30. E. J. Frieble, et al. Effect of low dose rate irradiation on doped silica core optical fibers. *App. Opt.* 23:4202–4208 (1984).
31. J. B. Haber, et al. Assessment of radiation induced loss for AT&T fiber-optic transmission systems in the terestrial environment. *IEEE Journ. Lightwave Tech.* 6:150–154 (1988).
32. D. Cunningham, M. Noel, D. Hanson, and L. Kazofsky. IEEE802.3z worst case link model. see http://www.ieee802. org/3/z/public/presentations/ mar1997/DCwpaper.pdf
33. IEEE 802.3ae 10G Ethernet optical link budget spreadsheet available from http://www.ieee802. org/3/ae/public/ adhoc/serial_pmd/documents/
34. ANSI T1.646-1995, Broadband ISDN-Physical Layer Specification For User-Network Interfaces, Appendix B.
35. G. D. Brown, Bandwidth and rise time calculations for digital multimode fiber-optic data links. *Journal of Lightwave Technology* 10, no. 5: 672–678 (1992, May).
36. J. L. Gimlett, and N. K. Cheung. Dispersion penalty analysis for LED/ single-mode fiber transmission systems. *Journal of Lightwave Technology.* LT-4, no.9:1381–1392 (1986, September).
37. P. Govind, Agrawal, P. J. Anthony, and T. M. Shen. Dispersion penalty for 1.3-mm lightwave systems with multimode semiconductor lasers. *Journal of Lightwave Technology* 6, no. 5:620–625 (1988, May).
38. R. J. S. Bates, D. M. Kuchta, and K. P. Jackson. Improved multimode fiber link BER calculations due to modal noise and non self-pulsating laser diodes. *Optical and Quantum Electronics* 27:203–224 (1995).
39. R. G. Smith, and S. D. Personick. Receiver design for optical communication systems. In *Topics in Applied Physics*, Vol. 39, *Semiconductor Devices for Optical Communications*, H. Kressel (ed.). Berlin: Springer-Verlag (1982).
40. A. Gumaste, and T. Anthony. DWDM network designs and engineering solutions. Indianapolis, Ind.: Cisco Press (2003).

41. J. W. Seeser, Current topics in fiber-optic communications. http://www.comsoc.org/stl/presentations%5COGSA%20presentation.ppt

42. P. C. Becker, N. A. Olssen, and J.R. Simpson. *Erbium-doped fiber amplifiers: fundamentals and technology.* New York: Academic Press (1999).

ADDITIONAL REFERENCE MATERIAL:

D. Hanson and D. Cunningham, http://www.ieee802.org/3/10G_study/public/email_attach/All_1250 .xls

Petrich, Methodologies for Jitter Specification, Rev 10.0, ftp://ftp.t1 1.org/t11/pub/fc/jitter_meth/99-151v2.pdf

D. Hanson, D. Cunningham, Dawe, http://www.ieee802.org/3/10G_study/public/email_attach/All_1250v2.xls

D. Hanson, D. Cunningham, P. Dawe, D. Dolfi, http://www.ieee802.org/3/1 0G_study/public/email_attach/3pmd046.xls

D. Dolfi, http://www.ieee802.org/3/10G_study/public/email_attach/new_isi.pdf

D. Cunningham and D. Lane, *Gigabit Ethernet networking.* Macmillan Technical Publishing, ISBN 1-57870-062-0

P. Pepeljugoski, M. Marsland, D. Williamson, http://www.ieee802.org/3/ae/public/mar00/pepeljugoski_1_0300.pdf

P. Dawe, http://www.ieee802.org/3/ae/public/mar00/dawe_1_0300.pdf

D. Cunningham, M. Nowell, D. Hanson, Proposed worst case link model for optical physical media dependent specification development. http://www.ieee802.org/3/z/public/presentations/jan1997/dc_model.pdf

M. Nowell, D. Cunningham, D. Hanson, L. Kazovsky. 2000. Evaluation of Gb/s laser based fibre LAN links: Review of the Gigabit Ethernet model. *Optical and Quantum Electronics.* 32:169–192.

Gair D. Brown, 1992, May. Bandwidth and rise time calculations for digital multimode fiber-optic data links. *JLT* 10, no. 5:672–678.

P. Dawe and D. Dolfi, http://www.ieee802.org/3/ae/public/jul00/dawe_1_0700.pdf P. Dawe, D. Dolfi, P. Pepeljugoski, D. Hanson, http://www.ieee802.org/3/ae/public/sep00/dawe_1_0900.pdf

References on Reflection Noise: Fröjdh and Öhlen, http://www.ieee802.org/3/ae/public/mar01/ohlen_1_0301.pdf

P. Pepeljugoski and P. Öhlen, http://www.ieee802.org/3/ae/public/mar01/pepeljugoski_1_0301.pdf

P. Pepeljugoski and G. Sefler, http://www.ieee802.org/3/ae/public/mar01/pepeljugoski_2_0301.pdf

K. Fröjdh and P. Öhlen, http://www.ieee802.org/3/ae/public/jan01/frojdh_1_0101.pdf

P. Pepeljugoski, http://www.ieee802.org/3/ae/public/adhoc/serial_pmd/documents/interferometric_ noise3a.xls

P. Pepeljugoski, http://www.ieee802.org/3/ae/public/adhoc/serial_pmd/documents/useful_IN_ formulas.pdf

K. Fröjdh and P. Öhlen, http://www.ieee802.org/3/ae/public/adhoc/serial_pmd/documents/interferometric_noise3.pdf

K. Fröjdh, http://www.ieee802.org/3/ae/public/adhoc/serial_pmd/documents/interferometric_noise3 .xls

G. Sefler and P. Pepejugoski, Interferometric noise penalty in 10 Gb/s LAN links, ECOC 2001 paper We.B.3.3

Chapter 6.4

6.4

ROADMs in network systems

Kaminow, Li and Willner

In less than a decade, the state of the art in fiber-optic transport systems has evolved from simple point-to-point chains of optically amplified fiber spans to massive networks with hundreds of optically amplified spans connecting transparent add/drop nodes spread over transcontinental distances. The primary driver for this transformation has been a remarkable improvement in cost, while its primary enabler has been the emergence of the reconfigurable optical add/drop multiplexer (ROADM) as a network element (NE).

This chapter begins with a brief description of how optical networks have progressed since their first deployments, and how ROADMs fit into this on-going evolution. In Section 6.4.2 the diverse nomenclature of ROADM technologies and architectures is reviewed and organized. Section 6.4.3 compares a network with ROADMs to three alternative architectures without ROADMs, to illustrate the economic advantages that ROADMs provide. Section 6.4.4 is a detailed analysis of the routing functionality offered by various types of ROADMs, while Section 6.4.5 discusses other features often included in ROADMs and requirements typical of many large carriers' networks. Section 6.4.6 focuses on switch design, including a brief summary of the underlying component technologies. Sections 6.4.7 and 6.4.8 discuss the design of ROADM transmission systems and the interplay between the ROADM and transmission performance. We conclude in Section 6.4.9 with a brief synopsis, and we identify some challenges remaining for ROADM-enabled networks to achieve their full potential.

6.4.1 ROADMs—A key component in the evolution of optical systems

Since optical transmission systems were first deployed there has been a constant push to improve system reach and capacity. A high-level overview of the evolution of optical transmission systems is shown in Figure 6.4.1. The development of wavelength-division multiplexing (WDM) and optical amplifiers, such as the erbium-doped fiber amplifier (EDFA), has led to cost-effective long-haul, high-capacity systems. Optical transmission has been realized over distances greater than the longest circuits in terrestrial networks, with a single-fiber capacity greater than the amount of traffic that currently needs to terminate at any single node [1].

These advances in transmission technology make it desirable for systems to have optical add/drop capability. Exploiting the inherent wavelength granularity of WDM, an optical add/drop multiplexer (OADM) allows some WDM channels (also referred to as wavelengths) to be dropped at a node, while the others traverse the same node without electronic regeneration. Previously, it was necessary to terminate line systems at each node served, and then regenerate the wavelength signals destined for other nodes. The ability to optically add/drop a fraction of a system's wavelengths at a node was first achieved using fixed OADMs. These were constructed from optical filters, and by enabling wavelengths to optically bypass nodes and eliminate unnecessary regeneration, they provided significant cost savings. However, because

Optical Fiber Telecommunications V B; ISBN: 9780123741721

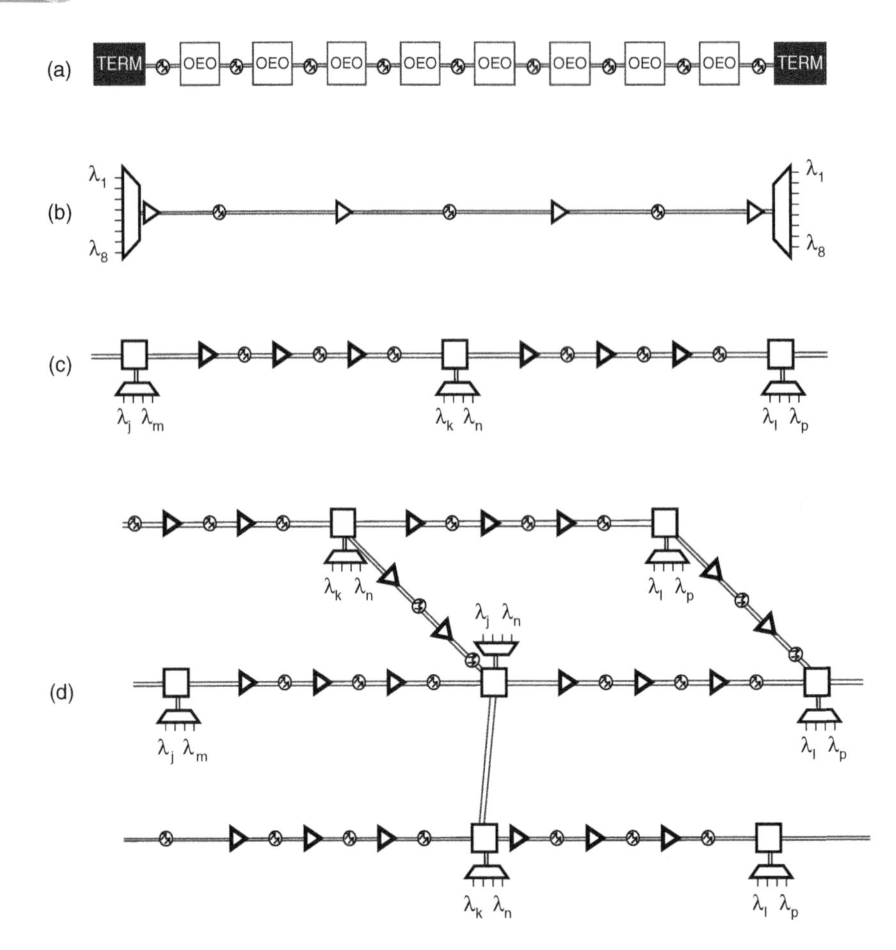

Figure 6.4.1 The evolution of optical transmission systems: (a) Early systems required full regeneration at each node; (b) WDM and the EDFA provide significant economies, enabling the reach and capacity of the systems to grow; (c) Further improvements in reach and capacity have driven the demand for OADMs, so that the same line system can serve intermediate nodes; (d) Switching in the optical domain will continue the evolution from the simple point-to-point systems (shown in b) to optical mesh networks (as shown in d).

traffic growth is inherently unpredictable, it is advantageous for the add/drop capability to be reconfigurable. Therefore, although fixed OADMs are usually lower in cost, ROADMs are supplanting them in all but the most cost-sensitive applications.

ROADMs provide many advantages beyond the savings achieved by optically bypassing nodes. In the future, multidegree ROADMs with adequate reconfiguration speeds may enable shared-mesh restoration at the optical layer [2]. Shared-mesh restoration significantly reduces the number of wavelength channels that must be installed as redundant protection circuits. ROADMs also provide operational advantages. Because ROADMs can be reconfigured remotely, they enable new wavelength channels to be installed by simply placing transponders at the end points, without needing to visit multiple intermediate sites. In addition to these cost-saving benefits, ROADMs will enable new services. For example, if transponders are preinstalled, then new circuits can be provided on-demand. The rapid network reconfiguration provided by ROADMs could also become an enabler of dynamic network services, such as switched video for IPTV. For all of these reasons, ROADMs will continue to have a significant effect on the design of optical networks.

6.4.2 Terminology—A ROADM is a network element

Generally, a ROADM is defined as an NE that permits the active selection of add and drop wavelengths within a WDM signal, while allowing the remaining wavelengths to be passed through transparently to other network nodes. Thus, the simplest ROADM will have two line ports (East and West) that connect to other nodes and one local port (add/drop) that connects to local transceivers. In today's networks, optical links are typically bidirectional, so each line port represents a pair of fibers. When using conventional local transceivers that can process only a single wavelength at a time, the number of

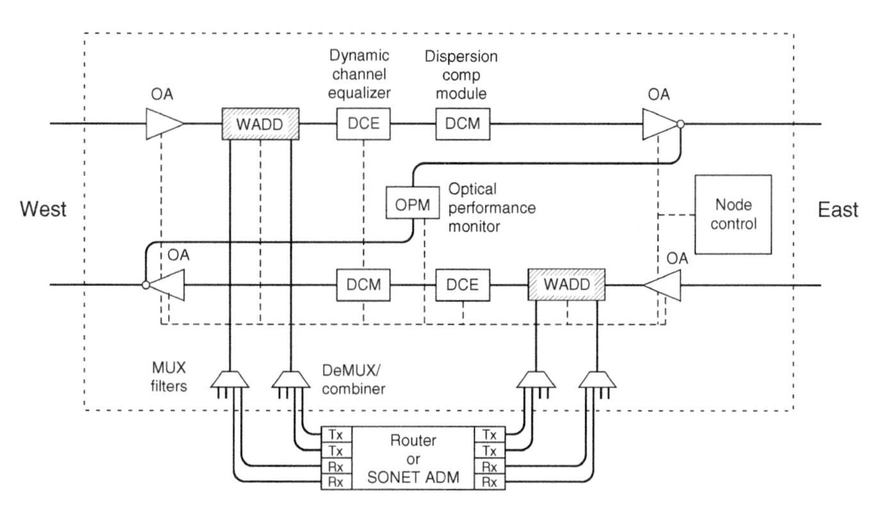

Figure 6.4.2 A two-degree ROADM connects the add/drop port at a node to the optical line. The WADD is a key element of the ROADM, capable of switching individual wavelengths between its input and output ports. Some ROADMs support additional functions, such as dynamic channel equalization.

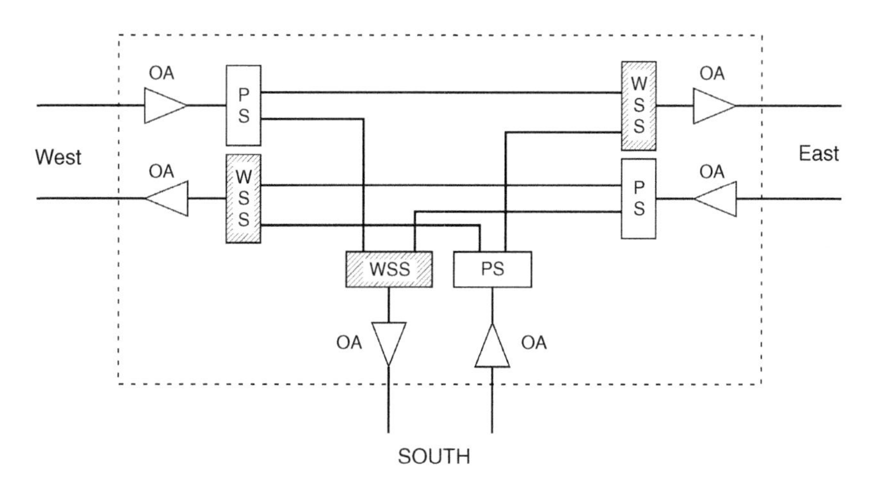

Figure 6.4.3 Simplified layout of a three-degree, broadcast-and-select type photonic crossconnect (PXC), made up of power splitters (PS), wavelength-selective switches (WSS), and optical amplifiers (OA). Ancillary subsystems have been omitted for clarity.

fibers in the add/drop port sets the maximum number of wavelengths that can be added or dropped at a given node. As shown in Figure 6.4.2, at the heart of the ROADM are wavelength add/drop devices (WADD) that perform the wavelength switching, but many other subsystems, such as amplifiers, performance monitors, and dispersion compensation modules, are needed to complete the NE. Depending on the particular technology chosen, wavelength multiplexing/demultiplexing, channel equalization or other functions may be integrated into the WADD module. Popular WADD types are outlined below, and include modules based on the two-fiber wavelength blocker (WB), the multifiber wavelength-selective switch (WSS), and the integrated DEMUX/switch array/MUX fabricated as a planar lightwave circuit (PLC).

Many of the elements shown in Figure 6.4.2 are also present in a photonic cross-connect (PXC, also sometimes referred to as a transparent crossconnect or an all-optical crossconnect[1]), defined here as an NE that interconnects WDM signals on multiple line fibers. An example of a PXC connecting three fiber routes is shown in Figure 6.4.3. Like the ROADM of Figure 6.4.2, the PXC of Figure 6.4.3 has three bidirectional ports. However, in the PXC case, each port consists of a single fiber pair, each handles WDM signals, and each maintains the full optical signal quality needed for propagation through a further cascade of fiber spans, PXCs, or

[1] To avoid confusion, the phrase "optical crossconnect" is not used in this chapter. It has been applied ambiguously in the literature to both electronic and all-optical switches.

ROADMs. (In a ROADM, it may be allowable for signals in the add/drop path to be degraded slightly, since full 3R regeneration immediately follows the drop process.)

In practical networks, locations suitable for a PXC usually need local add/drop capability as well, leading to the concept of a multidegree ROADM that combines both PXC and ROADM functions. In the language of Section 6.4.5.6, the *fiber degree* of a multidegree ROADM is equal to the number of line fiber pairs it supports.

Several other terms describing aspects of add/drop operation are worthy of note. A *full* ROADM is one that provides add/drop (de)multiplexing of any arbitrary combination of wavelengths supported by the system with no maximum, minimum, or grouping constraints. If a ROADM has access to only a subset of the wavelengths, or the choice of the first wavelength introduces constraints on other wavelengths to be dropped, it is called *partial* ROADM. The *drop fraction* of a ROADM is the maximum number of wavelengths that can be simultaneously dropped, divided by the total number of wavelengths in the WDM signal. (Typically, the analogously defined *add fraction* is equal to the *drop fraction.)* If a given add or drop fiber is capable of handling any wavelength, it is said to be *colorless*. If a given add or drop fiber can be set to address any of the line ports (e.g., East or West for a two-degree ROADM), it is said to be *steerable*. An NE is said to be *directionally separable* if there is no single failure that will cause loss of add/drop service to any two of its line ports. The path followed by a particular WDM channel from its source through various ROADMs and PXCs to its termination is denoted a *lightpath*.

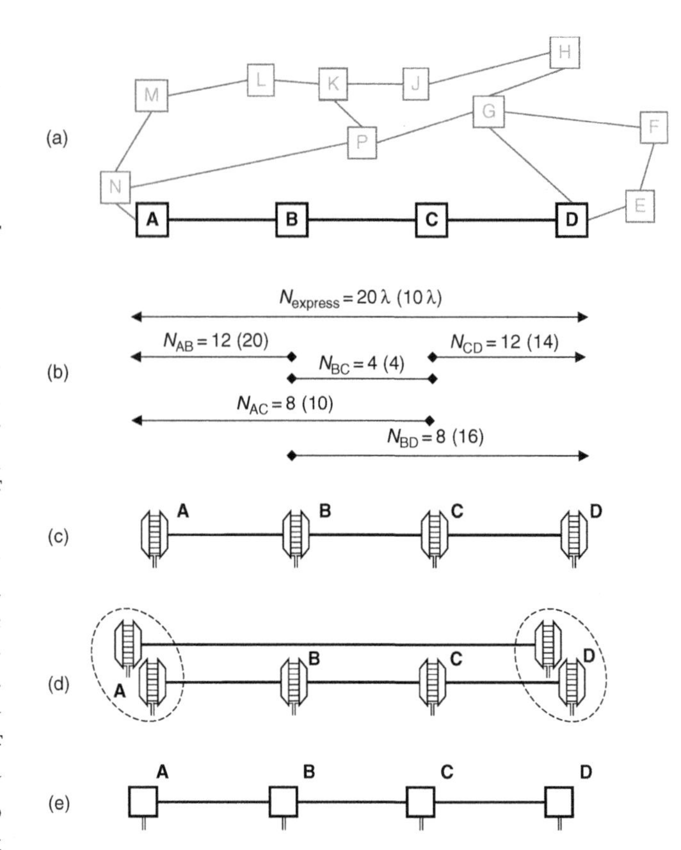

Figure 6.4.4 (a) Nodes A–D are a linear segment of a larger network. By examining how the number of transponders required varies as a function of architecture and traffic, one can develop a qualitative feel for the advantages of ROADMs. Two traffic scenarios (Cases I and II) and four different architectures are considered; (b) the two different traffic scenarios; (c) an architecture with full regeneration at each node; (d) a layered architecture that utilizes an express and local path; (e) two architectures—optical bypass can be achieved with either fixed OADMs or ROADMs at each node.

6.4.3 Simple comparison of four competing network architectures

The economic advantages of optical bypass have long been recognized [3]. Because it is impossible to predict future traffic requirements, realizing the full advantages of optical bypass demands OADMs that are reconfigurable.

Figure 6.4.4 illustrates the savings in transponders achieved by deploying ROADMs over various alternative network architectures. For simplicity, we consider only a linear segment of a larger network, consisting of four nodes denoted as A–D [as shown in Figure 6.4.4(a)]. We assume that a 40-wavelength transmission system is deployed to serve this network, and investigate two alternative end-of-life scenarios. The traffic demands for Case I and Case II (in parentheses) are shown in Figure 6.4.4(b). In both cases for every west-bound wavelength there is an additional east-bound wavelength provisioned for protection—both are included in the wavelength count shown in Figure 6.4.4(b).

Case I shows an end state in which 20 wavelengths express through this segment of the network. Case II (wavelength counts in parentheses) assumes that node B becomes a major traffic destination, with 20 wavelengths terminating there, and only 10 wavelengths express through the segment. A meticulous study would consider many cases, and take into account traffic growth from the beginning of life until the end, considering such factors as the relative costs of equipment, and how traffic churn affects the final state. While our illustrative examples do not represent a rigorous study, these two cases do illustrate the relative strengths of the various architectures. Our results are summarized in Table 6.4.1 and described in detail below.

One of the four competing architectures, shown in Figure 6.4.4(c), has full regeneration at each node. This architecture requires transponders at every node,

Table 6.4.1 The number of transponder pairs needed for each of the architectures in Figure 6.4.4.

# Transponder pairs	Case I (10 express)	Case II (10 express)
Full regeneration	120	120
Parallel systems	80	100
Fixed OADM (20 express)	80	Not served
Fixed OADM (10 express)	100	100
ROADM	64	74

regardless of the circuit's final destination. Figure 6.4.4(d) shows a network architecture that employs two transmission systems in parallel—one for express traffic and one for local traffic. This architecture requires twice the number of fibers be used, and roughly twice the number of amplifiers. Local traffic is regenerated at every node, while express traffic is transmitted over the express system, which only terminates at major nodes. The third architecture, shown in Figure 6.4.4(e), utilizes fixed OADMs. While fixed OADMs are less expensive than ROADMs, the efficiency of this architecture is dependent upon having a network with predictable and very stable traffic demand. If the network is designed assuming that ½ the traffic will be express, and the rest capable of dropping at each node, then this architecture will require the same number of transponders as the parallel express/local architecture shown in Figure 6.4.4(d). However, it will not be able to meet the traffic demands of Case II between the node A and the node B—and the network would need major work—either an additional transmission system would need to be deployed, or the preexisting network would need to be entirely reconfigured, which would interrupt all traffic on the route. Alternatively, the planners might have designed this fixed OADM architecture with many drops, to minimize the possibility that disruptive upgrades would be necessary. If the fixed OADMs were designed to express 10 wavelengths and to drop all others at each node, then both cases could be served by this architecture, but with less than optimal efficiency.

The final architecture schematically looks like Figure 6.4.4(e); however, it utilizes ROADMs, rather than fixed OADMs. By deploying ROADMs, this segment can meet traffic demands with the minimal number of transponders. Not only does this architecture require fewer transponders than any of the others, it can also be more easily maintained. When provisioning new wavelengths, only the end terminals for that circuit need to be visited, whereas the other architectures will frequently require that transponders be installed at midpoints.

The number of transponder pairs needed for each architecture is summarized in Table 6.4.1. This example was intentionally simplified—only a short segment of a network was considered, and only end-of-life traffic was modeled, without any of the complications that can be caused by wavelength blocking when traffic grows randomly. However, the trends illustrated in this example are in agreement with far more detailed studies that have taken traffic growth, and even relative equipment costs, into account [4, 5]. The example above only considered degree 2 ROADMs, as did the detailed studies just cited. When comparing higher-degree ROADMs to networks using only degree 2 ROADMs, the savings in transponder count are not as great, as more wavelength blocking is likely to occur—forcing the deployment of transponders for wavelength conversion [6]. However, the deployment of high-degree ROADMs offers additional advantages beyond transponder savings, including dynamic provisioning, and the possibility of providing mesh-based shared restoration at the optical layer [2].

6.4.4 Routing properties—Full flexibility is best

As a rule, the most desirable ROADMs are the ones with the greatest flexibility: i.e., full ROADMs with 100% add/drop fraction, and colorless, steerable add/drop fibers. However, these are also the most costly solutions, so less ideal options, such as partial ROADMs, have also been extensively investigated. In the analysis below, we will make use of the following parameters:

N = the number of wavelength channels supported by a WDM system,

k = maximum number of channels dropped by a WADD,

J = fiber degree of (i.e., number of line side fiber pairs supported by) a ROADM,

M = the number of input or output fibers provided by a WSS,

R = routing power.

Depending on their internal structure [7, 8] partial ROADMs may be limited to a simple contiguous band of wavelengths, or to a periodic comb of wavelengths, or they may have very complex wavelength constraints. Typically, the wavelength constraints in a partial ROADM are set by the WADDs that perform the wavelength switching. Feuer and Al-Salameh [9] introduced a figure of merit called routing power that provides a framework for comparisons of WADDs. The routing power describes the WADD's ability to establish network connectivity by counting how many distinct connection states [10] it supports. The routing power of a WADD is given by

$$R = \frac{\log(\text{no. of connection states supported by WADD})}{\log(\text{no. of connection states supported by fully flexible WADD})}. \quad (6.4.1)$$

A fully flexible four-fiber WADD without drop-and-continue function supports 2^N connection states, so we see that

$$R = \frac{\log(\text{no. of connection states supported by WADD})}{N\log(2)}. \quad (6.4.2)$$

Routing power ranges from zero (for fixed optical add/drop) to unity (for an ideal, fully flexible WADD), and it gives a fair representation of a ROADM's effectiveness in supporting networks with mesh-like traffic demands. Figure 6.4.5 shows a variety of partial WADD designs with a drop fraction of 0.25, together with the routing power of each. The full crossconnect type (Figure 6.4.5(a)) has no wavelength constraints except for the maximum drop fraction, so it has the highest routing power. The three banded designs (Figures 6.4.5(b–d)) allow access to only a fraction of the input wavelengths, so they have a low routing power. The tunable filter design of Figure 6.4.5(e) allows access to any wavelength, but not to two wavelengths in the same band, so it has an intermediate value of R.

The relationship of routing power to network value has been confirmed by simulations of traffic growth in a 32-wavelength, 8-node metro ring constructed from partial ROADMs [11]. In the simulations, ROADMs began with one partial WADD at each ring node. Randomly generated traffic demands (each representing a lightpath with 1 + 1 protection) were added to the ring sequentially. When a demand could not be satisfied by the existing ROADMs, an additional WADD module was added to one or both endpoint nodes to increase the drop fraction as needed. Accumulating the results of 1000 randomized trials, one obtains the average number of WADD modules required vs the number of traffic demands, as shown in Figure 6.4.6. Comparing ROADMs based on Figure 6.4.5(a), 6.4.5(b), and 6.4.5(e), it is clear that the low R of the banded WADD (Figure 6.4.5(b)) leads to a need for many more module installations, while the crossconnect WADD (Figure 6.4.5(a)) can satisfy the largest number of traffic demands. R can also be increased by using WADDs with a larger drop fraction, and Figure 6.4.7 shows the average number of demands satisfied per WADD, at full ring fill, for drop fraction of 0.125, 0.25, and 0.5, for all the three WADD types. Whether R is increased by adding drop fibers to the WADD or by changing the internal switch arrangement, there is a strong, roughly linear correlation between

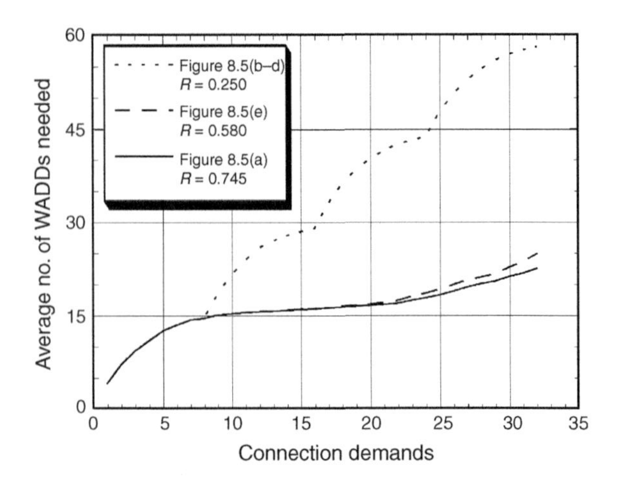

Figure 6.4.6 Average number of partial WADDs needed to satisfy traffic demands in an 8-node, 32-wavelength ring network with uniform random demands. For designs with a high routing power, fewer WADDs are needed.

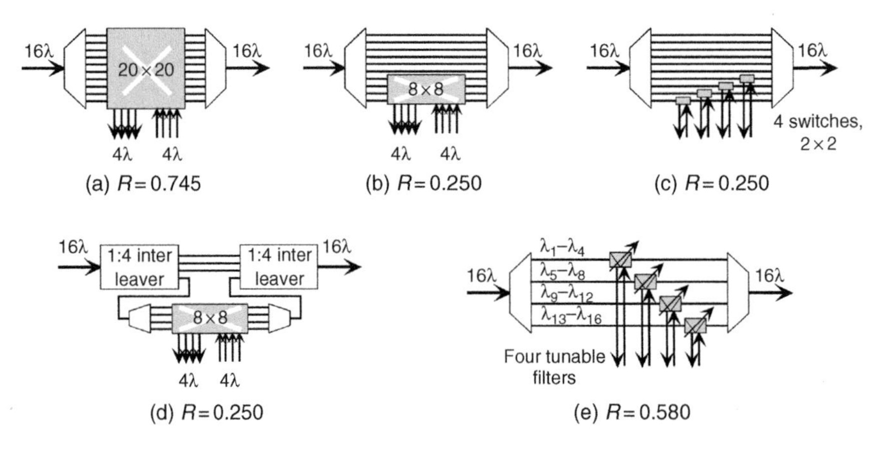

Figure 6.4.5 Five partial WADD designs with the same add/drop fraction (k/N = 25%), but different wavelength constraints.

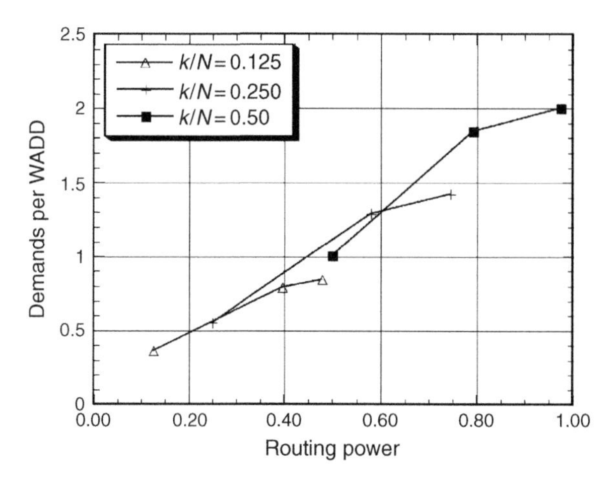

Figure 6.4.7 Demands satisfied per WADD as a function of the WADD's routing power in the network of Figure 6.4.6 with fully filled links. k/N is the drop fraction of the WADD.

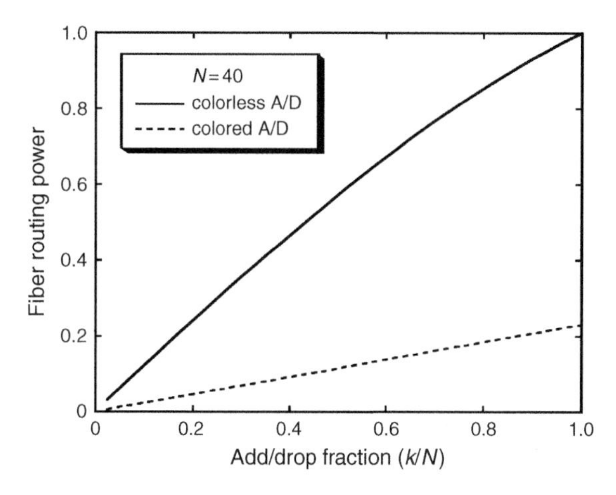

Figure 6.4.8 Fiber routing power for colored and colorless WADD designs.

demands satisfied and routing power, confirming the usefulness of R as a figure of merit for WADDs.

It is worth noting that routing power, which effectively assigns equal value to all possible network connection states, is most appropriate for networks with mesh-like (i.e., uniform) traffic demands. In networks with highly structured traffic patterns, such as hubbed rings, WADDs with lower R may be satisfactory. For example, in an access ring with a single hub, only connections with one endpoint at the hub are needed, and a WADD based on a three-fiber tunable filter with variable bandpass width has been proposed [12].

The WADD designs of Figures 6.4.5(b) and 6.4.5(c) illustrate the difference between colored and colorless add and drop fibers. In the colored case (Figure 6.4.5(c)), each add/drop fiber pair is locked to a single wavelength. Even if the transponder attached to it is tunable, service cannot be switched to a new wavelength without manual reconnection to a new fiber pair. This makes pre-need deployment of transponders less practical, and prohibits applications such as bandwidth-on-demand and $1:N$ protection against transponder failures. The routing power can be adapted to compare colored and colorless operation by treating each add/drop fiber pair as a separate port and applying the general definition of Eqn (6.4.1). The result is the fiber routing power R_f, plotted in Figure 6.4.8 as a function of add/drop fraction. Figure 6.4.8 shows that the advantage of colorless operation is most dramatic for WADDs that permit full add/drop.

To understand the role of directional steering of transponders, it is necessary to move up from the WADD to the level of the complete ROADM. With a ROADM design like that of Figure 6.4.2, each transponder is dedicated to a single line fiber pair, limiting its use for mesh protection or bandwidth-on-demand applications. An alternative approach is to begin with the PXC of Figure 6.4.3 and add a wavelength MUX/DMUX pair to one fiber pair to transform that fiber pair into an add/drop port. This creates a two-degree ROADM with one steerable add/drop port: transponders attached to such a ROADM can serve either the East or West direction. The design of Figure 6.4.3 is readily scalable, as shown in Figure 6.4.9, up to the limit set by the splitter loss or by the number of fibers supported by the WSS. The routing power of the complete ROADM, R_R, is derived from Eqn (6.4.1) by counting the add/ drop connection possibilities associated with different directions, as well as different wavelengths. (We have not counted the line interconnection possibilities of the multidegree ROADM. Because we require that all possible line interconnections be supported, they will cancel out. We have also not included optical multicast connections, though that is a possible extension of the concept.) Figure 6.4.10 compares the ROADM routing power of an 8-degree, 40-wavelength ROADM with k add/drop fiber pairs, in a steerable, colorless design, a non-steerable, colorless design, and a nonsteerable, colored design. Although the results demonstrate the expected correspondence between routing power and ROADM flexibility, one should keep in mind that network design studies validating the use of routing power for colorless and steerable ROADMs have not yet been carried out.

The discussion so far has excluded any consideration of optical multicast, the process of splitting a WDM channel and sending it to two or more destinations. Broadcast is the traditional method of distributing one-way services such as cable television, and some consider it a valuable element of future optical access as well [13, 14]. Multicast is an important tool for optimizing IP networks, and it promises to become more so as IP video and other streaming media continue to

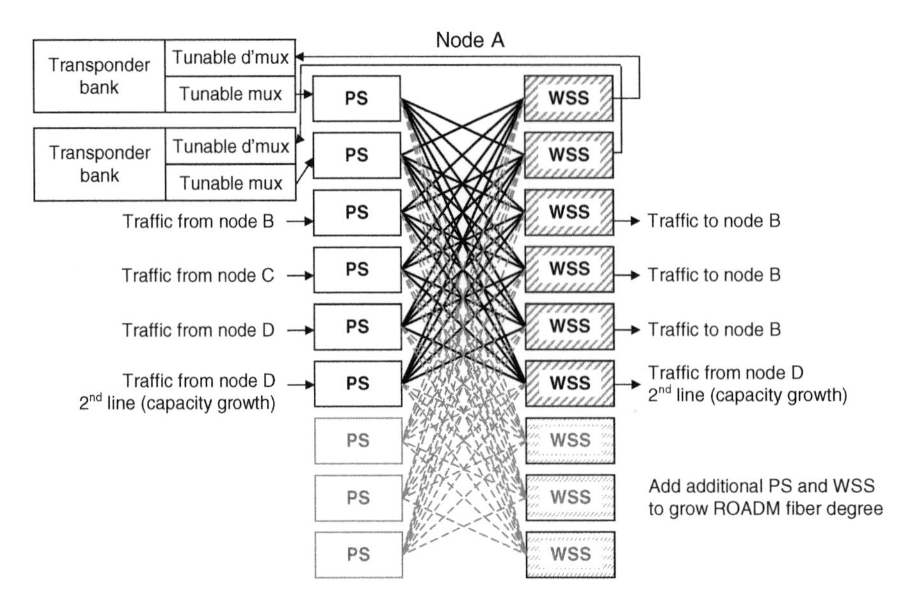

Figure 6.4.9 This multidegree ROADM supporting colorless, steerable add/drop is assembled from 1 × 8 power splitters and 8 × 1 wavelength-selective switches (WSSs). This node can grow in-service to degree 9 by adding additional power splitters and WSS.

expand. Some ROADMs, typically those with power splitters at the drop branch, support optical multicast easily. The multidegree ROADM of Figure 6.4.9 is one example. Multicast-capable 1 × M WSS have also been demonstrated [15], but these WSS are not yet able to broadcast to all M ports at the same time. For a two-degree OADM, optical multicast is equivalent to a drop-and-continue function, and it has been demonstrated with both fixed OADMs [16] and ROADMs [13]. In small networks where wavelength reuse is considered dispensable, an entire ring can be operated in broadcast and select mode [13]. In larger mesh networks, multicast operation alters the paradigm of bidirectional links that usually pertains to optical transmission, so it may require significant rethinking of network management schemes.

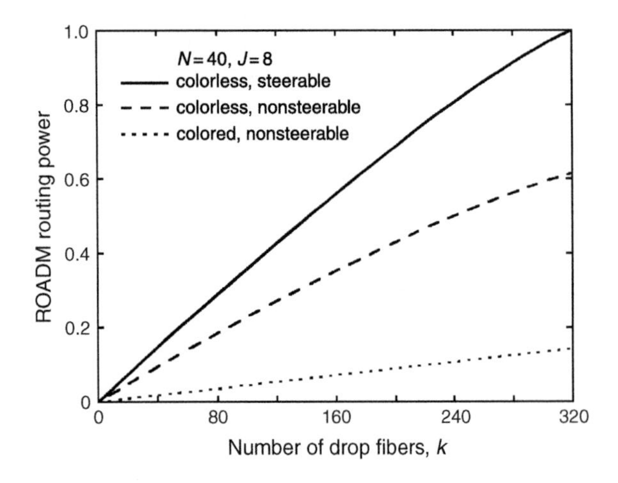

Figure 6.4.10 Effect of colorless and/or steerable add/drop capability on the fiber routing power of a 40-wavelength ROADM of fiber degree $J = 8$.

6.4.5 Additional attributes—Rounding out the picture

Section 6.4.4 has described the routing function of the ROADM, but other basic characteristics and attractive features also exist. The most basic are often required because their absence would adversely affect network reliability. Some features are necessary for the ROADM to be cost-effective for a particular application, while others enhance the transmission performance of the optical network. In this section we outline various additional attributes of a ROADM, and discuss what drives carriers to request, or even require, these features.

6.4.5.1 Hitless operation

Hitless operation is typically a "must-have" requirement. In most transmission system applications, a ROADM must be capable of switching one wavelength without inducing any errors on the other wavelengths [17]. It has been suggested that forward error correction (FEC) could handle the short burst of errors that might be created in such circumstances; however, this is usually unacceptable for two reasons. First, the performance enhancement provided by FEC is often relied upon by the transmission system itself, and it might not be capable of handling additional errors. Second, FEC is optimized for dealing with uncorrelated errors, rather than bursts of errors as might be caused by switching an adjacent channel in a nonhitless ROADM. Note that it is possible to design a hitless ROADM using a nonhitless WADD by using other elements to block a wavelength before switching.

6.4.5.2 Directional separability

Separability is a design principle meant to insure that a single component failure will not affect both the working and protection paths of any circuit. In ring-based Synchronous Optical NETwork (SONET) systems, the key precept is East-West separability: SONET add/drop multiplexers (ADMs) are constructed so that no single point of failure will disrupt both East-facing and West-facing traffic. This principle is applied even to redundant node controllers and power supply systems. Note that *East-facing* means traffic dropped *from* the East and added *to* the East—it is not the same as *Eastbound*. The principle of East–West separability can be easily extended to degree 2 ROADMs. In a multidegree ROADM this concept expands to directional separability—the requirement that no single point of failure disrupts local add/drop traffic on more than one of its line fiber pairs. This definition assures that subtending SONET or similar equipment will be able to perform client-level protection regardless of which direction is used for the protection path. Achieving directional separability is discussed in more detail in Section 6.4.6.

6.4.5.3 Modular deployment

Carriers wish to deploy systems in the most cost-effective manner possible. Today, it is far more cost-effective to initially deploy the minimal amount of equipment that can smoothly evolve to meet future needs, rather than to deploy a fully loaded system configuration from the very beginning. Currently (and for the foreseeable future), transponders make up the dominant cost of a fully loaded optical communication system. If a full set of transponders were included in the initial deployment, then a substantial cost would be incurred before the network had sufficient traffic to support the expense. Therefore, systems are routinely designed to permit incremental deployment of transponders on an as-needed basis. Similar considerations also apply to multiplexers, although the economic drivers are not as strong. In general, modular growth will be supported whenever the additional cost and complication of upgrading to higher capacity in the future is small compared to the financial impact of a full equipment deployment at startup.

This pay-as-you-grow approach should be designed into ROADMs, so that the network itself can grow in a cost-effective manner. Traditional networks grow by adding and interconnecting stand-alone line systems, incurring substantial cost and complexity. By using ROADMs that allow for modular deployment of additional ports, network growth can benefit from both the equipment and operational efficiencies of integrating line systems as they are needed into a seamless network (see Section 6.4.5.4). Because networks are deployed over the course of years, carriers prefer to be able to grow the nodes of the network from terminals or amplifiers, into degree 2 ROADMs, and eventually into multidegree ROADMs [18]. This not only allows the expense to be spread out over years, it also enables the network designers to respond to unforeseen traffic growth patterns.

6.4.5.4 High degree for span relief

As multidegree ROADMs become more common in backbone and metro networks, equipment builders and network operators must decide on the maximum number of line ports (degree) they will support. NE cost is expected to grow with the degree of the node, and a route-level view of national and continental-scale backbone networks would suggest that a degree higher than four is rare. However, as noted in Ref. [19], there are several reasons why the number of fiber pairs (fiber degree) needed might substantially exceed the number of intersecting routes (route degree).

Continuing growth of traffic (often in directions unpredictable at the time of the original system layout) will eventually require the lighting of additional fiber pairs (overlay) on one or more routes. If these fiber pairs are not connected to the rest of the network via the ROADM, then they will be isolated from the existing optical domain, and many of the advantages provided by transparent optical networking will be lost [20]. Traffic traversing these overlay fibers would need to be dropped, regenerated, and added at the node before it could travel farther in the network. Unless steerable add functionality were available, such overlay traffic would also lose the directional reconfigurability enjoyed by the original traffic. With a ROADM architecture capable of serving a sufficient number of fiber pairs, this isolation of new capacity can be avoided, and the network will be tolerant of uncertain traffic forecasts. Even with a ROADM design capable of modular deployment, as described in Section 6.4.5.3, the maximum degree of the ROADM is usually determined at the time it is initially deployed. The ability to expand the ROADM to serve a large number of fiber pairs must be designed into the ROADM from the beginning.

Additional line ports may also be needed to provide steerability for local add/ drop. As noted in Section 6.4.4, each steerable transponder bank consumes one fiber pair.

Thus, it is not unreasonable to expect the fiber degree to be at least twice the route degree. Since current WSS-based designs are manageable for a fiber degree of 8–10, many network plans are aiming at multidegree ROADMs in this range. To date, there is no widely accepted strategy for dealing with exceptional cases requiring growth to a higher degree (e.g., a four-route intersection with overlay on three routes and 100% drop fraction with full steerability). This could present a fruitful opportunity for future research.

6.4.5.5 Network scaling—How many is enough?

A key parameter in the design of wavelength-routed networks is the number of ROADMs to be cascaded by any given lightpath. Insertion of a ROADM necessarily involves some impairment of the transmitted signal, such as signal-to-noise degradation due to loss, signal distortion due to channel spectrum narrowing, and polarization-dependent loss (see Section 6.4.7.2). The ROADM design should balance the cost of improved performance against the cost of additional regeneration given the maximum number of cascaded nodes required by the network. The maximum number of cascaded ROADMs may be different for national or continental backbone networks, which traverse long distances, than it is for metro/regional core or feeder networks, which must connect many relatively dense access locations. Expectations for metro systems are also influenced by today's SONET capability of up to 16 cascaded nodes. A maximum cascade of 8-10 ROADM nodes is typically adequate for today's continental backbones, while interconnected metro rings may be asked to support up to 16 or more cascaded ROADM nodes [11]. This places stringent demands on the ROADM implementation, as discussed in Section 6.4.7.

6.4.5.6 Channel conditioning

One beneficial feature of ROADMs is that the WADD is often capable of more than simply transmitting or blocking a wavelength. For example, many WADDs can also act as a channel-specific variable optical attenuators (VOAs). This enables the ROADM to perform channel power equalization. Channel power equalization is particularly useful in networks where not all signals travel the same lightpath, and in long-reach systems that require correction of accumulated gain ripple. This aspect is discussed in detail in Sections 6.4.7 and 6.4.8. Channel power equalization is such a useful feature, that many WADDs have been redesigned to incorporate VOAs.

To perform channel power equalization, a ROADM must be able to accurately measure the channel powers, typically with better than 0.5 dB accuracy. A channel monitor provides such additional measurement capability. More advanced types of optical performance monitors [21] may provide channel monitoring capability plus other measures of signal quality such as the optical signal-to-noise ratio or the polarization mode dispersion. At a minimum, channel monitoring must be provided on the line at some point after the channel power equalization device to provide feedback for that component. Between 1% and 10% of the line power might be tapped for this purpose. Since power equalization is performed infrequently, a single monitor may potentially be shared among power equalization components in a multidegree or multidirection ROADM node as shown in Figure 6.4.2. In systems with distributed Raman amplification, channel monitoring may also be desirable at the input to the ROADM (output of the transmission span) or output of the dispersion compensating module (DCM) to optimize the Raman pumps for flat channel gain [22].

The technical challenges associated with the channel monitoring function should not be underestimated [21]. An accurate power reading is needed under all operating conditions of the transparent network. This potentially includes monitoring signals with different modulation formats and bit rates, large channel to channel power variations, and signal spectral shaping over different distances and node pass-through. The accumulated optical noise can also impact the accuracy of the signal power measurement [21].

In addition to providing feedback for power equalization, a channel monitoring device can also provide channel path telemetry and identification/discovery capabilities. Channel telemetry is used to trace the path of different WDM channels through the network. Telemetry information can be either explicit or implicit. Explicit telemetry (also called lightpath labeling) involves the use of a channel tag that is carried by the channel along its propagation path and provides an unambiguous indication of channel presence at a ROADM. Most commonly these tags are sinusoidal modulation tones placed on the channel power [23], although more recently an all-digital label coding scheme was introduced [24, 25]. The digital label scheme obviates the need for extra optical modulators and eliminates linear crosstalk between the label and the data information, though it adds a slight overhead (~2%) to the line data rate. In contrast, implicit telemetry information is transferred between nodes through conventional signaling (e.g., using the optical supervisory channel) and then cross-checked with a total power or channel power measurement. Thus, the node is told by the physical layer management software which channels should be present and the ROADM then confirms their presence by the power reading from the channel monitor. Additional information from a sophisticated channel monitor, such as modulation format or bit rate, can enhance the reliability of this implicit technique, as it provides additional cross-referencing of the correct channel presence.

This telemetry function is a natural adjunct of the wavelength-routing capability built into ROADM networks, needed to verify the accuracy of the routing and to diagnose and locate routing faults. With explicit telemetry, label reading is a local process implemented within each node, while the implicit type involves correlation of information from many nodes. Implicit path trace can be implemented with technology that is already present in the network, but it may be unable to detect certain hardware or software flaws that cause a ROADM to report its status incorrectly. In addition, amplified spontaneous emission (ASE) noise can be spectrally shaped by the ROADM nodes such that over long distances the accumulated ASE may acquire the spectral signature of a low power channel. Both types of telemetry are susceptible to some of the same challenges that complicate channel monitoring (see above), due to signal impairments and the diversity of spectra possible in transparent systems.

6.4.5.7 Switching speed

The switching speed required from the ROADMs will depend strongly on the kinds of network services they are expected to provide. The initial application of photonic (optically transparent) networks is circuit provisioning, a process which has historically taken weeks or months to complete. Even with the expectation that photonic networking will streamline the provisioning process, a ROADM switching time of seconds or even minutes is quite acceptable for present purposes. Dynamic wavelength services, in which a customer turns up additional bandwidth on an on-demand basis, are under active investigation with an eye to unlocking additional revenue streams for network providers. Although this kind of service might set more stringent limits on the ROADM switching time, the primary obstacles to implementation come from software, network management (including the physical layer), and operational issues. In addition, finding the right economic model to support the preinstallation of transponders before they are actually used presents a challenge. Implementation of dynamic wavelength services might drive ROADM switching times down to ~ 1 s, which should be accessible to most of today's leading component technology families. The next step in ROADM speed might be driven by burst or flow switching—since a data burst duration might be as brief as a few seconds, ROADM speeds of ~ 100 ms would be adequate. Some have suggested that millisecond wavelength services may be desirable [26], which is challenging for some technologies. An alternative driver for improved ROADM speed would be shared wavelength protection/restoration strategies. Since SONET has set a benchmark of 50 ms for the complete decision and switching process, ROADM speeds of ~10ms or less would be desirable for this application. Although shared protection can offer substantial efficiencies compared to 1 + 1 protection, network reconfiguration on these time scales presents many challenges beyond the ROADM switching, including the need for colorless and steerable add/ drops, control of transient phenomena in the optical amplifier chains along each affected lightpath, and strategies to manage control message latency and path computation/selection delays in complex mesh networks.

6.4.6 ROADM/WADD architecture—Thinking inside the box

Many WADD designs suitable for use in the ROADM framework shown in Figure 6.4.2 can be grouped into two classes: the parallel WADD, in which an assembly of space switches is sandwiched between the DMUX and MUX components (Figure 6.4.5(a), 6.4.5(b), or 6.4.5(c)); and the serial WADD composed of cascaded tunable filters (Figure 6.4.11). However, years of active ingenuity have also produced proposals or demonstrations of a wide variety of hybrid designs and unique reconfigurable components. Rather than try to catalog them all here, we will concentrate on three common designs that are commercially available.

6.4.6.1 Common designs

In the earliest commercial ROADMs, the classic parallel WADD of Figure 6.4.5(c) was implemented using arrayed waveguide gratings for DMUX and MUX, bracketing an array of 1 × 2 or 2 × 2 switches. Although parallel WADDs assembled from discrete components have been deployed, the discrete approach has been superseded by more highly integrated designs. WADDs based on PLC technology have been quite successful in systems with moderate wavelength counts, and PLC WADDs have been demonstrated with over 40

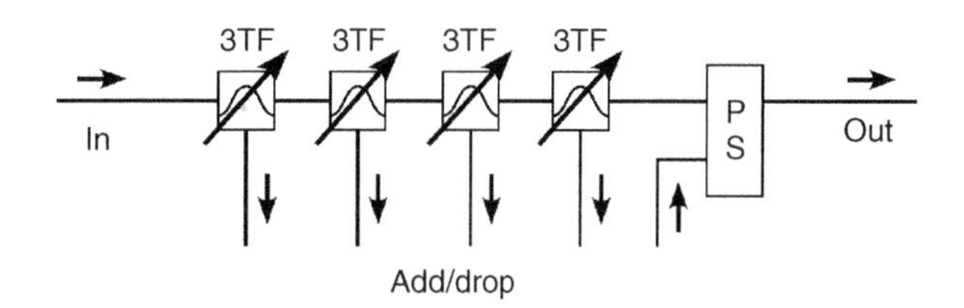

Figure 6.4.11 Serial WADD assembled from three-port tunable filters (3TF) and a wavelength-insensitive power splitter/combiner (PS).

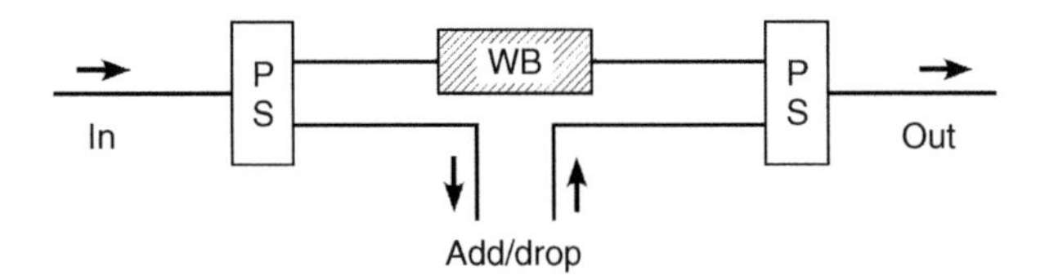

Figure 6.4.12 In this widely deployed WADD design, the wavelength blocker serves to block individual WDM channels to permit wavelength reuse. WB is a wavelength blocker and PS is a wavelength-independent power splitter/combiner.

wavelengths. These designs have built-in MUX/DMUX of the add/drop channels, which is convenient for ROADMs, but inconvenient for PXC operation such as ring interconnect. Due to insertion loss and passband narrowing, these early ROADMs had limited cascadability, and the discrete-component implementations were bulky. Per-channel equalization has been achieved by incorporating an array of VOAs in series with the switches. Colorless or steerable add/drops would require the addition of an external matrix switch in the local port. Extension of this design to a PXC or multidegree ROADM would involve a large increase in the number and complexity of components, so practical implementation of either integrated or discrete versions may be challenging.

To improve cascadability and cost in systems with 40 or more wavelengths, broadcast-and-select WADDs based on WBs have been extensively deployed in core networks. The WB is a component with one input fiber and one output fiber, capable of independently blocking or transmitting wavelength channels in any arbitrary pattern. It is assembled with wavelength-independent power couplers to construct a WADD, as shown in Figure 6.4.12 (see, e.g., Ref. [27]). The typical WB is a free-beam optical device actuated by an array of MEMS or liquid crystal (LC) cells. The add and drop fibers carry WDM signals, so external DMUX/MUX subsystems are needed for local add/drop. Per-channel equalization of express wavelengths is achieved by partial blocking, and colorless add/drop is possible with appropriate choices of DMUX and MUX. Building a PXC or multidegree ROADM with WB-based WADDs is possible, but the number of WBs needed scales as $J(J-1)$, where J is the fiber degree of the ROADM, making this an unattractive approach for ROADMs with $J > 3$.

To enable a smooth upgrade to multidegree ROADMs (see Section 6.4.5.3), WADDs based on WSSs are gaining in popularity [28]. The WSS is a component with three or more fibers that can be set to establish independent connections for each wavelength, from any input fiber to any output fiber. In a typical WSS, micro electromechanical systems (MEMS), or LC elements are used to steer a free-space beam for each wavelength toward its chosen connection. Although the free-space optics are intrinsically reciprocal, practical considerations such as beam control and back reflection mean that devices are usually personalized at the factory as either $1 \times M$ (multi-output) or $M \times 1$ (multi-input) versions. Either version, or both, may be used in a WADD, as shown in Figure 6.4.13. Values of M up to 10 and N up to 80 are commercially available, and per-channel equalization is provided. Since each WADD provides multiple drop fibers and multiple add fibers, construction of a multidegree ROADM is straightforward (at least from a hardware perspective) as shown in Figure 6.4.9. Incremental growth of the fiber degree J up to a maximum of $J = M + 1$ is also possible, using only J or 2J WSS modules (the additional degree is possible by eliminating loop back capability, so that signals input on one fiber pair can only be output on other fiber pairs). Steerable add/drop is automatically achieved by the design of Figure 6.4.9, in which an add/drop fiber pair is interchangeable with a line side pair. However, as fiber pairs are dedicated to steerable add/drop, the maximum fiber degree of the node is decreased proportionately. Thus, to guarantee a steerable 100% add/drop for all directions, one must reduce the maximum value of J to $(M + 1)/2$. Possible responses to this limit include: (a) develop WSS modules with increased M; (b) provide power taps on each line fiber pair to provide some nonsteerable add/drop capability; (c) adopt a more complex multidegree ROADM for high-degree nodes; or (d) design the network so that 100% add/drop is not needed at locations with high fiber degree.

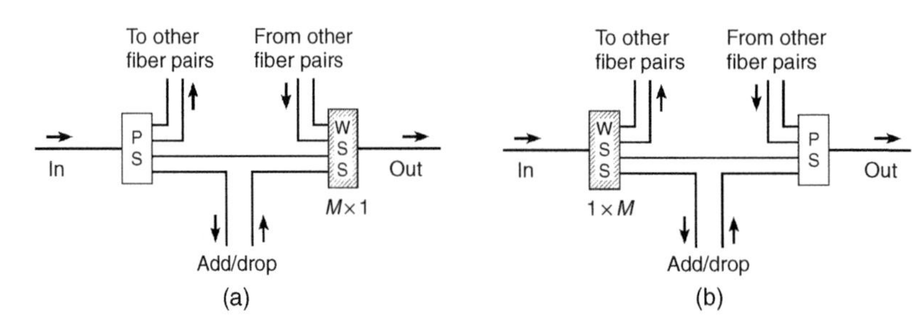

Figure 6.4.13 Two WADD designs based on wavelength-selective switches (WSSs). Only variant (a) is East–West separable under all WSS failure modes.

6.4.6.2 Designs for colorless Add/Drop

Although colorless add/drop capability is highly desirable, cost-effective means for implementing the flexible MUX/DMUX are still under debate. For the WADD of Figure 6.4.5(c), which effectively incorporates fixed MUX/DMUX into the WADD itself, the only option is an external matrix switch. A two-degree ROADM based on Figure 6.4.5(c) would need four switches of dimension $N \times N$ to achieve 100% add/drop that is colorless but nonsteerable. If the add/drop is required to be steerable as well, two matrix switches of dimension $2N \times 2N$ are required. With recent progress that has reduced the cost and optical loss of such matrix switches, this could become a realistic option, at least for systems with $N \leq 40$. As noted above, this WADD is not attractive for multidegree ROADMs.

WADDs based on WBs or WSSs require external MUX/DMUX modules attached to their add/drop fibers. We first consider the drop side under the assumption that 100% add/drop must be supported. One early proposal was to use a $1 \times N$ power splitter followed by tunable filters, as shown in Figure 6.4.14(a). This approach is attractive for its simplicity and its pay-as-you-grow economics (see Section 6.4.5.3). If the first stage of deployment includes only a single dropped wavelength, only a single tunable filter need be installed. In practice, however, the large loss associated with the splitting process demands extra amplifiers in the DMUX chain, raising the cost and failure rate of the complete system and degrading the signal-to-noise ratio at the receiver. The performance demanded from the tunable filters is quite high and the cost at increasing drop fractions quickly becomes higher than other alternatives.

Colorless drop for WB-based and WSS-based ROADMs can also be realized by following fixed DMUXs with matrix switches, as shown in Figure 6.4.14(b). To enable 100% colorless add/drop in a ROADM of fiber degree J, 2J matrix switches of dimension $N \times N$ are needed. Steerability may be supported by enlarging the matrix switches to dimension $JN \times JN$ (only two switches are then needed), or by consuming some fiber degrees of the ROADM for local add/drop, as discussed in connection with Figure 6.4.9.

Figure 6.4.14(c) illustrates a third approach to colorless drop, using $1 \times M$ WSSs as configurable DMUXs. Since M is usually less than N, a cascaded structure is needed to support 100% drop. For $M^2 \geq N$, a two-layer cascade is adequate. Depending on the split ratio N/M, the higher layer of the cascade could be a power coupler or a WSS. Assuming a two-layer cascade of WSSs, a ROADM serving $J/2$ line fiber pairs with colorless, steerable, 100% drop would employ $(J/2) \times (1 + N/M)$ WSS modules, showing the powerful incentive to develop WSS modules with higher values of M, especially when the wavelength count N is large.

Many variants and combinations of the three approaches outlined in Figure 6.4.14 have been devised, but no solution has yet achieved commercial dominance. Multiplexing structures for colorless add function are generally analogous to the demultiplexing structures, but there is a fundamental difference. Whereas the color of the light detected is determined in the drop demultiplexers, the wavelength of the light transmitted is determined in the transponder. Therefore, in systems with few WDM channels, it may be practical to implement a power combiner without tunable filters for the add function.

6.4.6.3 Separability and failure modes

As mentioned in Section 6.4.5.2, management of equipment failures is an essential part of NE design. A requirement that ROADMs have directional separability has far-reaching consequences on how ROADM elements are partitioned onto cards, since all elements on a given card will be disrupted when the card is replaced. For example, if a booster amplifier drives an East-facing fiber, it cannot be packaged on a card with a preamplifier that receives traffic from the West. To assure separability,

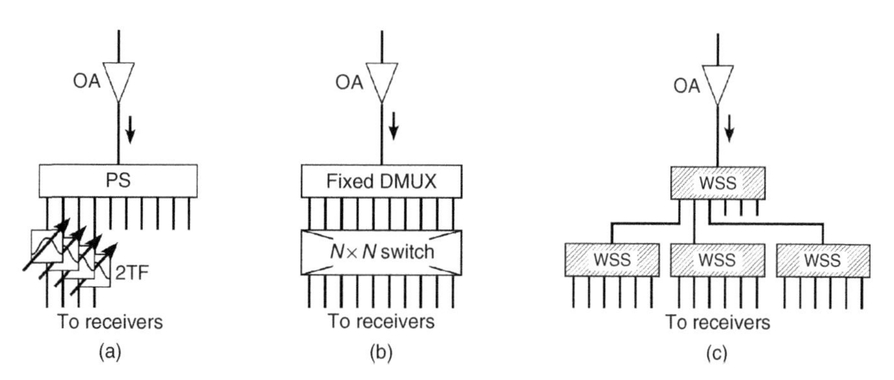

Figure 6.4.14 Three demultiplexing structures that achieve colorless operation of drop fibers. The labels 2TF, PS, OA, WSS represent two-port tunable filters, power splitters, optical amplifiers, and wavelength-selective switches, respectively.

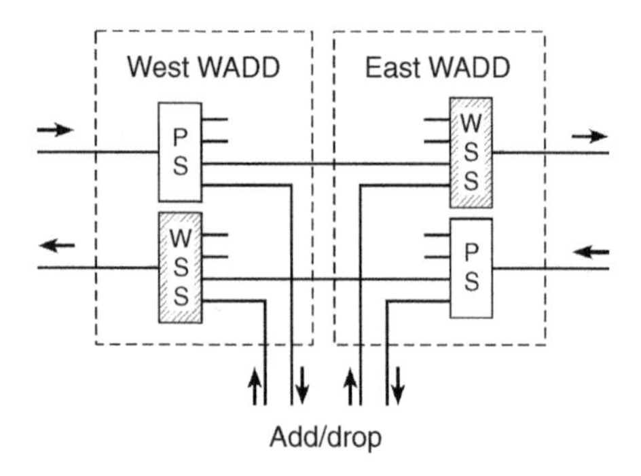

Figure 6.4.15 A typical scheme of partitioning WADD elements on separate line cards to achieve East–West separability.

the WADD subsystems discussed above must be divided among two cards. A typical arrangement is shown in Figure 6.4.15. Somewhat surprisingly, the two WADD configurations of Figure 6.4.13 show different separability properties. Configuration (a) can be East-West separable (with appropriate partitioning), since no failure of the WSS can disrupt both West-facing drop and East-facing add. Configuration (b), on the other hand, is not fully East-West separable, since a WSS that transmits when it should be blocking a wavelength will disrupt the East-facing add (due to optical interference), even as it fails to drop the West-facing channel(s). If this "fail-to-white" condition and the related "fail-to-gray" state (i.e., some light is transmitted in the failed mode) are rare enough to be negligible, this conditional separability may be acceptable. This example highlights the importance of understanding failure modes of these new optical subsystems.

The WADD partitioning of Figure 6.4.15 can be applied to the multidegree ROADM design of Figure 6.4.9 to obtain a directionally separable solution. This simple definition of directional separability does present a problem for a shared transponder bank that uses steerable add/drop to serve multiple line fiber pairs. In this case it is wise to consider what drives the separability requirement—that working and protection paths are not both vulnerable to the same single point of failure. This can be achieved by taking care when provisioning circuits to insure that the working and protection paths of any circuit are on separate transponder banks. Unfortunately, this solution may complicate the implementation of shared-mesh protection.

6.4.6.4 Underlying WADD technologies

Although ROADMs have been built from discrete components, the cost, size, and performance targets of high-volume deployment will demand technologies capable of a high degree of functional integration. For ROADMs of today and the near future, such integration is provided by PLC, MEMS and LC technologies [28]. The PLCs of greatest interest for ROADMs are made up of index-guided waveguides fabricated on silica or related glassy materials. Passive devices such as fixed waveguide DMUXs and power splitters built from PLCs offer excellent accuracy, reliability, and fiber coupling. Tuning and switching functions are typically integrated as thermo-optic elements driven by microheaters integrated onto the PLC. A wavelength multiplexer with an integrated array of VOAs (often called a VMUX) is an example of a successful PLC product. In very large-scale integration, as needed for a 1×9 WSS, thermo-optically controlled PLCs can have problems with total power dissipation, thermal crosstalk among different elements, and polarization dependence. The basic element of LC technology is a variable optical phase delay. In combination with crossed polarizers, a switch/attenuator is realized. If a diffraction grating or PLC DMUX is used to spatially separate the wavelength channels of a WDM signal, an array of LC switch/attenuator elements can be used to construct a WB, and in fact such WBs have achieved significant commercial success. Although the LC switch is intrinsically a two-mode device, multiple LC phase-delay elements per wavelength can be used to achieve diffractive steering, enabling the LC-based WSS [15]. MEMS was the original technology used to build WSSs, and it is still the technology platform of most WSS vendors. An array of beam-steering mirrors, one per wavelength channel, is used to direct each wavelength to its appropriate input and output fibers. Diffractive steering with MEMS is also possible, but has been less popular to date. As of this writing, it is not clear whether the marketplace will ultimately prefer ROADMs based on LC technology, MEMS technology, or a hybrid of the two.

6.4.7 ROADM transmission system design

The design of ROADM-based fiber-optic transmission systems shares many common elements with traditional point-to-point transmission system design. Balancing the degrading effects of noise and nonlinear phenomena is still a fundamental driver for system performance. However, in ROADM systems the bandwidth management flexibility is another dimension that is traded off against longer reach. For example, the loss associated with internal ROADM elements can be significant and require additional amplification that leads to a corresponding reduction in the noise-limited system reach. The ROADMs can also add new forms of signal impairment such as bandwidth narrowing penalties associated with passes through multiple cascaded ROADMs. Along with new penalties, ROADMs can add complexity to the system design. For example, while point-to-point systems can be designed for a specific desired reach, ROADM systems must accommodate WDM channels with different reach requirements, as well as channels with overlapping lightpaths that start and end at different nodes. In this section we describe the design elements shared by most ROADM-based transmission systems. In Section 6.4.9 we explore the additional challenges that arise when mesh networks are considered.

6.4.7.1 Designing for reach diversity

The ability to launch traffic from any ROADM node and drop traffic at any distance within the maximum reach capability creates a paradigm shift in transmission system design. Whereas previously the figure of merit for optical transport was the distance-bandwidth product alone, in ROADM systems, the bandwidth management flexibility must be taken into account.

Traditionally, the system dispersion map, which describes the accumulated dispersion as a function of transmission distance, was designed to enable optimized transmission to a specified distance. A dispersion map can include some combination of dispersion precompensation at the channel add location, in-line dispersion compensation and dispersion postcompensation [29]. In some cases, system designers may intentionally introduce or allow large accumulated dispersion during transmission, to avoid nonlinear effects that are enhanced by the propagation of well-formed bit patterns and long dispersion walk-off distances [30]. While this can enable good performance for the design distance, additional signal conditioning may be required to drop a signal at an intermediate distance and the performance will not be optimized for that distance. Several strategies have been proposed to address the problem of add anywhere, drop at any distance dispersion mapping, which we divide into three categories: (1) optimization for shared singly periodic dispersion maps; (2) return to zero and doubly periodic dispersion maps; and (3) per-channel dispersion management.

A singly periodic dispersion map (Figure 6.4.16(a)) is characterized by a constant residual dispersion per span (RDPS). Although optimized transmission can be achieved by using different RDPS values for different transmission distances, fixed RDPS values that are optimized for all distances up to the maximum reach have also been identified [31]. Different combinations of pre- and postcompensation can be incorporated to facilitate a single RDPS. The RDPS can be chosen such that the positive dispersion walk-off of the RDPS balances the nonlinear shaping due to self-phase modulation along the transmission path. This soliton-like dispersion map design is effective for nonreturn to zero formats as well as return to zero formats that can be configured for dispersion-managed soliton propagation, and has been demonstrated in a variety of ROADM experiments and field trials [22, 32, 33].

In general, doubly periodic dispersion maps (Figure 6.4.16(b)) provide an extra degree of freedom that can be exploited to achieve improved performance over singly periodic maps, albeit at the cost of greater complexity. A doubly periodic map will use a constant RDPS through some number of nodes followed by compensation to a new value and the pattern repeats, thus accommodating a wider range of transmission constraints. However, a common approach is to use a large RDPS at amplifier repeater sites and then bring the accumulated dispersion back down to a target value at each ROADM node. Returning the dispersion to zero at each

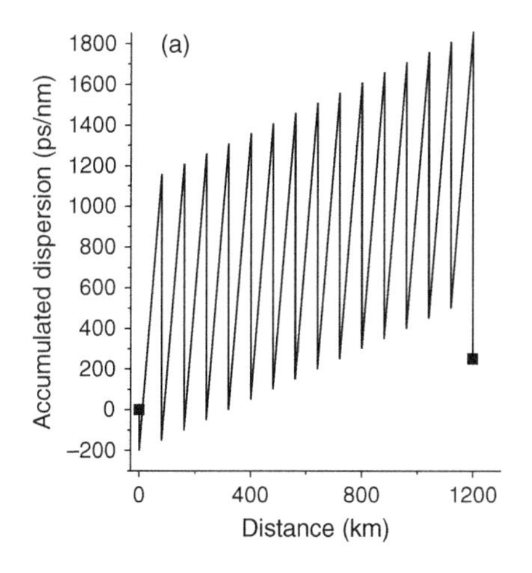

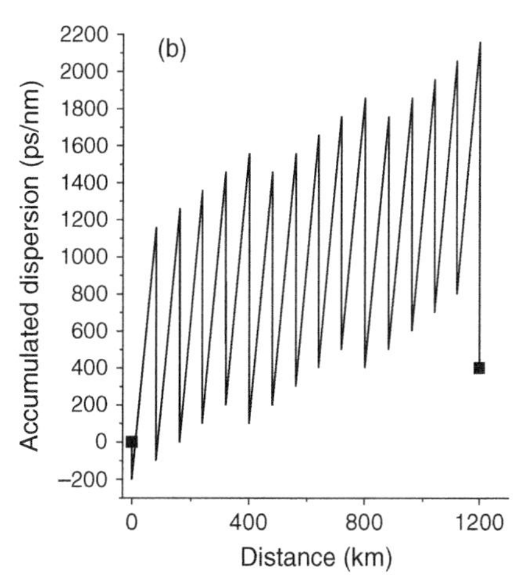

Figure 6.4.16 Examples of singly and doubly periodic dispersion maps for use in ROADM systems; (a) singly periodic dispersion map with −200ps/nm precompensation, 50ps/nm RDPS, and −300ps/nm postcompensation and (b) doubly periodic map with −200ps/nm precompensation, 100ps/nm RDPS, −300ps/nm postcompensation, and an additional −200ps/nm after every five spans.

ROADM is attractive because each channel will have the same nominal dispersion at the add/drop location. Interchannel nonlinear interactions, however, are increased because the bits are re-formed and re-aligned. This effect is particularly problematic for the extreme case of returning the dispersion to zero (or any fixed value) after every span. Several methods have been proposed to reduce such nonlinear crosstalk by de-correlating neighboring channels. For interleaver-based ROADM architectures (as shown in Figure 6.4.5(d)), this de-correlation may come largely for free due to the inevitable path-length differences among even and odd interleaver branches [34].

Per-channel dispersion compensation can be viewed as an alternative to dispersion maps or used in combination with a dispersion map. Much attention has been focused on compensation at the optical transceivers. Both electronic and optical dispersion compensation have been implemented in commercial optical receiver modules and are considered extensively elsewhere (Chapter 18 in Volume A). A given technology will have a maximum amount of accumulated dispersion that can be compensated. When used without other compensation devices, this maximum range will limit the size of network that can be accommodated for a given transmission format and rate. When used with a dispersion map, per-channel dispersion compensation at the receiver will increase the tolerance to dispersion variations (see Section 6.4.8.1). Likewise, per-channel dispersion trimming along the transmission path can further mitigate accumulated dispersion penalties, although at the expense of additional stand-alone single-channel compensation elements. However, integrating dispersion compensation into the WADD devices along with the power leveling capability would allow tuning at each ROADM node (see Section 6.4.7.2). Per-channel dispersion management also can be facilitated by modifying the optical transmitters. In one such technique distortion is applied to the transmitted waveform such that the waveform will be undistorted at the receiver [35]. This approach requires communication between the source and the destination nodes, and since the predistortion is path specific, the range of distances accommodated for drop and continue ROADM applications will be limited. Another approach is to apply a controlled chirp to the transmitted signal, which has an effect similar to using dispersion compensating fiber for precompensation [36]. As with postcompensation, these methods can be used with a dispersion map to provide greater tolerance to dispersion variations. Another alternative is to modify the modulation format and bit rate to achieve greater dispersion tolerance [37] (Chapter 2 in this volume). Recently, coherent receivers have attracted renewed attention because of their potential to provide electric field re-construction enabling one to correct for dispersion, as well as other sources of signal distortion.

Other dispersion management techniques, such as the use of nonlinear optical phase conjugators, have been proposed and are discussed elsewhere [38]. The most commonly used dispersion management method for transmission beyond a few hundred kilometers is the singly periodic dispersion map. Return-to-zero dispersion maps at every ROADM or at every span are sometimes used for simplicity, although at the price of reducing the transmission reach and power budget. More complex doubly periodic maps have generated research interest, but they generally do not provide sufficient improvement over singly periodic maps to justify their added complexity. Recently, per-channel dispersion management has been employed in 40 Gb/s transponders for transmission over in-line compensated links [22] and is commercially available at 10 Gb/s both for relaxing dispersion map engineering rules, e.g., Ref. [36] and providing transmission without broadband dispersion compensation [39].

6.4.7.2 ROADM cascade penalties

Signals transmitted through multiple ROADMs will accumulate additional penalties due to the ROADM transmission characteristics. The optical signal-to-noise ratio (OSNR) will be degraded due to optical losses and associated amplifier noise in the ROADM, which can be characterized by an effective noise figure. The effective noise figure is defined for each input port to output port path (including add and drop ports) as the noise figure of an equivalent amplifier used to replace the ROADM for that lightpath, with the same gain and noise performance. A multidegree ROADM core, as shown in Figure 6.4.9, will have 5–10 dB loss associated with the WSS and another 6–9 dB from the passive drop split. In addition, power leveling in the WSS requires an average loss of at least ½ of the maximum peak-to-peak channel power ripple, 2–4 dB. This leads to a total loss in the range 13–23 dB, not unlike the loss of a fiber span.

Wavelength multiplexing and demultiplexing for both the wavelength add/ drop and switching functions incur optical filtering. The details of the optical passband can impact the ROADM transmission performance. The passband must be sufficiently wide to accommodate the signal bandwidth over the full range of operating conditions, including the effects of spectral broadening due to fiber nonlinearities. Cascaded filtering results in a narrowing of the passband [28]. Figure 6.4.17 shows the passband narrowing of a WSS placed in a recirculating loop. Broadband ASE noise was incident on the input to the loop and subsequently shaped by each pass through the WSS. In Figure 6.4.17, the passband ripple growth is proportional to the number of ROADMs traversed because the filter shaping is identical with each

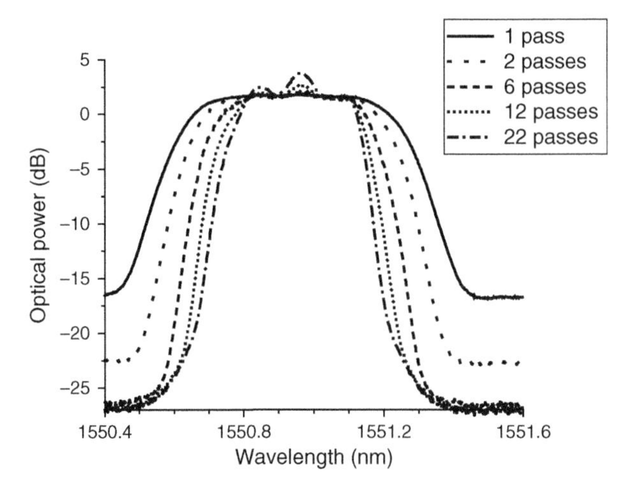

Figure 6.4.17 Filter narrowing due to multiple passes through a wavelength-selective switch (WSS). The noise floor is due to an amplifier after the WSS and the slight broadening below −20 dB is due to the accumulated ASE from many passes.

pass. For most practical ROADM systems, the passband ripple will grow more slowly because the filter shape will vary for different ROADMs. In general, with transmission through multiple ROADMs the signal bandwidth will tend to broaden due to nonlinearities while the cascaded ROADM bandwidth narrows. The resulting transmission penalty is a complicated function of the interaction between the ROADM filtering and the signal spectral shaping. It will depend on the details of the system design, such as modulation format and dispersion map, as well as the filter characteristics of the ROADMs. System studies have shown both positive and negative impact from cascaded filtering [40–42]. In addition to the amplitude response, the group-delay ripple, polarization-dependent loss, and polarization mode dispersion need to be characterized over the full passband.

Channel crosstalk is another parameter that must be carefully controlled in ROADM design. Crosstalk can be separated into in-band and out-of-band, as well as coherent and incoherent. In-band crosstalk is taken relative to the receiver bandwidth and includes any corresponding beat terms that fall within this bandwidth. In-band crosstalk can result in signal–signal beating on the one bits, in general inducing the largest penalties [43]. This beating is also polarization and signal quality dependent and will be worse if the interfering signal is a coherent copy of the original signal. In contrast, out-of-band crosstalk contributes a uniform background noise due to the total CW light detected. Primary sources of crosstalk in ROADM systems (Figure 6.4.18) include in-band interference due to channel blocking leakage, neighboring channel interference due to WDM channel filter leakage, and in-band multipath interference. The worst-case crosstalk transmission penalty will typically grow linearly with the number of occurrences at low crosstalk levels. However, the penalty depends on many factors including the characteristics of the signal and interferers (bit rate, modulation format, extinction ratio, OSNR, etc.) [44–46]. Therefore, one characterizes the worst-case penalty $p_{1\mathrm{pass}}(\varepsilon)$ as a function of the crosstalk power to signal power ratio ε, and then scales ε by the worst-case number of occurrences expected within the network, N, to determine the estimated penalty for transmission margin allocation: $P_{\mathrm{crosstalk}} = 10\log[p_{1\mathrm{pass}}(N\varepsilon)]$. Usually $P_{\mathrm{crosstalk}}$ is chosen to be a small value and this result is inverted to determine the required crosstalk rejection ratio ε for a given $P_{\mathrm{crosstalk}}$ and N. In some systems, the precise dependence of the penalty with N may need to be determined through careful measurements. With on the order of 10 interferers, less than 1 dB penalty is typically obtained for crosstalk rejection ratios in the range of $10^{-3}-10^{-4}$.

In-band multiple interferers due to channel blocking leakage is often the limiting case for ROADM crosstalk, particularly in mesh systems. When a channel is first added to the system, upstream traffic at that wavelength must be blocked. At each subsequent node, traffic at the same wavelength incident from a different input port, also must be blocked to express the channel of interest. Thus, for each ROADM that a signal cascades through there may potentially be traffic on the same wavelength that is blocked and thus a potential source of crosstalk if the ROADM blocking is not adequate. On the other hand, encroachment from neighboring channels is more of an issue related to the filter function. For the tight channel spacing, steep filter skirts will reduce crosstalk while minimizing the filter narrowing. Multipath interference (MPI) is a general problem in transmission systems. For the mesh systems, one has the additional source shown in Figure 6.4.18(c) in which different mesh

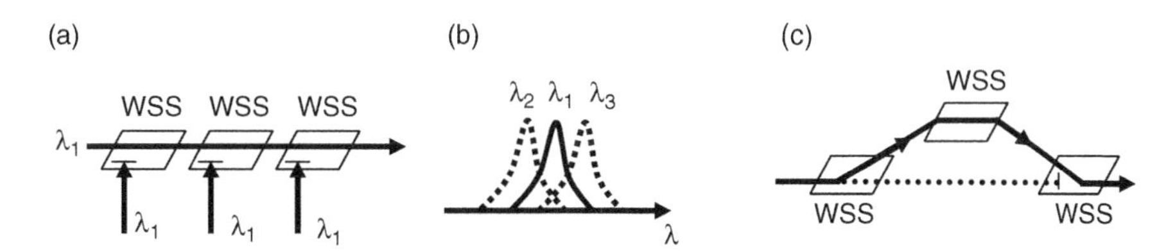

Figure 6.4.18 Sources of crosstalk in ROADM systems: (a) in-band multiple interferers, (b) WDM neighboring channel filter leakage, and (c) in-band multiple path interference.

lightpaths can lead to MPI, although this tends to be an infrequent source that is usually avoided due to the strong blocking required for the multiple interferer problem. MPI can also exist within ROADM components. Stray light scatter in the optical switching and MUX/DMUX elements can lead to MPI effects, which will occur with each ROADM pass and can be coherent. The crosstalk rejection of ROADM components is tested across the full channel passband. Often, the rejection is not uniform across the passband and the worst case may not reflect the channel performance because it is not present over the full passband. An appropriate weighting function can be incorporated to help relax the rejection requirements in these cases [47]. The behavior of crosstalk with channel attenuation must also be understood, since WBs and WSSs often provide a channel equalization function (see Section 6.4.5.5). In particular, some WSS designs can develop undesirable passband sidelobes when attenuated, e.g., due to diffraction effects. If present, these sidelobes must be accounted for in the system design. Ideally, the crosstalk rejection of ROADM components would be tested over all possible device configurations, but the large number of configurations for a WSS may render it impractical to characterize every device shipped in this way.

A promising approach for managing the limitations of the ROADM passband is to use WADDs that support channel passband adaptation. Recently, WSS devices have been introduced that can allow for variable channel spacing and passband width [15]. This capability could be particularly beneficial for supporting networks with mixed bit-rate and/or modulation-format signals. The WSS passband can be adapted to the width required for the different signal bandwidths or possibly even for different transmission distance/ROADM pass-through requirements. Furthermore, adaptation of the dispersion and optical loss across the passband has been demonstrated [48, 49], allowing for functionality such as dispersion compensation and optical signal equalization within the WADD.

Table 6.4.2 Sample margin allocation table for a ROADM transmission system (10.7 Gb/s NRZ-OOK, 2000 km, 15 ROADMs).

Line	Parameter	Value or penalty (dB)
1	Back-to-back required OSNR for BER $= 10^{-3}$	10.0
2	Transmitter/Receiver mismatch between vendors	0.5
3	Nonlinear transmission penalty (single channel)	1.0
4	Interchannel crosstalk penalty (WDM)	0.5
5	Polarization mode dispersion penalty	1.0
6	Dispersion slope penalty	0.5
7	Penalty due to channel power ripple and tilt	0.5
8	Polarization-dependent loss (including ROADM PDL) penalty	0.5
9	ROADM in-band crosstalk penalty	0.2
10	ROADM optical filtering penalty	0.2
11	ROADM network dispersion walk-off/error penalty	1.0
12	ROADM power control inaccuracy	0.5
13	Aging and repairs	1.5
14	Total margin allocation	7.9
15	Required EOL minimum received OSNR	17.9

6.4.7.3 ROADM system margin calculations and planning tools

Transmission margin calculations for ROADM systems will take into account additional penalties due to the ROADM transmission characteristics, and penalties associated with reconfiguration and transparency. Table 6.4.2 is a sample margin allocation table for a WDM 10.7 Gb/s non-return-to-zero, on-off keyed (NRZ-OOK) ROADM system with a maximum transmission distance of 2000 km and up to 15 ROADM passes. This table relates the received OSNR requirement for a system to the OSNR required in a back-to-back case. Note that the specific values and penalties will depend sensitively on the details of the system design and the numbers used here do not correspond to any particular system, but rather indicate possible values. Penalties on lines 2–8 are common to most transmission systems; whereas the penalties on lines 9–12 are unique to ROADMs. An overview of margin allocation for common transmission penalties can be found in Ref. [50]. Starting from the back-to-back transceiver required OSNR, the transmission penalties and margins are applied to determine the minimum received OSNR for the channel at its drop location. If the predicted end of life (EOL) delivered OSNR of a channel along a particular lightpath falls below the EOL minimum required OSNR on line 15, then the channel will need regeneration at an intermediate site to be provisioned along that connection.

If the penalties vary significantly between different network configurations, then multiple margin tables can be used to cover the different cases. For example, separate margin allocation will typically be provided for 10 and 40 Gb/s transmission cases. Separate margins may also be used for different fiber types or channel spacing. The penalties will correspond to the maximum value for transmission up to a particular distance and number of ROADM passes. Clearly a wide range of different distances and ROADM passes could be specified. However, including a broad range of system configurations in the margin allocation may unnecessarily increase the margins. The use of a planning tool can help to tighten the margins by calculating unique penalties for the different configurations.

Network planning tools have gained increased attention in ROADM systems to automate the complexity associated with the system engineering rules. These tools are used to optimally select equipment and also provide wavelength assignment and routing to minimize regeneration and cost. This includes incorporating differences in the margin allocation for each channel along its transmission path. These differences can be implemented as unique fixed values as in Table 6.4.2 or using algorithms to determine the accumulated penalty along the channel path. The planning tool can also be used to calculate the OSNR along the channel path to estimate whether the channel will reach its drop location and identify regenerator locations. Recently, there has been much interest in defining quality of transmission functions to estimate channel performance in planning tools [51]. These algorithms can be derived from detailed off-line simulation results covering the range of channel performance expected in the system and often accumulated over many months. The algorithms should then be validated against simulations and laboratory measurements of specific test cases [52]. Planning tools may include a suite of physical layer channel management and system configuration modules that in some cases intersect with network management software [53].

6.4.7.4 WDM channel power management

Perhaps the most significant impact of ROADMs on transmission system design is related to the WDM channel management aspects. In opaque networks, the channels are terminated at every node and the optical powers at the outputs of a node are generated by new transmitters and therefore decoupled from the optical power at the node inputs. The optical transparency engendered by ROADMs provides a continuous optical connection between the node inputs and outputs to enable selected channels to transparently bypass the node. This can potentially create optical coupling through the entire network. The system design must manage this coupling to prevent effects adverse to the system performance. A combination of architectural choices together with appropriate power control mechanisms can ensure stable system operation.

6.4.7.4.1 Steady-state channel power control

Transmission control can be segregated into steady-state and transient modes. Steady-state algorithms maintain the state of the system at or close to the optimum transmission parameters and manage planned transitions between such states, e.g., when new channels are provisioned. The steady-state control in ROADM systems is similar to the transmission control used in traditional point-to-point systems. Channel power levels are maintained near their designed targets and amplifier pump settings adjusted for minimum noise figure. The use of ROADMs, however, also introduces some unique requirements on the steady-state system control.

The power of WDM channels will evolve along the transmission path due to both time- and wavelength-dependent variations in the transmission spectra of the system components and fiber plant. Deviation of the channel power from the target value will lead to a penalty that increases with distance. Furthermore, the receiver will tolerate a range of input powers beyond which signal penalties will accrue. For short transmission distances, the receiver power tolerance will limit the allowable power error. Beyond a certain distance, the transmission impairments will dominate. In point-to-point links, deviations in WDM channel power are often characterized by a total power or average channel power error and channel power ripple and tilt. In ROADM systems there will also be a channel group power error, corresponding to groups of channels that originate from different node inputs and add ports. Changes in the total output power of an amplifier on a ROADM add path, e.g., will result in a power error unique to that group of add channels. Channel power interactions due to amplifier characteristics or fiber nonlinearities (notably stimulated Raman scattering) may cause this error to be transferred to other groups of channels.

Much of the channel power error can be accommodated by tuning the amplifier settings. These steady-state amplifier adjustments in ROADM systems are complicated due to the variability of the channel assignment. In point-to-point systems, wavelength channels can be turned up in a controlled pattern to minimize the impact on amplifier performance and simplify the amplifier control. In ROADM networks, the growth pattern implemented on a particular add port will be constrained by the wavelengths of channels already present from upstream nodes. Since the network must accommodate the unpredictable growth patterns of local traffic, it becomes important to maintain the target power levels and amplifier noise figures in the presence of arbitrary channel configurations. This can be a challenging task as the maximum number of channels supported by the system increases. Arbitrary channel configurations also complicate the detection of a loss of signal condition. Monitor photodiodes must be able to distinguish between the power of a single channel vs the total ASE noise power with no channels present. As the number of channels increases, allowing for a single channel at any location in the spectrum and allowing for the worst-case ASE growth can make discrimination difficult. Blocking unoccupied upstream channels will help to minimize ASE accumulation; however, one must still account for the maximum number of high loss spans between ROADMs.

Optical amplifiers in high-capacity ROADM systems are operated with some level of gain saturation. EDFAs are strongly saturated with gain compression similar to or greater than the range of channel loading [54]. Raman amplifiers will be mildly saturated, such that the gain will change <5 dB for the maximum variation in the number of input channels, even up to >100 channels [55]. Therefore, in each case, the amplifier must be adjusted to maintain a constant channel output power as the number of the input channels is varied. The channel power can be regulated by using either constant gain or constant output power control. In ROADM systems, amplifiers are usually operated with constant gain to allow for potentially large unplanned variations in channel loading, including those due to power transient events (see the section on Transient Control below). Constant power can be used if the unplanned changes in channel powers are within the system power tolerance or if other techniques are used to minimize the impact of gain saturation, such as using saturating lasers [56] or gain clamping [57]. Even with constant gain control, however, an amplifier will exhibit a constant power response until the action of the control system can adapt to a rapid input power change.

In ROADM systems, the amplifier control is complicated by the wavelength dependence of the channel loading. Here we focus on EDFA control, which is predominantly used in today's optical communication systems. With a single-channel growth pattern, e.g., adding channels from the shortest to the longest wavelength, and a flat input channel power spectrum, the gain and total output power for a single EDFA stage will come close to a one-to-one relationship with the pump power, simplifying the control design. However, for arbitrary channel loading the required pump power becomes a complicated function of the channel spectrum. In the example of Figure 6.4.19, the control process gain, defined as the pump power to total output power transfer function, is mapped out varying the EDFA input power by changing the number of channels at the input from 32 to 1, either starting from the long-wavelength end or from the short-wavelength end [58]. This single pump input stage amplifier is designed to operate with a 10.4 dB gain to yield a flat output spectrum using the gain-flattening filter. These results illustrate the wavelength dependence of the control gain and the amplifier gain saturation. For a given pump power, the control process gain will vary linearly with input power (on a linear scale) following a slope equal to the signal gain divided by the pump power, if the amplifier is linear. The fact that the amplifier control process gain is not constant indicates that nonlinear control techniques are necessary for optimal performance and the nonlinear parameters must account for differences due to different channel configurations. Such methods for characterizing amplifier behavior take into account the channel loading dependence. Besides the control process gain, another important parameter is the gain saturation measured as a function of channel loading, referred to as the channel gain saturation [59]. This quantity is more relevant to ROADM channel power dynamics than the amplifier gain saturation, which is typically measured as a function of total input power for a single reference channel configuration [60].

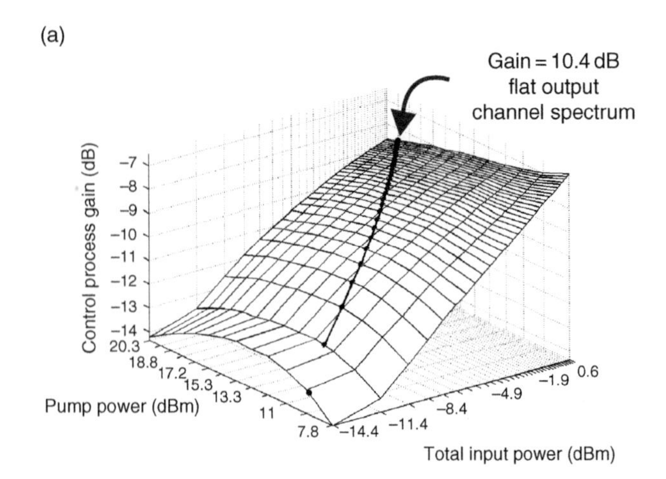

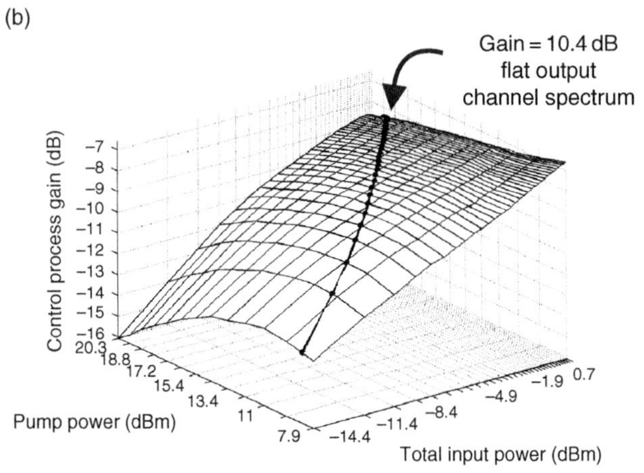

Figure 6.4.19 Control process gain ($k = P_{out}/P_{pump}$) for the input stage of a 32-channel WDM line amplifier (a) dropping channels from long to short wavelength and (b) from short to long wavelength; each input power grid line corresponds to a particular number of channels: −14.4 dBm corresponds to one channel, −11.4 corresponds to two channels, etc.

For a given mode of amplifier operation, constant gain or constant power, there will be different forms of channel power coupling that will impact the steady-state system control. For ROADM systems such coupling is particularly important with respect to the power control of channel groups associated with different mesh paths. These coupling mechanisms were first studied in constant power operation, typical of point-to-point systems. In the case of ideal, complete saturation, the total output power is a constant and equal to the product of the gain and the sum of the per-channel input powers. Assuming a flat gain spectrum with gain G, the total output power can be written in terms of the total input power P_{in}: $P_T = GP_1 + GP_2 = G[(1 - \eta)P_{in} + \eta P_{in}]$, where η is the fraction of input power (or channel number if the input channel powers are equal) initially in the second of two groups of channels, represented by total group input powers P_1 and P_2. If the first group of channels is adjusted in power by a factor f, P_T will remain fixed and the total group output power in channel group 2 will become [61]:

$$\begin{aligned} P_{o2(dB)} &= 10\log(G'\eta P_{in}) \\ &= 10\log(\eta P_T) - 10\log[\eta + (1-\eta)f], \end{aligned} \quad (6.4.3)$$

due to the change in $G \rightarrow G'$ required to maintain constant P_T. For the case that one channel group is completely dropped, $f = 0$ and $P_{o2(dB)}$ is increased by the usual factor of 10 logη. The impact of such channel power coupling in constant power networks is illustrated in Figure 6.4.20. In the ideal case and neglecting tilt, this effect is modified by the addition or deletion of channel groups, but does not vary with the number of amplifiers. As new groups of channels interact with groups carrying a deviation, higher products of the channel power/number ratio will appear and we refer to these successive interactions as being high order. Examples of first-, second-, and third-order channel power interactions are shown in Figure 6.4.20. From Figure 6.4.20 it is evident that the channel group power deviations decrease with higher order coupling. If more than two channel groups are involved, then additional features can be observed. Power coupling will only occur if the total power at the input to an amplifier changes. Thus, for case (b), if group G1 was not dropped at node B, then group G3 would be unaffected. The power deviation on group G1 was compensated by group G2 and the total input power to amplifier number three would remain unchanged. This effect can lead to situations in which a power error shared by groups of channels can propagate through

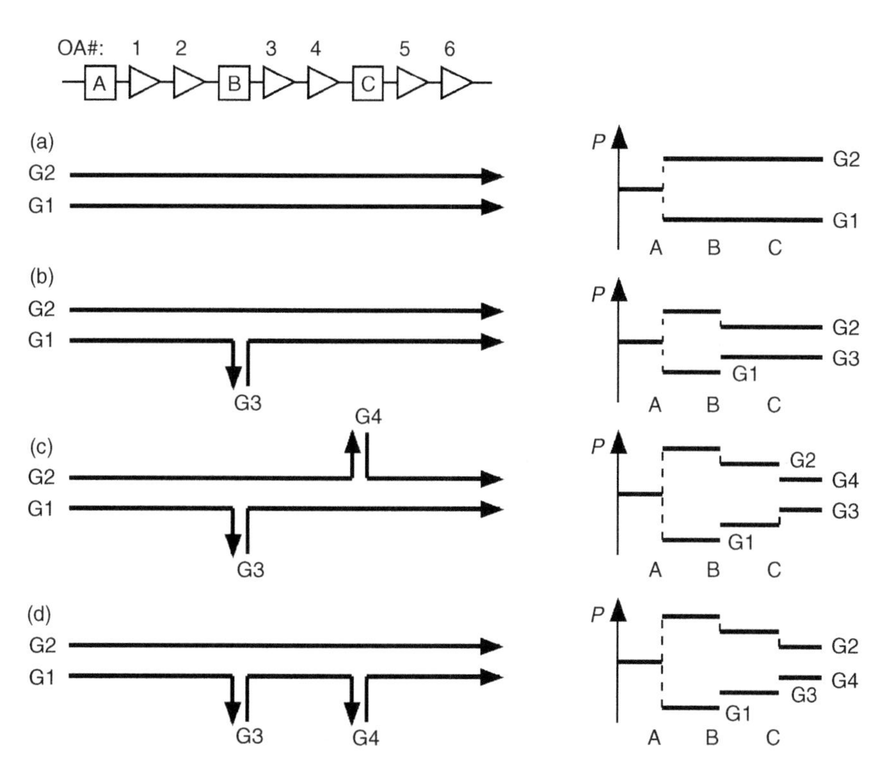

Figure 6.4.20 Channel power deviations for constant power amplified ROADM systems. As channel group G1 is reduced in power, (a) positive excursion occurs on neighboring channels G2 through a first-order interaction; (b) second-order interaction creates negative excursion in G3; (c) third-order interaction creates positive excursion in group G4; (d) first-, second-, and third-order interactions with groups G2, G3, and G4; the plots on the right-hand side illustrate the group power deviations as a function of distance corresponding to the add/drop patterns.

a network without impacting other channel groups until an affected group reaches its drop location. Furthermore, dropping groups of channels from a set that are influenced by a particular interaction, will result in a larger deviation for the remainder of the channels in the set, because the channel fraction η will decrease. Finally, it is important to note that if all of the channels at the input to an amplifier experience the same power deviation, then the constant power condition of the amplifier will remove the deviation and reset the channels to their correct power levels.

While the simple configurations in Figure 6.4.20 are useful to understand how groups of channels interact, in practice, very complex channel configurations may exist and numerical simulations are required to evaluate the dynamics. Figure 6.4.21 shows results from a simulation of a 64 ROADM node linear chain interconnected by single transmission spans (modeled as a simple loss element) and ideal constant power amplifiers [61]. Channel groups with a mean of four WDM channels were added and dropped between different ROADM nodes up to a maximum of 128 WDM channels per link. The channel groups were randomly distributed in wavelength, location, and distance such that the mean number of hops was 9 and the full-width at half-maximum was 2.5. A 3 dB power drop was applied to all of the add channels at a given node and the resulting power deviation ($\Delta P/P$) of the other channels as a function of hops from the event location is plotted after averaging over results from events occurring at every node. The deviation falls off rapidly with distance beyond the mean demand length as expected from interactions of the type illustrated in Figure 6.4.20. Here the absolute value of the deviation is plotted and the sharp dips correspond to sign changes as seen in Figure 6.4.20(c) such that the periodicity of the full oscillation is equal to the mean demand length. These features soften as the channel distance distribution

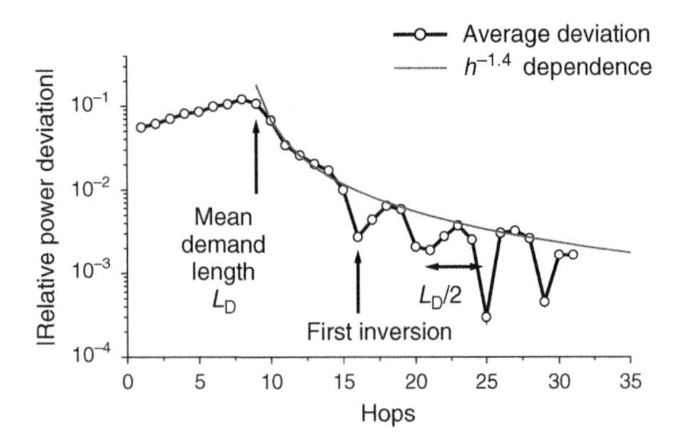

Figure 6.4.21 Simulated power deviations in a constant power amplified 64-node degree 2 ROADM system as a function of hop distance from the initial event location (this figure may be seen in color on the included CD-ROM).

broadens. Similar behavior was found in simulations for which the ideal amplifiers were replaced by EDFAs operating in deep saturation typical of transmission systems [61].

Although constant gain control was shown to mitigate the channel power coupling in early constant power experiments, gain ripple and tilt can give rise to coupling in constant gain amplification systems [62]. The impact of gain ripple and tilt can be understood by considering two groups of channels occupying different portions of the amplifier gain spectrum. Groups with total group input powers P_1 and P_2 experience different gains G_1 and G_2, respectively. A constant gain controlled amplifier maintains the total power gain such that $G = (G_1P_1 + G_2P_2)/(P_1 + P_2)$ is constant. For $G_1 \neq G2$, cross-coupling will occur when P_1 and P_2 are adjusted independently. For $G_1 > G_2$, an increase in P_1 will result in a change in the amplifier's total output power that is greater than the change in the total input power. As a result, the amplifier gain must be reduced to maintain constant total power gain. This gain change impacts both the channel groups and can result in additional gain tilt or ripple. We can write the individual group gains as $G_{1,2} = fG_{A0}g_{1,2}t_{1,2}$, where G_{A0} is the nominal amplifier total power gain, $g_{1,2}$ is the nominal gain ripple and tilt, $t_{1,2}$ is the gain dependent tilt, and f is the change required to maintain constant total power gain G. For an EDFA, the tilt is a function of the gain through the amplifier inversion level, and therefore it will also change, potentially exacerbating or mitigating the effect through the factor $t_{1,2}$. Unlike the channel power deviations in constant power amplifiers, the resulting errors in G_1 and G_2 grow linearly in cascade, modified by amplifier saturation and fiber nonlinearities.

Other mechanisms can also contribute to channel power coupling. Depending on the power and the number of channels (i.e., range of channel wavelengths), short-wavelength channels transfer power to longer wavelength channels through stimulated Raman scattering in the fiber. Also, there can be errors in the amplifier automatic gain control algorithm due to electronic noise and power measurement inaccuracy (usually only a potential problem in systems that require a large power adjustment dynamic range), optical noise (ASE) that contributes to the total power [63], spectral hole burning, and other nonlinearities in the amplifier [64–67]. Figure 6.4.22 illustrates how gain error in the automatic gain control can potentially evolve over distance and time when compensating for power changes on one channel group (G1) relative to another channel group (G2).

In addition to modifying amplifier settings, the steady-state control algorithm will tune control devices that often have much longer response times than the amplifiers. For example, the channel power attenuation in the WADD device may require one second or longer to

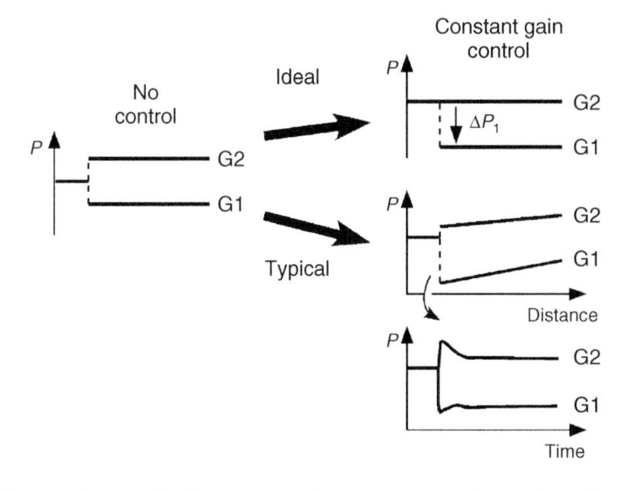

Figure 6.4.22 Ideal constant gain control of a saturated amplifier provides linear amplification and maintains the output power of the channels in group G2 at their target when channel group G1 falls below its target power by ΔP_1 (dB). In practice, both the channel groups will experience a finite gain error that varies over distance and time.

adjust a full set of channels. Furthermore, the steady-state control algorithm must propagate from node to node around the network. In part because of the channel power coupling mechanisms mentioned above, the adjustments made at one node can influence adjustments at other nodes. Therefore the settling time for a steady-state control algorithm can be quite long, depending on the size of the system. For this reason, a separate fast control mechanism is often implemented.

6.4.7.4.2 Transient Control

A fast control algorithm is implemented in ROADM systems to handle rapid and potentially large variations in channel power, e.g., due to fiber cuts. This form of transient control will suppress the channel performance degradation that may occur between the time of the fault and the time required for the steady-state control to adapt. In EDFA systems, transient control is usually implemented through a constant gain amplifier fast control algorithm. The speed required of the amplifier control is determined by the minimum rise time of the channel power excursions during transient events. If a set of channels loading an amplifier is rapidly removed, the remaining or surviving channels will increase in power due to the amplifier saturation as described above [68]. In Figure 6.4.23(a), the power excursion experienced by a surviving channel due to a 16–8 channel drop is shown for a cascade of 6 amplifiers [69]. Figure 6.4.23(b) shows maximum derivative of the rising edge of the excursion as a function of the number of amplifiers in the cascade. It is well known that the slope of the rising edge of the EDFA gain transient response will increase linearly in cascade according to [70]:

$$\dot{G}_n(0^+) \sim -\sum_{j=1}^{n} \Delta p_j^{in} g_j(0), \tag{6.4.4}$$

where G_n is the logarithmic gain of the nth amplifier, Δp_j^{in} is the change in input power for amplifier j, and $g_j(0)$ is the linear amplifier gain immediately prior to the event. Notice, however, in Figure 6.4.23(a) that the size of the excursion also varies in the cascade; this is due to channel power tilt and ripple. The same power coupling and gain errors discussed above for the steady control will also apply for the transient control. For these experiments there was no transmission fiber. If transmission fiber and DCF were included, the response would be further modified by the nonlinear interactions in the fiber.

The transient response in Raman amplifiers has several strong differences from the EDFA response. Figure 6.4.24 shows the uncontrolled transient response of an all-Raman amplified recirculating loop for surviving channels at the long- and short-wavelength ends of the spectrum [71]. The power excursions have been normalized to the final power after the event and correspond

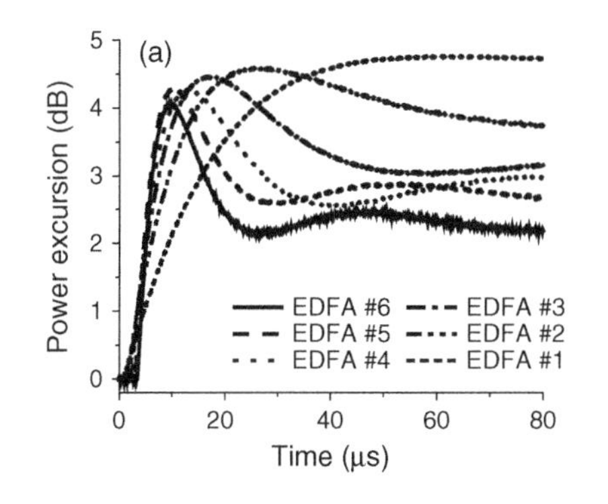

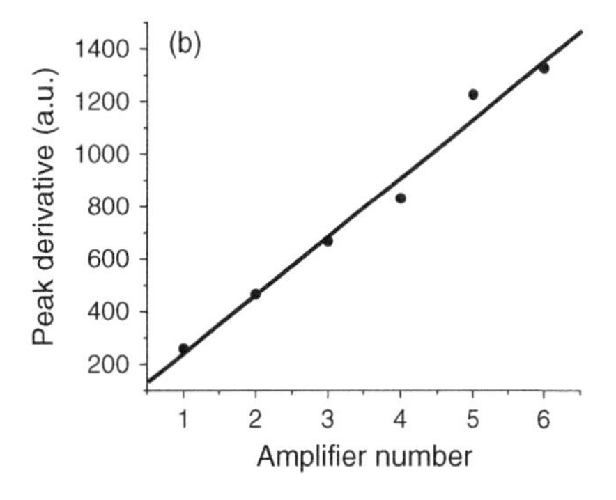

Figure 6.4.23 (a) Channel power excursion in an uncontrolled EDFA cascade experienced by 1 of 8 surviving channels after a drop from 16 to 8 channels. (b) Peak slope of rising edge grows linearly with number of amplifiers.

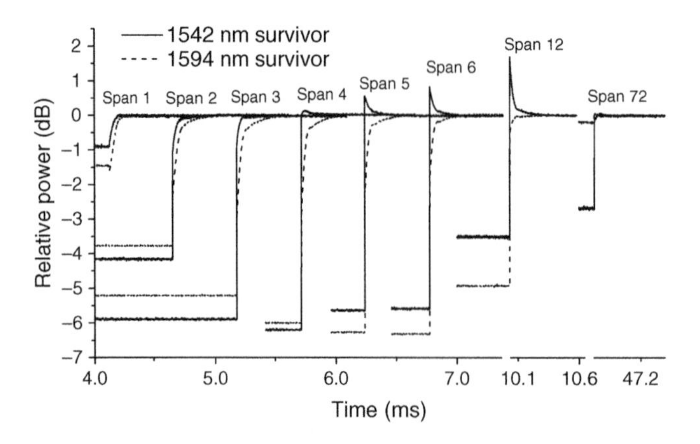

Figure 6.4.24 Surviving channel power excursions in an uncontrolled all-Raman amplified recirculating loop system following a 109–16 channel drop with 50 GHz spaced channels.

to the input of the indicated span number. Each transmission span consists of co-propagating and counter-propagating multipump amplifiers in the span and counter-propagating multipump amplifiers in the DCM fiber. The maximum power excursion is not reached until the input to the fourth span. Each of the three line amplifier types is characterized by a different characteristic response time. The DCM and span counter-pumped amplifiers create the double exponential rise evident on the curves from the long-wavelength channel at spans 3 and 4. The co-propagating pump and Raman channel interactions act on nanosecond time scales and result in the sharp edges. In the case of the short-wavelength channels, the co-propagating pump and channel Raman interactions add to create an overshoot, whereas the opposite is true for the long-wavelength channels. Beyond span 6, the growth of ASE noise reduces the power change and creates channel tilt.

Due to the large power excursions shown in Figures 6.4.23 and 6.4.24, amplifiers in ROADM systems are typically operated with constant gain control for fast power transient mitigation [72]. Both the Raman and the EDFA amplifiers are commercially available with fast pump power control to maintain constant gain. Feed-forward pump control was shown to be particularly effective in EDFAs to suppress the initial overshoot response [73]. Other EDFA techniques include replacing the lost signal power with power from additional control signals or counter-propagating signals and gain clamping methods [56, 57]. As noted above, spectral hole burning in EDFAs, which results in channel wavelength- and power-dependent gain variations, was shown to lead to significant gain error in long distance transmission [64–66]. EDFA amplifier control does not correct tilt variations due to interchannel Raman interactions in the transmission fiber. Fast tilt control techniques have been investigated to correct for Raman tilt errors [67, 74]. A variety of techniques have also been proposed for transient control in Raman amplifiers [75, 76], including multipump algorithms [77].

The responses in Figures 6.4.23 and 6.4.24 correspond to a simple amplifier cascade and not in general to the response of a network with ROADMs, in which channels are added and dropped all along the transmission path. As with the steady-state power errors discussed above, transient effects may be transferred from one channel group to another, persisting to other portions of the network in which the channel loading did not change [69]. In principle, such transient effects could propagate beyond the reach of any single lightpath. For example, the EDFA cascade used to obtain the results in Figure 6.4.23 was modified such that after the first two amplifiers in the cascade, all of channels that are cut in the transient event are dropped at a ROADM between amplifiers 2 and 3. However, the surviving channels continue and a new set of eight channels is added, maintaining amplifiers 3-6 at constant channel loading during the transient event (as in Figure 6.4.20(b) for the case that group G1 is cut upstream). However, the positive power excursion of the surviving channels stimulates a negative power excursion in the channels that add at the ROADM (Figure 6.4.25). Several features of this response are important for transient control in ROADM systems: (1) the rising edge of the original surviving channels no longer steepen, (2) the magnitude of the power excursion is reduced, being shared with the add channels to maintain constant power, and (3) the falling edge of the add channels steepens; however, it is bounded by the maximum slope of the rising edge of the surviving channels. Feature (3) is also true for the rising edge of the surviving channels relative to the falling edge of the cut channels. The consequence of (1) and (3) is that the control response time need only be as fast as the longest path shared by two independent channel groups in a transient event. No further steepening will occur. From (2) it is seen that the magnitude

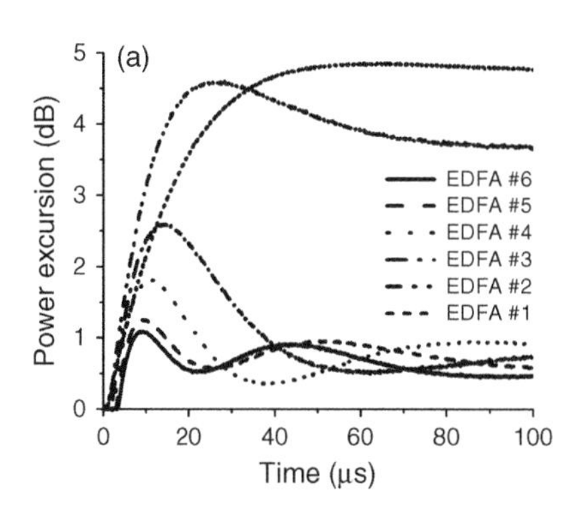

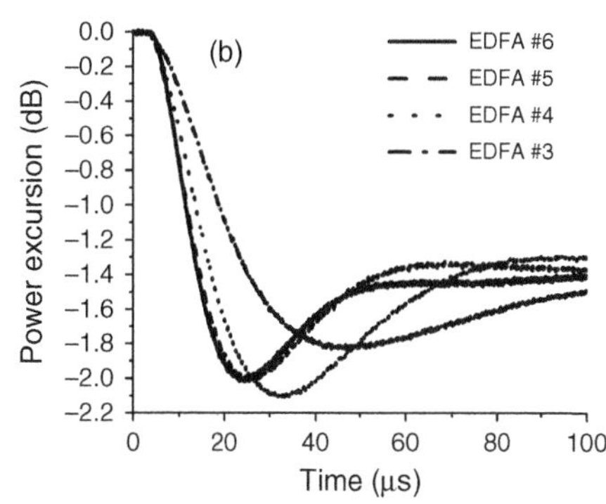

Figure 6.4.25 (a) Power excursion in an uncontrolled EDFA cascade experienced by 1 of 8 surviving channels following a 16–8 channel drop with channel reconfiguration after the second amplifier (as in Figure 6.4.20b). (b) Power excursion on 1 of 8 channels added after second amplifier due to second-order channel power interaction.

of the transient event will damp out in ROADM systems with a constant power constraint, both simplifying the power control and mitigating the propagation of power excursions.

6.4.8 ROADM networks

The use of ROADMs in mesh configurations requires additional considerations due to the complexity of optically transparent mesh networks. These issues include mesh dispersion map walk-off, wavelength routing, and mesh power control.

6.4.8.1 Optically transparent mesh transmission issues

Transmission in optically transparent mesh systems is a new topic of research. Much of the work to date has involved straight-line transmission experiments with mesh NEs such as ROADMs introduced to quantify the additional element penalties [27, 78]. Due to their size, mesh systems are difficult to study both from a computational [79] and a hardware point of view. Conventional recirculating loop experiments are well suited for the study of point-to-point systems. To study mesh-related effects, transmission experiments have been carried out using multiple interconnected recirculating loops [80, 81]. Another approach has been to use large laboratory experiments or field trials [22, 72, 82]. Recently, the issue of mesh dispersion map walk-off was studied using a dynamically reconfigurable recirculating loop experiment and is discussed below as an example of the types issues currently under investigation [83].

Dispersion map design will usually incorporate some allowance for deviations from the map. As mentioned in Section 6.4.7.1, these deviations can result from span dispersion measurement uncertainty or error and from systematic errors due to the dispersion slope across the band and granularity of available DCMs. In addition to these errors, in mesh systems a dispersion walk-off can occur due to intersecting line systems at nodes with degree greater than 2. Consider, e.g., a degree 3 ROADM for which the channels at one input have a span compensation error of 3 km because the closest available DCM value corresponds to spans 3 km longer than the span on the input. Suppose that the channels at the other input port experience a −4 km under compensation, again because of the DCM granularity. If the ROADM has only the first input port, then the compensation on the span at the ROADM output would be chosen for a target 3 km shorter than the actual span to correct for the compensation error on the previous span. Notice that this introduces a one time −3 km error on the add channels at the ROADM, even for a degree 2 ROADM. However, the channels on the second input port would experience a −7 km error because of the original error at the input plus the compensation for the other ROADM input port. This error can continue to grow with each ROADM for which the path of interest is not chosen for compensation at the ROADM output. Therefore, this walk-off must be included in the transmission penalties for a mesh design and will depend on the size of the DCM value granularity.

Experimentally verifying the system performance under conditions of DCM granularity can be challenging since many ROADMs would be required for a full system implementation. Recently, a dynamically reconfigurable recirculating loop was used to quantify transmission performance in the presence of compensation granularity error and walk-off [83]. Fast switches were placed at the output of each span so that different spools of fiber could be selected for the end of each span, varying the span length, while maintaining constant DCMs at mid-stage

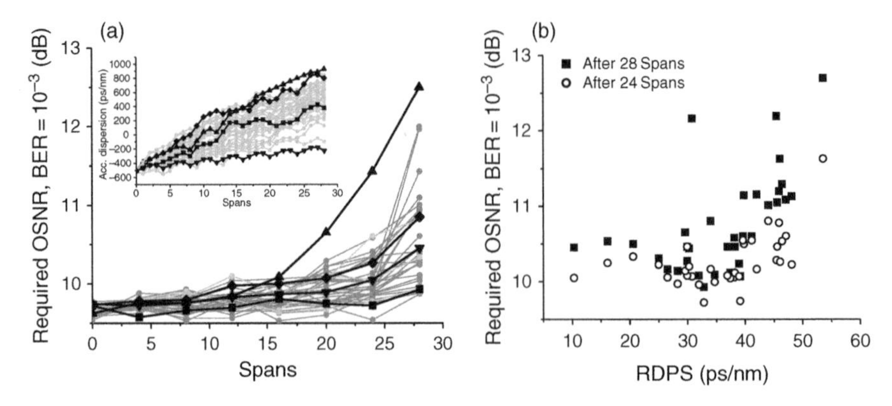

Figure 6.4.26 (a) Required OSNR for WDM channel at 1551 nm as a function of distance for 34 different dispersion maps (inset) realized using a dynamically reconfigurable recirculating loop. (b) Corresponding required OSNR as a function of RDPS.

on each of the amplifiers at the span output. The switches were operated synchronous with the recirculating loop loading switches such that 34 different dispersion maps could be realized. Figure 6.4.26(a) shows the maps and corresponding required OSNR for a BER of 10^{-3}. Figure 6.4.26(b) shows the corresponding required OSNR as a function of the mean RDPS over 24 and 28 spans, with the mean taken over 28 spans.

6.4.8.2 Wavelength routing and assignment

Wavelength routing and assignment in mesh networks is a challenging optimization problem [84]. A full account of the issues and literature is beyond the scope of this chapter. Because of the computational complexity of the problem, much of the optimization is reserved for predeployment system configuration planning. This need has led to the emergence of network planning tools (Section 6.4.7.3), which are distinct from network management and control software. Planning tools can be viewed as a collection of software packages that are needed for the development and deployment of ROADM systems. At a minimum, they will configure the channel wavelengths and routing in a network deployment, taking into account the system engineering rules and traffic matrices. More advanced tools will include data layer traffic engineering, equipment selection, physical layer transmission performance simulation, and maintenance/trouble shooting features. An important asset of planning tools is that they can incorporate known system performance data or measurements for a particular deployment and use that information to aid in the network optimization. In particular, whereas point-to-point systems need to provide margins that account for all channels traversing the longest reach, planning tools allow one to incorporate transmission-distance-dependent rules. This provides the potential to use impairment-constrained wavelength routing, which is a challenging optimization problem.

6.4.8.3 Optically transparent mesh channel management

Due to power coupling mechanisms in ROADM systems, mesh networks have the potential to form undesirable feedback paths around closed loops. The worst case involves direct optical power coupling at a single wavelength around a closed loop. Such closed loops can cause control mechanisms to run away and lead to optical lasing effects if enough amplification is available. Wavelength routing rules, ROADM control algorithms, and in many cases the physical switch design should prevent a single wavelength channel from being routed around a closed loop. Due to lightpath diversity requirements for protection, however, channels may be forced to take circuitous routes that could travel much of the distance around a loop. This will not lead to direct power feedback that would create optical closed loops, but care must be taken to avoid forming complete loops in these cases due to single switch failures or during provisioning and reconfiguration. A well-designed physical layer system control algorithm will include safeguards to prevent the possibility of forming closed loops, as well as channel collisions during normal operation. Likewise, ASE noise, both out-of-band and in-band, must be blocked. WSSs or blockers should be kept in the blocked position unless explicitly provisioned to create a planned path.

In addition to direct optical power coupling at a single wavelength, other power coupling mechanisms considered in Section 6.4.7.4 should be taken into account. Yoo et al. [85] demonstrated that the WDM channel power coupling that results from constant power amplifier operation can result in feedback and instability for ROADM mesh systems. In this case, no channel travels around the

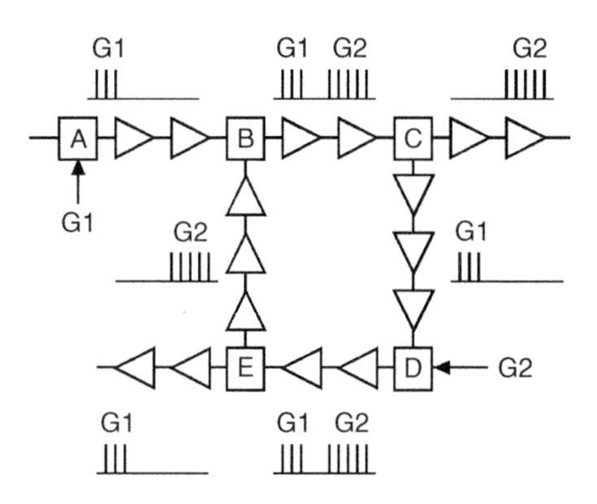

Figure 6.4.27 Mesh ROADM system with a closed loop. Three channels in group G1 add at node A and follow path A–B–C–D–E–; five channels in group G2 add at node D and follow path D–E–B–C–. Simultaneous adjustments at ROADMs B and E can couple through channel interactions due to overlapping paths.

complete loop. However, a change in power for one channel can influence the power level of other channels that may travel the remaining distance around the loop to complete the feedback path. This situation is illustrated in Figure 6.4.27. Channel groups G1 and G2 add at nodes A and D, respectively, and travel partially around the ring formed by nodes B, C, D, E. The groups do not take the shortest routes, which may occur, e.g., when a channel wavelength is already used by another channel in the link that provides the shorter route (this problem is referred to as wavelength blocking). Other groups of channels may also be present in this example, helping to facilitate the power coupling. However, as shown in Section 6.4.7.4, the amplitude of the power excursion is damped with the number of power coupling events and in proportion to the channel number or power ratio of the two groups of channels involved in the transfer. Therefore, the requirement that the amplitude of the feedback around a closed loop be greater than 1 is generally not met. A channel group with a large number of channels can transfer much of its power deviations to a smaller group, however, but the smaller group will not be effective at coupling the deviations back on the large group to complete a closed loop. Thus channel groups of similar size generally give the largest coupling around a ring. Another important factor is the channel power control algorithms that are used [86]. If the channel power controls operate simultaneously and independently, then this can lead to instability [85]. This does not occur if the power adjustments by the controllers are sequenced. While independent power control at each node is attractive in scaling the network, it leaves open the potential for instability.

For constant gain amplification, which is more common in ROADM systems, channel power coupling will occur through gain ripple and tilt, as well as fiber nonlinearities (see Section 6.4.7.4). Again, proper sequencing of the channel power control in different nodes is important to ensure stability. Figure 6.4.28(a) shows the oscillations that can occur for simultaneously adjusting nodes [62]. In this case, three WSS ROADM nodes similar to the architecture in Figure 6.4.13(a) were assembled in a ring configuration with four transmission spans. Groups of eight channels were placed at either end of the EDFA gain spectrum. The control algorithms were simultaneously stepwise executed and the power tilt between the two channel groups was measured after each iteration of a control cycle. The outputs of two ROADMs are shown to have opposite tilt that oscillates between positive and negative swings with each adjustment. These oscillations grew from a small initial tilt error. In Figure 6.4.28(b), the simultaneous control stepping was stopped and instead each WSS was adjusted through three control iterations before moving sequentially to the

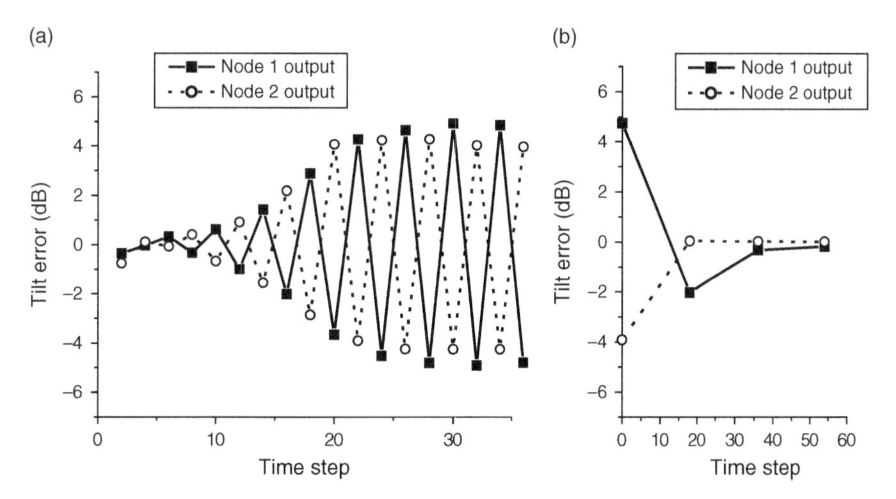

Figure 6.4.28 (a) Channel tilt oscillations created by simultaneous WSS channel power leveling in a ring configuration with constant gain amplification. (b) Tilt error correction when sequenced WSS channel power leveling is used.

next ROADM. The resulting tilt error is shown after all three ROADMs have been adjusted. After just two adjustments around the loop, the tilt error is negligible.

The use of node-sequenced channel power tuning avoids instabilities due to competing channel power adjustments. Sequencing through all nodes in a network, however, does not scale well. The time required to adjust the network can become large. The dynamic domains technique uses the channel paths in the network to define domains between which strong channel power coupling will not exist. These domains are allowed to adjust simultaneously allowing for scalable control [62].

6.4.9 Conclusions

In this chapter, we have reviewed the motivations, subsystem technologies, and network architectures underlying ROADMs and ROADM-based optical networks. Simple examples have been used to show the economic value of ROADMs. A variety of ROADM designs has been assessed and compared in terms of wavelength routing capability and other features, including modular growth, directional separability, and maximum fiber degree. An overview of the most successful current implementations has been presented, with an eye toward identifying appropriate network niches for each. The complex interplay between ROADM properties and optical transmission has also been explored, including a detailed discussion of static and dynamic channel power control. Finally, issues specific to optical networks containing ROADMs, such as dispersion map walk-off error and optically transparent mesh channel management, have been addressed.

ROADMs enable an automated and transparent network capable of rapid reconfiguration. To fully realize this vision within the growing global communication fabric, transmission systems must be capable of dealing with continual changes, including power transients and varying transmission conditions. Network management systems must solve complex problems in routing and wavelength blocking, path verification, and more as the photonic layer assumes some of the tasks previously handled by higher layers. Going forward, we can expect network operators to make more and more use of the capabilities of ROADMs, both to meet the growing traffic demand and to offer new services, such as providing entire wavelengths on demand. Advanced ROADM functionality, such as colorless add/drop ports, steerable transponders, and adaptive passbands, will be increasingly sought after, as will new and better solutions for signaling, network management, and mesh transmission. By meeting these challenges, the optical R&D community will help address the world's need for flexible, economical, and scalable networks.

Acknowledgments

The authors would like to acknowledge the help and support of their colleagues at AT&T, Bell Laboratories, and the Alcatel-Lucent Optical Networks Business division, and to extend special thanks to Martin Birk, S. Chandrasekhar, Randy Giles, Dah-Min Huang, Peter Magill, Patrick Mock, and Chris White.

References

[1] A. H. Gnauck and P. J. Winzer, "Optical phase-shift-keyed transmission," *J. Lightw. Technol.*, 23, 115–130 (2005) and references therein.

[2] J. Strand and A. Chiu, "Realizing the advantages of optical reconfigurability and restoration with integrated optical cross-connects," *J. Lightw. Technol.*, 21, 2871–2882 (2003).

[3] J. M. Simmons, E. L. Goldstein, and A. A. M. Saleh, "Quantifying the benefit of wavelength add–drop in WDM rings with distance-independent and dependent traffic," *J. Lightw. Technol.*, 17, 48–57 (1999).

[4] S. Chaudhuri and E. Goldstein, "On the value of optical-layer reconfigurability in IP-over-WDM lightwave networks," *IEEE Photon. Technol. Lett.*, 12, 1097–1099 (2000).

[5] A. F. Wallace, "Ultra Long-Haul DWDM: Network Economics," in *Proc. OFC.*, paper TuT1 (2001).

[6] A. Chiu and C. Yu, "Economic Benefits of Transparent OXC Networks as Compared to Long Systems with OADMs," in *Proc. OFC.*, paper WQ2 (2003).

[7] C. F. Lam, N. J. Frigo, and M. D. Feuer, "A taxonomical consideration of optical add/drop multiplexers," *Photonic Netw. Commun.*, 3, 327–333 (2001).

[8] J. Ford, "Micromechanical Wavelength Add/Drop Switching: From Device to Network Architecture," in *"Proc. OFC*, pp. 253–255 (2003).

[9] M. D. Feuer and D. Al-Salameh, "Routing Power: A Metric for Reconfigurable Wavelength Add/Drops," in *"Proc. OFC*, pp. 156–158 (2002).

[10] C. R. Giles and M. Spector, "The wavelength add/drop multiplexer for lightwave communication networks," *Bell Labs Technical Journal*, 4, 207–229 (1999).

[11] P. Mock and M. D. Feuer, private communication.

[12] T. Strasser, "ROADMs turn economical for edge networks," *Lightwave Mag.*, 24(7) (2007).

[13] C. F. Lam, "Next Generation Metropolitan Area WDM Optical Networks," in *Proc. APOC.*, paper 6354–42 (2006).

[14] E. B. Basch, R. Egorov, S. Gringeri, and S. Elby, "Architectural tradeoffs for reconfigurable dense wavelength--division multiplexing systems," *IEEE J. Sel Top. Quantum Electron.*, 12, 615–626 (2006).

[15] G. Baxter, S. Frisken, D. Abakoumov et al., "Highly Programmable Wavelength Selective Switch Based on Liquid Crystal on Silicon Switching Elements," in *Proc. OFC*, paper OTuF2 (2006).

[16] S. L. Woodward, M. D. Feuer, C. F. Lam et al., "Broadcasting a Single Wavelength Over a WDM Network," in *Proc. OFC.*, paper WBB2 (2001).

[17] C. F. Lam, M. Boroditsky, B. Desai, and N. J. Frigo, "A novel dynamic crosstalk characterization technique for 3-D photonic crossconnects," *IEEE Photon. Technol. Lett.*, 15, 141–143 (2003).

[18] T. Afferton, quoted in P. A. Bonenfant, and M. L. Jones, "OFC 2003 workshop on wavelength selective switching based optical networks," *J. Lightw. Technol.*, 22, 305–309 (2004).

[19] S. L. Woodward, M. D. Feuer, J. Calvitti et al., "A High-Degree Photonic Cross-Connect for Transparent Networking, Flexible Provisioning & Capacity Growth," in *Proc. ECOC*, paper Th1.2.2 (2006).

[20] Dah-Min Huang, private communication.

[21] D. C. Kilper, R. Bach, D. J. Blumenthal et al., "Optical performance monitoring," *J. Lightw. Technol.*, 22, 294–304 (2004).

[22] D. A. Fishman, W. A. Thompson, and L. Vallone, "*LambdaXtreme*®" transport system: R&D of a high capacity system for low cost, ultra long haul DWDM transport," *Bell Labs Technical Journal*, 11, 27–53 (2006).

[23] For a general review of the use of pilot tones in WDM systems see, H. C. Ji et al., "Optical performance monitoring based on pilot tones for WDM network applications," *J. Opt. Netwk.*, 3, 510–533 (2004).

[24] M. D. Feuer and V. A. Vaishampayan, "Demonstration of an in-Band Auxiliary Channel for Path Trace in Photonic Networks," in *Proc. OFC.*, paper OWB6 (2005).

[25] M. D. Feuer and V. A. Vaishampayan, "Rejection of interlabel crosstalk in a digital lightpath labeling system with low-cost all-wavelength receivers," *J. Lightw. Technol.*, 24, 1121–1128 (2006).

[26] A. A. M. Saleh and J. M. Simmons, "Evolution toward the next-generation core optical network," *J. Lightw. Technol.*, 24, 3303–3321 (2006).

[27] I. Tomkos, M. Vasilyev, J.-K. Rhee et al., "80 × 10.7 Gb/s Ultra-Long-Haul (+4200km) DWDM Network with Reconfigurable 'Broadcast & Select' OADMs," in *Proc. OFC*, paper FC1 (2002).

[28] D. T. Neilson, C. R. Doerr, D. M. Marom et al., "Wavelength selective switching for optical bandwidth management," *Bell Labs Technical Journal*, 11, 105–128 (2006).

[29] X. Xiao, S. Gao, Y. Tian, and C. Yang, "Analytical optimization of the net residual dispersion in SPM-limited dispersion managed systems," *J. Lightw. Technol.*, 24, 2038–2044 (2006); including a discussion of recent map design considerations and references therein.

[30] J.-R. Essiambre, G. Raybon, and B. Mikkelsen, "Pseudo-linear transmission of high-speed TDM signals: 40 and 160 Gb/s," in *Optical Fiber Telecommunications IV B – Systems and Impairments* (I. Kaminow and T. Li, eds), New York: Academic, pp. 232–304 (2002).

[31] N. Hanik, A. Ehrhardt, A. Gladisch et al., "Extension of all-optical network-transparent domains based on normalized transmission sections," *J. Lightw. Technol.*, 22, 1439–1453 (2004).

[32] D. F. Grosz, A. Agarwal, S. Banerjee et al., "All-Raman ultralong-haul single-wideband DWDM transmission systems with OADM capability," *J. Lightw. Technol.*, 22, 423–432 (2004).

[33] A. R. Pratt, P. Harper, S. B. Alleston et al., "5,745km DWDM Transcontinental Field Trial Using 10Gbit/s Dispersion Managed Solitons and Dynamic Gain Equalization," in *Proc. OFC.*, paper PD26 (2003).

[34] C. Xie, "A doubly periodic dispersion map for ultralong-haul 10- and 40-Gb/s hybrid DWDM optical mesh *networks,*" *IEEE Photon. Technol. Lett.*, 17, 1091–1093 (2005).

[35] J. McNicol, M. O'Sullivan, K. Roberts et al., "Electrical Domain Compensation of Optical Dispersion" in *Proc. OFC.*, paper OThJ3 (2005).

[36] S. Chandrasekhar, D. C. Kilper, X. Zheng et al., "Evaluation of Chirp-Managed Lasers in a Dispersion Managed DWDM Transmission Over 24 spans," in *Proc. OFC.*, paper OthL7 (2007).

[37] P. J. Winzer and R.-J. Essiambre, "Advanced modulation formats for high capacity optical transport networks," *J. Lightw. Technol.*, 24, 4711–4728 (2006); see also Chapter 2 in this volume.

[38] R.-D. Li, P. Kumar, and W. L. Kath, "Dispersion compensation with phase-sensitive optical amplifiers," *J. Lightw. Technol.*, 12, 541–549 (1994).

[39] M. Birk, X. Zhou, M. Boroditsky et al. , "WDM Technical Trial with Complete Electronic Dispersion Compensation," in *Proc. ECOC*, paper Th2.5.6 (2006).

[40] H. K. Kim, S. Chandrasekhar, M. Spector, and L. Buhl, "Error-free operation of large scale (16 nodes cascaded over total distance of 600 km) optical network based on dense WDM (15 100GHz-spaced l0Gbit/s/channel) Channels," *Electron. Lett.*, 36, 654–655 (2000).

[41] E. Pincemin et al., "Robustness of 40Gb/s ASK modulation formats in the practical system infrastructure," *Opt. Express*, 14, 12049–12062 (2006).

[42] J. Wiesenfeld, L. D. Garrett, M. Shtaif et al., "Effects of DGE Channel Bandwidth on Nonlinear ULH Systems," in *Proc. OFC.*, paper OWA2 (2005).

[43] E. L. Goldstein and L. Eskildsen, "Scaling limitations in transparent optical networks due to low-level crosstalk," *IEEE Photon. Technol. Lett.*, 7, 93–94 (1995).

[44] F. Liu, C. J. Rasmussen, and R. S. J. Pedersen, "Experimental verification of a new model describing the influence of incomplete signal extinction ratio on the sensitivity degradation due to multiple interferometric crosstalk," *IEEE Photon. Technol. Lett.*, 11, 137–139 (1999).

[45] H. K. Kim and S. Chandrasekhar, "Dependence of in-band crosstalk penalty on the signal quality in optical network systems," *IEEE Photon. Technol. Lett.*, 12, 1273–1274 (2000).

[46] P. J. Winzer, M. Pfennigbauer, and R. -J. Essiambre, "Coherent crosstalk in ultradense WDM systems," *J. Lightw. Technol.*, 23, 1734–1744 (2005).

[47] T. Zami, B. Lavigne, E. Balmefrezol et al., "Comparative Study of Crosstalk Created in 50GHz-Spaced

Wavelength Selective Switch for Various Modulation Formats at 43Gbit/s," in *Proc. ECOC*, paper We3.P.81 (2006).

[48] R. Ryf, Y. Su, L. Moeller et al., "Wavelength blocking filter with flexible data rates and channel spacing," *J. Lightw. Technol.*, 23, 54–60 (2005).

[49] M. A. F. Roelens, J. Bolger, G. Baxter et al., "Tunable Dispersion Trimming in Dynamic Wavelength Processor at 80Gbit/s per Channel," in *Proc. OFC.*, paper PDP44 (2007).

[50] ITU–Series Recommendation—Supplement 39, 2006 ITU-T G-Series Recommendation—Supplement 39, "Optical system design and engineering considerations," *International Telecommunication Union* (2006).

[51] D. Penninckx and C. Perret, "New physical analysis of 10Gb/s transparent optical networks," *IEEE Photon. Technol. Lett.*, 15, 778–780 (2003).

[52] S. Herbst, H. Lucken, C. Furst et al., "Routing Criterion for XPM-Limited Transmission in Transparent Optical Networks," in *Proc. ECOC* (2003).

[53] S. A. Taegar, D. A. Fishman, and D. L. Correa, "Engineering and Planning Tool for an Ultra-Long Haul Optical Mesh Transport System," in *Proc. NFOEC.*, paper NTuF4 (2006).

[54] E. L. Goldstein, L. Eskildsen, C. Lin, and R. E. Tench, "Multiwavelength propagation in lightwave systems with strongly inverted amplifiers," *IEEE Photon. Technol. Lett.*, 6, 266–269 (1994).

[55] M. Karasek and M. Menif, "Channel addition/removal response in raman fiber amplifiers: modeling and experimentation," *J. Lightw. Technol.*, 20, 1680–1687 (2002).

[56] A. K. Srivastava, J. L. Zyskind, Y. Sun et al., "Fast-link control protection of surviving channels in multiwavelength optical networks," *IEEE Photon. Technol. Lett.*, 9, 1667–1669 (1997).

[57] D. H. Richards and J. L. Jackel, "A theoretical investigation of dynamic all-optical gain control in multichannel EDFA's and EDFA cascades," *IEEEJ. Sel. Top. Quantum Electron.*, 3,1027–1036 (1997).

[58] D. C. Kilper, M. Waldow, W. Etter, and C. Xie, "Control Process Gain of Erbium Doped Fiber Amplifiers with Wavelength Division Multiplexed Signals," in *Proc. OAA* (2006).

[59] P. Vorreau, D. C. Kilper, and C. White, "Gain saturation spectrum of backward-pumped broadband Raman amplifiers," *IEEE Photon. Technol. Lett.*, 17, 1405–1407 (2005).

[60] N. Takahashi, T. Hirono, H. Akashi et al., "An output power stabilized erbium-doped fiber amplifier with automatic gain control," *IEEE J. Sel Top. Quantum Electron.*, 3, 1019–1026 (1997).

[61] D. C. Kilper and C. A. White, "Fundamental Saturated Amplifier Dynamics in Transparent Networks," in *Proc. ECOC*, paper We4.P.96 (2005).

[62] D. C. Kilper, C. A. White, and S. Chandrasekhar, "Control of Channel Power Instabilities in Transparent Networks Using Scalable Mesh Scheduling," in *Proc. OFC.*, paper PDP11 (2007).

[63] C. Simonneau and V. Leng, "Impact of Spectral Hole Burning and ASE in Constant Gain EDFA with Input Channel Count Variation," in *Proc. ECOC*, paper We3.P.9 (2006).

[64] H. Nakaji and M. Shigematsu, "Wavelength dependence of dynamic gain fluctuation in a highspeed automatic gain controlled erbium-doped fiber amplifier," *IEEE Photon. Technol. Lett.*, 15, 203 (2003).

[65] C. Fuerst et al. "Impact of Spectral Hole Burning and Raman Effect in Transparent Optical Networks," in *Proc. ECOC*, paper Tu.4.2.5 (2003).

[66] M. Bolshtyanski, "Spectral hole burning in erbium doped fiber amplifiers," *J. Lightw. Technol.* 21, 1032–1038 (2003).

[67] P. Krummrich and M. Birk, "Experimental investigation of compensation of Raman induced power transients from WDM channel interactions," *IEEE Photon. Technol. Lett.*, 17, 1094–1096 (2005).

[68] Y. Sun, A. K. Srivastava, J. L. Zyskind et al., "Fast power transients in WDM optical networks with cascaded EDFAs," *Electron. Lett.*, 33, 313–314 (1997).

[69] D. C. Kilper, S. Chandrasekhar, and C. A. White, "Transient Gain Dynamics of Cascaded Erbium Doped Fiber Amplifiers with Re-Configured Channel Loading," in *Proc. OFC.*, paper OtuK6 (2006).

[70] A. Bononi and L. Rusch, "Doped fiber amplifier dynamics: a system perspective," *J. Lightw. Technol.*, 16, 945–956 (1998).

[71] A. R. Grant and D. C. Kilper, "Signal Transient Propagation in an All Raman Amplified System," in *Proc. OFC.*, paper ThT3 (2004).

[72] N. Madamopoulos, D. C. Friedman, I. Tomkos, and A. Boskovic, "Study of the performance of a transparent and reconfigurable metropolitan area network," *J. Lightw. Technol.*, 20, 937–945 (2002).

[73] C. Tian and S. Kinoshita, "Analysis and control of transient dynamics of EDFA pumped by 1480- and 980-nm lasers," *J. Lightw. Technol.*, 21, 1728 (2003).

[74] K. Motoshima, N. Suzuki, K. Shimizu et al., "A channel-number insensitive erbium-doped fiber amplifier with automatic gain and power regulation function," *J. Lightw. Technol.*, 19, 1759–1767 (2001).

[75] C. J. Chen and W. S. Wong, "Transient effects in saturated Raman amplifiers," *Electron. Lett.*, 37, 371–372 (2001).

[76] Y. Sugaya, S. Muro, Y. Sato, and E. Ishikawa, "Suppression Method of Transient Power Response of Raman Amplifier Caused by Channel Adddrop," in *Proc. ECOC*, paper 5.2. 3 (2002).

[77] X. Zhou, M. Feuer, and M. Birk, "Sub-microsecond transient control for a forward pumped Raman amplifier," *IEEE Photon. Technol. Lett.*, 17, 2059–2061 (2005).

[78] G. Raybon, S. Chandrasekhar, A. H. Gnauck et al., "Experimental Investigation of Long-Haul Transport at 42.7-Gb/s Through Concatenated Optical Add/Drop Nodes," in *Proc. OFC*, paper ThE4 (2002).

[79] C. Checkuri, P. Claisse, R.-J. Essiambre et al., "Design tools for transparent optical networks," *Bell Labs Technical Journal*, 11, 129–143 (2006).

[80] P. Peloso, D. Penninckx, M. Prunaire, and L. Noirie, "Optical transparency of a heterogeneous pan-European network," *J. Lightw. Technol.*, 22, 242–248 (2004).

[81] L. M. Gleeson et al., in *Proc. ECOC*, Postdeadline paper Th.4.4.7 (2003).

[82] A. R. Pratt, B. Charbonnier, P. Harper et al., "40 × 10.7 Gbit/s DWDM Transmission Over a Meshed ULH Network with Dynamically Re-configurable Optical Cross Connects," in *Proc. OFC.*, paper PD9 (2003).

[83] D. Kilper, S. Chandrasekhar, E. Burrows et al., "Local Dispersion Map Deviations in Metro-Regional Transmission Investigated Using

a Dynamically Re-configurable Re-circulating Loop," in *Proc. OFC*, paper OThL5 (2007).

[84] H. Zang, J. P. Jue, and B. Mukherjee, "A Review of Routing and Wavelength Assignment Approaches for Wavelength-Routed Optical WDM Networks," *Opt. Netw. Mag.*, 1 (2000).

[85] S. B. J. Yoo, W. Xin, L. D. Garrett et al., "Observation of prolonged power transients in a reconfigurable multiwavelength network and their suppression by gain-clamping of optical amplifiers," *IEEE Photon. Technol. Lett.*, 10, 1659–1661 (1989).

[86] D. Gorinevsky and G. Farber, "System analysis of power transients in advanced WDM networks," *J. Lightw. Technol.*, 22, 2245–2255 (2004).

Chapter 6.5

6.5

Fiber-based broadband access technology

Kaminow, Li and Willner

Abstract

After more than 20 years of research and development, a combination of technological, regulatory, and competitive forces are finally bringing fiber-based broadband access to commercial fruition. Three main approaches, hybrid fiber coax, fiber to the cabinet, and fiber to the home, are each vying for a leading position in the industry, and each has significant future potential to grow customers and increase bandwidth and associated service offerings. Further technical advances and cost reductions will be adopted, eventually bringing performance levels and bandwidth to Gb/s rates when user demand warrants while keeping service costs affordable.

6.5.1 Introduction

The use of fiber-optic technology in telecommunications systems has grown over the past 25 years, since its introduction as a transmission media for linking metropolitan central offices in 1980. In the decade after that, long-distance networks deployed fiber-based systems extensively, followed by significant construction of metropolitan interoffice network infrastructure. During that time the reliability and availability of transport systems improved dramatically due largely to the low failure rates of the technology. As a result, researchers and developers dreamed of a time when fiberoptic technology could be economically applied in access networks [1] to replace the copper-based systems extending from central offices to residences and businesses (see Figure 6.5.1 for a diagram [2] of such scenarios). A perfect example of this visionary dream was articulated by Paul Shumate and Richard Snelling in an *IEEE Communications Magazine* article published in 1989 [3], where they predicted that a fiber-to-the-home (FTTH) solution could be at economic parity with copper-based solutions by 1995 if a carrier were to deploy 1–3 million lines (see Figure 6.5.2 for a replication of their prediction chart). They went on to explain that it could be possible to convert the entire US network to FTTH over a 25-year period, producing a broadband network for the Information Age. They further noted that there were many details to be worked out, including standardization, network interfaces, powering at the customer end, video capability, and network evolution. This work was pioneering and prophetic, as we now know today.

But what those authors, and others with them, did not foresee was the dramatic effect of the telecommunications consent decree on competition in the United States, the dampening effect of the federal communications commission (FCC) policy allowing competing carriers to lease infrastructure at cost or below, the delaying effect of the telecom "bubble" bursting [4] (Figure 6.5.3), and the bandwidth growth driver of the Internet when used for business and entertainment.

Now, in Asia, North America, Europe, and globally, we are seeing significant capital spending on fiber-based broadband access infrastructure [5, 6]. Some of the earliest spending was in Japan, where government funding and initiatives helped to push carriers toward fiber-based access network technologies. Soon after, in the United States, spending was driven by three events: the FCC has made an historic policy decision in 2003 to free fiber-based solutions from competitive regulations, cable TV (CATV) operators have increased telecommunications

Optical Fiber Telecommunications V B; ISBN: 9780123741721

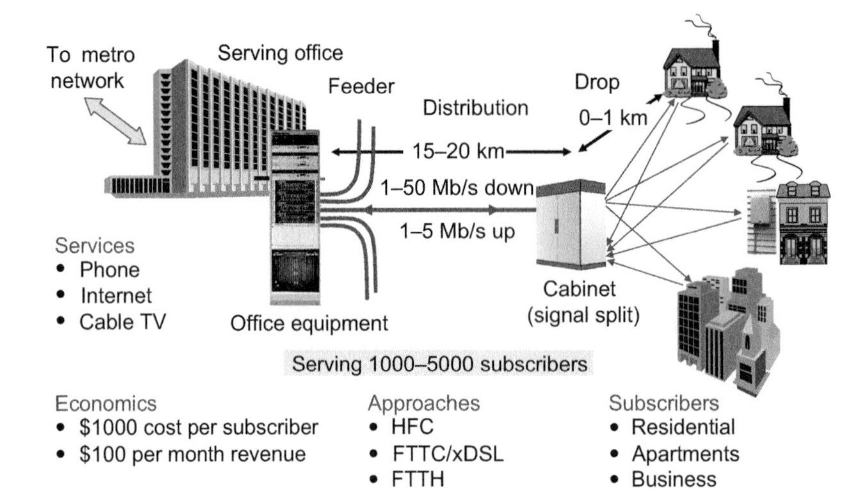

Figure 6.5.1 FTTx access network reference models.

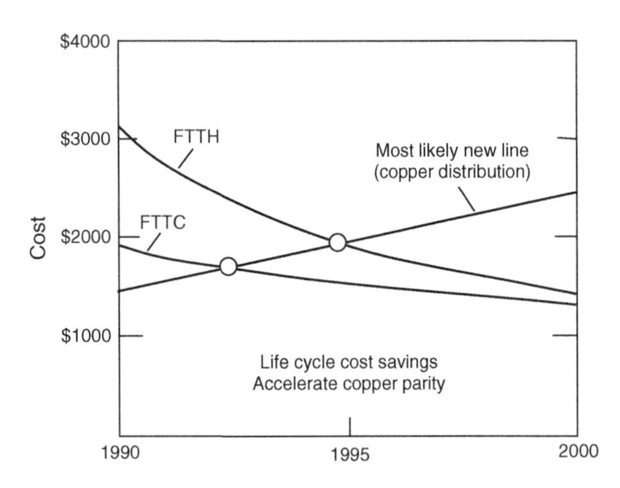

Figure 6.5.2 FTTH vs copper introduction cross-over date (C. 1991 Shumate).

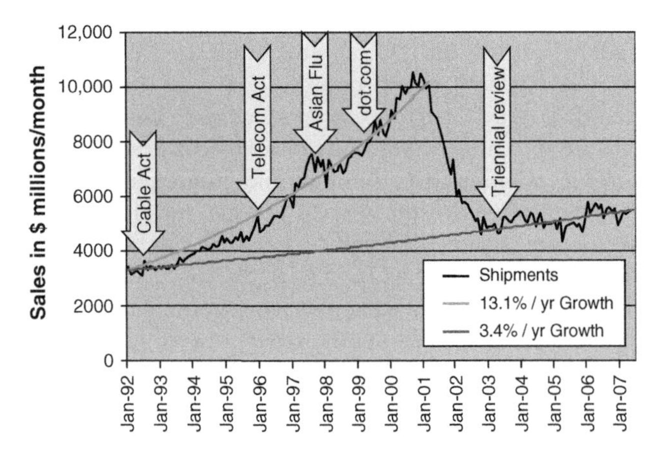

Figure 6.5.3 Telecom bubble—telecom equipment shipped in the United States vs time.

industry competition by offering high data rate Internet service and telephone service over the Internet, and users have grown accustomed to using the Internet for sharing large digital files containing e-mail messages, photos, music, video, and news clips.

In response to these drivers, the three major local exchange carriers in the United States (Verizon, BellSouth, and SBC) issued a common request for information to broadband access system suppliers, trusting that the resulting volume manufacturing incentives would bring the cost of deploying such systems in line with associated service revenues. Now it appears that this risky move was justified, because in North America we have seen about 8 million fiber-based broadband access lines deployed to homes since then, with more than 1.5 million customers taking the services being offered. Globally there are, by the end of 2007, about 11 million fiber-based access customers. The overall result of this is lowered costs for the technology and its installation, significant technological advancements in the equipment and installation methods, and improved service offerings to customers at affordable rates.

In alignment with this trend, European carriers are now committing funds to fiber-based access networks as well, and in a few years China will ramp up spending for this purpose.

This chapter will focus on the fiber-based approaches to broadband access worldwide, including some of the drivers for deployment, the architectural options, the capital and operational costs, the technological advances, and the future potential of these systems. Three variants of fiber-based broadband access, collectively called FTTx (fiber to the x) in this chapter, have emerged as particularly important. They are hybrid-fiber-coax (HFC) systems, fiber-to-the-cabinet (FTTC) systems, and FTTH systems.

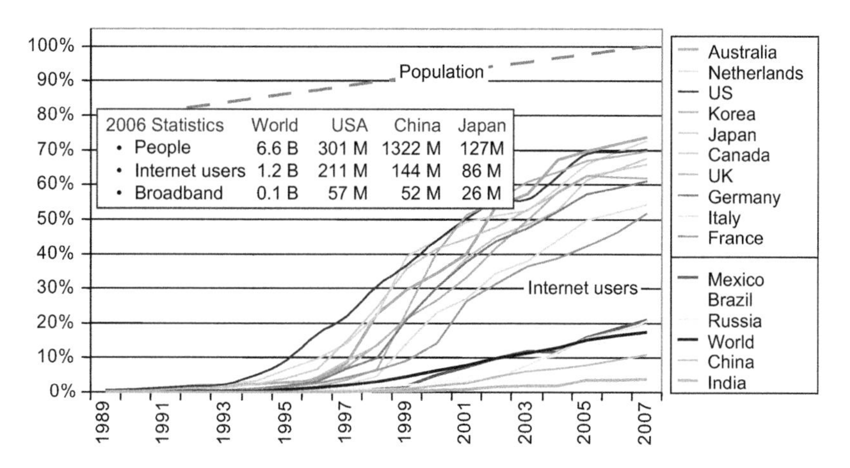

Figure 6.5.4 Internet usage worldwide.

6.5.2 User demographics

One of the most dramatic and unexpected drivers for fiber-based broadband access has been the explosive growth of the internet and its associated applications. As early as April 1993, when the Mosaic browser became available, the public began to use the internet for transfer of text messages, photos, and data files. Building on that initial application, the number of Internet users has grown to 211 million in the United States alone, and represents about 70% of the population now, while globally there are about 1.2 billion internet user representing about 18% of the world population [7, 8] (Figure 6.5.4). Not only has the number of users grown, but individual usage has expanded exponentially, until today we routinely exchange music, photos, videos, and software files as large as 10's of megabytes each.

A close look at the bandwidth available to access the Internet shows that it was gated by the technology being offered [9–11] (Figure 6.5.5), with new ones adopted in cycles of roughly 5–6 years each [2] (Figure 6.5.6). For example, the earliest users bought phone modems, initially at 9.6 kb/s, then increasing in performance to 14.4 kb/s, 28kb/s, and finally 56 kb/s [10]. Typically, a group of early users (indicated as Initial Users in Figure 6.5.6, representing the top 25% of users) adopt the new technology first; the mainstream follows, and finally everybody else (Slow Adopters in Figure 6.5.6, representing the bottom 25% of users) takes advantage of the technology. While the entire adoption cycle takes 8–10 years from introduction to 50% adoption, newer technology is introduced before even half of the users have the old in place. By the time the Slow Adopters begin to use one approach, another newer approach is being offered and adopted by a new set of Initial Users. The cycle of technology adoption by Initial Users, Mainstream Users, and Slow Adopters has experienced two full cycles now, and is into its third cycle: phone modems, then cable modems and asymmetric digital subscriber line (aDSL) modems, and now fiber-based access.

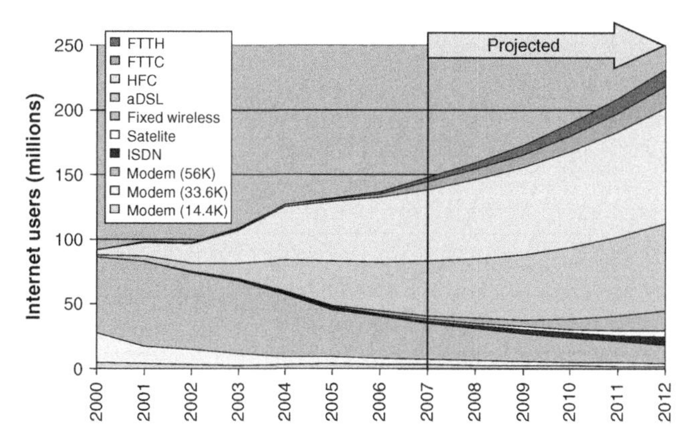

Figure 6.5.5 US residential access technology adoption over time (this figure may be seen in color on the included CD-ROM).

In Figure 6.5.6, phone modems were introduced in 1993 and by 2001 the overall average user bandwidth exceeded 56 kb/s. Cable modems, introduced in late 1997, were adopted similarly with average user bandwidth exceeding the 1 Mb/s cable modem bandwidth by 2006. Now that FTTH has been introduced, with primary service offerings ranging from 5 to 30 Mb/s, initial users are beginning to subscribe to this technology, especially at 10–15 Mb/s, which will take several years to reach the 25% penetration level and a few more years to achieve the mainstream. We can expect that by 2010 a significant number (perhaps a few million) of those users will be looking for something more, very likely 100 Mb/s service or higher. It is interesting to note that each of these technology cycles has resulted in bandwidth offerings about an order of magnitude larger than the previous, even though the price of the service offerings has only doubled for each cycle.

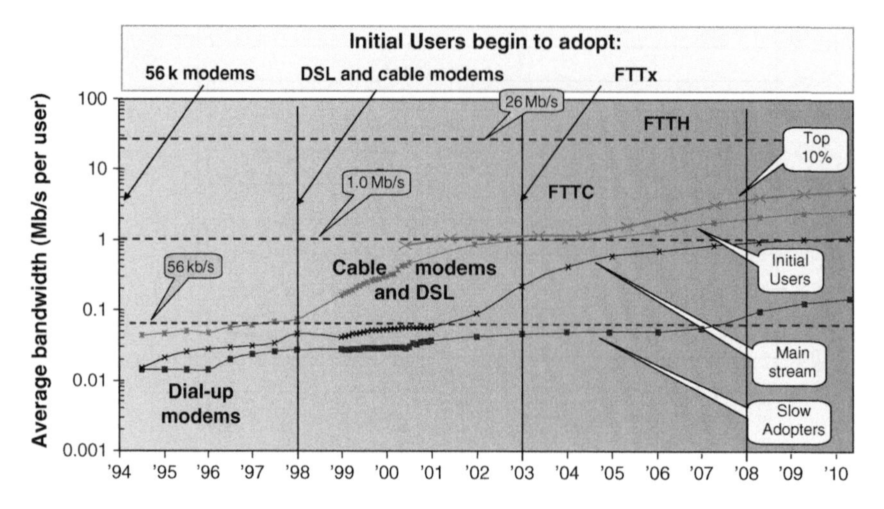

Figure 6.5.6 US user downstream connection speed trend and projection over time.

If we extend this trend into the future, we can expect at least two more cycles of technology adoption by users (Figure 6.5.7). One of these would offer 100 Mb/s service, projected by this data to have a few million subscribers by 2010, and another higher service offering of 1 Gb/s Ethernet projected to be penetrating the customer base by 2016. The earlier cycle that utilized cable modems and DSL technologies was characterized by asymmetric traffic, in which the downstream signals from the service provider to the user employed a much higher data rate than the upstream signals—for both technological and service reasons—because subscribers were more likely to view content than to generate content. The recent and future cycles, though, need to support nearly symmetric traffic because subscribers desire to share high-resolution images, large files, and video recorder clips with other users—which requires upstream data rates that are similar to downstream rates.

Beyond service offerings of 1 Gb/s with symmetric traffic, it is difficult now to envision what might be the next advance, because completely unimagined applications, spurred by the increased bandwidth offerings, are very likely to arise in the interim. But it is interesting to note that the raw bandwidth of high-definition TV (HDTV) signals for large-screen displays is nearly 10 Gb/s, leading to the expectation that even higher access speeds could be of interest in the very long term.

Technology	Per user	Start	50%	
Phone modems	<100 kb/s	1993	2001	History
Cable modems	1 Mb/s	1998	2006	History
FTTx approaches	10 Mb/s	2004	2012	History
Next generation fiber technology	100 Mb/s	2010	2018	Forecast
Big broadband technology	1 Gb/s	2016	2024	Forecast

Figure 6.5.7 Projection of user demand for bandwidth, showing a Gb/s target eventually.

Against this backdrop of continued bandwidth enhancements stands the much less variable location of the users. For example, residential users in metropolitan areas in the United States are typically located within 5 km of the central offices that serve them, and in suburban and rural areas this range extends at the greatest to about 10 and 20 km, respectively [12, 13] (Figure 6.5.8). In Asia and Europe, where the population density is higher, the users are typically closer to the central offices that serve them than they are in the United States. This density of users is not changing much over decades because it is determined by the cultural and socio-economic norms of suitable living arrangements. So the technologies that carriers deploy need to be capable of serving customers within this rather fixed range, while being able to be upgraded to provide ever higher user bandwidth. In practice, carriers have tended to provide options that offer cost or performance advantages in shorter-reach,

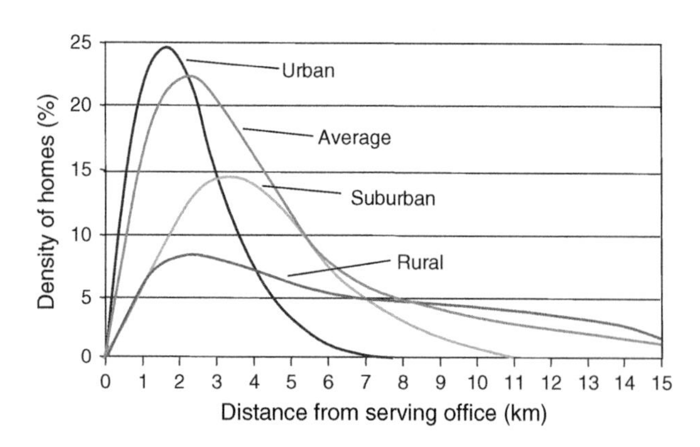

Figure 6.5.8 Distribution of homes from a serving office in the United States.

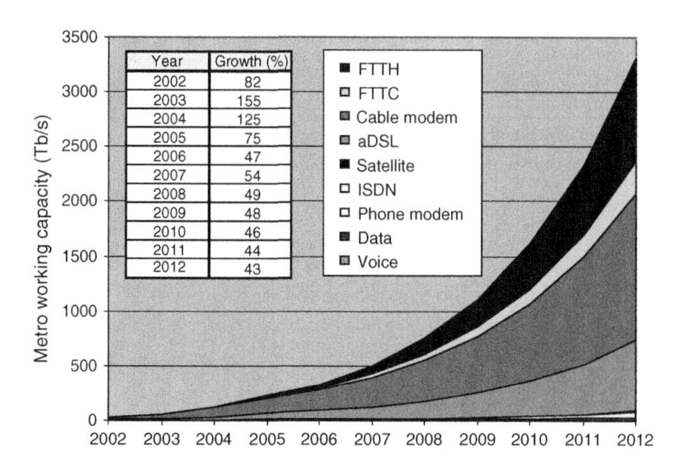

Figure 6.5.9 US Metro Internet traffic growth over time and in the future, with FTTx factored in (this figure may be seen in color on the included CD-ROM).

higher density metropolitan locations compared to suburban and rural areas. But until recently they have paid little attention to the looming issue of extending the user bandwidth to Gb/s rates.

Ultimately, the introduction of FTTx technologies and their subsequent adoption by mainstream users will have an effect on the traffic load offered to the metropolitan networks that serve the users [14]. This means that the metropolitan interoffice network demands will be advanced by a couple of years compared to a scenario where such technology is not introduced (see Figure 6.5.9). Such an effect is just now beginning to be felt by service providers as a result of their introduction of FTTx technologies.

6.5.3 Regulatory policy

Still, with all of these drivers, applications and technology advances available to carriers, they were reluctant to invest in new broadband access infrastructure until recently. In the United States, the 1996 Telecom Act [15] and the regulatory policy of the FCC stipulated that new infrastructure was regulated and had to be offered to competitors at a price set by strict policy rules. This discouraged new network builds, because investors feared that their expenditures would be exploited by competitors at costs lower than their own investments would support. Fortunately, this roadblock was removed to a great extent in February 2003, when the FCC adopted new rules for local phone carriers relating to broadband access networks, and issued a notice of proposed rule-making intended to give relief to major carriers of this burden by allowing new FTTx infrastructure to be built without being required to be offered to competitors at regulated prices. This was followed in August 2003 by the actual decree that formalized the FCC policy decision [16]. During this period, SBC, BellSouth, and Verizon issued a common Request For Proposals for FTTx broadband systems, offering encouragement to system houses to design and supply standard FTTx systems. While the initial FCC decision applied to FTTH architectures, follow-up clarifications and decisions related to the 2003 FCC ruling gave adequate relief to some FTTC architectures as well. Somewhat earlier, the Japanese and Korean governments instituted initiatives to help drive their carriers to deploy FTTx technologies, and recently European carriers are finding ways to overcome regulatory issues in the countries they serve. At the same time, municipal governments, independent contractors, and housing builders began to offer FTTx systems as part of their economic development packages. With regulatory relief in the United States, the number of fiber-based access homes passed in North America grew to 8 million in 4 years. In the same time the number of homes connected grew to nearly 1.5 million [17, 18], representing a penetration of about 19% of the homes passed (Figure 6.5.10). This growth is expected to continue to nearly 40 million homes passed by 2012, with a 44% penetration of homes connected.

Worldwide the number of homes connected has grown in this same period to about 11 million [19–24] and this is expected to grow to 110 million by 2012 (Figure 6.5.11). Initially Japan, with its national push, was the leading country to deploy FTTx technology, followed by the United States as a result of the regulatory policy changes that took place there, and then by Europe and finally in a few years by China.

Looking toward the future, there are several global policy issues in active debate today that may impact continued broadband deployment. These include video franchise reform to streamline competitive entry, municipal broadband deployment allowing local governments to deploy and operate broadband networks, and a universal service fund ensuring fair deployment to all communities regardless of socioeconomic status. It is too early to tell, but the outcome of these policy decisions could encourage further spending by carriers if it is favorable to them.

6.5.4 Network architectures

In the past 15 years, global service providers have gradually pushed fiber technologies farther out from the serving offices and closer to the users. This trend started with the deployment of Subscriber Loop Carrier systems, which simply replaced multiple aging copper lines with fiber transport for the first few km from the serving office. Later, beginning in the early 1990s, CATV

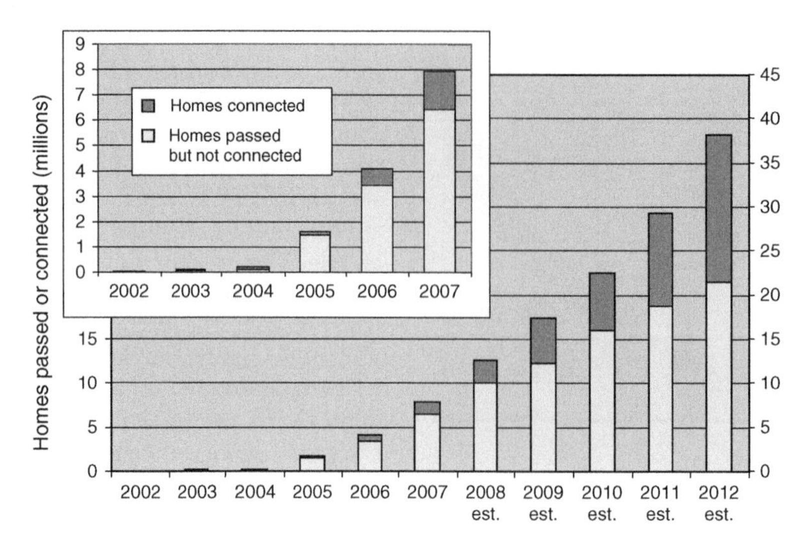

Figure 6.5.10 FTTx deployments in North America.

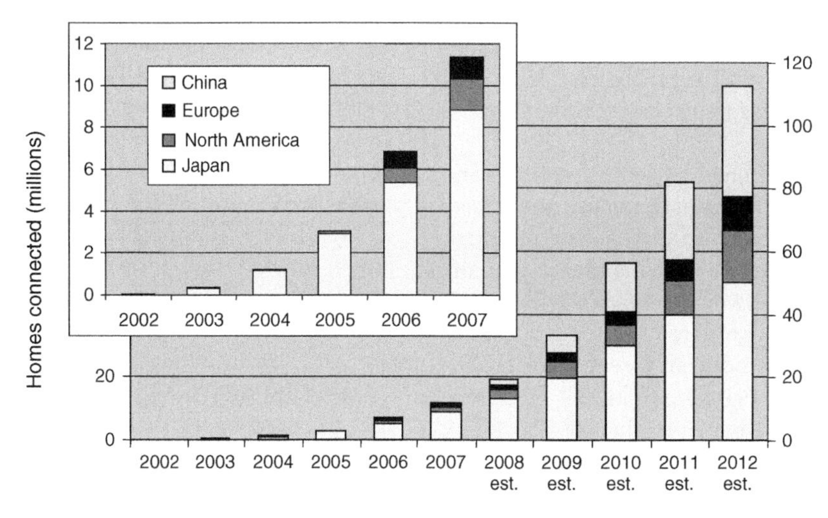

Figure 6.5.11 FTTx deployments globally.

operators began to enhance their broadcast TV infrastructure by deploying HFC systems [25], which brought fiber out to within a kilometer or so of the users. Beginning in 1998, BellSouth began to use fiber-to-the-curb systems whenever they needed to refurbish aging copper plant or had new housing start installations, bringing fiber out to within 150 m of the users. These systems are a variant of FTTC systems, with the cabinet placed close to the home at the curb. FTTC is also a hybrid approach, as it uses digital subscriber line (DSL) technology over the existing copper twisted pair infrastructure in the drop portion of the network [26]. In 2002, Japan began deploying FTTH systems, using a point-to-point architecture. Shortly after that, in 2004, Verizon began deploying FTTH systems using a passive optical network (PON) architecture in the USA, pushing fiber all the way to the user. Verizon generalized the approach, calling their system fiber-to-the-premises (FTTP) to allow for the possibility of serving multiple dwelling unit structures (duplex homes and apartments) as well as businesses with the same fiber-based approach. AT&T is now deploying fiber to the node (FTTN) in the United States, a variant of FTTC that brings fiber to within 1.0 km of the user. For the hybrid approaches, longer distances from the termination of the fiber plant to the user have an adverse effect on the bandwidth that can be delivered to the user over copper twisted pairs or coaxial cable.

The network architectures of each of these three variants of fiber-based broadband access are very similar (Figure 6.5.12). They each have office terminal equipment, large fiber count feeder plant, medium fiber count distribution plant, a drop cable, and customer premises equipment. The three variants differ in a very important respect: the transmission media used for the drop portion of the network. In HFC the drop is coaxial cable, in

FTTC the drop is twisted wire pairs, and in FTTH the drop is fiber. They also differ in another important respect: HFC and FTTC require active devices and powering in the outside plant, while FTTH has a totally passive outside plant that requires no power. These variants contribute to differences in the possible user service offerings, as well as differences in the cost structure for deployment and operations.

There are three basic services that each of these architectures is expected to deliver: telephone service, entertainment video service, and high-speed Internet data service. These represent the so-called triple play in FTTx architectures. Since entertainment video has historically been offered using NTSC (National Television System Committee) analog video standards, these systems typically strive to support both conventional analog video channels as well as digital data channels. In HFC systems, both analog and digital channels are multiplexed onto the same optical carrier; in FTTH systems, the analog and digital signals are multiplexed onto different optical carriers (different wavelengths); and in FTTC systems only digital data channels are supported. Digital TV (DTV) services are normally delivered on the analog channels, while video-on-demand (VoD) is handled by the high-speed data channel. Since FTTC has no analog support, it can not simultaneously carry all of the entertainment channels, but is limited to offering VoD services and a limited number of DTV channels.

FTTC differs from HFC and FTTH in that it must seek to deliver entertainment video services in a digital format. In the long run, this distinction about how entertainment video is supported (either analog or digital) may disappear, since there is a significant trend to shift to digital formats, such as DTV and HDTV. Such a transition to digital television is mandated in the United States by the FCC for over-the-air broadcast TV signals to be

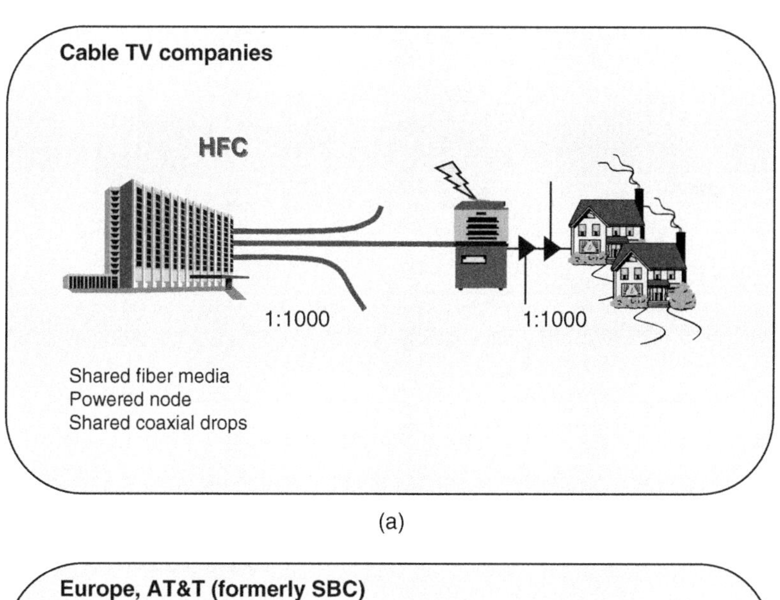

(a)

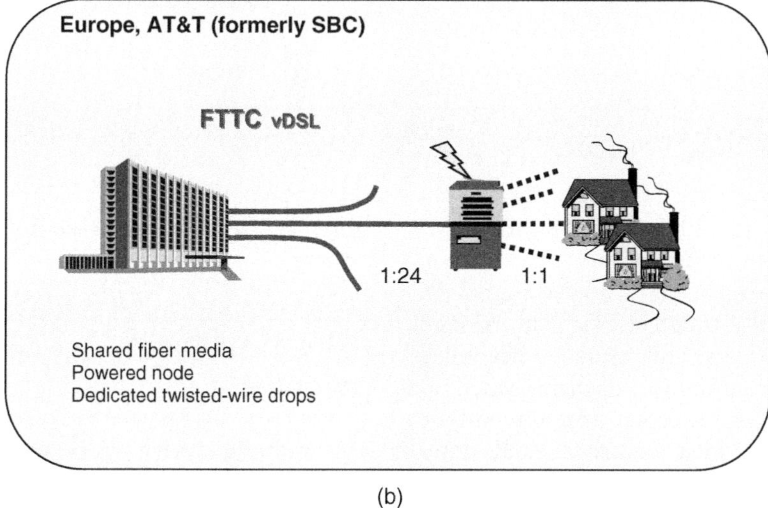

(b)

Figure 6.5.12 Broadband access network architectures for (a) HFC, (b) FTTC/xDSL.

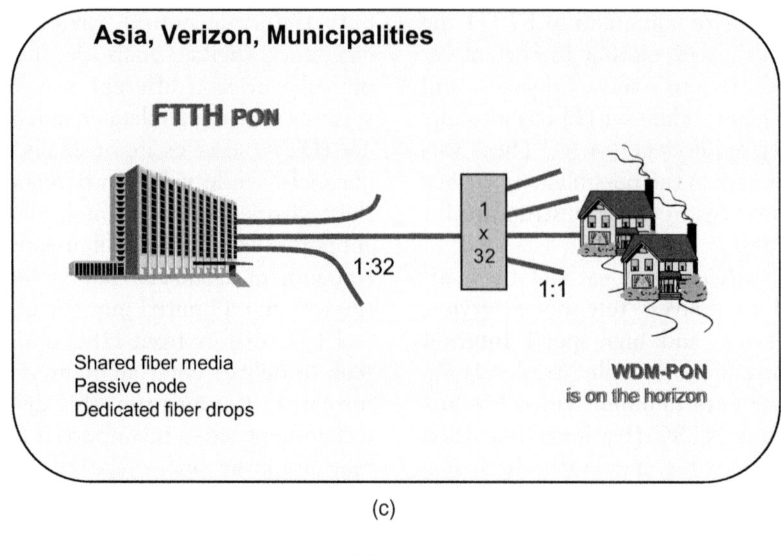

(c)

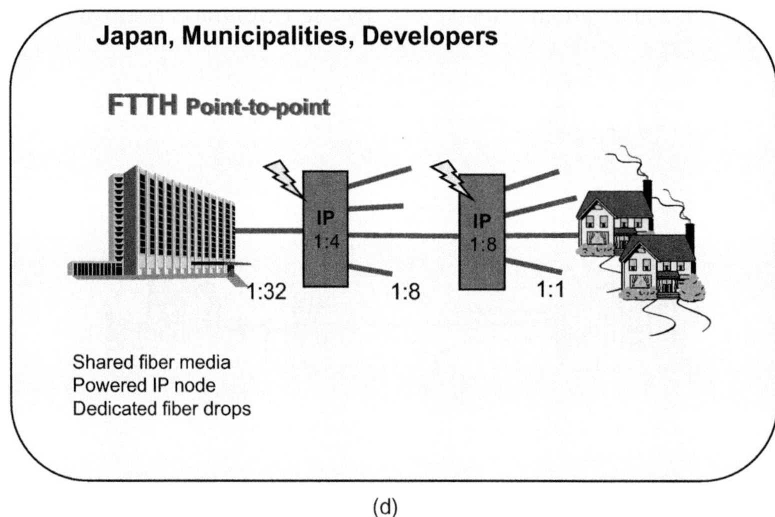

(d)

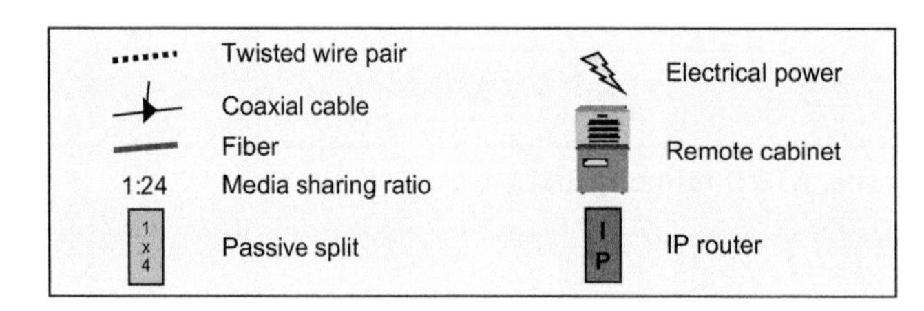

Figure 6.5.12 (c) FTTH PON, and (d) FTTH Pt-Pt.

completed by February 17, 2009 [27], but for video signals carried in a closed medium (fiber, coaxial cable, twisted pair) there is no equivalent mandated timetable. Other countries have similar timetables for conversion of TV signals to HDTV formats. Still, users are becoming increasingly aware of the advantages of digital video, as they become accustomed to digital video monitors, Video iPODs™, Digital Video Recorders, and other similar visual display appliances. Consequently, it is very likely that by the end of this decade the most common format for video signals will be digital formats that can store and retrieve video content from digital storage devices.

6.5.4.1 HFC network architecture and its potential

The multiple service operators (MSOs) have deployed HFC networks to more than 40 million customers to date. These architectures include fiber-based transport from a service office to a powered node, which typically supports 500–1000 homes. The modulation format on the transport is subcarrier multiplexed (SCM) analog TV channels on a 1550 nm optical carrier. Since the TV signals are analog in nature, the transmitter must be exceptionally linear to avoid second- and third-order intermodulation impairments among the multiplexed TV channels, and the optical receiver power must be high to avoid noise impairments. This means that high-power erbium-doped fiber amplifier (EDFAs) are used at the serving office, to boost the analog signals to high power and to enable splitting the analog signals to serve many nodes. Every node receives the same set of analog TV channels. In addition, each individual node receives its own channel that is dedicated to Internet service for that node. This means that the unique Internet data must be multiplexed onto the Internet channel for its particular node. Since all users of the node get the same signal, they are all sharing the Internet channel, so none of them can benefit from the full Internet channel bandwidth. At the node, all of the channels are amplified and split to feed to each subscriber associated with that node. Each subscriber receives the same analog signal as every other subscriber fed by that node, and their cable modem must filter out the TV channels they want to watch and select the Internet packets destined for their account. Cable modems sold today separate the Internet channel from the TV channels and deliver 42 Mb/s of Internet data downstream shared by all users, in compliance with a Data Over Cable Interface Specification (DOCSIS) 2.0 standard. In the upstream direction, Internet data from each subscriber is multiplexed onto a single upstream channel, which in DOCSIS 2.0 is 30 Mb/s shared by all users of the node. The upstream channel is limited by the coaxial amplifiers in the distribution plant, and this is the main limitation of HFC systems for high-speed Internet service.

Over time, there has been a progression of standards for HFC networks, with the DOCSIS cable modem standard being the controlling factor for user bandwidth. The trend has been to increase available downstream user bandwidth, then to increase upstream user bandwidth, then to use more channels for Internet access, and eventually to remove coaxial amplifiers from the distribution plant to allow more channels in the upstream direction. This progression of standardization is intended to increase user bandwidth for Internet service, and that trend is illustrated in Figure 6.5.13. Note that to achieve such increases, it is necessary to limit the number of customers that a node supports, and correspondingly to move the nodes closer to the user. Eventually, when the drop distances are short enough to remove all amplifiers, and when the node supports 25 users, an HFC network architecture should be able to deliver Gb/s data rate services to fulfill the demands of future users.

6.5.4.2 FTTC network architecture and its potential

BellSouth has deployed FTTC to more than 1.3 million customers by 2006, AT&T has a plan to deploy FTTN widely to their customers, and BT in the United Kingdom offers this architecture to its customers. These architectures, originally standardized by ANSI (American National Standards Institute) and FSAN (Full Service Access Network Group) with the local exchange carriers input and later standardized by International Telecommunications Union (ITU), use GbE transport from a serving office to a remote cabinet. The optical signal formats and system are simple single-channel digital systems at 1310 nm, with either time-division multiplexed (TDM) or packet multiplexing of individual

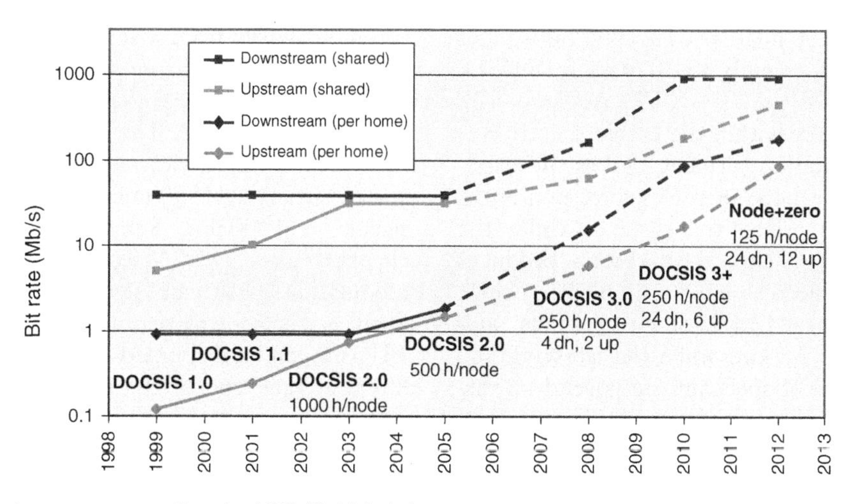

Figure 6.5.13 Standards progress over time for HFC (CableLabs).

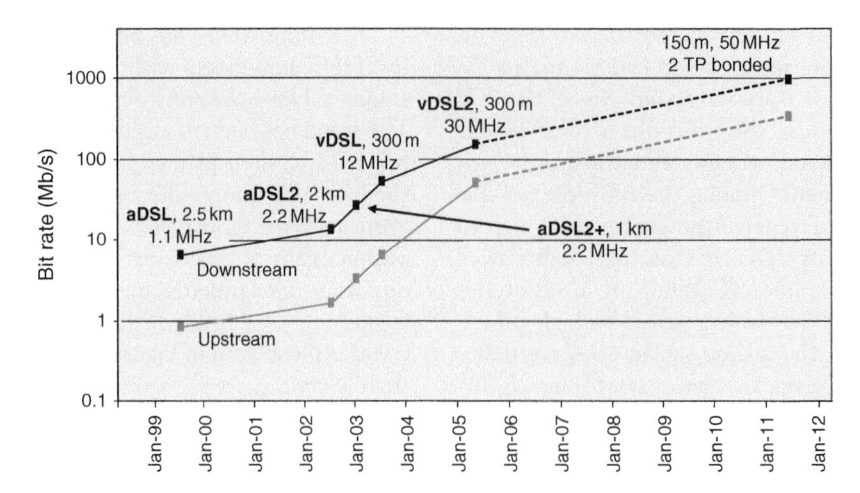

Figure 6.5.14 Standards progress over time for xDSL (ITU-T).

subscriber's information at the office and corresponding electronic demultiplexing at the remote cabinet in the downstream direction. Normally a single fiber supports the downstream multiplexed signals of 24 subscribers, and at the remote cabinet these are separated and remodulated using DSL formats that can be applied to the individual twisted wire pair drops. A similar arrangement is used in the upstream direction, using a second fiber with electronic multiplexers at the remote cabinet and corresponding demultiplexers at the office.

Over time, the standards and commercial DSL products have evolved to support ever higher user bandwidths. These standards prescribe a discrete multitone (DMT) modulation format, which uses a large number of 4kHz-spaced orthogonal subcarriers in the band, each modulated at a low rate, which allows the system to cope with severe impairments that may be introduced by the twisted-pair copper medium [28]. Early DSL systems used 1.1 MHz of spectrum on the copper wires to achieve a bit rate of 364 kb/s, but this spectrum use has been increased to 2.2 MHz for aDSL2, then to 12 MHz for very high-speed Digital Subscriber Line (vDSL) and most recently to 30 MHz for vDSL2 technology.

In each case, the noise within the usable spectrum is the limiting transmission impairment, and as the bandwidth increases the total noise increases correspondingly. The noise limitation translates to a distance limit—the distance for which the signal-to-noise ratio produces satisfactory error rates decreases as the usable bandwidth increases. So DSL using 1.1 MHz of spectrum can support drop distances of 2.5 km, while the latest vDSL2 standard using 30 MHz of spectrum supports drop distances of only 300 m. Consequently, the FTTN system of AT&T, with drop distance of 1.0 km can probably support a user bandwidth of about 25 Mb/s, while the FTTC system of BellSouth (now known as AT&T Southeast), with a drop distance of 300 m can probably support a vDSL2 user bandwidth of about 100 Mb/s.

A summary of the standards evolution is depicted in Figure 6.5.14. In the future, if DSL standards should move to an even higher utilization of copper spectrum of 50 MHz, and if service providers are willing to dedicate two twisted pairs per user and maintain drop distances less than 150 m, it is conceivable that DSL technology can support 1 Gb/s data rate services to individual users.

6.5.4.3 FTTH network architecture and its potential

Verizon began deploying FTTH (they use the term FTTP to include homes, multidwelling units, and businesses) in May 2004 and by July 2007 had passed more than 7 million homes and had an estimated 1.5 million subscribers. Carriers in Asia and Europe, as well as many municipalities globally deploy this architecture as well. The FTTH systems use a PON architecture, transporting common signals from a serving office to multiple users with a 1:32 optical power split at a passive cabinet [29], and then a fiber drop to a network interface unit on the outside of the house. The analog and digital signals are carried on different wavelengths, with the downstream analog signals at 1550 nm and the downstream digital signals at 1490 nm. Remember, though, that analog channels can carry DTV signals via quadrature amplitude modulation. Upstream signals are carried in the same fiber as the downstream signals at a wavelength of 1310 nm, and are coupled to the fiber through coupling filters at each end of the network. The upstream data signals are multiplexed together using time-division multiple access (TDMA) methods, and each user is assigned one or more unique timeslots. A key problem in PON networks is that each user's upstream signal will arrive at a variable time determined by the distance the user is from the serving office and the transit time of the signal in fiber, potentially causing signal contention at the point where the upstream TDMA signals are to be multiplexed. This is handled by an autoranging synchronization signal that prompts each user when to transmit so that their signal arrives at the multiplex point in the correct timeslot.

Initially, the PON standards addressed FTTH systems with a downstream data rate of 622 Mb/s and an upstream data rate of 155 Mb/s using the asynchronous transfer mode (ATM) format (A-PON), but later this was enhanced to include the analog channel for broadband access (B-PON). Eventually the ATM approach was extended to 2.5 Gb/s downstream and 622 Mb/s upstream (G-PON), and another approach using GbEthernet without an analog overlay was standardized as well (GE-PON). With the ATM format, the user has a well-defined and guaranteed bandwidth and quality of service, while with the Ethernet format the user shares the full bandwidth on a best-effort basis. The ATM-based PON systems are favored by US carriers, since the ATM format is compliant with their legacy transport systems, while the Ethernet format requires other capital-intensive changes to how they build their legacy networks. The user bandwidth characteristics of these various options for FTTH systems are summarized in Figure 6.5.15, and it is clear that user bandwidths of 100 Mb/s are quite reasonable today, and these can very likely move to Gb/s data rates [30] by making use of the statistical multiplexing inherent in Ethernet protocols.

While each of the FTTx architectures can be extended to provide significantly more bandwidth per user, even up to Gb/s rates per user, inside the home there is still the issue of how to distribute those signals to multiple rooms. The FTTH and FTTC systems provide the user an interface at the side of the home and the HFC systems provide a cable modem interface inside the home, but from a user perspective the signals must reach each PC, Internet appliance, display, and entertainment system in the household for the bandwidth to be useful. In Japan and in China, the FTTH system are installed with the user interface inside the home, allowing this interface to have substantially less environmental stress from temperature and the elements, and offering the possibility of tailoring the interface to the specific needs of the user. In-home networking has been identified as a key problem in the recent introduction of FTTP in the United States and Japan, and has been a lingering issue for HFC installations with multiple PCs. While this is not strictly the carriers issue to address, they will increasingly be in the position to suggest solutions to the homeowner in order to gain the maximum adoption of their service offerings. Fortunately, each of these system interfaces can be equipped with wireless (Wi-Fi, wireless-fidelity) base stations which today are capable of distributing several Mb/s of data anywhere inside a typical home. But as FTTx bandwidth increases over time, the demands of in-home networking capability will grow correspondingly to bandwidths of Gb/s. While the standards activity for Wi-Fi is on track to keep pace with the introduction of FTTx technologies (Figure 6.5.16), another option is to deploy specially designed fiber [31] in structured cabling arrangements for in-home networks.

6.5.5 Capital investment

While any one of these FTTx approaches can satisfy the current demands and affordability of the current market, they each require substantial capital investment to deploy widely. By way of example, imagine that the estimated 110 million homes is served by fiber-based access in 2012 as shown in Figure 6.5.11, that this represents

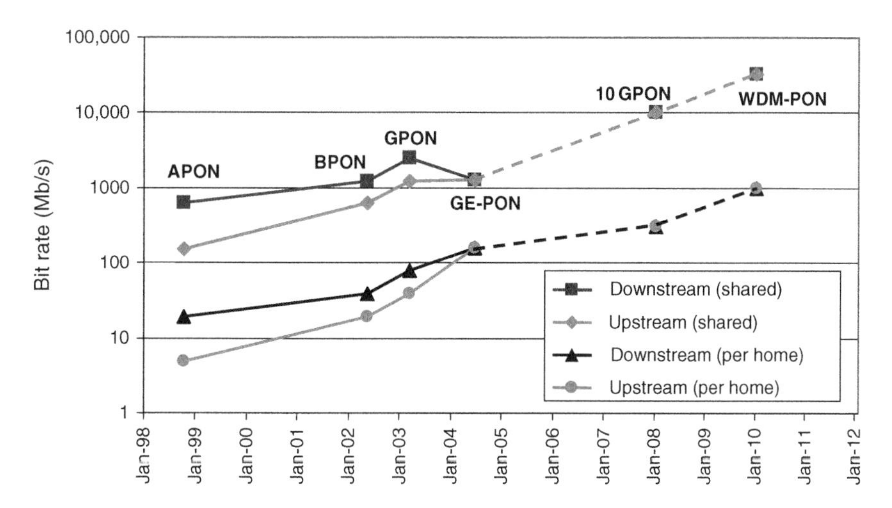

Figure 6.5.15 Standards progress over time for PON (ITU-T).

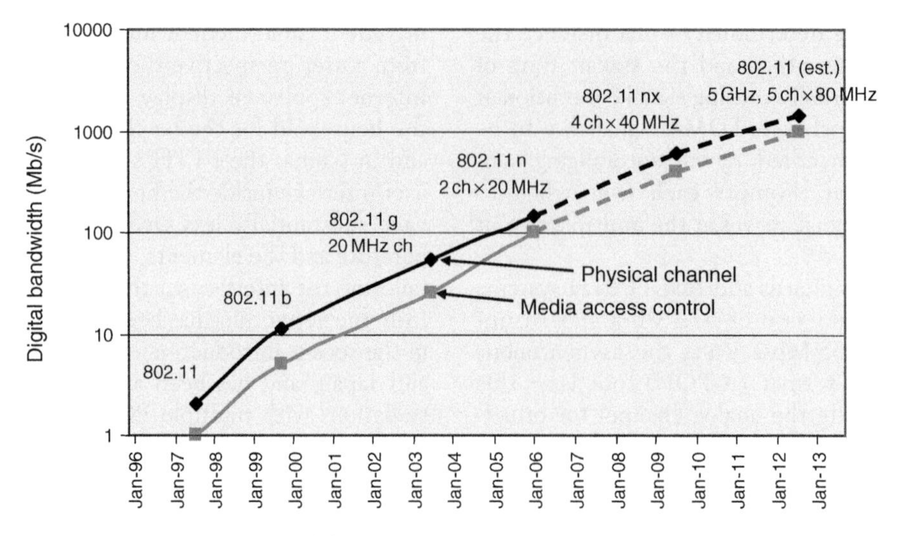

Figure 6.5.16 Standards progress over time for Wi-Fi (IEEE).

a penetration of homes passed of 30%, and that each home passed could be provided for as little as US $1000 per household. This would require about US $370 billion in cumulative capital expenditures over a 5-year period. This represents a combined annual capital spending for fiber-based access related systems of about US $75 billion by all carriers globally. But 110 million homes is only about 5% of those available in the developing countries, so it will require a 20-year or more infrastructure commitment to bring fiber to all homes worldwide. A comparison of the FTTx approaches indicates that US $1000 per home is a target that is eventually achievable, with the more expensive approaches providing more capability in the long run.

A comparative assessment of the capital costs for FTTx systems is illustrated in Figure 6.5.17, where it is assumed that 100% of homes passed are taking service, that there is a mix of 70% aerial and 30% buried plant, that the density of users is typical of a combination of urban and suburban customers, and that all of the equipment, installation, project, and subscriber costs are included in the comparison and averaged over all customers. At the lower end, FTTC/vDSL requires the smallest capital expenditures because it capitalizes on the existing telephone copper wire drop plant, but it also provides the most restrictive service options because it does not support analog TV programming. In the mid-range, HFC capital expenditures are higher than for FTTC/vDSL, and the extra expenditure buys the capability for analog TV programming, although HFC systems offer shared and therefore somewhat limited bandwidth for high-speed Internet services. At the upper end of capital expenditures are the FTTH systems, which provide for analog TV services as well as the

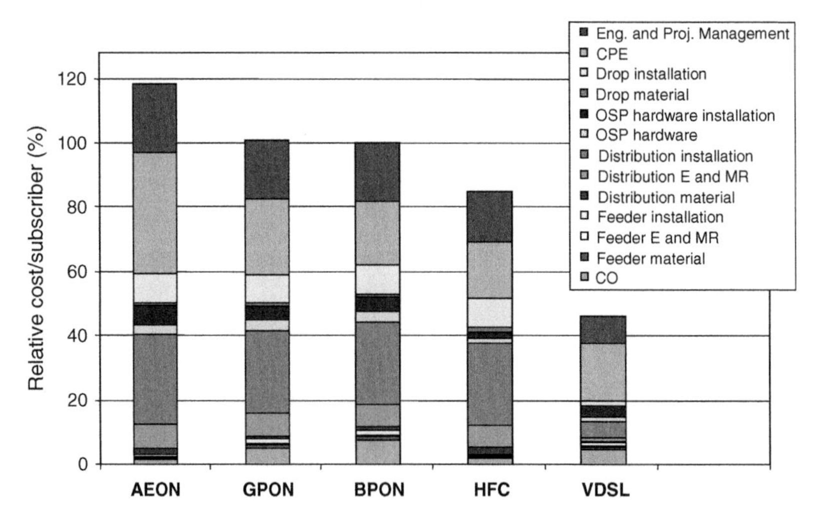

Figure 6.5.17 Comparison of relative capital costs per subscriber of HFC, FTTC/xDSL, and FTTH options, with B-PON arbitrarily chosen as a reference.

highest potential digital Internet bandwidth. Because the MSO and ILECs (Incumbent Local Exchange Carriers) already have coaxial cable and twisted-pair drop cables in place, those two options represent lower spending and medium-term risk positions for satisfying future bandwidth demand growth. On the other hand, FTTH requires new infrastructure all the way to the home, making it the most capital intensive but also offering a better longer-term risk for satisfying future growing bandwidth demands.

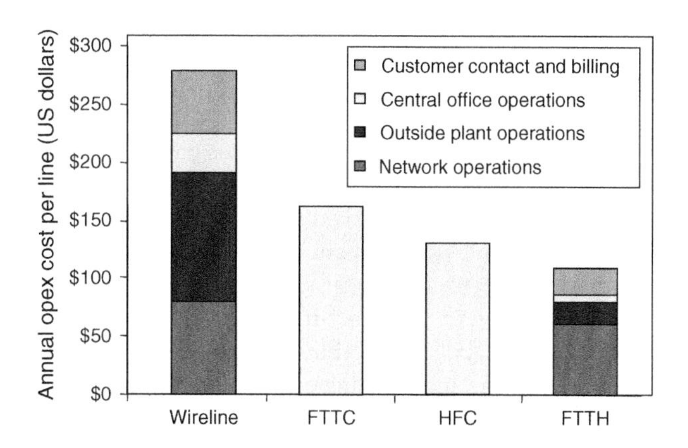

Figure 6.5.18 Comparison of operations costs per subscriber for FTT.

6.5.6 Operational savings

Given that FTTx deployments consume so much capital, a carrier is bound to ask: "Why would it benefit my company to take the risk to offer such services?" Of course, the answer is twofold: there are new service-related revenues to reap, and there are also operational cost savings associated with fiber-based systems. Taking a lesson from long-haul and metro transport systems, carriers have learned that fiber systems have fewer failures, higher availability, are more reliable [32], and have more capacity than copper-based systems. So FTTx solutions have a strong historical track-record on which to rely for both good reliability and high bandwidth at long distances. In addition, passive plant is especially attractive, because the number of active devices can be made much smaller in the outside plant where the parts are less accessible, and because passive devices in the outside plant are highly reliable [33]. For this reason alone, fiber-based systems are more attractive than copper-based systems that require periodic amplification and signal shaping. Further, new operational savings can be built into new infrastructure that take into account the ability to use sophisticated computer algorithms to enable faster and more efficient service provisioning, churn, administration, and easier fault location. All of these network operations and "back office" functions add up to significant annual operational savings compared to the way things are done today.

As an example, a comparison between FTTH annual operations expenditures relative to today's copper-based wireline technology [34, 35] is illustrated in Figure 6.5.18. While details of the customer contact and billing, central office, outside plant and network operations costs have been analyzed for both FTTH and today's technology [34], only the overall operations costs have been estimated for HFC and FTTC solutions [35]. This information indicates that more than US $150 per year per subscriber can be saved in operational expenditures for FTTH compared to the labor-intensive wireline operations activities associated with customer service requests, as well as provisioning, terminating, and re-provisioning service. Over a 7-year period, this nearly pays for the installation expenses, without even taking service revenue into account. A similar, but less dramatic, scenario is the case for HFC and FTTC solutions as well, but their reliance on active plant prevents them from reaping the benefits of a fully passive outside plant.

6.5.7 Technological advancements

Over the last decade, system suppliers have worked at reducing the cost of equipment, the first installed costs of active and passive cable plant and the operations costs of FTTx systems. Both HFC and FTTC solutions have been deployed for most of the decade, and are relatively simple optically, consisting of point-to-point fiber transmission followed by more complex electronic splitting in a cabinet near the customers. Consequently, a major cost reduction driver has been electronics advances of IC integration. For HFC, this has resulted from the DOCSIS 1.0 and 2.0 standards, which allows IC designers to build special-purpose ICs to a common standard and increase the IC volumes dramatically. In a similar vein, the xDSL standardization process has allowed aDSL and vDSL to be implemented in standard IC designs, so that the most recent vDSL2 IC designs are capable of supporting all previous standards by firmware control. As a result of the standardization and volume manufacture of chip sets for HFC and FTTC, the OEM chip costs have been reduced, leading to lower first installed costs for carriers. In both cases, these IC advances also reduce the serving office, outside plant, and customer equipment size, power consumption, and number of active parts in the system. These changes tend to reduce the installation costs as well as the ongoing operations costs of these systems. We can expect similar IC advances in FTTH systems as well when the annual deployment rates are large enough to justify custom IC development costs for FTTH systems—thus bringing down the cost of FTTH equipment in like manner.

6.5.7.1 Optical transmitter and receiver technology advances

For all three FTTx options, the photonics parts costs have been driven down by volume manufacturing. A prime example of this is the triplexer that is used in FTTH systems to receive downstream digital and analog signals and transmit upstream digital signals. Only a few years ago triplexers were manufactured using packaged photodiodes, packaged lasers, thin-film filter components, and bulk coupling and splitting optics—resulting in labor-intensive packaging costs and a triplexer price of US $350 per unit. Now, with volume demands for triplexers at a few million per year, the manufacturing approaches have turned to integrated solutions, where automated pick-and-place can be applied. For example, planar lightwave circuits [36–39] are used to form the coupling waveguides and wavelength filters, photodiode, and laser chips are flip-chip mounted directly on the PLC substrate, and fiber alignment is done by direct coupling via precision grooves in the PLC substrate [40] (Figure 6.5.19). These approaches bring the advantage that they can reliably achieve fiber alignment tolerances and good fiber coupling efficiency with a controlled waveguide transition design to the PLC waveguide, by using the precision and yield associated with IC fabrication processes [41]. The result of all these technology improvements is a triplexer that can be offered at a price of US $50 or lower to system suppliers. Since the triplexer is a dedicated part (one per subscriber), its cost reduction has a dollar-for-dollar impact on the cost of the overall system. The cost reduction of this single component alone has made a significant impact on lowering the per subscriber cost of FTTH systems.

6.5.7.2 Cable management technology advances

Installation of optical fiber cable, cabinets and other passive plant components in the feeder, distribution, and drop portions of the outside plant can account for more than half of the cost of installation of an FTTx system. This cost component has received a lot of attention to simplify installation techniques, to speed deployment, and to improve labor efficiency and effectiveness. One way to accomplish this is to integrate much of the time-consuming cable management parts together in the factory [42], where volume manufacturing methods and sophisticated assembly equipment can be easily supported. This means that the distribution cable, branch entry points, and cable terminations must be uniquely designed for a specific location and assembled under controlled conditions in the factory. Since each location has a different set of physical measurements along its route from distribution cable to branch entry point to splitter cabinet, this requires careful network planning and advance engineering effort by the carrier. In any installation, whether using field installed terminations or terminations installed at the factory, it is necessary to engineer and plan the network for a specific location. But when you integrate the cable assembly in the factory, the engineering measurements must be more accurate because there is no chance to make a final cable length adjustment in the field by cutting the cable shorter.

For cable assemblies that are integrated in the factory, the carrier's craft personnel are sent into the field prior to ordering the outside plant products, where they use precise laser rangefinders, measuring wheels, and pull-tapes to determine the distances of each cable, branch point, and termination on the specific route. These measurements are provided to a few inches tolerance for the cable manufacturer via an online configuration manager, which generates a bill of materials automatically. Then each cable management assembly is custom-manufactured to those demanding specifications and delivered to the installation site—ready to roll off the reel and fasten in place—without any field-related cutting of cables or splicing of cable branch entry points, splice points, or closures [43, 44]. Figure 6.5.20 shows photos of aerial cable terminal closure and branch points

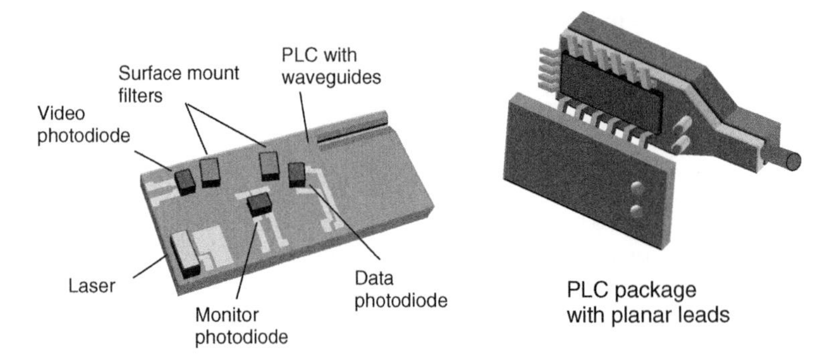

Figure 6.5.19 Sketch of triplexer solution enabling volume manufacturing.

Figure 6.5.20 Factory installed aerial cable branch point and fiber drop terminal closure (photos).

being installed, and Figure 6.5.21 shows photos of drop cable terminations and drop cable with robust connectors manufactured for this method. With this technique, installation times can be cut from nearly 2 days to less than 3 hours, and the entire process takes fewer truck rolls, fewer tools, and requires much less time working overhead in a bucket. To further ease installation, and to assure a minimum of future maintenance events, the splitter cabinets, enclosures, branch points, terminals, and drop cables are all fitted with connectors at the factory, making them waterproof and environmentally robust while minimizing splicing in the field.

6.5.7.3 Outside plant cabinet technology advances

In a similar fashion, the cabinets that house the splitters are factory assembled and internally preconnected according to an engineering design, so that they can be set

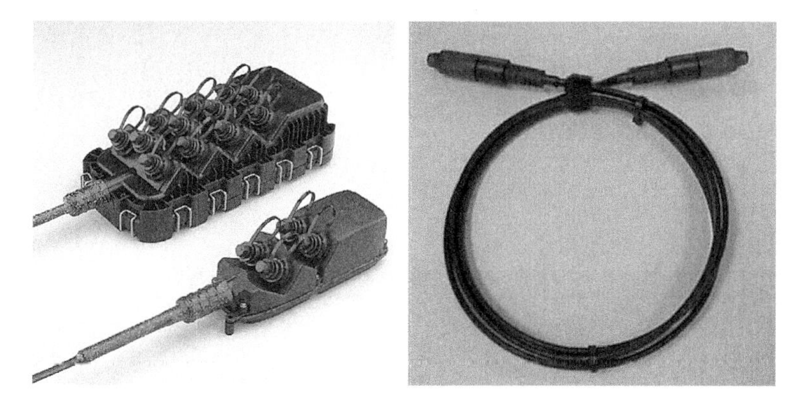

Figure 6.5.21 Fiber drop terminals and drop cable with robust connectors installed at the factory (photos).

in place and attached to their dedicated cables without further craft involvement interior to the cabinet. The initial designs of the cabinets were rather massive, requiring several laborers and a crane to unload and place them. Subsequent technology advances allowed the cabinets to be miniaturized so that they can be lifted by hand and set in place by an individual [45]. An enabling step toward this objective was to improve the bend tolerance of the fiber [19, 46, 47] used in the 1:32 splitters, from 75 mm bending radius to 30 mm bending radius. This was combined with a reduction in the fiber jumper diameter from 2.9 to 2.0 mm, to allow the 1:32 splitters [33] to be reduced in size from 191 mm x 131 mm to 125 mm x 63 mm on the long dimensions, because the excess fiber coiled inside the splitter package could be correspondingly reduced in diameter. Overall this resulted in a 78% reduction in the spatial volume taken up by the splitters (Figure 6.5.22). At the same time, the layout of the cabinet was improved to make it more craft friendly. The splitter size reduction together with the improved cabinet layout then enabled the splitter cabinets to be redesigned to allow a smaller footprint, resulting in a smaller cabinet with less metal, making it much lighter. An example of the cabinet size reduction that resulted from improved fiber bend tolerance is shown in Figure 6.5.23, where it is clear that the cabinet volume was reduced by more than a factor of four. Such size reductions have diverse impact on costs; ranging from reduced space to store inventory prior to installation, reduced shipping fees, reduced need for heavy equipment to install the cabinets, and reduced labor during cabinet installation. In addition, the smaller cabinets are less intrusive to the environment, can be more easily placed in a wider variety of locations than the bigger cabinets and experience less vandalism.

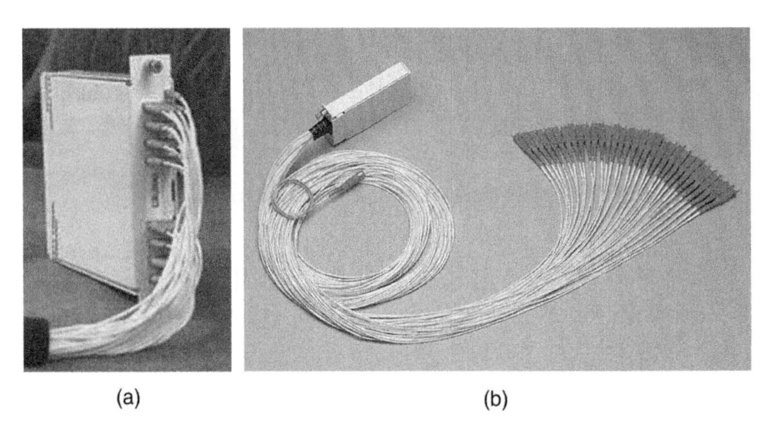

Figure 6.5.22 Splitter size reduction associated with bend tolerant fiber (a) before and (b) after size reduction (photos).

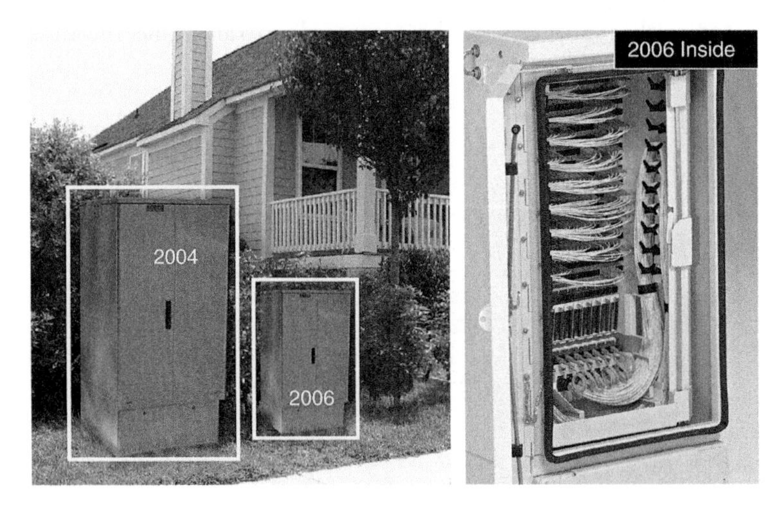

Figure 6.5.23 Cabinet size reduction associated with bend tolerant fiber, smaller cable diameter, and smaller splitters (photos).

6.5.7.4 Fiber performance technology advances

There is another important technology development that is beginning to play a role in reducing the overall FTTx system cost. This is the discovery that the stimulated Brillouin scattering (SBS) threshold in fiber can be increased, allowing higher launch power for analog signals without introducing more system impairments [48, 49]. To begin with, in both HFC and FTTH systems, the analog signal loss budget is the limiting factor in determining the reach from the serving office out to the active optoelectronics (see Figure 6.5.24). The loss budget is set at the subscriber end by the receiver noise limitations at a level of −5 dBm to −10 dBm, depending on the system under consideration. Then to get a high enough loss budget for reasonable reach, the launch power has to be very high—in excess of +15 dBm. With standard single-mode fiber (meeting G.652 standards), increasing the launch power above about +17 dBm for the analog signals introduces significant impairments due to the reflected SBS power, which emphasizes all of the key analog impairments [50]. The received noise is increased (carrier-to-noise ratio (CNR) is reduced) as a result of laser phase-to-intensity conversion by the SBS, and intermodulation distortions (CSO and composite triple beat (CTB)) are introduced in the modulated carrier because the optical carrier wave is above the SBS threshold and attenuated relative to the modulated sidebands which are below the SBS threshold. Fortunately, this SBS threshold can be controlled to some extent by the design of the fiber profile, which can be optimized to reduce the interaction between the acoustic wave and the optical field in the waveguide and ameliorate the cause of SBS [51, 52] (see Figure 6.5.25). A suitable fiber design can produce Brillouin gain spectra that are only about 35–40% as strong as the Brillouin gain spectra in standard single-mode fiber [50, 53] (see Figure 6.5.26).

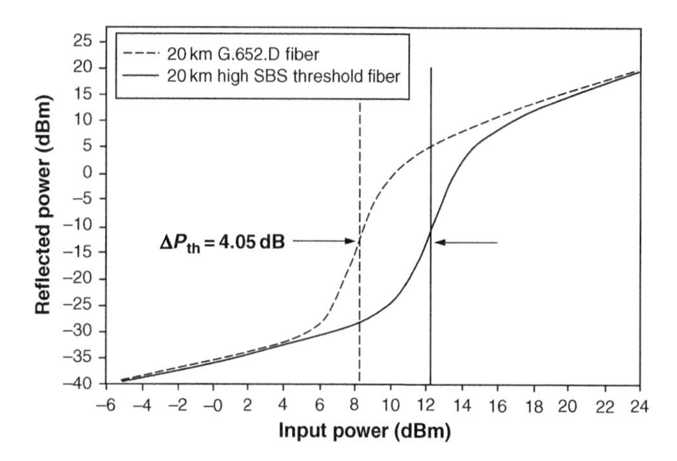

Figure 6.5.25 Comparison of fiber meeting G.652 standards and high-SBS threshold fiber performance.

The result is a fiber that can accommodate about 4 dB higher launch power, while still retaining the same analog impairments seen in standard single-mode fiber. Since the fiber design itself avoids excess SBS impairments, cheaper video transmitters can be used because they don't have to compensate for that impairment.

The impact of the high-threshold SBS fiber is to provide relief to the system in several potential ways. In HFC systems, it is possible to launch higher power and place the node farther from the serving office. In FTTH

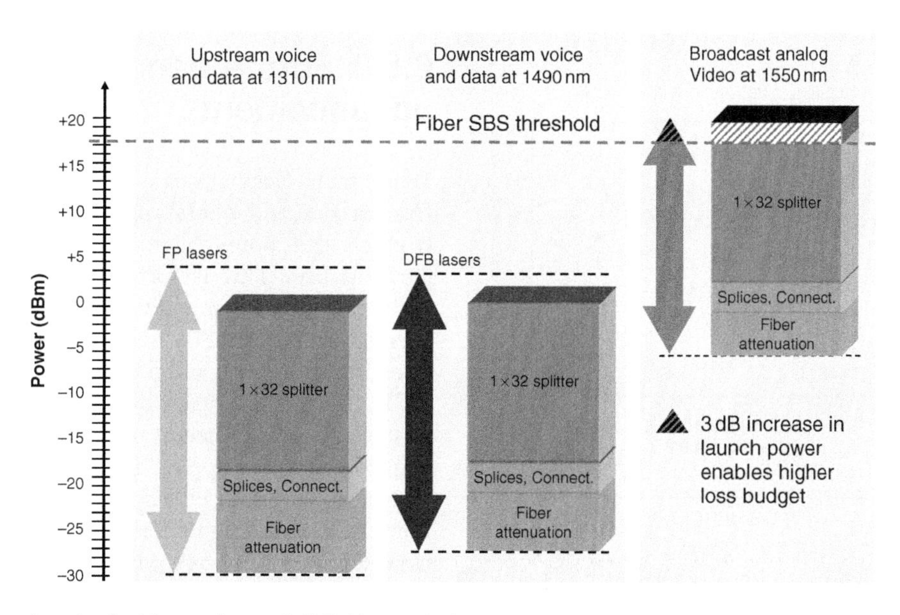

Figure 6.5.24 System loss budget for analog and digital transmission.

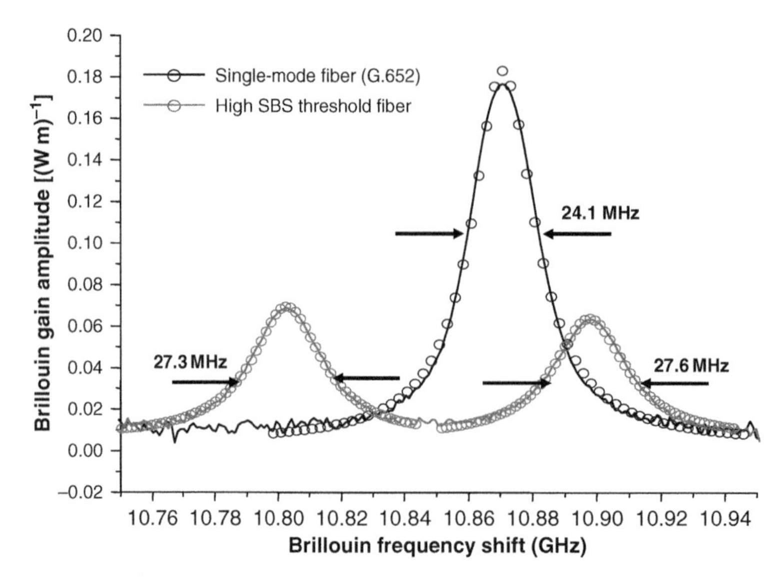

Figure 6.5.26 Fiber theory for improving SBS threshold.

systems, the higher launch power enables twice the splitting with a corresponding increase in cost sharing of key system components [54], or alternatively an increased reach of 10 km or so which enables more subscribers to be accessible to the serving office. Since this fiber performance has to be designed into the system, the impact of this improvement in fiber attributes is only now beginning to be felt in the industry. But laboratory measurements confirm that the system improvements suggested by the Brillouin gain spectrum reductions can in fact be realized (Figure 6.5.27). Since the improved fiber is completely compatible with standard single-mode fiber, achieving similar or better attenuation, fiber coupling, and splicing attributes, these system cost savings can be passed on to a large extent to system suppliers, to the carriers and ultimately to the subscribers as lower access fees.

Taken together, all of these technological advancements have made a steady impact on the costs of FTTx systems. For example, a progression over time of the system costs for FTTH approaches is shown in Figure 6.5.28, where the system cost has declined about 15% per year during the 1990s and about 20% per year between 2000 and 2004 [35], to a point where the cost is about US $1300 per subscriber in 2006. With continued deployment, increasing volumes, improved system range and split ratio [13, 55], and competitive pricing the cost per subscriber will likely continue moving downward toward the US $1000 target.

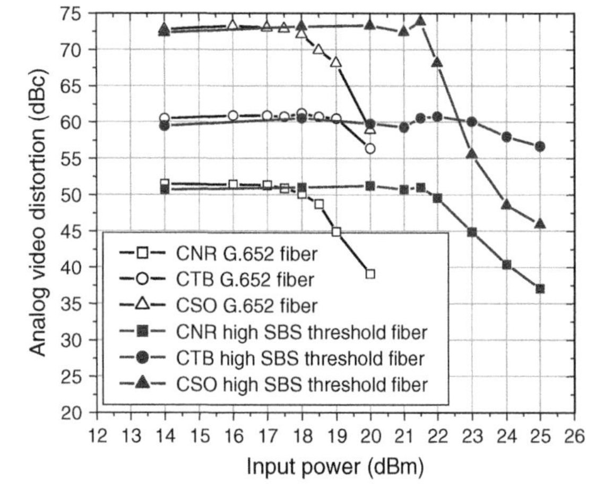

Figure 6.5.27 Analog impairments with fiber meeting G.652 standards and high-SBS threshold fiber.

6.5.8 Future bandwidth advancements

While technology advancements are helping to bring infrastructure and operations costs down, they also introduce the potential for increased user bandwidth performance. This is fortuitous, as the cycle of user technology adoption points toward a 10-fold increase in user bandwidth by 2010 to 100 Mb/s per user [56] for a few million users, and 5 years later to 1 Gb/s per user [57] with a similar user base. These changes in the industry and in user needs are going to drive the technologies that carriers deploy. In fact, the trends in standard products for HFC, FTTC, and FTTH all show the potential for achieving upgrades to 1 Gb/s per user by the time such needs develop, but this requires pushing fiber closer to the home than is currently the case for FTTC and HFC networks. So those service providers will have

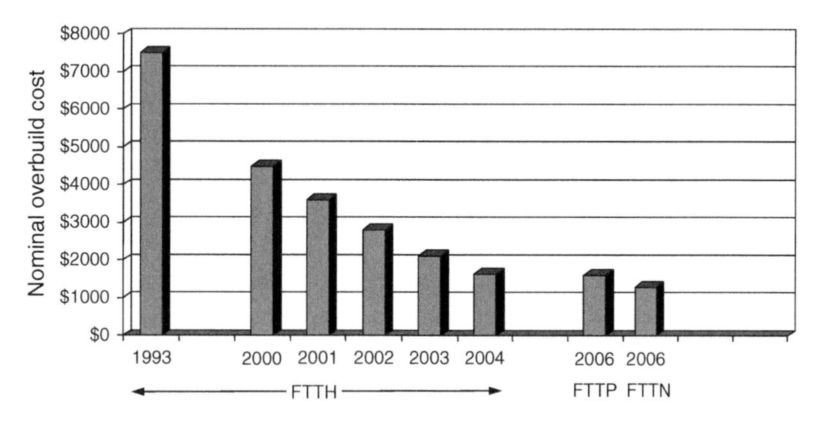

Figure 6.5.28 Cost decline of FTTx solutions with time.

to consider carefully their evolution and capital spending plans to assure they can continue to meet the competitive needs of users in the long term.

Of course, it is difficult to envision exactly how 1 Gb/s per user could be useful a decade from now. But we have seen technology transform our social and entertainment behavior in the past. Phonograph records gave way to CDs, which have much higher fidelity for music, and letters gave way to e-mail, which is faster and can go to multiple recipients. Film gave way to digital images, which enable instant proofing and sharing with friends and family. Black and white TV gave way to color TV, and now analog TV is being replaced by digital TV, which can enable interactive viewing and game playing. Movie theaters are giving way to DVD players with wide screens, surround sound, and home theater arrangements. Each of these technical advances has ushered in new ways to view and share information and entertainment content, leading to unexpected demands for higher bandwidth in our daily lives. As the trend continues, there will inevitably be more sharing of digital images, digital music files, digital movie files, digital home movie clips, digital news feeds, with the consumer electronics industry fueling the trend by offering a whole host of innovative communications and entertainment appliances catering to our growing wants and needs. Ultimately, the increased capability and fidelity of sharing and displaying images and movies will lead to the adoption of 1 Gb/s data rates being delivered to our homes and distributed throughout the household, most likely with capabilities for symmetric traffic flows in both directions.

With some imagination, and imperfect knowledge that the historical trends presented here project into the future, it is possible to envision the following scenario playing out over the next two decades.

- *Applications:* all industrial and entertainment content is digital, and able to be delivered to (and from) homes and businesses at Gb/s rates, where it is stored for on-demand use at the subscriber's convenience, and viewed on large- screen digital displays.
- *Technology:* fiber-based systems are used exclusively for broadband access, with many homes and business connected directly to fiber and the remainder connected from cabinets within 150 m of their building.
- *Deployment:* carriers offer integrated multimedia service packages that include entertainment (music and movies), high-speed Internet, cellular communications, as well as voice services. They also offer in-home networking using wireless and structured cabling options, so subscribers can effectively utilize their bandwidth.
- *Economics:* the capital expenditures for broadband access have been largely recovered and the business cases are positive because of reduced operations costs, allowing carriers to compete on price and bandwidth, and to expand infrastructure to support heavy use of Gb/s access rates.

By the time that 1 Gb/s per user is needed and affordable, much of the broadband access infrastructure will already have been installed and in service in the world. Because of this, upgrade strategies are of key importance for how to evolve from today's architectures to those that support Gb/s per user capacity 10–15 years from now. Even beyond this time frame, WDM-PON (wavelength-division multiplexed-PON) and 10G-PON (10 Gb/s-PON) approaches could enable even higher capacity, and complicate the evolution strategy and planning further [58]. While FTTH systems can accommodate such change by replacing or upgrading the transmission equipment at the ends of the access network, the FTTC and HFC systems may require labor-intensive changes in the drop plant as well. These are complex techno-economic issues and worthy of study and experimental investigation right now, although the issues will linger so that there is time to work through the possibilities to arrive at sound solutions before it is too late.

6.5.9 Summary

After more than 20 years of research and development, a combination of technological, regulatory, and competitive forces are finally bringing fiber-based broadband access to commercial fruition. Three main approaches, HFC, FTTC/vDSL, and FTTH, are each vying for a leading position in the industry, and each has significant future potential to grow customers and increase bandwidth and associated service offerings. No matter which approach wins, or even if all three remain important, construction of the infrastructure needed to serve the entire global network will take one or two decades to complete, because the capital requirements are enormous. During this time it is almost certain that further technical advances and cost reductions will be adopted, bringing performance levels and bandwidth ever higher and keeping service costs affordable. Ultimately, the potential for Gb/s access speeds is on the horizon, and is a target that can be reached economically, when user demand warrants it, through evolution of the infrastructure that is being deployed today.

Acknowledgments

The author would like to acknowledge the very broad base of work at Corning Incorporated, from which significant content of this paper is drawn. Contributors to that base of work include colleagues from Corning Science and Technology, Corning Optical Fiber, and Corning Cable Systems organizations. In particular, the author would like to thank John Igel, Robert Whitman, Mark Vaughn, Boh Ruffin, and Scott Bickham for specific contributions to this work. In addition, the author would like to thank Paul Shumate and Tingye Li for sharing their perspective on the historical trends for fiber to the home.

List of Acronyms

ANSI	American National Standards Institute, standardizing T1E1 for DSL formats
ATM	Asynchronos Transfer Mode, a signal format for combining digital signals
CATV	Cable TV, a means to provide TV via coaxial cable to homes
CNR	Carrier-to-Noise-Ratio, which is the RF carrier strength relative to the noise
CSO	Composite Second Order, an analog impairment due to non-linear effects
CTB	Composite Triple Beat, an analog impairment due to non-linear effects
DMT	Discrete Multi-Tone modulation format, the line code used by aDSL and vDSL systems
DOCSIS	Data Over Cable Interface Specification, a standard to describe cable modems
aDSL, aDSL2	Asymmetric Digital Subscriber Line, a standard for providing digital signals over copper wires used with FTTC systems
vDSL, vDSL2	Very high speed Digital Subscriber Line, a standard for providing digital signals over copper wires used with FTTC systems
DTV	Digital TV, provides digital television with resolution comparable to analog TV
EDFA	Erbium Doped Fiber Amplifier, provides gain in the 1550 nm band
FCC	Federal Communications Commission, a US regulatory body for communications
FSAN	Full Service Access Network Group, broadband access network standards forum
FTTC	Fiber to the cabinet, which brings fiber to a powered cabinet near the subscribers
FTTH	Fiber to the Home, which brings fiber all the way to the side of the home
FTTN	Fiber to the Node, AT&Ts name for FTTC with the cabinet being at a node up to 1.0 km from the subscribers
FTTP	Fiber to the Premises, Verizon's name for expanded FTTH systems to include fiber to business and multiple dwelling units
FTTx	Fiber to the X, describes fiber-based access systems, such as HFC, FTTC, FTTH
G.652	Standard single mode fiber, meeting the ITU-T standard named G.652
GbE	1.0 Gb/s Ethernet, a standard for Ethernet connections
HDTV	High Definition TV, provides high resolution digital television
HFC	Hybrid Fiber Coax, which brings fiber to a powered cabinet near the subscribers
ILEC	Incumbent Local Exchange Carriers, offer local telephone services
ITU	International Telecommunications Union, a standards body for communications
MSO	Multiple Service Operators, companies offering cable TV, internet and phone service
NTSC	National Television System Committee, analog television standard body
10G-PON	10 Gb/s – PON, a system with downstream data rates of 10 Gb/s
PON	Passive Optical Network, a method of providing passive splitter and handling upstream congestion by auto-ranging used with FTTH systems
A-PON	ATM-PON, a passive system using the ATM signal format

B-PON	Broadband-PON, a passive system using ATM signals and an analog overlay
G-PON	Gb/s PON, a system with Gb/s data rates and ATM signal formats
GE-PON	Gigabit Ethernet-PON, a system using GbE for downstream signals
WDM-PON	Wavelength Division Multiplexed-PON, a system using multiple wavelengths, one for each subscriber
SBS	Stimulated Brillouin Scattering, due to an interaction of acoustic and optical waves
SCM	Sub-Carrier Multiplexed, a means to combine multiple analog signals by interleaving RF carriers, each of which are modulated with analog signals
TDM	Time Division Multiplexed, combines signals by interleaving digital streams
TDMA	Time Division Multiple Access, combines the signals from many users into one
Wi-Fi	Wireless-Fidelity, system for broadcasting internet signals inside homes

References

[1] R. E. Wagner, J. I. Igel, R. Whitman et al., "Fiber-based broadband-access deployment in the United States," *J. Lightw. Technol.*, 24(12), 4526 (December 2006).

[2] R. E. Wagner, "Broadband Access Network Options," OIDA Workshop on Broadband Access for the Home and Small Business, Palo Alto, April 22–23 (2003).

[3] P. W. Shumate and R. K. Snelling, "Evolution of fiber in the residential loop plant," *IEEE Commun. Mag.*, 29(3), 68–74 (March 1991).

[4] US Census Bureau, "Manufacturing, Mining and Construction Statistics" Manufacturers' Shipments Historic Timeseries, http://www.census.gov/indicator/www/m3/hist/naicshist.htm (August 29, 2005).

[5] R. L. Whitman, "Global FTTH Risks and Rewards," FTTH Council Conference, Las Vegas (October 3–6, 2005).

[6] J. B. Bourgeouis, "FTTx in the US: Can we see the Light?" FTTH Council paper, http://www.ftthcouncil.org/documents/763530.pdf (November 2005).

[7] "Internet Users (ITU estimates)," United Nations Statistics Division, Millennium Indicators Database, http://millenniumindicators.un.org/unsd/mdg/Handlers/ExportHandler.ashx?Type=Excel& Series=608 (March 10, 2005).

[8] "World Internet Usage and Population Statistics," Miniwatts Marketing Group, http://www.internetworldstats.com/stats.htm (June 30, 2007).

[9] Nielsons NetRatings, "NetView Usage Metrics," http://www.nielsen-netratings.com/news.jsp?section=dat_to&country=us, published monthly.

[10] C. Kehoe, J. Pitkow, K. Sutton et al., GVU Tenth World Wide Web User Survey, Georgia Institute of Technology, www.cc.gatech.edu/gvu/user_surveys (May 14, 1999).

[11] M. J. Flanigan, *"TIA 's 2006 Telecommunications Market Review and Forecast,"* Telecommunications Industry Association, Arlington, VA (February 2006).

[12] M. D. Vaughn, "FTTH Cost Modeling: Impact of Split Location and Value of Added Splits, Extended Reach and Equipment Consolidation," 2003 FTTH Council Conference, Session B06, New Orleans (October 7–9, 2003).

[13] M. D. Vaughn, Member, IEEE, David Kozischek, David Meis, Aleksandra Boskovic, and Richard E. Wagner, "Value of reach and split ratio increase in FTTH Access networks," *J. Lightw. Technol.*, 22, 2617–2622 (November 2004).

[14] M. D. Vaughn, M. C. Meow, and R. E. Wagner, "A bottom-up traffic demand model for LH and metro optical networks," *OFC 2003 Technical Digest*, Paper MF111, Atlanta (March 23–28, 2003).

[15] Telecommunications Act of 1996, Pub. L. No. 104–104, 110 Stat. 56.

[16] M. H. Dortch, Secretary FCC, "Review of the Section 251 Unbundling Obligations of Incumbent Local Exchange Carriers" FCC 03-36, *Report and Order and Order on Remand and Further Notice of Proposed Rulemaking* in CC Docket No. 01-338, adopted February 20 (released August 21, 2003).

[17] M. Render, "FTTH/FTTP Update" 2005 FTTH Council Conference, Render, Vanderslice and Associates Research, www.ftthcouncil.org/documents/732751.pdf, Las Vegas (October 3–6, 2005).

[18] M. Render, "Fiber to the Home: Advanced Broadband 2007 Volume One," Render, Vanderslice and Associates Research, http://www.rvallc.com/ftth_reports.aspx (June 2007).

[19] J. George and S. Mettler, "FTTP market and drivers in the USA," ITU-T Study Group 6, TD 227A1 (GEN/6), Geneva (May 14–18, 2007).

[20] G. Finnie, "Heavy Reading European FTTH Market report," FTTH Council Europe Annual Conference, Barcelona (February 7–8, 2007).

[21] D. van der Woude, "An overview of Fiber to the Home and Fiber backbone projects," FTTH Council Europe Annual Conference, Barcelona (February 7–8, 2007).

[22] R. Montagne, "FTTH deployment dynamics: Overview and French case," IDATE Consulting and Research, FTTH Council Europe Annual Conference, Barcelona (February 7–8, 2007).

[23] X. Zhuang, S. Huiping, Xuemengchi, and W. Hong, "FTTH in China," Ministry of Information Industry of China, ITU-T Study Group 6, TD 227A1 Rev.1 (GEN/6), Geneva (May 14–18, 2007).

[24] M. Coppet, J. F. Hillier, J. C. Hodulik et al., "Global Communications," UBS Investment Research Q-series (July 10, 2007).

[25] W. Ciciora, J. Farmer, D. Large, and M. Adams, "Chapter 18–Architectural Elements and Examples," *Modern Cable Television Technology,* Second Edition, Morgan Kufmann Publishers in an Imprint of Elsevier, San Francisco, pp. 733–766 (2004).

[26] M. D. Nava and C. Del-Toso, "A Short Overview of theVDSL System

Requirements," *IEEE Commun. Mag.* (2002).

[27] W. F. Caton, Acting Secretary FCC, "Advanced Television Systems and Their Impact Upon the Existing Television Broadcast Service," FCC 97–115, 6th Report And Order on MM Docket No. 87–268 (adopted April 3, 1997).

[28] P. Eriksson and B. Odenhammar, "VDSL2: Next important broadband technology," http://www.ericsson.com/ericsson/corpinfo/publications/review/2006_01/06.shtml, Ericcson Review, Issue No. 1 (2006).

[29] D. Meis et al., "Centralized vs Distributed Splitting in Passive Optical Networks," NFOEC'05, Paper NWI4, Anaheim (March 6–11, 2005).

[30] D. Faulkner, "Recent Developments in PON Systems Standards in ITU-T," SPIE Optics East 2005, Paper 6012–27, Tutorial IV, Boston (October 23–26, 2005).

[31] K. Nakajima, K. Hogari, J. Zhou et al., "Hole-assisted fiber design for small bending and splice losses," *IEEE Photon. Technol. Lett.*, 15, 1737–1739 (2003).

[32] See, for example, transport availability standards requirements listed in Bellcore GR-418-CORE (downtime minutes per year), ETSI standard EN 300 416 (outages per year), and ITU-T standard G.828 (errored seconds per year).

[33] C. Saravanos, "Reliability Performance of Optical Components in the Outside Plant Environment," 2004 FTTH Council Conference, www.ftthcouncil.org/documents/212550.pdf, Orlando (October 4–6, 2004).

[34] J. Halpern, G. Garceau, and S. Thomas, "Fiber-to-the-Premises: Revolutionizing the Bell's Telecom Networks," Bernstein Research & Telcordia Research Study, http://www.telcordia.com/products/fttp/bernstein-telcordia.pdf (May 20, 2004).

[35] P. Garvey, "Making Cents of it All - Costs/Key Drivers for Profitable FTTH," FTTH Council Conference, Las Vegas (October 3–6, 2005).

[36] M. Pearson, S. Bidnyk, A. Balakrishnan, and M. Gao, "PLC Platform for Low-Cost Optical Access Components," *Technical Digest IEEE LEOS 2005,* Paper WX1, Sidney (October 23–27, 2005).

[37] Y. Nakanishi, H. Hirota, K. Watanabe et al., "PLC-based WDM Transceiver with Modular Structure using Chip-Scale-Packaged OE-Devices," *Technical Digest IEEE LEOS 2005,* Paper WX2, Sidney (October 23–27, 2005).

[38] W. Chen, K. B. Little, W. Chen et al. "Compact, Low cost Chip Scale Triplexer WDM Filters," *OFC 2006 Technical Digest,* Paper PDP12, Anaheim (March 5–10, 2006).

[39] H. Sasaki, M. Uekawa, Y. Maeno et al., "A low-cost micro-BOSA using Si microlens integrated on Si optical bench for PON application," *OFC 2006 Technical Digest,* Paper OWL6, Anaheim (March 5–10, 2006).

[40] D. W. Vernooy, J. S. Paslaski, H. A. Blauvelt et al., "Alignment-Insensitive Coupling for PLC-Based Surface Mount Photonics," *IEEE PTL,* 16(1), 269–271 (2004).

[41] Wenhua Lin, and T. Smith, "Silicon Opto-Electronic Integrated Circuits: Bringing the Excellence of Silicon into Optical Communications – The Key to Large Scale Integration" http://www.kotura.com/SOEIC_Technology.pdf (February 2004).

[42] M. Turner, "Moving backwards in the OSP: Improving the FTTH distribution segment helps advanced services move forward," OutsidePlant Magazine, Article EVO-634 (January 2006).

[43] D. Meis, "Get big savings in time and money," Broadband Properties Magazine, p. 20 (August 2005).

[44] V. O'Byrne, D. Kokkinos, D. Meis et al., "UPC vs APC Connector Performance in Passive Optical Networks," NFOEC'05, Paper NTuF3, Anaheim (March 6–11, 2005).

[45] A. Woodfin, "Access makes the Parts Grow Stronger," FTTH Council Conference, Las Vegas (October 3–6, 2005).

[46] K. Himeno, S. Matsuo, N. Guan, and A. Wada, "Low-Bending-Loss Single-Mode Fibers for Fiber-to-the-Home," *JLT,* 23(11), November 3494–3499 (2005).

[47] G.S. Glaesemann, M.J. Winningham, and S.R. Bickham, "Single-Mode Fiber for High Power Applications with Small Bend Radii," in Proc. *SPIE Photonics Europe,* paper 6193–23, Strasbourg (April 3–7, 2006).

[48] A. Woodfin, A. B. Ruffin, and J. Painter, "Advances in optical fiber technology for analog transport: technical advantages and recent deployment experience," National Cable and Telecommunications Association, NCTA Technical Papers, 44th Edition, pp. 93–100, Chicago (June 8–11, 2003).

[49] A. Woodfin, J. Painter, and A. Boh Ruffin, "Stimulated Brillouin Scattering Suppression with Alternate Fiber Types - Analysis and Deployment Experience," NFOEC'03, *Technical Digest,* 4, (1265), Orlando (September 7–11, 2003).

[50] A. B. Ruffin, F. Annunziata, S. Bickham et al., "Passive Stimulated Brillouin Scattering Suppression for Broadband Passive Optical Networks," *Technical Digest LEOS'04,* Paper ThE2, Puerto Rico (November 7–11, 2004).

[51] A. Kobyakov, S. Kumar, D. Q. Chowdhury et al., "Design concept for optical fibers with enhanced SBS threshold," *Opt. Express,* 13, 5338–5346 (July 2005).

[52] M.-J. Li, X. Chen, J. Wang et al., "Fiber Designs for Reducing Stimulated Brillouin Scattering," *OFC 2006 Technical Digest,* Paper OTuA4, Anaheim (March 5–10, 2006).

[53] A. B. Ruffin, Ming-Jun Li, Xin Chen et al., "Brillouin gain analysis for fibers with different refractive indices," *Opt. Lett.,* 30(23), 3123–3125 (December 1, 2005).

[54] M. D. Vaughn, A. B. Ruffin, A. Kobyakov et al., "Techno-economic study of the value of high stimulated Brillouin scattering threshold single-mode fiber utilization in fiber-to-the-home access networks," *JON,* 5(1) (January 4, 2006).

[55] J. Lepley, M. Thakur, I. Tsalamanis et al., "VDSL transmission over a fiber extended-access network," *J. Opt. Netw.,* 4, 517–523 (2005).

[56] R. Hunt, "Reforming Telecom Policy for the Big Broadband Era," New America Foundation article, http://www.newamerica.net/index.cfm?pg=publications&SecID=30&Al-lOf=2003 (December 19, 2003).

[57] IEEE-USA Committee on Communications and Information Policy, "FTTH Policy: The Case for Ubiquitous Gigabit Open-Access Nets," *Broadband Properties Magazine,* p. 12 (August 2005).

[58] R. Heron, "FTTx Architecture Transition Strategies," *NFOEC'2007,* Paper NThF, Anaheim (March 25–29, 2007).

Chapter 6.6

6.6

Metropolitan networks

Kaminow, Li and Willner

Abstract

This chapter summarizes the innovation in network architectures and optical transport that has enabled metropolitan networks to meet the diverse service needs of enterprise and residential applications, and cost-effectively scale to hundreds of Gb/s of capacity, and hundreds of kilometers of reach. A converged metro network, where IP/Ethernet services and traditional time-division multiplexed (TDM) traffic operate over a common intelligent wavelength-division multiplexed (WDM) transport layer, has become the most appropriate architecture for significantly reduced network operational cost. At the same time, advanced technology, and system-level intelligence have improved the deployment and manageability of WDM transport. The most important application drivers, system advancements, and associated technology innovations in metropolitan optical networks are being reviewed.

6.6.1 Introduction and definitions

This chapter discusses the evolution of optical metropolitan networks. We start from the evolution of services over the past several years and next few years, and drill down into increasing details about the implementation of the solution. To understand why the network is evolving the way it is and how it will continue to evolve, one has to first understand how services are evolving from simple point-to-point transport services to sophisticated packet services for video and other applications. This is covered in Section 6.6.2. The services, in turn, drive the architecture of the entire network, and this is covered in the Section 6.6.3. Once the architecture is defined, we are ready to delve into the implications of the architecture on the physical layer as described in the Section 6.6.4, while Section 6.6.5 discusses network automation tools required for successful design, deployment, and operation. We summarize the chapter in Section 6.6.6 and provide an outlook into the future of the network in Section 6.6.7.

Figure 6.6.1 depicts several network layers based on their packet functionality: access to customers, aggregation of traffic from various access points into larger central offices, the edge of the packet layer, and the core of the network. The table below the figure represents the most common technologies per layer, from a transport perspective (L0-L1 typically) and a packet perspective (L2-L3). A different segmentation is mostly based on geographical reach: access, metro, regional, and long-haul networks. While aggregation networks often correspond to metro/regional networks and core networks are often long-haul, this is not always the case: regional service providers (SPs) often run a metro core network, and access networks in sparsely populated areas often cover regional distances. In the rest of the section, we focus on the geographic segmentation as we find it more meaningful for dense wavelength-division multiplexing (DWDM) technology.

Access networks are typically classified by reaches below 50 km. Access networks are commonly deployed in a ring-based architecture to provide protection against fiber cuts, and are historically SONET/SDH (Synchronous Optical NETwork/Synchronous digital hierarchy), and more recently coarse wavelength division multiplexed (CWDM) systems. A 10 nm spacing of channel wavelengths in a CWDM system drives down the cost of

Optical Fiber Telecommunications V B; ISBN: 9780123741721

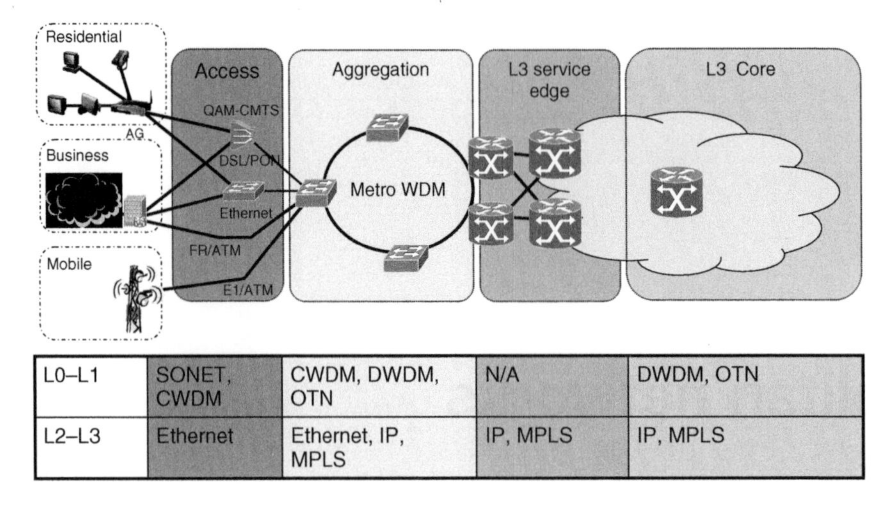

	Access	Aggregation	L3 service edge	L3 Core
L0–L1	SONET, CWDM	CWDM, DWDM, OTN	N/A	DWDM, OTN
L2–L3	Ethernet	Ethernet, IP, MPLS	IP, MPLS	IP, MPLS

Fig. 6.6.1 Context for metro networking within the entire network and technologies typically deployed.

pluggable transceivers by eliminating component cooling requirements, and simplifies optical filter design and manufacture. These systems also tend not to have optical amplification. Optical channel data rates in the access network today are predominantly at or below 2.48 Gb/s, with 4 and 10 Gb/s gaining some recent deployments.

Metro systems may be classified by reaches typically below ~300km, with currently typical node traffic capacities of one to several 10's Gb/s. Given the reach, number of nodes and optical add/drop granularity, Metro networks are typically equipped with optical amplifiers and support DWDM with wavelength channel spacing of 0.8 nm (100 GHz). The lasers now require cooling, but can still be operated without active wavelength locking. The majority of channels in deployment operate at 2.48 Gb/s, but 10 Gb/s data rates are gaining in market share, and higher data rates are starting to see some spot deployments.

Regional systems are classified by reaches on the order of 600 km and below, and long haul is anything above 600 km. Traffic capacities are on the order of multiple 100's of Gb/s. These networks are always equipped with optical amplifiers. Data rates of 10 Gb/s are prevalent for these systems, with channel spacing as low as 0.2 nm (25 GHz), though 0.4 nm (50 GHz) is much more widely deployed. Active wavelength locking is mandatory for such tight channel spacing, with data rates of 40 Gb/s seeing deployments, and 100 Gb/s being pursued within standards bodies and industry development groups.

6.6.2 Metro network applications and services

SPs have traditionally relied on different networks to address different consumer and enterprise market needs. POTS, TDM/PSTN, and to some extend ISDN, have typically served the voice-dominated consumer applications, while Frame Relay, asynchronous transfer mode (ATM) and TDM leased-line networks have served the more data-intensive enterprise applications. At the same time, video has been mostly distributed on separate, extensively analog, networks. Internet access, while representing a significant departure from the 64-kB/s voice access lines, has been relatively lightly used—mainly carrying content limited by human participation— such as e-mail and web access. In recent years, the paradigm has undergone a dramatic shift toward streaming and peer-to-peer applications, driving significant growth in the utilization of access lines as well as the core, as it is less dependent on the presence of a human being at the computer to drive the utilization of the network.

Whether the cause is, the spread of cell phones or voice over IP (VoIP), SPs have seen a steady decline in revenues from traditional telephony and a steep increase in voice-originated and other packet traffic. As a result, SPs are trying to find new revenue streams from residential customers via a strategy commonly referred to as "triple (or even quadruple) play": providing voice, video, and Internet over a common packet infrastructure. This infrastructure requires significant investment in upgrading residential access, as well as back-end systems to create functions and generate content that will increase customer loyalty. The primary example for such applications is video—including high-definition broadcast and on-demand content. In an ideal SP world, customers will receive all their video needs from a network that is engineered around video delivery. Such a network is shown in Figure 6.6.2.

However in many geographic locations, particularly in the USA, SPs are facing tough competition from cable providers, who are much more experienced in delivering video content and are also building their own triple play

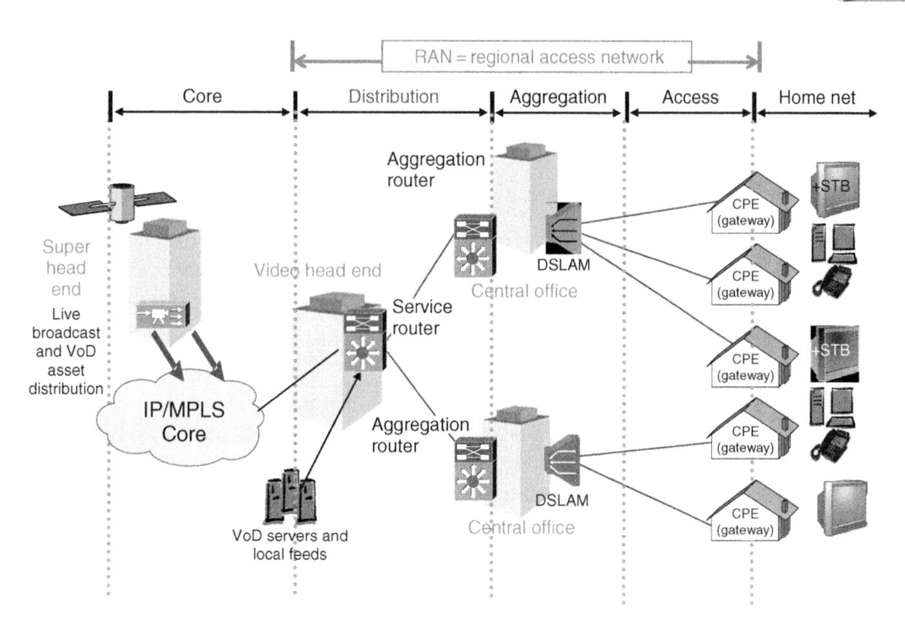

Fig. 6.6.2 Video over broadband architecture overview.

networks over a coax cable infrastructure that is less bandwidth-constrained than the twisted-pairs SPs own. This, in turn, drives SPs to revamp the actual access medium to fiber (to the home, curb, or neighborhood). Perhaps the most serious competition, for both SPs and cable providers, comes from companies who are providing more innovation over the Internet. These "over-the-top" providers use the high-speed Internet access that is part of triple play to deliver video, music, photo sharing, peer-to-peer, virtual communities, multiparty gaming, and many other services without facing the access infrastructure costs. This competition for the hearts and pockets of consumers is driven by application-level innovation delivered over the Internet Protocol (IP), and therefore the transport network should be optimized for either IP or for Ethernet, which is its closely related lower-layer packet transport mechanism.

Enterprise services are experiencing an equally phenomenal growth. The wide-scale adoption of e-commerce, data warehousing, business continuance, server consolidation, application hosting, and supply chain management applications, has fueled significant bandwidth growth in the enterprise network connectivity and storage needs. The business critical nature of most of these applications also calls for uninterrupted and unconstrained connectivity of employees and customers. To best support this, most enterprises have upgraded their networks, replacing ATM, Frame Relay, and TDM private lines, with a ubiquitous Ethernet (GE and 10GE) transport. In addition, regulatory requirements in the financial and insurance industries increased significantly the bandwidth needed to support disaster recovery. The large financial firms became the early adopters of enterprise WDM metro networks, driven by the need to support very high-bandwidth storage applications for disaster recovery such as asynchronous and synchronous data replication over metro/regional distances. An overview of storage-related services can be found in Figure 6.6.3.

To increase the value of the service and the resulting revenue per bit, carriers are looking for ways to provide higher level connectivity—beyond simple point-to-point connections between a pair of Ethernet ports. This includes multiplexed services—in which multiple services may be delivered over the same port— distinguished by a "virtual LAN" (VLAN) tag, and handled differently inside the network. It also includes point-to-multipoint and even multipoint-to-multipoint services, in which the network appears to the user switches and routers as a distributed Ethernet switch. These services are realized

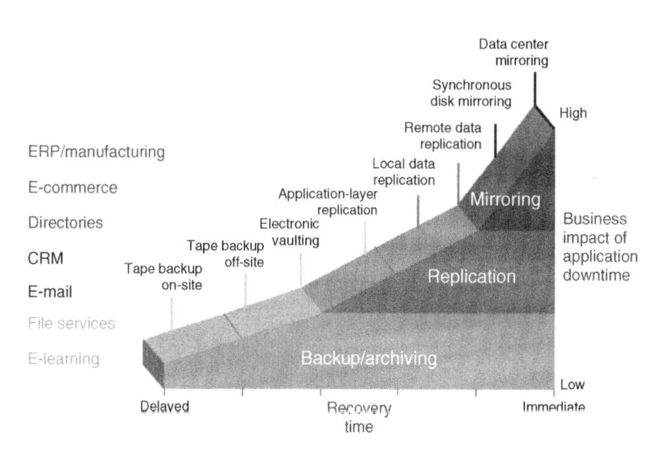

Fig. 6.6.3 Mapping business continuance solutions.

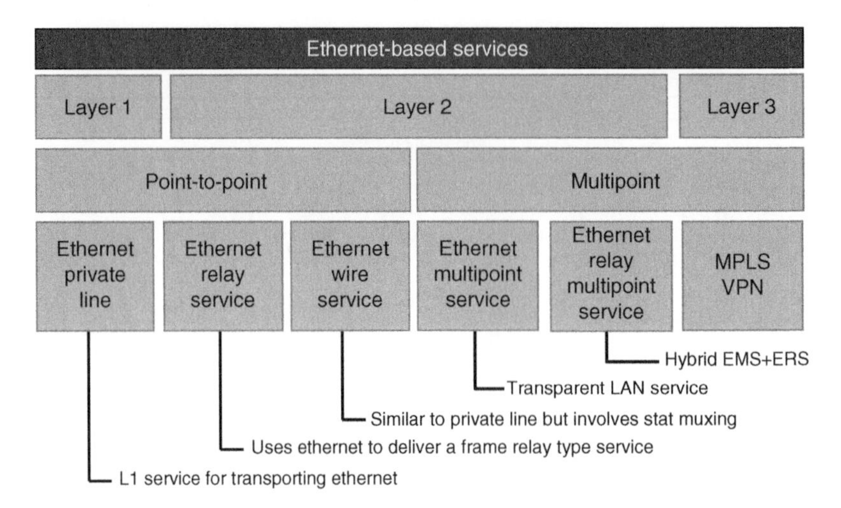

Fig. 6.6.4 Overview of Ethernet-based services (this figure may be seen in color on the included CD-ROM).

via various Layer 2 and even Layer 3 mechanisms. A summary of the various Metro Ethernet services can be found in Figure 6.6.4.

6.6.3 Evolution of metro network architectures

6.6.3.1 Network architecture drivers

In an environment of an ever increasing richness of services, at progressively higher bit rates, but with a price point that must grow slower (or even decline) than their required bandwidth, service providers must build very efficient networks— with the lowest possible Capital expenditure (CAPEX), as well as low operational costs (OPEX).

CAPEX can be optimized by the following means:

- Allowing for as much bandwidth oversubscription as possible, while still respecting the quality of service (QoS) customers are expecting. This cannot be achieved by connections with fixed preallocated bandwidth such as SONET private lines, and drives toward the adoption of packet technologies in the access and aggregation layers;
- Convergence of multiple per-service layers into a unified network. This allows for a better utilization of the network and for better economies of scale, as bandwidth can freely move from one application to another;
- Service flexibility. As new services are introduced, the existing infrastructure must be able to support them, even when the service deviates from the original design of the network;
- Hardware modularity. Hardware must be designed and deployed in a manner that can accept new technologies as they become available, without requiring complete new overbuilds.

OPEX can be optimized by the following means:

- Convergence of layers also helps reduce OPEX as operators do not have to be trained on diverse network elements and distinctive network management capabilities, but rather on one technology. This also allows for more efficient use of the resources as the same operators can work across the entire network instead of working in a "silo" dedicated to one service;
- Increasing the level of automation. Yesterday's transport systems required manual intervention for any change in connection bandwidth and endpoints. Moving to a network that adapts automatically to changes in traffic pattern and bandwidth allow reduction in manual work and cost.

Another related consideration is a barrier to entry and speed of deployment of a new service. Clearly new services will be introduced at an ever increasing rate, sometimes without a clear understanding of their commercial viability. Therefore, it is critical that they can be introduced with minor changes to the existing gear, without requiring a large capital and operational investment that goes with a new infrastructure. This can only be achieved with a converged network that is flexible enough.

So how does a carrier meet these requirements? By converging on a small number of very flexible and cost effective network technologies. Specifically, the following technologies have a good track record of meeting these needs:

- DWDM transport: since photonic systems are less sensitive to protocol, format, and even bit rate, they are able to meet diverse needs over the same

fiber infrastructure with tremendous scaling properties;
- Optical transport network (OTN) standard describes a digital wrapper tech nology for providing a unified way to transport both synchronous (i.e., SONET, SDH) and asynchronous (i.e., Ethernet) protocols, and provide a unified way to manage a diverse services infrastructure [1, 2];
- Ethernet: this technology has morphed over a large number of years to support Enterprise services as well as effective transport for TCP/IP, pre dominantly in metro aggregation networks;
- TCP/IP: has proven resilient to the mind-boggling changes that the Internet has gone through, and is the basis for much of the application level innovation.

Metropolitan area networks (MANs) have been the most appropriate initial convergence points for multiservice architectures. The significant growth of applications with extensive metropolitan networking requirements has placed increased emphasis on the scalability of multiservice MANs [3]. MAN architecture has thus evolved; Internetworking multiple "access" traffic-collector fiber rings in "logical" star or mesh, through a larger "regional" network [4, 5]. A converged metro network architecture of Ethernet/IP and high-speed OTN services operating over a common intelligent WDM layer, reduces CAPEX, and even more importantly OPEX, by enabling easier deployment and manageability of services.

6.6.3.2 Metro optical transport convergence

WDM has been acknowledged early as the most promising technology for scalable metro networks [5, 6]. Its initial deployment, however, remained rather limited, addressing mostly fiber exhaust applications. Metro optical transport is particularly sensitive to the initial cost of the deployed systems, and the CAPEX cost of WDM technologies had been prohibitively high. Attempts to control CAPEX through the use of systems with coarse WDM channel spacing (i.e., 10 or 20 nm) has met with only limited success, as system costs are still set by the transponders, and total unregenerated optical reach and total system capacity is limited. Attempts to increase system capacity by introducing denser channel spacing (i.e., DWDM with 100 GHz spacing) while still avoiding expensive optical amplifiers, resulted in systems with a severe limitation on the number of accessible nodes and total reach.

However, metro networks have critical requirements for service flexibility, and operational simplicity. Moreover, metro WDM cost has to account not only for a fully deployed network, but also for its ability to scale with the amount of deployed bandwidth; as most metro networks do not employ all (or most) WDM channels at the initial deployment phase, but rather "light" unused channels only when actually needed to serve (often unpredictable) network growth. As a result, network operators delayed WDM deployment in their metro networks until these problems were solved with mature technologies from a stable supply base.

At the same time, the evolution of the SONET/SDH transport standards, enabled a successful generation of systems that supported efficient bandwidth provisioning, addressing most of the initial MAN needs, leveraging the advancements in electronics, and 2.5 Gb/s (STM-16) and 10 Gb/s transport (STM-64) [7]. These "next-generation" SONET/SDH systems further allowed improved packet-based transport over the existing time-division multiplexed (TDM) infrastructure, based on data encapsulation and transport protocols (GFP, VCAT, LCAS). Packet-aware service provisioning enabled Ethernet "virtual" private network (VPN) over a common service provider MAN. The initial rate-limited best-effort Ethernet service architectures, evolved to offer QoS guarantees for Ethernet, as well as IP services (like VOIP), and packet-aware ring architecture, like the resilient packet ring (RPR) IEEE 802.17 standard, provided bandwidth spatial reuse. Such Layer-2, and eventually Layer-3, intelligent multiservice provisioning has enabled significant statistical multiplexing gains, enhancing network scalability. Table 6.6.1 illustrates an example of a network with VC-4 granularity that serves a VPN with 4 gigabit Ethernet (GE) sites, six additional point-to-point GE connections, and a storage area network with 2GE and two fiber channel (FC) services. A "purely" optical solution would require at least six STM-64 rings that, even after leveraging VCAT, would be at least 75% full. An advanced multilayer implementation that employs packet level aggregation and QoS in conjunction with VCAT (ML + VCAT) could be based on just four STM-64 rings, each with less than 40% of capacity utilization, saving more than 50% in network capacity. This new generation of multiservice platforms allowed, for the first time, different services to be deployed over a common network infrastructure, instead of separate networks, improving network operations.

VCAT Only	Required VC4:	ML + VCAT
91	Ring 1	30
84	Ring 2	30
49	Ring 3	26
21	Ring 4	8
245	Total VC4	94
(6) 75% full	STM64 Rings	(4) 40% Ml

As traffic needs grew beyond 10 Gb/s per fiber, however, WDM transport became the best alternative for network scalability. To this end, multi-service systems evolved to "incorporate" WDM interfaces that connect them directly onto metro fibers, thus eliminating client optoelectronic (OEO) conversions and costs. The integration of WDM interfaces in the service platforms also changed the traditional "service demarcation point" in the network architecture. This seemingly straightforward convergence of the transport and service layers has introduced additional requirements for improved manageability in the WDM transport. The introduction of many different "wavelength services" amplified the value for "open" WDM architectures that provide robust and flexible transport. In this sense, a converged, flexible WDM metro transport architecture that supports all the different services with the lowest possible OPEX, leveraging elaborate planning and operational tools, and enabling standards-based interoperability, has become increasingly important.

A related but somewhat opposite trend is the integration of increased service layer functionality into the DWDM layer: as packet processors and other service handling mechanisms have become more compact and less expensive, transponders in the DWDM system are no longer restricted to converting client signals to WDM, but have taken on the task of multiplexing services into a wavelength, switching these services to their destination—potentially adding new services along the path, and the related management and control functions. Thus, ADM, MSPP and Ethernet switch platforms "on a blade" have been introduced, replacing small ADMs and small Ethernet switches that would be managed separately from the DWDM layer by devices that are fully integrated into the DWDM layer. These devices typically have only a few WDM interfaces and are limited in size, and they can be logically interconnected in rings over the WDM layer. An example of such a device and its usage in the network can be found in Figure 6.6.5.

Figure 6.6.5(a) shows a conceptual drawing of a DWDM shelf with 3 ADM on a blade cards, each terminating a number of client interfaces and a single WDM wavelength each. These concepts hold for Ethernet-based cards as well.

Figure 6.6.5(b) shows a typical use of these cards on a physical ring topology. Each rectangle represents an ADM on a blade card and the color codes represent rings of ADMs on a blade. In some cases, these cards are concatenated in the same site to terminate a higher amount of traffic.

Figure 6.6.5(c) shows how these cards can be deployed over a physical mesh topology.

Another example is fiber channel (FC) "port extenders" which adapt FC over long distances by "spoofing" acknowledgements from the remote device toward the local device, thereby allowing the local device to increase its throughput without waiting a round trip delay for the remote device to acknowledge the messages. Another value such devices provide is that they

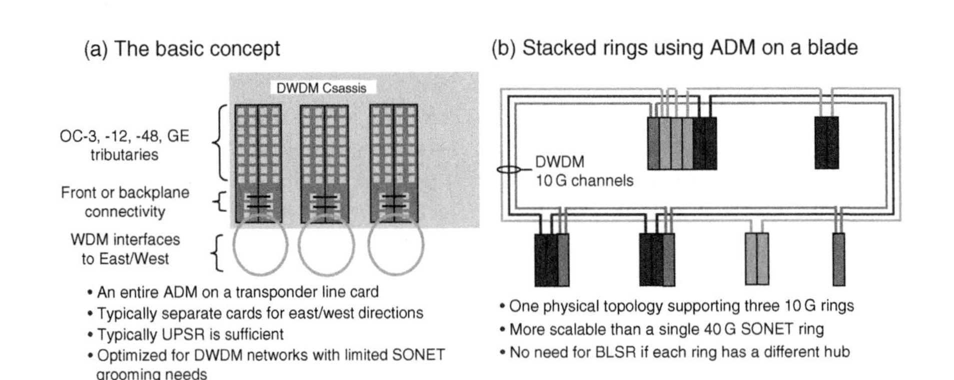

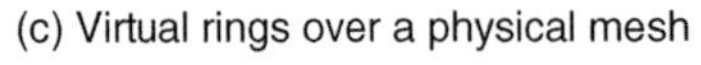

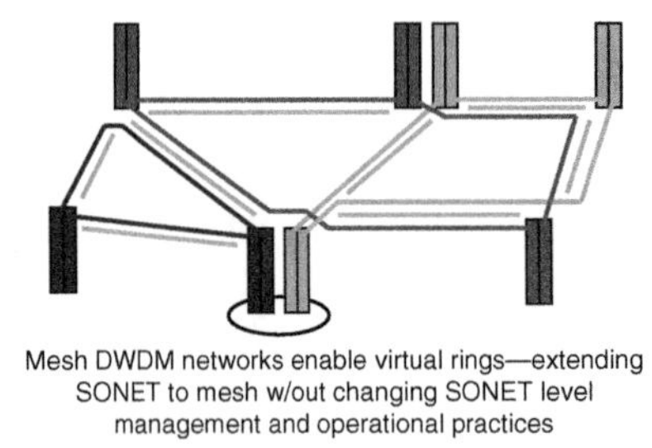

(a) The basic concept
DWDM Csassis
OC-3, -12, -48, GE tributaries

• An entire ADM on a transponder line card
• Typically separate cards for east/west directions
• Typically UPSR is sufficient

(b) Stacked rings using ADM on a blade

10 G channels
• One physical topology supporting three 10 G rings
• More scalable than a single 40 G SONET ring
• No need for BLSR if each ring has a different hub

Mesh DWDM networks enable virtual rings—extending SONET to mesh w/out changing SONET level management and operational practices

Fig. 6.6.5 The ADM on a blade concept.

give the SP visibility into application level issues and therefore enhance the SPs ability to troubleshoot the system.

Eventually, a new generation of metro-optimized WDM transport (often referred to as multiservice transport platforms or MSTPs) has contributed significantly to the recent progress in MAN WDM deployments. This WDM transport enables elaborate optical add-drop multiplexing (OADM) architectures that transparently interconnect the different MAN nodes. Moreover, such MSTP WDM systems have scaled cost-effectively to hundreds of Gb/s, and to hundreds of kilometers, and have significantly enhanced ease of deployment and operation, by automated control and integrated management of the optical transport layer [8].

6.6.3.3 Network survivability

Due to the critical nature and volume of information carried over the network, carriers must ensure that network failures do not result in a loss of customer data. There are a number of schemes that may be implemented, with a general characteristic of providing redundancy in physical transmission route and equipment. At a high level there are two approaches to survivability: (1) protection in the optical layer and (2) protection in the client layer. While optical layer protection provides lower cost, it does not protect against all failures and cannot differentiate between traffic that requires protection and traffic that does not. Therefore, typically, intelligent clients such as routers are in charge of protecting their own traffic over unprotected wavelengths, while less intelligent clients rely on optical layer protection. In the rest of this section, we first focus on optical layer protection and then move to client layer protection.

A common optical layer protection scheme arranges nodes into a physical ring topology, such that a connection between any pair of nodes can take one of two possible physical routes. Should one physical route experience a breakdown in fiber or equipment, an automatic protection switch (APS) is executed.

Figure 6.6.6 shows a four-node ring arrangement that is used to clarify various protection schemes encountered in metro networks, i.e., ULSR, UPSR, BLSR, BPSR. First letter indicates data flow direction around the ring, such that Unidirectional (U) implies that Node 1 communicates to Node 2 in a clockwise (CW) direction, and Node 2 also communicates to Node 1 in the same CW direction passing via Nodes 3 and 4. In this case, the counter-clockwise (CCW) direction serves a protection function. Bidirectional (B) implies that Node 1 communicates to Node 2 in a CW direction, while Node 2 communicates to Node 1 in

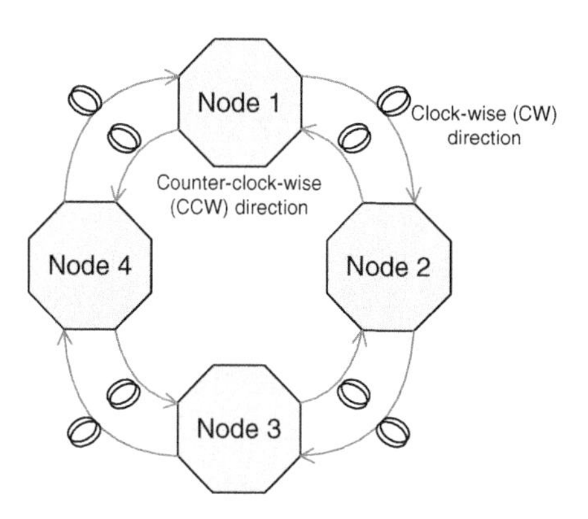

Fig. 6.6.6 Two-fiber (2F) optical ring architecture.

a CCW direction. In this case, the other ring portion serves a protection function. Second letter refers to the whether protection is done at a Line (L) or Path (P) level. SR refers to the basic switched-ring network architecture.

While the terminology varies, the above schemes are generally applicable to both SONET/SDH, DWDM, and optical transport network (OTN) protection (see G.872). The schemes deployed vary based on ease of implementation and changes in demand patterns: in SONET/SDH networks, the most common protection is UPSR followed by BLSR for some core networks. In the DWDM layer, simple 1 + 1 protection is typically implemented, which is equivalent to BPSR in the above notation.

It should be noted that the outlined protection approach inherently doubles the overall network bandwidth requirements relative to actual demand load. Unidirectional case allocates one fiber fully to work data, and one to protection data; bidirectional case allocates each fiber's bandwidth to work/protect in a 50%/50% split.

Figure 6.6.6 provides a very high-level view of the metro network ring architecture. Actual protection switching within a node can also be done in many different ways. For example, Figure 6.6.7 shows a few that see common implementation, in a general order of increasing cost. Figure 6.6.7(b) shows an implementation that protects against fiber cuts only, but minimizes hardware requirements. Figure 6.6.7(c) shows an implementation that both protects against fiber cuts and provides transport hardware redundancy, but requires only a single connection to the client equipment. Finally, Figure 6.6.7(d) shows that protection may be implemented at the electronic router/switch level, while increasing the required size of the electronic fabrics. All of these approaches fall into a general category of 1 + 1 protection schemes, i.e., each work demand has a dedicated corresponding protection demand through a geographically disjoint route, and all approaches are capable

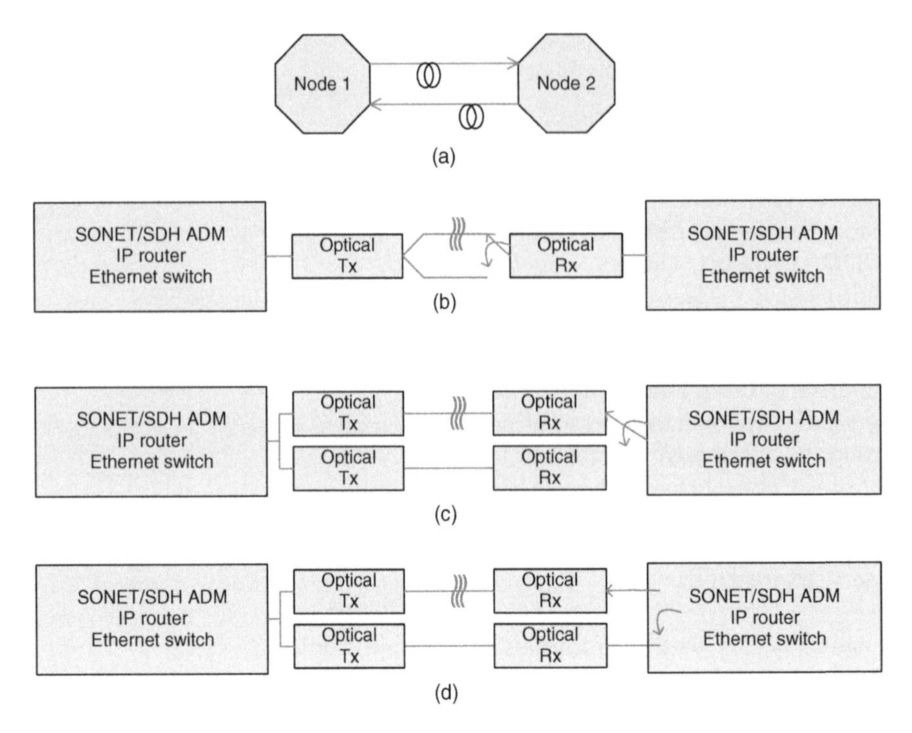

Fig. 6.6 7 Protection switch options within a Node: (a) Generic node 2F connection, (b) Line-side optical switch, (c) client-side optical switch, (d) Line Terminating equipment switch.

of providing protection within a 50 ms SONET/SDH requirement.

All of the above approaches require an effective doubling of the network capacity, while providing protection only against a single route failure. The demands placed on the networks continue to grow in geographic extent, number of interconnected nodes, and in an overall network demand load [9]. As the size grows, a single ring implementation may not be practical from reliability and bandwidth capacity perspectives. The network may still be partitioned into a set of ring connections, with ring-to-ring interconnections. A single ring-to-ring connection will look like a single point of failure in a network, and rings may need to be joined at multiple points: physically the network starts to look like a mesh arrangement of nodes, as shown in Figure 6.6.8.

The rich physical connectivity of a mesh network is obvious as a node-to-node demand connection may take many diverse routes through the network. This diverse connectivity offers an opportunity to significantly improve network's utilization efficiency. Recall that a fully protected ring-based network required dedicated doubling of its capacity relative to the actual bandwidth. A shared protection scheme allocates protection only after a failure has occurred. Thus, assuming that a network suffers only a small number of simultaneous failures, and that a rich physical network connectivity affords several route choices for the protection capacity, each optical route needs to carry only an incremental amount of excess protection bandwidth [10, 11] reduced by a factor of $1/(d\text{-}1)$, where d is the average number of diverse routes connected to nodes. In addition to reduced cost, another benefit of this approach is an ability to gracefully handle multiple network failures. The actual capacity is consumed only after a failure is detected. However, the trade-off is a substantially more complicated restoration algorithm that now requires both network resources and time to compute and configure a protection route [12].

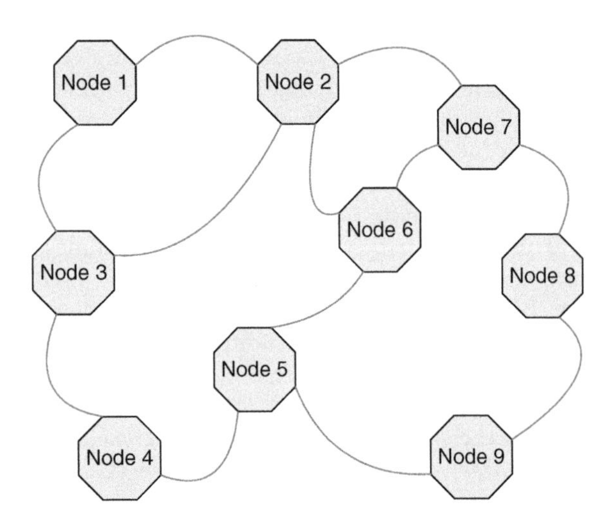

Fig. 6.6.8 Network with a large number of interconnected nodes takes on a "mesh" appearance.

Further, re-routing is done via electronic layer, not optical one, given today's status of optical technology.

As the service layer moves to packet-based devices, new protection mechanisms are being considered. Examples include MPLS fast reroute (FRR), RPR, as well as Ethernet convergence. Out of these mechanisms, RPR is the closest to SONET protection, in that it is mainly confined to rings and loops around the ring in the event of a failure. However, since RPR uses statistical multiplexing, it is able to drop traffic that has low priority depending on the actual amount of high priority traffic currently in the network, whereas the optical layer was limited to protecting the total working bandwidth irrespective of the actual usage of the bandwidth. This added flexibility allows SPs to offer a large number of services, while with SONET the only service that was available was a fully protected service (99.999%). It is worth noting that some SPs tried to also offer a preemptible service using protection bandwidth in SONET, but since this bandwidth was frequently preempted, the service was only useful for niche applications.

MPLS offers a more flexible mechanism that is not restricted to rings, based on a working path and a predefined protection path. Again, thanks to statistical multiplexing, the bandwidth along the protection path does not have to be reserved to a particular working connection, but rather is used by the traffic that requires protection based on priorities. While the packet level mechanism is simple, this scheme does require planning to ensure protection bandwidth is not oversubscribed to a point that the service level agreement cannot be guaranteed.

Finally, Ethernet also offers a convergence mechanism that ensures packet can be forwarded along a new spanning tree should the original spanning tree fail. However, this mechanism is typically slow and does not scale to larger networks.

Given these protection mechanisms, what is the role of the WDM layer in protecting traffic? Quite a bit of research has been performed on how WDM protection can coexist with service layer protection and how the layers coordinate and benefit from the respective protection mechanisms [13, 14]. However in reality, most SPs prefer to keep protection to one layer for simplicity. Naturally, when protection does exist in the service layer, it is more beneficial to use it, as it covers failure modes that are unrecoverable in the optical layer (such as an interface failing on the service box). As discussed, often times service level protection is more efficient driving to an overall lower cost of protection even as service layer equipment is more expensive than WDM equipment [15]. This leaves narrow room for protection at the WDM layer: typically for point-to-point applications that do not have their own protection mechanism, such as SAN applications.

6.6.4 WDM network physical building blocks

6.6.4.1 Client service interfaces

Metro networks generally see traffic that has already gone through several levels of aggregation multiplexing, and equipment interconnections are done at 1 Gb/s data rate and above. By definition, these are meant to connect equipment from a wide variety of manufacturers, and several international standards (as well as industry-wide Multi-Sources Agreements) have been developed [16, 17] that cover optical, mechanical, electrical, thermal, etc. aspects. Given the required high data rate and the physical connectivity length, metro equipment client interconnections are almost exclusively optical.

Early systems had interface hardware built from discrete components. A highly beneficial aspect of developing and adhering to standards is an ability of multiple vendors to provide competitive interchangeable solutions. Over the last several years, the optoelectronics industry has seen a tremendous amount of client interface development, and a near complete transition from custom-made interfaces to multi-sourced, hot-pluggable modules. The interfaces have evolved [18] from GBICs offered at 1 Gb/s data rate [19] to SFP for multirate application up to 2.5 Gb/s data rate [20] to XFP at 10 Gb/s [21]. The economics of manufacturing are such that it is frequently simpler to produce a more sophisticated component and use it across multiple applications. Further, providing a single electrical socket on the equipment allows the interfaces to be reconfigured simply by plugging in a different client module, with same interfaces being able to support SONET/SDH, Ethernet, FC, etc. applications. Figure 6.6.9 shows a comparison of relative physical form factors for a variety of pluggable interfaces targeting 10 Gb/s data rate, and the rapid relative size reduction over the course of a few years.

6.6.4.2 WDM network-side optical interfaces

The requirements on the WDM (network-side) of the optical system are more stringent than on the client interfaces. While client interfaces need connectivity over a relatively short distances, with 90% falling within 10 km distance, network side interfaces need to cover distances of hundreds of km and many demands are multiplexed on the same fiber using WDM.

The optical wavelength of client side interfaces has a wide tolerance range and uncooled lasers are most often used. The WDM side, due to multichannel requirement, has lagged the client side interface in size development.

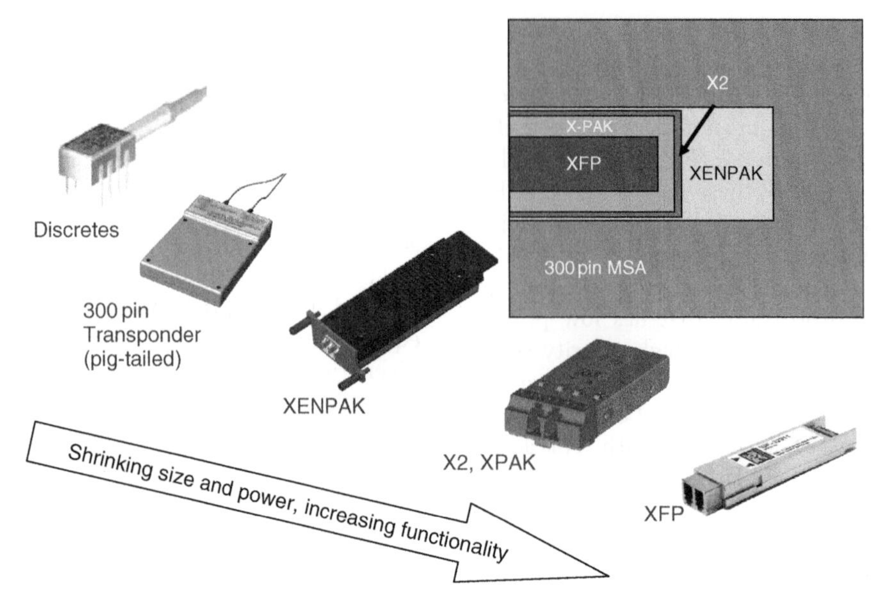

Fig. 6.6.9 Evolution of client size interface form factors (this figure may be seen in color on the included CD-ROM).

Initial network side pluggable modules were developed in GBIC form-factor for 2.5 Gb/s application, and used uncooled lasers for CWDM with relaxed 10 nm wave separation. More recently, 2.5 Gb/s interfaces with DWDM (100 GHz) channel spacing were implemented in SFP form factor, too. As networks evolved to support higher bandwidth services, 10 Gb/s network side interfaces were implemented in 300-pin MSA form factors [22]. Subsequently, MSA modules evolved to support 50 GHz channel spacing with full tunability across all C-band wavelengths. Tunable laser technologies have being increasingly employed in Metro WDM systems to reduce inventory cost, and improve operations [16]. The choice of the appropriate transmitter technology is particularly important, as its cost usually dominates the total cost of a fully deployed transport system [23]. At the same time, much smaller XFP packages supporting 10 Gb/s WDM interface at 100 GHz channel spacing were developed using lasers without wavelength locking. Such next-generation pluggable transmitters are very important, not only for their enhanced performance or lower cost, but also for more easily integrating into the different service platforms further simplifying the network architecture and thus reducing the overall network cost. At such high data rates, however, optical performance, predominantly dispersion-tolerant (chirp-minimized) modulation, becomes also important. Current development efforts are pursing fully C-band tunable 50 GHz WDM interfaces in XFP form factor. Higher data rate 40 Gb/s interfaces currently require a larger package, but are following the same general trajectory of rapidly decreasing size and increasing capability.

In addition to the extremely rapid advances in the optical technology developments, electronics technology is also providing increased performance, and reduced size and power consumption. Of particular interest are the field programmable gate array (FPGA) technology and the FEC technology. Metro networks are generally called on to support a rich variety of services, such as SONET/SDH, Ethernet, and FC. The protocols and framing formats, overhead, and performance monitoring parameters are very different, while intrinsic data rates may be quite close. Since client interfaces are pluggable, it is highly desirable to provide a software-configurable, flexible electrical processing interface such that a single hardware circuit pack can be field-reconfigured to support different services. FPGA elegantly fulfills such a role, providing high gate count, and low-power and high-speed capability with 65 nm CMOS geometries available in 2007, and 45 nm CMOS geometries expected to be available in 2009–2010 time frame. At the same time, FEC and increasingly enhanced FEC enable much more flexible transmission performance.

6.6.4.3 Modulation formats for metro networks

Non-return-to-zero on-off keying (NRZ-OOK) is arguably the simplest modulation format to implement for WDM network signal transmission. NRZ-OOK is the format with the widest deployed base of commercial systems, given that excellent propagation characteristics can be achieved with quality implementations having good control over rise/fall times, limited waveform distortions, and high optical extinction ratio. The longer (300 km) demand reach requirements imply that both fiber dispersion and loss become quite important. The

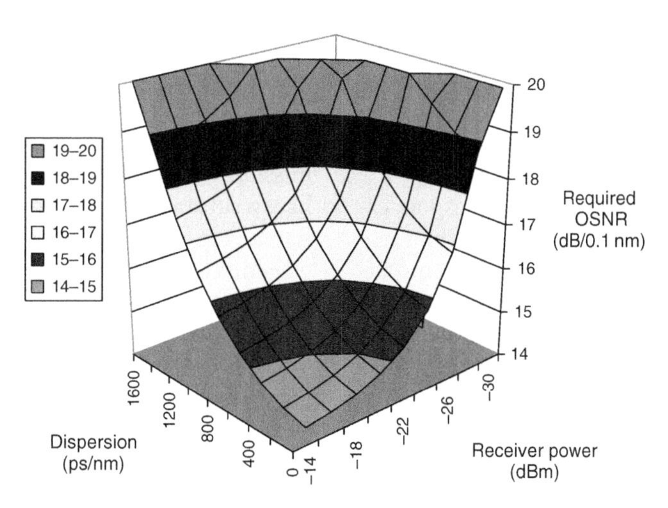

Fig. 6.6.10 Target OSNR required to achieve a BER ~10^{-7} for a typical 10-Gb/s NRZ-OOK implementation, which assumes a receiver sensitivity of -28 dBm at 10^{-9} BER, and an OSNR sensitivity of 10 dB/0.1 nm at 10^{-3} BER.

intrinsic dispersion tolerance is determined by the modulation format and data rate. For example, NRZ-OOK at 2.5 Gb/s has an intrinsic dispersion tolerance of ~ 17,000 ps/nm, corresponding to ~ 1000 km of NDSF fiber. The fact that this is well above reach "sufficiency" for Metro networks permits an engineering trade for a lower-cost, lower-quality implementation. Relaxing transmitter chirp control can lower costs substantially, while still allowing for a dispersion tolerance in 1600 ps/nm to 2400 ps/nm range (i.e., 140 km of NDSF fiber).

Intrinsic optical transceiver characteristics are set by launch power, dispersion tolerance, receiver sensitivity and ASE tolerance. A network must satisfy this multidimensional demand simultaneously and preferably with a single modular implementation. Metro networks geographic characteristics and traffic demands cover a broad range from a few tens of kilometers up to 200 km. Such wide range of networks implies that there is no generic target characteristics: networks may be limited by any combination of the above mechanisms, depending on demand reach length, number of nodes, fiber characteristics, etc. Even within the same network some demands may be power-limited, while others may be limited by dispersion, while others may be limited by ASE noise. A desire to minimized network hardware costs, while still supporting the demands, poses a challenging optimization problem.

Unamplified links have two dominant limiting characteristics, and performance is typically expressed in terms of receiver power penalty as a function of dispersion. Noise determined by receiver electronics is independent of input signal power, and affects both "0" and "1" signal levels equally. System performance must be kept above a threshold line defined by a minimum receiver power required to achieve desired bit error rate (BER) performance at a set dispersion. Optically amplified system noise is primarily determined by the beating between signal and ASE components, and impacts primarily "1" level for OOK modulation formats [24, 25]. Its functional form is different from a direct power penalty, and systems need to consider three characteristics: received power, dispersion and OSNR. Amplified systems performance can be expressed in terms of two-dimensional power and dispersion surface, and shown as a target OSNR required to achieve a certain BER, with an example shown in Figure 6.6.10.

Optically amplified, ASE-limited systems exhibit an inverse relationship between required target OSNR and unregenerated optical reach. Thus, a 1-dB increase in required OSNR produces a corresponding reduction in unregenerated reach, and can easily cross the network performance threshold.

6.6.4.4 Handling group velocity dispersion

As service demands pushed network transport rates to 10 Gb/s, the requirement to cover an identical network geographic extent remained unchanged. Current 10 Gb/s transmission is still most frequently done with NRZ-OOK format, which intrinsic dispersion tolerance is 16 times smaller than equivalent 2.5 Gb/s (i.e., ~1200 ps/nm for unchirped versions). 10 Gb/s data rate crosses the threshold of dispersion tolerance for many Metro networks, and some form of dispersion compensation is required. In-line dispersion compensating fiber (DCF) is the most commonly deployed technology. DCF is very reliable and completely passive, has a spectrally transparent pass band compatible with any format and channel spacing, and can be made to compensate dispersion across the full spectral range for most deployed transmission fiber types. The disadvantages are added insertion loss that is most conveniently "hidden" in the optical amplifier midstage, nonlinear effects requiring controlled optical input power, both of which degrade overall link noise figure. Existing networks cannot be easily retro-fitted with DCFs without significant common equipment disruption.

New ways of handling dispersion based on transponder-based technologies are advantageous for seamless network upgrades. Particularly, transmitter-side modulation formats [26] or receiver-side electronic distortion compensation (EDC) are attractive, if they can be made economically viable. However, it is instructive to note some options that have been specifically applied to metro networks: prechirped NRZ-OOK modulation, duobinary modulation, and receiver-side EDC. With both prechirped and duobinary modulation formats optimal dispersion is shifted to higher values, but performance may be degraded for other parameters, such as ASE tolerance or overall dispersion window. Receiver side equalization, whether electronic or optical, is a technology that introduces additional cost, power consumption, and board space. A decision to use transceiver-based dispersion compensation is not simple or universal, though EDC is finding good adoption especially on the client interfaces.

Modulation formats may mitigate some of the dispersion problems. However, the desire to upgrade existing field-deployed networks to support higher channel data rates transport is impeded by several considerations: (1) higher rate signals require OSNR increase of 3 dB for each rate doubling (assuming constant format and FEC), (2) receiver optical power sensitivity increasing by 3 dB for each data rate doubling, (3) dispersion tolerance decreasing by 6 dB for each data rate doubling, (5) in-line optical filtering, and (4) system software upgrades.

6.6.4.5 Optical amplifiers

Optical signal loss in metro networks accumulates from three components: transmission fiber, optical components embedded in the frequent add/drop nodes, and passive fiber segment connections. Metro networks are generally deployed in very active environments, and while not very long, they are subject to frequent mechanical disturbances and breaks. Metro network fiber accumulates passive loss due to frequent repairs and splicing.

There are three main optical amplifier choices available for overcoming the loss: erbium-doped fiber amplifier (EDFA), semiconductor optical amplifier (SOA), and Raman amplifier. Of these, EDFA provides a cost-effective solution to overcome loss in the network with good performance. SOA amplifiers have the advantage of wide amplification bandwidth, but have more limited output power, susceptibility to interchannel crosstalk issues, and high noise figure. Distributed Raman amplifiers' primary benefit is in reducing effective amplifier spacing which lowers the overall link noise figure. Metro nodes are already closely spaced and "distributed" benefit is small. Distributed Raman also relies on transmission fiber quality, and passive losses and reflections have a substantially deleterious effect [29], making use of Raman amplifiers in metro quite rare.

Amplifier response to optical transients is as critical a characteristic as gain, noise figure, output power, and spectral flatness. Rich traffic connectivity patterns require a high level of immunity to possible optical breaks. Figure 6.6.11 shows an example whereby a break between Node 1 and Node 2 causes a loss of optical Work 1-2 and Protect 2-3 channels. Protect 2-3 channels are optically coupled to Protect 1-2 channels, and any disruption on Protect 1-2 channels will cause a corresponding loss of connectivity between Node 1 and Node 2.

Table 6.6.2 below shows a table of typical effects that can lead to EDFA transients and associated time constants. The same physical properties that make EDFA excellent for multichannel transmission with negligible crosstalk also make it hard to suppress optical transients. Techniques based on optical reservoir channels suffer from having to add extra optical hardware, and the fact that EDFA dynamics are spectrally dependent, i.e., losing channels at short wavelengths does not have the same effect as adding a reservoir channel at long wavelengths. The same is true with gain clamping via lasing [30]. The most cost-effective strategy, and one that has seen actual field deployment, is using optical tap power monitors on the amplifier input and output ports, with electronic feedback to the pump lasers. The electronic feedback loop can be made quite fast, but pump laser electrical bandwidth and intrinsic EDFA gain medium dynamics limit possible control speed. A simple addition of

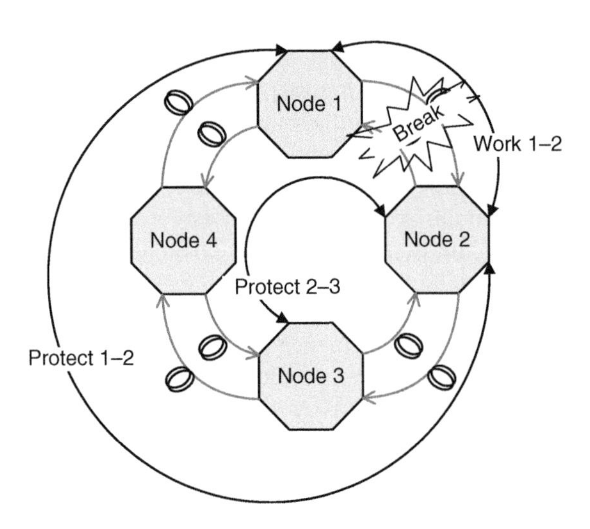

Fig. 6.6.11 Optical interaction between optical work and protect routes.

Table 6.6.2 Table of possible optical transient time constants, in order of increasing speed.

3 mm jacketed fiber slow bending	500 ms
3 mm jacketed fiber caught in shelf cover	100 ms
3 mm jacketed fiber fast bending	7 ms
3 mm jacketed fiber wire stripper cut	5 ms
Connector E2000 fast unplug	2 ms
Bare fiber wire stripper cut	200 μs
Bare fiber knife chop cut	100 μs
Bare fiber (break at splice using tensile force)	2 μs

a controlled attenuator is insufficient to guarantee flat spectral output under different channel load.

Figure 6.6.12 shows an example measurement of an electronically stabilized EDFA amplifier transient response to an optical step function, with step function fall time as the parameter. Several temporal regions can be identified. First, there is an optical power increase due to a redistribution of optical power into surviving channels, with a rise time corresponding to the optical channel power loss fall time. Second, the pump power is rapidly reduced by the electronic feedback control, and the channel power recovers with a time constant set by the control loop dynamics. Third, depending on the parameters of the control loop and their interplay with optical dynamics, there may be some amount of transient undershoot and possibly ringing. Finally, a steady state is reached that is likely to have some finite error in the channel power due to electronic errors, due to finite broadband ASE power, etc.

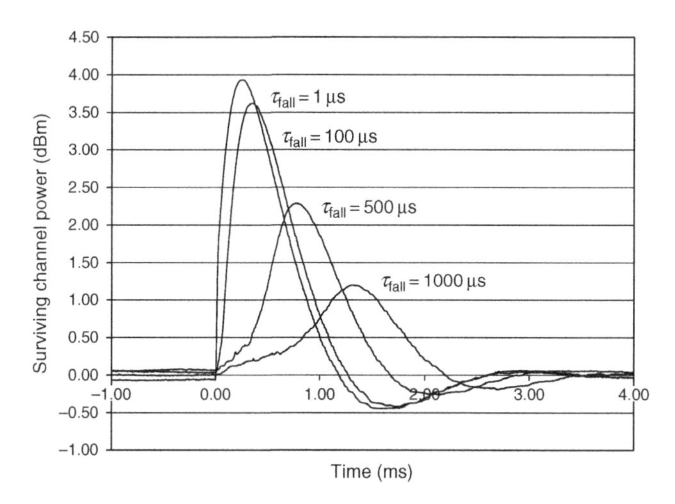

Fig. 6.6.12 Example amplifier optical transient response, with optical channel loss fall time as a parameter.

The transient are shown for a single amplifier, while real systems employ many cascaded amplifiers in a route. Each subsequent amplifier will see not only the original step-like loss of the optical signal, but will also experience the accumulated effects of all of the preceding amplifiers. Thus, control loop parameters have to be developed and verified on amplifier cascades, in addition to individual modules [31].

It is fundamentally not possible to completely prevent and eliminate optical transients, and these impact optical channel performance in several ways. Increasing optical power may lead to nonlinear fiber effects and corresponding distortions in the received signal. Low power leads to a decrease in optical SNR. Further, power may exceed the dynamic range of the receivers, and may interact with the dynamic response of the receiver electrical amplification and decision threshold mechanisms. These combined effects may lead to burst errors in the optical channel. Protection switching mechanisms need to be designed with corresponding hold-off times to prevent such events from triggering unnecessary switching.

6.6.4.6 Optical add/drop nodes

WDM multiplexing and connectivity provides some of the functionality previously supported at the SONET layer, such as multiplexing and circuit provisioning. As discussed previously and shown in Figure 6.6.11, networks nodes may be arranged on an optical fiber ring. However, demand connections may be such that an optical channel bypasses a node and stays in the optical domain. OADMs accomplish this function. Most of the metro networks were deployed with OADMs implemented with fixed optical spectral filters. The filters are positioned within the optical path along the ring to both drop and add signals of pre-determined wavelengths. This provides a very cost-effective access to the optical spectrum, minimizes signal insertion loss, and allows for possible wavelength reuse in different segments of the optical ring or mesh network. In a sense, predeployed optical filters associate a specific destination wavelength range address with each node. Optical filters themselves may be based on a wide variety of technologies, and are beyond the scope of this chapter.

Fixed filter based nodes are cost-effective, have low optical loss, and are simple to design and deploy. However, they pose an operational challenge due to the intrinsic lack of reconfigurability. Each node must have a predesigned amount of capacity (i.e., wavelength address space) associated with it and cannot be changed without significant traffic interruptions. For example, some portions of the network may experience more than expected capacity growth and may require additional spectral allocations, which would be impossible to achieve without inserting additional filters into the common signal path, thereby interrupting traffic. Others portion of the network may lag expected growth, and will thus strand the bandwidth by removing unused spectrum from being accessible to express paths. Network deployment with fixed filters require significant foresight into the expected capacity growth, and several studies have addressed the question of what happens when actual demands deviate from the expectations [32], with as much as half of the overall network capacity possibly being inaccessible. One of the frequently proposed techniques to overcome such wavelength blocking limits is the use of strategically placed wavelength converters in the network [33, 34], but is quite expensive in terms of additional hardware that must be either predeployed or require as-needed service field trips.

An alternative to deploying wavelength converters is to provide dynamic reconfigurability that is generally associated with electrical switching directly at the optical layer. Developments in optical technology have allowed a new level of functionality to be brought to the OADMs, and fall under a general term of reconfigurable OADM (ROADM). It should be pointed out that while ROADMs substantially reduce wavelength blocking probability [35, 36], they cannot completely eliminate it, especially if all wavelengths remain static after assignment. Some amount of wavelength conversion or dynamic wavelength retuning may be required.

A variety of optical ROADM node architectures can be considered, depending on the particular goals of the network designer [37, 38]. These architectures can be subdivided into three broad categories. The first category can be described as a space-switch-based architecture surrounded by MUX/DEMUX elements. An example for a Degree 2 node is shown in Figure 6.6.13(a). With the current state of the art, the implementation is done with integrated MUX/DEMUX, add/drop switch, and direction-switch elements. The channelized aspect of the architecture introduces optical filtering effects into the express path, thereby limiting bit rate and channel spacing transparency. Increasing the connectivity degree of the node requires a change in the configuration by adding either a new level of integration of additional MUX/DEMUX and switch elements, or externally interconnecting smaller building blocks of Figure 6.6.13(a) with additional external switching.

The second category can be described as broadcast and select architecture, and is shown in Figure 6.6.13(b). The architecture relies on an integrated wavelength blocker (WB), which can provide arbitrarily selectable pass and stop bands with continuous spectrum and single-channel resolution. The express path continuous-spectrum attribute reduces channel filtering effects, improves cascadability, and permits a high level of bit rate and channel spacing transparency. One disadvantage of the broadcast and select architecture is its requirement of a separate MUX/DEMUX structure to handle local add/drop traffic, which adds cost and complexity. A second disadvantage is that scaling to a higher Degree N node interconnect requires N x (N–1) WB blocks, with a Degree 4 node requiring 12 WB blocks.

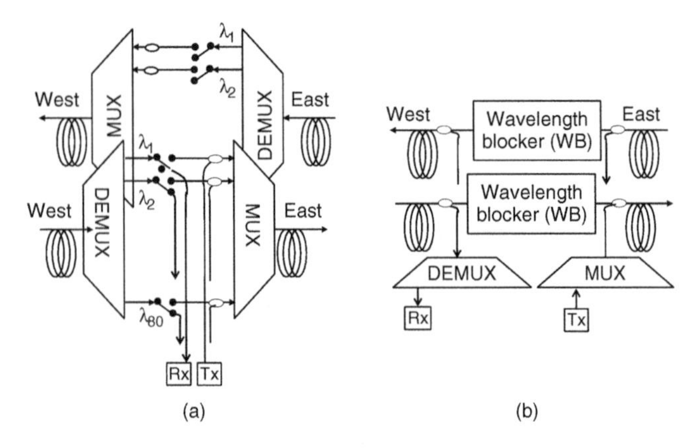

Fig. 6.6.13 (a) Degree 2 space-based OADM; (b) Degree 2 broadcast-and-select OADM.

More recently, a third ROADM architecture has been introduced that attempts to combine a high level of integration and express transparency associated with a broadcast and select architecture, with an integrated MUX/DEMUX functionality [39]. The architecture, shown in Figure 6.6.14, is based on an integrated multi-wavelength, multiport switch (MWS) and is particularly attractive for metro-type applications that are susceptible to frequent traffic and node churn, but have moderate bandwidth requirements. MWS elements have several output ports (in the 4-9 range) and can selectably direct any combination of wavelengths to any output ports.

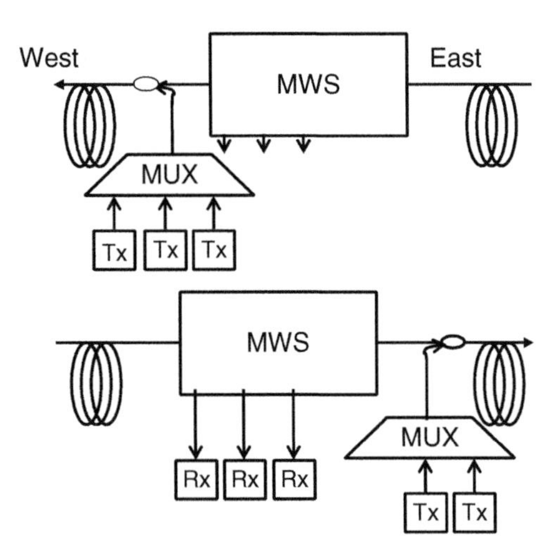

Fig. 6.6.14 Reconfigurable OADM Architecture.

A receiver can be connected directly to the MWS if only a small number of add/drop wavelengths is required, or a second level of DEMUX can be implemented to increase the add/drop capacity. The second level itself can be based on a low-cost fixed architecture, or a more complex wavelength-tunable one.

Table 6.6.3 shows a comparison table of possible space switch and MWS-based architectures, and associate trade-offs.

Recently, the optoelectronics industry has evolved a new direction toward higher levels of functional integration, which has the potential to reduce the cost of OE and EO conversion. Possible availability of such low-cost optical interface components has prompted a re-evaluation of the existing trade-offs between optical and electrical switching, and proposals for a "Digital ROADM" have been presented. A Digital ROADM is more accurately termed a reconfigurable electrical add/drop multiplexer. It is an electrical switching fabric very similar in functionality to the legacy SONET/SDH add/drop Multiplexers or to an Ethernet switch, but updated to operate on signals compliant to the recently developed OTN framing standard. Highly integrated photonics indeed hold a promise of reduced costs; unfortunately, optical interfaces comprise only a small fraction of the overall node cost. The impact of electronically processing every digital bit stream holds both advantages of being able to completely regenerate wavelength signals, monitor bit stream quality, and improve their grooming efficiency, as well as detrimental aspects of substantially increased electronic power consumption and mechanical footprint. Table 6.6.4 captures the more salient comparison points, and argues that metro networks still preferentially benefit from optical OADM rather than electrical one.

In summary, ROADM has introduced network design flexibility, and automated and scalable link engineering, both critical to the success of metro WDM architectures. Wavelength-level add/drop and pass-through, with automated reconfigurability (ROADM) at each service node, is also only operationally robust solution for WDM deployments that support the uncertain (often unpredictable) future traffic patterns in MANs that scale to hundred of Gb/s. ROADM network flexibility also provides the ability to set up a wavelength connection without visiting any intermediate sites, thus minimizing the risk of erroneous service disruptions during network upgrades. In the context of the present analysis, it is also useful to identify and distinguish between two main functional characteristics of ROADMs at the network level: (1) ROADM solutions allow for switching of each individual wavelength between the WDM ingress and egress, potentially among more that one fiber facility; (2) More elaborate solutions could also allow extraction or insertion of any client interfaces to any wavelength of any fiber. This latter solution has been often proposed, in combination with predeployed tunable transmitters and/or receivers, to realize advanced network automation. Such a network, with the addition of a GMPLS control plane, enables dynamic bandwidth provisioning, and fast shared optical layer mesh protection. Current network deployments, however, are primarily interested in the ROADM functionality and the most cost-effective-related technologies (rather than in the most advanced solution that would meet any conceivable future need, irrespective of price). In this sense, the technologies captured in the above table, are currently the main focus of network deployments, as they have sufficient functionality to meet most customer needs, and are the most mature and thus cost-effective technologies [40].

Table 6.6.3 Comparison of three optical ROADM architectures.

Parameter	Space switched	MWS + MWS DEMUX	MWS + fixed DEMUX
Cost	Independent of # A/D channel	Same up to 8 ch, 5 x at 40 ch	Same up to 8 ch, 2 x at 40 ch
Optical channel monitor	Built-in	Usually external	Usually external
Channel power equalization	Yes	Yes	Yes
Multidegree	Highly integration dependent. Difficult in-service growth	Up to Degree 8 (in service)	Up to Degree 8 (in service)
0.5 dB passband	~40 GHz	~50 GHz	~50 GHz
Express loss	<12 dB	<13 dB	<13 dB
Flexibility	Add/drop ports have fixed wavelengths	Drop ports are fully tunable. Adds are wavelength independent	Drop ports are commonly fixed wavelengths. Adds are wavelength independent

Table 6.6.4 Comparisons of reconfigurable electrical and optical add/drop nodes.

Parameter	Electrical ADM	Optical ADM
Optical mux/demux	First level external optical Second level integrated within Rx	First level MWS switch Second level fixed filter or MWS
Channel monitor	Full electrical PM	Analog optical wavelength power
Channel regeneration	Yes, but not critical for metro	Only power equalization, OK for metro
Multidegree	Requires highly sophisticated switching fabric with full nonblocking interconnect capability	MWS provide direct degree interconnects, but wavelength blocking may exist
Optical line amps	Requires Rx-side and Tx-side OLA to deal with MUX/DEMUX loss.	Requires Rx-side and Tx-side OLA to deal with MUX/DEMUX loss
Per-channel power consumption	~50-100 W per 10 Gbps data stream (estimated average from published Ethernet and OTN switch fabrics)	~2.5 W per 10 Gbps wavelength (<100 W for OLA + MWS)
Granularity	Provides subwavelength switching capability	Wavelength-level switching capability
Subrate channel flexibility	Subrate channels electrically multiplexed to wavelength	Subrate channels electrically multiplexed to wavelength.
Super-rate channels	Must be inverse multiplexed across several wavelengths. For example, 40 Gbps services occupy 4 x 10 Gbps λ's, and may cross integrate part boundary	May be inverse multiplexed. Or may used new modulation format technology for direct transport over existing line system
Total system capacity	Fixed on Day 1 install	Can grow as new XCVR technology is introduced to populate unfilled spectrum, i.e., improved FEC, modulation format, equalization
Relative cost of a high capacity ADM node	Assuming OEO interfaces are low-cost, high-capacity electrical FEC, framing and switching fabric expected to dominate costs	OEO interfaces maybe relatively higher, but optical switch fabric is much lower cost than comparable electronic one

6.6.5 Network automation

Network design tools are mandatory for proper design, deployment, and operation of an optical WDM network, especially when large degree of reconfigurability is deployed. Network planning process focuses an optimized network design for efficient capacity utilization and optimum network performance for a given service load. Such design and planning tools must combine a set of functions that span multiple layers of the network operation, from a definition of user demands, to service aggregation and demand routing, and finally to the physical transport layer. Higher (logical) network layers need not worry about the physics of light when calculating paths across a network, with a possible exception of physical latency associated with the connection. However, the physical layer must include important optical propagation effects when deciding demand routing and capacity load attributes.

A typical network design process, graphically illustrated in Figure 6.6.15, includes several steps. First, a set of "Input" parameters is formed by a combination of logical user service demands and by an abstracted layer describing known physical connectivity and limitations. The inputs may also include a description of existing network configuration, which may be automatically uploaded from field-deployed hardware, or may be a completely new installation. Second, service demands are aggregated into optical wavelengths, considering actual electronic hardware limitations, user-defined constraints, and required network performance. Third, aggregated wavelength demands and a refined definition of the physical layer (i.e., definitions of specific fiber types, lengths, optical losses) are provided as an input to drive physical network design process. The result of this multi-step process is a complete description of the network hardware, with a detailed description of the Bill of Material, deployment process documents and drawings, as well an estimate of the network performance characteristics.

The actual software implementation may partition the overall process into relatively independent modules, with each step performed sequentially. An alternative is to provide coupling between each step to allow a higher

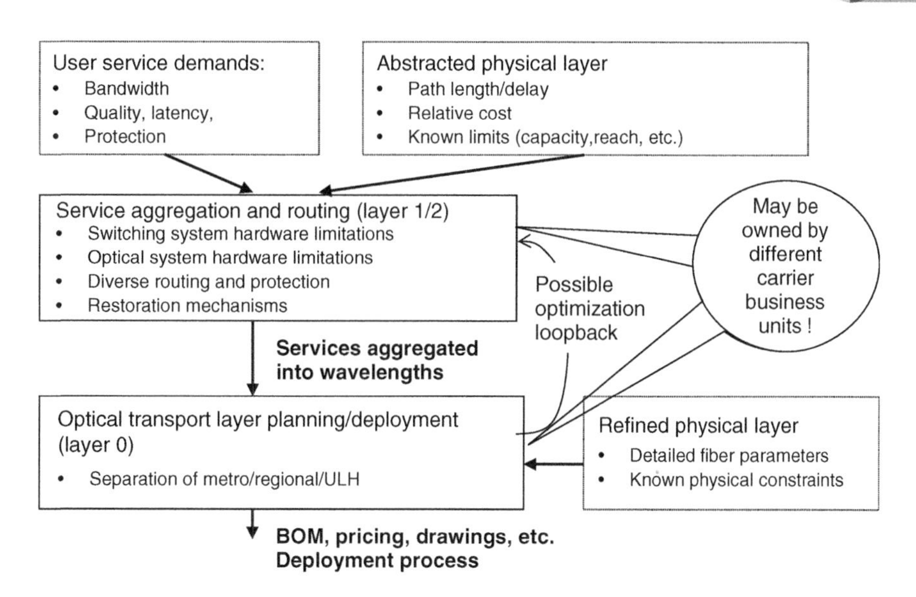

Fig. 6.6.15 Network planning and design process.

level of network optimization and refinement. The software itself may be an off-line design/planning tool driven by a user interested in targeting substantial network configuration changes or upgrades. A similar process and software may also be used for handling new service demand requests that may arrive to the automatically switched optical networks, with the process triggered when a single new service demand arrives to the network and must be satisfied within the shortest amount of time and within constraints of null or minimal hardware change.

The geography of the metro networks is frequently characterized by a wide variability in traffic-generating node separation, which may range from sub-km range for nearly co-located customers to 100 km, and possibly longer. Premise space and power availability, traffic add/drop capacity requirements are also quite variable. Networks with such diversity benefit from a highly modular system transmission design, whereby components such as add/drop filters, optical amplifiers, dispersion compensation modules, and other signal conditioning elements are independent of each other and are deployed on as needed basis. Especially considering the case of closely spaced nodes, the decision to deploy optical amplifiers and dispersion compensation modules cannot be made based on a purely "local, nearest node" basis. The presence of optically transparent degree 3 and higher nodes complicate the configuration process even further by optically coupling multiple network segments and even coupling directions. For example, a purely linear bi-directional system has independent optical propagation for the individual directions. However, a T-branch network geometry couples East-to-West and West-to-East directions, since both share a common Southbound path. The configuration and optimization of such networks cannot be made based on simplified engineering rules in networks that are mostly focused on the cost of the solution. Fortunately, the field of optimization algorithms is very advanced [41] and can be leveraged directly to solving the network problem. The algorithms may be additionally fine-tuned to target specific carrier requirements, such as preferentially focusing on lowest initial cost, highest network flexibility, best capacity scalability, or some other parameters defined by the carrier.

The operational aspects of reconfigurable optical networks require that same algorithms used for network design and planning be applied in-service. Whenever a network receives a new demand request, it must rapidly assess its current operational state and equipment availability, determine if the requested demand can be satisfied either directly or with some dynamic reconfiguration, and consider both logical connectivity and physical layer impairments with the newly proposed wavelength assignment and routing. This is a nontrivial problem requiring a large number of complex computations in near real time, and is currently seeing increasing research interest [42].

6.6.6 Summary

In this chapter, we discussed innovations in network architectures and optical transport that enable metropolitan networks to cost-effectively scale to hundreds Gb/s of capacity, and to hundreds of kilometers of reach, meeting the diverse service needs of enterprise and residential applications. A converged metro network, where Ethernet/IP services, along with the traditional TDM traffic, operate over an intelligent WDM transport layer is increasingly becoming the most attractive architecture addressing the primary need of network operators for significantly improved capital and operational network cost. At the same time, the optical layer of this converged network has to introduce intelligence, and leverage advanced technology in order to significantly improve the deployment and manageability of WDM transport. We reviewed the most important operational advancements, and the technologies that cost-effectively enhance the network flexibility, and advance the proliferation of WDM transport in multi-service metro networks.

This chapter has identified the two main trends in the transport layer of the Metro optical networks. First, there is a preference to use standards-based approaches at all layers of the network. Second, high network flexibility as demanded by the extremely rapid evolving market dynamics. Standards-based approaches allow carriers and equipment manufacturers to leverage a wide base of industry-wide efforts. Flexibility, when provided at low incremental costs, allows carriers to be "wrong" at initial network deployment, but still rapidly adapt to the ever changing and evolving markets. Software definable, flexible client interfaces allow a single network to support a rich variety of existing services, as well as to allow real-time changes as older protocols are removed and new ones are added to the same hardware. At the same time, new unanticipated services and protocol developments can be readily added without affecting installed hardware base. A different approach to the same end is a protocol-agnostic open WDM layer that allows alien wavelengths that carry traffic directly from service-optimized and integrated WDM interfaces on client platforms to operate over a common WDM infrastructure. Manufacturing advances and high volumes have made wide-band wavelength tunability a reality for the Metro space, where a single part number could be produced in extremely high quantities to cover all applications. In addition, wavelength tunability offers the promise of reconfigurability with little or no capital outlay for separate optical switch components. Fixed wavelength filters may serve as an effective "optical address" to which a tunable transmitter may be tuned to establish a particular traffic pattern. Tunable optical filters would have a similar application. Reconfigurable optical add/drop multiplexers bring a level of flexibility and optical transparency to optical switching and routing. Again, unanticipated network growth patterns, new services, and evolving channel data rates can be easily added. As new transceiver technology is developed, overall network capacity may even be increased substantially beyond initial design parameters without requiring new overbuilds. Finally, evolving sophistication of the network planning and design tools lets carriers minimize their CAPEX costs, while at the same time retaining highly desirable flexible OPEX characteristics.

6.6.7 Future outlook

From a transport perspective, DWDM is the main growth technology for metro networks as required bitrates on a fiber—in particular in the presence of video services—outpace the ability of single-wavelength transmission technologies to deliver the bandwidth cost-effectively. Moreover, photonic switching technologies allow for the electrical layer to scale more moderately, as much of the traffic can bypass electronics in most sites.

At the same time, Ethernet has become much more mature and robust, and is incorporating various mechanisms that will allow it to scale more gracefully—including fault detection mechanisms, hierarchical addressing schemes (such as 802.1 ah and 802.1ad), as well as carrier class switch implementations. In fact, Ethernet, in its various incarnations, is gaining popularity as a replacement of SONET/SDH for sub-wavelength transport.

Finally, the IP layer has clearly become the convergence layer of most services—both for the residential and enterprise markets. IP—in particular MPLS Pseudo-wires, has also been used increasingly as a mechanism to converge legacy technologies such as FR and ATM over IP and as a back-haul mechanism to aggregate them into the router. Different approaches exist in terms of role of Ethernet vs IP in the metro. Some carriers see the value of Layer 3 intelligence as close to the customer as possible. This allows for efficient multicast, deep packet inspection and security mechanisms that guarantee that a malicious user will have a minimal impact on the network. Other carriers are trying to centralize Layer 3 functions as much as possible, claiming that this reduces cost and allows for more efficient management. Whatever the right mix of Ethernet and IP may be, it is clear that the network will benefit from tighter integration of the packet layer and the optical layer.

Efficient packet transport based on WDM modules in routers and switches have recently enabled the first such "converged" deployment in the emerging IP-video MAN and WAN architectures [43]. The advents in optical switching and transmission technologies discussed in Section 6.6.3, further allow a flexible optical infrastructure that efficiently transmits and manages the "optical" bandwidth, enabling advanced network architectures to leverage the IP-WDM convergence, and to realize the associated CAPEX and OPEX savings. These architectures would ideally be based on open "WDM" solutions (like the ones described above), so that the same WDM infrastructure can also continue to support traditional TDM traffic, as well as wavelength services or other emerging applications that may not converge over the IP network, e.g., high-speed (4 or 10 Gb/s) fiber channel. Such open WDM architectures further need to offer performance guarantees for the different types of wavelength services, including "alien" wavelengths, support mesh network configurations, and also benefit from coordinated management, and network control.

Acknowledgments

The authors would like to acknowledge many colleagues at Cisco and Ciena.

REFERENCES

[1] ITU G.709 standard

[2] A. McCormick, "Service-enabled networks aid convergence," *Lightwave* (May 2007).

[3] Special Issue on "Metro & access networks," *J. Lightw. Technol.*, 22(11) (November 2004).

[4] L. Paraschis, "Advancements in metro optical network architectures," *SPIE*, 5626(45) (2004).

[5] N. Ghani et al., "Metropolitan optical networks," in *Optical Fiber Telecommunications IV B: Systems and Impairments*, (I. P. Kaminow and T. Li, eds), Academic Press, pp. 329-403, Chapter 14 (2002).

[6] A. Saleh et al., "Architectural principles of ... ," *J. Lightw. Technol.*, 17(10), 2431-2448 (December 1999).

[7] D. Cavendish, "Evolution of optical transport technologies ... ," *IEEE Comm.*, 38(6) (2002).

[8] M. Macchi, et al., "Design and Demonstration of a Reconfigurable Metro-Regional WDM 32 x 10 Gb/s System Scaling beyond 500 km of G. 652 Fiber," in *Optical Fiber Communication Technical Digest Series, Conference Edition, 2004*, Paper OTuP2, *Optical Society of America*.

[9] M. J. O'Mahony, et al., "Future optical networks," *J. Lightw. Technol.*, 24(12), 4684-4696 (December 2006).

[10] S. Ramamurthy, et al., "Survivable WDM mesh networks," *J. Lihgtw. Technol.*, 21(4), 870-883 (April 2003).

[11] N. Mallick, "Protection Capacity Savings due to an End-To-End Shared Backup Path Restoration in Optical Mesh Networks," in Proc. *OFC 2006*, paper NThB2 (March 2006).

[12] M. Bhardwaj et al., "Simulation and modeling of the restoration performance of path based restoration schemes in planar mesh networks," *J. Opt. Netw.*, 5(12), 967-984 (December 2006).

[13] P. Demeester et al., "Resilience in multilayer networks," *IEEE Communications Magazine*, 37(8), 70-76 (August 1999).

[14] J. Manchester, P. Bonenfant, and C. Newton, "The evolution of network survivability," *IEEE Communications Magazine*, 37(8), 44-51 (August 1999).

[15] R. Batchellor and O. Gerstel, "Protection in Core Packet Networks: Layer 1 or Layer 3?" in proceedings *ECOC Conference*, Paper Tu3.6.4 (Sept 2006).

[16] L. Paraschis, O. Gerstel, and R. Ramaswami, "Tunable Lasers Applications in Metropolitan Networks," in *Optical Fiber Communication Technical Digest Series, Conference Edition, 2004*. Paper MF 105. *Optical Society of America*.

[17] ITU-T Rec. G.957, "Optical interfaces for equipment and systems relating to the Synchronous Digital Hierarchy," (1995).

[18] http://www.schelto.com/

[19] ftp://ftp.seagate.com/sff/INF-8053.PDF

[20] ftp://ftp.seagate.com/sff/INF-8074.PDF

[21] http://www.xfpmsa.org/cgi-bin/home.cgi

[22] 300 pin Multi Source Agreement web site: http://300pinmsa.org/

[23] L. Paraschis, "Innovations for Cost-effective 10 Gb/S Optical Transport in Metro Networks," Invited Presentation in *LEOS 2004: The 17th Annual Meeting of the IEEE Lasers and Electro-Optics Society. November 2004*, Paper WU1).

[24] S. Norimatsu et al., "Accurate Q-factor estimation of optically amplified systems in the presense of waveform distortion," *J. Lightw. Technol.*, 20(1), 19-27 (January 2002).

[25] J. D. Downie, "Relationship of Q penalty to eye-closure penalty for NRZ and RZ signals with signal-dependent noise," *J. Lightw. Technol.*, 23(6), 2031-2038 (June 2005).

[26] A. Tzanakaki, "Performance study of modulation formats for 10 Gb/s WDM metropolitan area networks," *IEEE Photon. Technol. Lett.*, 16(7), 1769-1771 (July 2004).

[27] Q. Yu and A. Shanbhag, "Electronic data processing for error and dispersion compensation," *J. Lightwave Techn.*, 24(12), 4514-4525 (December 2006).

[28] T. Nielsen and S. Chandresekhar, *J. Lightw. Technol.*, 23(1), 131 (January 2005).

[29] S. K. Kim et al., "Distributed fiber Raman amplifiers with localized loss," *J. Lightw. Technol*, 21(5), 1286-1293 (May 2003).

[30] M. Karasek and J. A. Valles, "Analysis of channel addition/removal response in all-optical gain-controlled cascade of Erbium-doped fiber amplifiers," *J. Lightw. Technol.*, 16(10), 1795-1803 (October 1998).

[31] S. Pachnicke, et al., "Electronic EDFA Gain Control for the Suppression of Transient Gain Dynamics in Long-Haul Transmission Systems," *OFC*, paper JWA15 (March 2007).

[32] A. Sridharan, "Blocking in all-optical networks," *IEEE/ACM Trans. On Netw.*, 12(2), 384-397 (April 2004).

[33] C. C. Sue, "Wavelength routing with spare reconfiguration for all-optical WDM networks," *J. Lightwave Technol.*, 23(6), 1991-2000 (June 2005).

[34] O. Gerstel, et al., "Worst-case analysis of dynamic wavelength allocation in optical networks," *IEEE/ACM Trans. On Netw.*, 7(6), 833-845 (December 1999).

[35] J. Wagener, et al., "Characterization of the economic impact of stranded bandwidth in fixed OADM relative to ROADM networks," *OFC*, paper OThM6 (March 2006).

[36] H. Zhu, "Online connection provisioning in metro optical WDM networks using reconfigurable OADMs," *J. Lightw. Technol.*, 23(10), 2893-2901 (October 2005).

[37] S. Okamoto, A. Watanabe, and K. Sato, "Optical path cross-connect node architectures for photonic transport networks," *J. Lightw. Technol.*, 14(6), 1410-1422 (June 1996).

[38] E. Iannone and R. Sabella, "Optical path technologies: a comparison among different cross-connect architectures," *J. Lightw. Technol.*, 14(10), 2184-2196 (October 1996).

[39] M. Fuller, "RFP activity boosts ROADM development," *Lightwave*, April 2004. http://lw.pennnet.com/Articles/Article_Display.cfm?Section=ARTCL8ARTICLE_ID=203231&VERSION_NUM=1

[40] C. Ferrari, et al., "Flexible and Reconfigurable Metro-Regional WDM Scaling beyond 32x10Gb/s and 1000km of G.652 Fiber," in *LEOS 2005: The 18th Annual Meeting of the IEEE Lasers and Electro-Optics Society*, TuG5 (November 2005).

[41] Optimization Software Guide by Jorge J. More and Stephen J. Wright SIAM Publications (1993).

[42] G. Markidis, "Impairment-constraint-based routing in ultralong-haul optical networks with 2R regeneration," *Photon. Techn. Lett.*, 19(6), 420-422 (March 2007).

[43] O. Gerstel, M. Tatipamula, K. Ahuja et al., "Optimizing core networks for IP Transport," *IEC Annual Review of Communications*, 59 (2006).

Index

PHYSICAL CONSTANTS IN SI UNITS

Absolute zero temperature	$-273.3°C$
Acceleration due to gravity, g	$9.807\ m\ s^{-2}$
Avogadro's number, N_A	6.022×10^{23}
Base of natural logarithms, e	2.718
Boltsmann's constant, k	$1.381 \times 10^{-23}\ JK^{-1}$
Faraday's constant, k	$9.648 \times 10^{4}\ C\ mol^{-1}$
Gas constant, R	$8.314\ J\ mol^{-1}\ K^{-1}$
Permeabillity of vacuum, μ_0	$1.257 \times 10^{-6}\ H\ m^{-1}$
Permittivity of vacuum, ε_0	$8.854 \times 10^{-12}\ F\ m^{-1}$
Planck's constant, h	$6.626 \times 10^{-34}\ J\ s^{-1}$
Velocity of light in vacuum, c	$2.998 \times 10^{8}\ m\ s^{-1}$
Volume of perfect gas at STP	$22.41 \times 10^{-3}\ m^3\ mol^{-1}$

CONVERSION OF UNITS

Angle, θ	1 rad	$57.30°$
Density, ρ	$1\ lb\ ft^{-3}$	$16.03\ kg\ m^{-3}$
Diffusion Coefficient, D	$1\ cm^3\ s^{-1}$	$1.0 \times 10^{-4}\ m^2\ s^{-1}$
Force, F	1 kgf	9.807 N
	1 lbf	4.448 N
	1 dyne	$1.0 \times 10^{-5}\ N$
Length, L	1 ft	304.8 mm
	1 inch	25.40 mm
	1 Å	0.1 nm
Mass, M	1 tonne	1000 kg
	1 short ton	908 kg
	1 long ton	1107 kg
	1 lb mass	0.454 kg
Specific Heat, Cp	$1\ cal\ gal^{-1}\ °C$	$4.188\ kJ\ kg^{-1}\ °C$
	$Btu\ lb^{-1}\ °F$	$4.187\ kJ\ kg^{-1}\ °C$
Stress Intensity, K_{IC}	$1\ ksi\sqrt{in}$	$1.10\ MN\ m^{-3/2}$
Surface Energy, γ	$1\ erg\ cm^{-2}$	$1\ mJ\ m^{-2}$
Temperature, T	1 °F	0.556 K
Thermal Conducitivity, λ	$1\ cal\ s^{-1}\ cm\ °C$	$418.8\ W\ m^{-1}\ °C$
	$1\ Btu\ h^{-1}\ ft\ °F$	$1.731\ W\ m^{-1}\ °C$
Volume, V	1 Imperial gall	$4.546 \times 10^{-3}\ m^3$
	1 US gall	$3.785 \times 10^{-3}\ m^3$
Viscosity, η	1 poise	$0.1\ N\ s\ m^{-2}$
	1 lb ft s	$0.1517\ N\ s\ m^{-2}$

CONVERSION OF UNITS—STRESS AND PRESSURE*

	MN m^{-2}	dyn cm^{-2}	lb in^{-2}	kgf mm^{-2}	bar	long ton in^{-2}
MN m^{-2}	1	10^7	1.45×10^2	0.102	10	6.48×10^{-2}
dyn cm^{-2}	10^{-7}	1	1.45×10^{-5}	1.02×10^{-8}	10^{-8}	6.48×10^{-9}
lb in^{-2}	6.89×10^{-3}	6.89×10^4	1	703×10^{-4}	6.89×10^{-2}	4.46×10^{-4}
kgf mm^{-2}	9.81	9.81×10^7	1.42×10^3	1	98.1	63.5×10^{-2}
bar	0.10	10^6	14.48	1.02×10^{-2}	1	6.48×10^{-3}
long ton in^{-2}	15.44	1.54×10^8	2.24×10^3	1.54	1.54×10^2	1

CONVERSION OF UNITS—ENERGY*

	J	erg	cal	eV	Btu	ft lbf
J	1	10^7	0.239	6.24×10^{18}	9.48×10^{-4}	0.738
erg	10^{-7}	1	2.39×10^{-8}	6.24×10^{11}	9.48×10^{-11}	7.38×10^{-8}
cal	4.19	4.19×10^7	1	2.61×10^{19}	3.97×10^{-3}	3.09
eV	1.60×10^{-19}	1.60×10^{-12}	3.38×10^{-20}	1	1.52×10^{-22}	1.18×10^{-19}
Btu	1.06×10^3	1.06×10^{10}	2.52×10^2	6.59×10^{21}	1	7.78×10^{-2}
ft lbf	1.36	1.36×10^7	0.324	8.46×10^{18}	1.29×10^{-3}	1

CONVERSION OF UNITS—POWER*

	kW (kj s^{-1})	erg s^{-1}	hp	ft lbf $^{s-1}$
kW (kJ s^{-1})	1	10^{-10}	1.34	7.38×10^2
erg s^{-1}	10^{-10}	1	1.34×10^{-10}	7.38×10^{-8}
hp	7.46×10^{-1}	7.46×10^9	1	15.50×10^2
ft lbf s^{-1}	1.36×10^{-3}	1.36×10^7	1.82×10^{-3}	1

* To convert row unit to column unit, multiply by the number at the column-row intersection, thus 1 MN m^{-2} =10 bar

Printed and bound by CPI Group (UK) Ltd, Croydon, CR0 4YY
08/05/2025
01864928-0003